# AN INTRODUCTION TO STATISTICAL METHODS AND DATA ANALYSIS

**Lyman Ott**
Merrell-National Laboratories

**Duxbury Press**
North Scituate, Massachusetts

**To my wife, Sally**
**and two children,**
**Curtis and Kathryn**

**Library of Congress Cataloging in Publication Data**

Ott, Lyman.
An Introduction to statistical methods and data analysis.

Includes index.
1. Mathematical statistics. I. Title.
QA276.077 519.5 77-4722
ISBN 0-87872-134-7

**Duxbury Press**
A Division of Wadsworth Publishing Company, Inc.



*Introduction to Statistical Methods and Data Analysis* was edited and prepared for composition by Carol Beal. Interior design was provided by Richard Spencer. The cover was designed by Joseph Landry.

L.C. Cat. Card No.: 77-4722
ISBN 0-87872-134-7
Printed in the United States of America
3 4 5 6 7 8 9 — 81 80 79

# Contents

# Preface

The demands placed on professional statisticians working in government, teaching, and private industry have changed greatly in recent years. First, more and more people are becoming aware of the important role that statistics plays in the design and analysis of experiments in all areas of scientific endeavor. Research statisticians must respond to these needs by focusing on research problems that ultimately solve practical problems posed in other disciplines. Consulting statisticians can respond to these changing demands and needs by remaining abreast of recent research developments so that they provide scientists with appropriate up-to-date statistical analyses. And professors of statistics must respond to this increased awareness by offering scientists from many different disciplines a meaningful initial exposure to statistics that develops a recognition of the role of statistics in experimentation and provides appropriate analytical tools for solving a wide variety of current practical problems.

Second, the field of statistics has been changed by the availability and accessibility of high-speed computing facilities and portable electronic calculators. No longer must the student of statistics spend endless hours doing his or her calculations by hand to solve a classroom problem.

In their statistical course work, students must be exposed to computer software packages and other aids to computations so that they may function efficiently in our present environment. We, as teachers of statistics, must accept this challenge and treat the introduction of computing assistance as a blessing rather than a plague. It is up to us to make certain that our students understand how to compute a solution *and* understand how to obtain the same solution on a computer.

In an attempt to address the changes in direction and emphases of statistics, I have written a text that I hope will appeal to scientists from many different disciplines as a first exposure to statistical methods and data analysis. The text should serve also as a reference for research workers involved with performing statistical calculations or interpreting the results of statistics.

In addition to the standard statistical methods, the following features are included in the text:

- Solutions to selected examples illustrated by some of the more common statistical software packages (BMDP, MINITAB, SAS, SPSS, etc.).
- Starred exercises particularly appropriate for a computer solution.
- More than 100 worked examples from a wide variety of disciplines, such as biology, agriculture, engineering, business and economics, sociology, education, medicine and health, psychology, and environmental studies.
- An introductory section for each chapter that motivates the material to be presented and relates it to previous work.

- A summary section for each chapter.
- New topics motivated in terms of previously encountered problems or in terms of practical problems yet to be discussed.
- Use of statistical models as a unifying theme between regression and analysis of variance.
- An up-to-date chapter on multiple comparisons.
- A discussion of hypothesis testing that utilizes the concept of level of significance as well as the more standard decision-theoretic concepts.
- A separate chapter on experimental design.
- An up-to-date presentation of the elements of sample survey designs.
- A modern easy-to-understand discussion of covariance analysis.
- A discussion of control charts.
- A discussion of some quick portable statistical techniques.
- Matrices presented in an easily understood manner that can be comprehended by those having a background in high school algebra.
- The binomial distribution presented as a special case of the multinomial.
- A discussion of multivariate categorical data analysis.
- Solutions to almost all of the more than 300 exercises.

This text can be used for either a one-term or a one-year course for students having a background in high school algebra. No knowledge of statistics is required; however, those students who have had an introductory course in probability and statistics will find that the material in the first few chapters provides a good review of their previous course work.

We begin by answering the question, "What is statistics?" and then introduce descriptive statistics. Probability, which lays the basis for inferential statistics, is presented in an easily understood fashion based on the relative frequency concept of probability. Rather than discussing specific discrete random variables and their probability distributions, we move quickly into a discussion of sampling distributions and the calculation of areas (probabilities) associated with the normal curve.

We introduce inferential statistics with a discussion of estimation (point and interval) and hypothesis testing. The usual large- and small-sample tests for $\mu$ and $\mu_1 - \mu_2$ are presented, along with a discussion of the sample-size problem.

We then launch into a discussion of models that provides the basis for most of the material in the remainder of the text. By introducing models at this point, students become familiar with writing a model for an experimental situation and are able to phrase statistical estimation or testing problems in terms of model parameters. Having covered or reviewed this basic material, many options are available for developing a course to suit your needs. To assist you in adapting the textbook material to either a one-term course or a one-year sequence, the list that follows matches textbook chapters with corresponding prerequisite material needed for adequate comprehension.

| CHAPTER | PREREQUISITE MATERIAL |
|---|---|
| 1 | None |
| 2 | None |
| 3 | Chapter 2 |
| 4 | Chapters 2, 3 |
| 5 | Chapters 2-4 |
| 6 | None |
| 7 | Chapter 6 |
| 8 | Chapters 2-7 |
| 9 | Chapters 2-5 |
| 10 | Chapters 2-5 |
| 11 | Chapters 2-5, 9 |
| 12 | Chapters 2-5 |
| 13 | Chapters 2-5, 9, 12 |
| 14 | Chapters 2-5, 9, 12, 13 |
| 15 | Chapters 2-5, 9, 12-14 |
| 16 | Chapters 2-10, 12-15 |
| 17 | Chapters 2-9, 12-16 |
| 18 | Chapters 2-6, 9, 12-15 |
| 19 | Chapters 2-9, 12-17 |
| 20 | Chapters 2-9, 12-16 |
| 21 | Chapters 2-5, 11, 12 |
| 22 | Chapters 2-5 |

I am deeply indebted to the SAS Institute, Inc., Duxbury Press, the University of California Press, and the McGraw-Hill Book Company and Norman H. Nie* for allowing me to illustrate computer solutions to examples and exercises using SAS, MINITAB, BMDP and SPSS, respectively.

In addition, I wish to express my appreciation to friends and colleagues who have made many constructive criticisms and suggestions at various stages of revision. Particular thanks are due to Harvey Arnold, Oakland University; Christopher Barry, University of Florida; Franklin A. Graybill, Colorado State University; David Macky, California State University, San Diego; and Roger Pfaffenberger, University of Maryland who served as reviewers. Also thanks are due A. Hald, E. S. Pearson, H. O. Hartley, R. E. Kirk, the *Biometrika* Trustees, the Chemical Rubber Company, Lederle Laboratories, the Editors of the *Annals of Mathematical Statistics,* D. B. Duncan, R. A. Waller, the Editors of *Biometrics,* the Editors of *JASA*, and the American Society of Testing and Materials for permission to reprint tables, and my typists Mrs. Ruth Campbell, Mrs. Jeanette Beach, and Mrs. Ellen Evans. And finally I must acknowledge the love, support, and forbearance of my wife Sally and two children Curtis and Kathryn during the long evenings and weekends of work over the past several years.

# What Is Statistics? 1

## Introduction 1.1

Almost everyone is confronted with statistics in our day-to-day living, but few people have much of a notion about the discipline of statistics. A commonly held opinion is that statistics deals with data tabulations and summarizations in the form of batting averages, divorce rates, incidences of lung cancer for a given sector of the United States, and, in general, the tedious and somewhat boring description of our world by using numbers. While the field of statistics does utilize numbers and statisticians do attempt to describe phenomena in our world, modern day statistics deals primarily with statistical inference.

statistical inference

The development and testing of the Salk vaccine for protection against poliomyelitis (polio) provide an excellent example of how statistics can be used in reaching a decision about a practical problem. Most parents and children of the years prior to 1954 can recall the panic brought on by the outbreak of cases of polio during the summer months. Nowadays, similar although less severe concern arises when we learn of additional agents that appear to be carcinogenic.

Although relatively few children fell victim to polio each year, the pattern of its outbreak was unpredictable, and it caused great concern because of the possibility of paralysis or even death. The fact that very few of today's youth have even heard of polio demonstrates the great success of the Salk vaccine and the testing program that preceded its release on the market.

How do researchers determine whether a drug product is effective in the control of a specific disease? Following elaborate laboratory testing, a drug product will eventually be introduced into human beings under the close surveillance of highly trained investigators. If the product passes these initial tests in humans, standard practice requires further testing on a larger scale in humans. These investigations (often called clinical trials) are then used to determine the effectiveness of the product.

For example, after consultations with statisticians, it was decided that 400,000 children should be included in the Salk vaccine clinical trial begun in 1954, with the children randomly assigned so that half would receive the vaccine and the other half would receive an inactive solution that simply looked like the vaccine (called a placebo). When the study was over, less than 200 cases of polio were reported, but more than three times as many cases occurred in the placebo group. Was this evidence sufficient to say that the vaccine was effective? Is it likely that we could run another trial of the same magnitude and have a vaccine group with as many or more polio cases than the placebo group? These are some of the questions that statistics can help to answer.

The development of the Salk vaccine is not an isolated case of the use of statistics in the testing and development of new drug products. In recent years the Food and Drug Administration has placed stringent requirements on scientists to establish the effectiveness of proposed drug products. Statistics has played an important role in the development of birth control pills, rubella vaccines, chemotherapeutic agents for the treatment of cancer, and many other preparations.

A second example of the use of statistics is found in the predictions or estimations made about the total eligible work force that is unemployed. An accurate estimate is an important measure of social well-being. Clearly it would be impossible to locate and interview each and every person to ascertain his or her work status. And even if we could locate all persons, the unemployment situation would change while we were completing our survey. Thus it is both more realistic and more efficient to obtain information from a *representative sample* of the work force to make an estimate of the percentage of unemployed persons in the entire eligible work force.

sample

Our final example of the use of statistics relates to the problem of urban traffic congestion. One aspect of this problem, investigated by the Federal Highway Administration, is how to merge automobiles successfully onto high-speed areas of interstate highways. To study this problem an automatic merging system was installed on the entrance to I-75 at Tampa, Florida. Through the use of a series of display lights, drivers are told whether or not they are traveling at an appropriate speed to merge successfully into the existing traffic on the highway. Prior to the installation of the system, investigators measured the stress levels of many drivers merging into the highway during the 4–6 P.M. rush hour period. Similar testing on a sample of 50 drivers was conducted after the installation of the system. Suppose, for the purpose of illustration, the average stress level prior to installation of the system was 8.2 (as measured on a 10-point scale), and the sample of 50 drivers (after installation) had an average stress level of 7.6. Can this and other sample information be used to conclude that the merging system lowers the average stress level?

In each of these three examples, there are certain commonalities. First, each example involved making observations of a phenomenon that could not be predicted with certainty in advance. Indeed, results of repeated measurements are likely to bob about in a haphazard, or random, manner. For example, we could not predict with certainty in advance whether a child administered the polio vaccine would contract polio. Similarly, it would be impossible to predict with certainty in advance whether a member of the work force was currently employed or what the stress level of a driver would be under the new merging system on I-75.

observations

random

Second, each example involved *sampling*. A sample of 400,000 children was used in the Salk clinical trial, a sample of the labor force was used to estimate the percentage unemployed, and a sample of drivers was used to examine the postinstallation stress levels.

sampling

Third, each example involved the collection of data or measurements, one measurement corresponding to each element in the sample. Obviously, we would obtain one measurement (stress level) for each of the 50 drivers included in the sample in the third example. For the first and second examples, the observations on elements in the samples are qualitative (described by kind, or quality) rather than quantitative (described by numbers). However, we could make an arbitrary assignment of a measurement to each qualitative category as another means of identification. Thus in the first example, we could assign the measurement 1 to each child in the sample who contracted polio and 0 to each child who did not. Similarly, in the second example, we could assign the measurement 1 to each person who is employed and 0 to each person unemployed. Now all three examples involve a collection of measurements, one measurement for each element in the sample.

collection of data

Finally, in each example we had a common objective, namely, a statistical inference. Using the information obtained from a sample, we tried to make an inference about a much larger set of measurements called the *population*.

inference
population

For the first example the statisticians wished to make an inference about the proportion of children who would contract polio after being administered the vaccine. The population of interest here is the set of 1s and 0s corresponding to all children who might receive the vaccine. In this case the population is conceptual rather than real since we can only imagine the set of all children who might receive the vaccine.

The population of interest in the second example is the set of 1s and 0s corresponding to employed and unemployed members of the eligible work force. In contrast to the conceptual population of the first example, this population is real and does in fact exist. Our objective is to estimate the percentage of unemployed members (0s) in the population.

The population associated with the third example is the set of stress levels associated with all persons who have used or may use the automatic merge system. Note again that the population is conceptual, and we wish to reach a decision about whether the mean stress level for the population is lower than 8.2.

Combining these common characteristics, we see that statistical problems involve the measurement of phenomena which we cannot predict with certainty in advance, sampling, the collection of measurements, and inference. Information is obtained by experimentation and is employed to make an inference about a large set of measurements called the population.

**Definition 1.1** A *population* is the set of all measurements of interest to the experimenter.

**Definition 1.2** A *sample* is a subset of measurements selected from the population.

**Definition 1.3** The *objective of statistical inference* is to make an inference about a population based on information contained in a sample.

# 1.2 What Do Statisticians Do?

design

Statisticians, both in consulting and in research, devote their time to two major areas. The first concerns the acquisition of the sample data. Sample surveys and experiments cost money and they yield information—numbers on sheets of paper. By varying the survey or experimental procedure—how we select the data and how many observations we take from each source—we can vary the cost, quality, and quantity of information in the experiment. Rather simple modifications in the data selection procedure can reduce the cost of the sample to one hundredth or less of the cost of conventional sampling procedures. Thus statisticians study various methods for designing sample surveys and experiments and attempt to find the method that will yield a specified amount of information at a minimum cost.

choose appropriate method

The second task facing statisticians is to select the appropriate method of inference for a given sample survey or experimental design. Some of these methods are good, some are bad, and some seem to be best for most occasions. It is the statistician's job to choose the appropriate method for a given situation.

measure goodness

The preceding discussion leads to the most important contribution of statistics to science and business. Anyone can devise a method to make inferences based on the sample data. The major contribution of statistics is in evaluating the "goodness" of the inference. In other words, when predicting, we seek an upper limit to the error of prediction. Or in making a decision concerning a characteristic of the population, we wish to know the probability of reaching a correct conclusion.

To summarize, statisticians first design surveys and experiments to minimize the cost of obtaining a specified quantity of information. Second, they seek the best method for making an inference for a given sampling situation. Finally, statisticians measure the goodness of their inference.

# 2 Data Description

## 2.1 Introduction

data description

The field of statistics can be divided into two major branches, descriptive statistics and inferential statistics. In both branches we will be working with a set of measurements. For situations where data description is our major objective, the set of measurements available to us is frequently the entire population. For example, suppose that we wish to describe the distribution of annual incomes for all families registered in the 1970 census. Since these data are recorded and are available on computer tapes, we do not need to obtain a random sample from the population; the complete set of measurements is at our disposal. Our major problem is in organizing, summarizing, and describing these data.

statistical inference

In situations where we are concerned with statistical inference, a sample is usually the only set of measurements available to us. We use information in the sample to draw conclusions about the population from which the sample was drawn. For example, the fear of contracting botulism has spurred officials of canning companies to maintain rigid quality control standards. In particular, they may be interested in determining the proportion of cans out of the total production that are improperly sealed or damaged in transit to retail outlets. Obviously it is impossible for them to inspect all cans, but they can obtain a random sample of the company's production and use the proportion of defective or damaged cans in the sample to *estimate* the actual proportion of improperly sealed or defective cans in the entire production.

In this particular example description of the sample data is an important step leading towards the estimate (inference) which we make. Thus no matter what our objective—statistical inference or data description—we must first describe the set of measurements at our disposal.

There are two ways to describe a set of measurements. We can use either graphical techniques or numerical descriptive techniques. Section 2.2 is concerned with graphical description of data. In sections 2.3 and 2.4 we discuss numerical techniques for describing a set of measurements.

## 2.2 Graphical Techniques for Describing Data

When repeated measurements are obtained on a variable, the data must first be organized, prior to presentation, by using one of several graphical techniques. As a general rule, the data should be arranged in such a way that each observation can fall into *one and only one* category of the variable. This procedure eliminates any ambiguity that might arise in placing observations into categories and aids in the interpretation of the data.

guideline for organizing data

For example, if we are trying to categorize beef according to the qualitative variable "beef quality," appropriate categories of the variable might be as listed in the table that is shown below.

| *Quality of Beef* |
|---|
| prime |
| choice |
| good |
| standard |

Assuming these categories are clearly defined and assuming a rater is properly trained, all cuts of beef could be placed into one and only one category of the variable. However, if we are asking economists to respond to a questionnaire concerning their reactions to a new wage and price freeze, the categories listed on page 8 would not suffice.

| *Reactions of Economists to a Wage and Price Freeze* |
|---|
| too much too soon |
| long overdue |
| places the farmer at a disadvantage |
| will adversely affect the balance of payments |
| in the long run the freeze will hurt the consumer |
| other (please specify) |

Clearly several answers might adequately reflect a single economist's reactions to the controls. Hence the data from the qualitative variable "reaction to new controls" could not be organized according to our guidelines.

Having organized the data according to the guideline suggested, there are several ways to graphically display the data. The first and simplest graphical procedure for data organized in this manner is the *pie chart* or *circle chart.* It is used to display the percentage of the total number of measurements falling into each of the categories of the variable by partitioning a circle (much as one might slice a pie). For example, the data of table 2.1 represent the per capita

pie chart

**Table 2.1** Per capita consumption of new metals

| *Metal* | *Pounds* | *Percentage* |
|---|---|---|
| iron and steel | 1200 | 89.6 |
| copper, zinc, and lead | 55 | 4.1 |
| aluminum | 50 | 3.7 |
| other metals | 35 | 2.6 |

*Source:* Data from *U.S. News and World Report,* 30 June 1973, p. 58.

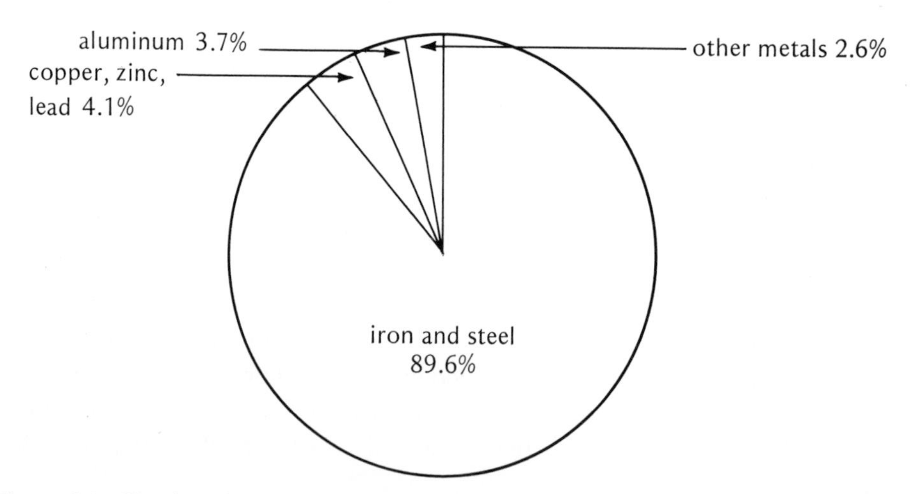


**Figure 2.1** Pie chart for per capita metal consumption, 1970

consumption of new metals in 1970. These data can be displayed in a pie chart, as shown in figure 2.1. One quick glance at figure 2.1 shows the overwhelming per capita consumption of iron and steel.

Similar displays in figure 2.2 give the percentage distributions of the world production and consumption of petroleum from 1966 to 1970. Again a very

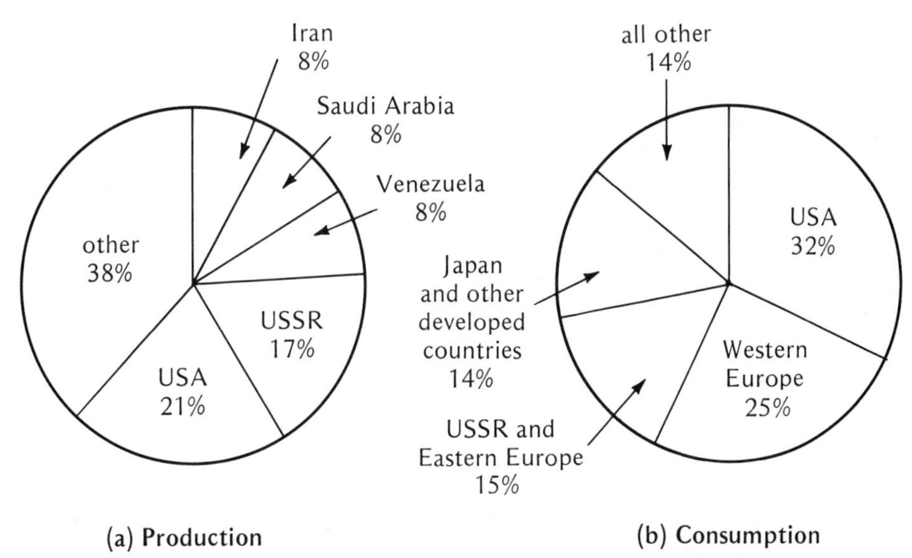


**Figure 2.2** World production and consumption of petroleum, 1966–70

*Source:* Data from *Business Week,* 30 June 1973, p. 56.

quick perusal of figure 2.2 gives the impression that the major users of petroleum do not necessarily produce as much as they consume.

In summary, the pie chart or circle chart can be used to display percentages associated with each category of the variable. The following guidelines should help you to obtain clarity of presentation in your pie charts.

**Guidelines for Constructing Pie Charts**

1. Choose a small number of categories for the variable, preferably around 5 or 6. Too many categories make the pie chart difficult to interpret.
2. Whenever possible, construct the pie chart so that percentages are in either ascending or descending order (see figures 2.1 and 2.2).

bar chart

The second graphical technique for data organized according to the recommended guideline is the *bar chart* or *bar graph.* Figure 2.3 displays four racial-ethnic populations in the United States for 1970, by sex.

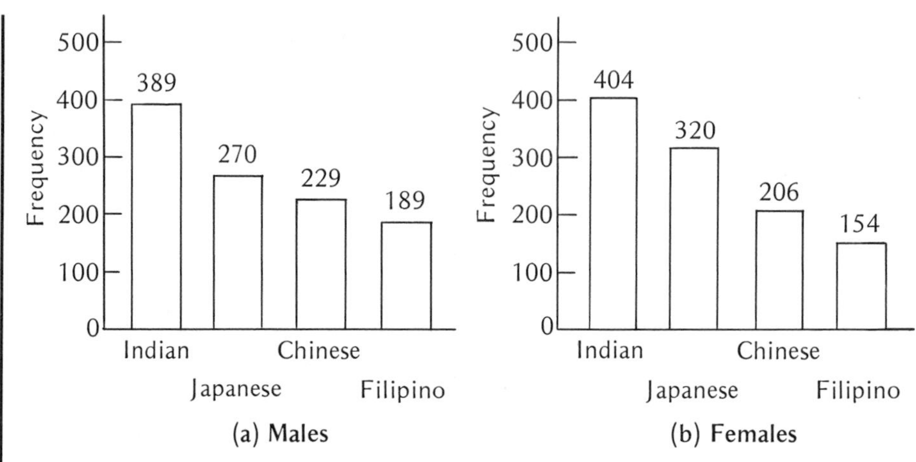


**Figure 2.3** Four racial-ethnic populations (in thousands) in the United States in 1970, by sex

Bar charts are relatively easy to construct if you use the guidelines given in the box.

**Guidelines for Constructing Bar Charts**

1. Label frequencies along the vertical axis and categories of the qualitative variable along the horizontal axis.
2. Construct a rectangle over each category of the variable with a height equal to the frequency (number of observations) in the category.
3. Leave a space between each category on the horizontal axis for clarity of presentation.

There are several variations of the bar chart, which you may have seen in published form. Sometimes the roles of the axes are reversed so that the rectangles are horizontal, as in figure 2.4.

frequency histogram
relative frequency histogram

The final two graphical techniques that we will discuss in this text are the *frequency histogram* and the *relative frequency histogram*. Both of these procedures are applicable only to quantitative data. As with the previous graphical techniques, we must organize the data prior to constructing a graph.

Consider the following kind of measurement: weight gain for each of 100 baby chickens fed on a new diet and observed over a period of 8 weeks. These data are recorded in table 2.2.

In trying to describe the set of measurements recorded in table 2.2, we note that the largest weight gain is 4.9 and the smallest is 3.6. But although we might examine the table very closely, it is difficult to describe how the measurements

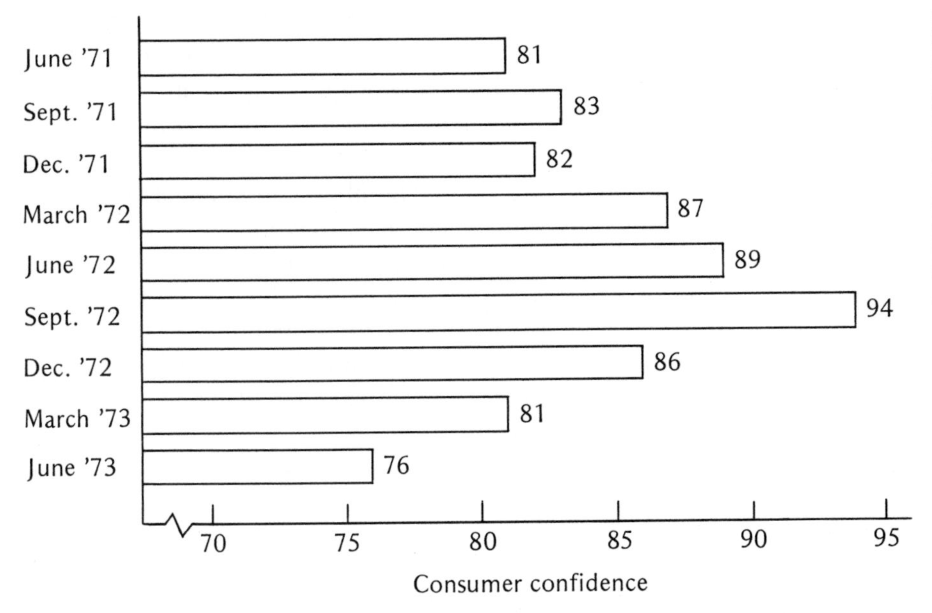


**Figure 2.4** Consumer confidence index towards the economy, 1971–73

*Source:* Data from *Time,* 9 July 1973, p. 56.

are situated along the interval from 3.6 to 4.9. Are most of the measurements near 3.6, near 4.9, or are they evenly distributed along the interval? To answer these questions we summarize the data in a *frequency table.*

frequency table

To construct a frequency table, we begin by dividing the range from 3.6 to 4.9 into an arbitrary number of subintervals called *class intervals.* The number of subintervals chosen depends on the number of measurements in the set, but we generally recommend using from 5 to 20 class intervals. The more data we have, the larger the number of classes we require. The guidelines given in the box can be used for constructing the appropriate class intervals.

class intervals

**Table 2.2** Weight gains for baby chickens (in grams)

| | | | | | | | | | |
|---|---|---|---|---|---|---|---|---|---|
| 3.7 | 4.2 | 4.4 | 4.4 | 4.3 | 4.2 | 4.4 | 4.8 | 4.9 | 4.4 |
| 4.2 | 3.8 | 4.2 | 4.4 | 4.6 | 3.9 | 4.3 | 4.5 | 4.8 | 3.9 |
| 4.7 | 4.2 | 4.2 | 4.8 | 4.5 | 3.6 | 4.1 | 4.3 | 3.9 | 4.2 |
| 4.0 | 4.2 | 4.0 | 4.5 | 4.4 | 4.1 | 4.0 | 4.0 | 3.8 | 4.6 |
| 4.9 | 3.8 | 4.3 | 4.3 | 3.9 | 3.8 | 4.7 | 3.9 | 4.0 | 4.2 |
| 4.3 | 4.7 | 4.1 | 4.0 | 4.6 | 4.4 | 4.6 | 4.4 | 4.9 | 4.4 |
| 4.0 | 3.9 | 4.5 | 4.3 | 3.8 | 4.1 | 4.3 | 4.2 | 4.5 | 4.4 |
| 4.2 | 4.7 | 3.8 | 4.5 | 4.0 | 4.2 | 4.1 | 4.0 | 4.7 | 4.1 |
| 4.7 | 4.1 | 4.8 | 4.1 | 4.3 | 4.7 | 4.2 | 4.1 | 4.4 | 4.8 |
| 4.1 | 4.9 | 4.3 | 4.4 | 4.4 | 4.3 | 4.6 | 4.5 | 4.6 | 4.0 |

**Guidelines for Constructing Class Intervals**

1. Divide the *range* of the measurements (the difference between the largest and the smallest measurements) by the approximate number of class intervals desired. Generally we will wish to have from 5 to 20 class intervals.
2. After dividing the range by the desired number of subintervals, round the resulting number to a convenient (easy to work with) unit. This unit represents a common width for the class intervals.
3. Choose the first class interval so that it contains the smallest measurement. It is also advisable to choose a starting point for the first interval so that no measurement falls on a point of division between two subintervals. This eliminates any ambiguity in placing measurements into the class intervals.

For the data in table 2.2, the range is

$$\text{range} = 4.9 - 3.6 = 1.3$$

Let us suppose that we would like to have approximately 10 subintervals. Dividing the range by 10 and rounding to a convenient unit, we have $1.3/10 = .13 \approx .1$. Thus the class interval width is .1.

It is convenient to choose the first interval to be 3.55 to 3.65, the second to be 3.65 to 3.75, and so on. Note that the smallest measurement, 3.6, falls in the first interval and that no measurement falls on the endpoint of a class interval (see table 2.3).

**Table 2.3** Frequency table for the baby chick data

| *Class* | *Class Interval* | *Frequency, $f_i$* | *Relative Frequency, $f_i/n$* |
|---|---|---|---|
| 1 | 3.55–3.65 | 1 | .01 |
| 2 | 3.65–3.75 | 1 | .01 |
| 3 | 3.75–3.85 | 6 | .06 |
| 4 | 3.85–3.95 | 6 | .06 |
| 5 | 3.95–4.05 | 10 | .10 |
| 6 | 4.05–4.15 | 10 | .10 |
| 7 | 4.15–4.25 | 13 | .13 |
| 8 | 4.25–4.35 | 11 | .11 |
| 9 | 4.35–4.45 | 13 | .13 |
| 10 | 4.45–4.55 | 7 | .07 |
| 11 | 4.55–4.65 | 6 | .06 |
| 12 | 4.65–4.75 | 7 | .07 |
| 13 | 4.75–4.85 | 5 | .05 |
| 14 | 4.85–4.95 | 4 | .04 |
| *Totals* | | $n = 100$ | 1.00 |

Having determined the class interval, we then construct a frequency table for the data. The first column labels the classes by number and the second column indicates the class intervals. We then examine the 100 measurements of table 2.2, keeping a tally of the number of measurements falling in each interval. The number of measurements falling in a given class interval is called the *class frequency*. These data are recorded in the third column of the frequency table (see table 2.3).

class frequency

The *relative frequency* of a class is defined to be the frequency of the class divided by the total number of measurements in the set (total frequency). Thus if we let $f_i$ denote the frequency for class $i$ and $n$ denote the total number of measurements, the relative frequency for class $i$ is $f_i/n$. The relative frequencies for all the classes are listed in the fourth column of table 2.3.

relative frequency

The data of table 2.2 have been organized into a frequency table, which can now be used to construct a frequency histogram or a relative frequency histogram. To construct a frequency histogram, we draw two axes, a horizontal axis labeled with the class intervals and a vertical axis labeled with the frequencies. We then construct a rectangle over each class interval with a height equal to the number of measurements falling in a given subinterval. The frequency histogram for the data of table 2.3 is shown in figure 2.5.

The relative frequency histogram is constructed in much the same way as a frequency histogram. In the relative frequency histogram, however, the vertical

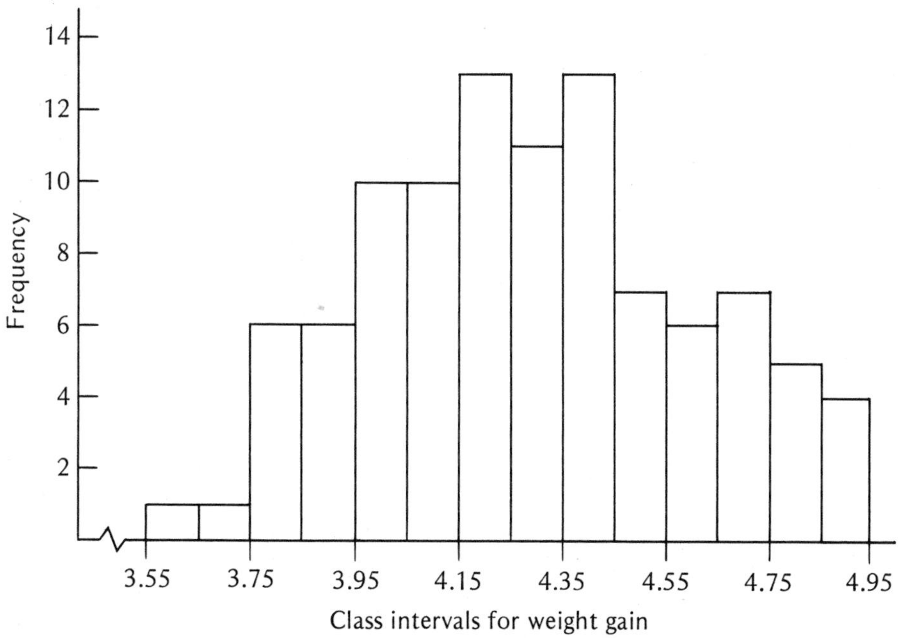

**Figure 2.5** Frequency histogram for the baby chick data of table 2.3

axis is labeled as relative frequency, and a rectangle is constructed over each class interval with a height equal to the class relative frequency, the fourth column of table 2.3. The relative frequency histogram for the data of table 2.3 is shown in figure 2.6. Clearly, the two histograms of figures 2.5 and 2.6 are of

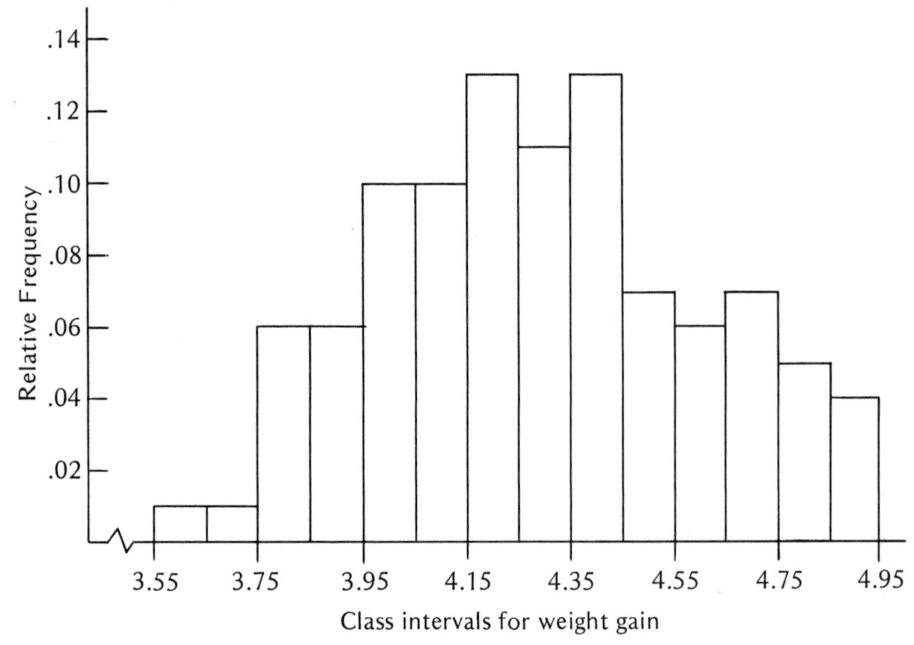


**Figure 2.6** Relative frequency histogram for the baby chick data

the same shape and would be identical if the vertical axes were equivalent. We shall frequently refer to either one as simply a histogram.

There are several comments regarding histograms that should be made here. First, the histogram is the most important graphical technique we will present because of the role it plays in statistical inference, a subject we will discuss in later chapters. Second, if we had an extremely large set of measurements, and if we constructed a histogram using many class intervals, each with a very narrow width, the histogram for the set of measurements would be, for all practical purposes, a smooth curve. Third, the fraction of the total number of measurements in an interval is equal to the fraction of the total area under the histogram over that interval. For example, if we consider the interval 3.75 to 4.35 for the baby chick data of table 2.3, there are exactly 56 measurements in that interval. Thus .56, the fraction of the total number of sample measurements falling in that interval, is equal to the fraction of the total area under the histogram over that interval, as indicated in figure 2.7.

area under histogram

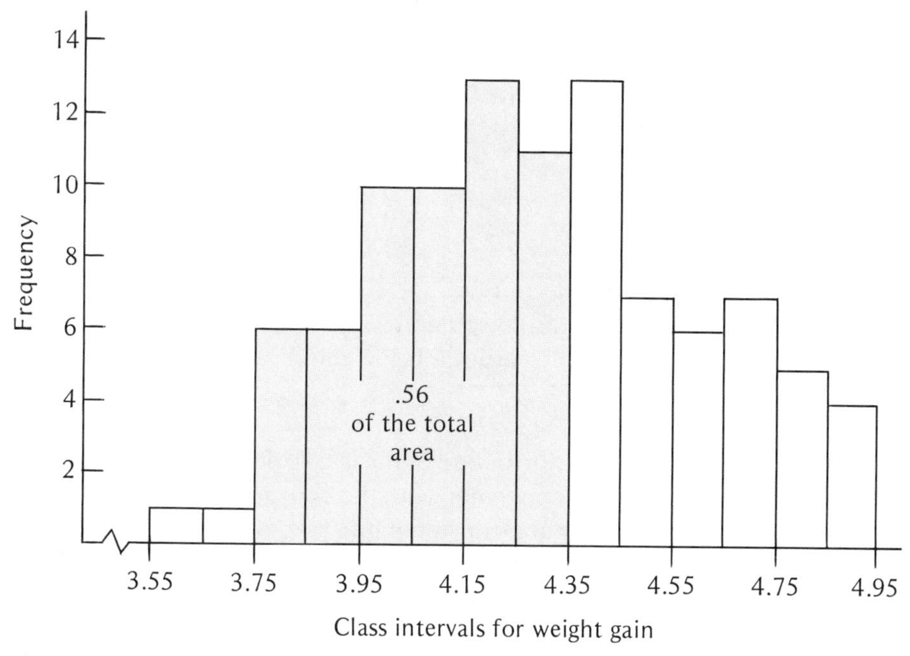


**Figure 2.7** Fraction of measurements in the interval 3.75 to 4.35 for the baby chick data

Finally, if a single measurement is selected at random from the set of sample measurements, the chance, or probability, that it lies in a particular interval is equal to the fraction of the total number of sample measurements falling in that interval. This same fraction will be used to estimate the probability that a measurement randomly selected from the population lies in the interval of interest. For example, from the sample data of table 2.3, the chance or probability of selecting a baby chicken with a weight gain in the interval 3.75 to 4.35 is .56. This value, .56, is an approximation to the probability of selecting a measurement in the interval 3.75 to 4.35 for the population of all weight gains for baby chickens.

probability

# Exercises

(Note: Throughout the text an asterisk (*) before an exercise indicates that the exercise involves a computer solution or is well-suited for computer solution.)

2.1. University officials periodically review the distribution of undergraduate majors within the colleges of the university to help determine a fair

allocation of resources to departments within the colleges. At one such review the following data were obtained:

| *College* | *Number of Majors* |
|---|---|
| agriculture | 1,500 |
| arts and sciences | 11,000 |
| business administration | 7,000 |
| education | 2,000 |
| engineering | 5,000 |

a. Construct a pie chart for these data.
b. Use the same data to construct a bar graph.

$ 2.2. During the inflation-recession period of the late sixties and early seventies, everyone was particularly concerned with the consumer price index (CPI). The percentage increase in the consumer price index over the previous 12 months is given for selected dates in the data below. (The data were recorded in December of the given year.)

| *Year* | 1969 | 1970 | 1971 | 1972 | 1973 | 1974 |
|---|---|---|---|---|---|---|
| *CPI* | 6.1 | 5.5 | 3.4 | 3.4 | 8.8 | 12.5 |

a. Would a pie chart be an appropriate graphical method for describing these data? Explain.
b. Construct a bar graph for the six 12-month periods.

2.3. The regulations of the board of health in a particular state specify that the fluoride level must not exceed 1.5 parts per million (ppm). The 25 measurements given below represent the fluoride levels for a sample of 25 days. Although fluoride levels are measured more than once per day, these data represent the early morning readings for the 25 days sampled.

| | | | | |
|---|---|---|---|---|
| .75 | .86 | .84 | .85 | .97 |
| .94 | .89 | .84 | .83 | .89 |
| .88 | .78 | .77 | .76 | .82 |
| .72 | .92 | 1.05 | .94 | .83 |
| .81 | .85 | .97 | .93 | .79 |

a. Determine the range of the measurements.
b. Dividing the range by 7, the number of subintervals selected, and rounding, we have a class interval width of .05. Using .705 as the lower limit of the first interval, construct a frequency histogram.
c. Compute relative frequencies for each class interval and construct a

relative frequency histogram. Note that the frequency and relative frequency histograms for these data have the same shape.

d. If one of these 25 days were selected at random, what is the chance (probability) that the fluoride reading will be greater than .90 ppm? Guess (predict) what proportion of days in the coming year will have a fluoride reading greater than .90 ppm.

*2.4. Use an available computer program to generate a frequency histogram for the data in table 2.2 (p. 11). A sample program from Minitab (1976) is shown here for purposes of illustration only. Appropriate system cards specific to your computer center would precede this Minitab program.

```
SET THE FOLLOWING DATA INTO COL C5

3.7     4.1     4.2     4.4
4.2     4.5     3.9     4.2
4.7     3.8     3.6     4.0
4.0     4.8     4.1     4.1
 :       :       :       :
4.0     4.3     4.0     4.8
4.3     4.4     3.9     4.0

NOTE C1 TO C4 GIVE WEIGHT GAINS FOR THE 100 CHICKS
HISTOGRAM OF COL C5
STOP
```

*2.5. Some programs allow you to choose the class width and starting point for the first class. Other programs (such as Minitab) choose their own width and starting point automatically. If you do not like the intervals chosen, use some computer packages to rank the data. You can then set up your own frequency histogram. If you have such a program at your disposal, rank the data of table 2.2. A Minitab program is shown below for purposes of illustration.

```
SET THE FOLLOWING DATA INTO COL C5

3.7     4.1     4.2     4.4
4.2     4.5     3.9     4.2
4.7     3.8     3.6     4.0
4.0     4.8     4.1     4.1
 :       :       :       :
4.0     4.3     4.0     4.8
4.3     4.4     3.9     4.0

ORDER C5 PUT ORDERED NUMBERS IN C6
PRINT C5 AND C6
STOP
```

# 2.3 Numerical Descriptive Measures for Describing Data—Central Tendency

Numerical descriptive measures are commonly used to convey a mental image of pictures, objects, and other phenomena. There are two main reasons for this. The first reason is that graphical descriptive measures are inappropriate for statistical inference since it is a difficult task to describe the similarity of a sample frequency histogram and the corresponding population frequency histogram. The second reason for using numerical descriptive measures is one of expediency—we never seem to carry the appropriate graphs or histograms with us, and hence we must resort to our powers of verbal communication to convey the appropriate picture. For example, nearly every adult understands the meaning attached to the expression "36–24–36" when trying to convey a mental picture of a particular female and would have no difficulty in creating a mental picture of her. So it is with statistics. We seek several numbers, called *numerical descriptive measures,* that will create a mental picture of the frequency distribution for a set of measurements.

central tendency
variability

The two most common numerical descriptive measures are measures of *central tendency* and measures of *variability*. That is, we seek to describe the center of the distribution of measurements and also how the measurements vary about the center of the distribution. We will draw a distinction between numerical descriptive measures for a population—called *parameters*—and numerical descriptive measures for a sample—called *statistics*. In problems requiring statistical inference, we will not be able to calculate values for various parameters, but we will be able to compute corresponding statistics from the sample and use these quantities to estimate the corresponding population parameters.

parameter
statistic

In this section we will consider various measures of central tendency. These will be followed in section 2.4 by a discussion of measures of variability.

mode

The first measure of central tendency we consider is the *mode*.

---

**Definition 2.1**

The *mode* of a set of measurements is defined to be the measurement that occurs most often (with the highest frequency).

---

We illustrate the use and determination of the mode in an example.

**EXAMPLE 2.1**

A research team consisting of an anthropologist and a sociologist analyzed the family structure of a large Hutterite community (a religious commune of the northwestern United States). One task involved counting the number of children in each family. The data given below represent a portion of the total number of Hutterite families.

Determine the modal number of children per family.

*Number of Children in 25 Hutterite Families*

| | | | | |
|---|---|---|---|---|
| 7 | 10 | 8 | 11 | 9 |
| 9 | 9 | 8 | 9 | 8 |
| 9 | 9 | 9 | 8 | 9 |
| 8 | 8 | 9 | 10 | 11 |
| 10 | 7 | 10 | 9 | 7 |

**SOLUTION**

First, we arrange the measurements in order from the smallest to the largest:

7, 7, 7, 8, 8, 8, 8, 8, 8, 9, 9, 9, 9, 9, 9, 9, 9, 9, 9, 10, 10, 10, 10, 11, 11

It is clear from these data that 9 is the measurement that occurs most often, and hence 9 is the mode.

Identification of the mode for example 2.1 was quite easy because we were able to arrange the measurements in ascending order and count the number of times each measurement occurred. Similarly, when dealing with *grouped data*—data presented in the form of a frequency table—we can easily determine the mode. Since we would not know the actual measurements, but only how many measurements fall into each interval, we define the mode in this case to be the midpoint of the class interval with the highest frequency.

grouped data

grouped data mode

The mode is also commonly used as a measure of popularity which reflects central tendency or opinion. For example, we might talk about the most preferred stock, a most preferred model of washing machine, or the most popular candidate. In each case we would be referring to the mode of the distribution.

It should be noted that some distributions have more than one measurement that occurs with the highest frequency. Thus we might encounter bimodal, trimodal, and so on, distributions.

The second measure of central tendency we consider is the *median*.

median

**Definition 2.2**

The *median* of a set of measurements is defined to be the middle value when the measurements are arranged in order of magnitude.

The median is most often used to measure the midpoint of a large set of measurements. For example, we may read about the median wage increase won by union members, the median age of persons receiving social security, and the median weight of cattle prior to slaughter during a given month. Each of these situations involves a large set of measurements, and the median would reflect the central value of the data.

However, we may use the definition of median for small sets of measurements by using the following convention. The median for an even number of measurements will be the average of the two midpoint values when the measurements are arranged in order of magnitude. When there are an odd number of measurements, the median is still the midpoint value.

### EXAMPLE 2.2

Each of 10 children in the second grade was given a reading aptitude test. The scores were as shown below.

Determine the median test score.

95, 86, 78, 90, 62, 73, 89, 92, 84, 76

### SOLUTION

We must first arrange the scores in order of magnitude.

62, 73, 76, 78, 84, 86, 89, 90, 92, 95

Since there is an even number of measurements, the median is the average of the two midpoint scores.

$$\text{median} = \frac{84 + 86}{2} = 85$$

### EXAMPLE 2.3

An experiment was conducted to measure the effectiveness of a new procedure for pruning grapes. Each of 13 men was assigned the task of pruning an acre of grapes. The productivity, measured in man-hours/acre, is recorded for each man.

4.4, 4.9, 4.2, 4.4, 4.8, 4.9, 4.8, 4.5, 4.3, 4.8, 4.7, 4.4, 4.2

Determine the mode and median productivity for the group.

### SOLUTION

First arrange the measurements in order of magnitude:

4.2, 4.2, 4.3, 4.4, 4.4, 4.4, 4.5, 4.7, 4.8, 4.8, 4.8, 4.9, 4.9

For these data we have two measurements appearing three times each, hence the data are bimodal with modes of 4.4 and 4.8. The median for the odd number of measurements is the midpoint score, 4.5.

grouped data median

The median for *grouped data* is slightly more difficult to compute. Since the exact value of the measurements is unknown, we know only that the median occurs in a particular class interval, but we do not know where to locate the median within the interval. Let

$L$ = lower class limit of the interval that contains the median

$n$ = total frequency

$cf_b$ = the sum of frequencies (cumulative frequency) for all classes before the median class

$f_m$ = frequency of the class interval containing the median

$w$ = interval width

Then for grouped data,

$$\text{median} = L + \frac{w}{f_m}(.5n - cf_b)$$

We illustrate how to find the median for grouped data in an example.

**EXAMPLE 2.4**

Table 2.4 is the frequency table for the baby chick data of table 2.3. Compute the median weight gain for these data.

**Table 2.4** Frequency table for the baby chick data, example 2.4

| *Class Interval* | $f_i$ | *Cumulative* $f_i$ | $f_i/n$ | *Cumulative* $f_i/n$ |
|---|---|---|---|---|
| 3.55–3.65 | 1 | 1 | .01 | .01 |
| 3.65–3.75 | 1 | 2 | .01 | .02 |
| 3.75–3.85 | 6 | 8 | .06 | .08 |
| 3.85–3.95 | 6 | 14 | .06 | .14 |
| 3.95–4.05 | 10 | 24 | .10 | .24 |
| 4.05–4.15 | 10 | 34 | .10 | .34 |
| 4.15–4.25 | 13 | 47 | .13 | .47 |
| 4.25–4.35 | 11 | 58 | .11 | .58 |
| 4.35–4.45 | 13 | 71 | .13 | .71 |
| 4.45–4.55 | 7 | 78 | .07 | .78 |
| 4.55–4.65 | 6 | 84 | .06 | .84 |
| 4.65–4.75 | 7 | 91 | .07 | .91 |
| 4.75–4.85 | 5 | 96 | .05 | .96 |
| 4.85–4.95 | 4 | 100 | .04 | 1.00 |
| *Totals* | $n$ = 100 | | 1.00 | |

**SOLUTION**

Let the cumulative relative frequency for class $j$ equal the sum of the relative frequencies for class 1 through class $j$. To determine the interval that contains the median, we must find the first interval for which the cumulative relative frequency exceeds .50. This interval will be the interval containing the median. For these data, the interval from 4.25 to 4.35 is the first interval for which the cumulative relative frequency exceeds .50, as shown in table 2.4, column 5. So this interval contains the median. Then

$$L = 4.25 \qquad f_m = 11$$

$$n = 100 \qquad w = .1$$

$$cf_b = 47$$

and

$$\text{median} = L + \frac{w}{f_m}(.5n - cf_b) = 4.25 + \frac{.1}{11}(50 - 47) = 4.28$$

mean

The third, and last, measure of central tendency we will discuss in this text is the arithmetic mean, known simply as the *mean*.

---

**Definition 2.3**

The *arithmetic mean,* or *mean,* of a set of measurements is defined to be the sum of the measurements divided by the total number of measurements.

---

Quite often when people talk about an "average," they are referring to the mean.

$\mu$

$\bar{y}$

Because of the important role that the mean will play in statistical inference in later chapters, we give special symbols to the population mean and the sample mean. The *population mean* is denoted by the Greek letter $\mu$ (read "mu"), and the *sample mean* is denoted by the symbol $\bar{y}$ (read "y-bar"). *We will use the sample mean $\bar{y}$ to make inferences about the corresponding population parameter $\mu$.*

We can now give a computational formula for the sample mean. If we let $y_1, y_2, \ldots, y_n$ denote the measurements observed in a sample of size $n$, then the sample mean $\bar{y}$ can be written as

$$\bar{y} = \frac{\sum_{i=1}^{n} y_i}{n}$$

where the symbol appearing in the numerator,

$$\sum_{i=1}^{n}$$

is the notation used to designate a sum of $n$ measurements, $i = 1, 2, \ldots, n$. Thus we can write a sum of measurements as

$$\sum_{i=1}^{n} y_i = y_1 + y_2 + \cdots + y_n$$

**EXAMPLE 2.5**

A representative sample of $n = 15$ overdue accounts in a large department store yields the following amounts due:

| | | |
|---|---|---|
| \$55.20 | \$ 4.88 | \$271.95 |
| 18.06 | 180.29 | 365.29 |
| 28.16 | 399.11 | 807.80 |
| 44.14 | 97.47 | 9.98 |
| 61.61 | 56.89 | 82.73 |

a. Determine the mean amount due for the 15 accounts sampled.
b. If there are a total of 150 overdue accounts, use the sample mean to predict the total amount overdue for all 150 accounts.

**SOLUTION**

a. The sample mean is computed as follows:

$$\bar{y} = \frac{\sum_{i=1}^{15} y_i}{15} = \frac{55.20 + 18.06 + \cdots + 82.73}{15} = \frac{2483.56}{15} = \$165.57$$

b. From part a we found that the 15 accounts sampled averaged \$165.57 overdue. Using this information, we would predict, or estimate, the total amount overdue for the 150 accounts to be 150(165.57) = \$24,835.50.

grouped data mean

The sample mean formula for grouped data is only slightly more complicated than the formula just presented for ungrouped data. Since we do not know the individual sample measurements, but only the interval to which a measurement is assigned, this formula will be an approximation to the actual sample mean. Hence when the sample measurements are known, the formula for ungrouped data should be used. We will use the same symbol $\bar{y}$ to designate the sample mean for grouped data. If there are $k$ class intervals and

$y_i$ = midpoint of the $i$th class interval

$f_i$ = frequency associated with the $i$th class interval

and $n$ is the total number of measurements, then

$$\bar{y} = \frac{\sum_{i=1}^{k} f_i y_i}{n}$$

**EXAMPLE 2.6**

The data of example 2.4 are reproduced here in table 2.5, along with two additional columns, $y_i$ and $f_i y_i$, that will be helpful in computing the mean. Compute the sample mean for this set of grouped data.

**Table 2.5** Baby chick data, example 2.6

| *Class Interval* | $f_i$ | $y_i$ | $f_i y_i$ |
|---|---|---|---|
| 3.55–3.65 | 1 | 3.6 | 3.6 |
| 3.65–3.75 | 1 | 3.7 | 3.7 |
| 3.75–3.85 | 6 | 3.8 | 22.8 |
| 3.85–3.95 | 6 | 3.9 | 23.4 |
| 3.95–4.05 | 10 | 4.0 | 40.0 |
| 4.05–4.15 | 10 | 4.1 | 41.0 |
| 4.15–4.25 | 13 | 4.2 | 54.6 |
| 4.25–4.35 | 11 | 4.3 | 47.3 |
| 4.35–4.45 | 13 | 4.4 | 57.2 |
| 4.45–4.55 | 7 | 4.5 | 31.5 |
| 4.55–4.65 | 6 | 4.6 | 27.6 |
| 4.65–4.75 | 7 | 4.7 | 32.9 |
| 4.75–4.85 | 5 | 4.8 | 24.0 |
| 4.85–4.95 | 4 | 4.9 | 19.6 |
| *Totals* | 100 | | 429.2 |

**SOLUTION**

Adding the entries in the $f_i y_i$ column and substituting into the formula, we find the sample mean to be

$$\bar{y} = \frac{\sum_{i=1}^{14} f_i y_i}{100} = \frac{429.2}{100} = 4.29$$

We have discussed three measures of central tendency and illustrated the computation of each. However, we will deal almost exclusively with the sample mean as a measure of central tendency. The reason for this is quite simple. The sample mean is an important statistic for making inferences, as we will see in later chapters. Thus, although the sample mean serves as a descriptive measure for the sample data, we will see later that it usually also provides an accurate estimate of the mean of the population from which the sample was drawn.

applications

As noted earlier, the mode is used when we wish to know the most frequently observed value of $y$. It tends to be a measure of popularity and provides a measure of central tendency for qualitative data where the mean and median are not applicable.

In some situations the median provides more information about the center of a distribution than the mean. For example, suppose we are interested in describing the distribution of ages for bridegrooms in a particular city over a given year. It would be easy to sample the marriage licenses on file with a justice of the peace. We would undoubtedly find many grooms who were in their late teens or early twenties, but some grooms would be much older. The median, which reflects the center of the sample, would therefore be a better indicator than the mean of the midpoint of the distribution of bridegroom ages because the mean age would be inflated by the ages of the older bridegrooms.

For a bell-shaped distribution of measurements, the values of the mean, the median, and the mode are identical [see figure 2.8(a)]. If the distribution of measurements is skewed to the left (negative skew), the mean $\mu$ is the smallest of the three measures while the mode $M_o$ is the largest [see figure 2.8(b)].

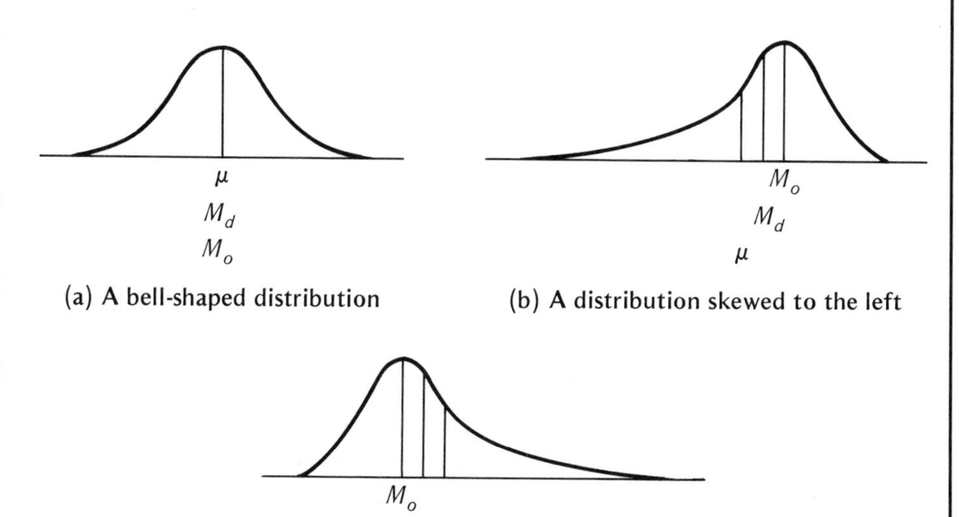


**Figure 2.8** Relationships among the mean $\mu$, the median $M_d$, and the mode $M_o$

When the distribution of measurements is skewed to the right (positive skew), $\mu$ is the largest of the three values and $M_0$ is the smallest [see figure 2.8(c)]. In each of these skewed cases, the median is the more central value, and it is for this reason that the median is a widely used measure for locating the center of a distribution.

Measures of central tendency do not provide a complete mental picture of the frequency distribution for a set of measurements. In addition to determining the center of the distribution, we must have some measure of the spread of the data. In the next section we discuss measures of variability, or dispersion.

# Exercises

2.6. Federal officials keep extensive records of disease rates to monitor trends in occurrences of serious diseases. One such record concerns the incidence rate of encephalitis per 1,000,000 US population, that is, the number of occurrences for every 1,000,000 people in the United States, during a given year. The data are given below.

| *Year* | *Encephalitis Rate* |
|---|---|
| 1945 | 5.6 |
| 1950 | 7.5 |
| 1955 | 13.1 |
| 1960 | 13.0 |
| 1965 | 3.9 |
| 1970 | 9.5 |

a. Find the median encephalitis rate.
b. Compute the mean encephalitis rate. Should the mean and the median be equal? Explain.

2.7. Nitrogen is a limiting factor in the yield of many different plants. In particular, the yield of apple trees is directly related to the nitrogen content of apple tree leaves and must be carefully monitored to protect the trees in an orchard. Research has shown that the nitrogen content should be approximately 2.5% for best yield results. (It should be noted that some researchers report their results in parts per million. Hence 1% would be equivalent to 10,000 ppm.)

To determine the nitrogen content of trees in an orchard, the growing tips of 150 leaves are clipped from trees throughout the orchard. These leaves are ground to form one composite sample, which the researcher assays for percentage of nitrogen. Composite samples obtained from a random sample of 36 orchards throughout the state gave the following nitrogen contents:

| | | | | | |
|---|---|---|---|---|---|
| 2.0968 | 2.8220 | 2.1739 | 1.9928 | 2.2194 | 3.0926 |
| 2.4685 | 2.5198 | 2.7983 | 2.0961 | 2.9216 | 2.1997 |
| 1.7486 | 2.7741 | 2.8241 | 2.6691 | 3.0521 | 2.9263 |
| 2.9367 | 1.9762 | 2.3821 | 2.6456 | 2.7678 | 1.8488 |
| 1.6850 | 2.7043 | 2.6814 | 2.0596 | 2.3597 | 2.2783 |
| 2.7507 | 2.4259 | 2.3936 | 2.5464 | 1.8049 | 1.9629 |

a. Round each of these measurements to the nearest hundredth. (Use the convention that 5 is rounded up.)
b. Determine the sample mode for the rounded data.
c. Determine the sample median for the rounded data.
d. Determine the sample mean for the rounded data.

2.8. Refer to the data of example 2.5. Since the sample mean is greater than 10 of the 15 observations, suggest a more appropriate measure of central tendency. Compute its value. How does the distribution of amounts for overdue accounts appear to be skewed? $

*2.9. Use a computer program to compute the maximum, minimum, sample mean, and sample median for the data (unrounded) shown in exercise 2.5 (table 2.2, p. 11). A Minitab program is shown for purposes of illustration.

```
SET THE FOLLOWING DATA INTO COL C5

3.7    4.1    4.2    4.4
4.2    4.5    3.9    4.2
4.7    3.8    3.6    4.0
4.0    4.8    4.1    4.1
 :      :      :      :
4.0    4.3    4.0    4.8
4.3    4.4    3.9    4.0


NOTE C1 TO C4 GIVE WEIGHT GAINS FOR THE 100 CHICKS
MAXIMUM OF C5, PUT IN C6
MINIMUM OF C5, PUT IN C7
AVERAGE OF C5, PUT IN C8
MEDIAN OF C5, PUT IN C9
PRINT COLS C5, C6, C7, C8, C9
STOP
```

| VARIABLE NO. NAME | MEAN | STANDARD DEVIATION | ST. ERR. OF MEAN | COEFF. OF VARIATION | SMALLEST VALUE | SMALLEST Z-SCORE | LARGEST VALUE | LARGEST Z-SCORE | RANGE | TOTAL FREQUENCY |
|---|---|---|---|---|---|---|---|---|---|---|
| 1 RATE | 425.395 | 183.536 | 19.3464 | 0.43145 | 91.000 | −1.82196 | 949.000 | 2.85288 | 858.000 | 90 |

*2.10. The computer printout (BMDP 1975) above provides many descriptive statistics for the violent crime rates shown below.

| *Violent Crime Rates* | | | | | | | | | |
|---|---|---|---|---|---|---|---|---|---|
| 707 | 467 | 383 | 672 | 607 | 175 | 325 | 538 | 313 | 433 |
| 372 | 466 | 535 | 396 | 949 | 143 | 311 | 283 | 468 | 217 |
| 703 | 643 | 367 | 418 | 513 | 281 | 358 | 524 | 518 | 396 |
| 411 | 675 | 461 | 146 | 259 | 226 | 185 | 226 | 676 | 590 |
| 609 | 517 | 354 | 165 | 91 | 585 | 293 | 679 | 385 | 448 |
| 703 | 348 | 650 | 431 | 609 | 654 | 331 | 624 | 290 | 309 |
| 372 | 360 | 504 | 290 | 264 | 518 | 507 | 890 | 400 | 572 |
| 518 | 285 | 253 | 793 | 96 | 120 | 217 | 419 | 688 | 379 |
| 594 | 189 | 331 | 341 | 268 | 545 | 355 | 352 | 305 | 510 |

a. Identify the sample size and sample mean.
b. Identify the largest and smallest measurements.

# 2.4 Numerical Descriptive Measures for Describing Data—Variability

variability

The need for some measure of variability is illustrated in the relative frequency histograms of figure 2.9. All the histograms have the same mean but each has a different spread, or *variability,* about the mean. For purposes of illustration, we have shown the histograms as smooth curves.

range

The simplest measure of data variation is the *range.* Recall that we alluded to the range in section 2.2. We now present its definition.

**Definition 2.4**

The *range* of a set of measurements is defined to be the difference between the largest and the smallest measurements of the set.

**EXAMPLE 2.7**

Determine the range of the 15 overdue accounts of example 2.5 (p. 23).

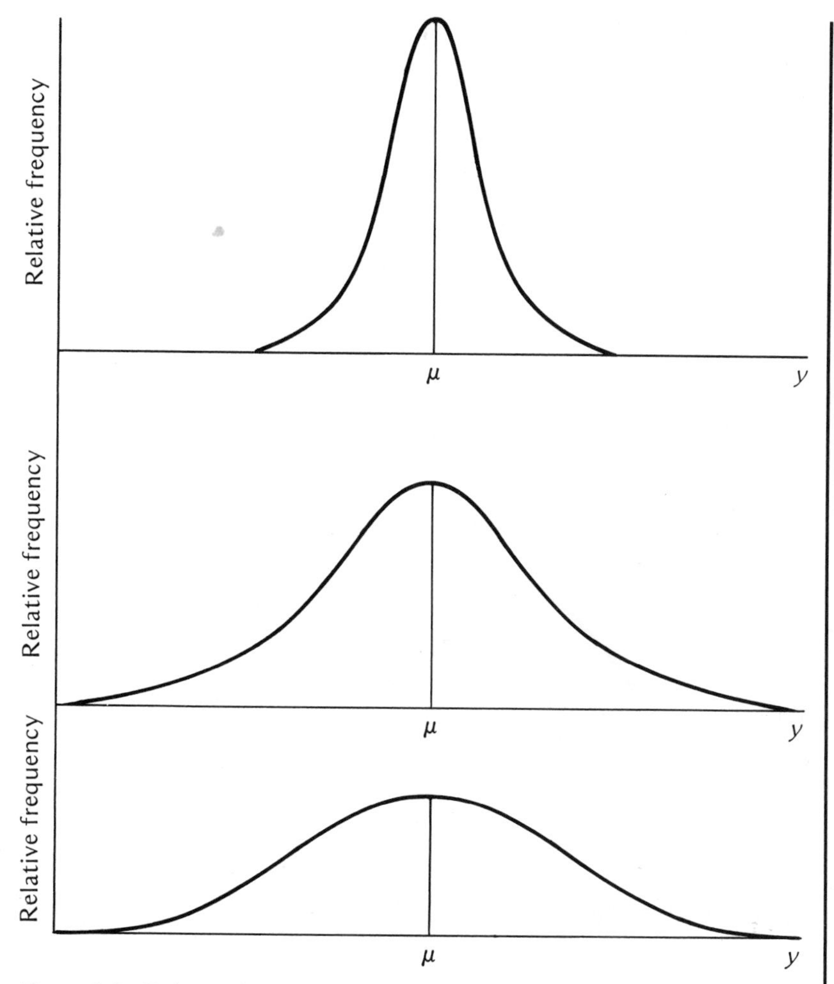


**Figure 2.9** Relative frequency histograms with different variabilities but the same mean

**SOLUTION**

The smallest measurement is \$4.88 and the largest is \$807.80. Hence the range is

807.80 − 4.88 = \$802.92

For grouped data, since we do not know the individual measurements, the range is taken to be the difference between the upper limit of the last interval and the lower limit of the first interval.

grouped data range

Although the range is easy to compute, it gives very little information about the variability or dispersion of the data about the mean. In figure 2.9 the middle and the bottom distributions have the same mean and the same range yet they differ substantially in their variability about the mean. What we seek is a measure of variability that is more sensitive to the piling up of data about the mean.

percentile

A second measure of variability involves the use of *percentiles*.

**Definition 2.5**

The *pth percentile* of a set of $n$ measurements arranged in order of magnitude is that value that has $p$% of the measurements below it and $(100 - p)$% above it.

For example, the 60th percentile of a set of measurements is illustrated in figure 2.10.

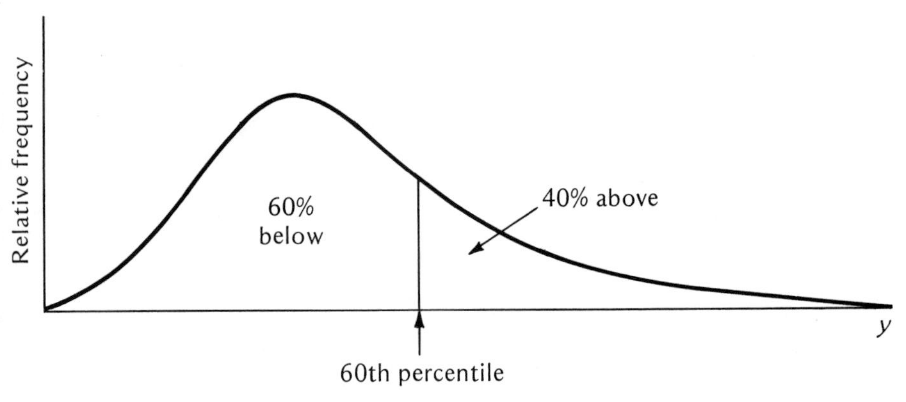


**Figure 2.10** The 60th percentile of a set of measurements

Percentiles are frequently used to describe the results of achievement test scores and the ranking of a person in comparison to the rest of the people taking an examination. Specific percentiles of interest are the 25th, 50th, and 75th percentiles, often called the lower quartile, the middle quartile (median), and the upper quartile, respectively (see figure 2.11).

grouped data percentile

Having learned how to compute the median—50th percentile—for grouped data, other percentiles follow immediately. Let

$P$ = percentile of interest

$L$ = lower limit of the class interval that includes percentile of interest

$n$ = total frequency

$cf_b$ = cumulative frequency for all class intervals before percentile class

$f_p$ = frequency of the class interval that includes percentile of interest

$w$ = interval width

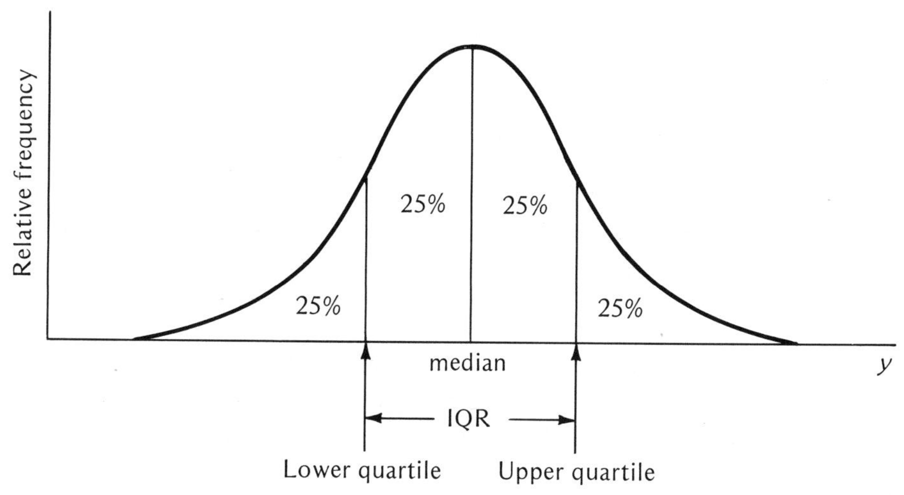


**Figure 2.11** Quartiles of a distribution

Then, for example, the 65th percentile for a set of grouped data would be computed using the formula

$$P = L + \frac{w}{f_p}(.65n - cf_b)$$

To determine $L$, $f_p$, and $cf_b$, begin with the lowest interval and find the first interval for which the cumulative relative frequency exceeds .65. This interval would contain the 65th percentile.

**EXAMPLE 2.8**

Refer to the data of table 2.4 (p. 21). Compute the 90th percentile.

**SOLUTION**

Since the twelfth interval is the first interval for which the cumulative relative frequency exceeds .90, then we have

$L = 4.65$

$n = 100$

$cf_b = 84$

$f_{90} = 7$

$w = .1$

Thus the 90th percentile is

$$P_{90} = 4.65 + \frac{.1}{7}[.9(100) - 84] = 4.74$$

This figure means that 90% of the measurements lie below this value and 10% lie above it.

interquartile range

The second measure of variability, the interquartile range, is now defined.

---

**Definition 2.6**

The *interquartile range* (IQR) of a set of measurements is defined to be the difference between the upper and lower quartiles (see figure 2.11).

---

The interquartile range, although more sensitive to data pileup about the midpoint than the range, is still not sufficient for our purposes. In particular, the IQR can be used for comparing the variability of two sets of measurements, but not much useful information is gained from the IQR in interpreting the variability of a single set of measurements.

We seek now a sensitive measure of variability, not only for comparing the variabilities of two sets of measurements, but also for interpreting the variability of a single set of measurements. To do this we work with the *deviation* $y - \bar{y}$ of a measurement $y$ from its mean $\bar{y}$.

To illustrate, suppose we have the set of 5 sample measurements $y_1 = 68$, $y_2 = 67$, $y_3 = 66$, $y_4 = 63$, and $y_5 = 61$, which represent the percentages of registered voters in 5 cities who exercised their right to vote at least once during the past year. These measurements are shown in the dot diagram of figure 2.12. Each measurement is located by a dot above the horizontal axis of

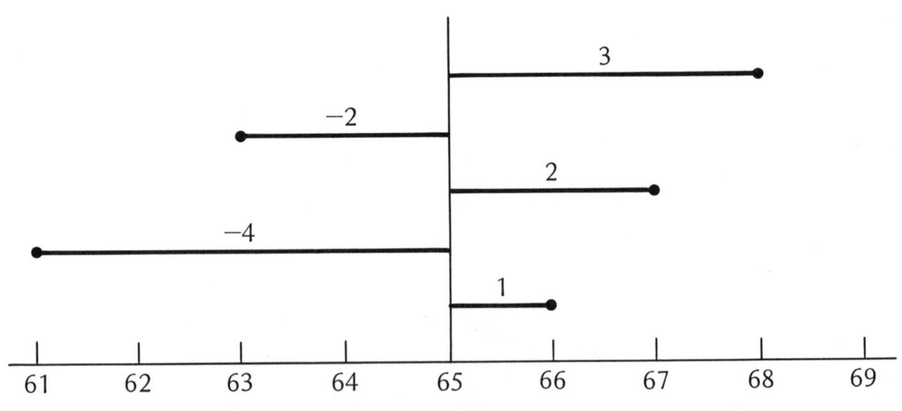


**Figure 2.12** Dot diagram of the percentages of registered voters in 5 cities

the diagram. We use the sample mean

$$\bar{y} = \frac{\sum_{i=1}^{5} y_i}{n} = \frac{325}{5} = 65$$

to locate the center of the set. We construct horizontal lines in figure 2.12 to represent the deviations of the sample measurements from their mean. The deviations of the measurements are computed by using the formula $y - \bar{y}$. The five measurements and their deviations are shown in figure 2.12.

A data set with very little variability will have most of the measurements located near the center of the distribution. Deviations from the mean of a more variable set of measurements would be relatively large.

Many different measures of variability can be constructed by using the deviations $y - \bar{y}$. A first thought would be to use the mean deviation, but this will always equal zero, as it does for our example. A second possibility would be to ignore the minus signs and compute the average of the absolute values. However, a more easily interpreted function of the deviations involves the sum of the squared deviations of the measurements from their mean. This measure is called the *variance*.

variance

---

**Definition 2.7**

The *variance* of a set of $n$ measurements $y_1, y_2, \ldots, y_n$ with mean $\bar{y}$ is the sum of the squared deviations divided by $n - 1$.

$$\frac{\sum_{i=1}^{n} (y_i - \bar{y})^2}{n - 1}$$

---

As with the sample and population means, we have special symbols to denote the sample and population variances. The symbol $s^2$ represents the sample variance, and the corresponding population variance is denoted by the symbol $\sigma^2$.

$s^2$

$\sigma^2$

Many statisticians define the sample variance to be the average of the squared deviations, $\Sigma (y - \bar{y})^2/n$. However, this quantity tends to underestimate the corresponding population variance $\sigma^2$. By using the divisor $n - 1$, we get a better estimate of $\sigma^2$, and so this is the divisor we will use in our definition of the sample variance.

The sample variance can be determined by using the following formula:

$$s^2 = \frac{\sum_{i=1}^{n} (y_i - \bar{y})^2}{n - 1}$$

standard deviation

Another useful measure of variability, the *standard deviation,* involves the square root of the variance.

**Definition 2.8**

The *standard deviation* of a set of measurements is defined to be the positive square root of the variance.

$s$
$\sigma$

We then have $s$ denoting the sample standard deviation and $\sigma$ denoting the corresponding population standard deviation.

**EXAMPLE 2.9**

The time between an electric light stimulus and a bar press to avoid a shock was noted for each of 5 conditioned rats. Use the data below to compute the sample variance and standard deviation.

shock avoidance times (in seconds): 5, 4, 3, 1, 3

**SOLUTION**

The deviation and the squared deviations are shown below. The sample mean $\bar{y}$ is 3.2.

| | $y_i$ | $y_i - \bar{y}$ | $(y_i - \bar{y})^2$ |
|---|---|---|---|
| | 5 | 1.8 | 3.24 |
| | 4 | .8 | 0.64 |
| | 3 | − .2 | .04 |
| | 1 | −2.2 | 4.84 |
| | 3 | − .2 | .04 |
| *Totals* | 16 | 0 | 8.80 |

Using the total of the squared deviations column, we find the sample variance to be

$$s^2 = \frac{\sum_{i=1}^{5}(y_i - \bar{y})^2}{4} = \frac{8.80}{4} = 2.2$$

and the sample standard deviation to be

$s = \sqrt{2.2} = 1.48$

(Note: You can refer to table 6 in the appendix or use any of a number of different electronic calculators for determining square roots.)

Computation of the quantities $s^2$ and $s$ is sometimes simplified using the following algebraic identity:

$$\sum_{i=1}^{n}(y_i - \bar{y})^2 = \sum_{i=1}^{n} y_i^2 - \frac{\left(\sum_{i=1}^{n} y_i\right)^2}{n}$$

Hence we have the shortcut formula for $s^2$ (and $s$) given in the box.

**Shortcut Formula for $s^2$ and $s$**

$$s^2 = \frac{1}{n-1}\left[\sum_{i=1}^{n} y_i^2 - \frac{\left(\sum_{i=1}^{n} y_i\right)^2}{n}\right] \quad \text{and} \quad s = \sqrt{s^2}$$

**EXAMPLE 2.10**

Use the data of example 2.9 to compute the sample variance using the shortcut formula.

**SOLUTION**

It is convenient to construct the following table when we do not have a calculator to perform the calculations:

| | $y_i$ | $y_i^2$ |
|---|---|---|
| | 5 | 25 |
| | 4 | 16 |
| | 3 | 9 |
| | 1 | 1 |
| | 3 | 9 |
| *Totals* | 16 | 60 |

Using the totals from the table, we have

$$s^2 = \frac{1}{4}\left[60 - \frac{(16)^2}{5}\right] = \frac{1}{4}[60 - 51.2] = 2.2$$

which is exactly the result we obtained in example 2.9.

Either of the formulas for calculating $s^2$ (and hence $s$) is appropriate and both will yield the same result. However, the shortcut formula is particularly appropriate for use with an electronic calculator, where we are able to simultaneously accumulate sums and sums of squares of measurements. Some electronic desk calculators even compute $s^2$ (or $s$) directly and hence, for those fortunate enough to have access to such equipment, the question of which formula to use is purely academic.

grouped data formula

We can make a simple modification of our shortcut formula for the sample variance so that it applies to grouped data. Recall that in computing the sample mean for grouped data, we let $y_i$ and $f_i$ denote the midpoint and frequency, respectively, for the $i$th class interval. With this notation the sample variance for grouped data is

$$s^2 = \frac{1}{n-1}\left[\sum_{i=1}^{k} f_i y_i^2 - \frac{\left(\sum_{i=1}^{k} f_i y_i\right)^2}{n}\right]$$

The sample standard deviation is $\sqrt{s^2}$.

**EXAMPLE 2.11**

Refer to the baby chick data from table 2.5 of example 2.6 (p. 24). Calculate the sample variance and standard deviation for these data.

**SOLUTION**

In addition to the calculations in table 2.5, we also need the calculations for $f_i y_i^2$. These calculations, formed by multiplying corresponding elements in the $y_i$ and $f_i y_i$ columns, are shown in the listing below.

$f_i y_i^2$: 12.96, 13.69, 86.64, 91.26, 160.00, 168.10, 229.32, 203.39,

251.68, 141.75, 126.96, 154.63, 115.20, 96.04

The sum of the $f_i y_i^2$ calculations is 1851.62. Using this total and the total of $f_i y_i$ in table 2.5, we can determine $s^2$ and $s$.

$$s^2 = \frac{1}{n-1}\left[\sum_{i=1}^{k} f_i y_i^2 - \frac{\left(\sum_{i=1}^{k} f_i y_i\right)^2}{n}\right]$$

$$= \frac{1}{99}\left[1851.62 - \frac{(429.2)^2}{100}\right] = \frac{9.49}{99} = .10$$

$$s = \sqrt{.10} = .32$$

We have now discussed several measures of variability, each of which can be used to compare the variabilities of two or more sets of measurements. The standard deviation is particularly appealing for two reasons: (1) we can compare the variabilities of *two or more* sets of data using the standard deviation, and (2) we can also use the results of the rule that follows to interpret the standard deviation of a *single* set of measurements. The wide applicability of this rule leads us to call it the *Empirical Rule*.

Empirical Rule

**Empirical Rule**

Given a set of $n$ measurements possessing a mound-shaped histogram, then the interval

$\bar{y} \pm s$ contains approximately 68% of the measurements;

$\bar{y} \pm 2s$ contains approximately 95% of the measurements;

$\bar{y} \pm 3s$ contains approximately all the measurements.

**EXAMPLE 2.12**

A sample of 20 days throughout the previous year indicated that the average price per pound wholesale for steers at a particular stockyard was \$.61, with a standard deviation of \$.07. If the histogram for the measurements is mound-shaped, describe the variability of the data using the Empirical Rule.

**SOLUTION**

Applying the Empirical Rule, the interval

$.61 \pm .07$ or \$.54 to \$.68

contains approximately 68% of the measurements. The interval

$.61 \pm .14$ or \$.47 to \$.75

contains approximately 95% of the measurements. The interval

$.61 \pm .21$ or \$.40 to \$.82

contains approximately all the measurements.

The results of the Empirical Rule enable us to obtain a check of the calculation of the sample standard deviation $s$. The Empirical Rule states that approximately 95% of the measurements lie in the interval $\bar{y} \pm 2s$. The length of this interval is therefore $4s$. Since the range of the measurements is approximately $4s$, we obtain an approximate value for $s$ by dividing the range by 4.

approximating $s$

$$\text{approximate value of } s = \frac{\text{range}}{4}$$

Some might wonder why we did not equate the range to $6s$ since the interval $\bar{y} \pm 3s$ should contain almost all the measurements. This procedure would yield an approximate value for $s$ which is smaller than the one obtained by the procedure above. If we are going to make an error (as we're bound to do with any approximation), it is better to overestimate the sample standard deviation so that we are not lead to believe there is less variability than may be the case.

**EXAMPLE 2.13**

The following data represent the percentages of family income allocated to groceries for a sample of 30 shoppers:

| | | | | | |
|---|---|---|---|---|---|
| 26 | 28 | 30 | 37 | 33 | 30 |
| 29 | 39 | 49 | 31 | 38 | 36 |
| 33 | 24 | 34 | 40 | 29 | 41 |
| 40 | 29 | 35 | 44 | 32 | 45 |
| 35 | 26 | 42 | 36 | 37 | 35 |

For these data $\Sigma_{i=1}^{30} y_i = 1043$ and $\Sigma_{i=1}^{30} y_i^2 = 37{,}331$.

Compute the mean, variance, and standard deviation of the percentage of income spent on food. Check your calculation of $s$.

**SOLUTION**

The sample mean is

$$\bar{y} = \frac{\sum_{i=1}^{30} y_i}{30} = \frac{1043}{30} = 34.77$$

The corresponding sample variance and standard deviation are

$$s^2 = \frac{1}{n-1}\left[\sum_{i=1}^{30} y_i^2 - \frac{\left(\sum_{i=1}^{30} y_i\right)^2}{n}\right]$$

$$= \frac{1}{29}[37{,}331 - 36{,}261.63] = \frac{1069.37}{29} = 36.87$$

$$s = \sqrt{36.87} = 6.07$$

We can check our calculation of $s$ by using the range approximation. The largest measurement is 49 and the smallest is 24. Hence an approximate value of $s$ is

$$s \approx \frac{\text{range}}{4} = \frac{49 - 24}{4} = 6.25$$

Note how close the approximation is to our computed value.

While there will not always be such close agreement as found in example 2.13, the range approximation provides a useful and quick check on the calculation of $s$.

# Coding to Simplify Calculations | 2.5

Data are frequently coded to simplify the calculations of $\bar{y}$ and $s^2$ (and $s$). For example, it is much easier to calculate the mean of the 5 measurements $-.1, .2, .1, 0$, and $.2$ than of the 5 measurements 99.9, 100.2, 100.1, 100.0, and 100.2. The first set of measurements was obtained by subtracting 100 from each measurement in the second set. Similarly, we might wish to simplify a set of measurements by multiplying by a constant. It is easier to work with the set 3, 1, 4, 6, 4, 2 than with the set .003, .001, .004, .006, .004, .002. The first set was obtained by multiplying each element of the second set by 1000.

coding

Data are generally *coded* by performing one or both of the following operations: subtracting (or adding) a constant $m$ from each measurement, or multiplying (or dividing) each measurement by a constant $k$.

How are the mean and standard deviation of the coded measurements related to the mean and standard deviation of the sample measurements $y_1, y_2, \ldots, y_n$? The theorem given in the box answers this question.

**Theorem 2.1**

Let $y_1, y_2, \ldots, y_n$ be $n$ measurements with sample mean $\bar{y}$ and sample standard deviation $s$. Then we have the following:

1. If we subtract a constant $m$ from each measurement, the mean and standard deviation for the coded measurements will be

$$\bar{y}_c = \bar{y} - m \qquad \text{and} \qquad s_c = s$$

2. If we multiply each measurement by a positive constant $k$, the mean and standard deviation for the coded data will be

$$\bar{y}_c = k\bar{y} \qquad \text{and} \qquad s_c = ks$$

**EXAMPLE 2.14**

Use the results of theorem 2.1 to compute the mean and standard deviation of the measurements .003, .001, .004, .006, .004, .002 by multiplying each by 1000.

**SOLUTION**

Multiplying each measurement by 1000, we have the coded data 3, 1, 4, 6, 4, 2. For the data, then,

$$\bar{y}_c = \frac{\sum y_c}{6} = \frac{20}{6} = 3.33$$

Similarly, with $\Sigma y_c = 20$ and $\Sigma y_c^2 = 82$, we have

$$s_c^2 = \frac{1}{5}\left[82 - \frac{(20)^2}{6}\right] = 3.07 \qquad \text{and} \qquad s_c = 1.75$$

Then applying the results of theorem 2.1, with $k = 1000$,

$$\bar{y} = \frac{3.33}{1000} = .00333 \qquad \text{and} \qquad s = \frac{1.75}{1000} = .00175$$

Note in example 2.14 that we used the symbol $\Sigma$ to denote a sum of terms in general. For simplicity of notation we omitted the subscripts on both the $\Sigma$ and the variables. Thus the notation $\Sigma y_c$ is equivalent to the notation

$$\sum_{i=1}^{n} (y_c)_i$$

and, in general, the notation $\Sigma y$ is an abbreviation for the notation

$$\sum_{i=1}^{n} y_i$$

coding grouped data

We can also code grouped data to simplify the calculations of $\bar{y}$ and $s$.

---

**Rules for Coding Grouped Data**

1. Based upon inspection, select the class interval that you think is likely to contain the mean. Let $m$ denote the midpoint of this interval (note: the selection of $m$ is not critical).
2. Code the interval midpoints as

$$y_c = \frac{y - m}{w}$$

   where $w$ is the interval width and $y$ is the midpoint (uncoded units).
3. Compute $\bar{y}_c$ and $s_c$ in the usual way for grouped data.
4. $\bar{y} = w\bar{y}_c + m$ and $s = ws_c$.

---

**EXAMPLE 2.15**

Use the frequency distribution of crime rates (violent crimes per 100,000 inhabitants) for a sample of 90 cities, shown in table 2.6, to compute $\bar{y}$ and $s$.

**SOLUTION**

Suppose we think the mean lies in the seventh interval from the bottom; then $m = 449.5$. Step 2 of the coding procedure is given in column 4 of

**Table 2.6** Frequency distribution data of crime rates for 90 cities, example 2.15

| *Class Interval* | *Midpoint, y* | *Frequency, f* | $y_c = (y - m)/w$ | $fy_c$ | $fy_c^2$ |
|---|---|---|---|---|---|
| 899.5–959.5 | 929.5 | 1 | 8 | 8 | 64 |
| 839.5–899.5 | 869.5 | 1 | 7 | 7 | 49 |
| 779.5–839.5 | 809.5 | 1 | 6 | 6 | 36 |
| 719.5–779.5 | 749.5 | 0 | 5 | 0 | 0 |
| 659.5–719.5 | 689.5 | 8 | 4 | 32 | 128 |
| 599.5–659.5 | 629.5 | 7 | 3 | 21 | 63 |
| 539.5–599.5 | 569.5 | 5 | 2 | 10 | 20 |
| 479.5–539.5 | 509.5 | 10 | 1 | 10 | 10 |
| 419.5–479.5 | 449.5 | 7 | 0 | 0 | 0 |
| 359.5–419.5 | 389.5 | 13 | −1 | −13 | 13 |
| 299.5–359.5 | 329.5 | 13 | −2 | −26 | 52 |
| 239.5–299.5 | 269.5 | 10 | −3 | −30 | 90 |
| 179.5–239.5 | 209.5 | 6 | −4 | −24 | 96 |
| 119.5–179.5 | 149.5 | 6 | −5 | −30 | 150 |
| 59.5–119.5 | 89.5 | 2 | −6 | −12 | 72 |
| *Totals* | | 90 | | −41 | 843 |

table 2.6 using $w = 60$. From column 5, the mean in coded units is

$$\bar{y}_c = \frac{\sum fy_c}{n} = \frac{-41}{90} = -.46$$

The mean for the original units is

$$\bar{y} = w\bar{y}_c + m = 60(-.46) + 449.5 = 421.9$$

The sample variance (in coded units) can be found by using columns 5 and 6 of the table.

$$s_c^2 = \frac{1}{n-1}\left[\sum fy_c^2 - \frac{\left(\sum fy_c\right)^2}{n}\right]$$

$$= \frac{1}{89}\left[843 - \frac{(-41)^2}{90}\right] = \frac{1}{89}(843 - 18.68) = 9.26$$

Then the standard deviation (in coded units) is

$$s_c = \sqrt{9.26} = 3.04$$

The standard deviation for the original crime rate data is

$$s = ws_c = 60(3.04) = 182.40$$

You may feel that the additional steps required in coding do not really simplify things, especially when most of us have access to an electronic calculator. However, the mere fact that the numbers are smaller and simpler to manipulate makes the coding worthwhile. And, more importantly, by working with smaller numbers, we are probably less prone to make arithmetic errors.

## 2.6 Summary

This chapter was concerned with graphical and numerical description of data. The pie chart and the bar graph are particularly appropriate for graphically displaying data obtained from a qualitative variable. The frequency and relative frequency histograms are graphical techniques applicable only to quantitative data.

Numerical descriptive measures of data are used to convey a mental image of the distribution of measurements. Measures of central tendency include the mode, the median, and the arithmetic mean. Measures of variability include the range, the interquartile range, the variance, and the standard deviation of a set of measurements. While some disciplines emphasize different measures, we will use the mean and the standard deviation of a set of measurements as the primary numerical measures of central tendency and variability, respectively. One explanation for our choice is that not only can we compare variabilities of *two* sets of measurements using the standard deviation of each but we can also interpret the variability of a *single* set of measurements using the mean, the standard deviation, and the Empirical Rule.

## Exercises

2.11. To practice using the shortcut method to calculate $s^2$ and $s$, consider a small number of sample measurements, say 0, 1, 2, 4, 4.

a. Verify that $\Sigma_{i=1}^{5} y_i = 11$ and $\Sigma_{i=1}^{5} y_i^2 = 37$.

b. Use the results of part a to compute $s^2$ and $s$ with the shortcut formula.

2.12. The rounded nitrogen contents for the 36 composite apple leaf samples of exercise 2.7 are presented below.

| | | | | | |
|---|---|---|---|---|---|
| 2.10 | 2.82 | 2.17 | 1.99 | 2.22 | 3.09 |
| 2.47 | 2.52 | 2.80 | 2.10 | 2.92 | 2.20 |
| 1.75 | 2.77 | 2.82 | 2.67 | 3.05 | 2.93 |
| 2.94 | 1.98 | 2.38 | 2.65 | 2.77 | 1.85 |
| 1.69 | 2.70 | 2.68 | 2.06 | 2.36 | 2.28 |
| 2.75 | 2.43 | 2.39 | 2.55 | 1.80 | 1.96 |

a. Use the shortcut formula to compute $s^2$ and $s$. You can verify that

$$\sum_{i=1}^{36} y_i = 87.61 \qquad \text{and} \qquad \sum_{i=1}^{36} y_i^2 = 218.7297$$

b. Use the range approximation to check your calculation of $s$.

c. To increase your confidence in the Empirical Rule, construct the intervals $\bar{y} \pm s$, $\bar{y} \pm 2s$, and $\bar{y} \pm 3s$. Count the number of rounded nitrogen content readings falling in each of the three intervals. Convert these numbers to percentages and compare your results to the Empirical Rule.

2.13. The College of Dentistry at the University of Florida has made a commitment to develop its entire curriculum around the use of self-paced instructional materials such as video tapes, slide tapes, syllabi, and so on. It is hoped that each student will proceed at a pace commensurate with his or her ability and that the instructional staff will have more free time for personal consultation in student-faculty interaction. One such instructional module was developed and tested on the first 50 students proceeding through the curriculum. The measurements below represent the number of hours it took these students to complete the required modular material.

| | | | | | | | | | |
|---|---|---|---|---|---|---|---|---|---|
| 16 | 8 | 33 | 21 | 34 | 17 | 12 | 14 | 27 | 6 |
| 33 | 25 | 16 | 7 | 15 | 18 | 25 | 29 | 19 | 27 |
| 5 | 12 | 29 | 22 | 14 | 25 | 21 | 17 | 9 | 4 |
| 12 | 15 | 13 | 11 | 6 | 9 | 26 | 5 | 16 | 5 |
| 9 | 11 | 5 | 4 | 5 | 23 | 21 | 10 | 17 | 15 |

a. Calculate the mode, the median, and the mean for these recorded completion times.

b. Guess the value of $s$.

c. Compute $s$ by using the shortcut formula and compare your answer to that of part b.

d. Would you expect the Empirical Rule to adequately describe the variability of these data? Explain.

2.14. Refer to the data of examples 2.6 and 2.11. We previously computed the sample mean and standard deviation to be 4.29 and .32, respectively. Use the coding procedures of section 2.5 to compute $\bar{y}$ and $s$. Proceed assuming that you think the eighth class interval contains the mean.

2.15. Repeat exercise 2.14 by using $m = 4.4$, the midpoint of the ninth interval.

2.16. Repeat exercise 2.14 by using $m = 4.5$, the midpoint of the tenth interval. Are your answers to exercises 2.14, 2.15, and 2.16 identical? If not, check your calculations.

2.17. A study was conducted to determine urine flow of sheep (in milliliters/minute) when infused intravenously with the antidiuretic hormone ADH. The urine flows of ten sheep are recorded below.

0.7, 0.5, 0.5, 0.6, 0.5, 0.4, 0.3, 0.9, 1.2, 0.9

a. Determine the mean, the median, and the mode for these sample data.
b. Suppose that the largest measurement is 6.8 rather than 1.2. How does this affect the mean, the median, and the mode?

2.18. Refer to exercise 2.17.
a. Compute the range and the sample standard deviation.
b. Check your calculation of $s$ using the range approximation.
c. How are the range and standard deviation affected if the largest measurement is 6.8 rather than 1.2?

2.19. Refer to exercise 2.17. Code the data by multiplying each measurement by 10. Compute the sample mean and standard deviation for the original set using the coded values.

2.20. A random sample of 90 standard metropolitan statistical areas (SMSA) was studied to obtain information on murder rates. The murder rate (number of murders per 100,000 people) was recorded, and these data are summarized in the frequency table displayed below.

| *Class Interval* | $f_i$ | *Class Interval* | $f_i$ |
|---|---|---|---|
| −.5–1.5 | 2 | 13.5–15.5 | 9 |
| 1.5–3.5 | 18 | 15.5–17.5 | 4 |
| 3.5–5.5 | 15 | 17.5–19.5 | 2 |
| 5.5–7.5 | 13 | 19.5–21.5 | 1 |
| 7.5–9.5 | 9 | 21.5–23.5 | 1 |
| 9.5–11.5 | 8 | 23.5–25.5 | 1 |
| 11.5–13.5 | 7 | | |

Construct a relative frequency histogram for these data.

2.21. Refer to the data of exercise 2.20.
a. Compute the sample median and the mode.
b. Compute the sample mean.

c. Which measure of central tendency would you use to describe the center of the distribution of murder rates?

2.22. Refer to the data of exercise 2.20.

a. Compute the interquartile range.

b. Compute the sample standard deviation.

2.23. Refer to the data of exercise 2.20. If you did not employ coding to compute $\bar{y}$ and $s$ in exercises 2.21 and 2.22, use the midpoint of the fifth interval as $m$. Code the sample data to compute $\bar{y}$ and $s$. Compare your answers to those of exercises 2.21 and 2.22.

2.24. Every 20 minutes a sample of 10 transistors was drawn from the outgoing product on a production line and tested. The data are summarized below for the first 500 samples of 10 measurements.

Construct a relative frequency distribution depicting the interquartile range. (Note: $y_i$ in the table is the number of defectives in a sample of 10.)

| $y_i$ | 0 | 1 | 2 | 3 | 4 | 5 | 6 | 7 | 8 | 9 | 10 |
|---|---|---|---|---|---|---|---|---|---|---|---|
| $f_i$ | 170 | 185 | 75 | 25 | 15 | 10 | 8 | 5 | 4 | 2 | 1 |

2.25. Refer to exercise 2.24.

a. Determine the sample median and the mode.

b. Calculate the sample mean.

c. Based on the mean, the median, and the mode, how is the distribution skewed?

2.26. Refer to the data of exercise 2.24.

a. Code the sample data to compute the sample standard deviation.

b. Can the Empirical Rule be used to describe this set of measurements?

2.27. Use the frequency table below to find the sample mean and the standard deviation for the self-esteem scores of a random sample of $C^+$ students. Use the original units.

| Score | Frequency | Score | Frequency |
|---|---|---|---|
| 10–11 | 2 | 22–23 | 13 |
| 12–13 | 16 | 24–25 | 12 |
| 14–15 | 12 | 26–27 | 2 |
| 16–17 | 16 | 28–29 | 4 |
| 18–19 | 27 | 30–31 | 1 |
| 20–21 | 37 | 32–33 | 1 |

2.28. Repeat exercise 2.27, but use the coding steps with $m = 20.5$.

2.29. Repeat exercise 2.27, but use the coding steps with $m = 24.5$. Are your answers to exercises 2.27, 2.28, and 2.29 the same? If not, check your calculations.

2.30. Refer to example 2.9 (p. 34). Find the sample mean and the standard deviation by using the coding steps of section 2.5. Compare your answers to those of example 2.9.

*2.31. Refer to the computer output shown in exercise 2.10 (p. 28).
a. Identify the standard deviation.
b. Compute $s/\sqrt{n}$. This quantity is often called the standard error of the mean. This quantity is shown in the output, but we will delay a discussion of it until chapter 3.

*2.32. Descriptive statistics have been computed for the sample data shown below. Use the BMDP computer printout that follows to give an answer to these items:
a. Identify the mean and sample size.
b. Verify the computer results for part a.
c. Give the sample standard deviation.
d. Give the sample median (refer to the data).

*Median Values of Homes*

| | | | | | |
|---|---|---|---|---|---|
| 20,700 | 13,600 | 14,400 | 12,800 | 10,500 | 16,300 |
| 12,800 | 17,500 | 13,000 | 13,000 | 14,600 | 16,800 |
| 16,400 | 17,600 | 11,400 | 14,600 | 11,100 | 22,400 |
| 16,400 | 13,800 | 12,400 | 15,000 | 12,000 | 21,900 |
| 20,600 | 19,800 | 12,100 | 12,200 | 16,300 | 18,500 |
| 18,200 | 17,500 | 15,100 | 12,600 | 14,000 | 15,400 |
| 25,500 | 17,300 | 17,100 | 12,500 | 14,100 | 23,100 |
| 18,900 | 14,500 | 21,300 | 12,200 | 15,300 | 22,700 |
| 22,500 | 18,000 | 18,700 | 11,200 | 17,300 | 35,100 |
| 23,400 | 13,900 | 17,100 | 12,000 | 13,000 | 11,300 |

| VARIABLE NO. NAME | MEAN | STANDARD DEVIATION | ST. ERR. OF MEAN | COEFF. OF VARIATION | SMALLEST VALUE | SMALLEST Z-SCORE |
|---|---|---|---|---|---|---|
| 1 Y | 16354.938 | 4436.070 | 572.6941 | 0.27124 | 10500.000 | −1.31985 |

| LARGEST VALUE | LARGEST Z-SCORE | RANGE | TOTAL FREQUENCY |
|---|---|---|---|
| 35100.000 | 4.22560 | 24600.000 | 60 |

# References


Barr, A.J.; Goodnight, J.H.; Sall, S.P.; and Helwig, J.T. 1976. *A user's guide to SAS 76*. Raleigh, N.C.: SAS Institute, Inc.

Dixon, W. J., ed. 1975. *BMDP, biomedical computer programs*. Berkeley: University of California Press.

Mendenhall, W.; Ott, L.; and Larson, R. 1974. *Statistics: A tool for the social sciences*. N. Scituate, Mass.: Duxbury Press. Chapters 3 and 4.

Ryan, T.A.; Joiner, B.L.; and Ryan, B.F. 1976. *Minitab student handbook*. N. Scituate, Mass.: Duxbury Press.

Service, J. 1972. *A user's guide to the statistical analysis system*. Raleigh, N.C.: Student Supply Stores, North Carolina State University.

# 3 The Normal Probability Distribution

## 3.1 Introduction

normal curve

Many variables of interest, including several statistics to be discussed in later sections and chapters, have mound-shaped frequency distributions that can be approximated by using a *normal curve*. For example, the distribution of total scores on the Brief Psychiatric Rating Scale for outpatients having a current history of repeated aggressive acts would be mound-shaped. Other practical examples of mound-shaped distributions are social perceptiveness scores of preschool children selected from a particular socioeconomic background, psychomotor retardation scores for patients with circular-type manic-depressive illness, milk yields for cattle of a particular breed, and perceived anxiety scores for residents of a community. Each of these mound-shaped distributions could be approximated with a normal curve.

Since the normal distribution has been well tabulated, areas under a normal curve—which correspond to probabilities—can be used to approximate probabilities associated with the variables of interest in our experimentation. Thus the normal random variable and its associated distribution will play an important role in statistical inference.

normal probability distribution

The relative frequency histogram for the normal random variable, called the *normal probability distribution,* is a smooth bell-shaped curve. Figure 3.1 shows a normal curve. If we let $y$ represent the normal random variable, then the height of the probability distribution for a specific value of $y$ is represented by $f(y)$.

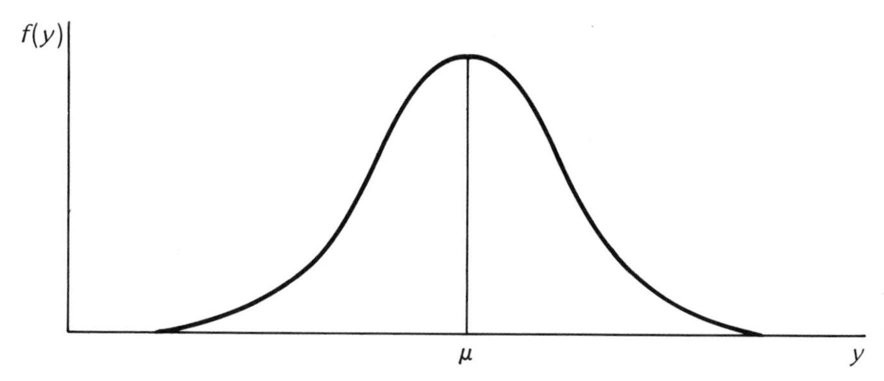


**Figure 3.1** A normal curve with mean $\mu$ and standard deviation $\sigma$

As we see from figure 3.1, the normal probability distribution is bell-shaped and symmetrical about the mean $\mu$. Although the normal random variable $y$ may theoretically assume values from $-\infty$ to $+\infty$, we know from the Empirical Rule that approximately all the measurements are within 3 standard deviations ($3\sigma$) of $\mu$. Hence this theoretical restriction does not weaken the usefulness of the normal random variable.

Before we begin our study of the normal probability distribution, we present a brief discussion of random sampling.

# Random Sampling 3.2

So far in the text we have alluded to "random" or "representative" samples. What is the importance of random sampling? We must know how the sample was selected so that we can determine probabilities associated with various sample outcomes. The probabilities of samples selected *in a random manner* can be determined and we can use these probabilities to make inferences about the population from which the sample was drawn. Nonrandom samples have unknown probabilities associated with them and hence cannot be used for purposes of statistical inference.

random sample

We turn now to a definition of a *random sample* of $n$ measurements selected from a population containing $N$ measurements ($N > n$).

**Definition 3.1**

A sample of $n$ measurements selected from a population is said to be a *random sample* if every different sample of size $n$ from the population has an equal probability of being selected.

**EXAMPLE 3.1**

Let us suppose that a population consists of the 6 measurements 1, 2, 3, 4, 5, 7. List all possible different samples of 2 measurements that could be selected from the population. Give the probability associated with each sample in a random sample of $n = 2$ measurements selected from the population.

**SOLUTION**

All possible samples are listed below.

| *Sample* | *Measurements* |
|---|---|
| 1 | 1,2 |
| 2 | 1,3 |
| 3 | 1,4 |
| 4 | 1,5 |
| 5 | 1,7 |
| 6 | 2,3 |
| 7 | 2,4 |
| 8 | 2,5 |
| 9 | 2,7 |
| 10 | 3,4 |
| 11 | 3,5 |
| 12 | 3,7 |
| 13 | 4,5 |
| 14 | 4,7 |
| 15 | 5,7 |

Now let us suppose that we draw a single sample of $n = 2$ measurements from the 15 possible samples of two measurements. The sample selected is called a random sample if every sample had an equal probability ($\frac{1}{15}$) of being selected.

random number table

It is rather unlikely that we would ever achieve a truly random sample, because the probabilities of selection will not always be exactly equal. But we do the best we can. One of the simplest and most reliable ways to select a random sample of $n$ measurements from a population is to use a table of random numbers (see table 8 in the appendix). Random number tables are constructed in such a way that no matter where you start in the table and no matter what direction you move, the digits occur randomly and with equal probability. Thus if we wished to choose a random sample of $n = 10$ measurements from a population containing 100 measurements, we could label the measurements in the population from 0 to 99 (or 1 to 100). Then by referring to table 8 in the appendix and choosing a random starting point, the next 10 two-digit numbers going across the page would indicate the labels of the particular measurements to be included in the random sample. Similarly, by moving up or down the page, we would also obtain a random sample.

**EXAMPLE 3.2**

A small community consists of 850 families. We wish to obtain a random sample of 20 families to ascertain public acceptance of a wage and price freeze. Refer to table 8 in the appendix to determine which families should be sampled.

**SOLUTION**

Assuming that some list of all families in the community is available (such as a telephone directory), we could label the families from 0 to 849 (or, equivalently, from 1 to 850). Then referring to table 8, the appendix, we choose a starting point. Suppose we have decided to start at line 1, column 3. Going down the page we will choose the first 20 three-digit numbers between 000 and 849. From table 8, we have

| | | | |
|---|---|---|---|
| 015 | 110 | 482 | 333 |
| 255 | 564 | 526 | 463 |
| 225 | 054 | 710 | 337 |
| 062 | 636 | 518 | 224 |
| 818 | 533 | 524 | 055 |

These 20 numbers identify the 20 families that are to be included in our sample.

It should be noted that a telephone directory is not always the best source list for names, especially in surveys related to economics or politics. In the 1936 presidential election, Alfred Landon was predicted (incorrectly) to beat Franklin D. Roosevelt based on the results of a telephone survey. Unfortunately, since not all people had telephones, the survey was biased because it related only to the opinions of those who could be reached by a telephone. While the problem is not so serious now, tax records, residential utility records, or other lists might provide a more accurate source of names upon which to base a survey than a telephone directory.

Having defined mound-shaped distributions and random sampling, in the next section we turn to the question, "Why do many variables possess relative frequency histograms that are mound-shaped?"

# Exercises

3.1. A psychologist was interested in studying women who are in the process of obtaining a divorce to determine whether there are significant attitudinal changes after the divorce has been finalized. Existing records from the geographic area in question show that 798 couples have recently filed for divorce. If a sample of 25 women is needed for the study, use table 8 of the

appendix to determine which women should be asked to participate in the study. (Hint: Begin in column 2, row 1, and proceed down.)

3.2. Refer to exercise 3.1. As is the case in most surveys, not all persons chosen for a study will agree to participate. Suppose that 5 of the 25 women selected refuse to participate. Determine 5 more women to be included in the study.

*3.3. Suppose you have been asked to run a public opinion poll related to an upcoming election. There are 1000 registered voters in a specific precinct and you wish to obtain a random sample of 50 persons. Use a computer program to indicate which individuals are to be included in the sample. A Minitab program is shown below for purposes of illustration. (Note: We are assuming that there is a list of the 1000 voters, with the numbers 1 to 1000 corresponding to people on the list.)

```
IRANDOM 50 INTEGERS BETWEEN 1 AND 1000, PUT IN C1
STOP
```

# 3.3 The Central Limit Theorem

As indicated previously, the relative frequency histograms plotted for many sets of measurements will be mound-shaped. A very plausible explanation for this phenomenon is offered by a series of theorems in mathematical statistics called central limit theorems. We will discuss one such theorem (without proof).

---

**Theorem 3.1**

If random samples of $n$ measurements are repeatedly drawn from a population with a finite mean $\mu$ and a standard deviation $\sigma$, then, when $n$ is large, the relative frequency histogram for the sample means (calculated from the repeated samples) will be approximately normal (bell-shaped) with mean $\mu$ and standard deviation $\sigma/\sqrt{n}$. (Note: The approximation becomes more precise as $n$ increases.)

---

Figure 3.2 illustrates theorem 3.1. Figure 3.2(a) shows the distribution of the original measurement $y$ from which the samples are to be drawn. No specific shape need be assigned to the distribution of $y$. All we know is that its mean is $\mu$ and its variance is $\sigma^2$. Figure 3.2(b) illustrates the relative frequency histo-

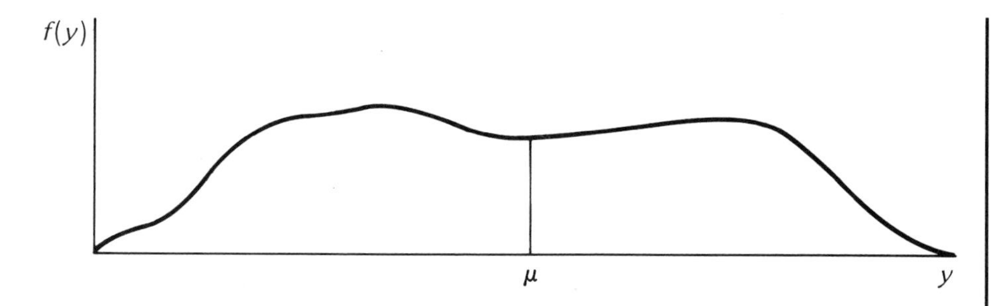


(a) Probability distribution of $y$, with mean $\mu$ and standard deviation $\sigma$

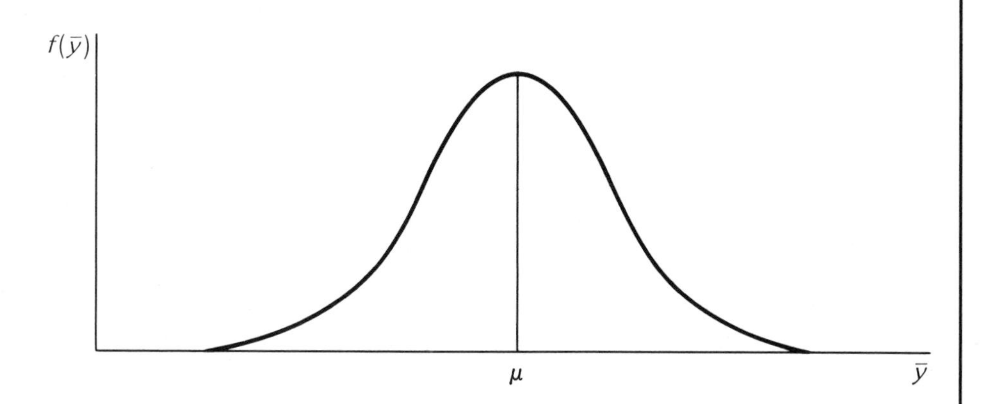


(b) Sampling distribution of $\bar{y}$, with mean $\mu$ and standard deviation $\sigma/\sqrt{n}$

**Figure 3.2** The probability distribution of $y$ and the sampling distribution of $\bar{y}$

sampling distribution for $\bar{y}$

sampling distribution

gram, called the *sampling distribution,* for the sample mean $\bar{y}$. Repeated samples of size $n$ are to be drawn from the population illustrated in figure 3.2(a). For each sample drawn we compute the sample mean $\bar{y}$. If we were to continue this process over and over again, finally plotting the relative frequency histogram for the sample means, it would appear as in figure 3.2(b). The mean for the sampling distribution of $\bar{y}$ is $\mu$, the same as that for the original $y$ measurements, and the standard deviation of the sampling distribution (also called the standard error of the mean) is equal to the standard deviation of the $y$ measurements divided by $\sqrt{n}$.

standard error std. dev./$\sqrt{n}$

We will be able to use the information we have concerning the distribution of $\bar{y}$ to make inferences about the parameter $\mu$. For example, we know by the Empirical Rule that approximately 95% of the $\bar{y}$'s will be within 2 standard deviations ($2\sigma/\sqrt{n}$) of their mean. In a given sample where we calculate a single sample mean $\bar{y}$, we would expect our calculated $\bar{y}$ to be within $2\sigma/\sqrt{n}$ of $\mu$ (see figure 3.3). Hence not only can we use $\bar{y}$ to estimate $\mu$ but also we can

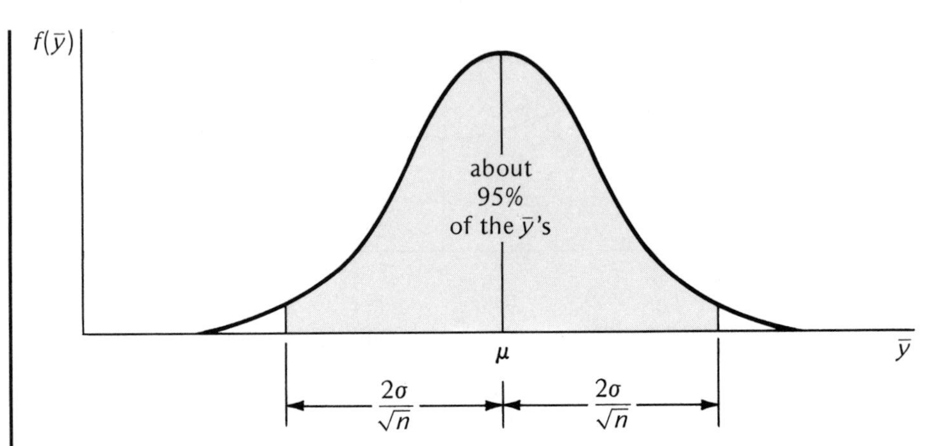


**Figure 3.3** Application of the empirical rule to a sampling distribution

say how close to $\mu$ we expect our estimate to be—that is, we can provide a *measure of goodness* for our estimate.

We can use the table of random numbers to illustrate the Central Limit Theorem which we applied to sample means. Table 3.1 lists the murder rates (number of murders per 100,000 people) associated with 90 metropolitan areas throughout the United States in the North, the South, and the West in 1971. Although it is unrealistic to assume that these 90 murder rates represent a population of measurements, we will, for illustration purposes, take the 90 measurements of table 3.1 as the population of interest.

To examine the sampling distribution of the sample mean $\bar{y}$, we randomly draw 50 samples of $n = 5$ measurements each. For example, numbering the cities from 1 to 90, we refer to table 8, the appendix, and proceed down column 1, row 1, to obtain 5 two-digit numbers. These numbers, 10, 22, 24, 42, and 37, designate the numbers of the 5 cities to be included in the first sample, as shown below.

| *City Number* | *City* | *Murder Rate* |
|---|---|---|
| 10 | Dallas | 18 |
| 22 | Nashville | 13 |
| 24 | Orlando | 11 |
| 42 | Kalamazoo | 4 |
| 37 | Cleveland | 14 |
| | | 60 |

We then compute $\bar{y}_1$, the mean for the first sample of $n = 5$ measurements, to be $\bar{y}_1 = 60/5 = 12$. This same procedure is repeated 49 more times to obtain 50

**Table 3.1** Murder rates (per 100,000 inhabitants) in 1971 for 90 cities in the United States in the North, the South, and the West

| *South* | *Rate* | *North* | *Rate* | *West* | *Rate* |
|---|---|---|---|---|---|
| Atlanta | 20 | Albany, NY | 3 | Bakersfield | 8 |
| Augusta, GA | 22 | Allentown | 2 | Boise | 5 |
| Baton Rouge | 10 | Atlantic City | 5 | Colorado Springs | 5 |
| Beaumont, TX | 10 | Canton, OH | 3 | Denver | 8 |
| Birmingham | 14 | Chicago | 13 | Eugene, OR | 4 |
| Charlotte, NC | 25 | Cincinnati | 6 | Fresno | 8 |
| Chattanooga | 15 | Cleveland | 14 | Honolulu | 4 |
| Columbia, SC | 13 | Detroit | 15 | Kansas City, MO | 13 |
| Corpus Christi | 13 | Evansville | 7 | Lawton, OK | 6 |
| Dallas | 18 | Grand Rapids | 3 | Los Angeles | 9 |
| El Paso | 4 | Johnstown, PA | 2 | Modesto, CA | 2 |
| Fort Lauderdale | 14 | Kalamazoo | 4 | Oklahoma City | 6 |
| Greensboro, NC | 11 | Kenosha, WI | 2 | Oxnard, CA | 2 |
| Houston | 17 | Lancaster, PA | 2 | Pueblo, CO | 3 |
| Jackson, MS | 16 | Lansing | 3 | Sacramento | 6 |
| Knoxville | 8 | Lima, OH | 3 | St. Louis | 15 |
| Lexington, KY | 13 | Madison, WI | 2 | Salinas, KS | 6 |
| Lynchburg, VA | 18 | Mansfield, OH | 7 | Salt Lake City | 4 |
| Macon, GA | 13 | Milwaukee | 4 | San Bernardino | 6 |
| Miami | 16 | Newark, NJ | 10 | San Francisco | 8 |
| Monroe, LA | 15 | Paterson, NJ | 3 | San Jose | 2 |
| Nashville | 13 | Philadelphia | 9 | Seattle | 4 |
| Newport News | 8 | Pittsfield, MA | 1 | Sioux City | 3 |
| Orlando | 11 | Racine, WI | 5 | Spokane | 1 |
| Richmond, VA | 15 | Rockford, IL | 4 | Stockton, CA | 9 |
| Roanoke | 10 | South Bend | 6 | Tacoma | 6 |
| Shreveport, LA | 15 | Springfield, IL | 2 | Topeka | 2 |
| Washington, D.C. | 11 | Syracuse | 4 | Tucson | 17 |
| Wichita Falls, TX | 6 | Vineland, NJ | 10 | Vallejo, CA | 4 |
| Wilmington | 7 | Youngstown | 7 | Waterloo, IA | 4 |

*Source: Uniform Crime Reports for the United States: 1970* (Washington, D.C.: Department of Justice), pp. 78–94.

sample means, each based on $n = 5$ measurements. Our results are listed below.

| | | | | |
|---|---|---|---|---|
| 12.0 | 8.8 | 11.4 | 5.0 | 8.8 |
| 7.0 | 12.4 | 15.0 | 8.6 | 8.6 |
| 14.2 | 7.4 | 10.2 | 8.4 | 7.2 |
| 11.2 | 5.4 | 7.8 | 9.6 | 10.2 |
| 9.2 | 8.8 | 6.2 | 4.4 | 9.0 |
| 10.0 | 6.4 | 6.0 | 10.4 | 8.4 |
| 9.0 | 7.0 | 11.0 | 11.4 | 8.6 |
| 9.8 | 8.2 | 8.2 | 7.4 | 5.4 |
| 7.4 | 10.0 | 13.2 | 7.4 | 8.4 |
| 7.0 | 7.8 | 7.6 | 4.2 | 9.6 |

To see how the sampling distribution of the sample mean $\bar{y}$ relates to the original distribution for the 90 murder rates, we can construct histograms for both sets of measurements. These are displayed in figure 3.4.

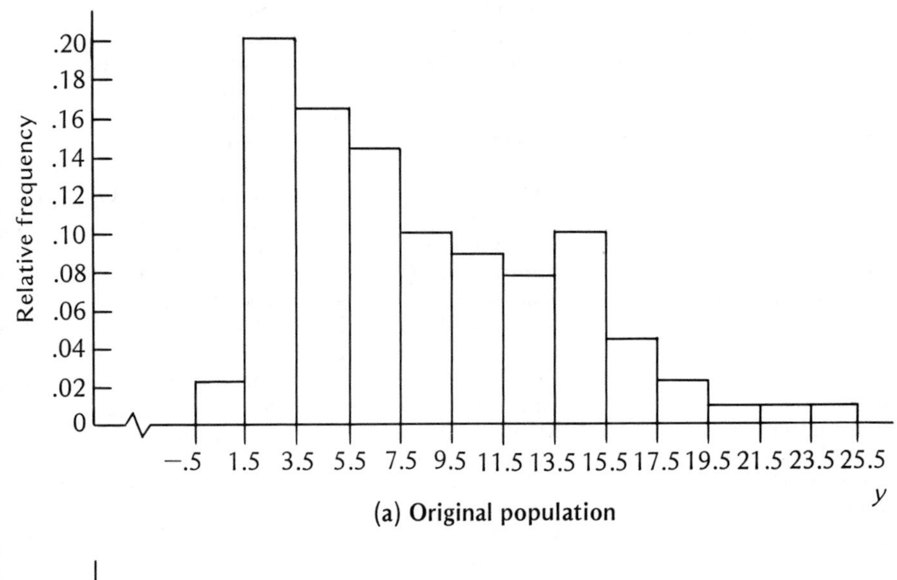


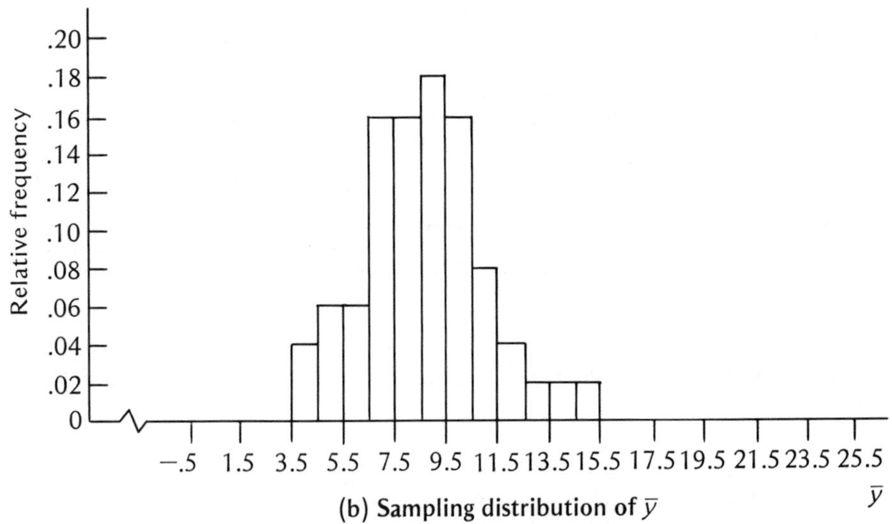


**Figure 3.4** Relative frequency histograms for the original population and for the sampling distribution of $\bar{y}$ for the murder rate data

As can be seen from figure 3.4, the original population is certainly not mound-shaped or symmetrical. In fact, the distribution is skewed to the right (tails off to the right). But even with a small number of samples (50) and a small number of measurements per sample (5), the sampling distribution of $\bar{y}$

is beginning to look bell-shaped, with most of the sample means grouped closely about the mean of the population (which in this case can be computed and is $\mu = 8.24$).

The illustration we have presented could have been made more convincing by assuming a much larger population and by taking more samples. The important point to remember is that in repeated sampling, $\bar{y}$ will be approximately normally distributed with mean $\mu$ and standard deviation $\sigma/\sqrt{n}$. The approximation will be more precise as $n$, the sample size for each sample, increases. Thus the frequency histogram for $\bar{y}$ in our example would have been even more bell-shaped if $n$ had been 10 rather than 5, or 15 rather 10, and so on.

We can use the results of coding, presented in chapter 2, to extend the Central Limit Theorem to the sample sum $\Sigma_{i=1}^{n} y_i$. If repeated samples of size $n$ are drawn from a population, and if, for each sample drawn, we compute $\Sigma_{i=1}^{n} y_i = n\bar{y}$, we have, in essense, coded the sample means for each sample by multiplying by $n$. Applying our coding results from chapter 2 to the results of the Central Limit Theorem, the mean and the standard deviation for $\Sigma_{i=1}^{n} y_i$ will be, respectively, $n\mu$ and $n\sigma/\sqrt{n} = \sqrt{n}\,\sigma$. This result is illustrated in figure 3.5.

sampling distribution for $\Sigma y$

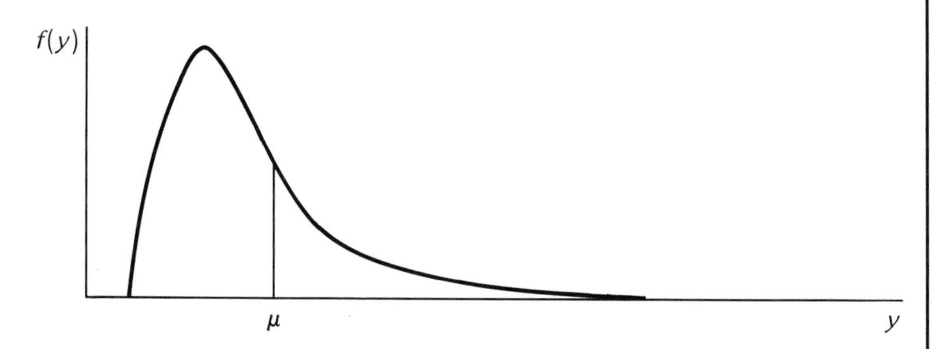


**(a) Probability distribution of $y$, with mean $\mu$ and standard deviation $\sigma$**

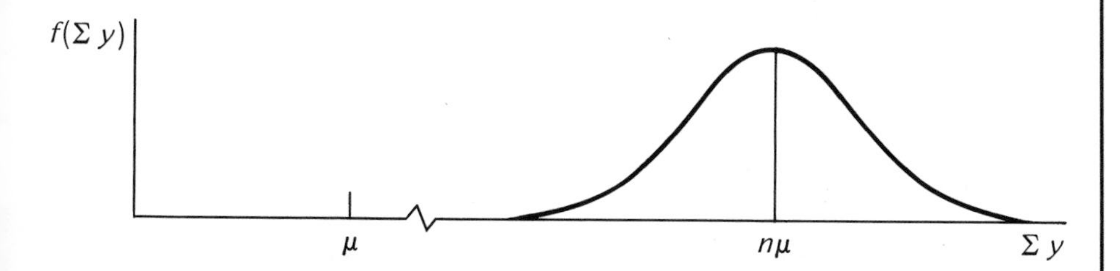


**(b) Sampling distribution $\sum_{i=1}^{n} y_i$, with mean $n\mu$ and standard deviation $\sqrt{n}\sigma$**

**Figure 3.5** Probability distribution of $y$ and of the sample sum $\Sigma_{i=1}^{n} y_i$

Many of the statistics that we will encounter in later chapters will be either sums or averages of variables. Hence we will employ the Central Limit Theorem discussed in this chapter to specify their sampling distributions.

## Exercises

3.4. A random sample of 16 measurements is drawn from a population with a mean of 60 and a standard deviation of 5. Describe the sampling distribution of $\bar{y}$, the sample mean. Within what interval would you expect $\bar{y}$ to lie approximately 95% of the time?

3.5. Refer to exercise 3.4. Describe the sampling distribution for the sample sum $\Sigma_{i=1}^{16} y_i$. Is it unlikely (improbable) that $\Sigma_{i=1}^{16} y_i$ would be more than 70 units away from 960? Explain.

*3.6. In exercise 3.4, a random sample of 16 observations was to be selected from a population with $\mu = 60$ and $\sigma = 5$. If we assume that the original population of measurements is normal, use a computer program to simulate the distribution of $\bar{y}$ based on 40 sample means. A Minitab program is shown below for purposes of illustration.

```
NRANDOM 16 OBSN, MU = 60, SIGMA = 5, PUT IN C1
AVERAGE THE OBSERVATIONS IN C1
```

This set of 2 instructions would be repeated 39 more times. The complete set of 80 instructions would be entered simultaneously followed by a STOP instruction card.

# 3.4 Areas—Probabilities—Associated with the Normal Curve

The Empirical Rule has given us some information about probabilities associated with a normal random variable. In particular, we know that if we select a measurement at random from a population of measurements that possesses a mound-shaped distribution, the probability is approximately .68 that the

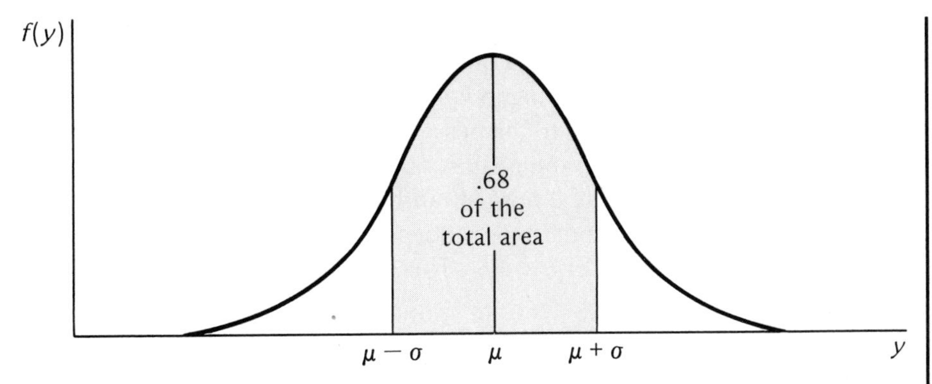


**Figure 3.6** Area under a normal curve within one standard deviation of the mean

measurement will lie within one standard deviation of its mean (see figure 3.6). Similarly, we know that the probability is approximately .95 that a value will lie in the interval $\mu \pm 2\sigma$. What we don't know, however, is the probability that the measurement will be within 1.65 standard deviations of its mean, or within 2.58 standard deviations of its mean. The procedure we are going to discuss in this section will enable us to calculate the probability that a measurement falls within any distance of the mean $\mu$ for a normal curve.

Since there are many different normal curves (depending on the parameters $\mu$ and $\sigma^2$), it might seem to be an impossible task to tabulate areas (probabilities) for all normal curves, especially if each curve requires a separate table. Fortunately, this is not the case. By specifying the probability that a variable $y$ lies within a certain number of standard deviations of its mean (just as we did in using the Empirical Rule), we need only one table of probabilities.

area under a normal curve

Table 1 in the appendix gives the area under a normal curve from the mean $\mu$ to a point $z$ standard deviations ($z\sigma$) to the right of $\mu$ (see figure 3.7). Because of

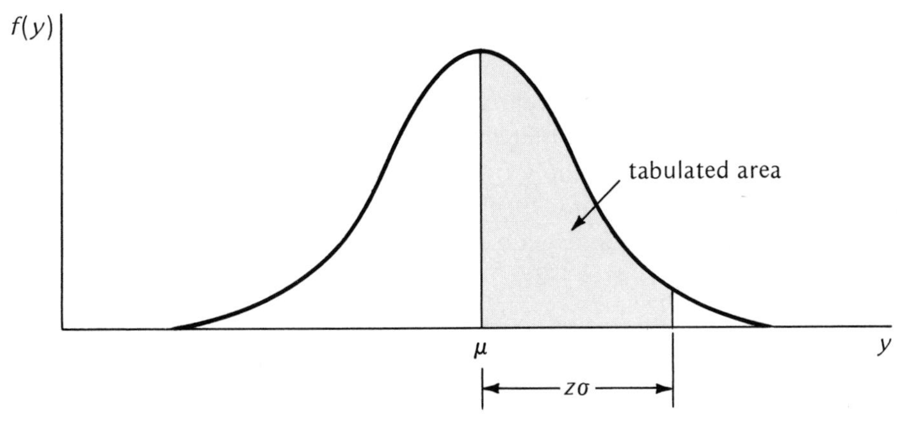


**Figure 3.7** Area under the normal curve, as given in table 1, the appendix

z standard deviations

the symmetry of the normal probability distribution, this would be the same as the area between the mean and a point $z$ standard deviations to the left of $\mu$.

The area shown by the shading in figure 3.7 is the probability listed in table 1 in the appendix. Values of $z$ to the nearest tenth are listed along the left-hand column of table 1, with $z$ to the nearest hundredth along the top of the table. Thus to find the probability a normal random variable will lie in the interval from $\mu$ to a point 1.65 standard deviations above the mean, we look up the table entry corresponding to $z = 1.65$. This probability is .4505 (see figure 3.8).

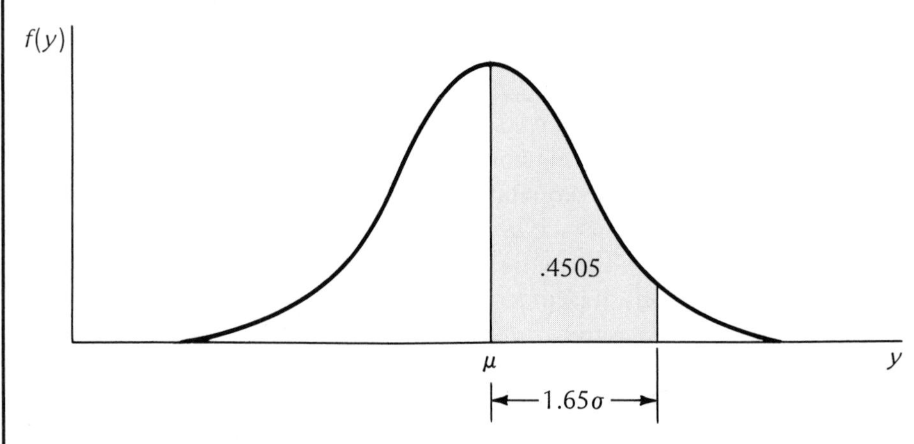


**Figure 3.8** Area under a normal curve from $\mu$ to a point 1.65 standard deviations above the mean

z score

To determine the probability that a measurement will fall in the interval from $\mu$ to some value $y$ to the right of $\mu$, we first calculate the number of standard deviations that $y$ lies away from the mean by using the formula

formula for z

$$z = \frac{y - \mu}{\sigma}$$

The value of $z$ computed using this formula is sometimes referred to as the $z$ *score* associated with the $y$-value. Using the computed value of $z$, we determine the appropriate probability by using table 1, the appendix. Note that we are merely coding the value $y$ by subtracting $\mu$ and dividing by $\sigma$. (In other words, $y = z\sigma + \mu$.) Figure 3.9 illustrates the values of $z$ corresponding to specific values of $y$. Thus a value of $y$ 2 standard deviations below (to the left of $\mu$) corresponds to $z = -2$.

**EXAMPLE 3.3**

Consider a normal distribution with $\mu = 20$ and $\sigma = 2$. Determine the probability that a measurement will be in the interval from 20 to 23.

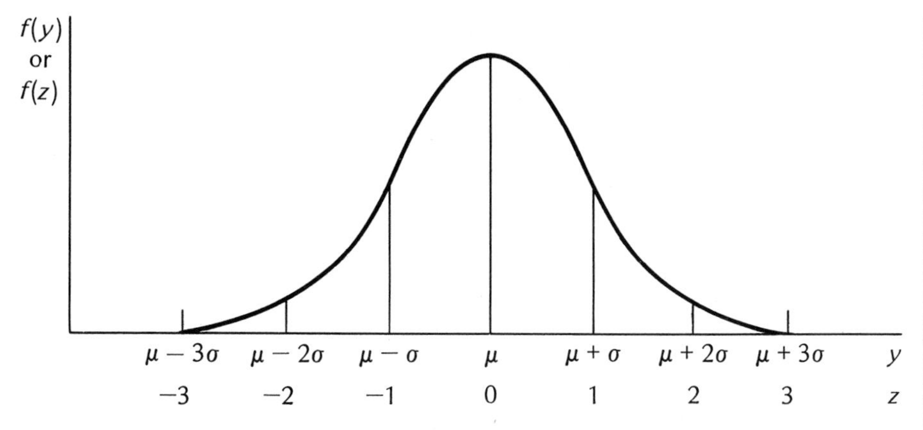


**Figure 3.9** Relationship between specific values of $y$ and $z = (y - \mu)/\sigma$

**SOLUTION**

When first working problems of this type, it might be a good idea to draw a picture so that you can see the area in question. We have done this in figure 3.10.

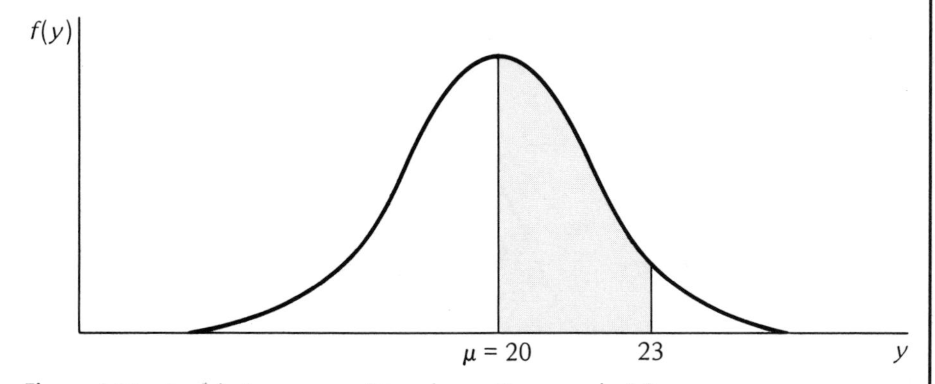


**Figure 3.10** Area between $\mu = 20$ and $y = 23$, example 3.3

To determine the area under the curve from $\mu = 20$ to the value $y = 23$, we first calculate the number of standard deviations $y = 23$ lies away from the mean.

$$z = \frac{y - \mu}{\sigma} = \frac{23 - 20}{2} = 1.5$$

Thus $y = 23$ lies 1.5 standard deviations above $\mu = 20$. Referring to table 1, the appendix, we find the area corresponding to $z = 1.5$ to be .4332. This is the probability that a measurement falls in the interval from 20 to 23.

Similarly, when finding the probability that a measurement lies in the interval from $\mu$ to some value of $y$ to the left of the mean, we again compute $z$, using

$$z = \frac{y - \mu}{\sigma}$$

negative $z$

The computed value of $z$ will be negative but we ignore the negative sign and again refer to table 1, the appendix, for the appropriate probability.

### EXAMPLE 3.4

For the normal distribution of example 3.3, with $\mu = 20$ and $\sigma = 2$, find the probability that $y$ will lie in the interval from 16 to 20.

### SOLUTION

In determining the area between 16 and 20, we use

$$z = \frac{y - \mu}{\sigma} = \frac{16 - 20}{2} = -2$$

Ignoring the negative sign, we find the appropriate area from table 1 to be .4772. Thus .4772 is the probability that a measurement falls in the interval from 16 to 20. The area is shown in figure 3.11.

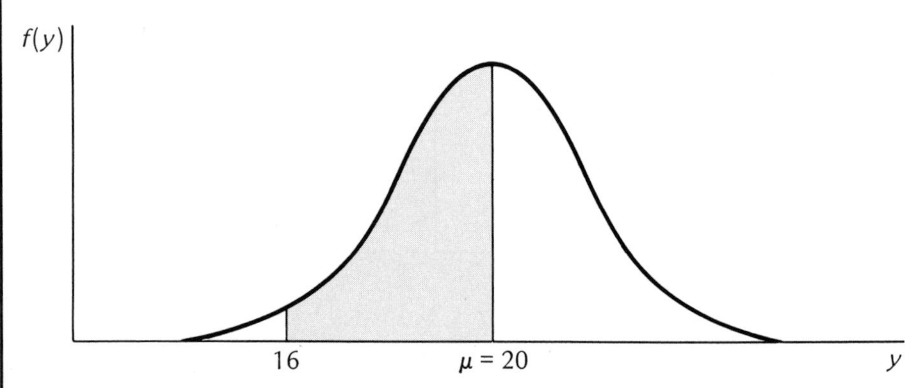


**Figure 3.11** Area between $y = 16$ and $\mu = 20$, example 3.4

### EXAMPLE 3.5

Total yearly milk productions for cows in a herd of guernsey cows are assumed to be normally distributed with $\mu = 70$ pounds and $\sigma = 13$ pounds. What is the probability that the milk production for a cow chosen at random will lie in the interval from 60 pounds to 90 pounds? What is the probability that the milk production for the randomly selected cow will exceed 90 pounds in a given year?

### SOLUTION

We begin by drawing a figure to picture the area we are looking for (figure 3.12).

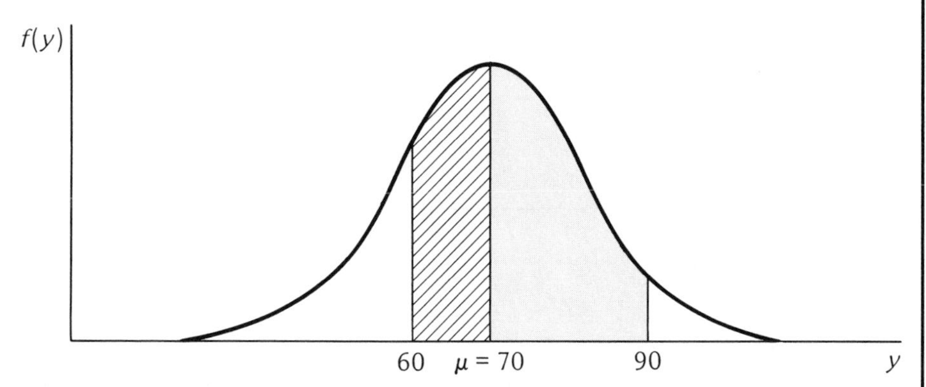


**Figure 3.12** Areas between 60 and 70 and between 70 and 90, example 3.5

To answer the first question, we must compute two areas, the area between 60 and 70 and the area between 70 and 90. The value $y = 60$ corresponds to

$$z = \frac{y - \mu}{\sigma} = \frac{60 - 70}{13} = -.77$$

Referring to table 1, the area between $y = 60$ and $\mu = 70$ is .2794.

The value $y = 90$ corresponds to

$$z = \frac{y - \mu}{\sigma} = \frac{90 - 70}{13} = 1.54$$

So the tabulated area between 70 and 90 is .4382. Adding these two areas we find that the probability that a cow's milk production will lie in the interval from 60 pounds to 90 pounds is .7176.

The probability that the cow's milk production will exceed 90 pounds can be computed using the second area computed above. Because of the symmetry of a normal curve, we know that the total area under the curve to the right of $\mu$ is .5 (similarly, the total area to the left of $\mu$ is .5). We computed the area for the interval from 70 to 90 to be .4382. Subtracting this value from .5, we know that the probability of exceeding 90 is $.5 - .4382 = .0618$ (See figure 3.13 on the next page).

### EXAMPLE 3.6

Annual incomes for career service employees at a large university are approximately normally distributed with a mean of \$6200 and a standard deviation of \$900. Find the probability that an employee chosen at random will have an annual income less than \$5000; an income greater than \$7000.

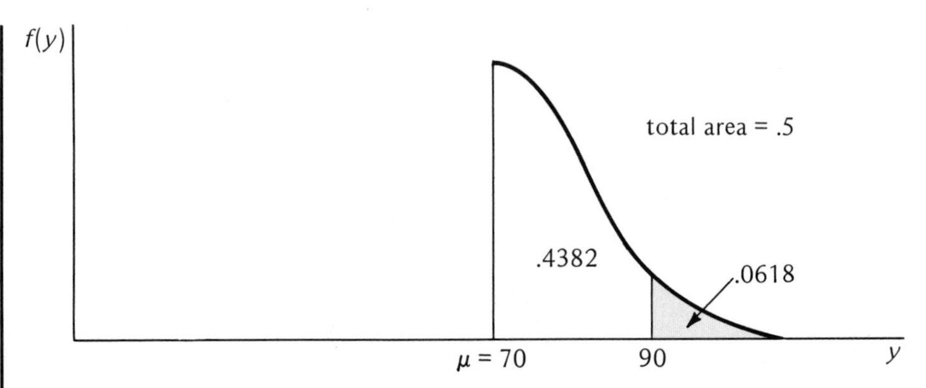


**Figure 3.13** Area above $y = 90$ for $\mu = 70$ and $\sigma = 13$

## SOLUTION

First we draw a figure showing the areas in question (figure 3.14).
Now we must determine the area between 5000 and 6200.

$$z = \frac{y - \mu}{\sigma} = \frac{5000 - 6200}{900} = \frac{-1200}{900} = -1.33$$

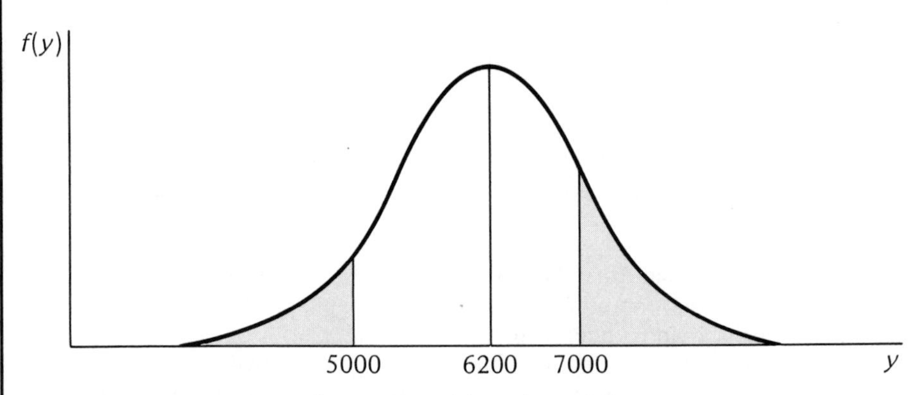


**Figure 3.14** Areas greater than 7000 and less than 5000 for $\mu = 6200$ and $\sigma = 900$, example 3.6

The area between the mean of a normal distribution and a value 1.33 standard deviations to the left of the mean is, from table 1, the appendix, .4082. Hence the probability of observing an annual income less than \$5000 is

$$.5 - .4082 = .0918$$

Similarly, to compute the probability of observing a salary above \$7000, we determine the area between 6200 and 7000.

$$z = \frac{y - \mu}{\sigma} = \frac{7000 - 6200}{900} = .89$$

The area corresponding to $z = .89$ is .3133. Hence the desired probability is

$$.5 - .3133 = .1867$$

**EXAMPLE 3.7**

From income tax returns in the previous year, it has been found that for a given income classification, the amount of money owed to the government over and above the amount paid in the estimated tax vouchers for the first three payments is approximately normally distributed with a mean of \$530 and a standard deviation of \$205. Find the 75th percentile for this distribution of measurements.

percentile

**SOLUTION**

The 75th percentile is, by definition, the value of $y$ such that 75% of the measurements are below it and 25% above it, as shown in figure 3.15. By

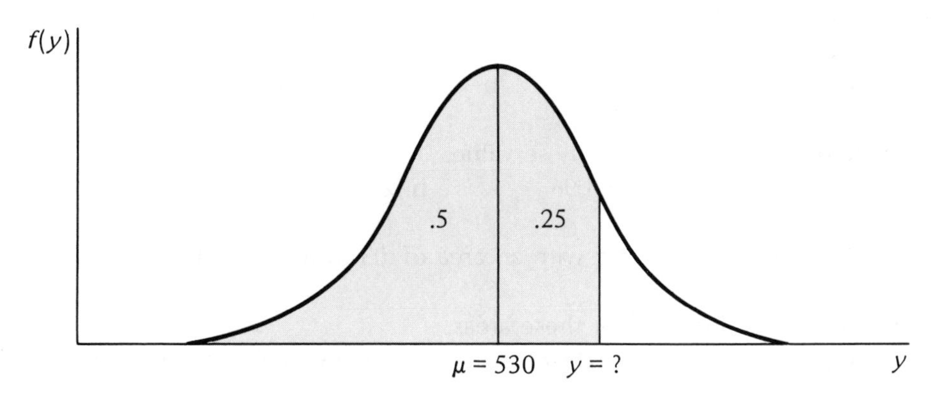


**Figure 3.15** Area under the normal curve for the 75th percentile, example 3.7

referring to table 1, the appendix, the value of $z$ corresponding to an area of .25 to the right of $\mu = 530$ is about .67. We find this value of $z$ by first looking for an area in table 1 that is close to .25. Once we find that, we look across to the $z$-value and to the top of the table for the hundredths in the $z$-value. For an area of .25, $z$ is about .67. Substituting into $z = (y - \mu)/\sigma$ and solving for $y$, we have $y = 667.35$. This value is the 75th percentile of the distribution of owed money for this tax classification.

# Summary 3.5

Many of the statistics computed from a random sample of measurements selected from the population of interest will be sums or averages. Using the Central Limit Theorem of this chapter, we know that the sampling distributions for these statistics will be normal, and with table 1, the appendix, we can determine the probability that a certain statistic falls in a given interval. Using this known probability, we will be able to make statistical inferences.

Do not be concerned now about how these inferences will be made. All you need to be aware of for the moment is that we will know the form of many sampling distributions. Because these distributions will be normal and because we can compute probabilities associated with a normal random variable, we will be able to determine probabilities associated with these statistics.

# Exercises

3.7. Use table 1, the appendix, to calculate the area under the curve between these values.

a. $z = 0$ and $z = 1.5$ b. $z = 0$ and $z = 1.8$

3.8. Repeat exercise 3.7 for these values.

a. $z = -1.96$ and $z = 1.96$ b. $z = -2.33$ and $z = 2.33$

3.9. What is the value of $z$ with an area of .05 to its right? To its left?

3.10. Find the value of $z$ for these areas.

a. an area .01 to the right of $z$ b. an area .10 to the left of $z$

3.11. Find the probability of observing a value of $z$ greater than these values.

a. 1.96 b. 2.21 c. 2.86 d. 0.73

3.12. Find the probability of observing a value of $z$ less than these values.

a. $-1.20$ b. $-2.62$ c. 1.84 d. 2.17

$ 3.13. Records maintained by the office of budget in a particular state indicate that the amount of time elapsed between the submission of travel vouchers and the final reimbursement of funds has approximately a normal distribution with a mean of 39 days and a standard deviation of 6 days.

a. What is the probability that the elapsed time between submission and reimbursement will exceed 50 days?

b. If you had a travel voucher submitted more than 55 days ago, what might you conclude?

3.14. Psychomotor retardation scores for a large group of manic-depressive patients were found to be approximately normal with a mean of 930 and a standard deviation of 130.

a. What fraction of the patients scored between 800 and 1100?

b. Less than 800?

c. Greater than 1200?

3.15. Refer to exercise 3.14.

a. Find the 90th percentile for the distribution of manic-depressive scores. (Hint: Solve for $y$ in the expression $z = (y - \mu)/\sigma$, where $z$ is the number of standard deviations the 90th percentile lies above the mean $\mu$.)

b. Find the interquartile range.

3.16. Federal resources have been tentatively approved for funding the construction of an outpatient clinic. But in order for the designers to present plans for a facility that will handle patient load requirements while still staying within a limited budget, a study of patient demand was made. From studying a similar facility in the area, it was found that the distribution of the number of patients requiring hospitalization during a week could be approximated by a normal distribution with a mean of 125 and a standard deviation of 32.

a. Use the Empirical Rule to describe the distribution of $y$, the number of patients requesting service in a week.

b. If the facility was built with a 160-patient capacity, what fraction of the weeks might the clinic be unable to handle the demand?

3.17. Refer to exercise 3.16. What size facility should be built so that the probability of the patient load exceeding the clinic capacity is .05? .01?

3.18. The distribution of the milk fat percentages for holstein cattle in a particular state during the 1960s was approximately normal with a mean of 3.7 and a standard deviation of .3.

a. What percentage of the holsteins had a milk fat percentage less than 3?

b. Greater than 4.5?

3.19. Refer to exercise 3.18.

a. Find the limits within which 90% of the milk fat percentages fell.

b. Compute the 95th percentile for the distribution of milk fat percentages.

3.20. Refer to exercise 3.18. Suppose a random sample of $n = 25$ holsteins is selected from the population of holstein cattle in the state.
a. Describe the distribution of $\bar{y}$, the mean milk fat percentage for the sample of 25 cattle.
b. Compare the distribution of $\bar{y}$ in part a to that for a distribution of $\bar{y}$ from a sample of 100 holsteins.
c. What is the probability that the sample mean milk fat percentage would exceed 4 in part a?

3.21. Random samples of size 20 are repeatedly drawn from a normal distribution with a mean of 65 and a standard deviation of 8.
a. Describe the sampling distribution for $\bar{y}$.
b. What fraction of the sample mean should be in the interval from 60 to 72?

3.22. Refer to exercise 3.21.
a. Describe the sampling distribution of the sample sum $\Sigma_{i=1}^{20} y_i$.
b. Locate the 25th and 75th percentiles for the sampling distribution of $\Sigma_{i=1}^{20} y_i$.

3.23. The breaking strengths for one-foot-square samples of a particular synthetic fabric are approximately normally distributed with a mean of 2250 pounds per square inch (psi) and a standard deviation of 10.2 psi.
a. Find the probability of selecting a one-foot-square sample of material at random which upon testing would have a breaking strength in excess of 2265 psi.
b. Describe the sampling distribution for $\bar{y}$ based on random samples of 15 one-foot sections.

3.24. Refer to exercise 3.23. Suppose that a new synthetic fabric has been developed which may have a different mean breaking strength. A random sample of 15 one-foot sections is obtained and each section is tested for breaking strength. If we assume that the population standard deviation for the new fabric is identical to that for the old fabric (exercise 3.23), give the standard deviation for the sampling distribution of $\bar{y}$ using the new fabric.

3.25. Refer to exercise 3.23. Suppose that the mean breaking strength for the sample of 15 one-foot sections of the new synthetic fabric is 2268. What is the probability of observing a value of $\bar{y}$ equal to or greater than 2268, assuming that the mean breaking strength for the new fabric is 2250, the same as that for the old.

3.26. Refer to exercise 3.25. Based on your answer in exercise 3.25, do you believe the new fabric has the same mean breaking strength as the old? (Assume $\sigma = 10.2$.)

*3.27. In figure 3.4 (p. 56) we visually inspected the relative frequency histogram for 50 sample means, each based on $n = 5$ measurements, and noted its bell shape. Another way to determine whether a set of measurements is bell-shaped (normal) is to construct a plot of the sample data on *probability paper*. This plot is called a *probability plot*. If the probability plot is approximately a straight line, we say the measurements were selected from a normal population. Use a computer program to construct a probability plot for the 50 sample means. A Minitab program is shown below for purposes of illustration. (Note: This program will display the histogram and the probability plot.)

```
SET THE FOLLOWING DATA INTO COL C6

12.0      8.8      11.4      5.0      8.8
 7.0     12.4      15.0      8.6      8.6
  ⋮        ⋮         ⋮         ⋮        ⋮
 7.0      7.8       7.6      4.2      9.6

HISTOGRAM OF COL C6
NSCORES OF C6 PUT IN C7
PLOT C6 VS C7
STOP
```

# References


Mendenhall, W., and Ott, L. 1976. *Understanding statistics*. 2d ed. N. Scituate, Mass.: Duxbury Press. Chapter 6.

Mendenhall, W.; Ott, L; and Larson, R. 1974. *Statistics: A tool for the social sciences*. N. Scituate, Mass.: Duxbury Press. Chapter 5.

Ryan, T.A.; Joiner, B.L.; and Ryan, B.F. 1976. *Minitab student handbook*. N. Scituate, Mass.: Duxbury Press.

# 4 Inferences About a Population Mean: Large-Sample Results

## 4.1 Introduction

We have stated previously that the objective of inferential statistics is to make inferences about population parameters based on information contained in a sample. However, our inferences can be phrased in two different ways. Consider the following situations.

In a preliminary study of servicemen returning home from a tour of duty overseas, we wish to administer a political awareness questionnaire to a sample of returnees. Because of the ever-changing political situation, very little previous information is available to help us in estimating the average response for returnees now or in the near future. However, the sample information could be used to *estimate* (predict) the mean political awareness of all returnees, and this estimate could then be used to project a proper format for future debriefing sessions.

In an experimental situation, we are concerned about the effectiveness of a new drug to control a particular species of worms in the stomach of cattle. Let us suppose that a standard drug exists and that after a fixed period of treatment on the standard, the average infestation level has been shown to be $\mu = 10.3$. If the new drug is indeed effective, it must have a mean infestation level after treatment at least as low as that for the standard. Thus we might wish to *test the research hypothesis* that the mean infestation level for the test drug is less than 10.3. Note that in this situation we are not asking, "What is the mean infestation level after treatment?" We are asking a more pointed question, "Is the mean infestation level lower than 10.3?"

These two examples illustrate the two different inference-making procedures we can use: *estimation* or *test of hypothesis*. The two procedures are contrasted by the following questions. In estimating or predicting a population parameter, we are asking the question, "What is the value of the population parameter?" This procedure reflects an experimental situation where we know very little about the parameter of interest. In testing an hypothesis we are asking the question, "Is the parameter equal to this specific value?" In this situation we do know something about the parameter of interest.

estimation

test of hypothesis

In this chapter we will consider these two inference-making procedures: estimation of a population mean $\mu$ and testing a statistical hypothesis about $\mu$.

## 4.2 Point Estimation of $\mu$

Estimation procedures can be divided into two types. We can calculate a single number, called a *point estimate*, from the sample data and use this value to estimate the parameter of interest. Or we can compute two numbers from the sample data and use the interval formed by the two numbers as an *interval estimate* of the parameter of interest. The distinction between the two is quite simple. In the first case we compute one number and infer that the parameter is that number. In the second case we compute two numbers and infer that the parameter lies in the interval between them. We will consider point estimation of a population mean in this section and interval estimation in section 4.3.

point estimate

interval estimate

Statistical inference-making procedures differ from ordinary inference-making procedures in that we not only make an inference but we also provide a measure of how good our inference is. For example, the value of $\bar{y}$ computed from the sample measurements will be a point estimate of $\mu$. This is our guessed value of $\mu$. However, since we know that there are many possible values of $\bar{y}$, and since it is quite likely that the observed value of $\bar{y}$ will not be exactly equal to $\mu$, we must say how close $\bar{y}$ should be to this parameter we are trying to estimate. To do this, we refer to the sampling distribution of $\bar{y}$ (discussed in section 3.3).

Recall that the Central Limit Theorem for $\bar{y}$ stated that if repeated samples of $n$ measurements each were drawn from a population with a finite mean $\mu$ and a standard deviation $\sigma$, then the sampling distribution of $\bar{y}$ would be approximately normal, with mean $\mu$ and standard deviation $\sigma/\sqrt{n}$. [Although a specific sample size requirement was not stated in chapter 3, in this chapter we will assume (without proof) that the sample size must be 30 or more in order for this Central Limit Theorem to hold.] Thus from knowledge of the sampling distribution of $\bar{y}$ and areas under the normal curve, we know, for example, that approximately 95% of the point estimates (sample means) calculated from repeated samples will be within $2\sigma_{\bar{y}} = 2\sigma/\sqrt{n}$ of $\mu$ (see figure 4.1). We can use this information to provide a measure of the goodness of our point estimate, as we will show in the following discussion.

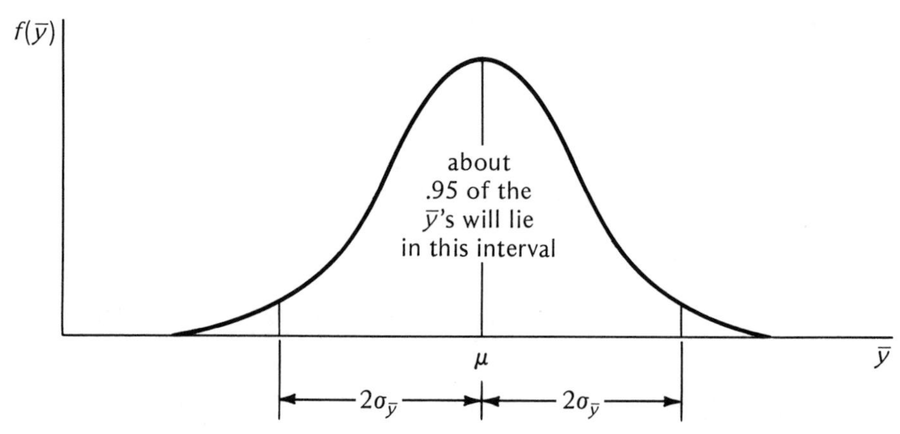


**Figure 4.1** Sampling distribution of $\bar{y}$

error of estimation

The *error of estimation* for a given point estimate is defined to be the difference $\bar{y} - \mu$, ignoring the sign of the result. That is, the error of estimation is the absolute value of the difference between what we think the parameter's value is and what it actually is. This quantity is designated by $|\bar{y} - \mu|$.

$$\text{error of estimation} = |\bar{y} - \mu|$$

bound on error

Because, by the Empirical Rule, the error of estimation will be less than $2\sigma_{\bar{y}}$ approximately 95% of the time, we call the quantity $2\sigma_{\bar{y}}$ a *bound on the error of estimation*. This quantity is a measure of how good our inference is. The smaller the bound on the error of estimation, the better the inference.

For a given point estimation problem, the point estimate is calculated along with a corresponding bound on the error of estimation. This procedure for estimating $\mu$ is summarized in the box.

**Point Estimation of a Population Mean $\mu$**

Point estimate of $\mu$: $\bar{y}$.

Bound on the error of estimation: $2\sigma_{\bar{y}} = 2\sigma/\sqrt{n}$.

Note: The sample size $n$ is assumed to be 30 or more.

**EXAMPLE 4.1**

A social worker is interested in estimating the average length of time spent outside of prison for first offenders who later commit a second crime and are sent to prison again. A random sample of $n = 150$ prison records in the county court house indicate that the average length of prison-free life between first and second offenses is 3.2 years, with a standard deviation of 1.1 years. Use the sample information to estimate $\mu$, the mean prison-free life between first and second offenses for all prisoners on record in the county court house. Place a bound on the error of estimation.

**SOLUTION**

The sample data indicate that $\bar{y} = 3.2$ and $s = 1.1$. Hence our point estimate of $\mu$ is 3.2. The corresponding bound on the error of estimation is given by

$$\text{bound on error} = 2\sigma_{\bar{y}} = \frac{2\sigma}{\sqrt{n}}$$

When $\sigma$ is unknown but $n \geq 30$, we can replace $\sigma$ by $s$ and obtain an *approximate* bound on the error.

$$2\sigma_{\bar{y}} \approx \frac{2s}{\sqrt{n}} = \frac{2(1.1)}{\sqrt{150}} = .18$$

Knowing that approximately 95% of the sample means calculated in repeated sampling would lie within two standard deviations of $\mu$, we feel fairly certain that $\bar{y} = 3.2$ is within a distance of .18 of the actual mean $\mu$.

# Exercises

4.1. The rust mite, a major pest of citrus in the state of Florida, punctures the cells of the leaves and fruit. Damage by rust mites is readily recognizable because the injured fruit will display a brownish (rust) color and be somewhat reduced in size depending on the severity of the attack. If the rust mites are not controlled, the affected groves will have a substantial reduction in both the fruit yield and the fruit quality. In either case the citrus grower suffers financially since his produce will be of a lower grade and sell for less on the fresh fruit market. This year more and more citrus growers have gone to a preventive program of maintenance spraying for rust mites. In evaluating the effectiveness of the program, a random sample of 60 10-acre plots, one each from 60 groves, showed an average yield of 850 boxes, with a standard deviation of 100 boxes. Give a point estimate of $\mu$, the average (10-acre) yield for all groves utilizing such a maintenance spraying program. Compute a bound on the error of estimation.

4.2. A study was conducted to examine the effect of a preparation of mosaic virus on tobacco leaves. In a random sample of $n = 32$ leaves, the mean number of lesions was 22, with a standard deviation of 3. Use these data to estimate the average number of lesions for leaves affected by a preparation of mosaic virus. Place a bound on the error of estimation.

4.3. We all remember being told, "Your fever has subsided and your temperature has returned to normal." What do we mean by the word "normal"? Most people use the benchmark of 98.6° Fahrenheit, but this does not apply to all people, only the "average" person. Without putting words into someone's mouth, we might define a person's normal temperature to be his or her average temperature when healthy. But even this definition is cloudy because there is variation in a person's temperature throughout the day. To determine a subject's normal temperature, we recorded his temperature for a random sample of 30 days. On each day selected for inclusion in the sample, the temperature reading was made at 7 A.M. The sample mean and standard deviation for these 30 readings were, respectively, 98.4 and .15. Assuming the subject was healthy on the days examined, use these data to estimate his 7A.M. "normal" temperature. Place a bound on the error of estimation.

## 4.3 Interval Estimation of $\mu$

We can use the point estimate $\bar{y}$ to form an interval estimate for the population mean $\mu$. From the Central Limit Theorem for the sample mean given in chapter 3, we know that for large $n$ (30 or more), $\bar{y}$ will be approximately normally distributed, with a mean $\mu$ and a standard deviation $\sigma_{\bar{y}}$. Then from our knowledge of the Empirical Rule and areas under a normal curve, we know that the interval $\mu \pm 2\sigma_{\bar{y}}$, or more precisely, the interval $\mu \pm 1.96\sigma_{\bar{y}}$, includes 95% of the $\bar{y}$'s in repeated sampling, as shown in figure 4.2.

Consider the interval $\bar{y} \pm 1.96\sigma_{\bar{y}}$. Any time $\bar{y}$ lies in the interval $\mu \pm 1.96\sigma_{\bar{y}}$, the interval $\bar{y} \pm 1.96\sigma_{\bar{y}}$ will contain the parameter $\mu$ (see figure 4.3), and this will occur with probability .95. The interval $\bar{y} \pm 1.96\sigma_{\bar{y}}$ represents an interval estimate of $\mu$.

We evaluate the goodness of an interval estimation procedure by examining the fraction of times in repeated sampling that interval estimates would

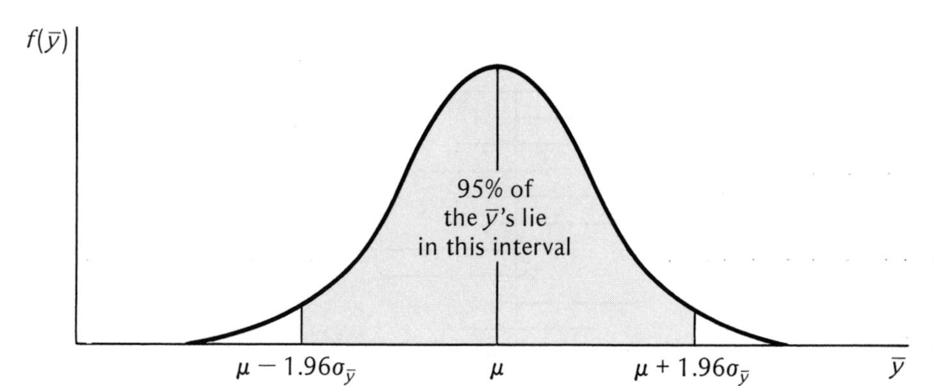


**Figure 4.2** Sampling distribution for $\bar{y}$

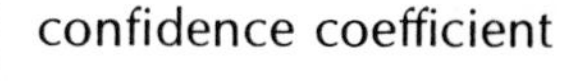
confidence coefficient

encompass the parameter to be estimated. This fraction, called the *confidence coefficient,* is .95 when using the formula $\bar{y} \pm 1.96\sigma_{\bar{y}}$. That is, 95% of the time in repeated sampling, intervals calculated using the formula $\bar{y} \pm 1.96\sigma_{\bar{y}}$ will contain the mean $\mu$.

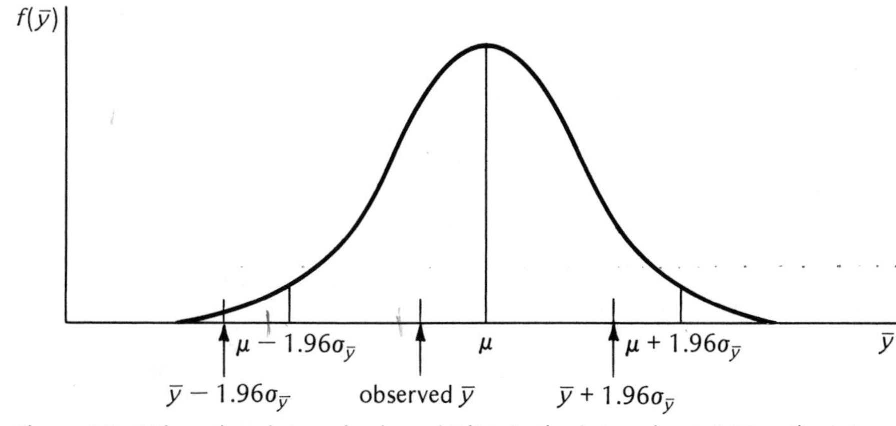


**Figure 4.3** When the observed value of $\bar{y}$ lies in the interval $\mu \pm 1.96\sigma_{\bar{y}}$, the interval $\bar{y} \pm 1.96\sigma_{\bar{y}}$ contains the parameter $\mu$

This idea is illustrated in figure 4.4. Twenty different samples were drawn from a population with mean $\mu$ and variance $\sigma^2$. For each sample an interval estimate was computed using the formula $\bar{y} \pm 1.96\sigma_{\bar{y}}$. Note that although the intervals bob about, most of them capture the parameter $\mu$. In fact, if we repeated the process of drawing samples and computing confidence intervals, 95% of the intervals so formed would contain $\mu$.

95% confidence interval

In a given experimental situation, we calculate only one such interval. This interval, called a *95% confidence interval,* represents an interval estimate of $\mu$.

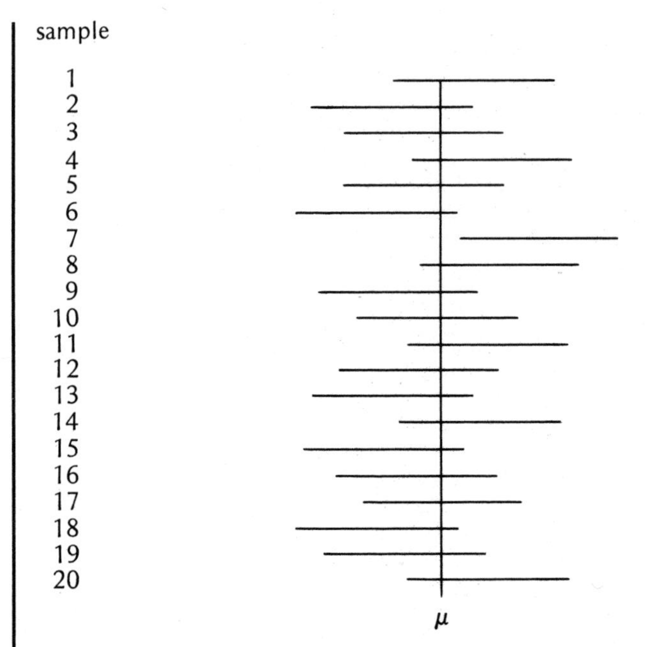


**Figure 4.4** Twenty interval estimates computed by using $\bar{y} \pm 1.96\sigma_{\bar{y}}$

### EXAMPLE 4.2

In a random sample of $n = 36$ parochial schools throughout the South, the average number of pupils per school was 379.2, with a standard deviation of 124. Use the sample to construct a 95% confidence interval for $\mu$, the mean number of pupils per school for all parochial schools in the South.

### SOLUTION

The sample data indicate that $\bar{y} = 379.2$ and $s = 124$. The appropriate 95% confidence interval is then computed by using the formula

$$\bar{y} \pm 1.96\sigma_{\bar{y}}$$

where $\sigma_{\bar{y}} = \sigma/\sqrt{n}$. In section 5.2 we will present a procedure for obtaining a confidence interval for $\mu$ when $\sigma$ is unknown. However, for all practical purposes, provided the sample size is 30 or more, we can substitute the sample standard deviation $s$ for $\sigma$ in the confidence interval formula. With $s$ replacing $\sigma$, our interval is

$$379.2 \pm \frac{1.96(124)}{\sqrt{36}} \quad \text{or} \quad 379.2 \pm 40.51$$

The interval from 338.69 to 419.71 forms a 95% confidence interval for $\mu$. In other words, we are 95% sure that the average number of pupils per school for parochial schools throughout the South lies between 338.69 and 419.71.

There are many different confidence intervals for $\mu$, depending on the confidence coefficient we choose. For example, the interval $\mu \pm 2.58\sigma_{\bar{y}}$ would

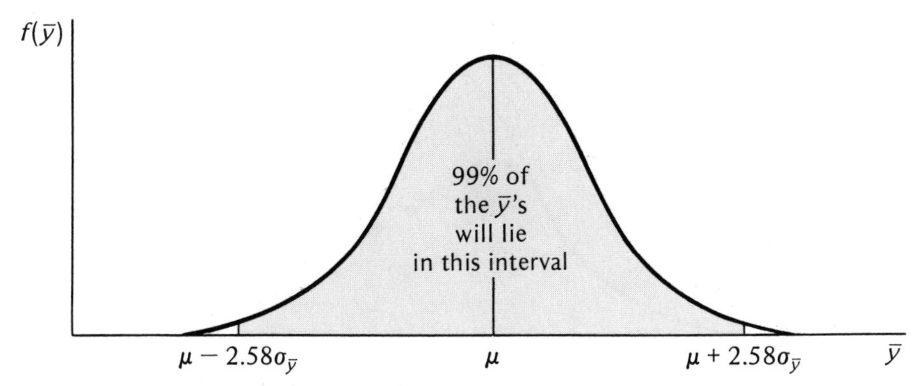


**Figure 4.5** Sampling distribution of $\bar{y}$

include 99% of the values of $\bar{y}$ in repeated sampling (see figure 4.5), and the interval $\bar{y} \pm 2.58\sigma_{\bar{y}}$ forms a 99% confidence interval for $\mu$.

99% confidence interval

We can state a general formula for a confidence interval for $\mu$ with a confidence coefficient of $(1 - \alpha)$, where $\alpha$ (Greek letter alpha) is between 0 and 1. For a specified value of $(1 - \alpha)$, a $100(1 - \alpha)\%$ confidence interval for $\mu$ is given by the formula in the box.

$(1 - \alpha)$ = confidence coefficient

**Confidence Interval for $\mu$**

$$\bar{y} \pm z_{\alpha/2}\sigma_{\bar{y}}$$

The quantity $z_{\alpha/2}$ is a value of $z$ having an area $\alpha/2$ to its right. In other words, at a distance of $z_{\alpha/2}$ standard deviations to the right of $\mu$, there is an area of $\alpha/2$ under the normal curve. Values of $z_{\alpha/2}$ can be obtained from table 1 in the appendix by looking up the $z$-value corresponding to an area of $(1 - \alpha)/2$ (see figure 4.6). Common values of the confidence coefficient $(1 - \alpha)$ and $z_{\alpha/2}$ are given in table 4.1.

$z_{\alpha/2}$

**Table 4.1** Common values of the confidence coefficient $(1 - \alpha)$ and the corresponding $z$-value, $z_{\alpha/2}$

| *Confidence Coefficient,* $(1 - \alpha)$ | *Area in Table 1,* $(1 - \alpha)/2$ | *Corresponding z-value,* $z_{\alpha/2}$ |
|---|---|---|
| .90 | .450 | 1.645 |
| .95 | .475 | 1.96 |
| .98 | .490 | 2.33 |
| .99 | .495 | 2.58 |

**EXAMPLE 4.3**

A forester is interested in estimating the average number of "count trees" per acre (trees larger than a specified size) on a 2000-acre plantation. He

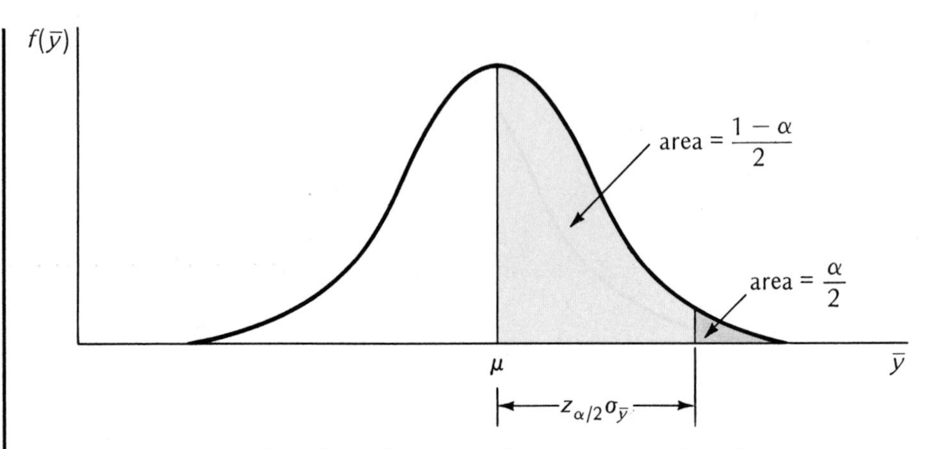


**Figure 4.6** Interpretation of $z_{\alpha/2}$ in the confidence interval formula

can then use this information to determine the total timber volume for trees in the plantation. A random sample of $n = 50$ one-acre plots was selected and examined. The average (mean) number of count trees per acre was found to be 27.3, with a standard deviation of 12.1. Use this information to construct a 99% confidence interval for $\mu$, the mean number of count trees per acre for the entire plantation.

**SOLUTION**

We use the general confidence interval with confidence coefficient equal to .99 and a $z_{\alpha/2}$-value equal to 2.58 (see table 4.1). Substituting into the formula $\bar{y} \pm 2.58\sigma_{\bar{y}}$ and replacing $\sigma$ by $s$ in $\sigma_{\bar{y}} = \sigma/\sqrt{n}$, we have

$$27.3 \pm \frac{2.58(12.1)}{\sqrt{50}}$$

This corresponds to the confidence interval $27.3 \pm 4.42$, that is, the interval from 22.88 to 31.72. Thus we are 99% sure that the average number of count trees per acre is between 22.88 and 31.72.

In section 4.2 we indicated that the goodness of a point estimate is measured by the magnitude of the bound on the error of estimation. For interval estimation the width of the confidence interval and the confidence coefficient measure the goodness of the inference. Obviously, for a given confidence coefficient, the smaller the width of the interval, the better the inference. The confidence coefficient must be specified by the experimenter, and it expresses how much assurance he or she wishes to place in whether the interval estimate encompasses the parameter of interest.

# Exercises

4.4. Refer to exercise 4.3 (p. 74). Use the sample data to construct a 95% confidence interval for $\mu$, the subject's average (normal) 7 A.M. temperature. How would your confidence interval change if a 98% confidence interval was required?

4.5. An experiment was conducted to examine the susceptibility of root stocks of a variety of lemon trees to a specific larva. Forty of the plants were subjected to the larvae and examined after a fixed period of time. The response of interest was the logarithm of the number of larvae per gram that was counted on each root stock. For these 40 plants the sample mean was 9.02 and the standard deviation was 1.12. Use these data to construct a 90% confidence interval for $\mu$, the mean susceptibility for the population of lemon tree root stocks from which the sample was drawn.

4.6. A mobility study was conducted among a random sample of 900 high school graduates of a particular state over the past ten years. For each of the persons sampled, the distance between the high school attended and the present permanent address was recorded. For these data $\bar{y} = 430$ miles and $s = 262$ miles. Using a 95% confidence interval, estimate the average number of miles between a person's high school and his or her present permanent address for high school graduates of the state over the past ten years.

4.7. Refer to exercise 4.2 (p. 74). Use the sample data to form a 99% confidence interval on $\mu$, the average number of lesions for tobacco leaves affected by a preparation of mosaic virus.

# Testing an Hypothesis About $\mu$ 4.4

The second type of inference-making procedure is a statistical test of an hypothesis. As with estimation procedures, we will make an inference about a population parameter, but here the inference will be of a different sort. In this section we will present a statistical test that will lead to an answer to the question, "Is the population mean equal to a specified value $\mu_0$?" For example, in studying the antipsychotic properties of an experimental compound, we might ask whether the average shock avoidance response for rats treated with a specific dose is 60, the same value that has been observed after extensive testing using a suitable standard drug.

statistical test

A *statistical test* is based on the concept of proof by contradiction and is composed of the five parts listed in the box.

**Five Parts of a Statistical Test**

1. Null hypothesis, denoted by $H_0$.
2. Research hypothesis (also called the alternative hypothesis), denoted by $H_a$.
3. Test statistic, denoted by T.S.
4. Rejection region, denoted by R.R.
5. Conclusion.

research hypothesis, $H_a$

null hypothesis, $H_0$

For example, in setting up a statistical test concerning the mean yield per acre (in bushels) for a particular variety of soybeans, we may be interested in the *research hypothesis* that the mean yield per acre $\mu$ is greater than 520 bushels, the average observed for farms throughout a state in the past several years. To verify the research hypothesis, we try to contradict another hypothesis, called the *null hypothesis,* that $\mu = 520$ (i.e., the highest average yield that is still contradictory to the research hypothesis).

test statistic, T.S.

Having stated the null and research hypotheses, we then obtain a random sample of one-acre yields from farms throughout the state and compute $\bar{y}$ and $s$, the sample mean and standard deviation, respectively. The decision to accept the null hypothesis or reject the null hypothesis in favor of the research hypothesis is based on a *test statistic* or decision maker computed from the sample data. A logical choice as a decision maker for $\mu$ would be $\bar{y}$ or some function of the sample mean.

rejection region, R.R.

If we choose $\bar{y}$ as the test statistic, we know that the sampling distribution of $\bar{y}$, assuming the null hypothesis is true, is approximately normal with mean $\mu = 520$. Values of $\bar{y}$ which are contradictory to the null hypothesis and are in favor of the research hypothesis will be those that lie in the upper tail of the distribution of $\bar{y}$. See figure 4.7. These contradictory values form a *rejection region* for our statistical test. If the observed value of $\bar{y}$ falls in the rejection region of figure 4.7, we would reject the null hypothesis that the mean yield per acre is $\mu = 520$ in favor of the research hypothesis that $\mu > 520$. Note that we are verifying the research hypothesis by contradicting the null hypothesis. If the observed value of $\bar{y}$ falls in the acceptance region rather than the rejection region, we accept the null hypothesis.

Before we give a precise location for the acceptance and rejection regions of figure 4.7, we should consider the two types of errors that can be made while performing a statistical test of an hypothesis. As with any two-way decision

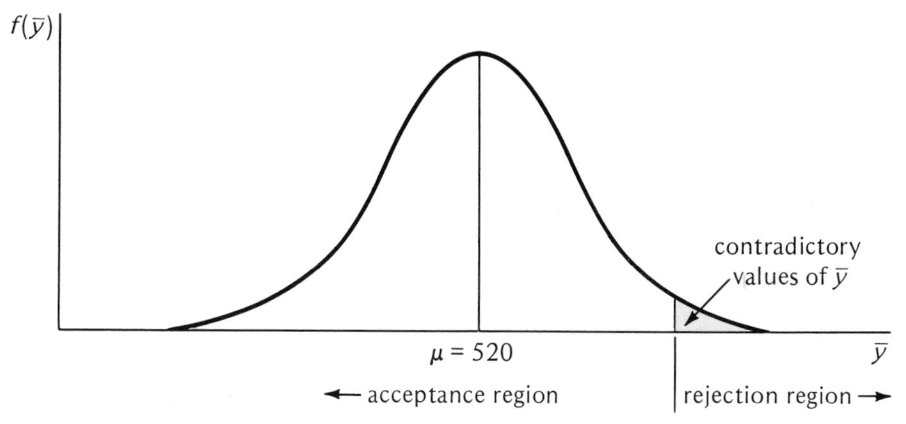


**Figure 4.7** Assuming $H_0$ is true, contradictory values of $\bar{y}$ are in the upper tail

process, we can make an error by falsely rejecting the null hypothesis or by falsely accepting the null hypothesis. We give these errors special names.

---

**Definition 4.1**
probability $\alpha$

A *type I error* is committed if we reject the null hypothesis when it is true. The probability of a type I error is denoted by the symbol $\alpha$.

---

**Definition 4.2**
probability $\beta$

A *type II error* is committed if we accept the null hypothesis when it is false and the research hypothesis is true. The probability of a type II error is denoted by the symbol $\beta$ (Greek letter beta).

---

The relationships of $\alpha$ and $\beta$ are shown in table 4.2.

**Table 4.2** Probability of committing a type I or a type II error

| | *Null Hypothesis* | |
|---|---|---|
| *Decision* | *True* | *False* |
| reject $H_0$ | $\alpha$ | correct |
| accept $H_0$ | correct | $\beta$ |

Although it would be desirable to determine the acceptance and rejection regions to minimize, simultaneously, both $\alpha$ and $\beta$, this is not possible. The probabilities associated with type I and type II errors are inversely related. For a fixed sample size $n$, as we change the rejection region to increase $\alpha$, then $\beta$ decreases, and vice versa.

To alleviate what appears to be an impossible bind, the experimenter specifies a tolerable probability for a type I error of the statistical test. Thus he may choose $\alpha$ to be .01, .05, .10, and so on. Specification of a value for $\alpha$ then locates the rejection region. Determination of the associated probability of a type II error is more complicated and will be delayed until later in the chapter.

Let us now see how the choice of $\alpha$ locates the rejection region. Returning to our soybean example, we will reject the null hypothesis for large values of the sample mean $\bar{y}$. Suppose we have decided to take a sample of $n = 36$ one-acre plots, and from these data we compute $\bar{y} = 573$ and $s = 124$. Can we conclude that the mean yield for all farms is above 520?

specifying $\alpha$

Before answering this question we must specify $\alpha$. If we are willing to take the risk that 1 time in 40 we would incorrectly reject the null hypothesis, then $\alpha = 1/40 = .025$. An appropriate rejection region can be specified for this value of $\alpha$ by referring to the sampling distribution of $\bar{y}$. Assuming the null hypothesis is true, then $\bar{y}$ is normally distributed, with $\mu = 520$ and $\sigma_{\bar{y}} \approx s/\sqrt{n} = 124/\sqrt{36} = 20.67$. Since the shaded area of figure 4.7 corresponds to $\alpha$, locating a rejection region with an area of .025 in the right-hand tail of the distribution of $\bar{y}$ would be equivalent to determining the value of $z$ that has an area .025 to its right. Referring to table 1, the appendix, this value of $z$ is 1.96. Thus the rejection region for our example would be located 1.96 standard deviations ($1.96\sigma_{\bar{y}}$) above the mean $\mu = 520$. If the observed value of $\bar{y}$ is greater than 1.96 standard deviations above $\mu = 520$, we reject the null hypothesis. This is shown in figure 4.8.

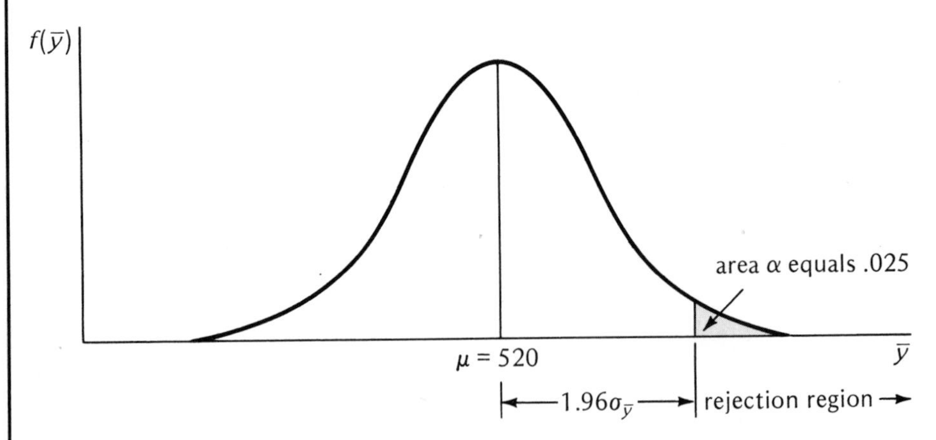


**Figure 4.8** Rejection region for the soybean example when $\alpha = .025$

**EXAMPLE 4.4**

Set up all the parts of a statistical test for the soybean example and use the sample data to reach a decision whether to accept or reject the null hypothesis. Set $\alpha = .025$.

**SOLUTION**

The five parts of the test are as follows:

$H_0$: $\mu = 520$.

$H_a$: $\mu > 520$.

T.S.: $\bar{y}$.

R.R.: For $\alpha = .025$, reject the null hypothesis if $\bar{y}$ lies more than 1.96 standard deviations above $\mu = 520$.

The computed value of $\bar{y}$ was 573. To determine the number of standard deviations that $\bar{y}$ lies above $\mu = 520$, we compute a z score for $\bar{y}$, using the formula

$$z = \frac{\bar{y} - \mu_0}{\sigma_{\bar{y}}}$$

where $\sigma_{\bar{y}}$ can be approximated by $s/\sqrt{n}$ (for $n \geq 30$). Substituting into the formula,

$$z = \frac{\bar{y} - \mu_0}{\sigma_{\bar{y}}} \approx \frac{573 - 520}{124/\sqrt{36}} = 2.56$$

Conclusion: Since the observed value of $\bar{y}$ lies more than 1.96 standard deviations above the hypothesized mean $\mu = 520$, we reject the null hypothesis in favor of the research hypothesis and conclude that the average soybean yield per acre is greater than 520.

one-tailed test

The statistical test we conducted in example 4.4 is called a *one-tailed test,* because the rejection region is located in only one tail of the distribution of $\bar{y}$.

If our research hypothesis is $H_a$: $\mu < 520$, small values of $\bar{y}$ would indicate rejection of the null hypothesis. This test would also be one-tailed but the rejection region would be located in the lower tail of the distribution of $\bar{y}$. Figure 4.9 displays the rejection region for the alternative hypothesis $H_a$: $\mu < 520$ when $\alpha = .025$.

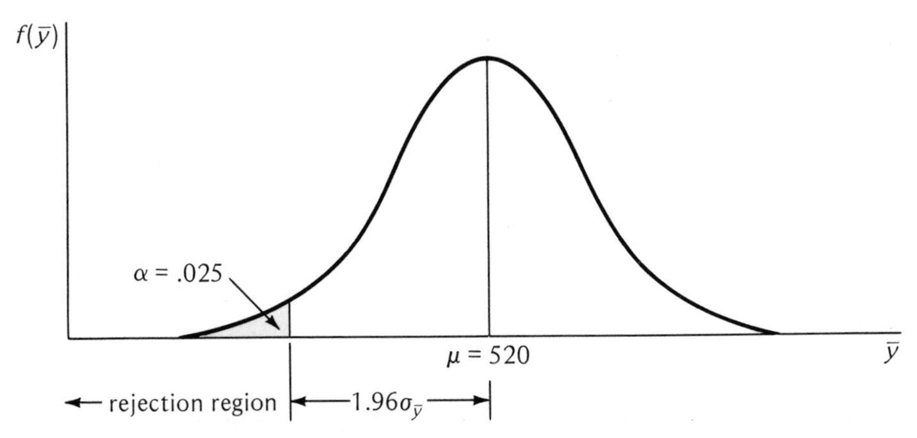


**Figure 4.9** Rejection region for $H_a$: $\mu < 520$ when $\alpha = .025$ for the soybean example

two-tailed test

We can formulate a *two-tailed test* for the research hypothesis $H_a: \mu \neq 520$, where we are interested in detecting whether the mean yield per acre of soybeans is greater or less than 520. Clearly both large and small values of $\bar{y}$ will contradict the null hypothesis, and we would locate the rejection region in both tails of the distribution of $\bar{y}$. A two-tailed rejection region for $H_a$: $\mu \neq 520$ and $\alpha = .05$ is shown in figure 4.10.

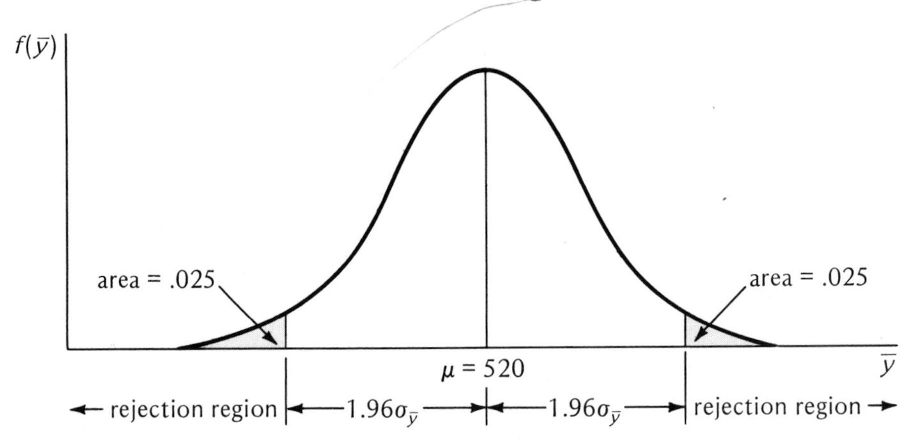


**Figure 4.10** Two-tailed rejection region for $H_a: \mu \neq 520$ when $\alpha = .05$ for the soybean example

## EXAMPLE 4.5

A corporation maintains a large fleet of company cars for its salespeople. To check the average number of miles driven per month per car, a random sample of $n = 40$ cars was examined. The mean and standard deviation for the sample were 2752 miles and 350 miles, respectively. Records for previous years indicate that the average number of miles driven per car per month was 2600. Use the sample data to test the research hypothesis that $\mu$ differs from 2600. Set $\alpha = .05$.

## SOLUTION

The research hypothesis for this statistical test is $H_a: \mu \neq 2600$; the null hypothesis is $H_0: \mu = 2600$. Using $\alpha = .05$, the two-tailed rejection region for this test is located as shown in figure 4.11.

To determine how many standard deviations our test statistic $\bar{y}$ lies away from $\mu = 2600$, we compute

$$z = \frac{\bar{y} - \mu}{\sigma/\sqrt{n}} \approx \frac{2752 - 2600}{350/\sqrt{40}} = 2.75$$

The observed value for $\bar{y}$ lies more than 1.96 standard deviations above the mean, hence we reject the null hypothesis in favor of the alternative $H_a$: $\mu \neq 2600$. Practically, since the computed value of $\bar{y}$ is greater than the hypothesized mean $\mu = 2600$, we conclude that the mean number of miles driven is greater than 2600.

**EXAMPLE 4.7**

Compute $\beta$ for the test in example 4.6 if the actual mean live weight of steers is 395.

**SOLUTION**

The research hypothesis for example 4.6 was $H_a$: $\mu > 380$. Using $\alpha = .01$ and the computing formula for $\beta$ with $\mu_0 = 380$ and $\mu_a = 395$, we have

$$\beta = P\left[z < z_{.01} - \frac{|\mu_0 - \mu_a|}{\sigma_{\bar{y}}}\right] = P\left[z < 2.33 - \frac{|380 - 395|}{35.2/\sqrt{50}}\right]$$

$$= P[z < 2.33 - 3.01] = P[z < -.68]$$

Referring to table 1, the appendix, the area corresponding to $z = .68$ is .2517. Hence $\beta = .5 - .2517 = .2483$.

Previously, when $\bar{y}$ did not fall in the rejection region, we concluded that there was insufficient evidence to reject $H_0$ because $\beta$ was unknown. Now when $\bar{y}$ falls in the acceptance region, we can compute $\beta$ corresponding to one (or more) alternative values for $\mu$ that appear reasonable in light of the experimental setting. Then provided we are willing to tolerate a probability of falsely accepting the null hypothesis equal to the computed value of $\beta$ for the alternative value(s) of $\mu$ considered, our decision is to accept the null hypothesis. Thus in example 4.7, if we are willing to risk a $\beta$ error of about .25 of falsely accepting the null hypothesis, we would then accept the null hypothesis $\mu = 380$.

**EXAMPLE 4.8**

Prospective salespeople for an encyclopedia company are now being offered a sales training program. Previous data indicate that the average number of sales per month for those who do not participate in the program is 33. To determine whether the training program is effective or not, a random sample of 35 new employees is given the sales training and then sent out into the field. One month later the mean and standard deviation for the number of sets of encyclopedias sold are 35 and 8.4, respectively. Do the data present sufficient evidence to indicate that the training program enhances sales? Use $\alpha = .05$.

**SOLUTION**

The five parts to our statistical test are as follows:

$H_0$: $\mu = 33$.

$H_a$: $\mu > 33$.

T.S.: $z = \dfrac{\bar{y} - \mu_0}{\sigma_{\bar{y}}} \approx \dfrac{35 - 33}{8.4/\sqrt{35}} = 1.41.$

R.R.: For $\alpha = .05$ we will reject the null hypothesis if $z > z_{.05} = 1.645$.

Conclusion: Since the observed value of $z$ does not fall into the rejection region, we reserve judgment on accepting $H_0$ until we calculate $\beta$. That is, we conclude that there is insufficient evidence to reject the null hypothesis that persons on the sales program have the same mean number of sales per month as those not under the program.

**EXAMPLE 4.9**

Refer to example 4.8. Suppose that the encyclopedia company thinks that the cost of financing the sales program will be offset by increased sales if those on the program average 38 sales per month. Compute $\beta$ for $\mu_a = 38$ and, based on the value of $\beta$, indicate whether you would accept the null hypothesis.

**SOLUTION**

Using the computational formula for $\beta$ with $\mu_0 = 33$, $\mu_a = 38$, and $\alpha = .05$, we have

$$\beta = P\left[z < z_{.05} - \frac{|\mu_0 - \mu_a|}{\sigma_{\bar{y}}}\right] = P\left[z < 1.645 - \frac{|33 - 38|}{8.4/\sqrt{35}}\right] = P[z < -1.88]$$

The area corresponding to $z = 1.88$ in table 1, the appendix, is .4699. Hence
$\beta = .5 - .4699 = .0301$

Because $\beta$ is relatively small, we accept the null hypothesis and conclude that the training program has not increased the average sales per month above the point where increased sales would offset the cost of the training program.

In sections 4.2 and 4.3 we discussed how we measure the goodness of point and interval estimates. The goodness of a statistical test can be measured by the magnitudes of the type I and type II errors, $\alpha$ and $\beta$. When $\alpha$ is preset at a tolerable level by the experimenter, $\beta$ is a function of the sample size for a fixed value of $\mu_a$. The larger the sample size, the more information we have concerning $\mu$ and hence the smaller the value of $\beta$. In chapter 9 we will consider the problem of designing an experiment for testing $H_0$: $\mu = \mu_0$ when $\alpha$ is specified and $\beta$ is preset for a fixed actual value $\mu_a$. This problem reduces to determining the sample size needed for the fixed values of $\alpha$ and $\beta$.

# 4.5 The Level of Significance of a Statistical Test

In the previous section we presented an introduction to hypothesis testing along rather traditional lines: we defined the parts of a statistical test along with the two types of errors and their associated probabilities $\alpha$ and $\beta$. In recent years many statisticians and users of statistics have objected to this decision-based approach to hypothesis testing. Rather than running a statistical test with a preset value of $\alpha$, it is argued that we should specify the null and alternative hypotheses, collect the sample data, and determine the weight of the evidence for rejecting the null hypothesis. This weight, given in terms of a probability, is called the *level of significance of the statistical test.*

level of significance

We illustrate the calculation of a level of significance with an example.

**EXAMPLE 4.10**

Refer to example 4.6 (p. 86). Rather than specifying a preset value for $\alpha$, determine the level of significance for the statistical test.

**SOLUTION**

The null and alternative hypotheses are

$H_0$: $\mu = 380$

$H_a$: $\mu > 380$

From the sample data, the computed value of the test statistic is

$$z = \frac{\bar{y} - 380}{s/\sqrt{n}} = \frac{390 - 380}{35.2/\sqrt{50}} = 2.01$$

The level of significance for this test (i.e., the weight of evidence for rejecting $H_0$) is the probability of observing a value of $\bar{y}$ greater than 390 assuming the null hypothesis is true. This value can be computed by using the $z$-value of the test statistic, 2.01, and referring to table 1, the appendix, to determine the probability of observing a $z$-value greater than 2.01. This probability, which is sometimes designated by the letter $p$, is seen to be $.5 - .4778 = .0222$. This value is shown by the shaded area in figure 4.13. Thus we would say that the level of significance for this test is .0222.

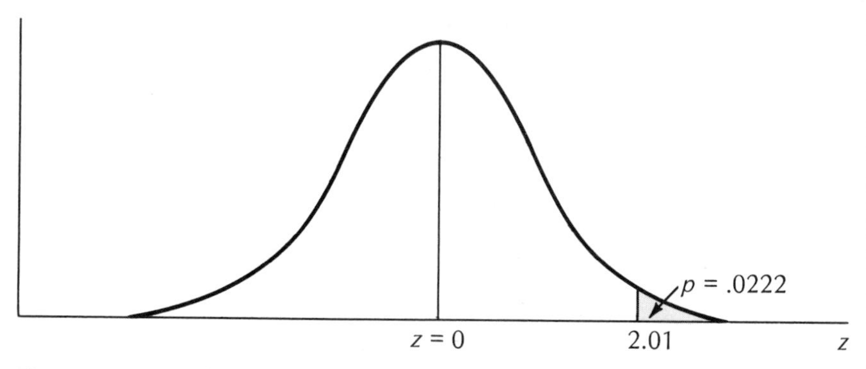


**Figure 4.13** Level of significance for example 4.10

As can be seen from example 4.10, the level of significance represents the probability of observing a sample outcome more contradictory to $H_0$ than the observed sample result. *The smaller the value of this probability, the heavier the weight of the sample evidence for rejecting $H_0$.* For example, a statistical test with a level of significance of $p = .01$ shows more evidence for the rejection of $H_0$ than does another statistical test with $p = .20$.

Suppose the null and alternative hypotheses in example 4.10 had been

$$H_0: \mu = 380$$

$$H_a: \mu < 380$$

and the computed value of $z$ had been $z = -2.01$. The level of significance would still be $p = .0222$ (see figure 4.14).

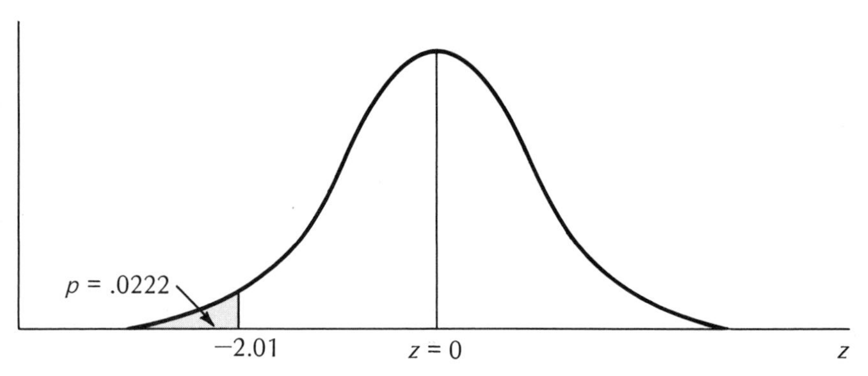


**Figure 4.14** Level of significance for $H_0$: $\mu = 380$, $H_a$: $\mu < 380$ and $z = -2.01$

*p*-value for one-tailed test

To summarize, the level of significance for a one-tailed test can be computed as follows:

$$p = P[z > |\text{computed } z|]$$

*p*-value for two-tailed test

For two-tailed tests (as determined by the form of $H_a$), we still compute the probability of obtaining a sample outcome more contradictory to $H_0$ than the observed result, but the level of significance is commonly taken to be twice this probability. That is, for a two-tailed test, the level of significance is

$$p = 2P[z > |\text{computed } z|]$$

**EXAMPLE 4.11**

Determine the level of significance for the data of example 4.6 if the null hypothesis and alternative hypothesis are

$$H_0: \mu = 380$$

$$H_a: \mu \neq 380$$

**SOLUTION**

The computed value of $z$ is 2.01. Since the probability of observing a value of $z$ greater than 2.01 is .0222, the level of significance for the two-tailed statistical test is $p = 2(.0222) = .0444$.

There is much to be said in favor of this approach to hypothesis testing. Rather than reaching a decision directly, the statistician (or person performing the statistical test) presents the experimenter with the weight of evidence for rejecting the null hypothesis. The experimenter can then draw his or her own conclusion. Some experimenters will reject a null hypothesis if $p = .10$; others will require $p < .05$ or $p < .01$ for rejecting the null hypothesis. The experimenter is left to make the decision based on what he or she feels is enough evidence to indicate rejection of the null hypothesis.

Many professional journals have followed this approach by reporting the results of a statistical test in terms of its level of significance. Thus we might read that a particular test was significant at the $p = .05$ level or perhaps the $p < .01$ level. By reporting results this way, the reader is left to draw his or her own conclusion.

One word of warning is given here. The $p = .05$ level of significance has become a magic level, and many seem to feel that a particular null hypothesis should not be rejected unless the test achieves the .05 level, or lower. This has resulted in part from attitudes held by people who for many years have used the decision-based approach with $\alpha$ preset at .05. Try not to fall into this trap when reading journal articles or reporting the results of your statistical tests. After all, statistical significance at a particular level does not dictate practical significance. Hence after determining the statistical level of significance of a test, the experimenter should always consider the practical significance of the finding.

Throughout the text we will conduct statistical tests from both the decision-based approach and from the level-of-significance approach to familiarize you with both avenues of thought.

## Summary 4.6

A population mean can be estimated using point or interval estimation. Which estimation procedure one chooses is a personal matter. However, most people use point estimation with a bound on the error of estimation when a crude statement of precision is required. The magnitude of the bound on the error of estimation measures the goodness of a point estimate. The smaller the bound, the better our estimate is. The flexibility of changing confidence coefficients allows us to be more precise when using interval estimation. The goodness of an interval estimate is given by the width of the interval and the confidence coefficient.

Following the traditional approach to hypothesis testing, a statistical test of an hypothesis about $\mu$ is composed of five parts: null hypothesis, research hypothesis, test statistic, rejection region, and conclusion. It employs the technique of proof by contradiction. We try to verify the research hypothesis by gathering information to contradict the null hypothesis $H_0$: $\mu = \mu_0$. As with any two-decision problem, there are two types of errors that can be committed, the rejection of $H_0$ when $H_0$ is true—a type I error—and the acceptance of $H_0$ when $H_0$ is false and some alternative is true—a type II error. The probabilities for these errors, designated by $\alpha$ and $\beta$, measure the goodness of the test procedure.

In this chapter we indicated that for a given sample size, $\alpha$ and $\beta$ are inversely related; as $\alpha$ is increased, $\beta$ is decreased, and vice versa. If we specify $n$ and $\alpha$ for a given test procedure, we can compute $\beta$ for alternative values of $\mu$. Sometimes we may wish to specify both $\alpha$ and $\beta$ *prior* to conducting the investigation. To do this, we determine the sample size required for the specific values of $\alpha$ and $\beta$. While this topic is beyond the scope of this chapter, we do address the problem in chapter 9.

Finally, we considered an alternative to the traditional decision-based approach for a statistical test of an hypothesis. Rather than relying on a preset level of $\alpha$, we compute the weight of evidence for rejecting the null hypothesis. This weight, expressed in terms of a probability, is called the level of significance for the test. Most professional journals summarize the results of a statistical test using the level of significance.

# Exercises

4.8. To test the effectiveness of a new spray for controlling rust mites, we would like to compare the average yield for treated groves with the average yield displayed for untreated groves in previous years. A random sample of 30 one-acre groves is chosen and sprayed according to a recommended schedule. The average yield for the sample of 30 groves was 830 boxes, with a standard deviation of 91. Yields from groves in the same area without rust mite maintenance spraying has averaged 760 boxes over previous years. Do these data present sufficient evidence to indicate that the mean yield for groves sprayed with the new preparation is higher than 760 boxes, the average over previous years without spraying? Is this a one-tailed or two-tailed test? Use $\alpha = .05$.

4.9. The board of health of a particular state was called to investigate claims that raw pollutants were being released into the river flowing past a small residential community. By applying financial pressure, the state was able to get the violating company to make major concessions towards the installation of a new water purification system. In the interim, different production systems were to be initiated to help reduce the pollution level of water entering the stream. To monitor the effect of the interim system, a random sample of 50 water specimens was taken throughout the month at a location downstream from the plant. If $\bar{y} = 5.0$ and $s = .70$, use the sample data to determine whether the mean dissolved oxygen count of the water (in ppm) is less than 5.2, the average reading at this location over the past year.

a. List the five parts of the statistical test, using $\alpha = .05$.

b. Conduct the statistical test and state your conclusion.

e

4.10. Refer to figure 4.12 (p. 88). Compute $\beta$ for (b) and (c) by using $H_0$: $\mu = 380$, $H_a$: $\mu > 380$. Recall that $n = 50$ and $s = 35.2$.

4.11. An automatic merge system has been installed at the entrance ramp to a major highway. Through the use of a series of pacer lights, a driver is told if she is traveling at the correct speed to successfully merge onto the highway. Prior to the installation of the system, investigators measured the stress level of many drivers merging onto the highway during rush hours and found the average stress level to be 8.2 on a 10-point scale. Similar testing on a random sample of 50 is to be conducted now that the automatic merge system has been installed. If $\bar{y} = 7.6$ and $s = 1.8$ for the 50 drivers sampled, conduct a statistical test of the research hypothesis that the average stress at peak hours for drivers under the new system is less than 8.2, the average stress level prior to the installation of the automatic merge system. Determine the level of significance of the statistical test. Interpret your findings.

4.12. The search for alternatives to oil for major sources of fuel and energy will inevitably bring about many environmental challenges. These challenges will require solutions to problems in such areas as strip mining and many others. Let us focus on one. If coal is considered as a major source of fuel and energy, we will have to consider ways to keep large amounts of sulfur dioxide ($SO_2$) and particulates from getting into the air. This is especially important at large government and industrial operations. There are several possibilities.

e

1. Build the smoke stack extremely high.
2. Remove the $SO_2$ and particulates from the coal prior to combustion.

3. Remove the $SO_2$ from the gases after the coal is burned but before the gases are released into the atmosphere. This is accomplished by using a scrubber.

Several scrubbers have been developed in recent years. Suppose that a new one has been constructed and is set for testing at a given power plant. Fifty samples are obtained at various times from gases emitted from the stack. The mean $SO_2$ emission is .13 lb per million Btu, with a standard deviation of .05 lb. Use the sample data to construct a statistical test of the null hypothesis $H_0$: $\mu = .145$, the average emission level for one of the more efficient scrubbers that has been developed. Choose an appropriate alternative hypothesis, with $\alpha = .05$.

4.13. Refer to exercise 4.6 (p. 79). Construct a 99% confidence interval for $\mu$, the average number of miles between a person's high school and his (or her) present permanent address for the state sample.

4.14. Refer to exercise 4.12. Rather than being interested in testing the research hypothesis that $\mu < .145$, the average emission level for one of the more efficient scrubbers, we may wish to estimate the mean emission level for the new scrubber. Use the sample data to construct a 99% confidence interval for $\mu$. Interpret your results.

4.15. As part of an overall evaluation of training methods, an experiment was conducted to determine the average exercise capacity of healthy male army inductees. To do this each male in a random sample of 35 healthy army inductees was allowed to exercise on a bicycle ergometer (a device for measuring work done by the muscles) under a fixed work load until he tired. Blood pressure, pulse rates, and other indicators were carefully monitored to ensure that no one's health was in danger. The exercise capacities (mean time, in minutes) for the 35 inductees are listed below.

| | | | | | | |
|---|---|---|---|---|---|---|
| 23 | 19 | 36 | 12 | 41 | 43 | 19 |
| 28 | 14 | 44 | 15 | 46 | 36 | 25 |
| 35 | 25 | 29 | 17 | 51 | 33 | 47 |
| 42 | 45 | 23 | 29 | 18 | 14 | 48 |
| 21 | 49 | 27 | 39 | 44 | 18 | 13 |

a. Use these data to construct a 95% confidence interval for $\mu$, the average exercise capacity for healthy male inductees. Interpret your findings.
b. How would your interval change using a 99% confidence interval?

4.16. Using the data of exercise 4.15, determine the number of sample observations that would be required to estimate $\mu$ to within 1 minute, using a

95% confidence interval. (Hint: Find the sample size for which $z_{\alpha/2}\, s/\sqrt{n} = 1.0$ when $s = 12.36$.)

4.17. Refer to the data of exercise 4.15. Suppose that the random sample of 35 inductees was selected from a large group of new army personnel being subjected to a new (and hopefully improved) physical fitness program. If previous testing with several thousand personnel over the past several years has shown an average exercise capacity of 29 minutes, run a statistical test for the research hypothesis that the average exercise capacity is improved for the new fitness program. Give the level of significance for the test. Interpret your findings.

4.18. Refer to exercise 4.17.

a. How would the research hypothesis change if we were interested in determining whether the new program is better or worse than the physical fitness program for inductees?

b. What is the level of significance for your test?

4.19. In a random sample of 40 hospitals from a list of hospitals with over 100 semiprivate beds, a researcher collected information on the proportion of persons whose bills are covered by a group policy under a major medical insurance carrier. The sample proportions are listed below.

| | | | | | | | |
|---|---|---|---|---|---|---|---|
| .67 | .74 | .68 | .63 | .91 | .81 | .79 | .73 |
| .82 | .93 | .92 | .59 | .90 | .75 | .76 | .88 |
| .85 | .90 | .77 | .51 | .67 | .67 | .92 | .72 |
| .69 | .73 | .71 | .76 | .84 | .74 | .54 | .79 |
| .71 | .75 | .70 | .82 | .93 | .83 | .58 | .84 |

Use the sample data to construct a 90% confidence interval on $\mu$, the average proportion of patients per hospital with group medical insurance coverage.

4.20. Refer to exercise 4.19. Use the same data to construct a 99% confidence interval.

4.21. Faculty members in a state university system who resign within 10 years of initial employment are entitled to receive the money paid into a retirement system, plus 4% per annum. Unfortunately, experience has shown that the state is pitifully slow in returning this money. Concerned about such a practice, a local teachers' organization decides to investigate. From a random sample of 50 employees who resigned from the state university system over the past 5 years, the average time between the termination date and reimbursement was 75 days, with a standard deviation of 15 days. Use the data to estimate the mean time to reimbursement, using a 95% confidence interval.

$ 4.22. Refer to exercise 4.21. After a confrontation with the teachers' union, the state has promised to make reimbursements within 60 days. Monitoring of the next 40 resignations yields an average of 58 days, with a standard deviation of 10 days. If we assume these 40 resignations represent a random sample of the state's future performance, estimate the mean reimbursement time, using a 99% confidence interval.

4.23. Refer to example 4.8 (p. 89). Compute $\beta$ for $\mu_a = 40$. What would be your conclusion based on the magnitude of $\beta$?

4.24. Refer to exercise 4.23. Using the values of $\beta$ computed for $\mu_a = 38$ and $\mu_a = 40$, calculate the probability of a type II error for several other values of $\mu_a$ in order to construct a graph of $\beta$ against $\mu_a$.

*4.25. Refer to the data of exercise 4.15. It can be shown that the sample standard deviation is 12.36. Use a computer program to construct a 90% confidence interval for $\mu$. A Minitab program is shown here for illustration purposes.

```
SET THE FOLLOWING DATA INTO COL C6

23      25      27      46      14
28      45      12      51      18
 :       :       :       :       :
14      23      41      33      13

ZINTERVAL 90 PERCENT CONFIDENCE, SIGMA = 12.36, DATA IN C7
STOP
```

*4.26. Use the Minitab output shown below to respond to the statements that follow concerning the confidence interval for $\mu$, using the data of exercise 4.19.

```
— — SET DATA INTO C1
COLUMN C1
COUNT 40

0.67000    0.73000    0.59000    0.93000    0.92000
0.82000    0.75000    0.51000    0.81000    0.54000
0.85000    0.68000    0.76000    0.75000    0.58000
0.69000    0.92000    0.82000    0.67000    0.73000
0.71000    0.77000    0.91000    0.74000    0.88000
0.74000    0.71000    0.90000    0.83000    0.72000
0.93000    0.70000    0.67000    0.79000    0.79000
0.90000    0.63000    0.84000    0.76000    0.84000
```

```
— — ZINTERVAL 98 PERCENT CONFIDENCE, SIGMA = 0.10861, DATA IN C1
     C1          N = 40          MEAN = 0.76200          ST.DEV. = 0.10861
     THE ASSUMED SIGMA = 0.10861
     A 98.00 PERCENT C.I. FOR MU IS: (0.72044, 0.80356)
— — STOP
```

a. Identify the sample mean for the data shown in the computer output.
b. What is the confidence coefficient for the confidence interval shown?
c. Give the confidence limits and interpret the interval estimate.

4.27. A random sample of 40 inner-city birth rates showed an average of 35 per thousand, with a standard deviation of 6.3. Estimate the mean inner-city birth rate. Use a 95% confidence interval.

4.28. A random sample of 30 standard metropolitan statistical areas (SMSAs) was selected and the ratio (per 1000) of registered voters to the total number of persons 18 years and over was recorded in each area. Use the data below to test the research hypothesis that $\mu$, the average ratio (per 1000), is different from 675, last year's average ratio. Give the level of significance for your test.

| | | | | | |
|---|---|---|---|---|---|
| 802 | 497 | 653 | 600 | 729 | 812 |
| 751 | 730 | 635 | 605 | 760 | 681 |
| 807 | 747 | 728 | 561 | 696 | 710 |
| 641 | 848 | 672 | 740 | 818 | 725 |
| 694 | 854 | 674 | 683 | 695 | 803 |

# References


Mendenhall, W., and Ott, L. 1976. *Understanding statistics.* 2d ed. N. Scituate, Mass.: Duxbury Press.

Ryan, T.A.; Joiner, B.L.; and Ryan, B.F. 1976. *Minitab student handbook.* N. Scituate, Mass.: Duxbury Press.

# 5 Inferences: Small-Sample Results

## 5.1 Introduction

The estimation and test procedures about $\mu$ in chapter 4 were based on the assumption that we would obtain a random sample of 30 or more observations. Sometimes it is not possible to obtain so large a sample size. For example, in determining the blood level of persons suffering from a very rare disease, it may be impossible to obtain a random sample of 30 or more observations at a given time.

W. S. Gosset faced a similar problem around the turn of the century when, as a chemist for Guinness Breweries, he was asked to make judgments on the mean quality of various brews. As might be expected, he was not supplied with large sample sizes to reach his conclusions.

Gosset felt that, for small sample sizes, when he used the test statistic

$$z = \frac{\bar{y} - \mu_0}{\sigma/\sqrt{n}}$$

with $\sigma$ replaced by $s$, he was falsely rejecting the null hypothesis $H_0: \mu = \mu_0$ at a much higher rate than that specified by $\alpha$. He became intrigued by this problem and set out to derive the distribution and percentage points of the test statistic

$$\frac{\bar{y} - \mu_0}{s/\sqrt{n}}$$

for $n < 30$.

For example, suppose an experimenter sets $\alpha$ at a nominal level, say .05. Then he expects to falsely reject the null hypothesis approximately 1 time in 20. However, Gosset proved that the actual probability of a type I error for this test was somewhat higher than the nominal level designated by $\alpha$. The results of his study were published under the pen name Student, because it was against company policy for him to publish his results in his own name at that time. The quantity

$$\frac{\bar{y} - \mu_0}{s/\sqrt{n}}$$

is called the $t$ statistic and its distribution is called the *Student's t distribution* or, simply, Student's $t$ (see figure 5.1).

Student's $t$

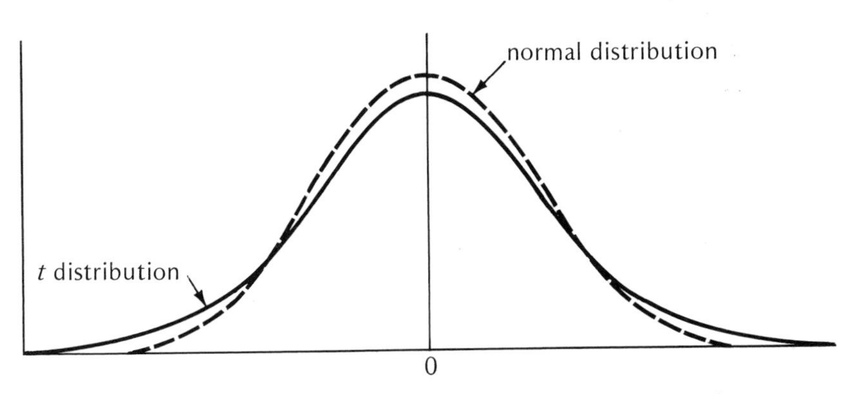


**Figure 5.1** A $t$ distribution with a normal distribution superimposed

Although the quantity

$$\frac{\bar{y} - \mu_0}{s/\sqrt{n}}$$

will possess a $t$ distribution only when the sample is selected from a normal population, the $t$ distribution provides a reasonable approximation to the distribution of

$$\frac{\bar{y} - \mu_0}{s/\sqrt{n}}$$

when the sample is selected from a population with a mound-shaped distribution.* We summarize the properties of $t$ in the box.

---

**Properties of Student's $t$ Distribution**

degrees of freedom, df

1. The $t$ distribution, like that of $z$, is symmetrical about 0.
2. The $t$ distribution is more variable than the $z$ distribution (see figure 5.1).
3. There are many different $t$ distributions. We specify a particular one by a parameter called the *degrees of freedom* (df). Thus we specify

$$t = \frac{\bar{y} - \mu_0}{s/\sqrt{n}} \qquad \text{df} = n - 1$$

4. As $n$ (or equivalently df) increases, the distribution of $t$ approaches the distribution of $z$.

---

Everyone seems to want to know what df really means. There is no *valid* explanation; df is a parameter (a numerical descriptive measure) of the $t$ distribution. As we stated, df $= n - 1$ for this $t$ distribution. The larger (or smaller) $n$, the larger (or smaller) df. As $n$ (and hence df) gets large, the $t$ distribution approaches the $z$ distribution.

$t_a$

Because of the symmetry of $t$, only upper-tail percentage points (probabilities or areas) of the distribution of $t$ have been tabulated; these appear in table 2 of the appendix. The degrees of freedom (df) are listed along the left-hand column of the page. An entry in the table specifies a value of $t$, say $t_a$, such that

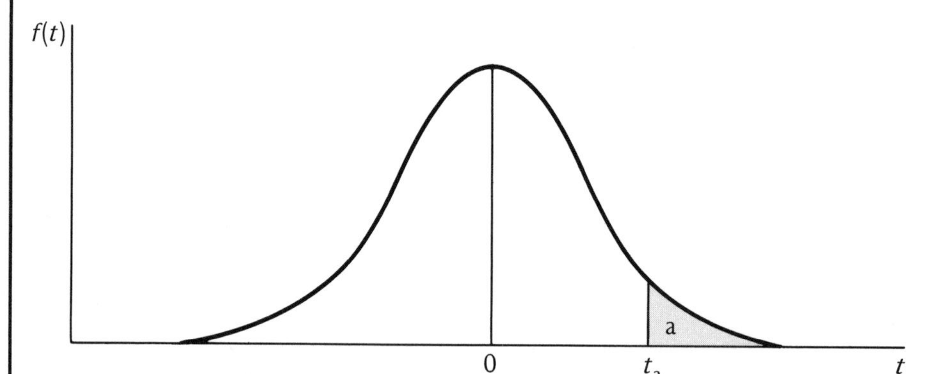


**Figure 5.2** Illustration of area tabulated in table 2 of the appendix for the $t$ distribution

*Actually, symmetry of the distribution for the population is not the major concern. Rather, we need to avoid working with heavy-tailed distributions (peaked distributions with a good deal of the area in the tails). More information concerning measures of symmetry (or skewness) and heaviness of tails is given in chapter 21.

an area "a" lies to its right. Various values of "a" appear across the top of the page. Thus, for example, with df = 7, the value of $t$ with an area .05 to its right is 1.895 (found in the a = .05 column and df = 7 row).

# 5.2 Small-Sample Inferences About a Population Mean $\mu$

We can use the $t$ distribution to make inferences about a population mean $\mu$. The sample test concerning $\mu$ is summarized in the box. The only difference between the large-sample test of chapter 4 and the small-sample test given here is that $t$ replaces $z$.

small-sample test, $\mu$

**Small-Sample Test About $\mu$**

$H_0$: $\mu = \mu_0$.

$H_a$: 1. $\mu > \mu_0$.
2. $\mu < \mu_0$.
3. $\mu \neq \mu_0$.

T.S.: $t = \dfrac{\bar{y} - \mu_0}{s/\sqrt{n}}$.

R.R.: For a probability $\alpha$ of a type I error and df $= n - 1$,

1. reject $H_0$ if $t > t_\alpha$;
2. reject $H_0$ if $t < -t_\alpha$;
3. reject $H_0$ if $|t| > t_{\alpha/2}$.

Recall that "a" denotes the area in the tail of the $t$ distribution. For a one-tailed test with the probability of a type I error equal to $\alpha$, we locate the rejection region using the value from table 2, the appendix, for a $= \alpha$ and df $= n - 1$. But for a two-tailed test we would use the $t$-value from table 2 corresponding to a $= \alpha/2$ and df $= n - 1$.

Thus for a one-tailed test we reject the null hypothesis if the computed value of $t$ is greater than the $t$-value from table 2, the appendix, and a $= \alpha$ and df $= n - 1$. Similarly, for a two-tailed test we reject the null hypothesis if $|t|$ is greater than the $t$-value from table 2 for a $= \alpha/2$ and df $= n - 1$.

**EXAMPLE 5.1**

A tire company guarantees that a particular brand has a mean useful lifetime of 42,000 miles or more. A consumer test agency, wishing to verify this claim, observed $n = 10$ tires on a test wheel that simulated normal road conditions. The lifetimes (in thousands of miles) were as follows:

42, 36, 46, 43, 41, 35, 43, 45, 40, 39

Use these data to determine if there is sufficient evidence to contradict the manufacturer's claim. Set $\alpha = .05$.

## SOLUTION

The null and research hypotheses for this example are

$H_0$: $\mu = 42$

$H_a$: $\mu < 42$

Note that we are giving the manufacturer the benefit of the doubt by setting $\mu = 42$ for $H_0$.

Before setting up the test statistic and rejection region, we must first compute the sample mean and standard deviation. You can verify that

$$\sum_{i=1}^{10} y_i = 410 \quad \text{and} \quad \sum_{i=1}^{10} y_i^2 = 16{,}926$$

Then

$$\bar{y} = \frac{\sum_{i=1}^{10} y_i}{10} = \frac{410}{10} = 41$$

Similarly, substituting into the shortcut formula for $s^2$, we find

$$s^2 = \frac{1}{9}\left[\sum_{i=1}^{10} y_i^2 - \frac{\left(\sum_{i=1}^{10} y_i\right)^2}{10}\right] = 12.89$$

$$s = \sqrt{12.89} = 3.59$$

The test statistic, then, is

$$t = \frac{\bar{y} - \mu_0}{s/\sqrt{n}} = \frac{41 - 42}{3.59/\sqrt{10}} = -.88$$

The rejection region is

R.R.: Reject $H_0$ if $t < -t_{.05}$

From table 2, the appendix, the critical $t$-value with a $= .05$ and df $= 9$ is 1.833, so $-t_{.05}$ is $-1.833$. Since the observed value of $t$ is not less than $-1.833$, we have insufficient evidence to indicate that the mean lifetime of this brand of tires is less than 42,000 miles.

At this point someone might suggest calculating $\beta$, the probability of a type II error, to see if we can accept the manufacturer's claim. Unfortunately, this is a much more difficult task for a small-sample test than it is for the large-sample test and is beyond the scope of this text. (If you are interested in pursuing the topic, consult *Biometrika Tables for Statis-*

*ticians,* Volume I.) Our conclusion will be that there is insufficient evidence to reject the company's claim and we should continue sampling.

**EXAMPLE 5.2**

Refer to example 5.1. Rather than performing the statistical test with a preset $\alpha$ level, give the level of significance for the test.

**SOLUTION**

For the one-tailed lower-tail test, the computed $t$-value is $t = -.88$. Although we cannot compute the exact probability of observing a value of $t$ less than or equal to $-.88$, we see from table 2, the appendix, that the level of significance can be written as $p > .10$. Based on this probability, the experimenter would probably conclude that there was insufficient evidence to reject the null hypothesis. If you feel that the level of significance should be given more precisely, you could refer to more detailed tables of the $t$ distribution in the *Biometrika Tables for Statisticians*.

In addition to being able to run a statistical test for $\mu$ using small sample sizes, we can construct a small-sample confidence interval using $t$. The small-sample confidence interval for a population mean $\mu$, with confidence coefficient $(1 - \alpha)$, is identical to the corresponding large-sample confidence interval, with $z$ replaced by $t$.

**Small-Sample Confidence Interval for $\mu$**

$$\bar{y} \pm t_{\alpha/2}\frac{s}{\sqrt{n}}$$

Note: df $= n - 1$ and the confidence coefficient is $(1 - \alpha)$.

**EXAMPLE 5.3**

In a psychological depth perception test, a random sample of $n = 14$ airline pilots was asked to judge the distance between two markers at the other end of a laboratory. The sample data (recorded in feet) are listed below.

| | | | | | | |
|---|---|---|---|---|---|---|
| 2.7 | 2.4 | 1.9 | 2.6 | 2.4 | 1.9 | 2.3 |
| 2.2 | 2.5 | 2.3 | 1.8 | 2.5 | 2.0 | 2.2 |

Use the sample data to place a 95% confidence interval on $\mu$, the average recorded distance for this psychological test.

**SOLUTION**

Before setting up a 95% confidence interval on $\mu$, we must compute $\bar{y}$ and $s$. You can verify that

$$\sum_{i=1}^{14} y_i = 31.70 \quad \text{and} \quad \sum_{i=1}^{14} y_i^2 = 72.79$$

The sample mean, variance, and standard deviation are then

$$\bar{y} = \frac{\sum_{i=1}^{14} y_i}{14} = \frac{31.70}{14} = 2.26$$

$$s^2 = \frac{1}{13}\left[72.79 - \frac{(31.7)^2}{14}\right] = .078$$

$$s = \sqrt{.078} = .28$$

Referring to table 2, the appendix, the $t$-value corresponding to a = .025 and df = 13 is 2.160. Hence the 95% confidence interval is

$$\bar{y} \pm t_{\alpha/2}\frac{s}{\sqrt{n}} \qquad \text{or} \qquad 2.26 \pm \frac{2.160(.28)}{\sqrt{14}}$$

which is the interval 2.26 ± .16, or 2.10 to 2.42. Thus we are 95% confident that the interval from 2.10 to 2.42 will encompass the mean $\mu$.

# Exercises

5.1. A random sample of 10 students in a fourth-grade reading class were thoroughly tested to determine reading speed and reading comprehension. Based on a fixed-length standardized test reading passage, the following speeds (in minutes) and comprehension scores (based on a 100-point scale) were obtained.

| *Student* | 1 | 2 | 3 | 4 | 5 | 6 | 7 | 8 | 9 | 10 |
|---|---|---|---|---|---|---|---|---|---|---|
| *Reading Speed* | 5 | 7 | 15 | 12 | 8 | 7 | 10 | 11 | 13 | 9 |
| *Reading Comprehension* | 60 | 76 | 96 | 100 | 81 | 75 | 85 | 88 | 98 | 83 |

a. Use the reading speed data to place a 95% confidence interval on $\mu$, the average speed for all fourth-grade students in the large school from which the sample was drawn.
b. Interpret the interval estimate in part a.
c. How would your inference change by using a 98% confidence interval?

5.2. Refer to exercise 5.1. Using the reading comprehension data, test the research hypothesis that the mean for all fourth graders on the standardized examination is greater than 80, the statewide average for comparable students the previous year. Give the level of significance for your test. Interpret your findings.

5.3. Refer to exercise 5.2.

a. Set up all parts for a statistical test of the research hypothesis that the mean score for all fourth graders is different from 80, the statewide average the previous year.

b. Give the level of significance for this test.

*5.4. Refer to the data of exercise 5.1. Use a computer program to construct a 90% confidence interval for the mean total reading score (speed plus comprehension). A Minitab program is shown here for illustrative purposes.

```
READ C1, C2

 5      60
 7      76
15      96
 ⋮       ⋮
 9      83

ADD C1 TO C2, PUT IN C3
TINTERVAL WITH 90 PERCENT CONFIDENCE, DATA IN C3
PRINT C1, C2, C3
STOP
```

*5.5. The amount of sewage and industrial pollutants dumped into a body of water affects the health of the water by reducing the amount of dissolved oxygen available for aquatic life. Suppose that weekly readings are taken from the same location in a river over a two-month period. Use the summary data from the computer printout to conduct a statistical test of the research hypothesis that the mean dissolved oxygen content is less than 5.0 parts per million, a level some scientists feel is marginal for supplying enough dissolved oxygen for fish.

```
5.100000000
4.900000000
5.600000000
4.200000000
4.800000000
4.500000000
5.300000000
5.200000000
```

| | |
|---|---|
| 8.000000000 | sample size |
| 4.950000000 | $\bar{y}$ |
| .2028571428 | $s^2$ |
| .4503966505 | $s$ |

# 5.3 Control Charts for a Population Mean

skip

control chart

We can extend the notion of a confidence interval for $\mu$ to obtain a *control chart.* We, as consumers, are vitally interested in product quality. We expect product quality for a particular item to be both uniform from one time period to another, and we expect it to live up to the product description advertised by the manufacturer. For example, in buying paint from a paint store, we would expect different gallons of the same color to be uniform in color and we would expect the color to be identical to that advertised in the paint sample brochure. Similarly, the Food and Drug Administration (FDA) not only expects but also demands that drug products have uniform potency and meet the standards advertised by the pharmaceutical firm.

Consumers are not the only ones interested in product quality. Reputable manufacturers are also concerned that their products meet the standards they have claimed. If the quality of a product falls below the standards advertised by the company, then there is a risk that consumers will reject the product and buy from a competitor. Similarly, if the product quality drifts above the standards established by the company, then it would be in the company's interest to upgrade their advertising to reflect the increase in quality.

quality control

Quality control techniques have been developed to monitor the ongoing quality of a manufacturing process in an effort to maintain uniform quality or at least to detect when the product quality has shifted. We can monitor product quality of a production process by using a graph called a *control chart.* Thus we could graph the sample mean or sample range for samples collected over a period of time to monitor product quality.

Typically a control chart consists of three lines: a center line, an upper control line, and a lower control line. In a control chart for the mean, successive sample means would be plotted much as they appear in figure 5.3. The sample means are shown by the x's in figure 5.3. If one of the sample means falls outside either the upper or lower control lines, the process is judged to be out of control. That is, it appears that product quality has shifted. At this point company officials and production personnel would try to establish the cause of the shift and would initiate corrective changes in the production process.

center line

Establishment of the three control lines is quite simple. The center line (denoted by $\bar{y}_c$) represents the average of $k$ sample means, each based on $n$ observations. We generally recommend taking $k \geq 25$ and $n > 3$. These samples should be taken at some time when the process is judged to be under control. Then if we let $y_{ij}$ denote the $j$th observation in sample $i$ and

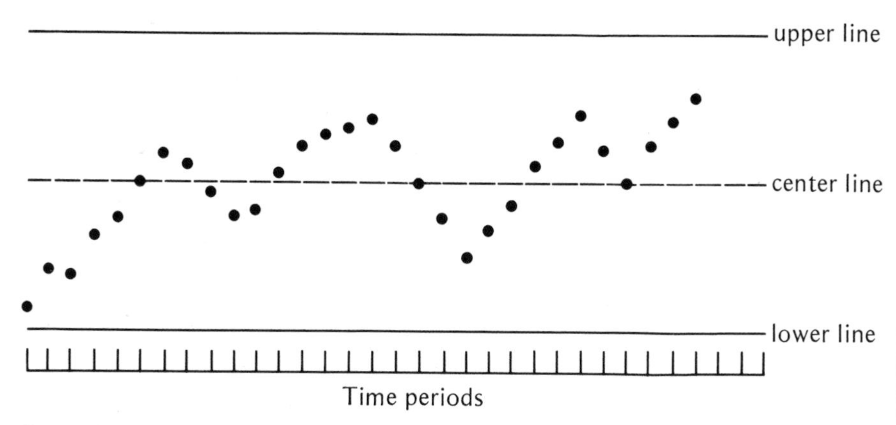


**Figure 5.3** Control chart for sample means

$\bar{y}_i = \Sigma_{j=1}^{n} y_{ij}/n$ denote the mean for sample $i$, the average of the $k$ sample means is

$\bar{y}_c$

$$\bar{y}_c = \frac{\sum_{i=1}^{k} \bar{y}_i}{k} = \frac{\sum_{i=1}^{k}\sum_{j=1}^{n} y_{ij}}{nk}$$

This summation notation is often abbreviated to

$$\bar{y}_c = \frac{\sum_{i} \bar{y}_i}{k} = \frac{\sum_{i}\sum_{j} y_{ij}}{nk}$$

UCL, LCL for mean quality

The upper control line (UCL) and the lower control line (LCL) are computed as follows:

$$\text{UCL} = \bar{y}_c + 3\frac{\sigma}{\sqrt{n}} \quad \text{and} \quad \text{LCL} = \bar{y}_c - 3\frac{\sigma}{\sqrt{n}}$$

From knowledge of the Empirical Rule, the interval $\bar{y}_c \pm (3\sigma/\sqrt{n})$ should contain nearly all the sample means $\bar{y}_i$ in repeated sampling. If a sample mean falls outside this interval, we have either observed an extremely unlikely event or the process quality has changed and $\bar{y}_c$ is no longer an accurate measure of the actual mean product quality. This latter conclusion is more realistic and is used to signal a manufacturing process out of control.

estimate of $\sigma$ for control chart

The standard deviation $\sigma$ in the formulas for the upper and lower control lines can be estimated either by using a pooled sample variance from the $k$ samples or, more quickly, by using the $k$ sample ranges. It is this latter procedure that we will employ. Letting $r_i$ denote the range for the $n$ sample measurements in sample $i$ and $\bar{r}$ the average of the $k$ sample ranges, we can estimate $\sigma$ by

$$\hat{\sigma} = \frac{\bar{r}}{d_n}$$

where $d_n$ is obtained from table 20 of the appendix. For example, suppose we have $k = 20$ different samples of $n = 7$ observations per sample and $\bar{r} = 5$. Then $d_7 = 2.704$ and $\hat{\sigma} = 5/2.704 = 1.849$.

**EXAMPLE 5.4**

A company that dyes rugs is interested in establishing a control chart for the mean color when dyeing solid colored rugs. Although maintaining uniform color is somewhat important for patterned or multicolored rugs, it is much more important for solid-colored rugs, where minor changes in solid colors are readily recognizable. Rug color quality can be monitored by taking readings on a colorimeter. Twenty-five samples of 5 measurements each from a rug being dyed red yielded the data listed in table 5.1. These data were obtained while the manager felt the process was in control.

Use the data of table 5.1 to construct a control chart for the mean colorimeter reading.

**SOLUTION**

From table 5.1 we have

$$\bar{y}_c = \frac{\sum_i \sum_j y_{ij}}{nk} = \frac{258.2}{5(25)} = 2.07$$

$$\bar{r} = \frac{\sum_i r_i}{k} = \frac{56.8}{25} = 2.27$$

From table 20, the appendix, we have $d_5 = 2.326$ and hence

$$\hat{\sigma} = \frac{\bar{r}}{d_n} = \frac{2.27}{2.326} = .98$$

The center line is then 2.07, with upper and lower control lines given by

$$\text{UCL} = \bar{y}_c + 3\frac{\hat{\sigma}}{\sqrt{n}} = 2.07 + \frac{3(.98)}{\sqrt{5}} = 3.38$$

$$\text{LCL} = \bar{y}_c - 3\frac{\hat{\sigma}}{\sqrt{n}} = 2.07 - \frac{3(.98)}{\sqrt{5}} = .76$$

**Table 5.1** Colorimeter readings for the 25 samples of example 5.4

| *Sample* | *Observation* | *Sample Sum* | *Sample Range* |
|---|---|---|---|
| 1 | 2.4, 1.8, 0.7, 1.0, 2.5 | 8.4 | 1.8 |
| 2 | 2.3, 3.0, 2.5, 1.2, 3.1 | 12.1 | 1.9 |
| 3 | 1.3, 1.2, 0.9, 1.2, 3.0 | 7.6 | 2.1 |
| 4 | 0.5, 2.2, 2.4, 1.5, 3.0 | 9.6 | 2.5 |
| 5 | 2.8, 1.9, 2.6, 1.3, 2.9 | 11.5 | 1.6 |
| 6 | 2.4, 3.1, 1.7, 3.3, 2.6 | 13.1 | 1.6 |
| 7 | 2.5, 2.9, 1.4, 4.0, 2.1 | 12.9 | 2.6 |
| 8 | 1.1, 2.9, 3.0, 1.4, 2.8 | 11.2 | 1.9 |
| 9 | 3.3, 2.2, 2.7, 2.8, 2.1 | 13.1 | 1.2 |
| 10 | 0.8, 4.2, 2.3, 1.4, 2.1 | 10.8 | 3.4 |
| 11 | 0.2, 2.6, 2.3, 0.7, 4.2 | 10.0 | 4.0 |
| 12 | 1.8, 1.6, 2.3, 2.1, 1.7 | 9.5 | .7 |
| 13 | 0.1, 3.9, 2.3, 1.4, 1.0 | 8.7 | 3.8 |
| 14 | 1.1, 3.1, 1.8, 0.9, 1.8 | 8.7 | 2.2 |
| 15 | 0.5, 0.9, 4.0, 2.2, 2.8 | 10.4 | 3.5 |
| 16 | 2.9, 3.3, 1.9, 3.1, 2.3 | 13.5 | 1.4 |
| 17 | 3.5, 2.0, 2.5, 2.0, 0.3 | 10.3 | 3.2 |
| 18 | 2.5, 2.1, 2.7, 1.7, 1.5 | 10.5 | 1.2 |
| 19 | 1.1, 3.9, 2.7, 1.2, 1.3 | 10.2 | 2.8 |
| 20 | 1.4, 2.0, 2.5, 4.2, 2.4 | 12.5 | 2.8 |
| 21 | 2.2, 1.9, 0.7, 1.3, 1.4 | 7.5 | 1.5 |
| 22 | 1.5, 1.5, 1.1, 2.3, 2.4 | 8.8 | 1.3 |
| 23 | 2.2, 1.3, 2.5, 1.9, 0.7 | 8.6 | 1.8 |
| 24 | 1.7, 0.1, 1.8, 0.7, 2.1 | 6.4 | 2.0 |
| 25 | 3.7, 1.5, 1.9, 0.6, 4.6 | 12.3 | 4.0 |
| *Totals* | | 258.2 | 56.8 |

As stated previously, an observation falling outside one of the control lines is a signal that something has changed. If $\sigma$ were known and the control lines were computed by using the known value of $\sigma$, a value outside a control line would suggest to us that $\bar{y}_c$ no longer represents the actual mean quality. Unfortunately, when $\sigma$ is unknown and must be estimated, a value outside one of the control lines could suggest a shift in the mean quality, an increase in $\sigma$, or both.

To protect ourselves, we should also keep a control chart on the product quality variability. Again, rather than working with the $k$ sample standard deviations, it is simpler to use the $k$ sample ranges. Upper and lower control lines for product quality variability are given by

UCL, LCL for product variability

$$\text{UCL} = D'_n\bar{r} \qquad \text{and} \qquad \text{LCL} = D_n\bar{r}$$

where $D'_n$ and $D_n$ are obtained from table 21 of the appendix. For example 5.4, $\bar{r} = 2.27$, and, based on $n = 5$ observations per sample, $D'_5 = 2.115$ and $D_5 = 0$.

Hence the control lines for the sample ranges are

$$\text{UCL} = 2.115(2.27) = 4.80 \qquad \text{and} \qquad \text{LCL} = 0$$

## Exercise

5.6. Refer to example 5.4. Graph the upper and lower control limits and the center line. Plot the sequence of sample means listed below to determine if and when the process is out of control. (Note: Each mean is based on 5 measurements.)

2.0, 1.9, 1.6, 1.5, 1.7, 1.8, 2.2, 2.1, 2.0, 2.3, 2.4, 2.7, 2.8, 2.9

## 5.4 Estimation and Tests Concerning the Difference Between Two Population Means

The inferences we have made so far have concerned a parameter from a single population. Quite often we are faced with an inference that concerns a comparison of more than one parameter from different populations. For example, we might wish to compare the mean corn crop yield for two different varieties of corn, the mean annual income for two ethnic groups, the mean nitrogen content of two different lakes, or the mean length of time between administration and eventual relief for two different antivertigo drugs.

two populations

common variance

In each of these situations, we will assume that we are sampling from two normal populations (1 and 2) with different means $\mu_1$ and $\mu_2$ but identical variances $\sigma^2$. *We then draw independent random samples of size $n_1$ and $n_2$.* The corresponding sample means and variances are $\bar{y}_i$ and $s_i^2$ ($i = 1$ or 2). Using the data from the two samples, we would like to make a comparison between the population means $\mu_1$ and $\mu_2$. In particular, we will estimate and test an hypothesis concerning the difference $\mu_1 - \mu_2$.

$\mu_1 - \mu_2$

A logical point estimate for the difference in population means is the sample difference $\bar{y}_1 - \bar{y}_2$. The form of the general small-sample confidence interval with confidence coefficient of $(1 - \alpha)$ is given in the box.

**Small-Sample Confidence Interval for $\mu_1 - \mu_2$**

$$(\bar{y}_1 - \bar{y}_2) \pm t_{\alpha/2} s \sqrt{\frac{1}{n_1} + \frac{1}{n_2}}$$

where

$$s = \sqrt{\frac{(n_1 - 1)s_1^2 + (n_2 - 1)s_2^2}{n_1 + n_2 - 2}}$$

and

$$\text{df} = n_1 + n_2 - 2$$

Note: Although we call this a small-sample procedure, it applies to any sample sizes $n_1$ and $n_2$.

The quantity $s$ in the confidence interval is an estimate of the standard deviation $\sigma$ for the two populations and is formed by combining information from the two samples. In fact, $s^2$ is a *weighted average* of the sample variances $s_1^2$ and $s_2^2$. For the special case where the sample sizes are the same ($n_1 = n_2$), the formula for $s^2$ reduces to $s^2 = (s_1^2 + s_2^2)/2$, the mean of the two sample variances. The degrees of freedom for the confidence interval are a combination of the degrees of freedom for the two samples; that is, $\text{df} = (n_1 - 1) + (n_2 - 1) = n_1 + n_2 - 2$.

$s^2$, a weighted average

Recall that we are assuming that the two populations from which we draw the samples are normal distributions with a common variance $\sigma^2$. If the confidence interval presented were valid only when these assumptions were met exactly, the estimation procedure would be of limited use. Fortunately, the confidence coefficient remains relatively stable if both distributions are mound-shaped and the sample sizes are approximately equal.

**EXAMPLE 5.5**

Company officials were concerned about the length of time a particular drug product retained its potency. A random sample, sample 1, of $n_1 = 10$ bottles of the product was drawn from the production line and analyzed for potency; a second sample, sample 2, of $n_2 = 10$ bottles was obtained and stored in a regulated environment for a period of one year.

The readings obtained from each sample are given in table 5.2.

**Table 5.2** Potency reading for two samples

| Sample 1 | | Sample 2 | |
|---|---|---|---|
| 10.2 | 10.6 | 9.8 | 9.7 |
| 10.5 | 10.7 | 9.6 | 9.5 |
| 10.3 | 10.2 | 10.1 | 9.6 |
| 10.8 | 10.0 | 10.2 | 9.8 |
| 9.8 | 10.6 | 10.1 | 9.9 |

Suppose we let $\mu_1$ denote the mean potency for all bottles that might be sampled coming off the production line and $\mu_2$ denote the mean potency for all bottles that may be retained for a period of one year. Estimate $\mu_1 - \mu_2$ by using a 95% confidence interval.

**SOLUTION**

The necessary sample calculations from the data of table 5.2 are presented below.

| Sample 1 | Sample 2 |
|---|---|
| $\sum_{j=1}^{10} y_{1j} = 103.7$ | $\sum_{j=1}^{10} y_{2j} = 98.3$ |
| $\sum_{j=1}^{10} y_{1j}^2 = 1076.31$ | $\sum_{j=1}^{10} y_{2j}^2 = 966.81$ |

Then

$$\bar{y}_1 = \frac{103.7}{10} = 10.37 \qquad \bar{y}_2 = \frac{98.3}{10} = 9.83$$

$$s_1^2 = \frac{1}{9}\left[1076.31 - \frac{(103.7)^2}{10}\right] = .105 \qquad s_2^2 = \frac{1}{9}\left[966.81 - \frac{(98.3)^2}{10}\right] = .058$$

The estimate of the common standard deviation $\sigma$ is

$$s = \sqrt{\frac{(n_1 - 1)s_1^2 + (n_2 - 1)s_2^2}{n_1 + n_2 - 2}} = \sqrt{\frac{9(.105) + 9(.058)}{18}}$$

which, for $n_1 = n_2 = 9$, reduces to

$$s = \sqrt{\frac{.105 + .058}{2}} = .285$$

The $t$-value based on df $= n_1 + n_2 - 2 = 18$ and a $= .025$ is 2.101. A 95% confidence interval for the difference in mean potencies is

$$(10.37 - 9.83) \pm 2.101(.285)\sqrt{\tfrac{1}{10} + \tfrac{1}{10}} \quad \text{or} \quad .54 \pm .268$$

We estimate that the difference in means, $\mu_1 - \mu_2$, lies in the interval .272 to .808.

We can also test an hypothesis about the difference between two population means. As with any test procedure, we begin by specifying a research hypothesis for the difference in population means. Thus we might, for example, specify that the difference $\mu_1 - \mu_2$ is greater than some value $D_0$ (note: $D_0$ could be zero). The entire test procedure is summarized in the box.

---

**A Small-Sample Statistical Test for $\mu_1 - \mu_2$**

$H_0$: $\mu_1 - \mu_2 = D_0$ ($D_0$ is specified).

$H_a$: 1. $\mu_1 - \mu_2 > D_0$.

2. $\mu_1 - \mu_2 < D_0$.

3. $\mu_1 - \mu_2 \neq D_0$.

T.S.: $t = \dfrac{(\bar{y}_1 - \bar{y}_2) - D_0}{s\sqrt{(1/n_1) + (1/n_2)}}$.

R.R.: For a type I error $\alpha$ and df $= n_1 + n_2 - 2$,

1. reject $H_0$ if $t > t_\alpha$;
2. reject $H_0$ if $t < t_\alpha$;
3. reject $H_0$ if $|t| > t_{\alpha/2}$.

Note: Although this procedure is called a small-sample test, it applies for any sample sizes $n_1$ and $n_2$.

---

**EXAMPLE 5.6**

An experiment was conducted to compare the mean number of tapeworms in the stomachs of sheep that had been treated for worms against the mean number in those that were untreated. A sample of 14 worm-infected lambs was randomly divided into two groups. Seven were injected with the drug and the remainder were left untreated. After a six-month period, the lambs were slaughtered and the following worm counts were recorded:

| | | | | | | | |
|---|---|---|---|---|---|---|---|
| *Drug Treated* | 18 | 43 | 28 | 50 | 16 | 32 | 13 |
| *Untreated* | 40 | 54 | 26 | 63 | 21 | 37 | 39 |

Test an hypothesis that there is no difference in the mean number of worms between treated and untreated lambs. Assume that the drug cannot increase the number of worms and hence use the alternative hypothesis that the mean for treated lambs is less than the mean for untreated lambs. Use $\alpha = .05$.

**SOLUTION**

The calculations for the samples of treated sheep and untreated sheep are summarized below.

| Drug Treated | Untreated |
|---|---|
| $\sum_{j=1}^{7} y_{1j} = 200$ | $\sum_{j=1}^{7} y_{2j} = 280$ |
| $\sum_{j=1}^{7} y_{1j}^2 = 6906$ | $\sum_{j=1}^{7} y_{2j}^2 = 12{,}492$ |
| $\bar{y}_1 = \dfrac{200}{7} = 28.57$ | $\bar{y}_2 = \dfrac{280}{7} = 40.0$ |

| Drug Treated | Untreated |
|---|---|
| $s_1^2 = \frac{1}{6}\left[6906 - \frac{(200)^2}{7}\right]$ | $s_2^2 = \frac{1}{6}\left[12{,}492 - \frac{(280)^2}{7}\right]$ |
| $= \frac{1}{6}[6906 - 5714.29]$ | $= \frac{1}{6}[12{,}492 - 11{,}200]$ |
| $= 198.62$ | $= 215.33$ |

The sample variances are combined to form an estimate of the common population standard deviation $\sigma$.

$$s = \sqrt{\frac{(n_1 - 1)s_1^2 + (n_2 - 1)s_2^2}{n_1 + n_2 - 2}} = \sqrt{\frac{6(198.62) + 6(215.33)}{12}} = 14.39$$

The test procedure for the research hypothesis that the treated sheep will have a mean infestation level ($\mu_1$) less than the mean level ($\mu_2$) for untreated sheep is as follows:

$H_0$: $\mu_1 - \mu_2 = 0$ (i.e., no difference in the mean infestation levels).

$H_a$: $\mu_1 - \mu_2 < 0$.

T.S.: $t = \frac{\bar{y}_1 - \bar{y}_2}{s\sqrt{(1/n_1) + (1/n_2)}} = \frac{28.57 - 40}{14.39\sqrt{\frac{1}{7} + \frac{1}{7}}} = -1.49.$

R.R.: For $\alpha = .05$, the critical $t$-value for a one-tailed test with $\text{df} = n_1 + n_2 - 2 = 12$ can be obtained from table 2, the appendix, using $a = .05$. We will reject $H_0$ if $t < -1.782$.

Conclusion: Since the observed value of $t$, $-1.49$, does not fall in the rejection region, we have insufficient evidence to reject the hypothesis that there is no difference in the mean number of worms in treated and untreated lambs.

The test procedures for comparing two population means presented in this chapter are based on the assumption that the two population variances, $\sigma_1^2$ and $\sigma_2^2$, are identical (but unknown). Although we will delay a discussion of how to test whether the population variances are identical (see chapter 12), we will present here a statistical test of $H_0$: $\mu_1 = \mu_2$ when $\sigma_1^2 \neq \sigma_2^2$.

$\sigma_1^2 \neq \sigma_2^2$

When samples are selected from populations with different variances, the test statistic

$$t = \frac{\bar{y} - \bar{y}_2}{s\sqrt{(1/n_1) + (1/n_2)}}$$

no longer possesses a $t$ distribution. However, Cochran (1964) showed that the quantity

$t'$

$$t' = \frac{\bar{y}_1 - \bar{y}_2}{\sqrt{(s_1^2/n_1) + (s_2^2/n_2)}}$$

can be approximated by a $t$ distribution under the situations outlined in the box.

**Approximating the Distribution of $t'$ with a $t$ Distribution When Testing $H_0$: $\mu_1 = \mu_2$**

1. When both sample sizes are the same (i.e., $n_1 = n_2 = n$), use the test statistic

$$t' = \frac{\bar{y}_1 - \bar{y}_2}{\sqrt{(s_1^2/n_1) + (s_2^2/n_2)}}$$

do not use this junk

The rejection region for $t'$ can be obtained from table 2 in the appendix for df $= n - 1$.

2. When $n_1 \neq n_2$, calculate the statistic

$$t' = \frac{\bar{y}_1 - \bar{y}_2}{\sqrt{(s_1^2/n_1) + (s_2^2/n_2)}}$$

Reject $H_0$ if

$$|t'| > \frac{(t_1 s_1^2/n_1) + (t_2 s_2^2/n_2)}{(s_1^2/n_1) + (s_2^2/n_2)}$$

where $t_i$ ($i = 1, 2$) is the critical $t$-value from table 2, the appendix, based on $n_i - 1$ degrees of freedom.

**EXAMPLE 5.7**

Refer to the experimental situation explained in example 5.6. Suppose the sample data were as shown below.

| | | | | | | | |
|---|---|---|---|---|---|---|---|
| *Drug Treated* | 5 | 13 | 18 | 6 | 4 | 2 | 15 |
| *Untreated* | 40 | 54 | 26 | 63 | 21 | 37 | 39 |

Test the research hypothesis $H_a$: $\mu_1 - \mu_2 < 0$ under the assumption that the two population variances are different. Use $\alpha = .05$.

**SOLUTION**

It is easy to verify that

$\bar{y}_1 = 9.00 \qquad \bar{y}_2 = 40.00$

$s_1^2 = 38.67 \qquad s_2^2 = 215.33$

Then the statistical test is set up as follows:

$H_0$: $\mu_1 - \mu_2 = 0$.

$H_a$: $\mu_1 - \mu_2 < 0$.

$$\text{T.S.: } t' = \frac{\bar{y}_1 - \bar{y}_2}{\sqrt{(s_1^2/n_1) + (s_2^2/n_2)}} = \frac{9 - 40}{\sqrt{(38.67/7) + (215.33/7)}} = -5.15.$$

R.R.: For $\alpha = .05$ reject $H_0$ if $t' < -1.943$, the $t$-value for a = .05 and df = 6.

Conclusion: Since the computed value of $t'$ is less than $-1.943$, we reject $H_0$ and conclude that the drug product reduces the average infestation level.

## 5.5 Summary

In this chapter we have considered small-sample inferences concerning a population mean and concerning the difference between two population means. The small-sample test and confidence interval for $\mu$ are identical to the large-sample results, with $t$ (based on $n - 1$ degrees of freedom) replacing $z$. Although not emphasized in our previous discussion, the use of $t$ in small-sample inferences about $\mu$ assumes that the original measurements ($y$-values) were drawn from a normal population. While this is an important assumption, the use of the $t$ statistic provides an adequate approximation when the data arise from a mound-shaped distribution.

The small-sample test and estimation procedures for comparing two population means also utilize a $t$ distribution based on $n_1 + n_2 - 2$ degrees of freedom. Here we assume independent random samples are drawn from two normal populations with unknown means but a common (unknown) variance $\sigma^2$. Again these assumptions can be relaxed somewhat, thus increasing the applicability of the procedures.

Finally we note that although the test and estimation procedures were developed specifically for small-sample situations, all sample sizes of one or more observations are permissible provided the assumptions are met.

## Exercises

e

5.7. Two different emission control devices were being tested to determine the average amount of nitric oxide being emitted by an automobile over a one-hour period of time. Twenty cars of the same model and year were selected for the study. Ten cars were randomly selected and equipped with a Type I emission control device, and the remaining cars were equipped with Type II devices. Each of the 20 cars was then monitored for a one-hour period to determine the amount of nitric oxide emitted.

Use the data on the next page to test the research hypothesis that the mean level of emission for Type I devices ($\mu_1$) is greater than the mean emission level for Type II devices ($\mu_2$). Use $\alpha = .01$.

| *Type I Device* | | *Type II Device* | |
|---|---|---|---|
| 1.35 | 1.28 | 1.01 | 0.96 |
| 1.16 | 1.21 | 0.98 | 0.99 |
| 1.23 | 1.25 | 0.95 | 0.98 |
| 1.20 | 1.17 | 1.02 | 1.01 |
| 1.32 | 1.19 | 1.05 | 1.02 |

5.8. It has been estimated that lead poisoning resulting from an unnatural craving (pica) for lead may affect as many as a quarter of a million children each year, causing them to suffer from severe, irreversible retardation. Explanations for why children voluntarily consume lead range from "improper parental supervision" to "a child's need to mouth objects." Some researchers, however, have been investigating whether this unnatural craving for lead has some nutritional explanation. One such study involved a comparison of a regular diet and a calcium deficient diet on the ingestion of a lead acetate solution in rats. Each rat in a group of 20 rats was randomly assigned to either an experimental or control group. Those in the control group received a normal diet; the experimental group received a calcium deficient diet. Each of the rats occupied a separate cage and was monitored to observe the quantity of a .15% lead acetate solution consumed during the study period. The sample results are summarized below.

| | | | | | | | | | | |
|---|---|---|---|---|---|---|---|---|---|---|
| *Control Group* | 5.4 | 6.2 | 3.1 | 3.8 | 6.5 | 5.8 | 6.4 | 4.5 | 4.9 | 4.0 |
| *Experimental Group* | 8.8 | 9.5 | 10.6 | 9.6 | 7.5 | 6.9 | 7.4 | 6.5 | 10.5 | 8.3 |

Run a test of the research hypothesis that the mean quantity of lead acetate consumed in the experimental group is greater than that for rats in the control group. Use $\alpha = .05$.

5.9. The results of a three-year study to examine the effect of ready-to-eat breakfast cereals on dental caries (decay) experience in adolescent children were recently reported by Rowe, Anderson, and Wanninger (1974). A sample of 375 adolescent children of both sexes from the Ann Arbor, Michigan, public schools was enrolled (after parental consent) in the study. Each of the participants was provided with toothpaste and boxes of different varieties of ready-to-eat cereals. Although these were brand name cereals, they were not packaged in their usual containers. Rather, each type of cereal was packaged in plain white seven-ounce boxes and labeled as wheat flakes, corn cereal, oat cereal, fruit-flavored corn puffs, corn puffs, cocoa-flavored cereal, and sugared oat cereal. Note that the last four varieties of cereal had been presweetened and the others had not.

Each of the children received a dental examination at the beginning of the study, twice during the study, and once at the end. The response of interest for us will be the incremental DMF surfaces, that is, the difference between the final (poststudy) and initial (prestudy) number of decayed, missing, and filled (DMF) tooth surfaces. Careful records for each participant were maintained throughout the three years, and at the end of the study, a person was classified as "noneater" if he or she had eaten less than 28 boxes of cereal throughout the study. All others were classified as "eaters." The incremental DMF surface readings have been summarized below for each group.

| | *Sample Size* | *Sample Mean* | *Sample Standard Deviation* |
|---|---|---|---|
| *Noneaters* | 73 | 6.41 | 5.62 |
| *Eaters* | 302 | 5.20 | 4.67 |

Use these data to test the research hypothesis that the mean incremental DMF surface for noneaters is larger than the corresponding mean for eaters. Give the level of significance for your test. Interpret your findings.

5.10. Refer to exercise 5.9. Although complete details of the original study have not been disclosed, critique the procedure that has been discussed.

5.11. The study of concentrations of atmospheric trace metals in isolated areas of the world has received considerable attention because of the concern that humans might somehow alter the climate of the earth by changing the amount and distribution of trace metals in the atmosphere. Consider a study at the South Pole, where at 10 different sampling periods throughout a two-month period, 10,000 standard cubic meters (scm) of air were obtained and analyzed for metal concentrations. The results associated with magnesium and europium are listed below. (Note: Magnesium results are in units of $10^{-9}$ g/scm; europium results are in units of $10^{-15}$ g/scm.)

| | *Sample Size* | *Sample Mean* | *Sample Standard Deviation* |
|---|---|---|---|
| *Magnesium* | 10 | 1.0 | 2.21 |
| *Europium* | 10 | 17.0 | 12.65 |

Construct a 99% confidence interval for the mean magnesium content in the atmosphere at the South Pole.

5.12. Refer to exercise 5.11. Construct a 90% confidence interval for the mean europium concentration in the atmosphere at the study site.

5.13. A study was conducted to compare the flavor of cheddar cheese made two different ways: with milk and with milk fat plus buttermilk solids. Cheese was processed in large vats, one vat for each of the different types of fat, and stored for a period of six months. A panel of 15 trained judges was employed to compare each of the cheeses. On the day of testing, the cheese was brought to room temperature. Each judge was presented with an unmarked square of each type of cheese and asked to taste the cheeses and judge them for cheddar flavor on a 0-to-8-point scale.

| None | Slight | Moderate | Pronounced* |
|---|---|---|---|
| 0 1 2 | 3 4 | 5 6 | 7 8 |

For each judge the score recorded was the difference between the rating on cheese made from milk and the rating on cheese made from milk fat supplemented with buttermilk solids. The sample data are summarized below.

sample size: 15 (number of scores)

sample mean: 1.3

sample standard deviation: 6.04

Use these data to test the null hypothesis that the mean score (mean difference in cheddar flavor ratings for the two cheeses) is zero against a two-sided alternative hypothesis. Give the level of significance of your test and interpret the results of this test.

5.14. Refer to exercise 5.13. Construct a 95% confidence interval for the mean difference in cheddar ratings for the two cheeses.

5.15. To determine the nutritional value of whey, a study was conducted on 16 dairy cattle. Eight cows were randomly assigned to a liquid regimen of water only (Group 1); the others received liquid whey only (Group 2). In addition, each animal was given 7.5 kg of grain per day and allowed to graze on hay at will. While no significant differences were observed between the groups in the dairy milk production gauges, such as milk

*Pronounced means almost identical to a typical cheddar flavor.

production and fat content of the milk, the following data on daily hay consumption (in kilograms/cow) were of interest:

| | | | | | | | | |
|---|---|---|---|---|---|---|---|---|
| *Group 1* | 15.1 | 14.9 | 14.8 | 14.2 | 13.1 | 12.8 | 15.5 | 15.9 |
| *Group 2* | 6.8 | 7.5 | 8.6 | 8.4 | 8.9 | 8.1 | 9.2 | 9.5 |

Use these sample data to test the research hypothesis that there is a difference in mean hay consumption for the two diets. Use $\alpha = .05$.

5.16. Refer to exercise 5.15. Construct a 90% confidence interval for the difference in mean hay consumption for cattle on the two diets.

5.17. Over the past decade or more there has been a steady decrease in the national average Scholastic Aptitude Test (SAT) score. Many parents, teachers, and administrators have been concerned about this trend and have sought means for halting the decline, at least at the local school level. To this end a group of 50 students (24 males and 26 females), matched as nearly as possible according to socioeconomic background, received parental permission to participate in a study to examine the effect of classroom atmosphere (strict or liberal) on student performance, as measured on a standardized achievement test score at the end of the year. The 50 students were randomly divided into two groups of 25 students each (12 males and 13 females), with Group I to study under a strict, closely regulated classroom atmosphere while Group II attended classes under a very permissive atmosphere. Since groups rotated to different subject matter classes throughout the day, it was possible to employ the same subject matter teachers for both groups. After nine months under this program, all students were given the same standardized test: the verbal test and the mathematics test. The results are given below.

| | *Group I* | | | *Group II* | | |
|---|---|---|---|---|---|---|
| | *Verbal* | *Math* | *Total* | *Verbal* | *Math* | *Total* |
| *Sample Mean* | 45.2 | 46.6 | 91.8 | 40.3 | 43.4 | 83.7 |
| *Sample Standard Deviation* | 6.0 | 4.7 | 8.1 | 5.1 | 5.9 | 8.4 |

a. Use the sample verbal data for the two groups to test the research hypothesis that the group with a more regimented classroom atmosphere will score higher, on the average, than the less restricted group. Use $\alpha = .05$.

b. Would your results change using a two-tailed test with $\alpha = .05$?

5.18. Refer to exercise 5.17. Using the verbal data, construct a 90% confidence interval on the difference in the mean scores for the two groups.

5.19. Using the mathematics test score data of exercise 5.17, construct a 99% confidence interval for the difference in mean responses between the two groups.

5.20. Refer to exercise 5.17. Using the math data, perform the same type of statistical test with $\alpha = .05$.

5.21. Refer to exercise 5.17. Some people might argue that the best measure of student performance using this standardized test is the sum of the verbal and mathematics scores. Use these data to test the research hypothesis that the two groups have different mean test scores. Use $\alpha = .01$.

5.22. Refer to exercise 5.21. Use the sample data in exercise 5.17 to construct a 98% confidence interval on the difference in mean test scores.

5.23. An industrial concern has experimented with several different mixtures of the four components magnesium, sodium nitrate, strontium nitrate, and binder which comprise a rocket propellant. The company has found that two mixtures in particular give higher flare illumination values than the others. Mixture 1 consists of a blend composed of the proportions .40, .10, .42, and .08 for the four components of the mixture; Mixture 2 consists of a blend using the proportions .60, .27, .10, and .03. Twenty different blends (10 of each mixture) are prepared and tested to obtain the flare illumination values. These data appear below (in units of 1000 candles).

| *Mixture 1* | 185 | 192 | 201 | 215 | 170 | 190 | 175 | 172 | 198 | 202 |
|---|---|---|---|---|---|---|---|---|---|---|
| *Mixture 2* | 221 | 210 | 215 | 202 | 204 | 196 | 225 | 230 | 214 | 217 |

Use the sample data to determine whether there is evidence to indicate that one of the mixtures has a higher mean flare illumination value than the other. Give the level of significance of the test and interpret your findings.

5.24. Refer to exercise 5.23. Instead of conducting a statistical test, use the sample data to answer the question, "What is the difference in mean flare illumination for the two mixtures?"

5.25. Refer to example 5.7 (p. 117). Suppose the seventh untreated animal died before the study was completed. Analyze the remaining observations to compare the two population means. Assume that $\sigma_1^2 \neq \sigma_2^2$. Give the level of significance for your test.

*5.26. A computer printout for the statistical test of the data in exercise 5.8 is shown at the bottom of the page.

a. Compare the computer results from BMDP (1975) to your calculations for exercise 5.8. [Note: Refer to T (POOLED) in the output.]

b. Give the value of the test statistic and the level of significance for a $t$ test of the research hypothesis that the experimental mean is greater than the control mean.

| | | | | | | | | | | |
|---|---|---|---|---|---|---|---|---|---|---|
| *Control Group* | 5.4 | 6.2 | 3.1 | 3.8 | 6.5 | 5.8 | 6.4 | 4.5 | 4.9 | 4.0 |
| *Experimental Group* | 8.8 | 9.5 | 10.6 | 9.6 | 7.5 | 6.9 | 7.4 | 6.5 | 10.5 | 8.3 |

5.27. A study of anxiety was conducted among residents of a southeastern metropolitan area. Each person selected for the study was asked to check a yes or a no for the presence of each of 12 anxiety symptoms. Anxiety scores ranged from 0 to 12, with higher scores related to higher perceived presence of anxiety symptoms. The results for a random sample of 50 residents, categorized by sex, are summarized below.

| | *Sample Size* | *Mean* | *Standard Deviation* |
|---|---|---|---|
| *Female* | 26 | 5.26 | 3.2 |
| *Male* | 24 | 7.02 | 3.9 |

Use these data to test the research hypothesis that the mean perceived anxiety score is different for males and females. Give the level of significance for your test.

```
FOR DIFFERENCES ON SINGLE VARIABLES

VARIABLE NUMBER  2 RESPONSE              GROUP     1 CONTROL  2 EXPTAL       1 CONTROL (N = 10)               2 EXPTAL (N = 10)
                                         MEAN        5.0600    8.5600
      STATISTICS            P VALUE  D. F.  STD DEV     1.1890    1.4714
                                         S.E.M.      0.3760    0.4653
T (SEPARATE)          -5.85  0.000  17.2  NUMBER    10         10
T (POOLED)            -5.85  0.000  18    MAXIMUM     6.5000   10.6000             H                                 X    X X
                                         MINIMUM     3.1000    6.5000       HHHHH HHHH                        XXX X X X X
F (FOR VARIANCES)      1.53  0.536   9, 9                                 MIN-------------------MAX  MIN-------------------MAX
                                                                              AN H = 1.0 CASES                 AN X = 1.0 CASES
```

# References


Anderson, M.J.; Lamb, R.C.; Mekelsen, C.H.; and Wiscombe, R.L. 1974. Feeding liquid whey to dairy cattle. *Journal of Dairy Science* 57.

Brownlee, K.A. 1965. *Statistical theory and methodology in science and engineering*. 2d ed. New York: Wiley.

Cochran, W.G. 1964. Approximate significance levels of the Behrens-Fisher test. *Biometrics* 20: 191–195.

Dixon, W.J., ed. 1975. *BMDP, biomedical computer programs*. Berkeley: University of California Press.

Dixon, W.J., and Massey, F.J., Jr. 1969. *Introduction to statistical analysis*. 3d ed. New York: McGraw-Hill.

Duncan, A.J. 1959. *Quality control and industrial statistics*. 2d ed. Homewood, Ill.: Richard D. Irwin, Inc. Chapter 21.

Mendenhall, W. 1975. *Introduction to probability and statistics*. 4th ed. N. Scituate, Mass.: Duxbury Press.

Ostle, B. 1963. *Statistics in research*. 2d ed. Ames, Iowa: Iowa State University Press.

Pearson, E.S., and Hartley, H.O. 1966. *Biometrika tables for statisticians*. 3d ed. Vol. I. London: Cambridge University Press.

Rowe, N.H.; Anderson, R.H.; and Wanninger, L.A. 1974. Effects of ready-to-eat breakfast cereals on dental caries experience in adolescent children: A three-year study. *Journal of Dental Research* 53: 33.

Ryan, T.A.; Joiner, B.L.; and Ryan, B.F. 1976. *Minitab student handbook*. N. Scituate, Mass.: Duxbury Press.

Snedecor, G.W., and Cochran, W.G. 1967. *Statistical methods*. 6th ed. Ames, Iowa: Iowa State University Press. Chapter 4.

Snowdon, C.T., and Sanderson, B.A. 1974. Lead pica produced in rats. *Science* 183: 92–94.

"Student." 1908. The probable error of a mean. *Biometrika* 6: 1–25.

# 6 Models Relating a Response to a Set of Quantitative Independent Variables

## 6.1 Introduction

In chapters 4 and 5 we considered estimation and the test of an hypothesis concerning a population mean or the difference between two population means. The problems we encountered were relatively simple and straightforward. Estimation of a population mean can become more involved, however, if the variable of interest, often called the *dependent variable,* is affected by one or more additional variables, called *independent variables.*

dependent variable
independent variable

For example, suppose we are interested in estimating the mean weight gain per month for steers fed on a particular variety of feed. The dependent variable, weight gain, could be affected by many variables—initial weight of the steer, amount of feed offered per day, protein content of the feed, water content of the feed, and so on. The problem of estimating the mean weight gain per month must now take into account the *levels,* or *settings,* of the independent variables.

levels, or settings

For example, we might wish to estimate the mean weight gain (in pounds per day) for steers fed a high-concentrate feed containing 15% protein, 10% water, and the rest carbohydrates. Here 15% is a setting, or level, of the independent variable "protein content of feed" and 10% is a setting of the independent variable "water content of feed." Similarly, we might wish to estimate, or predict, the mean weight gain for other combinations of settings of the independent variables. Thus estimating a population mean becomes a problem of estimating a population mean for each setting of the independent variables.

This estimation problem can be greatly simplified if we consider *models* relating a *response* (dependent variable) to a set of independent variables. In this chapter we consider the problem of model formulation; in succeeding chapters we will use these models to formulate estimates and tests of hypotheses about the mean of a dependent variable that is related to one or more independent variables.

model
response

These same models will be useful in many other prediction or estimation problems. For example, a biologist may wish to predict an animal's pulse rate based on the amount of a particular drug administered and the length of time since the drug's administration. A political scientist may wish to predict the outcome of an election (in terms of numbers of votes cast for a particular candidate) based on various socioeconomic factors and previous voting records of the population under study. A sociologist may wish to relate or predict the average number of prison-free years between first and second offenses for persons characterized by various variables, such as age, sex, number of years of schooling, IQ, and so on. A market analyst might want to predict the year-end sales for a company based on various economic indices. Each of these problems bears certain similarities and can be attacked using the models to be proposed in this chapter.

The simplest type of model relating a response $y$ to a single quantitative independent variable $x$ is given by the equation of a straight line:

equation of straight line

$$y = \beta_0 + \beta_1 x$$

where $\beta_0$ is the *y-intercept* (value of $y$ when $x = 0$) and $\beta_1$ is the *slope* of the straight line (unit change in $y$ for a unit change in $x$). For a given equation, $\beta_0$ and $\beta_1$ are constants. An equation of this form is called a *deterministic model* because there is no error in reading $y$. That is, for a given value of the independent variable $x$, we can predict $y$ exactly using the deterministic equation $y = \beta_0 + \beta_1 x$.

y-intercept $\beta_0$
slope $\beta_1$
deterministic model

Deterministic models need not be equations of straight lines. For example, a dependent variable $y$ might be related to $x$ in a *curvilinear* manner given by the deterministic equation

$$y = \beta_0 + \beta_1 x + \beta_2 x^2$$

This equation, for $\beta_0 = 1$, $\beta_1 = 1$, and $\beta_2 = .5$, is sketched in figure 6.1. Again we can predict $y$ exactly for any value of $x$ using the model relating $y$ to $x$. Thus when $x = 1$, we would predict $y$ to be

$$y = 1 + 1 + .5 = 2.5$$

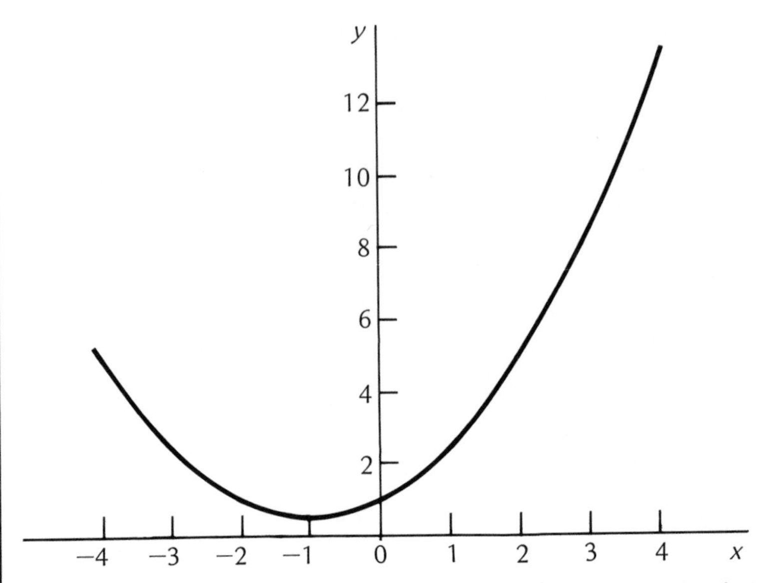


**Figure 6.1** Graph of the model $y = 1 + x + .5x^2$, a curvilinear relation

deterministic model

Although deterministic models are simple to use, they are unrealistic in many situations, since a dependent variable $y$ cannot always be adequately represented by a deterministic equation in one or more quantitative independent variables. Consider the data of table 6.1, which gives hospital expenses covered by an insurance carrier and the number of days of confinement for a random sample of $n = 10$ patients. In the table we let $y$ be the dependent variable (expense) and $x$ the independent variable (number of days of confinement). Suppose the insurance carrier and hospital administrators are interested in estimating the average hospital expense for a given number of days of confinement. Can we use a deterministic model for this problem?

**Table 6.1** Hospital expense data

| *Expense, y* | *Number of Days of Confinement, x* |
|---|---|
| $ 50 | 1 |
| 175 | 3 |
| 180 | 6 |
| 200 | 7 |
| 60 | 2 |
| 140 | 4 |
| 420 | 12 |
| 540 | 15 |
| 170 | 5 |
| 300 | 9 |

First, let us draw a picture of the data. The data of table 6.1 can be plotted by using a *scatter diagram*. In a scatter diagram we draw a vertical axis and a horizontal axis, labeled $y$ and $x$, respectively. The $n = 10$ data points are then plotted as shown in figure 6.2.

scatter diagram

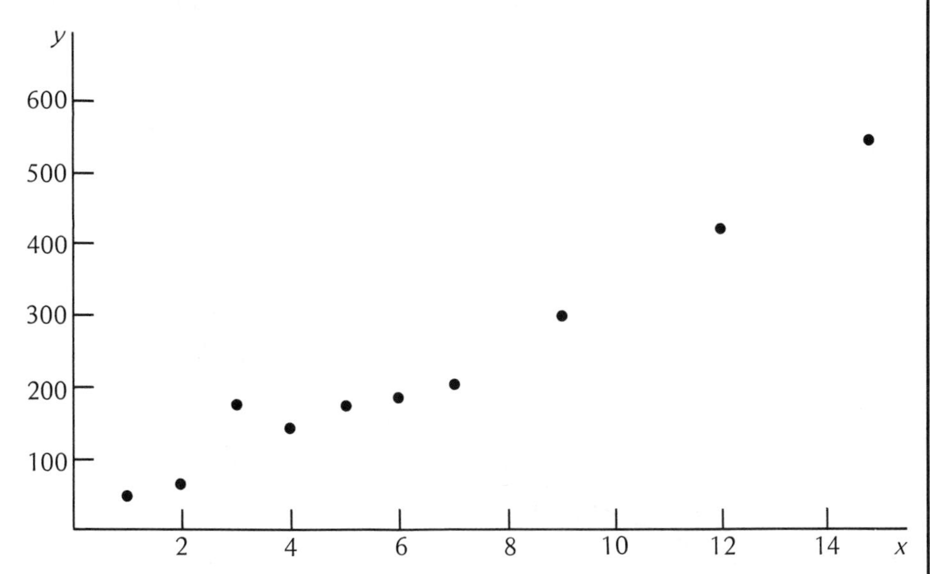


**Figure 6.2** Scatter diagram of the 10 data points for the hospital expense data of table 6.1

You can see from figure 6.2 that a straight line would adequately describe the trend in the data. However, we cannot predict $y$ *exactly* for a given value of $x$. Thus, for example, we could not predict the average value of $y$ for $x = 8$ days of confinement using a deterministic model of the form $y = \beta_0 + \beta_1 x$.

A model that allows for the possibility that the observations do not all lie on a straight line is the model

$$y = \beta_0 + \beta_1 x + \epsilon$$

where $\epsilon$ is a *random error*. In this model $\epsilon$ represents the difference between a measurement $y$ and a point on the line $\beta_0 + \beta_1 x$. We call this model a *probabilistic model*. Some of the properties of $\epsilon$ will be discussed now.

random error $\epsilon$
probabilistic model

One assumption made concerning the random error is that the average value of $\epsilon$ for a given value of $x$ is 0. Thus since $\beta_0$ and $\beta_1$ are constants, the *average* of $y$ (often called the *expected value* of $y$) for a fixed value of $x$ is $\beta_0 + \beta_1 x$. This line, denoted by

average value of $\epsilon = 0$ for fixed $x$
expected value of $y$

$$E(y) = \beta_0 + \beta_1 x$$

$E(y)$

is shown in figure 6.3. A point on the line denotes the average value of $y$ for the corresponding setting of $x$. The difference between a sample data point and the expected value of $y$ (a point on the line $\beta_0 + \beta_1 x$) is $\epsilon$. Thus the observed values of $y$ deviate above or below the line by a random amount $\epsilon$. The random errors associated with the 10 data points listed in table 6.1 are pictured here in figure 6.3.

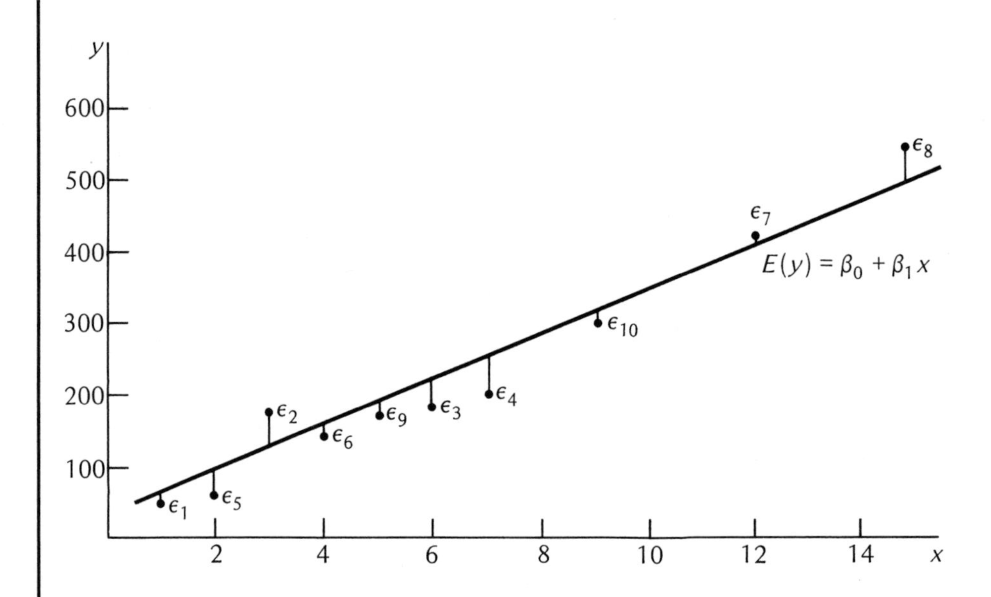


**Figure 6.3** A plot of $E(y) = \beta_0 + \beta_1 x$ for the hospital expense data of table 6.1

Unfortunately, since $\beta_0$ and $\beta_1$ are unknown parameters, we will never know the precise location of the line $E(y) = \beta_0 + \beta_1 x$. All we will have in a given experimental situation will be the $n$ data points. In succeeding chapters we will show how to use the sample information to construct estimates, $\hat{\beta}_0$ and $\hat{\beta}_1$, of the parameters $\beta_0$ and $\beta_1$ to be used in formulating an estimate of the line $E(y) = \beta_0 + \beta_1 x$. As with any estimation procedure, we will also provide a measure of the goodness of our estimation procedures. Appropriate confidence intervals and tests of hypotheses will also be discussed in the chapters that follow.

In the remainder of this chapter we will discuss various types of probabilistic models relating a response $y$ to one or more quantitative independent variables.

# Models Relating a Response to a Single Quantitative Independent Variable 6.2

The simplest type of probabilistic model relating the dependent variable $y$ to a quantitative independent variable $x$ is the one discussed in section 6.1:

$$y = \beta_0 + \beta_1 x + \epsilon$$

Under the assumption that the average value of $\epsilon$ (also called the *expected value* of $\epsilon$) for a given value of $x$ is $E(\epsilon) = 0$, this model indicates that the expected value of $y$ for a given value of $x$ is described by the straight line

expected value of $\epsilon$
$E(\epsilon) = 0$

$$E(y) = \beta_0 + \beta_1 x$$

Not all data sets are adequately described by a model whose expectation is a straight line. For example, consider the data of table 6.2, which gives the

**Table 6.2** Yield of 12 equal-sized plots of tomato plantings for different amounts of fertilizer

| *Plot* | *Yield, y (in bushels)* | *Amount of Fertilizer, x (in pounds per plot)* |
|---|---|---|
| 1 | 24 | 12 |
| 2 | 18 | 5 |
| 3 | 31 | 15 |
| 4 | 33 | 17 |
| 5 | 26 | 20 |
| 6 | 30 | 14 |
| 7 | 20 | 6 |
| 8 | 25 | 23 |
| 9 | 25 | 11 |
| 10 | 27 | 13 |
| 11 | 21 | 8 |
| 12 | 29 | 18 |

yields (in bushels) for 12 equal-sized plots planted in tomatoes for different levels of fertilization. It is evident from the scatter plot in figure 6.4 that a linear equation will not adequately represent the relationship between yield and the amount of fertilizer applied to the plot. The reason for this is that, whereas a modest amount of fertilizer may well enhance the crop yield, too much fertilizer can be destructive.

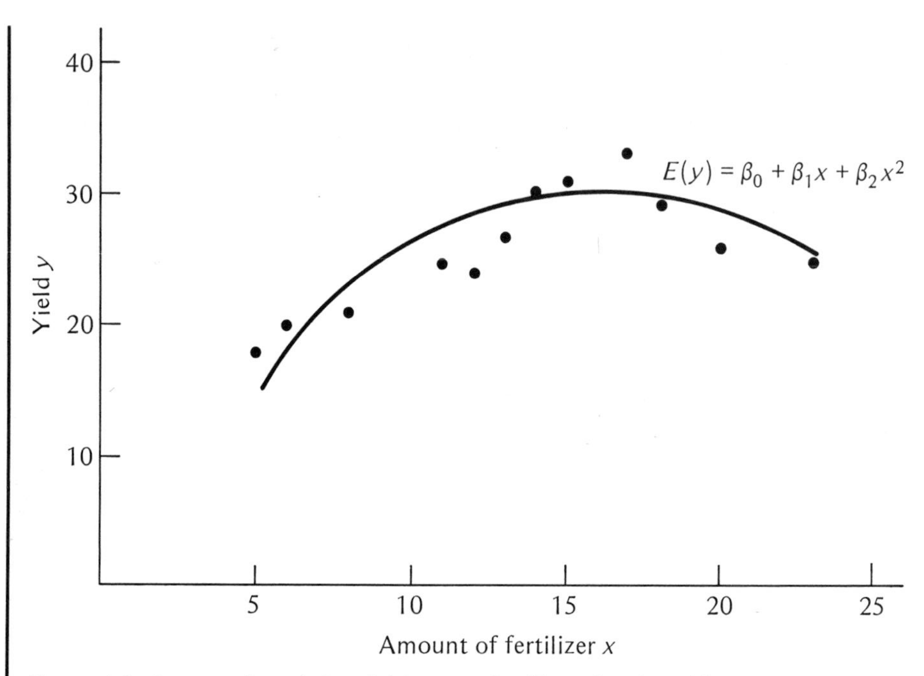


**Figure 6.4** Scatter plot of the yield versus fertilizer data in table 6.2

A model for this physical situation might be

$$y = \beta_0 + \beta_1 x + \beta_2 x^2 + \epsilon$$

Again with the assumption that $E(\epsilon) = 0$, the average value of $y$ for a given value of $x$ is

$$E(y) = \beta_0 + \beta_1 x + \beta_2 x^2$$

One such line is plotted in figure 6.4, superimposed on the data of table 6.2.

general probabilistic model, single variable

A *general polynomial probabilistic model* relating a dependent variable $y$ to a single independent variable $x$ is given by

$$y = \beta_0 + \beta_1 x + \beta_2 x^2 + \cdots + \beta_k x^p + \epsilon$$

with

$$E(y) = \beta_0 + \beta_1 x + \beta_2 x^2 + \cdots + \beta_k x^p$$

The choice of $p$ and hence the choice of an appropriate model will depend on the experimental situation.

# Models with Several Quantitative Independent Variables: General Linear Model 6.3

Not all probabilistic models relate a response $y$ to a single quantitative independent variable $x$. It is quite possible that more than one independent variable is related to the response and should be included in the probabilistic model. For example, in studying the yield of a tomato crop, several independent variables—amount of fertilizer ($x_1$), amount of water ($x_2$), hours of sunlight on clear days ($x_3$)—could all have an effect on yield. Hence to formulate a probabilistic model that adequately represents the yield of a tomato crop, we should include all these variables in a model. Typical probabilistic models relating a response $y$ to two quantitative independent variables, $x_1$ and $x_2$, are the following:

probabilistic model, two variables

$$y = \beta_0 + \beta_1 x_1 + \beta_2 x_2 + \epsilon$$

$$y = \beta_0 + \beta_1 x_1 + \beta_2 x_2 + \beta_3 x_1 x_2 + \epsilon$$

$$y = \beta_0 + \beta_1 x_1 + \beta_2 x_1^2 + \beta_3 x_2 + \beta_4 x_1 x_2 + \beta_5 x_1^2 x_2 + \epsilon$$

$$y = \beta_0 + \beta_1 x_1 + \beta_2 x_1^2 + \beta_3 x_2 + \beta_4 x_2^2 + \epsilon$$

$$y = \beta_0 + \beta_1 x_1 + \beta_2 x_1^2 + \beta_3 x_2 + \beta_4 x_2^2 + \beta_5 x_1 x_2 + \beta_6 x_1^2 x_2 + \beta_7 x_1 x_2^2 + \beta_8 x_1^2 x_2^2 + \epsilon$$

Several of the models are illustrated in figure 6.5.

Similarly, typical models relating $y$ to a set of three independent variables would be as follows:

three variables

$$y = \beta_0 + \beta_1 x_1 + \beta_2 x_2 + \beta_3 x_3 + \epsilon$$

$$y = \beta_0 + \beta_1 x_1 + \beta_2 x_2 + \beta_3 x_3 + \beta_4 x_1 x_2 + \beta_5 x_1 x_3 + \beta_6 x_2 x_3 + \epsilon$$

$$y = \beta_0 + \beta_1 x_1 + \beta_2 x_1^2 + \beta_3 x_2 + \beta_4 x_2^2 + \beta_5 x_3 + \beta_6 x_1 x_2 + \beta_7 x_1 x_3 + \beta_8 x_2 x_3 + \epsilon$$

$$y = \beta_0 + \beta_1 x_1 + \beta_2 x_1^2 + \beta_3 x_2 + \beta_4 x_2^2 + \beta_5 x_3 + \beta_6 x_3^2 + \epsilon$$

$$y = \beta_0 + \beta_1 x_1 + \beta_2 x_1^2 + \beta_3 x_2 + \beta_4 x_2^2 + \beta_5 x_3 + \beta_6 x_3^2 + \beta_7 x_1 x_2 + \beta_8 x_1 x_3 + \beta_9 x_2 x_3 + \beta_{10} x_1^2 x_2 + \beta_{11} x_1^2 x_3 + \beta_{12} x_1^2 x_2^2 + \epsilon$$

Again the particular choice of a model will depend on the actual experimental situation under study and the number of variables that affect the response. It

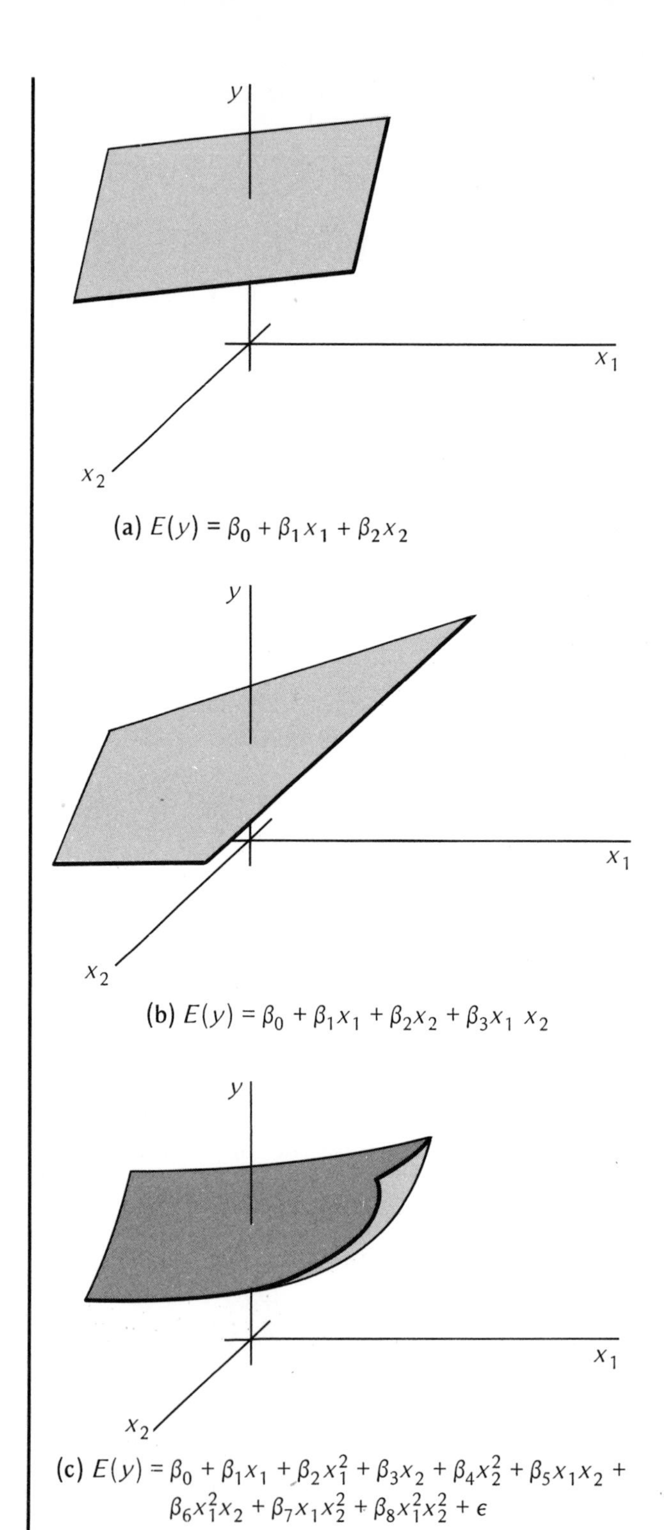


**Figure 6.5** Graphical illustrations of several models relating a response $y$ to two independent variables $x_1$ and $x_2$

suffices to say here that we will try to choose a model that best describes the relationship between the independent variables of interest and the response $y$.

We now present a model, called the *general linear model,* that relates a response to a set of $k$ ($k > 1$) independent variables.

general linear model

$$y = \beta_0 + \beta_1 x_1 + \beta_2 x_2 + \cdots + \beta_k x_k + \epsilon$$

where $x_1, x_2, \ldots, x_k$ are independent variables measured without error; $\beta_0, \beta_1, \beta_2, \ldots, \beta_k$ are unknown parameters; and $\epsilon$ is a random error term with $E(\epsilon) = 0$. All the probabilistic models discussed in this section and in section 6.2 are special cases of the general linear model. For example, the model

$$y = \beta_0 + \beta_1 x + \beta_2 x^2 + \beta_3 x^3 + \epsilon$$

is equivalent to the general linear model with $k = 3$, $x_1 = x$, $x_2 = x^2$, and $x_3 = x^3$. Similarly, the model

$$y = \beta_0 + \beta_1 x_1 + \beta_2 x_1^2 + \beta_3 x_2 + \beta_4 x_2^2 + \beta_5 x_1 x_2 + \beta_6 x_1^2 x_2 + \epsilon$$

is equivalent to the general linear model with $k = 6$ and

$$x_1 = x_1 \qquad x_3 = x_2 \qquad x_5 = x_1 x_2$$
$$x_2 = x_1^2 \qquad x_4 = x_2^2 \qquad x_6 = x_1^2 x_2$$

The fact that we have used the terms "general *linear* model" to describe a polynomial situation may seem confusing. We use the word "linear" in the general linear model to describe how the $\beta$'s are entered in the model, not how the independent variables are entered in the model. A general linear model is linear (in the usual algebraic sense) in the $\beta$'s. Thus the general linear model can be used to relate a dependent variable to one or more quantitative variables in any polynomial representation.

Succeeding chapters will deal with estimation and tests of hypotheses concerning individual parameters of the general linear model, as well as the average value,

$$E(y) = \beta_0 + \beta_1 x_1 + \beta_2 x_2 + \cdots + \beta_k x_k$$

In the next section we briefly discuss the classification of terms in the general linear model.

# 6.4 Classification of Terms in the General Linear Model

degree

pth-degree term

Individual terms in the general linear model are classified by their exponents. The *degree* of a term is given by the sum of the exponents for the independent variables appearing in the term. Thus any independent variable $x_i$ which appears in the general linear model as $x_i^p$ is called a *pth-degree term*. Hence $x_i$, $x_i^2$, $x_i^3$ are called first-, second-, and third-degree terms, respectively. Similarly, if two independent variables $x_i$ and $x_j$ appear together as $x_i^p x_j^q$, the term is called a $(p + q)$th-degree term. We would classify the terms $x_i x_j$, $x_i^2 x_j$, $x_i x_j^2$, $x_i^2 x_j^2$ as second-, third-, third-, and fourth-degree terms, respectively.

**EXAMPLE 6.1**

An experimenter feels that the probabilistic model

$$y = \beta_0 + \beta_1 x_1 + \beta_2 x_1^2 + \beta_3 x_2 + \beta_4 x_1 x_2 + \epsilon$$

adequately represents the relationship between a dependent variable $y$ and two independent variables $x_1$ and $x_2$. Identify the degree of all terms in the model containing one or more independent variables.

**SOLUTION**

Following the procedures just discussed for assigning degrees to terms, we have the following:

| *Term* | *Degree* |
|---|---|
| $\beta_1 x_1$ | first |
| $\beta_2 x_1^2$ | second |
| $\beta_3 x_2$ | first |
| $\beta_4 x_1 x_2$ | second |

first-order model

We just indicated that individual terms of the general linear model are classified by their exponents. We can also identify specific models by the types of terms that appear in the model. A *first-order model* is a general linear model that contains all possible first-degree terms in the independent variables.

**EXAMPLE 6.2**

Write a first-order probabilistic model relating a response $y$ to the independent variables $x_1$, $x_2$, and $x_3$.

**SOLUTION**

A first-order model in $x_1$, $x_2$, and $x_3$ includes all first-degree terms in these variables. The appropriate model is

$$y = \beta_0 + \beta_1 x_1 + \beta_2 x_2 + \beta_3 x_3 + \epsilon$$

Similarly, a *second-order model* is a general linear model that includes all possible first- and second-degree terms in the independent variables.

second-order model

**EXAMPLE 6.3**

Write a second-order model relating a dependent variable $y$ to the independent variables $x_1$, $x_2$, and $x_3$.

**SOLUTION**

The appropriate second-order model would include the terms of the corresponding first-order model (example 6.2) and all possible second-degree terms.

$$y = \beta_0 + \beta_1 x_1 + \beta_2 x_2 + \beta_3 x_3 + \beta_4 x_1^2 + \beta_5 x_2^2 + \beta_6 x_3^2 + \beta_7 x_1 x_2 + \beta_8 x_1 x_3 + \beta_9 x_2 x_3 + \epsilon$$

Higher-order models represent obvious extensions of the first- and second-order models. Several exercises at the end of this chapter offer you additional practice in formulating these models.

# Summary 6.5

As mentioned in the introduction to this chapter, inferences concerning a population mean can become quite involved. Hence we must develop techniques beyond the simplified results of chapters 4 and 5. In this chapter we presented the general linear model

$$y = \beta_0 + \beta_1 x_1 + \beta_2 x_2 + \cdots + \beta_k x_k + \epsilon$$

which relates the independent variables $x_1, x_2, \ldots, x_k$ to the dependent (response) variable $y$. The random error $\epsilon$ in the general linear model is assumed to have a mean equal to zero for any setting of the independent variables. Hence the average value of $y$ (expected value of $y$) for a given setting of the independent variables is

$$E(y) = \beta_0 + \beta_1 x_1 + \beta_2 x_2 + \cdots + \beta_k x_k$$

We showed that the general linear model can be used to represent polynomial models in one or more independent variables.

We classify terms in the general linear model by their exponents: the degree of a term is given by the sum of the exponents for the independent variables appearing in the term. Models can also be classified. We defined a first-order model to be a general linear model that contains all possible first-degree terms in the $k$ independent variables. Similarly, a second-order model contains all possible first- and second-degree terms in the $k$ independent variables.

Having discussed the formulation of the general linear model, we will turn in the next chapter to a procedure for developing an estimate of $E(y)$, the average value of $y$ when $y$ is related to a set of independent variables. This estimation procedure is called the method of least squares.

# Exercises

6.1. Suppose that a response $y$ is related to two independent variables $x_1$ and $x_2$.

a. Write a first-order probabilistic model.

b. Write three different probabilistic models relating $y$ to $x_1$ and $x_2$ using first- and second-degree terms.

6.2. Identify the degree for all terms in the model

$$y = \beta_0 + \beta_1 x_1 + \beta_2 x_1^2 + \beta_3 x_2 + \beta_4 x_1 x_2 + \beta_5 x_1^2 x_2 + \epsilon$$

6.3. Sketch the deterministic model

$$y = 1.5 + 1.0x^2$$

for $x$ in the range $-3 \leq x \leq 3$. (Hint: Substitute different values for $x$ into the model to determine corresponding values for $y$. Plot the $x$ and $y$ points.)

6.4. Sketch the deterministic model

$$y = 1.5 + 2.5x + 1.0x^2$$

for $x$ in the range $-3 \leq x \leq 3$.

6.5. Sketch the deterministic model

$$y = 1.5 - 2.5x + 1.0x^2$$

for $x$ in the range $-3 \leq x \leq 3$.

6.6. Compare the results of your sketches for exercises 6.3 through 6.5.

6.7. Distinguish between a deterministic model and a probabilistic model.

6.8. Consider the general linear model

$$y = \beta_0 + \beta_1 x_1 + \beta_2 x_1^2 + \beta_3 x_1^3 + \beta_4 x_1^4 + \epsilon$$

a. Specify the degree of each term in the general linear model.
b. What is the order of the model?

6.9. Write a second-order model relating a response $y$ to the independent variables $x_1$, $x_2$, and $x_3$.

6.10. Sketch the deterministic equation $y = 10 - 0.5x_1 - 1.0x_2$ on a graph with three axes: $y, x_1$, and $x_2$. [Hint: Substitute different combinations of $x_1$ and $x_2$ to determine $y$. Plot the $(x_1, x_2, y)$ points.]

6.11. Consider the general linear model

$$y = \beta_0 + \beta_1 x_1 + \beta_2 x_1^2 + \beta_3 x_2 + \beta_4 x_1 x_2 + \epsilon$$

a. Specify the degree of each term in the general linear model.
b. Is this a first- or second-order model? Explain.

6.12. Write a second-order probabilistic model relating a response $y$ to four independent variables $(x_1, x_2, x_3$, and $x_4)$.

6.13. Sketch plots for these deterministic models for $-1 \leq x \leq 1$:
a. $y = 1.2 + x$
b. $y = 1.2 + x + .4x^2$
c. $y = 1.2 + x + .4x^2 + .6x^3$

+ 6.14. Sketch a graph using the 5 data points given below.

| $x$ | 1 | 2 | 3 | 4 | 5 |
|---|---|---|---|---|---|
| $y$ | 5 | 10 | 14 | 21 | 26 |

6.15. Refer to the sketch in exercise 6.14. Write a probabilistic model relating $y$ to $x$.

6.16. Sketch a graph of the response $y$ as a function of the independent variable $x$ using the 5 data points given below.

| $x$ | 0 | $-1$ | 2 | 3 | $-2$ |
|---|---|---|---|---|---|
| $y$ | 15 | 12 | 31 | 50 | 11 |

6.17. Using the sketch drawn in exercise 6.16, indicate the form (without values for the $\beta$'s) of a general linear model relating the response $y$ to the independent variable $x$.

6.18. If a second-order model relating $y$ to $x$ has one peak (or, equivalently, one valley) when sketched, and a third-order model has one peak and one valley when sketched, how many peaks and valleys do you think a fourth-order model has when sketched? Sketch a typical fourth-order model relating a response $y$ to an independent variable $x$.

6.19. Earnings from a particular stock are listed below for the past seven years.

| *Year* | 1976 | 1975 | 1974 | 1973 | 1972 | 1971 | 1970 |
|---|---|---|---|---|---|---|---|
| *Earnings per Share* | 2.30 | 1.80 | 1.50 | 1.20 | 1.05 | 1.10 | 1.20 |

Sketch these 7 data points.

6.20. Refer to the sketch in exercise 6.19. Suggest an appropriate general linear model relating earnings per share to the independent variable "year."

# References

Mendenhall, W. 1968. *Introduction to linear models and the design and analysis of experiments.* Belmont, Calif.: Wadsworth.

# Obtaining a Prediction Equation Using the Method of Least Squares

# 7

## Introduction 7.1

In chapter 6 we presented the general linear model as a means of relating a response $y$ to a set of independent variables $x_1, x_2, \ldots, x_k$. For example, we might wish to predict the number of prison-free years between first and second offenses for persons characterized by independent variables such as age, sex, size of metropolitan area, and number of years of schooling. Under the assumption that the random error has expectation zero, the expected value of $y$ has the following form:

$$E(y) = \beta_0 + \beta_1 x_1 + \beta_2 x_2 + \cdots + \beta_k x_k$$

regression of $y$ on $x_1, x_2, \ldots, x_k$

This line, sometimes called the *regression* of $y$ on $x_1, x_2, \ldots, x_k$, is of concern to us in many experimental situations because it represents the average value of $y$ for any setting of the independent variables. As mentioned previously, the parameters $\beta_0, \beta_1, \beta_2, \ldots, \beta_k$ are unknown and hence we can never obtain $E(y)$. However, using the sample information, we can construct an estimate of $E(y)$ by using the equation

estimate of $E(y)$

$$\hat{y} = \hat{\beta}_0 + \hat{\beta}_1 x_1 + \hat{\beta}_2 x_2 + \cdots + \hat{\beta}_k x_k$$

where $\hat{\beta}_0, \hat{\beta}_1, \hat{\beta}_2, \ldots, \hat{\beta}_k$ are estimates of the corresponding unknown parameters $\beta_0, \beta_1, \beta_2, \ldots, \beta_k$. In this chapter we will be concerned with a procedure for obtaining estimates of the $\beta$'s in the equation for $E(y)$ from sample data. In chapter 8 we will be concerned with using these estimates in making inferences concerning any of the individual parameters $\beta_0, \beta_1, \beta_2, \ldots, \beta_k$ and also $E(y)$.

# 7.2 Linear Regression and the Method of Least Squares

The problem of obtaining estimates for parameters in the general linear model can be illustrated by using the probabilistic model

$$y = \beta_0 + \beta_1 x + \epsilon$$

linear regression

for the *linear regression*

$$E(y) = \beta_0 + \beta_1 x$$

Note that the linear regression is a special case of the regression of $y$ on $k$ independent variables. For the linear regression, $k = 1$.

There are many ways for determining an estimate of $E(y)$, which is represented by the equation

$$\hat{y} = \hat{\beta}_0 + \hat{\beta}_1 x$$

eyeball fitting

One procedure, called the eyeball-fitting technique, requires that we plot the data on a scatter diagram and then use a ruler to draw what we feel is the straight line that most accurately displays the linear trend of the data. Unfortunately, if each of us was given the same set of data, we might each come up with a different prediction equation.

method of least squares

residual, or error of prediction

The *method of least squares* is, in many respects, a formalization of the eyeball-fitting routine just discussed. If we let $\hat{y}$ denote the predicted value of $y$ for a given value of $x$, then the error of prediction (often called the *residual*) is $y - \hat{y}$, the difference between the actual value of $y$ and what we predict it to be. *The method of least squares chooses the prediction line $\hat{y} = \hat{\beta}_0 + \hat{\beta}_1 x$ that minimizes the sum of the squared errors of prediction $\Sigma (y - \hat{y})^2$ for all sample points.* We can denote the sum of the squared errors of prediction for the linear model $y = \beta_0 + \beta_1 x + \epsilon$ by

$$\sum (y - \widehat{y})^2 = \sum (y - \widehat{\beta}_0 - \widehat{\beta}_1 x)^2$$

Thus the method of least squares consists of finding those estimates $\widehat{\beta}_0$ and $\widehat{\beta}_1$ that minimize $\Sigma (y - \widehat{y})^2$.

While the procedure for deriving these estimates involves use of the calculus, we can summarize the results. The estimates, called *least squares estimates,* that minimize $\Sigma (y - \widehat{y})^2$ are computed as shown in the box.

least squares estimates

---

**Least Squares Estimates of $\beta_1$ and $\beta_0$**

$$\widehat{\beta}_1 = \frac{SS_{xy}}{SS_{xx}} \quad \text{and} \quad \widehat{\beta}_0 = \overline{y} - \widehat{\beta}_1 \overline{x}$$

where

$SS_{xx}$

$$SS_{xx} = \sum (x - \overline{x})^2 = \sum x^2 - \frac{\left(\sum x\right)^2}{n}$$

$SS_{xy}$

$$SS_{xy} = \sum (x - \overline{x})(y - \overline{y}) = \sum xy - \frac{\left(\sum x\right)\left(\sum y\right)}{n}$$

---

These ideas can probably be best understood by working an example.

**EXAMPLE 7.1**

In a random sample of $n = 9$ steers, the live weights and dressed weights were recorded. In table 7.1 we let $y$ denote the dressed weight (in hundreds of pounds) and $x$ denote the corresponding live weight (in hundreds of pounds). Use the sample data to obtain least squares estimates for the model

$$y = \beta_0 + \beta_1 x + \epsilon$$

**Table 7.1** Sample data for example 7.1; live weight ($x$) and dressed weight ($y$) of steers

| $x$ | $y$ |
|---|---|
| 4.2 | 2.8 |
| 3.8 | 2.5 |
| 4.8 | 3.1 |
| 3.4 | 2.1 |
| 4.5 | 2.9 |
| 4.6 | 2.8 |
| 4.3 | 2.6 |
| 3.7 | 2.4 |
| 3.9 | 2.5 |

## SOLUTION

When we do not have the use of a calculator, the least squares estimates can be computed fairly easily if we construct a summary table, such as that shown in table 7.2.

**Table 7.2** Summary table for the data of example 7.1

| | $x$ | $x^2$ | $y$ | $y^2$ | $xy$ |
|---|---|---|---|---|---|
| | 4.2 | 17.64 | 2.8 | 7.84 | 11.76 |
| | 3.8 | 14.44 | 2.5 | 6.25 | 9.50 |
| | 4.8 | 23.04 | 3.1 | 9.61 | 14.88 |
| | 3.4 | 11.56 | 2.1 | 4.41 | 7.14 |
| | 4.5 | 20.25 | 2.9 | 8.41 | 13.05 |
| | 4.6 | 21.16 | 2.8 | 7.84 | 12.88 |
| | 4.3 | 18.49 | 2.6 | 6.76 | 11.18 |
| | 3.7 | 13.69 | 2.4 | 5.76 | 8.88 |
| | 3.9 | 15.21 | 2.5 | 6.25 | 9.75 |
| *Totals* | 37.2 | 155.48 | 23.7 | 63.13 | 99.02 |

Using the computational formulas for $SS_{xx}$ and $SS_{xy}$, we have, from table 7.2,

$$SS_{xx} = \sum x^2 - \frac{\left(\sum x\right)^2}{n} = 155.48 - \frac{(37.2)^2}{9}$$

$$= 155.48 - 153.76 = 1.72$$

$$SS_{xy} = \sum xy - \frac{\left(\sum x\right)\left(\sum y\right)}{n} = 99.02 - \frac{(37.2)(23.7)}{9}$$

$$= 99.02 - 97.96 = 1.06$$

The least squares estimate for $\beta_1$ is

$$\hat{\beta}_1 = \frac{SS_{xy}}{SS_{xx}} = \frac{1.06}{1.72} = .616$$

The sample means $\bar{x}$ and $\bar{y}$ are

$$\bar{x} = \frac{\sum x}{n} = \frac{37.2}{9} = 4.133$$

$$\bar{y} = \frac{\sum y}{n} = \frac{23.7}{9} = 2.633$$

Substituting our calculated values into the formula for $\hat{\beta}_0$, we have

$$\hat{\beta}_0 = \bar{y} - \hat{\beta}_1\bar{x} = 2.633 - .616(4.133) = .087$$

The least squares equation for these data is

$$\hat{y} = .087 + .616x$$

It is plotted in figure 7.1, with the sample data superimposed. The predicted value of $y$ for any value of $x$ is given by a point on the line and can be computed by using the equation

$$\hat{y} = .087 + .616x$$

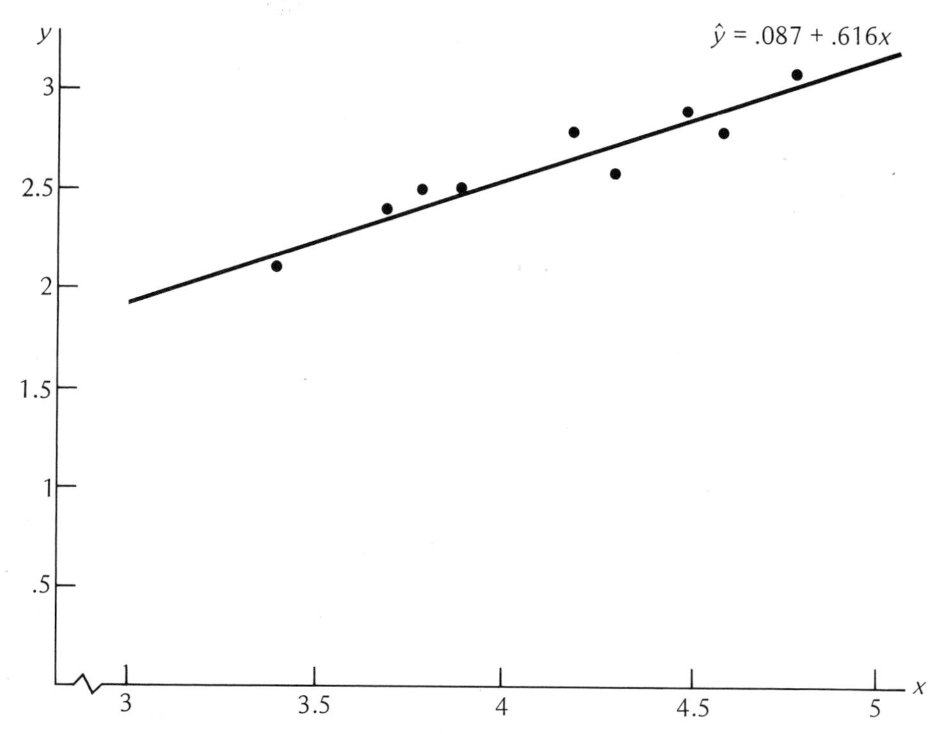


**Figure 7.1** Plot of the least squares equation for the data of example 7.1

The method of least squares can be applied to any experimental situation where the investigator would like to estimate parameters in the general linear model. For

$$y = \beta_0 + \beta_1 x_1 + \beta_2 x_2 + \cdots + \beta_k x_k + \epsilon$$

we try to find estimates $\hat{\beta}_0, \hat{\beta}_1, \hat{\beta}_2, \ldots, \hat{\beta}_k$ that minimize

$$\sum (y - \hat{y})^2 = \sum (y - \hat{\beta}_0 - \hat{\beta}_1 x_1 - \hat{\beta}_2 x_2 - \cdots - \hat{\beta}_k x_k)^2$$

Algebraically, it can be shown that from a least squares fit of a general linear model relating a response $y$ to $k$ independent variables, then

$$\sum (y - \bar{y})^2 = \sum (y - \hat{y})^2 + \sum (\hat{y} - \bar{y})^2$$

While the proof of this equality is beyond the scope of this text, we can obtain an intuitive understanding of this relationship by considering the following situation.

Suppose that we wish to fit a general linear model with $k = 0$, that is, the model

$$y = \beta_0 + \epsilon$$

In this model $\beta_0$ represents the population mean for the variable $y$, and, intuitively, we would estimate its value using the sample mean $\bar{y}$. (You can confirm this result by using the formula for the estimated intercept $\hat{\beta}_0$ in a linear model.) Since $\hat{y} = \bar{y}$ for this model, the sum of the squared errors of prediction is $\Sigma (y - \bar{y})^2$.

Now suppose the variable $y$ is related linearly to an independent variable $x$. From our previous work, we could fit the model $y = \beta_0 + \beta_1 x + \epsilon$ to obtain

$$\hat{y} = \hat{\beta}_0 + \hat{\beta}_1 x \quad \text{and} \quad \sum (y - \hat{y})^2$$

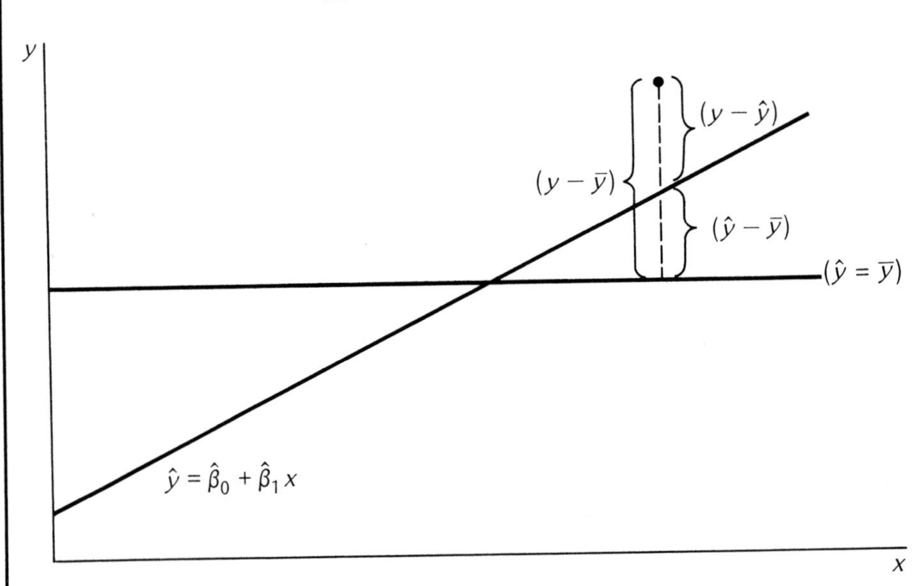


**Figure 7.2** Relationship between $\Sigma (y - \bar{y})^2$ and $\Sigma (y - \hat{y})^2$

In figure 7.2 we have presented two prediction equations, $\hat{y} = \bar{y}$ for the model $y = \beta_0 + \epsilon$ and $\hat{y} = \hat{\beta}_0 + \hat{\beta}_1 x$ for the model $y = \beta_0 + \beta_1 x + \epsilon$. Note that we can express the distance between an observation $y$ and the sample mean $\bar{y}$ as the sum of two components, $(\hat{y} - \bar{y})$ and $(y - \hat{y})$. The quantity $(\hat{y} - \bar{y})$ represents that portion of the overall distance which can be attributed to the independent variable $x$ (through the prediction equation $\hat{y} = \hat{\beta}_0 + \hat{\beta}_1 x$). The quantity $(y - \hat{y})$ represents that portion of the distance between $y$ and $\bar{y}$ which cannot be accounted for by the independent variable $x$ (and which we attribute to error). Combining this information for all sample observations, we can express the total variability in the sample measurements about the sample mean, $\Sigma (y - \bar{y})^2$, as the total of the sum of the squared deviations of the predicted values from $\bar{y}$, $\Sigma (\hat{y} - \bar{y})^2$, and the sum of the squared errors of prediction, $\Sigma (y - \hat{y})^2$. We illustrate these calculations with an example.

**EXAMPLE 7.2**

Consider the 5 data points listed in columns 1 and 2 of table 7.2. Fit the model

$$y = \beta_0 + \beta_1 x + \epsilon$$

Then verify that

$$\sum (y - \bar{y})^2 = \sum (y - \hat{y})^2 + \sum (\hat{y} - \bar{y})^2$$

**SOLUTION**

Applying the computational formulas of this section, it can be shown that

$\bar{y} = 7.6$ and $\hat{y} = 2.5405 + 1.5811x$

For each $x$-value we then compute $\hat{y}$ from the least squares prediction equation. We also compute the quantities $(y - \bar{y})$, $(y - \hat{y})$, and $(\hat{y} - \bar{y})$. These quantities are displayed in table 7.3.

**Table 7.3** Data and computations for example 7.2

| $x$ | $y$ | $\hat{y}$ | $y - \bar{y}$ | $y - \hat{y}$ | $\hat{y} - \bar{y}$ |
|---|---|---|---|---|---|
| 1 | 4 | 4.1216 | −3.6000 | −.1216 | −3.4784 |
| 2 | 6 | 5.7027 | −1.6000 | .2973 | −1.8973 |
| 3 | 7 | 7.2838 | −.6000 | −.2838 | −.3162 |
| 4 | 9 | 8.8649 | 1.4000 | .1351 | 1.2649 |
| 6 | 12 | 12.0271 | 4.4000 | −.0271 | 4.4271 |

From columns 4, 5, and 6 in the table, we have

$$\sum (y - \bar{y})^2 = 37.2000$$

$$\sum (y - \hat{y})^2 = .2027$$

$$\sum (\hat{y} - \bar{y})^2 = 36.9982$$

Note that, except for rounding errors,

$$\sum (y - \bar{y})^2 = \sum (y - \hat{y})^2 + \sum (\hat{y} - \bar{y})^2$$

In section 7.4 we will turn to the problem of determining least squares estimates for the parameters in a general linear model. Since there are no simple shortcut formulas, except for the special case where $k = 1$, we must introduce the concept of a matrix and show how we can use matrices to provide a simple expression for our least squares estimates. Thus an elementary introduction to matrices is presented in the next section, section 7.3. Those familiar with matrices may go directly to section 7.4 without loss of continuity. For additional reading see Graybill (1969) and Hohn (1958).

# Exercises

7.1. An experiment was conducted to examine the effect of different concentrations of pectin on the firmness of canned sweet potatoes. Three concentrations were to be used, 0%, 1.5%, and 3% pectin by weight. Six number-303-X-406 cans were packed with sweet potatoes in a 25% (by weight) sugar solution. Two cans were randomly assigned to each of the pectin concentrations with the appropriate percentage of pectin added to the sugar syrup. The cans were then sealed and placed in a 25°C environment for thirty days. At the end of the storage time, the cans were opened and a firmness determination made for the contents of each can. These data appear below.

| *Pectin Concentration* | 0% | 1.5% | 3.0% |
|---|---|---|---|
| *Firmness Reading* | 50.5, 46.8 | 62.3, 67.7 | 80.1, 79.2 |

a. Let $x$ denote the pectin concentration of a can and $y$ denote the firmness reading following the thirty days of storage at 25°C. Plot the sample data in a scatter diagram.
b. Obtain least squares estimates for the parameters in the model $y = \beta_0 + \beta_1 x + \epsilon$.

7.2. Refer to exercise 7.1. Predict the firmness for a can of sweet potatoes treated with a 1% concentration of pectin (by weight) after thirty days of storage at 25°C.

7.3. A study was conducted to examine the quality of fish after seven days in ice storage. Ten raw fish of the same kind and approximately the same size were caught and prepared for ice storage. Two of the fish were placed in storage immediately after being caught, 2 were placed in storage 3 hours after being caught, and 2 each were placed in storage at 6, 9, and 12 hours after being caught. Let $y$ denote a measurement of fish quality (on a 10-point scale) after the seven days of storage and $x$ denote the time after being caught that the fish were placed in ice packing. The sample data appear below.

| $y$ | 8.5 | 8.4 | 7.9 | 8.1 | 7.8 | 7.6 | 7.3 | 7.0 | 6.8 | 6.7 |
|---|---|---|---|---|---|---|---|---|---|---|
| $x$ | 0 | 0 | 3 | 3 | 6 | 6 | 9 | 9 | 12 | 12 |

a. Plot the sample data in a scatter diagram.
b. Use the method of least squares to obtain estimates of the parameters in the model $y = \beta_0 + \beta_1 x + \epsilon$.

7.4. Refer to example 7.2. Compute the least squares prediction equation. (Be sure to carry at least four decimal places.)

7.5. Refer to exercise 7.3. Predict the seven-day quality score of a fish placed in ice storage 10 hours after being caught. Would you be willing to predict a quality score for fish placed in storage 18 hours after being caught?

## Matrix Notation 7.3

We introduce the concept of a matrix here because matrices enable us to obtain a simple expression for least squares estimates of parameters in the general linear model. In section 7.2 we presented formulas to use in the special case where a response $y$ is linearly related to a single independent variable $x$. However, by using matrices, we will be able to obtain estimates for the parameters of the *general* linear model.

matrix

A *matrix* is defined to be a rectangular array of real numbers. We will indicate a particular matrix by a capital boldface letter. The real numbers of a matrix, called *elements*, appear in rows and columns, as indicated in figure 7.3.

elements

$$\mathbf{A} = \begin{bmatrix} 3 & 1 & 0 \\ 2 & 4 & 8 \\ 4 & 7 & 5 \end{bmatrix}$$

$$\mathbf{B} = \begin{bmatrix} 2 & -5 & 1 \\ 7 & 0 & 4 \end{bmatrix}$$

**Figure 7.3** Elements of a 3 × 3 matrix **A** and a 2 × 3 matrix **B**

dimension of matrix

Note that in addition to identifying a matrix by a capital boldface letter, we can also indicate the *dimension* of a matrix by specifying the number of rows and columns in the matrix. Thus a $3 \times 3$ (read "3 by 3") matrix contains 3 rows and 3 columns, a $2 \times 3$ matrix contains 2 rows and 3 columns, and a $1 \times 4$ matrix contains 1 row and 4 columns.

identity matrix

A matrix of considerable importance is the *identity matrix*. An identity matrix, denoted by **I**, is a square matrix (same number of rows and columns) whose diagonal elements, proceeding from the upper left to the lower right of the matrix, are 1, with all off-diagonal elements 0. Three identity matrices are shown in figure 7.4.

$$\mathbf{I} = \begin{bmatrix} 1 & 0 \\ 0 & 1 \end{bmatrix}$$

$$\mathbf{I} = \begin{bmatrix} 1 & 0 & 0 \\ 0 & 1 & 0 \\ 0 & 0 & 1 \end{bmatrix}$$

$$\mathbf{I} = \begin{bmatrix} 1 & 0 & 0 & 0 \\ 0 & 1 & 0 & 0 \\ 0 & 0 & 1 & 0 \\ 0 & 0 & 0 & 1 \end{bmatrix}$$

**Figure 7.4** $2 \times 2$, $3 \times 3$, and $4 \times 4$ identity matrices

As with other quantities used in statistics and mathematics, we will want to perform operations with matrices, such as addition, multiplication, and so on. Thus in the following discussions, we define the matrix operations we will need in our statistical work.

addition of matrices

*Two matrices,* **A** *and* **B,** *can be added only if they have the same dimensions.* For example, we could add two $3 \times 3$ matrices because both matrices have the same dimensions, but we could not add a $2 \times 3$ matrix and a $3 \times 3$ matrix. *The sum of two matrices,* **A** *and* **B,** *whose dimensions are the same is defined to be a new matrix formed by adding the corresponding elements of* **A** *and* **B.** This new matrix, **A** + **B,** has the same dimensions as **A** and **B.**

We illustrate matrix addition with some examples.

### EXAMPLE 7.3

Suppose a $2 \times 2$ matrix **A** and a $2 \times 2$ matrix **B** are as shown below.

$$\mathbf{A} = \begin{bmatrix} 1 & 3 \\ 2 & 5 \end{bmatrix} \quad \mathbf{B} = \begin{bmatrix} 4 & 1 \\ 3 & 7 \end{bmatrix}$$

Find the sum of the two matrices.

**SOLUTION**

Since the two matrices **A** and **B** have the same dimensions we can add them. The sum of the two matrices, denoted by **A** + **B**, is

$$\mathbf{A} + \mathbf{B} = \begin{bmatrix} 1 & 3 \\ 2 & 5 \end{bmatrix} + \begin{bmatrix} 4 & 1 \\ 3 & 7 \end{bmatrix} = \begin{bmatrix} (1+4) & (3+1) \\ (2+3) & (5+7) \end{bmatrix} = \begin{bmatrix} 5 & 4 \\ 5 & 12 \end{bmatrix}$$

**EXAMPLE 7.4**

For the 2 × 3 matrix **A** and 2 × 3 matrix **B** shown below, find **A** + **B**.

$$\mathbf{A} = \begin{bmatrix} 3 & 4 & -6 \\ 9 & 1 & 1 \end{bmatrix} \qquad \mathbf{B} = \begin{bmatrix} 1 & 7 & 4 \\ 0 & -9 & 5 \end{bmatrix}$$

**SOLUTION**

The sum of these two matrices is

$$\mathbf{A} + \mathbf{B} = \begin{bmatrix} (3+1) & (4+7) & (-6+4) \\ (9+0) & (1-9) & (1+5) \end{bmatrix} = \begin{bmatrix} 4 & 11 & -2 \\ 9 & -8 & 6 \end{bmatrix}$$

You can easily verify, using the previous example, that the addition of two matrices **A** and **B** is commutative; that is,

$$\mathbf{A} + \mathbf{B} = \mathbf{B} + \mathbf{A}$$

multiplication of matrices

*We can multiply a matrix* **A** *times a matrix* **B** *if the number of columns in* **A** *is equal to the number of rows in* **B.** For example, we could multiply a 2 × 3 matrix **A** times a 3 × 2 matrix **B** because the number of columns in **A** and the number of rows in **B** is the same, namely, 3. Similarly, we could multiply a 1 × 5 matrix **A** times a 5 × 4 matrix **B.**

The multiplication of matrices is somewhat complex. Basically, an element in the product matrix **AB** is found by multiplying each element in a row in **A** times each element in the corresponding column in **B** and adding the results. This procedure is best illustrated and understood by working some examples.

**EXAMPLE 7.5**

Let the 2 × 2 matrix **A** and the 2 × 1 matrix **B** be given as follows:

$$\mathbf{A} = \begin{bmatrix} 3 & 1 \\ 2 & 4 \end{bmatrix} \qquad \mathbf{B} = \begin{bmatrix} 1 \\ 2 \end{bmatrix}$$

Find the product **AB.**

**SOLUTION**

The first thing to note in the multiplication of two matrices **A** and **B** is that the resulting product **AB** will be a new matrix with dimensions given by

number of rows of **AB** = number of rows of **A**

number of columns of **AB** = number of columns of **B**

In our example

$$\underset{\underline{2}\times 2}{\mathbf{A}} \quad \underset{2\times \underline{1}}{\mathbf{B}}$$

will be a new 2 × 1 matrix. There will be two elements in the new matrix, one in the first row and first column and one in the second row and first column.

The element in the first row, first column, is found by multiplying the elements of the first *row* of **A** times the corresponding elements of the first *column* of **B** and adding the result:

$$\begin{bmatrix} 3 & 1 \\ 2 & 4 \end{bmatrix}\begin{bmatrix} 1 \\ 2 \end{bmatrix} = \begin{bmatrix} 3\cdot 1 + 1\cdot 2 \\ \phantom{0} \end{bmatrix} = \begin{bmatrix} 5 \\ \phantom{0} \end{bmatrix}$$

The element in the second row, first column, of the new matrix is found by multiplying the elements of the *second* row of **A** times the elements of the *first* column of **B** and adding the result.

$$\begin{bmatrix} 3 & 1 \\ 2 & 4 \end{bmatrix}\begin{bmatrix} 1 \\ 2 \end{bmatrix} = \begin{bmatrix} \phantom{0} \\ 2\cdot 1 + 4\cdot 2 \end{bmatrix} = \begin{bmatrix} \phantom{0} \\ 10 \end{bmatrix}$$

The product **AB** is then

$$\mathbf{AB} = \begin{bmatrix} 3 & 1 \\ 2 & 4 \end{bmatrix}\begin{bmatrix} 1 \\ 2 \end{bmatrix} = \begin{bmatrix} 5 \\ 10 \end{bmatrix}$$

## EXAMPLE 7.6

Let a 2 × 2 matrix **A** and a 2 × 3 matrix **B** be given as follows:

$$\mathbf{A} = \begin{bmatrix} 1 & 2 \\ 0 & 3 \end{bmatrix} \qquad \mathbf{B} = \begin{bmatrix} 4 & 4 & 0 \\ 8 & 3 & 2 \end{bmatrix}$$

a. Find **AB.**
b. Find **BA.**

## SOLUTION

a. Again we will illustrate this matrix multiplication in separate parts. First we know that the new matrix will be a 2 × 3 matrix

$$\underset{\underline{2}\times 2}{\mathbf{A}} \quad \underset{2\times \underline{3}}{\mathbf{B}}$$

The element in the first row, first column, of the new matrix [called the (1, 1) element] is obtained by multiplying the elements in the first row of **A** times corresponding elements in the first column of **B** and adding.

$$(1, 1)\text{ element} = \begin{bmatrix} 1 & 2 \\ 0 & 3 \end{bmatrix}\begin{bmatrix} 4 & 4 & 0 \\ 8 & 3 & 2 \end{bmatrix} = \begin{bmatrix} 1\cdot 4 + 2\cdot 8 & & \\ & & \end{bmatrix}$$

The element in the first row, second column [called the (1, 2) element], of the new matrix is formed by multiplying the elements of the first row of **A** and the second column of **B** and adding.

$$(1,2) \text{ element} = \begin{bmatrix} 1 & 2 \\ 0 & 3 \end{bmatrix} \begin{bmatrix} 4 & 4 & 0 \\ 8 & 3 & 2 \end{bmatrix} = \begin{bmatrix} & 1 \cdot 4 + 2 \cdot 3 & \\ & & \end{bmatrix}$$

The remaining elements are found in a similar manner.

$$(1,3) \text{ element} = \begin{bmatrix} 1 & 2 \\ 0 & 3 \end{bmatrix} \begin{bmatrix} 4 & 4 & 0 \\ 8 & 3 & 2 \end{bmatrix} = \begin{bmatrix} & & 1 \cdot 0 + 2 \cdot 2 \\ & & \end{bmatrix}$$

$$(2,1) \text{ element} = \begin{bmatrix} 1 & 2 \\ 0 & 3 \end{bmatrix} \begin{bmatrix} 4 & 4 & 0 \\ 8 & 3 & 2 \end{bmatrix} = \begin{bmatrix} & & \\ 0 \cdot 4 + 3 \cdot 8 & & \end{bmatrix}$$

$$(2,2) \text{ element} = \begin{bmatrix} 1 & 2 \\ 0 & 3 \end{bmatrix} \begin{bmatrix} 4 & 4 & 0 \\ 8 & 3 & 2 \end{bmatrix} = \begin{bmatrix} & & \\ & 0 \cdot 4 + 3 \cdot 3 & \end{bmatrix}$$

$$(2,3) \text{ element} = \begin{bmatrix} 1 & 2 \\ 0 & 3 \end{bmatrix} \begin{bmatrix} 4 & 4 & 0 \\ 8 & 3 & 2 \end{bmatrix} = \begin{bmatrix} & & \\ & & 0 \cdot 0 + 3 \cdot 2 \end{bmatrix}$$

Combining our results we have

$$\mathbf{AB} = \begin{bmatrix} 20 & 10 & 4 \\ 24 & 9 & 6 \end{bmatrix}$$

b. In order to multiply

$$\underset{2\times 3}{\mathbf{B}} \quad \text{times} \quad \underset{2\times 2}{\mathbf{A}}$$

the number of columns in **B** must equal the number of rows in **A.** Since this does not hold in this example, we cannot multiply **B** times **A.** *Note this also implies that* **BA** *is not necessarily equal to* **AB.**

The *transpose* of a matrix **A** is a new matrix, denoted by **A′** (called "*A* prime"), formed by interchanging the corresponding rows and columns of the original matrix **A.** Thus the first row of **A** becomes the first column of **A′,** the second row of **A** becomes the second column of **A′,** and so on.

transpose **A′**

**EXAMPLE 7.7**

Determine the transpose of the matrix

$$\mathbf{A} = \begin{bmatrix} 1 & 5 & 6 \\ 3 & 2 & 4 \end{bmatrix}$$

**SOLUTION**

We obtain the transpose of **A** by forming a new matrix, where the first and second rows of **A** become the first and second columns of the new matrix.

$$\mathbf{A}' = \begin{bmatrix} 1 & 3 \\ 5 & 2 \\ 6 & 4 \end{bmatrix}$$

Note that **A** had dimensions $2 \times 3$, while **A′** has the reverse dimensions, $3 \times 2$.

determinant

We can associate a number with every square matrix (i.e., the number of rows equals the number of columns), which we call the *determinant* of a matrix. The determinant of a $1 \times 1$ matrix, that is, a matrix with only one element, is the value of that element. Thus the determinant of the matrix $\mathbf{A} = [4]$ is 4. The computation of determinants will be illustrated for $2 \times 2$ and $3 \times 3$ matrices.

Let **A** be a $2 \times 2$ matrix with elements

$$\mathbf{A} = \begin{bmatrix} a_{11} & a_{12} \\ a_{21} & a_{22} \end{bmatrix}$$

The determinant of the matrix **A**, denoted by $|\mathbf{A}|$, is the quantity

$$|\mathbf{A}| = a_{11}a_{22} - a_{21}a_{12}$$

Note that the determinant of a $2 \times 2$ matrix is formed by multiplying elements along the diagonal of the matrix from upper left to lower right and then subtracting the product of the diagonal elements from lower left to upper right (see figure 7.5).

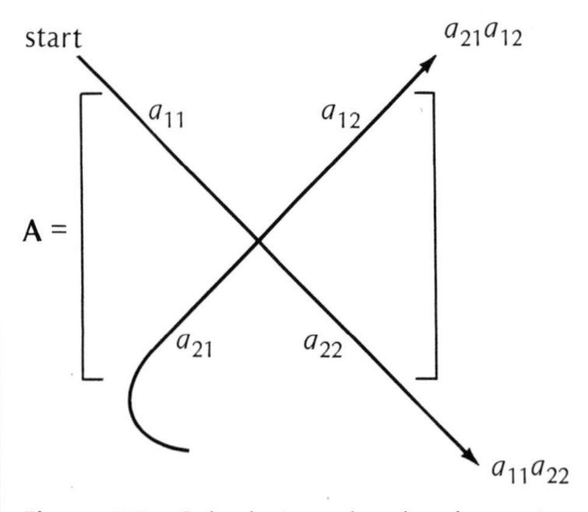


**Figure 7.5** Calculations for the determinant of a $2 \times 2$ matrix

**EXAMPLE 7.8**

Find the determinant of the matrix

$$\mathbf{A} = \begin{bmatrix} 2 & 3 \\ 1 & 5 \end{bmatrix}$$

**SOLUTION**

Using the computational formula, we have

$$|\mathbf{A}| = 2(5) - 1(3) = 10 - 3 = 7$$

The determinant of a 3 × 3 matrix can be computed in a similar way. Let

$$\mathbf{A} = \begin{bmatrix} a_{11} & a_{12} & a_{13} \\ a_{21} & a_{22} & a_{23} \\ a_{31} & a_{32} & a_{33} \end{bmatrix}$$

Then the determinant of this matrix is

$$|\mathbf{A}| = a_{11}a_{22}a_{33} + a_{21}a_{32}a_{13} + a_{31}a_{23}a_{12} - a_{13}a_{22}a_{31} - a_{23}a_{32}a_{11} - a_{33}a_{21}a_{12}$$

Although this computation is more difficult than that for the 2 × 2 matrix, it can be remembered easily using the procedure shown in figure 7.6.

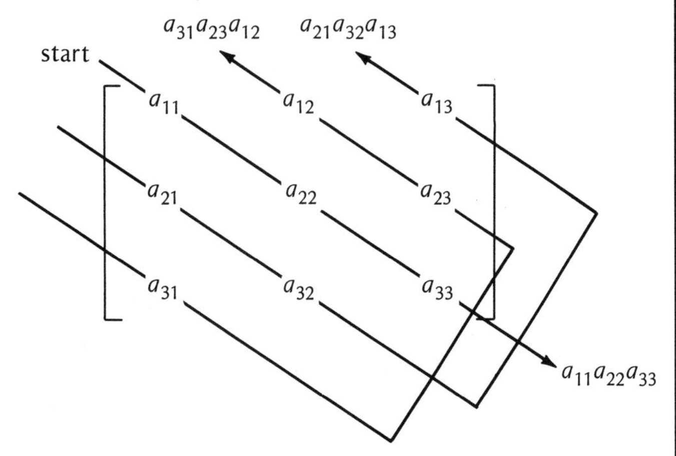


(a) **Computation of first three terms of |A|**

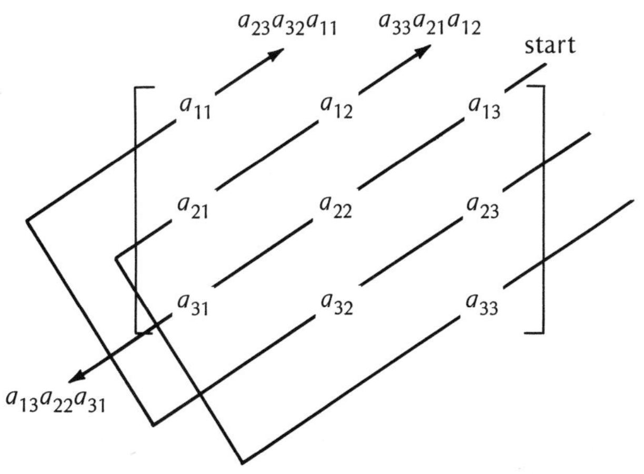


(b) **Computation of last three terms of |A|**

**Figure 7.6** Computation of $|\mathbf{A}|$ for a 3 × 3 matrix

**EXAMPLE 7.9**

Find the determinant of the matrix

$$\mathbf{A} = \begin{bmatrix} 1 & 3 & 1 \\ 2 & 1 & 1 \\ 1 & 3 & 3 \end{bmatrix}$$

**SOLUTION**

The first three terms of /**A**/, starting in the upper left-hand corner, are

$1(1)(3) = 3 \qquad 2(3)(1) = 6 \qquad 1(1)(3) = 3$

The last three terms are

$1(1)(1) = 1 \qquad 1(3)(1) = 3 \qquad 3(2)(3) = 18$

Combining, we have

$|\mathbf{A}| = 3 + 6 + 3 - 1 - 3 - 18 = -10$

Although there are computational formulas for calculating the determinant of a square matrix with more than three rows (and columns), the computational tricks for $2 \times 2$ and $3 \times 3$ matrices do not extend to higher-dimensional matrices. *When working with larger matrices, we will refer to a computer for a solution.*

cofactor of element

We can now use the definition of a determinant to define a new concept. The *cofactor* associated with the $(i, j)$ element of a square matrix **A** is defined to be $(-1)^{i+j}$ times the determinant of the matrix formed by deleting all the elements in row $i$ and column $j$ of the matrix **A.** We illustrate this idea with an example.

**EXAMPLE 7.10**

Find the cofactors of the (1, 2) element and the (2, 2) element of the matrix

$$\mathbf{A} = \begin{bmatrix} 1 & 3 & 1 \\ 2 & 1 & 1 \\ 1 & 3 & 3 \end{bmatrix}$$

**SOLUTION**

To find the cofactor of the (1, 2) element of **A,** we delete the first row and second column of **A.**

$$\begin{bmatrix} \not{1} & \not{3} & \not{1} \\ 2 & \not{1} & 1 \\ 1 & \not{3} & 3 \end{bmatrix}$$

The determinant of the matrix that remains is $2(3) - 1(1) = 6 - 1 = 5$, so the cofactor of the (1, 2) element is $(-1)^{1+2}5 = -5$.

To find the cofactor of the (2, 2) element of **A,** we first delete the second row and second column of the matrix.

$$\begin{bmatrix} 1 & \not{3} & 1 \\ \not{2} & \not{1} & \not{1} \\ 1 & \not{3} & 3 \end{bmatrix}$$

The determinant of the remaining matrix is $1(3) - 1(1) = 2$. Hence the cofactor of the (2, 2) element is $(-1)^{2+2}2 = 2$.

A *cofactor matrix* associated with a square matrix **A** is a new matrix formed by replacing each element by its cofactor.

cofactor matrix

**EXAMPLE 7.11**

Find the cofactor matrix associated with the matrix

$$\mathbf{A} = \begin{bmatrix} 1 & 3 & 1 \\ 2 & 1 & 1 \\ 1 & 3 & 3 \end{bmatrix}$$

**SOLUTION**

We have already found the cofactors associated with the (1, 2) and (2, 2) elements in example 7.10. In a similar way we can find the cofactors of the other elements. Table 7.4 summarizes these computations.

**Table 7.4** Computations for the cofactor matrix of **A** in example 7.11

| *Element (i, j)* | *Determinant After Deleting ith Row and jth Column* | *Cofactor of Element (i, j)* |
|---|---|---|
| (1, 1) | $1(3) - 3(1) = 0$ | $0$ |
| (1, 2) | see example 7.10 | $-5$ |
| (1, 3) | $2(3) - 1(1) = 5$ | $(-1)^{1+3}5 = 5$ |
| (2, 1) | $3(3) - 3(1) = 6$ | $(-1)^{2+1}6 = -6$ |
| (2, 2) | see example 7.10 | $2$ |
| (2, 3) | $1(3) - 1(3) = 0$ | $0$ |
| (3, 1) | $3(1) - 1(1) = 2$ | $(-1)^{3+1}2 = 2$ |
| (3, 2) | $1(1) - 2(1) = -1$ | $(-1)^{3+2}(-1) = 1$ |
| (3, 3) | $1(1) - 2(3) = -5$ | $(-1)^{3+3}(-5) = -5$ |

Hence the cofactor matrix is

$$\begin{bmatrix} 0 & -5 & 5 \\ -6 & 2 & 0 \\ 2 & 1 & -5 \end{bmatrix}$$

The final matrix concept that we present makes use of many of the previous operations and results we have discussed. The *inverse* of a square matrix **A**, denoted by $\mathbf{A}^{-1}$ (read "*A* inverse"), is a new matrix that has the property that both

inverse $\mathbf{A}^{-1}$

$$\mathbf{A}\mathbf{A}^{-1} = \mathbf{I} \quad \text{and} \quad \mathbf{A}^{-1}\mathbf{A} = \mathbf{I}$$

It should be noted that not all square matrices have an inverse; only those which have a nonzero determinant have inverses.

We will now show how we can apply our previous results to obtain the inverse of a $2 \times 2$ matrix and a $3 \times 3$ matrix.

---

**A Procedure for Obtaining the Inverse of a $2 \times 2$ or $3 \times 3$ Square Matrix A**

1. Find the determinant of **A.** If the determinant is nonzero, proceed; otherwise the inverse does not exist.
2. Find the cofactor matrix associated with **A.**
3. Find the transpose of the cofactor matrix.
4. Divide each element of this transposed matrix by $|\mathbf{A}|$. The resulting matrix is $\mathbf{A}^{-1}$.

Note: For matrices larger than $3 \times 3$, we will find the inverse of a matrix by using a computer package.

---

**EXAMPLE 7.12**

Find the inverse of the matrix

$$\mathbf{A} = \begin{bmatrix} 1 & 3 & 1 \\ 2 & 1 & 1 \\ 1 & 3 & 3 \end{bmatrix}$$

**SOLUTION**

Referring to examples 7.9 and 7.11 and following the steps for obtaining $\mathbf{A}^{-1}$, we have the following.

1. $|\mathbf{A}| = -10$.
2. The cofactor matrix of **A** is

$$\begin{bmatrix} 0 & -5 & 5 \\ -6 & 2 & 0 \\ 2 & 1 & -5 \end{bmatrix}$$

3. The transpose of the cofactor matrix is

$$\begin{bmatrix} 0 & -6 & 2 \\ -5 & 2 & 1 \\ 5 & 0 & -5 \end{bmatrix}$$

4. Dividing each element by $|\mathbf{A}| = -10$, we find the inverse of **A** to be

$$\mathbf{A}^{-1} = \begin{bmatrix} 0 & .6 & -.2 \\ .5 & -.2 & -.1 \\ -.5 & 0 & .5 \end{bmatrix}$$

Note that

$$AA^{-1} = \begin{bmatrix} 1 & 3 & 1 \\ 2 & 1 & 1 \\ 1 & 3 & 3 \end{bmatrix} \begin{bmatrix} 0 & .6 & -.2 \\ .5 & -.2 & -.1 \\ -.5 & 0 & .5 \end{bmatrix} = \begin{bmatrix} 1 & 0 & 0 \\ 0 & 1 & 0 \\ 0 & 0 & 1 \end{bmatrix}$$

$$A^{-1}A = \begin{bmatrix} 0 & .6 & -.2 \\ .5 & -.2 & -.1 \\ -.5 & 0 & .5 \end{bmatrix} \begin{bmatrix} 1 & 3 & 1 \\ 2 & 1 & 1 \\ 1 & 3 & 3 \end{bmatrix} = \begin{bmatrix} 1 & 0 & 0 \\ 0 & 1 & 0 \\ 0 & 0 & 1 \end{bmatrix}$$

Hence our calculations are correct.

**EXAMPLE 7.13**

Find the inverse of the matrix

$$\mathbf{A} = \begin{bmatrix} 5 & 0 & 0 \\ 0 & 6 & 0 \\ 0 & 0 & 2 \end{bmatrix}$$

**SOLUTION**

Although we could follow the general procedure for the inverse of a matrix, we can simplify our work when we have a *diagonal matrix,* that is, a matrix with nonzero elements on the diagonal from the upper left to the lower right and with all other elements being 0. The inverse of any diagonal matrix **A** is a matrix with each diagonal element equal to the inverse of the corresponding diagonal element of **A.** All other elements are zero. Thus we can immediately write $\mathbf{A}^{-1}$ as

diagonal matrix

$$\mathbf{A}^{-1} = \begin{bmatrix} \frac{1}{5} & 0 & 0 \\ 0 & \frac{1}{6} & 0 \\ 0 & 0 & \frac{1}{2} \end{bmatrix}$$

We have spent a good deal of time here learning about a few concepts in matrix algebra. We will use these matrix results in the next section to find least squares estimates for parameters in the general linear model.

# Exercises

7.6. Consider the two matrices

$$\mathbf{A} = \begin{bmatrix} 2 & 1 \\ 3 & 2 \end{bmatrix} \quad \mathbf{B} = \begin{bmatrix} 2 & 1 \\ 1 & 1 \end{bmatrix}$$

a. Compute **A + B.**
b. Verify that **A + B = B + A.**
c. Compute **AB.**

7.7. Refer to exercise 7.6.
a. Find the matrices $\mathbf{A}'$ and $\mathbf{B}'$. b. Compute $\mathbf{A}^{-1}$.

7.8. Refer to exercise 7.6. Compute $(\mathbf{AB})^{-1}$.

7.9. Consider the $3 \times 3$ matrix

$$\mathbf{A} = \begin{bmatrix} 3 & 0 & 2 \\ 0 & 2 & 0 \\ 2 & 0 & 2 \end{bmatrix}$$

a. Find $\mathbf{A}'$. b. Compute $|\mathbf{A}|$.

7.10. Refer to exercise 7.9. Find $\mathbf{A}^{-1}$. Verify that $\mathbf{AA}^{-1} = \mathbf{A}^{-1}\mathbf{A} = \mathbf{I}$.

7.11. Consider the matrix

$$\mathbf{A} = \begin{bmatrix} 1 & 1 \\ 1 & 2 \\ 1 & 3 \end{bmatrix}$$

a. Find $\mathbf{A}'$. b. Compute $\mathbf{A}'\mathbf{A}$.

7.12. Refer to exercise 7.11. Compute $(\mathbf{A}'\mathbf{A})^{-1}$.

7.13. Refer to exercises 7.11 and 7.12. Let

$$\mathbf{B} = \begin{bmatrix} 10 \\ 22 \end{bmatrix}$$

Compute $(\mathbf{A}'\mathbf{A})^{-1}\mathbf{B}$.

7.14. Consider the two matrices

$$\mathbf{A} = \begin{bmatrix} 1 & -1 & 1 \\ 2 & 0 & 1 \\ 3 & 1 & 3 \end{bmatrix}$$

$$\mathbf{B} = \begin{bmatrix} 1 & 0 & 0 \\ 0 & 1 & 0 \\ 0 & 0 & 1 \end{bmatrix}$$

a. Compute $\mathbf{A} + \mathbf{B}$. b. Compute $\mathbf{AB}$.

7.15. Refer to exercise 7.14.
a. Find $|\mathbf{A}|$. b. Compute $\mathbf{A}^{-1}$.

# Least Squares Solution to the General Linear Model 7.4

Recall that a model relating a response $y$ to a set of $k$ independent variables of the form

$$y = \beta_0 + \beta_1 x_1 + \beta_2 x_2 + \cdots + \beta_k x_k + \epsilon$$

has been called the general linear model. If a sample of $n$ ($n > k$) measurements are obtained for $n$ settings of the independent variables $x_1, x_2, \ldots, x_k$, we can write an individual observation as

$$y_i = \beta_0 + \beta_1 x_{i1} + \beta_2 x_{i2} + \cdots + \beta_k x_{ik} + \epsilon_i \qquad (i = 1, 2, \ldots, n)$$

where $x_{i1}, x_{i2}, \ldots, x_{ik}$ are the settings of the independent variables for the response $y_i$ and $\epsilon_i$ is the random error for the $i$th response.

The entire set of $n$ observations can be expressed in the general linear model using matrix notation. Let the $n \times 1$ matrix **Y**

$$\mathbf{Y} = \begin{bmatrix} y_1 \\ y_2 \\ \vdots \\ y_n \end{bmatrix}$$

be the matrix of observations, and let the $n \times (k + 1)$ matrix **X**

$$\mathbf{X} = \begin{bmatrix} 1 & x_{11} & x_{12} & \cdots & x_{1k} \\ 1 & x_{21} & x_{22} & \cdots & x_{2k} \\ \vdots & \vdots & \vdots & & \vdots \\ 1 & x_{n1} & x_{n2} & \cdots & x_{nk} \end{bmatrix}$$

be a matrix of settings for the independent variables augmented with a column of 1s. The first row of **X** contains a 1 and the settings on the $k$ independent variables for the first observation. Row two contains a 1 and corresponding settings on the independent variables for $y_2$. Similarly, the other rows contain settings for the remaining observations.

Let

$$\boldsymbol{\beta} = \begin{bmatrix} \beta_0 \\ \beta_1 \\ \beta_2 \\ \vdots \\ \beta_k \end{bmatrix}$$

be a $(k + 1) \times 1$ matrix containing the unknown parameters for the general linear model and let

$$\boldsymbol{\epsilon} = \begin{bmatrix} \epsilon_1 \\ \epsilon_2 \\ \vdots \\ \epsilon_n \end{bmatrix}$$

general linear model, matrix notation

be an $n \times 1$ matrix of errors associated with the $n$ observations. Then we can write the general linear model in matrix notation as

$$\begin{bmatrix} y_1 \\ y_2 \\ \vdots \\ y_n \end{bmatrix} = \begin{bmatrix} 1 & x_{11} & x_{12} & \cdots & x_{1k} \\ 1 & x_{21} & x_{22} & \cdots & x_{2k} \\ \vdots & \vdots & \vdots & & \vdots \\ 1 & x_{n1} & x_{n2} & \cdots & x_{nk} \end{bmatrix} \begin{bmatrix} \beta_0 \\ \beta_1 \\ \vdots \\ \beta_k \end{bmatrix} + \begin{bmatrix} \epsilon_1 \\ \epsilon_2 \\ \vdots \\ \epsilon_n \end{bmatrix}$$

or simply

$$\mathbf{Y} = \mathbf{X}\boldsymbol{\beta} + \boldsymbol{\epsilon}$$

Note that to obtain the equation for $y_1$, we multiply row one of **X** times the matrix of $\beta$'s and then add $\epsilon_1$ to obtain

$$\begin{bmatrix} y_1 \\ \\ \\ \end{bmatrix} = \begin{bmatrix} 1 & x_{11} & x_{12} & \cdots & x_{1k} \\ & & & & \\ & & & & \end{bmatrix} \begin{bmatrix} \beta_0 \\ \beta_1 \\ \vdots \\ \beta_k \end{bmatrix} + \begin{bmatrix} \epsilon_1 \\ \\ \\ \end{bmatrix}$$

or

$$y_1 = \beta_0 + \beta_1 x_{11} + \beta_2 x_{12} + \cdots + \beta_k x_{1k} + \epsilon_1$$

which is precisely as it was defined previously. In fact, any observation $y_i$ is obtained by multiplying the $i$th row of $\mathbf{X}$ times the matrix of $\beta$'s and then adding $\epsilon_i$.

If we let the matrix

$$\hat{\boldsymbol{\beta}} = \begin{bmatrix} \hat{\beta}_0 \\ \hat{\beta}_1 \\ \hat{\beta}_2 \\ \vdots \\ \hat{\beta}_k \end{bmatrix}$$

represent the matrix of least squares estimates for the parameters of the general linear model, then, provided the matrix $\mathbf{X'X}$ has an inverse, we can find these estimates using the matrix equation given in the box.

least squares estimates, general linear model

**Least Squares Estimates**

$$\hat{\boldsymbol{\beta}} = (\mathbf{X'X})^{-1}\mathbf{X'Y}$$

This matrix solution gives the set of parameter estimates $\hat{\beta}_0, \hat{\beta}_1, \ldots, \hat{\beta}_k$ in the general linear model that minimizes $\Sigma_{i=1}^{n}(y_i - \hat{y}_i)^2$ for the data collected.

**EXAMPLE 7.14**

Refer to example 7.1 (p. 143). We were trying to relate the dressed weight $y$ of a steer to its corresponding live weight $x$ by using the equation

$$y = \beta_0 + \beta_1 x + \epsilon$$

Find least squares estimates of the parameters $\beta_0$ and $\beta_1$, using the matrix approach just discussed.

**SOLUTION**

The model $y = \beta_0 + \beta_1 x + \epsilon$ can be considered a general linear model with $k = 1$ independent variable. Using the $n = 9$ sample observations, we can specify the following matrices:

$$\mathbf{Y} = \begin{bmatrix} 2.8 \\ 2.5 \\ 3.1 \\ 2.1 \\ 2.9 \\ 2.8 \\ 2.6 \\ 2.4 \\ 2.5 \end{bmatrix} \qquad \mathbf{X} = \begin{bmatrix} 1 & 4.2 \\ 1 & 3.8 \\ 1 & 4.8 \\ 1 & 3.4 \\ 1 & 4.5 \\ 1 & 4.6 \\ 1 & 4.3 \\ 1 & 3.7 \\ 1 & 3.9 \end{bmatrix}$$

Note that the second column of **X** gives the settings (live weights) corresponding to the observed dressed weights ($y$).

The transpose of **X** is

$$\mathbf{X}' = \begin{bmatrix} 1 & 1 & 1 & 1 & 1 & 1 & 1 & 1 & 1 \\ 4.2 & 3.8 & 4.8 & 3.4 & 4.5 & 4.6 & 4.3 & 3.7 & 3.9 \end{bmatrix}$$

Thus

$$\mathbf{X}'\mathbf{X} = \begin{bmatrix} 1 & 1 & 1 & 1 & 1 & 1 & 1 & 1 & 1 \\ 4.2 & 3.8 & 4.8 & 3.4 & 4.5 & 4.6 & 4.3 & 3.7 & 3.9 \end{bmatrix} \begin{bmatrix} 1 & 4.2 \\ 1 & 3.8 \\ 1 & 4.8 \\ 1 & 3.4 \\ 1 & 4.5 \\ 1 & 4.6 \\ 1 & 4.3 \\ 1 & 3.7 \\ 1 & 3.9 \end{bmatrix}$$

$$= \begin{bmatrix} 9 & 37.2 \\ 37.2 & 155.48 \end{bmatrix}$$

and

$$\mathbf{X}'\mathbf{Y} = \begin{bmatrix} 1 & 1 & 1 & 1 & 1 & 1 & 1 & 1 & 1 \\ 4.2 & 3.8 & 4.8 & 3.4 & 4.5 & 4.6 & 4.3 & 3.7 & 3.9 \end{bmatrix} \begin{bmatrix} 2.8 \\ 2.5 \\ 3.1 \\ 2.1 \\ 2.9 \\ 2.8 \\ 2.6 \\ 2.4 \\ 2.5 \end{bmatrix}$$

$$= \begin{bmatrix} 23.7 \\ 99.02 \end{bmatrix}$$

Before computing $\hat{\boldsymbol{\beta}}$ we must first obtain the inverse of the **X′X** matrix by following the procedure for obtaining an inverse (section 7.3). The inverse of **X′X** is found as follows:

1. The determinant of **X′X** is

   $$|\mathbf{X}'\mathbf{X}| = 9(155.48) - 37.2(37.2) = 1399.32 - 1383.84 = 15.48$$

2. The cofactors associated with the elements of **X′X** are given below.

| *Element* | *Determinant of Matrix After Deleting Row i and Col j* | *Cofactor* |
|---|---|---|
| (1, 1) | 155.48 | $(-1)^{1+1}155.48 = 155.48$ |
| (1, 2) | 37.2 | $(-1)^{1+2}37.2 = -37.2$ |
| (2, 1) | 37.2 | $(-1)^{2+1}37.2 = -37.2$ |
| (2, 2) | 9 | $(-1)^{2+2}9 = 9$ |

The cofactor matrix is

$$\begin{bmatrix} 155.48 & -37.2 \\ -37.2 & 9 \end{bmatrix}$$

3. Since the cofactor matrix is *symmetrical* about the diagonal, the transpose of the cofactor matrix is the same as the cofactor matrix.
4. Dividing each element of the transposed matrix by $|\mathbf{X'X}|$, we find $(\mathbf{X'X})^{-1}$ to be

$$(\mathbf{X'X})^{-1} = \begin{bmatrix} \dfrac{155.48}{15.48} & \dfrac{-37.2}{15.48} \\ \dfrac{-37.2}{15.48} & \dfrac{9}{15.48} \end{bmatrix} = \begin{bmatrix} 10.0439 & -2.4031 \\ -2.4031 & .5814 \end{bmatrix}$$

The least squares solution then is

$$\hat{\boldsymbol{\beta}} = (\mathbf{X'X})^{-1}\mathbf{X'Y}$$

$$\begin{bmatrix} \hat{\beta}_0 \\ \hat{\beta}_1 \end{bmatrix} = \begin{bmatrix} 10.0439 & -2.4031 \\ -2.4031 & .5814 \end{bmatrix} \begin{bmatrix} 23.7 \\ 99.02 \end{bmatrix} = \begin{bmatrix} .085 \\ .617 \end{bmatrix} \begin{matrix} \longleftarrow \hat{\beta}_0 \\ \longleftarrow \hat{\beta}_1 \end{matrix}$$

These estimates, except for minor rounding errors, are identical to those obtained in example 7.1 by using the algebraic shortcut formulas.

### EXAMPLE 7.15

A chemist is interested in determining the weight loss $y$ of a particular compound as a function of the amount of time the compound is exposed to the air and the humidity of the environment during exposure. The data in table 7.5 give the weight losses associated with $n = 12$ settings of the independent variables.

**Table 7.5** Weight loss, exposure time, and relative humidity data for example 7.15

| *Weight Loss, y (in pounds)* | *Exposure Time (in hours)* | *Relative Humidity* |
|---|---|---|
| 4.3 | 4 | .20 |
| 5.5 | 5 | .20 |
| 6.8 | 6 | .20 |
| 8.0 | 7 | .20 |
| 4.0 | 4 | .30 |
| 5.2 | 5 | .30 |
| 6.6 | 6 | .30 |
| 7.5 | 7 | .30 |
| 2.0 | 4 | .40 |
| 4.0 | 5 | .40 |
| 5.7 | 6 | .40 |
| 6.5 | 7 | .40 |

Find the least squares prediction equation for the model

$$y = \beta_0 + \beta_1 x_1 + \beta_2 x_2 + \epsilon$$

## SOLUTION

To simplify our calculations, let

$x_1$ = (exposure time − mean exposure time)

$x_2$ = (relative humidity − mean relative humidity)

The mean exposure time for the 12 sample observations is 5.5 and the mean relative humidity is .3. Hence using

$x_1$ = (exposure time − 5.5) and $x_2$ = (relative humidity − .3)

we have the **X** and **Y** matrices given below.

$$\mathbf{Y} = \begin{bmatrix} 4.3 \\ 5.5 \\ 6.8 \\ 8.0 \\ 4.0 \\ 5.2 \\ 6.6 \\ 7.5 \\ 2.0 \\ 4.0 \\ 5.7 \\ 6.5 \end{bmatrix} \qquad \mathbf{X} = \begin{bmatrix} 1 & -1.5 & -.1 \\ 1 & -.5 & -.1 \\ 1 & .5 & -.1 \\ 1 & 1.5 & -.1 \\ 1 & -1.5 & 0 \\ 1 & -.5 & 0 \\ 1 & .5 & 0 \\ 1 & 1.5 & 0 \\ 1 & -1.5 & .1 \\ 1 & -.5 & .1 \\ 1 & .5 & .1 \\ 1 & 1.5 & .1 \end{bmatrix}$$

Coding of the independent variables in this example makes the **X′X** matrix diagonal and simplifies the computation of $(\mathbf{X'X})^{-1}$.

$$\mathbf{X'X} = \begin{bmatrix} 1 & 1 & 1 & 1 & 1 & 1 & 1 & 1 & 1 & 1 & 1 & 1 \\ -1.5 & -.5 & .5 & 1.5 & -1.5 & -.5 & .5 & 1.5 & -1.5 & -.5 & .5 & 1.5 \\ -.1 & -.1 & -.1 & -.1 & 0 & 0 & 0 & 0 & .1 & .1 & .1 & .1 \end{bmatrix} \times \begin{bmatrix} 1 & -1.5 & -.1 \\ 1 & -.5 & -.1 \\ 1 & .5 & -.1 \\ 1 & 1.5 & -.1 \\ 1 & -1.5 & 0 \\ 1 & -.5 & 0 \\ 1 & .5 & 0 \\ 1 & 1.5 & 0 \\ 1 & -1.5 & .1 \\ 1 & -.5 & .1 \\ 1 & .5 & .1 \\ 1 & 1.5 & .1 \end{bmatrix}$$

$$= \begin{bmatrix} 12 & 0 & 0 \\ 0 & 15 & 0 \\ 0 & 0 & .08 \end{bmatrix}$$

and hence

$$(\mathbf{X'X})^{-1} = \begin{bmatrix} \frac{1}{12} & 0 & 0 \\ 0 & \frac{1}{15} & 0 \\ 0 & 0 & \frac{1}{.08} \end{bmatrix} = \begin{bmatrix} .083 & 0 & 0 \\ 0 & .067 & 0 \\ 0 & 0 & 12.5 \end{bmatrix}$$

The $\mathbf{X'Y}$ matrix is

$$\mathbf{X'Y} = \begin{bmatrix} 1 & 1 & 1 & 1 & 1 & 1 & 1 & 1 & 1 & 1 & 1 & 1 \\ -1.5 & -.5 & .5 & 1.5 & -1.5 & -.5 & .5 & 1.5 & -1.5 & -.5 & .5 & 1.5 \\ -.1 & -.1 & -.1 & -.1 & 0 & 0 & 0 & 0 & .1 & .1 & .1 & .1 \end{bmatrix} \times \begin{bmatrix} 4.3 \\ 5.5 \\ 6.8 \\ 8.0 \\ 4.0 \\ 5.2 \\ 6.6 \\ 7.5 \\ 2.0 \\ 4.0 \\ 5.7 \\ 6.5 \end{bmatrix}$$

$$= \begin{bmatrix} 66.10 \\ 19.75 \\ -.64 \end{bmatrix}$$

The least squares estimates of the parameters $\beta_0$, $\beta_1$, and $\beta_2$ are then found by

$$\hat{\boldsymbol{\beta}} = (\mathbf{X'X})^{-1}\mathbf{X'Y}$$

$$= \begin{bmatrix} .083 & 0 & 0 \\ 0 & .067 & 0 \\ 0 & 0 & 12.5 \end{bmatrix} \begin{bmatrix} 66.10 \\ 19.75 \\ -.64 \end{bmatrix} = \begin{bmatrix} 5.49 \\ 1.32 \\ -8.00 \end{bmatrix}$$

The corresponding prediction equation is

$$\hat{y} = 5.49 + 1.32x_1 - 8x_2$$

Note that if we wished to predict weight loss for 6.5 hours of exposure and a relative humidity of .35,

$$x_1 = (6.5 - 5.5) = 1$$

$$x_2 = (.35 - .3) = .05$$

Substituting these values into the least squares equation, our predicted weight loss for the compound is

$$\hat{y} = 5.49 + 1.32(1) - 8(.05) = 6.41$$

# Exercises

7.16. Refer to exercise 7.1 (p. 148).

a. Set up the **X** and **Y** matrices for the model

$$y = \beta_0 + \beta_1 x + \epsilon$$

b. Compute **X′X**, $(\mathbf{X'X})^{-1}$, and **X′Y**.

7.17. Refer to exercise 7.16. Obtain the least squares estimates for $\beta_0$ and $\beta_1$ using the matrix approach. Compare your answers to those obtained in exercise 7.1.

7.18. The Arab oil embargo of 1974 served to emphasize what many had been saying for some time, namely, that alternatives to automotive transportation will have to be developed in order to cope with steadily increasing demands for oil. Alternatives which have been proposed include expansion of mass transit facilities, development of effective, efficient pollution control devices for automobiles, and the introduction of legislation (with accompanying monetary backing) requiring automobile manufacturers to redesign their engines to improve the fuel economy and decrease the emissions from their automobiles.

One thermal pollution study considered the relationship between the weight of an automobile and the Btu per vehicle mile. Some sample data for 1971 and 1972 models are given below.

| *Weight, x* (*in* 1000 *lb*) | *Btu per Vehicle Mile, y* (*in* 1000*s*) |
|---|---|
| 1.8 | 4 |
| 2.6 | 5.2 |
| 4.2 | 8.5 |
| 5.0 | 11.6 |
| 4.8 | 10.1 |
| 3.4 | 6.3 |

a. Compute estimates of $\beta_0$ *and* $\beta_1$ by using the algebraic formulas of section 7.2.

b. Set up the **X, Y, X′X,** and **X′Y** matrices for these data.

7.19. Refer to exercise 7.18.

a. Obtain $(\mathbf{X'X})^{-1}$ and use the previous results to compute $\hat{\boldsymbol{\beta}}$, the matrix of least squares estimates. Compare your results to part a of exercise 7.18.

b. Predict the Btu pollution for a 1971–72 vehicle weighing 3000 pounds.

# A Computer Solution for Least Squares Estimates 7.5

Throughout the first six chapters of this text we have learned how to solve certain statistical problems by hand calculation. In addition, we have had specific exercises related to computer solutions for these problems. Although Minitab and BMDP were used in illustrating these programs and outputs, we were not trying to force you into using any one particular computer package. No software package (computer program) is universally accepted and the details of any one system become obsolete quickly.

In the remaining chapters we will use computer programs and outputs from several software packages to illustrate the techniques we will discuss. Keep in mind that we are not trying to stress one computer system over another. Our objective is to give you several options for obtaining computer solutions to our problems. The important point is that you learn to read computer outputs and that you learn to use the system that is available to you. Your professor will be of great assistance to you in this endeavor.

## Exercises

*7.20. The data from example 7.15 (p. 165) are reproduced here.

| *Weight Loss, y* | $x_1$ | $x_2$ |
|---|---|---|
| 4.3 | −1.5 | −.1 |
| 5.5 | −.5 | −.1 |
| 6.8 | .5 | −.1 |
| 8.0 | 1.5 | −.1 |
| 4.0 | −1.5 | 0 |
| 5.2 | −.5 | 0 |
| 6.6 | .5 | 0 |
| 7.5 | 1.5 | 0 |
| 2.0 | −1.5 | .1 |
| 4.0 | −.5 | .1 |
| 5.7 | .5 | .1 |
| 6.5 | 1.5 | .1 |

A computer output (SAS 1972) follows for a least squares fit of the model $y = \beta_0 + \beta_1 x_1 + \beta_2 x_2 + \epsilon$

a. Locate the $\mathbf{Y}$, $\mathbf{X'X}$, and $(\mathbf{X'X})^{-1}$ matrices. Compare your results to those of example 7.15.

b. Give the least squares prediction equation. Is it the same as that obtained in example 7.15?

```
STATISTICAL ANALYSIS SYSTEM

DATA WEIGHT;
INPUT Y 1-4 X1 6-9 X2 11-13;
CARDS;

  12 OBSERVATIONS IN DATA SET WEIGHT     3 VARIABLES

PROC PRINT;


OBS        Y          X1          X2

  1       4.3       -1.5         -.1
  2       5.5       -0.5         -.1
  3       6.8        0.5         -.1
  4       8.0        1.5         -.1
  5       4.0       -1.5          .0
  6       5.2       -0.5          .0
  7       6.6        0.5          .0
  8       7.5        1.5          .0
  9       2.0       -1.5          .1
 10       4.0       -0.5          .1
 11       5.7        0.5          .1
 12       6.5        1.5          .1


PROC REGR;
MODEL Y = X1 X2/ X I;
TITLE 'EXERCISE 7.20';


*************************************************************

PROC REGR  : EXERCISE 7.20
DATA SET   : WEIGHT            NUMBER OF VARIABLES = 3   NUMBER OF CLASSES = 0
VARIABLES  : Y X1 X2

*************************************************************


*************************************************************

MODEL          : EXERCISE 7.20
DEPENDENTS     : Y
INDEPENDENTS   : X1  X2

*************************************************************
```

EXERCISE 7.20
THE X′X MATRIX

| | INTERCPT | X1 | X2 |
|---|---|---|---|
| INTERCPT | 12.00000000 | 0.0 | 0.0 |
| X1 | 0.0 | 15.00000000 | 0.0 |
| X2 | 0.0 | 0.0 | 0.08000000 |

EXERCISE 7.20
THE X′X INVERSE MATRIX, RANK = 3

| | INTERCPT | X1 | X2 |
|---|---|---|---|
| INTERCPT | 0.08333333 | 0.0 | 0.0 |
| X1 | 0.0 | 0.06666667 | 0.0 |
| X2 | 0.0 | 0.0 | 12.50000000 |

EXERCISE 7.20

ANALYSIS OF VARIANCE TABLE, REGRESSION COEFFICIENTS, AND STATISTICS OF FIT FOR DEPENDENT VARIABLE Y

| SOURCE | DF | SUM OF SQUARES | MEAN SQUARE | F VALUE | PROB > F | R-SQUARE | C.V. |
|---|---|---|---|---|---|---|---|
| REGRESSION | 2 | 31.12416667 | 15.56208333 | 104.13290 | 0.0001 | 0.95857609 | 7.01810 |
| ERROR | 9 | 1.34500000 | 0.14944444 | | | STD DEV | Y MEAN |
| CORRECTED TOTAL | 11 | 32.46916667 | | | | 0.38658045 | 5.50833 |

| SOURCE | DF | SEQUENTIAL SS | F VALUE | PROB > F | PARTIAL SS | F VALUE | PROB > F |
|---|---|---|---|---|---|---|---|
| X1 | 1 | 26.00416667 | 174.00558 | 0.0001 | 26.00416667 | 174.00558 | 0.0001 |
| X2 | 1 | 5.12000000 | 34.26022 | 0.0002 | 5.12000000 | 34.26022 | 0.0002 |

| SOURCE | B VALUES | T FOR HO: B = 0 | PROB > \|T\| | STD ERR B | STD B VALUES |
|---|---|---|---|---|---|
| INTERCEPT | 5.50833333 | 49.35952 | 0.0001 | 0.11159616 | 0.0 |
| X1 | 1.31666667 | 13.19112 | 0.0001 | 0.09981464 | 0.89492347 |
| X2 | −8.00000000 | −5.85322 | 0.0002 | 1.36676829 | −0.39709956 |

*7.21. A BMDP (1975) solution to the data of exercise 7.20 follows. Compare your results.

BMDPIR—MULTIPLE LINEAR REGRESSION

HEALTH SCIENCES COMPUTING FACILITY
UNIVERSITY OF CALIFORNIA, LOS ANGELES

PROGRAM REVISED JULY 7, 1975
MANUAL DATE—1975

```
PROGRAM CONTROL INFORMATION

  PROBLEM TITLE IS 'REGRESSION ANALYSIS.'./
  INPUT VARIABLES = 3.
         FORMAT = '(3F4.1)'.
         CASE = 12./
  VARIABLE NAME = WGTLOSS,X1,X2./
  PRINT COVA.
        CORR.
        DATA./
  PLOT VARIABLE = X1,X2.
       PROBIT.
       PREP = X1,X2.
       RESID./
  REGRES DEPEND = WGT LOSS./
  END/

PROBLEM TITLE . . . . . . . . . . . REGRESSION ANALYSIS.
NUMBER OF VARIABLES TO READ IN . . . . . . . . . . . . .          3
NUMBER OF VARIABLES ADDED BY TRANSFORMATIONS .                     0
TOTAL NUMBER OF VARIABLES . . . . . . . . . . . . . . . .          3
NUMBER OF CASES TO READ IN . . . . . . . . . . . . . . .          12
CASE LABELING VARIABLES . . . . . . . . . . . . . . . . . . .
LIMITS AND MISSING VALUE CHECKED BEFORE TRANSFORMATIONS
BLANKS ARE . . . . . . . . . . . . . . . . . . . . . . . . .   ZEROS
INPUT UNIT NUMBER . . . . . . . . . . . . . . . . . . . .          5
REWIND INPUT UNIT PRIOR TO READING. . . .DATA . . .               NO

INPUT FORMAT
      (3F4.1)

VARIABLES TO BE USED
          1 WGTLOSS            2 X1            3 X2
REGRESSION INTERCEPT . . . . . . . . . . . . . . . . . . . . . .NON-ZERO
GROUPING VARIABLE . . . . . . . . . . . . . . . . . . . . . . .
WEIGHT VARIABLE. . . . . . . . . . . . . . . . . . . . . . . . .
PRINT COVARIANCE MATRIX . . . . . . . . . . . . . . . . .        YES
PRINT CORRELATION MATRIX. . . . . . . . . . . . . . . . .        YES
PRINT RESIDUALS . . . . . . . . . . . . . . . . . . . . . .      YES
PROBIT PLOT . . . . . . . . . . . . . . . . . . . . . . . .      YES

NUMBER OF CASES READ. . . . . . . . . . . . . . . . . . .         12

VARIABLE   MEAN STANDARD DEVIATION ST.  DEV/MEAN   MINIMUM    MAXIMUM
 1 WGTLOSS   5.50833     1.71806          0.31190   2.00000    8.00000
 2 X1        0.0         1.16775          0.0      -1.50000    1.50000
 3 X2        0.00000     0.08528 ***********       -0.10000    0.10000
  REGRESSION TITLE . . . . . . . . . . . . . . . .  REGRESSION ANALYSIS.
  DEPENDENT VARIABLE . . . . . . . . . . . . . . . .        1 WGTLOSS
  TOLERANCE . . . . . . . . . . . . . . . . . . . .   0.0100
ALL DATA CONSIDERED AS A SINGLE GROUP
MULTIPLE R          0.9791   STD. ERROR OF EST.   0.3866
MULTIPLE R-SQUARE   0.9586
```

```
ANALYSIS OF VARIANCE
                    SUM OF SQUARES   DF   MEAN SQUARE   F RATIO   P(TAIL)
  REGRESSION            31.124        2      15.562     104.129   0.00000
  RESIDUAL               1.345        9       0.149

                                           STD. REG
    VARIABLE   COEFFICIENT STD. ERROR   COEFF          T       P(2 TAIL)

      (CONSTANT      5.5083)
X1              2      1.317     0.100      0.895     13.191     0.0
X2              3     -8.000     1.367     -0.397     -5.853     0.000
```

*7.22. Using the sample data of example 7.15, the model

$$y = \beta_0 + \beta_1 x_1 + \beta_2 x_2 + \beta_3 x_2^2 + \beta_4 x_1 x_2 + \beta_5 x_1 x_2^2 + \epsilon$$

was fit. Refer to the computer output that follows.

a. Identify the **Y, X, (X′X)** and $(\mathbf{X'X})^{-1}$ matrices.
b. Give the prediction equation.
c. Predict weight loss for an exposure time of 4.5 hours and relative humidity of .25. (Hint: Recall that $x_1$ and $x_2$ have been coded.)

```
STATISTICAL ANALYSIS SYSTEM

DATA WEIGHT;
INPUT Y 1-4 X1 6-9 X2 11-13;
X1X2=X1 * X2;
X2SQ=X2 * * 2;
X1X2SQ=X1 * X2SQ;
CARDS;

      12 OBSERVATIONS IN DATA SET WEIGHT    6 VARIABLES

PROC PRINT;
```

| OBS | Y | X1 | X2 | X1X2 | X2SQ | X1X2SQ |
|---|---|---|---|---|---|---|
| 1 | 4.3 | −1.5 | −.1 | .15 | .01 | −.015 |
| 2 | 5.5 | −0.5 | −.1 | .05 | .01 | −.005 |
| 3 | 6.8 | 0.5 | −.1 | −.05 | .01 | .005 |
| 4 | 8.0 | 1.5 | −.1 | −.15 | .01 | .015 |
| 5 | 4.0 | −1.5 | .0 | .00 | .00 | .000 |
| 6 | 5.2 | −0.5 | .0 | .00 | .00 | .000 |
| 7 | 6.6 | 0.5 | .0 | .00 | .00 | .000 |
| 8 | 7.5 | 1.5 | .0 | .00 | .00 | .000 |
| 9 | 2.0 | −1.5 | .1 | −.15 | .01 | −.015 |
| 10 | 4.0 | −0.5 | .1 | −.05 | .01 | −.005 |
| 11 | 5.7 | 0.5 | .1 | .05 | .01 | .005 |
| 12 | 6.5 | 1.5 | .1 | .15 | .01 | .015 |

```
PROC REGR;
MODEL Y = X1 X2 X2SQ X1X2 X1X2SQ/ X I;
TITLE 'EXERCISE 7.22';
```

* * * * * * * * * * * * * * * * * * * * * * * * * * * * * * * * * * * * * * * * * * * * * * * * * * * *

PROC REGR : EXERCISE 7.22

DATA SET : WEIGHT NUMBER OF VARIABLES = 6 NUMBER OF CLASSES = 0

VARIABLES : Y X1 X2 X1X2 X2SQ X1X2SQ

* * * * * * * * * * * * * * * * * * * * * * * * * * * * * * * * * * * * * * * * * * * * * * * * * * * *

* * * * * * * * * * * * * * * * * * * * * * * * * * * * * * * * * * * * * * * * * * * * * * * * * * * *

MODEL : EXERCISE 7.22

DEPENDENTS : Y

INDEPENDENTS : X1 X2 X2SQ X1X2 X1X2SQ

* * * * * * * * * * * * * * * * * * * * * * * * * * * * * * * * * * * * * * * * * * * * * * * * * * * *

EXERCISE 7.22

THE X'X MATRIX

| | INTERCPT | X1 | X2 | X2SQ | X1X2 | X1X2SQ |
|---|---|---|---|---|---|---|
| INTERCPT | 12.00000000 | 0.0 | 0.0 | 0.08000000 | 0.0 | 0.0 |
| X1 | 0.0 | 15.00000000 | 0.0 | 0.0 | 0.00000000 | 0.10000000 |
| X2 | 0.0 | 0.0 | 0.08000000 | 0.00000000 | 0.0 | 0.0 |
| X2SQ | 0.08000000 | 0.0 | 0.00000000 | 0.00080000 | 0.0 | 0.0 |
| X1X2 | 0.0 | 0.00000000 | 0.0 | 0.0 | 0.10000000 | 0.00000000 |
| X1X2SQ | 0.0 | 0.10000000 | 0.0 | 0.0 | 0.00000000 | 0.00100000 |

EXERCISE 7.22

THE X'X INVERSE MATRIX, RANK = 6

| | INTERCPT | X1 | X2 | X2SQ | X1X2 | X1X2SQ |
|---|---|---|---|---|---|---|
| INTERCPT | 0.25000000 | 0.0 | 0.00000000 | −25.00000000 | 0.0 | 0.0 |
| X1 | 0.0 | 0.20000000 | 0.0 | 0.0 | 0.00000000 | −20.00000000 |

| | | | | | | |
|---|---|---|---|---|---|---|
| X2 | 0.00000000 | 0.0 | 12.50000000 | −0.00000000 | 0.0 | 0.0 |
| X2SQ | −25.00000000 | 0.0 | −0.00000000 | 3750.00000000 | 0.0 | 0.0 |
| X1X2 | 0.0 | 0.00000000 | 0.0 | 0.0 | 10.00000000 | −0.00000000 |
| X1X2SQ | 0.0 | −20.00000000 | 0.0 | 0.0 | −0.00000000 | 3000.00000000 |

EXERCISE 7.22

ANALYSIS OF VARIANCE TABLE, REGRESSION COEFFICIENTS, AND STATISTICS OF FIT FOR DEPENDENT VARIABLE Y

| SOURCE | DF | SUM OF SQUARES | MEAN SQUARE | F VALUE | PROB > F | R-SQUARE | C.V. |
|---|---|---|---|---|---|---|---|
| REGRESSION | 5 | 32.04216667 | 6.40843333 | 90.04824 | 0.0002 | 0.98684906 | 4.84304 % |
| ERROR | 6 | 0.42700000 | 0.07116667 | | | STD DEV | Y MEAN |
| CORRECTED TOTAL | 11 | 32.46916667 | | | | 0.26677081 | 5.50833 |

| SOURCE | DF | SEQUENTIAL SS | F VALUE | PROB > F | PARTIAL SS | F VALUE | PROB > F |
|---|---|---|---|---|---|---|---|
| X1 | 1 | 26.00416667 | 365.39813 | 0.0001 | 7.08050000 | 99.49180 | 0.0001 |
| X2 | 1 | 5.12000000 | 71.94379 | 0.0001 | 5.12000000 | 71.94379 | 0.0001 |
| X2SQ | 1 | 0.60166667 | 8.45433 | 0.0271 | 0.60166667 | 8.45433 | 0.0271 |
| X1X2 | 1 | 0.19600000 | 2.75410 | 0.1481 | 0.19600000 | 2.75410 | 0.1481 |
| X1X2SQ | 1 | 0.12033333 | 1.69087 | 0.2412 | 0.12033333 | 1.69087 | 0.2412 |

| SOURCE | B VALUES | T FOR HO: B = 0 | PROB > \|T\| | STD ERR B | STD B VALUES |
|---|---|---|---|---|---|
| INTERCEPT | 5.82500000 | 43.67044 | 0.0001 | 0.13338541 | 0.0 |
| X1 | 1.19000000 | 9.97456 | 0.0001 | 0.11930353 | 0.80882957 |
| X2 | −8.00000000 | −8.48197 | 0.0001 | 0.94317725 | −0.39709956 |
| X2SQ | −47.50000000 | −2.90763 | 0.0271 | 16.33630925 | −0.13612641 |
| X1X2 | 1.40000000 | 1.65955 | 0.1481 | 0.84360338 | 0.07769489 |
| X1X2SQ | 19.00000000 | 1.30033 | 0.2412 | 14.61163920 | 0.10544307 |

## Summary | 7.6

Letting $\widehat{y} = \widehat{\beta}_0 + \widehat{\beta}_1 x_1 + \widehat{\beta}_2 x_2 + \cdots + \widehat{\beta}_k x_k$ represent a predicted value of a response $y$, the method of least squares chooses the set of estimates $\widehat{\beta}_0, \widehat{\beta}_1, \widehat{\beta}_2, \ldots, \widehat{\beta}_k$ that minimizes the sum of the squared errors, $\Sigma_{i=1}^{n} (y_i - \widehat{y}_i)^2$. While these can be computed from algebraic formulas for simple models, we presented a procedure that utilizes matrices for obtaining least squares estimates for parameters in the general linear model. Perhaps the utility of the matrix procedure for obtaining least squares estimates may not be readily apparent after working example 7.15. The crucial point is that we can find the least squares estimates for *any* model that is written in the form of a general linear model using the equation

$$\hat{\beta} = (\mathbf{X}'\mathbf{X})^{-1}\mathbf{X}'\mathbf{Y}$$

provided that the **X** matrix is formed so that **X′X** has an inverse.

In this chapter we have presented the procedure for computation of the determinant of a matrix for a $2 \times 2$ matrix and a $3 \times 3$ matrix. Since we use the determinant of a matrix to find its inverse, we are restricted at this time to obtaining inverses for matrices of dimension $3 \times 3$ or less. The only exception that we have noted so far would be the inverse of a diagonal matrix. While you might feel we are severely restricted in applying these results by hand, we will see in succeeding chapters that we can, in fact, solve many different types of problems. In addition, we have a procedure that is readily adaptable to use on a computer.

# Exercises

*7.23. A pharmaceutical firm would like to obtain information on the relationship between the dose level of a drug product and its potency. To do this, each of 15 test tubes is inoculated with a virus culture and incubated for five days at 30° C. Three test tubes are randomly assigned to each of the 5 different dose levels to be investigated (2, 4, 8, 16, and 32 milligrams). Each tube was injected with only one dose level and the response of interest (a measure of the protective strength of the product against the virus culture) was obtained. The data are given below.

| *Dose Level* | *Response* |
|---|---|
| 2 | 5, 7, 3 |
| 4 | 10, 12, 14 |
| 8 | 15, 17, 18 |
| 16 | 20, 21, 19 |
| 32 | 23, 24, 29 |

a. Plot the data.
b. Fit both a linear and a quadratic model to these data.
c. Compare your results in part b to those obtained in the computer output that follows (SPSS 1975).

EXAMPLE OF REGRESSION MODELS

FILE NONAME

| VARIABLE | MEAN | STANDARD DEV | CASES |
|---|---|---|---|
| VAR01 | 12.4000 | 11.2935 | 15 |
| VAR02 | 15.8000 | 7.3892 | 15 |

EXAMPLE OF REGRESSION MODELS

FILE NONAME

CORRELATION COEFFICIENTS

A VALUE OF 99.00000 IS PRINTED
IF A COEFFICIENT CANNOT BE COMPUTED.

| | VAR01 | VAR02 |
|---|---|---|
| VAR01 | 1.00000 | 0.87923 |
| VAR02 | 0.87923 | 1.00000 |

EXAMPLE OF REGRESSION MODELS

FILE NONAME

* * * * * * * * * * * * * * * * * * * * * * * * * * * * * * MULTIPLE REGRESSION * * * * * * * * * * * * * * * * * * VARIABLE LIST 1
REGRESSION LIST 1

DEPENDENT VARIABLE.. VAR02

VARIABLE(S) ENTERED ON STEP NUMBER 1.. VAR01

| | | | | | | |
|---|---|---|---|---|---|---|
| MULTIPLE R | 0.87923 | ANALYSIS OF VARIANCE | DF | SUM OF SQUARES | MEAN SQUARE | F |
| R SQUARE | 0.77305 | REGRESSION | 1. | 590.91613 | 590.91613 | 44.28025 |
| ADJUSTED R SQUARE | 0.75559 | RESIDUAL | 13. | 173.48387 | 13.34491 | |
| STANDARD ERROR | 3.65307 | | | | | |

| ------------VARIABLES IN THE EQUATION------------- | | | | | -----VARIABLES NOT IN THE EQUATION------ | | | | |
|---|---|---|---|---|---|---|---|---|---|
| VARIABLE | B | BETA | STD ERROR B | F | VARIABLE | BETA IN | PARTIAL | TOLERANCE | F |
| VAR01 | 0.57527 | 0.87923 | 0.08645 | 44.280 | | | | | |
| (CONSTANT) | 8.66667 | | | | | | | | |

MAXIMUM STEP REACHED

```
EXAMPLE OF REGRESSION MODELS

FILE     NONAME

****************************** MULTIPLE REGRESSION ******************************

DEPENDENT VARIABLE: VAR02     FROM      VARIABLE LIST 1
                                      REGRESSION LIST 1

          OBSERVED     PREDICTED                        PLOT OF STANDARDIZED RESIDUAL
SEQNUM       VAR02         VAR02      RESIDUAL    -2.0      -1.0      0.0      1.0      2.0

     1    5.000000      9.817204     -4.817204                  *      I
     2    7.000000      9.817204     -2.817204                     *   I
     3    3.000000      9.817204     -6.817204                *        I
     4    10.00000      10.96774     -.9677419                        *I
     5    12.00000      10.96774      1.032258                         I*
     6    14.00000      10.96774      3.032258                         I   *
     7    15.00000      13.26882      1.731182                         I *
     8    17.00000      13.26882      3.731182                         I    *
     9    18.00000      13.26882      4.731182                         I      *
    10    20.00000      17.87096      2.129032                         I  *
    11    21.00000      17.87096      3.129032                         I   *
    12    19.00000      17.87096      1.129032                         I *
    13    23.00000      27.07526     -4.075269                   *     I
    14    24.00000      27.07526     -3.075269                     *   I
    15    29.00000      27.07526      1.924730                         I  *

DURBIN-WATSON TEST OF RESIDUAL DIFFERENCES COMPARED BY CASE ORDER (SEQNUM).

VARIABLE LIST    1,REGRESSION LIST    1.    DURBIN-WATSON TEST    0.77105

EXAMPLE OF REGRESSION MODELS

FILE     NONAME

VARIABLE         MEAN     STANDARD DEV     CASES

VAR01         12.4000          11.2935        15
VAR02         15.8000           7.3892        15
VAR03        272.8000         399.9234        15
```

EXAMPLE OF REGRESSION MODELS

FILE NONAME

CORRELATION COEFFICIENTS

A VALUE OF 99.00000 IS PRINTED
IF A COEFFICIENT CANNOT BE COMPUTED.

| | VAR01 | VAR02 | VAR03 |
|---|---|---|---|
| VAR01 | 1.00000 | 0.87923 | 0.97425 |
| VAR02 | 0.87923 | 1.00000 | 0.78234 |
| VAR03 | 0.97425 | 0.78234 | 1.00000 |

EXAMPLE OF REGRESSION MODELS

FILE NONAME

* * * * * * * * * * * * * * * * * * * * * * * * * * * MULTIPLE REGRESSION * * * * * * * * * * * * * * * * VARIABLE LIST 1
REGRESSION LIST 1

DEPENDENT VARIABLE.. VAR02

VARIABLE(S) ENTERED ON STEP NUMBER 1.. VAR01
VAR03

| | | ANALYSIS OF VARIANCE | DF | SUM OF SQUARES | MEAN SQUARE | F |
|---|---|---|---|---|---|---|
| MULTIPLE R | 0.93888 | | | | | |
| R SQUARE | 0.88150 | REGRESSION | 2. | 673.82062 | 336.91031 | 44.63404 |
| ADJUSTED R SQUARE | 0.86175 | RESIDUAL | 12. | 90.57938 | 7.54828 | |
| STANDARD ERROR | 2.74741 | | | | | |

-------------VARIABLES IN THE EQUATION-------------- -----VARIABLES NOT IN THE EQUATION------

| VARIABLE | B | BETA | STD ERROR B | F | VARIABLE | BETA IN | PARTIAL | TOLERANCE | F |
|---|---|---|---|---|---|---|---|---|---|
| VAR01 | 1.50633 | 2.30224 | 0.28836 | 27.287 | | | | | |
| VAR03 | −0.02699 | −1.46062 | 0.00814 | 10.983 | | | | | |
| (CONSTANT) | 4.48366 | | | | | | | | |

ALL VARIABLES ARE IN THE EQUATION

```
EXAMPLE OF REGRESSION MODELS

FILE     NONAME

******************************* MULTIPLE REGRESSION *******************************

DEPENDENT VARIABLE: VAR02        FROM        VARIABLE LIST 1
                                             REGRESSION LIST 1

                                                                  PLOT OF STANDARDIZED RESIDUAL
            OBSERVED       PREDICTED                           -2.0     -1.0     0.0     1.0     2.0
SEQNUM        VAR02          VAR02          RESIDUAL

    1       5.000000       7.388367       -2.388361                          *   I
    2       7.000000       7.388367       -.3883618                            * I
    3       3.000000       7.388367       -4.388361                        *     I
    4       10.00000       10.07717       -.7716632E-01                          *
    5       12.00000       10.07717        1.922833                              I   *
    6       14.00000       10.07717        3.922833                              I     *
    7       15.00000       14.80708        .1929159                              *
    8       17.00000       14.80708        2.192915                              I   *
    9       18.00000       14.80708        3.192915                              I     *
   10       20.00000       21.67615       -1.676154                          *   I
   11       21.00000       21.67615       -.6761543                            * I
   12       19.00000       21.67615       -2.676154                         *    I
   13       23.00000       25.05122       -2.051233                           *  I
   14       24.00000       25.05122       -1.051233                            * I
   15       29.00000       25.05122        3.948766                              I     *

DURBIN-WATSON TEST OF RESIDUAL DIFFERENCES COMPARED BY CASE ORDER (SEQNUM).

VARIABLE LIST   1, REGRESSION LIST   1. DUBRIN-WATSON TEST    1.33140
```

*7.24. Refer to the data of exercise 7.23. Many times a logarithmic transformation can be used on the dose levels to linearize the response with respect to the independent variable.

a. Refer to a set of log tables (see, for example, the Chemical Rubber Company tables) or an electronic calculator to obtain the logarithms of the 5 dose levels.

b. If $x_1$ denotes the log dose, fit the model

$$y = \beta_0 + \beta_1 x_1 + \epsilon$$

c. Compare your results in part b to those shown in the computer printout (SPSS 1975) that follows.

EXAMPLE OF REGRESSION MODELS

FILE NONAME

| VARIABLE | MEAN | STANDARD DEV | CASES |
|---|---|---|---|
| VAR02 | 15.8000 | 7.3892 | 15 |
| VAR04 | 0.9031 | 0.4407 | 15 |

EXAMPLE OF REGRESSION MODELS

FILE NONAME

CORRELATION COEFFICIENTS

A VALUE OF 99.00000 IS PRINTED
IF A COEFFICIENT CANNOT BE COMPUTED.

| | VAR02 | VAR04 |
|---|---|---|
| VAR02 | 1.00000 | 0.96412 |
| VAR04 | 0.96412 | 1.00000 |

EXAMPLE OF REGRESSION MODELS

FILE NONAME

* * * * * * * * * * * * * * * * * * * * * * * * * * * * * MULTIPLE REGRESSION * * * * * * * * * * * * * * * * VARIABLE LIST 1
REGRESSION LIST 1

DEPENDENT VARIABLE.. VAR02

VARIABLE(S) ENTERED ON STEP NUMBER 1.. VAR04

| | | ANALYSIS OF VARIANCE | DF | SUM OF SQUARES | MEAN SQUARE | F |
|---|---|---|---|---|---|---|
| MULTIPLE R | 0.96412 | | | | | |
| R SQUARE | 0.92953 | REGRESSION | 1. | 710.53334 | 710.53334 | 171.47775 |
| ADJUSTED R SQUARE | 0.92411 | RESIDUAL | 13. | 53.86666 | 4.14359 | |
| STANDARD ERROR | 2.03558 | | | | | |

------------VARIABLES IN THE EQUATION-------------

| VARIABLE | B | BETA | STD ERROR E | F |
|---|---|---|---|---|
| VAR04 | 16.16672 | 0.96412 | 1.23458 | 171.478 |
| (CONSTANT) | 1.20000 | | | |

-----VARIABLES NOT IN THE EQUATION------

| VARIABLE | BETA IN | PARTIAL | TOLERANCE | F |
|---|---|---|---|---|

MAXIMUM STEP REACHED

```
EXAMPLE OF REGRESSION MODELS

FILE     NONAME

* * * * * * * * * * * * * * * * * * * * * * * * * * * * * * MULTIPLE REGRESSION * * * * * * * * * * * * * * * * * * * * * * * * * * * * * *

DEPENDENT VARIABLE: VAR02     FROM        VARIABLE LIST 1
                                        REGRESSION LIST 1

              OBSERVED       PREDICTED                            PLOT OF STANDARDIZED RESIDUAL
SEQNUM          VAR02          VAR02        RESIDUAL      -2.0      -1.0      0.0      1.0      2.0

    1         5.000000       6.066669      -1.066667                         *I
    2         7.000000       6.066669       .9333332                          I*
    3         3.000000       6.066669      -3.066667                       *  I
    4         10.00000       10.93333      -.9333344                         *I
    5         12.00000       10.93333       1.066665                          I *
    6         14.00000       10.93333       3.066665                          I    *
    7         15.00000       15.80000      -.7999990                        * I
    8         17.00000       15.80000       1.200001                          I *
    9         18.00000       15.80000       2.200001                          I  *
   10         20.00000       20.66666      -.6666647                        * I
   11         21.00000       20.66666       .3333353                          *
   12         19.00000       20.66666      -1.666664                       *  I
   13         23.00000       25.53333      -2.533335                      *   I
   14         24.00000       25.53333      -1.533335                       *  I
   15         29.00000       25.53333       3.466664                          I    *

DURBIN-WATSON TEST OF RESIDUAL DIFFERENCES COMPARED BY CASE ORDER (SEQNUM).

VARIABLE LIST    1, REGRESSION LIST    1. DURBIN-WATSON TEST    1.71667
```

*7.25. The abrasive effect of a wear tester for experimental fabrics was tested on a particular fabric while run at 6 different machine speeds. Forty-eight identical five-inch-square pieces of fabric were cut, with 8 squares randomly assigned to each of the 6 machine speeds 100, 120, 140, 160, 180, and 200 revolutions per minute. The order of assignment of the squares to the machine was random, with each square tested for a three-minute period at the appropriate machine setting. The amount of wear was measured and recorded for each square. The data appear below.

| *Machine Speed (in rpm)* | *Wear* |
|---|---|
| 100 | 23.0, 23.5, 24.4, 25.2, 25.6, 26.1, 24.8, 25.6 |
| 120 | 26.7, 26.1, 25.8, 26.3, 27.2, 27.9, 28.3, 27.4 |
| 140 | 28.0, 28.4, 27.0, 28.8, 29.8, 29.4, 28.7, 29.3 |
| 160 | 32.7, 32.1, 31.9, 33.0, 33.5, 33.7, 34.0, 32.5 |
| 180 | 43.1, 41.7, 42.4, 42.1, 43.5, 43.8, 44.2, 43.6 |
| 200 | 54.2, 43.7, 53.1, 53.8, 55.6, 55.9, 54.7, 54.5 |

a. Plot the sample mean at each machine speed.
b. Fit a quadratic model to these data.

*7.26. Refer to the data of exercise 7.25. Suppose that another variable was controlled and that the first 4 squares at each speed were treated with a .2 concentration of protective coating, while the second 4 measurements were treated with a .4 concentration of the same coating. If $x_1$ denotes the machine speed and $x_2$ denotes the concentration of the protective coating, fit these models:

$$y = \beta_0 + \beta_1 x_1 + \beta_2 x_1^2 + \beta_3 x_2 + \epsilon$$

$$y = \beta_0 + \beta_1 x_1 + \beta_2 x_1^2 + \beta_3 x_2 + \beta_4 x_1 x_2 + \beta_5 x_1^2 x_2 + \epsilon$$

*7.27. A manufacturer of laundry detergent was interested in testing a new product prior to market release. One area of concern was the relationship between the height of the detergent suds in a washing machine as a function of the amount of detergent added and the degree of agitation in the wash cycle. For a standard size washing machine tub filled to the full level, random assignments of different agitation levels (measured in minutes) and amounts of detergent were made and tested on the washing machine. The data appear below.

$

| *Height, y* | *Agitation, $x_1$* | *Amount, $x_2$* |
|---|---|---|
| 28.1 | | 6 |
| 32.3 | | 7 |
| 34.8 | 1 | 8 |
| 38.2 | | 9 |
| 43.5 | | 10 |
| 60.3 | | 6 |
| 63.7 | | 7 |
| 65.4 | 2 | 8 |
| 69.2 | | 9 |
| 72.9 | | 10 |
| 88.2 | | 6 |
| 89.3 | | 7 |
| 94.1 | 3 | 8 |
| 95.7 | | 9 |
| 100.6 | | 10 |

Fit the following two models:

$$y = \beta_0 + \beta_1 x_1 + \beta_2 x_2 + \epsilon$$

$$y = \beta_0 + \beta_1 x_1 + \beta_2 x_2 + \beta_3 x_1 x_2 + \epsilon$$

*7.28. Refer to exercise 7.27. Fit the model

$$y = \beta_0 + \beta_1 x_1 + \beta_2 x_1^2 + \beta_3 x_2 + \beta_4 x_2^2 + \beta_5 x_1 x_2 + \beta_6 x_1 x_2^2 + \beta_7 x_1^2 x_2 + \beta_8 x_1^2 x_2^2 + \epsilon$$

Ψ 7.29. A psychologist is interested in examining the effects of loss of sleep on a person's ability to perform simple arithmetic tasks. To do this, prospective subjects are screened to obtain individuals whose daily sleep patterns were closely matched. From this group, 20 subjects are chosen. Each individual selected is randomly assigned to one of 5 groups, 4 individuals per group.

Group 1: 0 hours of sleep
Group 2: 2 hours of sleep
Group 3: 4 hours of sleep
Group 4: 6 hours of sleep
Group 5: 8 hours of sleep

All subjects are then placed on a standard routine for the next 24 hours.

The following day after breakfast, each individual is tested to determine the number of arithmetic additions done correctly in a 10-minute period. That evening the amount of sleep each person is allowed depends on the group to which he or she had been assigned. The following morning after breakfast, each person is again tested using a different, but equally difficult, set of additions.

Let the response of interest be the difference in the number of correct responses on the first test day minus the number correct on the second test day. Use the data below to respond to the statements that follow.

| *Group* | *Response, y* |
|---|---|
| 1 | 39, 33, 41, 40 |
| 2 | 25, 29, 34, 26 |
| 3 | 10, 18, 14, 17 |
| 4 | 4, 6, −1, 9 |
| 5 | −5, 0, −3, −8 |

a. Plot the sample data and use the plot to suggest a model.
b. Fit the suggested model.

7.30. Social adjustment and perceived self-image tests were administered to 6 ex-drug addicts. These data are shown on the next page.

a. Use these data to obtain a least squares fit for the linear regression line $E(y) = \beta_0 + \beta_1 x$.

| | | | | | | |
|---|---|---|---|---|---|---|
| *Perceived Self-image Score, x* | 35 | 23 | 42 | 18 | 31 | 45 |
| *Social Adjustment Score, y* | 55 | 37 | 61 | 28 | 52 | 70 |

b. Predict the social adjustment score for an ex-addict whose perceived self-image test score is 29.

*7.31. An experiment was conducted to determine the relationship between the amount of warping $y$ for a particular alloy and the temperature (in degrees Celsius) under which the experiment was conducted. The sample data appear below. Note that three observations were taken at each temperature setting.

| *Amount of Warping* | *Temperature (°C)* |
|---|---|
| 10, 13, 12 | 15 |
| 14, 12, 11 | 20 |
| 14, 12, 16 | 25 |
| 18, 19, 22 | 30 |
| 25, 21, 20 | 35 |
| 23, 25, 26 | 40 |
| 30, 31, 34 | 45 |
| 35, 33, 38 | 50 |

Use the computer output that follows to respond to these statements.

a. Plot the data to determine whether a linear or quadratic model appears more appropriate.
b. If a linear model is fit, indicate the prediction equation. Superimpose the prediction equation over the scatter diagram of $y$ versus $x$.
c. If a quadratic model is fit, identify the prediction equation. Superimpose the quadratic prediction equation on the scatter diagram. Which fit looks better, the linear or the quadratic?
d. Predict the amount of warping at a temperature of 27°C, using both the linear and the quadratic prediction equations.

```
STATISTICAL ANALYSIS SYSTEM

DATA WARPING;
INPUT Y 1-2 X 4-5;
XSQ = X * * 2;
CARDS;
```

24 OBSERVATIONS IN DATA SET WARPING 3 VARIABLES

PROC PRINT;

| OBS | Y | X | XSQ |
|---|---|---|---|
| 1 | 10 | 15 | 225 |
| 2 | 13 | 15 | 225 |
| 3 | 12 | 15 | 225 |
| 4 | 14 | 20 | 400 |
| 5 | 12 | 20 | 400 |
| 6 | 11 | 20 | 400 |
| 7 | 14 | 25 | 625 |
| 8 | 12 | 25 | 625 |
| 9 | 16 | 25 | 625 |
| 10 | 18 | 30 | 900 |
| 11 | 19 | 30 | 900 |
| 12 | 22 | 30 | 900 |
| 13 | 25 | 35 | 1225 |
| 14 | 21 | 35 | 1225 |
| 15 | 20 | 35 | 1225 |
| 16 | 23 | 40 | 1600 |
| 17 | 25 | 40 | 1600 |
| 18 | 26 | 40 | 1600 |
| 19 | 30 | 45 | 2025 |
| 20 | 31 | 45 | 2025 |
| 21 | 34 | 45 | 2025 |
| 22 | 35 | 50 | 2500 |
| 23 | 33 | 50 | 2500 |
| 24 | 38 | 50 | 2500 |

PROC REGR;
MODEL Y = X/ X I;
MODEL Y = X XSQ/ X I;
TITLE 'EXERCISE 7.31';

* * * * * * * * * * * * * * * * * * * * * * * * * * * * * * * * * * * * * * * * * * * * * * * * * * * * * * * *

MODEL : EXERCISE 7.31

DEPENDENTS : Y

INDEPENDENTS : X

* * * * * * * * * * * * * * * * * * * * * * * * * * * * * * * * * * * * * * * * * * * * * * * * * * * * * * * *

EXERCISE 7.31

THE X'X MATRIX

| | INTERCPT | X |
|---|---|---|
| INTERCPT | 24.00000000 | 780.00000000 |
| X | 780.00000000 | 28500.00000000 |

EXERCISE 7.31

THE X′X INVERSE MATRIX, RANK = 2

| | INTERCPT | X |
|---|---|---|
| INTERCPT | 0.37698413 | −0.01031746 |
| X | −0.01031746 | 0.00031746 |

EXERCISE 7.31

ANALYSIS OF VARIANCE TABLE, REGRESSION COEFFICIENTS, AND STATISTICS OF FIT FOR DEPENDENT VARIABLE Y

| SOURCE | DF | SUM OF SQUARES | MEAN SQUARE | F VALUE | PROB > F | R-SQUARE | C.V. |
|---|---|---|---|---|---|---|---|
| REGRESSION | 1 | 1571.62698413 | 1571.62698413 | 265.54614 | 0.0001 | 0.92349054 | 11.35933% |
| ERROR | 22 | 130.20634921 | 5.91847042 ← $\sigma^2$ | | | STD DEV | Y MEAN |
| CORRECTED TOTAL | 23 | 1701.83333333 | | | | 2.43279066 | 21.41667 |

| SOURCE | DF | SEQUENTIAL SS | F VALUE | PROB > F | PARTIAL SS | F VALUE | PROB > F |
|---|---|---|---|---|---|---|---|
| X | 1 | 1571.62698413 | 265.54614 | 0.0001 | 1571.62698413 | 265.54614 | 0.0001 |

| SOURCE | B VALUES | T FOR HO: B = 0 | PROB > \|T\| | STD ERR B | STD B VALUES |
|---|---|---|---|---|---|
| INTERCEPT | −1.53968254 | −1.03078 | 0.3138 | 1.49370995 | 0.0 |
| X | 0.70634921 | 16.29559 | 0.0001 | 0.04334604 | 0.96098415 |

* * * * * * * * * * * * * * * * * * * * * * * * * * * * * * * * * * * * * * * * * * * * * * * * * * * * * * *

MODEL : EXERCISE 7.31

DEPENDENTS : Y

INDEPENDENTS : X XSQ

* * * * * * * * * * * * * * * * * * * * * * * * * * * * * * * * * * * * * * * * * * * * * * * * * * * * * * *

EXERCISE 7.31

THE X′X MATRIX

| | INTERCPT | X | XSQ |
|---|---|---|---|
| INTERCPT | 24.00000000 | 780.00000000 | 28500.00000000 |
| X | 780.00000000 | 28500.00000000 | 1131000.00000000 |
| XSQ | 28500.00000000 | 1131000.00000000 | 47467500.00000000 |

EXERCISE 7.31

THE X'X INVERSE MATRIX, RANK = 3

| | INTERCPT | X | XSQ |
|---|---|---|---|
| INTERCPT | 3.09325397 | −0.20119048 | 0.00293651 |
| X | −0.20119048 | 0.01373016 | −0.00020635 |
| XSQ | 0.00293651 | −0.00020635 | 0.00000317 |

EXERCISE 7.31

ANALYSIS OF VARIANCE TABLE, REGRESSION COEFFICIENTS, AND STATISTICS OF FIT FOR DEPENDENT VARIABLE Y

| SOURCE | DF | SUM OF SQUARES | MEAN SQUARE | F VALUE | PROB > F | R-SQUARE | C.V. |
|---|---|---|---|---|---|---|---|
| REGRESSION | 2 | 1613.92063492 | 806.96031746 | 192.76131 | 0.0001 | 0.94834236 | 9.55354 % |
| ERROR | 21 | 87.91269841 | 4.18631897 | | | STD DEV | Y MEAN |
| CORRECTED TOTAL | 23 | 1701.83333333 | | | | 2.04604960 | 21.41667 |

| SOURCE | DF | SEQUENTIAL SS | F VALUE | PROB > F | PARTIAL SS | F VALUE | PROB > F |
|---|---|---|---|---|---|---|---|
| X | 1 | 1571.62698413 | 375.41979 | 0.0001 | 0.15969355 | 0.03815 | 0.8470 |
| XSQ | 1 | 42.29365079 | 10.10283 | 0.0045 | 42.29365079 | 10.10283 | 0.0045 |

| SOURCE | B VALUES | T FOR HO: B = 0 | PROB > \|T\| | STD ERR B | STD B VALUES |
|---|---|---|---|---|---|
| INTERCEPT | 9.17857143 | 2.55065 | 0.0186 | 3.59852022 | 0.0 |
| X | −0.04682540 | −0.19531 | 0.8470 | 0.23974742 | −0.06370569 |
| XSQ | 0.01158730 | 3.17849 | 0.0045 | 0.00364553 | 1.03674543 |

# References


Barr, A.J.; Goodnight, J.H.; Sall, J.P.; and Helwig, J.T. 1976. *A user's guide to SAS 76.* Raleigh, N.C.: SAS Institute, Inc.

Dixon, W.J., ed. 1975. *BMDP, biomedical computer programs.* Berkeley: University of California Press.

Draper, N.R., and Smith, H. 1966. *Applied regression analysis.* New York: Wiley.

Graybill, F.A. 1969. *Introduction to matrices with applications in statistics.* Belmont, Calif.: Wadsworth.

———. 1976. *Theory and application of the linear model.* N. Scituate, Mass.: Duxbury Press.

*Handbook of tables for probability and statistics.* 1966. Cleveland, Ohio: The Chemical Rubber Co.

Hohn, F.E. 1958. *Elementary matrix algebra.* New York: Macmillan.

Mendenhall, W. 1968. *Introduction to linear models and the design and analysis of experiments.* Belmont, Calif.: Wadsworth.

Nie, N.H.; Hull, C.H.; Jenkins, J.G.; Steinbrenner, K.; and Bent, D.H. 1975. *Statistical package for the social sciences.* 2d ed. New York: McGraw-Hill.

Pearson, E.S., and Hartley, H.O. 1966. *Biometrika tables for statisticians.* 3d ed. Vol. I. London: Cambridge University Press.

Service, J. 1972. *A user's guide to the statistical analysis system.* Raleigh, N.C.: Student Supply Stores, North Carolina State University.

Snedecor, G.W., and Cochran, W.G. 1967. *Statistical methods.* 6th ed. Ames, Iowa: Iowa State University Press.

# 8 Some Inferences Concerning the General Linear Model

## 8.1 Introduction

In chapter 7 we used matrices to find least squares estimates of parameters in the general linear model. We would now like to use these estimates to make inferences about the corresponding population parameters.

For example, suppose we assume that the model

$$y = \beta_0 + \beta_1 x + \beta_2 x^2 + \epsilon$$

adequately represents the relationship between the independent variable $x$, the amount of force applied to a one-foot section of steel, and the resulting increase in area $y$ of the steel sample. Applying the procedures of chapter 7, we obtain a random sample of $n$ observations and construct estimates of the model parameters by using

$$\hat{\boldsymbol{\beta}} = (\mathbf{X'X})^{-1}\mathbf{X'Y}$$

It might be of interest to check whether there is a curvilinear relationship between the amount of force applied $x$ and the increase in area $y$, as represented in our model. To do this, we could conduct a test of the null hypothesis $H_0$: $\beta_2 = 0$ against the alternative $H_a$: $\beta_2 \neq 0$. The sample estimate of $\beta_2$, $\hat{\beta}_2$, could be used to construct an appropriate test statistic for this test.

Another inference of practical significance would be to estimate the expected value of $y$, $E(y)$, for one or more values of $x$. Our estimate of $E(y)$ is, of course,

$$y = \widehat{\beta}_0 + \widehat{\beta}_1 x + \widehat{\beta}_2 x^2$$

but we have not yet discussed how to construct a confidence interval for $E(y)$.

In this chapter we will develop methods for making inferences about a single parameter $\beta$ in the general linear model and inferences concerning $E(y)$.

# Inferences About a Single Parameter in the General Linear Model 8.2

The general linear model in $k$ independent variables has the form

$$y_i = \beta_0 + \beta_1 x_{i1} + \beta_2 x_{i2} + \cdots + \beta_k x_{ik} + \epsilon_i \qquad (i = 1, 2, \ldots, n)$$

In previous discussions we have assumed that $\epsilon_i$, the random error associated with observation $i$, has expectation zero. We will expand our assumptions somewhat further to include the following:

assumptions for $\epsilon_i$

1. $\epsilon_1, \epsilon_2, \ldots, \epsilon_n$ are *independent* of each other; and
2. $\epsilon$, for a given setting of the independent variables $x_1, x_2, \ldots, x_k$, has mean zero and variance $\sigma^2$.

variance $v_{ii}\sigma^2$

Under these assumptions, it can be shown that the distribution of the least squares estimate $\widehat{\beta}_i$ has mean $\beta_i$ (the parameter estimated) and variance $v_{ii}\sigma^2$, where the constant $v_{ii}$ is the $i$th diagonal element of $(\mathbf{X}'\mathbf{X})^{-1}$:

$$(\mathbf{X}'\mathbf{X})^{-1} = \begin{bmatrix} v_{00} & v_{01} & v_{02} & \cdots & v_{0k} \\ v_{10} & v_{11} & v_{12} & \cdots & v_{1k} \\ v_{20} & v_{21} & v_{22} & \cdots & v_{2k} \\ \vdots & \vdots & \vdots & & \vdots \\ v_{k0} & v_{k1} & v_{k2} & \cdots & v_{kk} \end{bmatrix}$$

Note that the first diagonal element of $(\mathbf{X}'\mathbf{X})^{-1}$ is labeled $v_{00}$ to correspond to

$V(\beta_i)$

$\hat{\beta}_0$ and thus the variance of $\hat{\beta}_0$, denoted by $V(\hat{\beta}_0)$, is

$$V(\hat{\beta}_0) = v_{00}\sigma^2$$

Similarly, the variance of $\hat{\beta}_1$ is $V(\hat{\beta}_1) = v_{11}\sigma^2$.

**EXAMPLE 8.1**

In example 7.15 (p. 165) we examined the relationship between the weight loss of a compound and two independent variables, time of exposure and relative humidity of the room during exposure. The assumed model was

$$y = \beta_0 + \beta_1 x_1 + \beta_2 x_2 + \epsilon$$

for which we obtained the prediction equation

$$\hat{y} = 5.49 + 1.32x_1 - 8x_2$$

Determine the variances of the least squares estimates $\hat{\beta}_0$, $\hat{\beta}_1$, and $\hat{\beta}_2$.

**SOLUTION**

You will recall (p. 167) that the inverse matrix for these data was

$$(\mathbf{X}'\mathbf{X})^{-1} = \begin{bmatrix} .083 & 0 & 0 \\ 0 & .067 & 0 \\ 0 & 0 & 12.5 \end{bmatrix}$$

Hence

$$V(\hat{\beta}_0) = .083\sigma^2 \qquad V(\hat{\beta}_1) = .067\sigma^2 \qquad V(\hat{\beta}_2) = 12.50\sigma^2$$

estimate of $\sigma^2$

Before we can utilize the variances of the least squares estimates, we must obtain an estimate of $\sigma^2$ from the sample data. The general linear model, with

$$y = \beta_0 + \beta_1 x_1 + \beta_2 x_2 + \cdots + \beta_k x_k + \epsilon$$

residual sum of squares SSE

and $E(\epsilon) = 0$, has $\epsilon = y - E(y)$. But since we don't know $E(y)$, we must estimate its value using $\hat{y}$. Then the residuals (errors of prediction) can be used to form an estimate of $\sigma^2$. The quantity $\Sigma\,(y - \hat{y})^2$ is often referred to as either the *residual sum of squares* or as the *sum of squares for error* (SSE). Dividing SSE by $n - (k + 1)$ degrees of freedom, we obtain an estimate of $\sigma^2$. The degrees of freedom are easily computed as the sample size minus the number of parameters fit in the general linear model.

---

**Estimate of $\sigma^2$ Using the General Linear Model**

$$s^2 = \frac{\sum (y - \hat{y})^2}{n - (k + 1)} = \frac{\text{SSE}}{n - (k + 1)}$$

where $n$ is the sample size and $k$ is the number of independent variables in the model.

---

The matrix notation of chapter 7 can be used to provide a computational formula for SSE.

**Computational Formula for SSE**

$$\text{SSE} = \mathbf{Y}'\mathbf{Y} - \hat{\boldsymbol{\beta}}'\mathbf{X}'\mathbf{Y}$$

**EXAMPLE 8.2**

The yield per plot in bushels of corn was observed on $n = 10$ plots which had been fertilized in varying degrees. We let the independent variable $x_1$ denote (fertilizer − average amount of fertilizer applied). The data and the coded fertilizer values are recorded in table 8.1. Use the sample data of table 8.1 to obtain the least squares prediction equation $\hat{y}$ for the model $y = \beta_0 + \beta_1 x_1 + \epsilon$. Also, calculate an estimate of $\sigma^2$.

**Table 8.1** Corn yield data for example 8.2 [$x_1$ = (fertilizer − average fertilizer applied)]

| *Yield (in bushels)* | *Fertilizer (in pounds per plot)* | $x_1$ |
|---|---|---|
| 12 | 2 | −2 |
| 13 | 2 | −2 |
| 13 | 3 | −1 |
| 14 | 3 | −1 |
| 15 | 4 | 0 |
| 15 | 4 | 0 |
| 14 | 5 | 1 |
| 16 | 5 | 1 |
| 17 | 6 | 2 |
| 18 | 6 | 2 |

**SOLUTION**

For these data,

$$\mathbf{Y} = \begin{bmatrix} 12 \\ 13 \\ 13 \\ 14 \\ 15 \\ 15 \\ 14 \\ 16 \\ 17 \\ 18 \end{bmatrix} \qquad \mathbf{X} = \begin{bmatrix} 1 & -2 \\ 1 & -2 \\ 1 & -1 \\ 1 & -1 \\ 1 & 0 \\ 1 & 0 \\ 1 & 1 \\ 1 & 1 \\ 1 & 2 \\ 1 & 2 \end{bmatrix}$$

$$\mathbf{X}'\mathbf{X} = \begin{bmatrix} 10 & 0 \\ 0 & 20 \end{bmatrix} \qquad \mathbf{X}'\mathbf{Y} = \begin{bmatrix} 147 \\ 23 \end{bmatrix} \qquad (\mathbf{X}'\mathbf{X})^{-1} = \begin{bmatrix} .1 & 0 \\ 0 & .05 \end{bmatrix}$$

Thus we have that

$$\hat{\boldsymbol{\beta}} = (\mathbf{X}'\mathbf{X})^{-1}\mathbf{X}'\mathbf{Y} = \begin{bmatrix} .1 & 0 \\ 0 & .05 \end{bmatrix} \begin{bmatrix} 147 \\ 23 \end{bmatrix} = \begin{bmatrix} 14.7 \\ 1.15 \end{bmatrix}$$

The least squares prediction equation is then

$$\hat{y} = 14.7 + 1.15x_1$$

The estimate $s^2$ of $\sigma^2$ can be computed after finding SSE $= \mathbf{Y}'\mathbf{Y} - \hat{\boldsymbol{\beta}}'\mathbf{X}'\mathbf{Y}$.

$$\mathbf{Y}'\mathbf{Y} = [12 \quad 13 \quad 13 \quad 14 \quad 15 \quad 15 \quad 14 \quad 16 \quad 17 \quad 18] \begin{bmatrix} 12 \\ 13 \\ 13 \\ 14 \\ 15 \\ 15 \\ 14 \\ 16 \\ 17 \\ 18 \end{bmatrix} = \sum_{i=1}^{10} y_i^2 = 2193$$

$$\hat{\boldsymbol{\beta}}'\mathbf{X}'\mathbf{Y} = [14.7 \quad 1.15] \begin{bmatrix} 147 \\ 23 \end{bmatrix} = 2187.35$$

Combining we have

$$\text{SSE} = \mathbf{Y}'\mathbf{Y} - \hat{\boldsymbol{\beta}}'\mathbf{X}'\mathbf{Y} = 2193 - 2187.35 = 5.65$$

and, for $n = 10$ and $k = 1$,

$$s^2 = \frac{\text{SSE}}{n - (k+1)} = \frac{5.65}{8} = .71$$

If we assume, in addition to previous assumptions (p. 191), that the $\epsilon_i$'s of the general linear model are normally distributed, we can specify the following $100(1 - \alpha)\%$ confidence interval for a specific parameter of the general linear model.

---

**$100(1 - \alpha)\%$ Confidence Interval for $\beta_i$**

$$\hat{\beta}_i \pm t_{\alpha/2}\, s\, \sqrt{v_{ii}}$$

where

$$s^2 = \frac{\text{SSE}}{n - (k+1)}$$

and $t$ is based on df $= n - (k+1)$, the denominator of $s^2$.

---

**EXAMPLE 8.3**

Use the data of example 8.2 (p. 193) to place a 95% confidence interval on $\beta_1$.

**SOLUTION**

The sample calculations of example 8.2 yielded

$\hat{y} = 14.7 + 1.15x_1$

$s^2 = .71 \qquad \text{and} \qquad s = \sqrt{.71} = .84$

The critical $t$-value for a = .025 and df $= n - (k + 1) = 8$ is 2.306. Using

$$(\mathbf{X'X})^{-1} = \begin{bmatrix} v_{00} & v_{01} \\ v_{10} & v_{11} \end{bmatrix} = \begin{bmatrix} .1 & 0 \\ 0 & .05 \end{bmatrix}$$

we find $v_{11} = .05$. Substituting into the confidence interval formula for $\beta_1$, the appropriate interval is

$1.15 \pm 2.306(.84)\sqrt{.05} \qquad \text{or} \qquad 1.15 \pm .43$

That is, we estimate that the slope of the line relating $y$ (yield) to the independent variable $x$ (coded fertilizer levels) lies in the interval from .72 to 1.58.

**EXAMPLE 8.4**

The SAS computer output for the data of example 7.15 follows. Use the SAS output to give variances for the least squares estimates of $\beta_0$, $\beta_1$, and $\beta_2$ in the model

$y = \beta_0 + \beta_1 x_1 + \beta_2 x_2 + \epsilon$

Also give an estimate of $\sigma^2$.

**SOLUTION**

```
STATISTICAL ANALYSIS SYSTEM

DATA WEIGHT;
INPUT Y 1-4 X1 6-9 X2 11-13;
CARDS;


   12 OBSERVATIONS IN DATA SET WEIGHT     3 VARIABLES

PROC PRINT;


OBS        Y        X1        X2

  1       4.3      -1.5      -.1
  2       5.5      -0.5      -.1
  3       6.8       0.5      -.1
  4       8.0       1.5      -.1
  5       4.0      -1.5       .0
  6       5.2      -0.5       .0
  7       6.6       0.5       .0
  8       7.5       1.5       .0
```

```
  9      2.0     -1.5     .1
 10      4.0     -0.5     .1
 11      5.7      0.5     .1
 12      6.5      1.5     .1


PROC REGR;
MODEL Y = X1 X2/X 1;
TITLE 'EXAMPLE 8.4';

******************************************************************************

PROC REGR : EXAMPLE 8.4

DATA SET   : WEIGHT          NUMBER OF VARIABLES = 3    NUMBER OF CLASSES = 0

VARIABLES  : Y X1 X2

******************************************************************************

******************************************************************************

MODEL          : EXAMPLE 8.4

DEPENDENTS     : Y

INDEPENDENTS   : X1 X2

******************************************************************************

                                              EXAMPLE 8.4

                                            THE X'X MATRIX

                      INTERCPT              X1              X2

INTERCPT            12.00000000     0.0             0.0

X1                   0.0            15.00000000     0.0

X2                   0.0             0.0            0.08000000


                                                              EXAMPLE 8.4

                                          THE X'X INVERSE MATRIX, RANK = 3

                                    INTERCPT          X1              X2

INTERCPT                           0.08333333    0.0             0.0

X1                                 0.0           0.06666667      0.0

X2                                 0.0           0.0            12.50000000
```

EXAMPLE 8.4

ANALYSIS OF VARIANCE TABLE, REGRESSION COEFFICIENTS, AND STATISTICS OF FIT FOR DEPENDENT VARIABLE Y

| SOURCE | DF | SUM OF SQUARES | MEAN SQUARE | | F VALUE | PROB > F | R-SQUARE | C.V. |
|---|---|---|---|---|---|---|---|---|
| REGRESSION | 2 | 31.12416667 | 15.56208333 | | 104.13290 | 0.0001 | 0.95857609 | 7.01810 |
| ERROR | 9 | 1.34500000 | 0.14944444 | ⟵ | estimate of $\sigma^2$ | | | |
| | | | | | | | STD DEV | Y MEAN |
| CORRECTED TOTAL | 11 | 32.46916667 | | | | | | |
| | | | | | | | 0.38658045 | 5.50833 |

| SOURCE | DF | SEQUENTIAL SS | F VALUE | PROB > F | PARTIAL SS | F VALUE | PROB > F |
|---|---|---|---|---|---|---|---|
| X1 | 1 | 26.00416667 | 174.00558 | 0.0001 | 26.00416667 | 174.00558 | 0.0001 |
| X2 | 1 | 5.12000000 | 34.26022 | 0.0002 | 5.12000000 | 34.26022 | 0.0002 |

| SOURCE | B VALUES | T FOR HO: B = O | PROB > \|T\| | STD ERR B | STD B VALUES |
|---|---|---|---|---|---|
| INTERCEPT | 5.50833333 | 49.35952 | 0.0001 | 0.11159616 | 0.0 |
| X1 | 1.31666667 | 13.19112 | 0.0001 | 0.09981464 | 0.89492347 |
| X2 | −8.00000000 | −5.85322 | 0.0002 | 1.36676829 | −0.39709956 |

From the $(\mathbf{X'X})^{-1}$ matrix shown in the output, we obtain the diagonal elements

$$v_{00} = 0.08333333$$
$$v_{11} = 0.06666667$$
$$v_{22} = 12.50000000$$

The entry for the error source of variability in the MEAN SQUARE column of the analysis of variance table is $s^2$, an estimate of $\sigma^2$. This value is 0.14944444. Some people refer to this quantity as the mean square error (MSE). Using this estimate $s^2$, we calculate the estimated variances for $\hat{\beta}_0$, $\hat{\beta}_1$, and $\hat{\beta}_2$ to be, respectively,

$$0.08333333(0.14944444) = 0.01245370$$
$$0.06666667(0.14944444) = 0.00996296$$
$$12.50000000(0.14944444) = 1.86805550$$

An even easier way to obtain the estimated variances for the $\hat{\beta}$'s (or, equivalently, the estimated standard deviations) is to use the column labeled "STD ERR B." The entries for the intercept, $x_1$, and $x_2$ rows are the standard deviations for $\hat{\beta}_0$, $\hat{\beta}_1$, and $\hat{\beta}_2$, respectively. Note that these values are precisely the square roots of the three computed variances above.

In a similar way we can construct a test of an hypothesis concerning a model parameter $\beta_i$. Three different research hypotheses are presented along with the corresponding rejection regions. For a particular experimental situation, we must choose one of the specific alternatives.

**Test of an Hypothesis Concerning $\beta_i$**

---

$H_0$: $\beta_i = 0$.

$H_a$: 1. $\beta_i > 0$.
2. $\beta_i < 0$.
3. $\beta_i \neq 0$.

T.S.: $t = \dfrac{\hat{\beta}_i}{s\sqrt{v_{ii}}}$.

R.R.: For a given value of $\alpha$ and df $= n - (k + 1)$,

1. reject $H_0$ if $t > t_\alpha$;
2. reject $H_0$ if $t < -t_\alpha$;
3. reject $H_0$ if $t > |t_{\alpha/2}|$.

---

### EXAMPLE 8.5

Refer to the data of example 8.4 (p. 195); an experimenter was interested in determining the weight loss of a compound as a function of the amount of exposure time and the humidity of the environment during exposure. A sample of $n = 12$ observations was used to fit the model

$$y = \beta_0 + \beta_1 x_1 + \beta_2 x_2 + \epsilon$$

to obtain

$$\hat{y} = 5.50 + 1.32x_1 - 8.00x_2$$

Note that we computed the intercept to be 5.49 in example 7.15, but the computer output of example 8.4 shows the intercept to be 5.50. The difference is due to rounding errors in the calculations we did by hand.

Use these sample data to test the hypothesis that relative humidity has no linear effect on weight loss against the alternative that there is a negative linear effect; that is, as relative humidity increases, weight loss decreases.

### SOLUTION

We wish to test the following null and alternative hypotheses:

$H_0$: $\beta_2 = 0$

$H_a$: $\beta_2 < 0$

but before we do, we must compute $s^2$. Using the computer output from example 8.4, we see that the mean square error ($s^2$) is equal to .149 and hence $s = .386$.

Our statistical test is then

$H_0$: $\beta_2 = 0$.

$H_a$: $\beta_2 < 0$.

T.S.: $t = \dfrac{\hat{\beta}_2}{s\sqrt{v_{22}}} = \dfrac{-8}{.386\sqrt{12.5}} = -5.86$.

R.R.: For a one-tailed test with $\alpha = .05$ and df $= n - (k+1) = 9$, we reject $H_0$ if $t < -1.833$.

Conclusion: Since $t = -5.86$ is less than $-1.833$, we reject $H_0$: $\beta_2 = 0$ and conclude that $\beta_2 < 0$; that is, relative humidity does have a negative linear effect on weight loss of the compound.

## EXAMPLE 8.6

Refer to example 8.4. Use the SAS output to conduct a test of the null hypothesis $H_0$: $\beta_2 = 0$ against $H_a$: $\beta_2 < 0$. Give the level of significance for the test and interpret your findings.

## SOLUTION

Using the column of the analysis of variance table labeled "T FOR HO: B = O," we are given the value of the test statistic

$$t = \frac{\hat{\beta}_i}{s\sqrt{v_{ii}}}$$

for every parameter in the model. The analysis of variance portion of the computer output of example 8.4 is displayed here. The value of the test statistic for the null hypothesis $H_0$: $\beta_2 = 0$ is shown by the arrow. The value computed in example 8.5 $(-5.86)$ is different from the computer solution due to rounding errors in the hand calculations.

Using the entry from the next column of the output, labeled "PROB T," we are given the probability that $t$ is greater than the absolute value of the computed $t$, 5.853. From the computer output, the level of significance for the test is .0002. Since this probability is so small, we reject $H_0$ and conclude that $\beta_2$ is less than 0.

EXAMPLE 8.6

ANALYSIS OF VARIANCE TABLE, REGRESSION COEFFICIENTS, AND STATISTICS OF FIT FOR DEPENDENT VARIABLE Y

| SOURCE | DF | SUM OF SQUARES | MEAN SQUARE | F VALUE | PROB > F | R-SQUARE | C.V. |
|---|---|---|---|---|---|---|---|
| REGRESSION | 2 | 31.12416667 | 15.56208333 | 104.13290 | 0.0001 | 0.95857609 | 7.01810 |
| ERROR | 9 | 1.34500000 | 0.14944444 | | | STD DEV | Y MEAN |
| CORRECTED TOTAL | 11 | 32.46916667 | | | | 0.38658045 | 5.50833 |

| SOURCE | DF | SEQUENTIAL SS | F VALUE | PROB > F | PARTIAL SS | F VALUE | PROB > F |
|---|---|---|---|---|---|---|---|
| X1 | 1 | 26.00416667 | 174.00558 | 0.0001 | 26.00416667 | 174.00558 | 0.0001 |
| X2 | 1 | 5.12000000 | 34.26022 | 0.0002 | 5.12000000 | 34.26022 | 0.0002 |

| SOURCE | B VALUES | T FOR HO: B = 0 | PROB > \|T\| | STD ERR B | STD B VALUES |
|---|---|---|---|---|---|
| INTERCEPT | 5.50833333 | 49.35952 | 0.0001 | 0.11159616 | 0.0 |
| X1 | 1.31666667 | 13.19112 | 0.0001 | 0.09981464 | 0.89492347 |
| X2 | −8.00000000 | −5.85322 ⟵ | 0.0002 | 1.36676829 | −0.39709956 |

# Exercises

8.1. Recall in exercise 7.1 (p. 148) that we were interested in examining the relationship between the different concentrations of pectin (0%, 1.5%, and 3% by weight) on the firmness of canned sweet potatoes after storage in a controlled 25°C environment. The sample data for 6 cans are displayed again below.

| *y (firmness)* | 50.5 | 46.8 | 62.3 | 67.7 | 80.1 | 79.2 |
|---|---|---|---|---|---|---|
| *x (concentration of pectin)* | 0 | 0 | 1.5 | 1.5 | 3.0 | 3.0 |

a. Obtain the least squares estimates for the parameters in the model $y = \beta_0 + \beta_1 x + \epsilon$ by using the matrix approach. (Note: You may already have performed these calculations in exercise 7.1.)
b. Give the variance of $\widehat{\beta}_1$.
c. Obtain an estimate of $\sigma^2$.

8.2. Refer to exercise 8.1.
a. Give an estimate of the variance of $\widehat{\beta}_1$.
b. Perform a statistical test of the null hypothesis that there is no linear relationship between the concentration of pectin and the firmness of canned sweet potatoes after thirty days of storage at 25°C. Use $\alpha = .05$.

8.3. The extent of disease transmission can be affected greatly by the viability of infectious organisms suspended in the air. Because of the infectious nature of the disease under study, the viability of these organisms must be studied in an airtight chamber. One way to do this is to disperse an aerosol cloud, prepared from a solution containing the organisms, into the chamber. The biological recovery at any particular time is the percentage of the total number of organisms suspended in the aerosol that are viable. The data below give the biological recovery percentages computed from 13 different aerosol clouds. For each of the clouds, recovery percentages were determined at different times.

| *Cloud* | *Time, x (in minutes)* | *Biological Recovery (%)* |
|---|---|---|
| 1 | 0 | 70.6 |
| 2 | 5 | 52.0 |
| 3 | 10 | 33.4 |
| 4 | 15 | 22.0 |
| 5 | 20 | 18.3 |
| 6 | 25 | 15.1 |
| 7 | 30 | 13.0 |
| 8 | 35 | 10.0 |
| 9 | 40 | 9.1 |
| 10 | 45 | 8.3 |
| 11 | 50 | 7.9 |
| 12 | 55 | 7.7 |
| 13 | 60 | 7.7 |

a. Since the assumption of equal variance at different settings of the independent variable would probably not be satisfied for these data, some people have recommended using the logarithm of the biological recovery values in a general linear model. Give the logarithms (to the base 10) for the recovery percentages.
b. Plot the data in a scatter diagram with the vertical axis labeled $y$ (log units) and the horizontal axis labeled $x$ (time).

*8.4. Refer to exercise 8.3.
a. Using a matrix approach, fit the general linear model

$$y = \beta_0 + \beta_1 x + \beta_2 x^2 + \epsilon$$

to obtain $\hat{y}$.
b. Compute an estimate of $\sigma^2$.
c. Identify the standard deviations of $\hat{\beta}_0$, $\hat{\beta}_1$, and $\hat{\beta}_2$.

8.5. Refer to exercise 8.3. Conduct a test of the null hypothesis that $\beta_2 = 0$, that is, the log of the biological recovery percentage ($y$) is linearly related to time ($x$). Use $\alpha = .05$.

8.6. Refer to exercise 8.3. Place a 95% confidence interval on $\beta_0$, the mean log biological recovery percentage at time zero. (Note: $E(y) = \beta_0$ when $x = 0$.)

8.7. An experiment was conducted to examine the relationship between the weight gain of chickens whose diets had been supplemented by different amounts of the amino acid lysine and the amount of lysine ingested. Since the percentage of lysine is known, and we can monitor the amount of feed consumed, we can determine the amount of lysine eaten. A random sample of 12 two-week-old chickens was selected for the study. Each was caged separately and allowed to eat at will from feed composed of a base supplemented with lysine. The sample data summarizing weight gains and amounts of lysine eaten over the test period are given below. (In the data, $y$ represents weight gain, in grams, and $x$ represents the amount of lysine ingested, in grams.)

| *Chick* | $y$ | $x$ | *Chick* | $y$ | $x$ |
|---|---|---|---|---|---|
| 1 | 14.7 | .09 | 7 | 17.2 | .11 |
| 2 | 17.8 | .14 | 8 | 18.7 | .19 |
| 3 | 19.6 | .18 | 9 | 20.2 | .23 |
| 4 | 18.4 | .15 | 10 | 16.0 | .13 |
| 5 | 20.5 | .16 | 11 | 17.8 | .17 |
| 6 | 21.1 | .23 | 12 | 19.4 | .21 |

a. Plot the data in a scatter diagram.
b. Fit the general linear model $y = \beta_0 + \beta_1 x + \epsilon$ using the matrix approach.

8.8. Refer to exercise 8.7.
a. Identify the variance of $\hat{\beta}_1$.
b. Compute an estimate of $\sigma^2$.
c. Conduct a statistical test of the research hypothesis that for this diet preparation and length of study, there is a direct (positive) linear relationship between weight gain and the amount of lysine eaten.
d. Give an estimate of the standard deviation of $\hat{\beta}_1$.

8.9. Refer to exercises 8.7 and 8.8.
a. For this example would it make sense to give any physical interpretation to $\beta_0$? (Hint: The lysine was mixed in the feed.)
b. Consider an alternative model relating weight gain to amount of lysine ingested:

$$y = \beta_1 x + \epsilon$$

Fit the model (you can use a matrix approach). Compare the estimate of $\sigma^2$ for this model and the estimate obtained from exercise 8.8. How many degrees of freedom are available for estimating $\sigma^2$?

c. Which of the two models, $y = \beta_0 + \beta_1 x + \epsilon$ or $y = \beta_1 x + \epsilon$, appears to give a better fit to the sample data? (Plot the two prediction equations on a graph of the sample observations.)

# Inferences Concerning $E(y)$ 8.3

The methods of previous sections can be expanded to include inferences concerning the average value of $y$ for a given setting of the independent variables. For example, in evaluating the effects of different levels of advertising expenditure $x$ on sales $y$, it may be of interest to estimate the average sales per month for a given level of expenditure $x$. The estimate of $E(y)$ for a specific setting of $x_1, x_2, \ldots, x_k$ can be obtained by evaluating the prediction equation

$$\hat{y} = \hat{\beta}_0 + \hat{\beta}_1 x_1 + \hat{\beta}_2 x_2 + \cdots + \hat{\beta}_k x_k$$

at that setting. It can be shown that in repeated sampling at a particular setting of $x_1, x_2, \ldots, x_k$, the sampling distribution of $\hat{y}$ has a mean

mean $E(y)$

$$E(y) = \boldsymbol{\ell}'\boldsymbol{\beta} = \beta_0 + \beta_1 x_1 + \beta_2 x_2 + \cdots + \beta_k x_k$$

and a variance given by

variance $V(\hat{y})$

$$V(\hat{y}) = \boldsymbol{\ell}'(\mathbf{X}'\mathbf{X})^{-1}\boldsymbol{\ell}\sigma^2$$

where the matrix

$$\boldsymbol{\ell} = \begin{bmatrix} 1 \\ x_1 \\ x_2 \\ \vdots \\ x_k \end{bmatrix}$$

$\boldsymbol{\ell}$

displays the settings of the independent variables and where

$$\boldsymbol{\beta} = \begin{bmatrix} \beta_0 \\ \beta_1 \\ \beta_2 \\ \vdots \\ \beta_k \end{bmatrix}$$

Again assuming that the $\epsilon_i$'s are normally distributed, a $100(1-\alpha)\%$ confidence interval for $E(y)$ is given by the formula in the box.

**$100(1-\alpha)\%$ Confidence Interval for $E(y)$**

$$\hat{y} \pm t_{\alpha/2} s \sqrt{\boldsymbol{\ell}'(\mathbf{X}'\mathbf{X})^{-1}\boldsymbol{\ell}}$$

where

$$s^2 = \frac{\text{SSE}}{n-(k+1)}$$

and the $t$-value is based on $\text{df} = n-(k+1)$.

**EXAMPLE 8.7**

Use the data of example 8.2 (p. 193) to give a 90% confidence interval for the mean corn yield when 5 pounds of fertilizer are applied to a plot.

**SOLUTION**

The prediction equation in example 8.2 was

$\hat{y} = 14.7 + 1.15x_1$

where $x_1 =$ (fertilizer applied $-$ 4). For our example we need $x_1 = (5-4) = 1$, so that

$\hat{y} = 14.7 + 1.15(1) = 15.85$

The variance of $\hat{y}$ can be computed by using

$$\boldsymbol{\ell} = \begin{bmatrix} 1 \\ x_1 \end{bmatrix} = \begin{bmatrix} 1 \\ 1 \end{bmatrix} \qquad (\mathbf{X}'\mathbf{X})^{-1} = \begin{bmatrix} .1 & 0 \\ 0 & .05 \end{bmatrix}$$

$$s^2 = .71 \qquad s = .84$$

$$\boldsymbol{\ell}'(\mathbf{X}'\mathbf{X})^{-1}\boldsymbol{\ell} = [1 \quad 1]\begin{bmatrix} .1 & 0 \\ 0 & .05 \end{bmatrix}\begin{bmatrix} 1 \\ 1 \end{bmatrix} = [.1 \quad .05]\begin{bmatrix} 1 \\ 1 \end{bmatrix} = .15$$

The $t$-value in table 2 of the appendix for $a = .05$ and $\text{df} = n-(k+1) = 8$ is 1.86. Hence the appropriate confidence interval for the average corn yield per plot when 5 pounds of fertilizer are applied is

$15.85 \pm 1.86(.84)(\sqrt{.15})$ or $15.85 \pm .61$

that is, 15.24 to 16.46.

**EXAMPLE 8.8**

In example 8.7 we constructed a 90% confidence interval for the mean corn yield when 5 pounds of fertilizer are applied. Use the same sample data to construct a 90% confidence interval on $E(y)$ for any specific value of fertilizer in the range from 2 to 6. Graph your results.

### SOLUTION

Using the results from example 8.7, $\hat{y} = 14.7 + 1.15x_1$, $s = .84$,

$$\boldsymbol{\ell} = \begin{bmatrix} 1 \\ x_1 \end{bmatrix} \quad \text{and} \quad \boldsymbol{\ell}'(\mathbf{X}'\mathbf{X})^{-1}\boldsymbol{\ell} = [1 \quad x_1]\begin{bmatrix} .1 & 0 \\ 0 & .05 \end{bmatrix}\begin{bmatrix} 1 \\ x_1 \end{bmatrix} = .1 + .05x_1^2$$

Our 90% confidence interval for $E(y)$, then, is of the form $\hat{y} \pm 1.86(.84)\sqrt{.1 + .05x_1^2}$, and all we need to do is substitute a specific value of $x_1$ in this form to determine a confidence interval. Settings of $-2$, $-1$, 0, 1, and 2 correspond to 2, 3, 4, 5, and 6 pounds of fertilizer, respectively. For these values of $x_1$, the 90% confidence limits are given below.

| $x_1$ | *90% Confidence Interval* |
|---|---|
| $-2$ | 11.544 to 13.256 |
| $-1$ | 12.945 to 14.155 |
| 0 | 14.206 to 15.194 |
| 1 | 15.245 to 16.455 |
| 2 | 16.144 to 17.856 |

Plotting the endpoints of the confidence intervals and connecting the points, we get the general 90% confidence interval on $E(y)$ for any value of $x_1$ between $-2$ and $+2$. The graph in figure 8.1 displays 90% *confidence bands* for $E(y)$. Notice how the width of the confidence interval (the

confidence bands

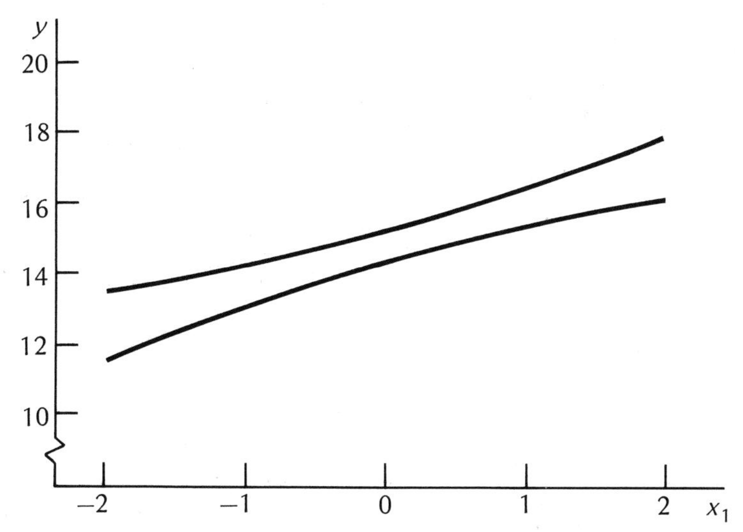


**Figure 8.1** 90% confidence band for $E(y)$ using the data of example 8.2

vertical distance between the two curves of the graph) varies for different values of $x_1$. $E(y)$ can be estimated more precisely for values of $x_1$ in the center of the experimental region. The widening of the gap between the bands at the extremities of the experimental region indicates that it would be unwise to extrapolate [try to estimate $E(y)$] beyond the region of experimentation.

A statistical test concerning $E(y)$ for a given setting of the independent variables can also be formulated; the test setup is shown in the box. Thus, for example, we might wish to test that the mean corn yield per plot is 16 when 6 pounds of fertilizer are applied.

**Test of an Hypothesis Concerning $E(y)$**

$H_0$: $E(y) = \mu_0$.

$H_a$: 1. $E(y) > \mu_0$.
2. $E(y) < \mu_0$.
3. $E(y) \neq \mu_0$.

T.S.: $t = \dfrac{\hat{y} - \mu_0}{s\sqrt{\boldsymbol{\ell}'(\mathbf{X}'\mathbf{X})^{-1}\boldsymbol{\ell}}}$.

R.R.: For a general value of $\alpha$ and df $= n - (k + 1)$,
1. reject $H_0$ if $t > t_\alpha$;
2. reject $H_0$ if $t < t_\alpha$;
3. reject $H_0$ if $t > |t_{\alpha/2}|$.

### EXAMPLE 8.9

Use the data of example 7.15 (p. 165) to test the hypothesis that the mean weight loss of a compound is 4.0 when exposed to drying for 4 hours in an exposure environment with a relative humidity of .2. Give the level of significance for your test. Interpret the test results.

### SOLUTION

Recall that

$x_1 =$ (exposure time $-$ 5.5)

$x_2 =$ (relative humidity $-$ .3)

For our example,

$x_1 = (4 - 5.5) = -1.5$

$x_2 = (.2 - .3) = -.1$

We have previously found that

$$(\mathbf{X}'\mathbf{X})^{-1} = \begin{bmatrix} .083 & 0 & 0 \\ 0 & .067 & 0 \\ 0 & 0 & 12.5 \end{bmatrix}$$

$$s^2 = .28 \qquad s = .53$$

$$\hat{y} = 5.50 + 1.32(-1.5) - 8(-.1) = 4.32$$

Using

$$\boldsymbol{\ell} = \begin{bmatrix} 1 \\ -1.5 \\ -.1 \end{bmatrix}$$

we have that

$$\boldsymbol{\ell}'(\mathbf{X}'\mathbf{X})^{-1}\boldsymbol{\ell} = [1 \; -1.5 \; -.1]\begin{bmatrix} .083 & 0 & 0 \\ 0 & .067 & 0 \\ 0 & 0 & 12.5 \end{bmatrix}\begin{bmatrix} 1 \\ -1.5 \\ -.1 \end{bmatrix} = .36$$

The appropriate statistical test is as follows:

$H_0$: $E(y) = 4$.

$H_a$: $E(y) \neq 4$.

$$\text{T.S.: } t = \frac{\hat{y} - \mu_0}{s\sqrt{\boldsymbol{\ell}'(\mathbf{X}'\mathbf{X})^{-1}\boldsymbol{\ell}}} = \frac{4.32 - 4}{.53\sqrt{.36}} = 1.01.$$

Using table 2 of the appendix, we see that for df = 9, the probability of observing a value of $t$ greater than 1.01 is more than .10. Hence the level of significance for the test is greater than .20. Since the level of significance is so large, we have insufficient evidence to indicate that the mean response differs from 4.

## Predicting $y$ for a Given Value of $x$ 8.4

In section 8.3 we were concerned with estimating the average value of $y$ for a given value of $x$. Suppose, however, that after obtaining a least squares prediction equation for the general linear model, an investigator would like to *predict* the actual value of $y$ (say the next measurement) for a given value of the independent variable x. Note that this problem differs from the problem discussed in the previous section in that we do not want to estimate the average value of $y$ for a given value of $x$, but rather we wish to predict what a particular observation will be for that same setting of $x$.

predict $y$

We still use the least squares equation $\hat{y}$ as our predictor, but the corresponding interval about the observation $y$ is called a *prediction interval*. (Prediction intervals are constructed about variables while confidence intervals are constructed about parameters.)

**General $100(1 - \alpha)\%$ Prediction Interval for $y$**

$$\hat{y} \pm t_{\alpha/2}\, s\sqrt{1 + \boldsymbol{\ell}'(\mathbf{X}'\mathbf{X})^{-1}\boldsymbol{\ell}}$$

where

$$s^2 = \frac{\text{SSE}}{n - (k + 1)}$$

and $t_{\alpha/2}$ is based on df $= n - (k + 1)$.

Note the similarity between the confidence interval for $E(y)$ and the prediction interval for the variable $y$. The only difference is that the above prediction interval has a 1 added to the quantity under the square root sign.

### EXAMPLE 8.10

Use the data of example 8.2 (p. 193) to predict the actual crop yield for a plot fertilized with 5 pounds of fertilizer. Place a 90% prediction interval about the actual value of $y$.

### SOLUTION

Using our previous work from example 8.7, the predicted value of $y$ (using $\hat{y}$) at $x_1 = 1$ is $\hat{y} = 15.85$. Also,

$$\boldsymbol{\ell}'(\mathbf{X}'\mathbf{X})^{-1}\boldsymbol{\ell} = .15$$

$$s^2 = .71 \qquad s = .84$$

The corresponding $t$-value for $\alpha = .05$ and df $= n - (k + 1) = 8$ is 1.86. Hence the 90% prediction interval is

$$15.85 \pm 1.86(.84)\sqrt{1 + .15}$$

or

$15.85 \pm 1.68$; that is, 14.17 to 17.53.

Note that the above interval is slightly wider than the corresponding interval for $E(y)$ of example 8.7. This is to be expected since here we are placing an interval about a quantity that may vary, while in example 8.7 we were placing an interval about $E(y)$, which cannot vary. Since both intervals are called 90% intervals, the prediction interval must be wider to have the same fraction of intervals (.90) covering $y$ in repeated sampling.

### EXAMPLE 8.11

a. Refer to the data of example 8.2 and construct a general 90% prediction interval for $y$ when $x_1$ takes the values $-2, -1, 0, 1$, and 2.
b. Graph your results to show a 90% prediction band for $y$ when $-2 \leq x_1 \leq 2$.
c. On the graph of part b, superimpose the graph of the 90% confidence band for $E(y)$.

### SOLUTION

a. From previous calculations, $\hat{y} = 14.7 + 1.15x_1$, $s = .84$,

$$\boldsymbol{\ell} = \begin{bmatrix} 1 \\ x_1 \end{bmatrix} \quad \text{and} \quad \boldsymbol{\ell}'(\mathbf{X}'\mathbf{X})^{-1}\boldsymbol{\ell} = .1 + .05x_1^2$$

Hence a general 90% prediction interval is of the form

$$\hat{y} \pm 1.86(.84)\sqrt{1.1 + .05x_1^2}$$

Substituting into this form the values $x_1 = -2, -1, 0, 1$, and 2, we have the intervals given on the next page.

| $x_1$ | *90% Prediction Interval* |
|---|---|
| −2 | 10.618 to 14.182 |
| −1 | 11.874 to 15.226 |
| 0 | 13.061 to 16.339 |
| 1 | 14.174 to 17.526 |
| 2 | 15.218 to 18.782 |

b., c. Plotting the endpoints of the prediction intervals and connecting the dots, we obtain the 90% prediction bands, shown by the solid lines in figure 8.2. The dotted lines indicate the corresponding 90% confidence

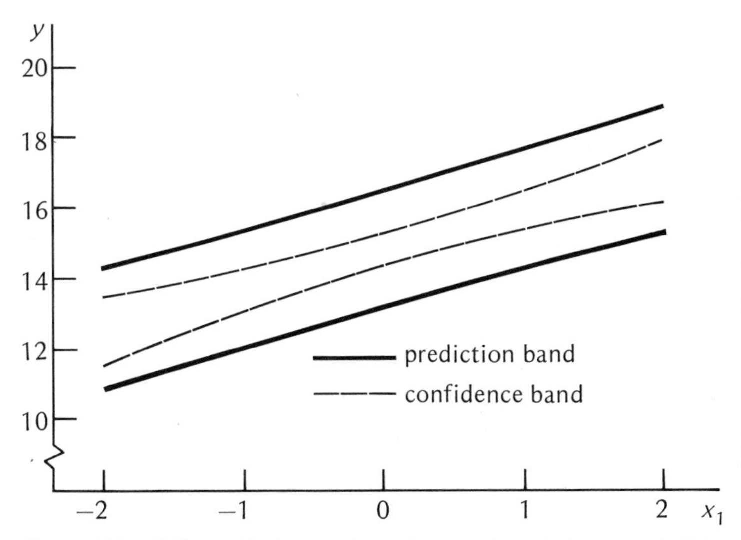


**Figure 8.2** 90% prediction and confidence bands for $y$ and $E(y)$, example 8.11

bands for $E(y)$. Notice that the prediction bands are wider than the corresponding confidence bands to allow for the fact that we are predicting the value of a random variable rather than estimating a parameter.

# Exercises

8.10. Refer to exercise 8.3 (p. 200). For the least squares equation

$$\hat{y} = \hat{\beta}_0 + \hat{\beta}_1 x + \hat{\beta}_2 x^2$$

estimate the mean log biological recovery percentage at 30 minutes, using a 95% confidence interval.

8.11. Refer to exercise 8.7 (p. 202). Estimate the mean weight gain for chickens fed on a diet supplemented with lysine if .19 grams of lysine were ingested over a study period of the same duration. Use a 95% confidence interval.

8.12. Refer to exercise 8.11. Construct a 95% prediction interval for the weight gain of a chick chosen at random and observed to ingest .19 grams of lysine. Compare your results to the confidence interval of exercise 8.11.

8.13. Using the data of exercise 8.10, construct a 95% prediction interval for the log biological recovery percentage at 30 minutes. Compare your result to the confidence interval on $E(y)$ of exercise 8.10.

*8.14. A portion of a computer output for the weight loss data of example 7.15 is shown here. Locate 95% confidence bands for $E(y)$.

```
STATISTICAL ANALYSIS SYSTEM

DATA WEIGHT;
INPUT Y 1-4 X1 6-9 X2 11-13;
CARDS;

   12 OBSERVATIONS IN DATA SET WEIGHT    3 VARIABLES

PROC PRINT;

OBS     Y      X1     X2

  1   4.3   -1.5    -.1
  2   5.5   -0.5    -.1
  3   6.8    0.5    -.1
  4   8.0    1.5    -.1
  5   4.0   -1.5     .0
  6   5.2   -0.5     .0
  7   6.6    0.5     .0
  8   7.5    1.5     .0
  9   2.0   -1.5     .1
 10   4.0   -0.5     .1
 11   5.7    0.5     .1
 12   6.5    1.5     .1

PROC REGR;
MODEL Y = X1 X2/ X I P CLM;
TITLE 'EXAMPLE 8.12';
```

```
**********************************************************************

PROC REGR : EXERCISE 8.14

DATA SET  : WEIGHT   NUMBER OF VARIABLES = 3   NUMBER OF CLASSES = 0

VARIABLES  : Y X1 X2

**********************************************************************
```

EXERCISE 8.14

ANALYSIS OF VARIANCE TABLE, REGRESSION COEFFICIENTS, AND STATISTICS OF FIT FOR DEPENDENT VARIABLE Y

| SOURCE | DF | SUM OF SQUARES | MEAN SQUARE | F VALUE | PROB > F | R-SQUARE | C.V. |
|---|---|---|---|---|---|---|---|
| REGRESSION | 2 | 31.12416667 | 15.56208333 | 104.13290 | 0.0001 | 0.95857609 | 7.01810 |
| ERROR | 9 | 1.34500000 | 0.14944444 | | | | |
| CORRECTED TOTAL | 11 | 32.46916667 | | | | STD DEV | Y MEAN |
| | | | | | | 0.38658045 | 5.50833 |

| SOURCE | DF | SEQUENTIAL SS | F VALUE | PROB > F | PARTIAL SS | F VALUE | PROB > F |
|---|---|---|---|---|---|---|---|
| X1 | 1 | 26.00416667 | 174.00558 | 0.0001 | 26.00416667 | 174.00558 | 0.0001 |
| X2 | 1 | 5.12000000 | 34.26022 | 0.0002 | 5.12000000 | 34.26022 | 0.0002 |

| SOURCE | B VALUES | T FOR HO: B = O | PROB > /T/ | STD ERR B | STD B VALUES |
|---|---|---|---|---|---|
| INTERCEPT | 5.50833333 | 49.35952 | 0.0001 | 0.11159616 | 0.0 |
| X1 | 1.31666667 | 13.19112 | 0.0001 | 0.09981464 | 0.89492347 |
| X2 | −8.00000000 | −5.85322 | 0.0002 | 1.36676829 | −0.39709956 |

| OBS NUMBER | OBSERVED VALUE | PREDICTED VALUE | RESIDUAL | LOWER 95% CL FOR MEAN | UPPER 95% CL FOR MEAN |
|---|---|---|---|---|---|
| 1 | 4.30000000 | 4.33333333 | −0.03333333 | 3.80984168 | 4.85682499 |
| 2 | 5.50000000 | 5.65000000 | −0.15000000 | 5.23518217 | 6.06481783 |
| 3 | 6.80000000 | 6.96666667 | −0.16666667 | 6.55184883 | 7.38148450 |
| 4 | 8.00000000 | 8.28333333 | −0.28333333 | 7.75984168 | 8.80682499 |
| 5 | 4.00000000 | 3.53333333 | 0.46666667 | 3.11090353 | 3.95576314 |
| 6 | 5.20000000 | 4.85000000 | 0.35000000 | 4.57345478 | 5.12654522 |
| 7 | 6.60000000 | 6.16666667 | 0.43333333 | 5.89012144 | 6.44321189 |
| 8 | 7.50000000 | 7.48333333 | 0.01666667 | 7.06090353 | 7.90576314 |
| 9 | 2.00000000 | 2.73333333 | −0.73333333 | 2.20984168 | 3.25682499 |
| 10 | 4.00000000 | 4.05000000 | −0.05000000 | 3.63518217 | 4.46481783 |
| 11 | 5.70000000 | 5.36666667 | 0.33333333 | 4.95184883 | 5.78148450 |
| 12 | 6.50000000 | 6.68333333 | −0.18333333 | 6.15984168 | 7.20682499 |

```
SUM OF RESIDUALS                                  = 0.00000000

SUM OF SQUARED RESIDUALS                          = 1.34500000

SUM OF SQUARED RESIDUALS - ERROR SS               = 0.00000000

FIRST ORDER AUTOCORRELATION OF RESIDUALS          = 0.16111679

DURBIN-WATSON D                                   = 1.65613383
```

# 8.5 The Calibration Problem: Predicting $x$ for a Given Value of $y$

Often in experimental situations we are interested in estimating the value of the independent variable corresponding to a measured value of the dependent variable. This problem will be illustrated for the case where the dependent variable $y$ is linearly related to an independent variable $x$.

Consider the calibration of an instrument that measures the flow rate of a chemical process. Let $x$ denote the actual flow rate and $y$ denote a reading on the calibrating instrument. In the calibration experiment the flow rate is controlled at $n$ levels $x_i$, and the corresponding instrument readings $y_i$ are observed. Suppose we assume a model of the form

$$y_i = \beta_0 + \beta_1 x_i + \epsilon_i$$

where the $\epsilon_i$'s are independent identically distributed normal random variables with mean zero and variance $\sigma^2$. Then using the $n$ data points $(x_i, y_i)$, we can obtain the least squares estimates $\widehat{\beta}_0$ and $\widehat{\beta}_1$ (see section 7.2). Sometime in the future the experimenter will be interested in estimating the flow rate $x$ from a particular instrument reading $y$.

The most commonly used estimate is found by replacing $\widehat{y}$ by $y$ and solving the least squares equation $\widehat{y} = \widehat{\beta}_0 + \widehat{\beta}_1 x$ for $x$:

$\widehat{x}$

$$\widehat{x} = \frac{y - \widehat{\beta}_0}{\widehat{\beta}_1}$$

Without going into any detail, expressions for the upper prediction limit (UPL) and the lower prediction limit (LPL) about the variable $x$ are given in the box.

**General 100(1 − α)% Prediction Interval for *x***

$$\text{UPL} = \bar{x} + \frac{1}{1-c^2}\left[(\hat{x}-\bar{x}) + \frac{t_{\alpha/2}s}{\hat{\beta}_1}\sqrt{\frac{n+1}{n}(1-c^2) + \frac{(\hat{x}-\bar{x})^2}{SS_{xx}}}\right]$$

$$\text{LPL} = \bar{x} + \frac{1}{1-c^2}\left[(\hat{x}-\bar{x}) - \frac{t_{\alpha/2}s}{\hat{\beta}_1}\sqrt{\frac{n+1}{n}(1-c^2) + \frac{(\hat{x}-\bar{x})^2}{SS_{xx}}}\right]$$

where

$$s^2 = \frac{\text{SSE}}{n-2} \qquad c^2 = \frac{t^2_{\alpha/2}s^2}{\hat{\beta}_1^2 SS_{xx}}$$

and $t_{\alpha/2}$ is based on $n - 2$ degrees of freedom.

$c^2$

It should be noted that since

$$t = \frac{\hat{\beta}_1}{s/\sqrt{SS_{xx}}}$$

is the test statistic for $H_0$: $\beta_1 = 0$, then $c = t_{\alpha/2}/t$. We will require that $|t| > t_{\alpha/2}$; that is, $\beta_1$ must be significantly different from zero. Then $c^2 < 1$ and $0 < (1 - c^2) < 1$. The greater the strength of the linear relationship between $x$ and $y$, the larger is the quantity $(1 - c^2)$, making the width of the prediction interval narrower. Note also that we will get a better prediction of $x$ when $\hat{x}$ is closer to the center of the experimental region, as measured by $\bar{x}$. Combining a prediction at an endpoint of the experimental region with a weak linear relationship between $x$ and $y$ ($t \approx t_{\alpha/2}$ and $c^2 < 1$) can create extremely wide limits for the prediction of $x$.

**EXAMPLE 8.12**

Let us suppose that an experimenter is interested in calibrating a flow rate meter. He fixes the process at $n = 10$ different flow rates $x$ and observes the corresponding meter readings $y$. Use the data given in table 8.2 to

**Table 8.2** Data for the calibration problem of example 8.12

| *Flow Rate, x* | *Instrument Reading, y* |
|---|---|
| 1 | 1.4 |
| 2 | 2.3 |
| 3 | 3.1 |
| 4 | 4.2 |
| 5 | 5.1 |
| 6 | 5.8 |
| 7 | 6.8 |
| 8 | 7.6 |
| 9 | 8.7 |
| 10 | 9.5 |

calibrate the flow rate instrument, assuming a model of the form $y = \beta_0 + \beta_1 x + \epsilon$.

**SOLUTION**

To obtain the calibration curve $\hat{y}$, we need the following calculations for the least squares estimates $\hat{\beta}_0$ and $\hat{\beta}_1$ (using the algebraic approach of chapter 7).

$$\sum_{i=1}^{10} x_i = 1 + 2 + \cdots + 10 = 55$$

$$\sum_{i=1}^{10} x_i^2 = 1^2 + 2^2 + \cdots + 10^2 = 385$$

$$\sum_{i=1}^{10} x_i y_i = 1(1.4) + 2(2.3) + \cdots + 10(9.5) = 374.1$$

$$\sum_{i=1}^{10} y_i = 1.4 + 2.3 + \cdots + 9.5 = 54.5$$

$$\sum_{i=1}^{10} y_i^2 = (1.4)^2 + (2.3)^2 + \cdots + (9.5)^2 = 364.09$$

Using these data, we find the following:

$$SS_{xx} = \sum x^2 - \frac{\left(\sum x\right)^2}{10} = 385 - \frac{(55)^2}{10} = 82.5$$

$$SS_{xy} = \sum xy - \frac{\left(\sum x\right)\left(\sum y\right)}{10} = 374.1 - \frac{(55)(54.5)}{10} = 74.35$$

$$\bar{x} = \frac{\sum x}{10} = 5.5 \qquad \bar{y} = \frac{\sum y}{10} = 5.45$$

Then we obtain*

$$\hat{\beta}_1 = \frac{SS_{xy}}{SS_{xx}} = \frac{74.35}{82.5} = .9012$$

$$\hat{\beta}_0 = \bar{y} - \hat{\beta}_1 \bar{x} = 5.45 - .9012(5.5) = .4934$$

The corresponding least squares prediction equation (here called the calibration curve) is

$$\hat{y} = .4934 + .9012x$$

*We are carrying more decimal places than usual in our hand calculations to avoid rounding errors.

The estimate of $x$ for a given value of $y$ is

$$\hat{x} = \frac{y - \hat{\beta}_0}{\hat{\beta}_1} = \frac{y - .4934}{.9012}$$

Substituting an instrument reading of $y = 4.0$, we estimate the actual flow rate to be

$$\hat{x} = \frac{4.0 - .4934}{.9012} = 3.8910$$

**EXAMPLE 8.13**

Use the data of example 8.12 to place a 95% prediction interval on $x$, the actual flow rate corresponding to an instrument reading of 4.0.

**SOLUTION**

Although we did not use the matrix approach to find the least squares estimates of $\beta_0$ and $\beta_1$, we can still apply the results of section 8.2 to find an estimate of $\sigma^2$. Recall that

$$\text{SSE} = \mathbf{Y'Y} - \hat{\boldsymbol{\beta}}'\mathbf{X'Y}$$

where

$$\mathbf{Y'Y} = \sum y^2 = 364.09 \qquad \text{and} \qquad \hat{\boldsymbol{\beta}} = \begin{bmatrix} .4934 \\ .9012 \end{bmatrix}$$

The $\mathbf{X}$ matrix for this example is

$$\mathbf{X} = \begin{bmatrix} 1 & 1 \\ 1 & 2 \\ 1 & 3 \\ 1 & 4 \\ 1 & 5 \\ 1 & 6 \\ 1 & 7 \\ 1 & 8 \\ 1 & 9 \\ 1 & 10 \end{bmatrix}$$

and $\mathbf{X'Y}$ is

$$\begin{bmatrix} 1 & 1 & 1 & 1 & 1 & 1 & 1 & 1 & 1 & 1 \\ 1 & 2 & 3 & 4 & 5 & 6 & 7 & 8 & 9 & 10 \end{bmatrix} \begin{bmatrix} 1.4 \\ 2.3 \\ 3.1 \\ 4.2 \\ 5.1 \\ 5.8 \\ 6.8 \\ 7.6 \\ 8.7 \\ 9.5 \end{bmatrix} = \begin{bmatrix} 54.5 \\ 374.1 \end{bmatrix}$$

The expression $\boldsymbol{\beta}'\mathbf{X}'\mathbf{Y}$ is

$$[.4934 \quad .9012]\begin{bmatrix} 54.5 \\ 374.1 \end{bmatrix} = 364.0292$$

and

$$\text{SSE} = \mathbf{Y}'\mathbf{Y} - \hat{\boldsymbol{\beta}}'\mathbf{X}'\mathbf{Y} = 364.09 - 364.0292 = .0608$$

The estimate of $\sigma^2$ for $k = 1$ is based on $n - (k + 1) = 8$ degrees of freedom.

$$s^2 = \frac{\text{SSE}}{n - (k + 1)} = \frac{.0608}{8} = .0076$$

$$s = .0872$$

For $\alpha = .05$, the $t$-value for df $= 8$ and a $= .05$ is 2.306.

$$c^2 = \frac{t^2_{\alpha/2}\, s^2}{\hat{\beta}_1^2 \text{SS}_{xx}} = \frac{(2.306)^2(.0076)}{(.9012)^2(82.5)} = .0006$$

and $1 - c^2 = .9994$. Then the upper and lower prediction limits for $x$ when $y = 4.0$ are as follows:

$$\text{UPL} = 5.5 + \frac{1}{.9994}\left[-1.6090 + \frac{2.306(.0872)}{.9012}\sqrt{\frac{11}{10}(.9994) + \frac{(-1.6090)^2}{82.5}}\right]$$

$$= 5.5 + \frac{1}{.9994}(-1.6090 + .2373) = 4.1274$$

$$\text{LPL} = 5.5 + \frac{1}{.9994}(-1.6090 - .2373) = 3.6526$$

Thus the 95% prediction limits for $x$ are 3.65 to 4.13. These limits are shown in figure 8.3.

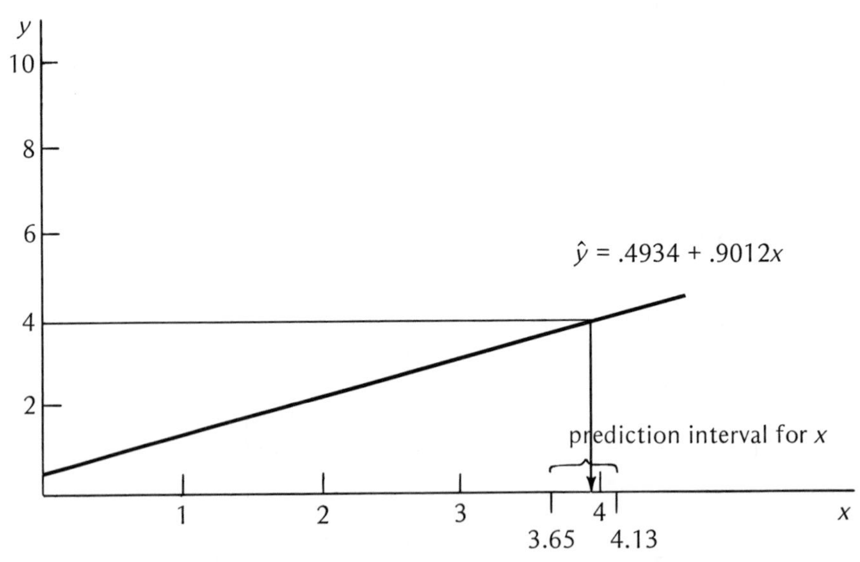


**Figure 8.3** 95% prediction interval for $x$ when $y = 4.0$, example 8.13

# Exercises

8.15. A particular forester has become adept at estimating the volume (in cubic feet) of trees on a particular site prior to a timber sale. Since his operation has now expanded, he would like to train another person to assist in estimating the cubic-foot volume of trees. He decides to calibrate his assistant's estimations of actual tree volume. The forester selects a random sample of trees soon to be felled. For each tree the assistant is to guess the cubic-foot volume $y$. In addition, the forester obtains the actual cubic-foot volume $x$ after the tree has been chopped down. From these data the forester obtains the calibration curve for the model

$$y = \beta_0 + \beta_1 x + \epsilon$$

Then in the near future he can use the calibration curve to correct the assistant's estimates of tree volumes. The sample data are summarized below.

| *Tree* | 1 | 2 | 3 | 4 | 5 | 6 | 7 | 8 | 9 | 10 |
|---|---|---|---|---|---|---|---|---|---|---|
| *Guessed Volume, y* | 12 | 14 | 8 | 12 | 17 | 16 | 14 | 14 | 15 | 17 |
| *Actual Volume, x* | 13 | 14 | 9 | 15 | 19 | 20 | 16 | 15 | 17 | 18 |

Fit the calibration curve using the method of least squares. Does the evidence indicate that the slope is significantly greater than zero? Use $\alpha = .05$.

8.16. Refer to exercise 8.15.

a. Predict the actual tree volume for a tree the assistant estimates to have a cubic-foot volume of 13.

b. Place a 95% prediction interval on $x$, the actual tree volume in part a.

8.17. A company was interested in calibrating an instrument to measure the consistency of the liquid in a paper mill. In the making of paper, small particles of wood fiber are conveyed in a liquid flow to the screening process, where the pulp is separated from the water, dried, and made into paper. Controlling the consistency of the fibers in the liquid flow represents an important step in maintaining the final quality of the paper. Let us assume that for the calibration of the consistency meter, we are able to control the consistency of a simulated laboratory liquid flow at various levels $x$ while monitoring the meter reading $y$. Assuming a model of the form

$$y = \beta_0 + \beta_1 x + \epsilon$$

use the sample data below to fit the model.

| Meter Reading, $y$ | 2.16 | 2.15 | 2.17 | 2.26 | 2.35 | 2.39 | 2.42 | 2.51 |
|---|---|---|---|---|---|---|---|---|
| Actual Consistency, $x$ | 2.16 | 2.17 | 2.18 | 2.20 | 2.22 | 2.24 | 2.26 | 2.28 |

Is there sufficient evidence to indicate that the slope is greater than zero? Use $\alpha = .05$.

*8.18. A computer output for the data of exercise 8.17 is shown here.
a. Compare the least squares estimates to those obtained in exercise 8.17.
b. Give the level of significance for a test of $H_0$: $\beta_1 = 0$.
c. Identify 95% prediction limits for the independent variable (actual consistency of the flow) when the observed meter reading is 2.20.

```
  2.16,2.16
? 2.17,2.15
? 2.18,2.17
? 2.20,2.26
? 2.22,2.35
? 2.24,2.39
? 2.26,2.42
? 2.28,2.51
WANT A LISTING OF THIS FILE (Y OR N) ? Y
WANT TRANSFORMATION OF INDEPENDENT VARIABLE
(0=NO, 1=NATURAL LOG, 2=COMMON LOG) ? 0
WANT TRANSFORMATION ON DEPENDENT VARIABLE
(0=NO, 1=NATURAL LOG, 2=COMMON LOG, 3=ARCSIN, 4=SQUARE ROOT) ? 0


SIMPLE LINEAR REGRESSION

        FILES:                      X                        Y

OBSERVATIONS:                  INDEPENDENT               DEPENDENT

                                 2.1600                   2.1600
                                 2.1700                   2.1500
                                 2.1800                   2.1700
                                 2.2000                   2.2600
                                 2.2200                   2.3500
                                 2.2400                   2.3900
                                 2.2600                   2.4200
                                 2.2800                   2.5100
```

| | INDEPENDENT | | DEPENDENT |
|---|---|---|---|
| NO. OBSERVATIONS | 8.0000 | | 8.0000 |
| ARITHMETIC MEANS | 2.2138 | | 2.3013 |
| STANDARD DEVIATION | 0.0437 | | 0.1361 |

| | Y-INTERCEPT | | SLOPE |
|---|---|---|---|
| ESTIMATE | −4.5054 | | 3.0747 |
| STANDARD DEVIATION | 0.4367 | | 0.1973 |
| T-STATISTIC--SIGNIFICANCE | −10.3159 | | 15.5877 |
| DEGREES OF FREEDOM | 6.0000 | | 6.0000 |
| PROBABILITY | 0.0000 | | 0.0000 |
| RESIDUAL SUM OF SQUARES | | 0.0031 | |
| RESIDUAL MEAN SQUARE | | 0.0005 | |

```
WANT TO ESTIMATE EFFECTIVE DOSE (Y OR N)? Y
ENTER NO. OF ESTIMATES? 1
ENTER Y VALUES
? 2.20
```

EFFECTIVE DOSE ESTIMATES

| Y VALUE | ESTIMATE | 95% CONFIDENCE INTERVAL |
|---|---|---|
| 2.2000 | 2.1808 | [2.1597, 2.2002] |

# 8.6 Correlation

In this section we will extend our study of the relationships between two or more variables. Not only might we like to predict the value of one variable (the dependent variable) based on information on one or more independent variables, as we have done in previous sections, but we might also wish to provide a measure of the strength of the relationship between these variables. This idea will be the topic of this section.

correlation coefficient

One measure of the strength of the relationship between two variables $x$ and $y$ is called the *coefficient of linear correlation,* or, simply, the *correlation coefficient.* Given $n$ pairs of observations $(x_i, y_i)$, we can compute the *sample correlation coefficient* $r$ as

sample correlation coefficient $r$

$$r = \frac{SS_{xy}}{\sqrt{SS_{xx}SS_{yy}}}$$

where

$SS_{yy}$

$$SS_{yy} = \sum y^2 - \frac{\left(\sum y\right)^2}{n}$$

You will immediately note the similarity between $r$ and the slope of the least squares equation

$$\hat{y} = \hat{\beta}_0 + \hat{\beta}_1 x$$

relating $y$ to $x$.

$$\hat{\beta}_1 = \frac{SS_{xy}}{SS_{xx}}$$

$$r = \sqrt{\frac{SS_{xx}}{SS_{yy}}}\,\hat{\beta}_1$$

For experimental situations where not all $x$'s and $y$'s are the same, both $SS_{xx}$ and $SS_{yy}$ are positive. Then $r$ and $\hat{\beta}_1$ have the same sign. Because of the relationship between $r$ and $\hat{\beta}_1$, the sample correlation coefficient $r$ measures the strength of the linear relationship between $x$ and $y$ and is used to estimate the corresponding population coefficient of linear correlation $\rho$.

---

**Properties of *r***

1. $r$ lies between $-1$ and $+1$. $r > 0$ indicates a positive linear relationship and $r < 0$ a negative linear relationship between $x$ and $y$. $r = 0$ indicates no linear relationship between $x$ and $y$. (See figure 8.4.)
2. $r^2$ gives the proportion of the total variability of the $y$-values that can be accounted for by the independent variable $x$.
3. Sample correlation coefficients are not additive. Thus if the sample correlation coefficient between $y$ and $x_1$ is $r_1$ and $r_2$ is the sample correlation coefficient between $y$ and another independent variable $x_2$, the proportion of the total variability accounted for by $x_1$ and $x_2$ is not $r_1^2 + r_2^2$. Indeed, the two independent variables $x_1$ and $x_2$ may be highly correlated and contribute the same information for predicting $y$.

---

While we will not give proofs for these properties, we will illustrate the second property of $r^2$ for the linear model $y = \beta_0 + \beta_1 x + \epsilon$. As noted in

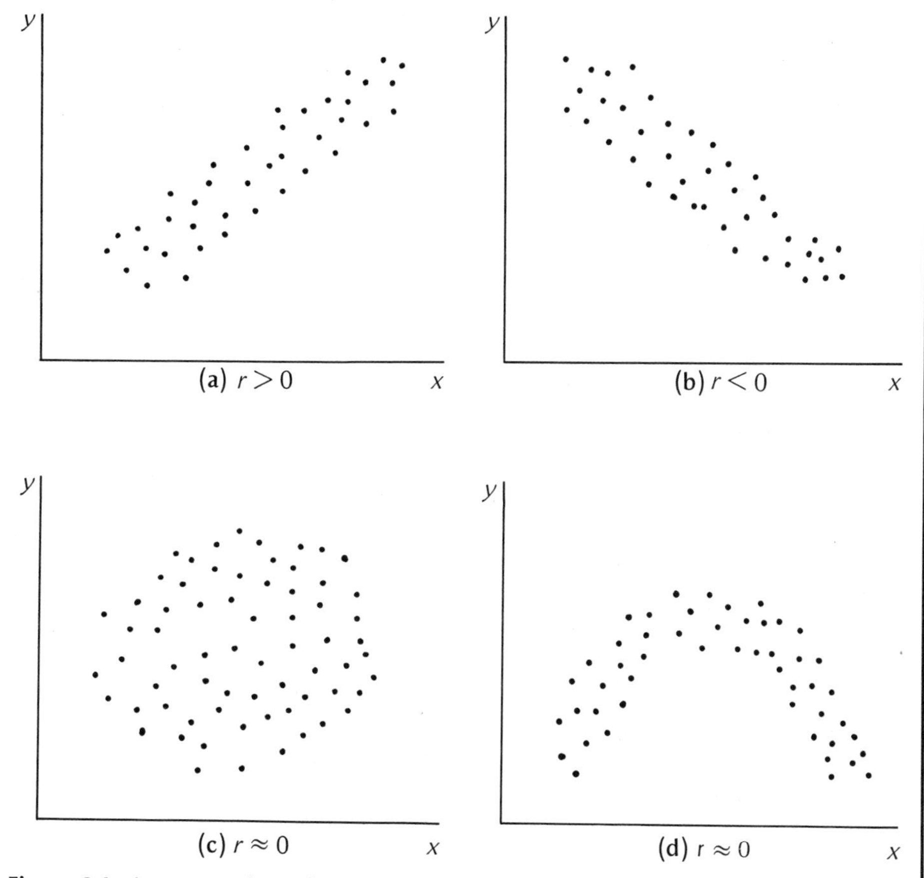


**Figure 8.4** Interpretation of $r$

chapter 7 the total variability of the $y$-values about their mean $\bar{y}$ can be expressed as

$$\sum (y - \bar{y})^2 = \sum (y - \hat{y})^2 + \sum (\hat{y} - \bar{y})^2$$

where $\Sigma\ (\hat{y} - \bar{y})^2$ is that portion of the total variability that can be accounted for by the independent variable $x$ and $\Sigma\ (y - \hat{y})$ is the sum of squares for error (SSE). Using a computational formula, we can rewrite the total sum of squares $SS_{yy}$ as

$$SS_{yy} = \sum (y - \bar{y})^2 = \sum y^2 - \frac{\left(\sum y\right)^2}{n}$$

Similarly, it can be shown that $\Sigma\,(y - \hat{y})^2$ for a linear model can be written as

$$\text{SSE} = \mathbf{Y'Y} - \boldsymbol{\beta}'\mathbf{X'Y} = \sum y^2 - \left[\frac{\left(\sum y\right)^2}{n} + \frac{\text{SS}_{xy}^2}{\text{SS}_{xx}}\right]$$

$$= \sum y^2 - \frac{\left(\sum y\right)^2}{n} - \frac{\text{SS}_{xy}^2}{\text{SS}_{xx}} = \text{SS}_{yy} - \frac{\text{SS}_{xy}^2}{\text{SS}_{xx}}$$

and by subtraction,

$$\sum (\hat{y} - \bar{y})^2 = \frac{\text{SS}_{xy}^2}{\text{SS}_{xx}}$$

Then expressing both SSE and $\Sigma\,(y - \bar{y})^2$ as a proportion of $\text{SS}_{yy} = \Sigma\,(y - \bar{y})^2$, we have

$r^2$

$$\frac{\text{SSE}}{\text{SS}_{yy}} = 1 - \frac{\text{SS}_{xy}^2}{\text{SS}_{xx}\text{SS}_{yy}} = 1 - r^2$$

$$\frac{\sum (\hat{y} - \bar{y})^2}{\text{SS}_{yy}} = \frac{\text{SS}_{xy}^2}{\text{SS}_{xx}\text{SS}_{yy}} = r^2$$

Thus $r^2$ represents that proportion of the total variability of the $y$-values that is accounted for by the independent variable $x$. Similarly, $1 - r^2$ represents that proportion of the total variability of the $y$-values that is not accounted for by the variable $x$.

**EXAMPLE 8.14**

Use the data of example 8.12 (p. 213) to calculate the coefficient of linear correlation between the actual flow rate $x$ and the instrument reading $y$.

**SOLUTION**

Recall from the solution to example 8.12 that $\text{SS}_{xx} = 82.5$ and $\text{SS}_{xy} = 74.35$. We can also quickly calculate $\text{SS}_{yy}$.

$$\text{SS}_{yy} = \sum y^2 - \frac{\left(\sum y\right)^2}{n} = 364.09 - \frac{(54.5)^2}{10} = 67.065$$

Combining, we have

$$r = \frac{SS_{xy}}{\sqrt{SS_{xx}SS_{yy}}} = \frac{74.35}{\sqrt{82.5(67.065)}} = .9996$$

That is, there appears to be a very strong positive linear relationship between the flow rate $x$ and the instrument reading $y$.

If we let $r$ denote the sample estimate of the corresponding population correlation coefficient $\rho$, we might wish to test the null hypothesis

$H_0$: $\rho = 0$ (i.e., no linear correlation between $x$ and $y$)

However, this hypothesis is equivalent to $H_0$: $\beta_1 = 0$ and hence can be tested using the methods of section 8.2.

As we mentioned previously, the sample correlation coefficients are not additive. If we assume a model of the form

$$y = \beta_0 + \beta_1 x_1 + \beta_2 x_2 + \cdots + \beta_k x_k + \epsilon$$

then using the sample data we can calculate the *coefficient of determination* $R^2$, which represents the proportion of the total variability among the $y$-values that is accounted for by the independent variables $x_1, x_2, \ldots, x_k$.

coefficient of determination $R^2$

---

**Computation of $R^2$**

$$R^2 = \frac{SS_{yy} - SSE}{SS_{yy}} \qquad 0 \leq R^2 \leq 1$$

where

$$SS_{yy} = \sum y^2 - \frac{\left(\sum y\right)^2}{n}$$

---

The square root $R$ of the coefficient of determination is designated as the *multiple correlation coefficient* and represents a measure of the strength of the relationship between $y$ and the independent variables $x_1, x_2, \ldots, x_k$. If we let

multiple correlation coefficient $R$

$$\hat{y} = \hat{\beta}_0 + \hat{\beta}_1 x_1 + \hat{\beta}_2 x_2 + \cdots + \hat{\beta}_k x_k$$

then the multiple correlation coefficient $R$ is merely the coefficient of linear correlation between the actual response $y$ and the predicted response $\hat{y}$. Thus from previous work we could compute $R$ using the formula

$$R = \frac{SS_{\hat{y}y}}{\sqrt{SS_{\hat{y}\hat{y}}SS_{yy}}}$$

where

$$SS_{\hat{y}y} = \sum \hat{y}y - \frac{\left(\sum \hat{y}\right)\left(\sum y\right)}{n}$$

$$SS_{\hat{y}\hat{y}} = \sum \hat{y}^2 - \frac{\left(\sum \hat{y}\right)^2}{n}$$

**EXAMPLE 8.15**

An experiment was conducted to investigate the effect of temperature and humidity on the yield of a production process. The sample data are summarized in matrix notation for the model

$$y = \beta_0 + \beta_1x_1 + \beta_2x_1^2 + \beta_3x_2 + \beta_4x_1x_2 + \epsilon$$

where $x_1$ and $x_2$ are coded values for temperature and humidity, respectively.

$$\mathbf{Y} = \begin{bmatrix} 65 \\ 78 \\ 52 \\ 70 \\ 77 \\ 83 \end{bmatrix} \quad \mathbf{X} = \begin{bmatrix} 1 & -1 & 1 & -1 & 1 \\ 1 & 0 & 0 & -1 & 0 \\ 1 & 1 & 1 & -1 & -1 \\ 1 & -1 & 1 & 1 & -1 \\ 1 & 0 & 0 & 1 & 0 \\ 1 & 1 & 1 & 1 & 1 \end{bmatrix}$$

$$\mathbf{X'X} = \begin{bmatrix} 6 & 0 & 4 & 0 & 0 \\ 0 & 4 & 0 & 0 & 0 \\ 4 & 0 & 4 & 0 & 0 \\ 0 & 0 & 0 & 6 & 0 \\ 0 & 0 & 0 & 0 & 4 \end{bmatrix}$$

Use these data to obtain the least squares prediction equation and then calculate $R^2$, the coefficient of determination.

**SOLUTION**

Applying the results of the exercises in chapter 7, we can obtain $(\mathbf{X'X})^{-1}$ by inverting the $3 \times 3$ and $2 \times 2$ matrices

$$\begin{bmatrix} 6 & 0 & 4 \\ 0 & 4 & 0 \\ 4 & 0 & 4 \end{bmatrix} \quad \text{and} \quad \begin{bmatrix} 6 & 0 \\ 0 & 4 \end{bmatrix}$$

Note that these two matrices are shown by the partitioning of **X′X** above. For the 3 × 3 matrix, the determinant is 32, and, after a few calculations, we find the cofactor matrix to be

$$\begin{bmatrix} 16 & 0 & -16 \\ 0 & 8 & 0 \\ -16 & 0 & 24 \end{bmatrix}$$

Hence the inverse of the 3 × 3 matrix is

$$\begin{bmatrix} \frac{1}{2} & 0 & -\frac{1}{2} \\ 0 & \frac{1}{4} & 0 \\ -\frac{1}{2} & 0 & \frac{3}{4} \end{bmatrix}$$

Similarly, we find the inverse of the 2 × 2 matrix to be

$$\begin{bmatrix} \frac{1}{6} & 0 \\ 0 & \frac{1}{4} \end{bmatrix}$$

Combining these we have

$$(\mathbf{X'X})^{-1} = \left[\begin{array}{ccc|cc} \frac{1}{2} & 0 & -\frac{1}{2} & & \\ 0 & \frac{1}{4} & 0 & & \\ -\frac{1}{2} & 0 & \frac{3}{4} & & \\ \hline & & & \frac{1}{6} & 0 \\ & & & 0 & \frac{1}{4} \end{array}\right]$$

The **X′Y** matrix is

$$\mathbf{X'Y} = \begin{bmatrix} 425 \\ 0 \\ 270 \\ 35 \\ 26 \end{bmatrix}$$

and hence the least squares estimates of $\beta_0, \ldots, \beta_4$ are
$\hat{\boldsymbol{\beta}} = (\mathbf{X'X})^{-1}\mathbf{X'Y}$

$$= \left[\begin{array}{ccc|cc} \frac{1}{2} & 0 & -\frac{1}{2} & & \\ 0 & \frac{1}{4} & 0 & & \\ -\frac{1}{2} & 0 & \frac{3}{4} & & \\ \hline & & & \frac{1}{6} & 0 \\ & & & 0 & \frac{1}{4} \end{array}\right] \begin{bmatrix} 425 \\ 0 \\ 270 \\ 35 \\ 26 \end{bmatrix}$$

$$= \begin{bmatrix} 77.5 \\ 0 \\ -10 \\ 5.83 \\ 6.50 \end{bmatrix}$$

To compute $R^2$ we must first calculate SSE and $SS_{yy}$. From the sample data,

$\mathbf{Y'Y} = \sum y^2 = 30{,}731$

$$\hat{\boldsymbol{\beta}}'\mathbf{X}'\mathbf{Y} = [77.5 \quad 0 \quad -10 \quad 5.83 \quad 6.5]\begin{bmatrix} 425 \\ 0 \\ 270 \\ 35 \\ 26 \end{bmatrix} = 30{,}610.55$$

Then

$$\text{SSE} = \mathbf{Y}'\mathbf{Y} - \hat{\boldsymbol{\beta}}'\mathbf{X}'\mathbf{Y} = 30{,}731 - 30{,}610.55 = 120.45$$

$$\text{SS}_{yy} = \sum y^2 - \frac{\left(\sum y\right)^2}{n} = 30{,}731 - \frac{(425)^2}{6}$$

$$= 30{,}731 - 30{,}104.17 = 626.83$$

Hence

$$R^2 = \frac{\text{SS}_{yy} - \text{SSE}}{\text{SS}_{yy}} = \frac{626.83 - 120.45}{626.83} = .81$$

That is, the independent variables $x_1$, $x_1^2$, $x_2$, and $x_1x_2$ have accounted for 81% of the total variability in the $y$-values.

In some experimental situations the investigator may wish to measure the strength of the relationship between the response $y$ and one independent variable while holding the other independent variables constant. For example, suppose we were investigating the effects of two independent variables

$x_1$ = index of new businesses

$x_2$ = (gross national product − inventories)

on the response

$y$ = net increase in the number of main telephones over a quarterly period

partial correlation

At some point in the investigation, we might want to measure the strength of the relationship between $y$ and $x_1$, adjusting for the other independent variable $x_2$. One such measure that allows us to do this is the *partial correlation* between $y$ and $x_1$. This is the correlation between $y$ and $x_1$ while holding $x_2$ constant, and it is denoted by $r_{y1.2}$. Similarly, we might wish to examine the partial correlation of $y$ and $x_2$ adjusted for $x_1$, denoted by $r_{y2.1}$. These quantities can be computed as shown in the box.

**Computation of Some Partial Correlations**

$$r_{y1.2} = \frac{r_{y1} - r_{y2}r_{12}}{\sqrt{(1 - r_{y2}^2)(1 - r_{12}^2)}}$$

where $r_{y1}$ is the sample correlation coefficient between $y$ and $x_1$ and $r_{12}$ is the sample correlation coefficient between $x_1$ and $x_2$. Similarly,

$$r_{y2.1} = \frac{r_{y2} - r_{y1}r_{12}}{\sqrt{(1 - r_{y1}^2)(1 - r_{12}^2)}}$$

**EXAMPLE 8.16**

Suppose that sample data for the business survey just discussed yield the following sample correlation coefficients:

$r_{y1} = .83 \qquad r_{y2} = .56 \qquad r_{12} = .67$

Determine the partial correlation coefficients $r_{y1.2}$ and $r_{y2.1}$.

**SOLUTION**

Using the computational formulas,

$$r_{y1.2} = \frac{r_{y1} - r_{y2}r_{12}}{\sqrt{(1 - r_{y2}^2)(1 - r_{12}^2)}} = \frac{.83 - (.56)(.67)}{\sqrt{[1 - (.56)^2][1 - (.67)^2]}}$$

$$= \frac{.83 - .375}{\sqrt{(.686)(.551)}} = \frac{.455}{.615} = .740$$

$$r_{y2.1} = \frac{r_{y2} - r_{y1}r_{12}}{\sqrt{(1 - r_{y1}^2)(1 - r_{12}^2)}} = \frac{.56 - (.83)(.67)}{\sqrt{[1 - (.83)^2][1 - (.67)^2]}}$$

$$= \frac{.56 - .556}{\sqrt{(.311)(.551)}} = \frac{.004}{.414} = .010$$

Thus for the sample data, the coefficient of linear correlation between $y$ and $x_1$ is .83 when $x_2$ is ignored and is .74 when controlling for levels of $x_2$. In either case there appears to be a positive linear relationship between $y$ and $x_1$. In contrast, there appears to be a positive linear relationship between $y$ and $x_2$ when $x_1$ is ignored but no linear relationship between $y$ and $x_2$ when we control for levels of $x_1$.

We might mention that the partial correlation coefficients just discussed can also be computed by fitting models. For example, to obtain the partial correlation coefficient between $y$ and $x_2$, adjusting for the independent variable $x_1$, we first fit the model

$$y = \beta_0 + \beta_1 x_1 + \epsilon$$

to obtain

$$\hat{y} = \hat{\beta}_0 + \hat{\beta}_1 x_1.$$

We also fit the model

$$x_2 = \alpha_0 + \alpha_1 x_1 + \varepsilon$$

to obtain the prediction equation

$$\hat{x}_2 = \hat{\alpha}_0 + \hat{\alpha}_1 x_1$$

If we compute the residuals $z_y = y - \hat{y}$ and $z_{x_2} = x_2 - \hat{x}_2$, the partial correlation between $y$ and $x_2$ after adjusting for $x_1$ is the sample correlation coefficient between the residuals $z_y$ and $z_{x_2}$. Thus using the formula on page 219,

$$r_{y2.1} = \frac{SS_{z_y z_{x2}}}{\sqrt{SS_{z_y z_y} SS_{z_{x2} z_{x2}}}}$$

We can extend these concepts to any number of independent variables. For example, to obtain the partial correlation between $y$ and $x_1$ adjusting for $x_2$ and $x_3$, $r_{y1.23}$, we would fit the models

$$y = \beta_0 + \beta_1 x_2 + \beta_2 x_3 + \varepsilon$$

and

$$x_1 = \alpha_0 + \alpha_1 x_2 + \alpha_2 x_3 + \varepsilon$$

If the residuals $z_y$ and $z_{x_1}$ are computed for these two models, the sample correlation coefficient between the residuals is the partial correlation coefficient between $y$ and $x_1$ adjusting for $x_2$ and $x_3$.

## 8.7 Summary

In this chapter we introduced some of the inferential techniques for dealing with the general linear model. First we discussed inferences concerning $\beta_i$, a parameter in the general linear model. These results made use of the fact that under certain assumptions (see p. 191) the variance of $\widehat{\beta}_i$ is $\sigma^2 v_{ii}$, where $v_{ii}$ is the $i$th diagonal element of $(\mathbf{X}'\mathbf{X})^{-1}$. An estimate of $\sigma^2$ was developed using SSE, the sum of the squared errors $y - \widehat{y}$. We also considered inferences concerning a linear function of the model parameters, namely, $E(y)$. A statistical test and an estimation procedure were presented.

The next important problem discussed in this chapter concerned the prediction of a future value of the response $y$ for given settings of the independent variables. Here we were not interested in estimating the average value of $y$ for fixed settings of $x_1, x_2, \ldots, x_k$ but rather in predicting a particular observation. The form of the prediction interval was similar to the confidence interval for $E(y)$ and used the estimate $y$.

While most estimation or prediction problems are directed towards inferences about $E(y)$ or $y$ for given settings of the independent variables, the calibration problem presents a situation for predicting the independent variable $x$ from the fitted model of a simple linear regression. The prediction of $x$ and an appropriate prediction interval were discussed.

We then presented the basic concepts involved in measuring the strength of the relationship between two or more variables. The coefficient of linear correlation $r$ can be computed from a set of $n$ pairs of observations $(x_i, y_i)$, and it measures the strength of the linear relationship between the two variables. The concept of correlation can be extended to more than two variables if $y$ is related to a set of $k$ independent variables in a general linear model. The coefficient of determination $R^2$ represents the proportion of the total variability in the $y$-values accounted for by the $k$ independent variables. The square root of $R^2$ is called the multiple correlation coefficient, but it can be viewed as the simple correlation coefficient between a response $y$ and the predicted response $\widehat{y}$ for the $n$ data points.

Finally, we examined the situation where the experimenter would like to measure the correlation between a response $y$ and an independent variable $x$ while holding other independent variables constant. Computational formulas for these partial correlation coefficients were presented and their use was discussed.

# Exercises

*8.19. Examine the computer output for the data of example 8.15.

a. Give the least squares estimates for the model

$$y = \beta_0 + \beta_1 x_1 + \beta_2 x_2 + \beta_3 x_1^2 + \beta_4 x_1 x_2 + \epsilon$$

b. Identify $R^2$ and compare it with the answer of example 8.15.

ANALYSIS OF VARIANCE TABLE, REGRESSION COEFFICIENTS, AND STATISTICS OF FIT FOR DEPENDENT VARIABLE Y

| SOURCE | DF | SUM OF SQUARES | MEAN SQUARE | F VALUE | PROB > F | R-SQUARE | C.V. |
|---|---|---|---|---|---|---|---|
| REGRESSION | 4 | 506.50000000 | 126.62500000 | 1.05229 | 0.3870 | 0.80802978 | 15.48657 |
| ERROR | 1 | 120.33333333 | 120.33333333 | | | STD DEV | Y MEAN |
| CORRECTED TOTAL | 5 | 626.83333333 | | | | 10.96965511 | 70.83333 |

| SOURCE | DF | SEQUENTIAL SS | F VALUE | PROB > F | PARTIAL SS | F VALUE | PROB > F |
|---|---|---|---|---|---|---|---|
| X1 | 1 | 0.0 | 0.0 | 1.0000 | 0.0 | 0.0 | 1.0000 |
| X1SQ | 1 | 133.33333333 | 1.10803 | 0.4837 | 133.33333333 | 1.10803 | 0.4837 |
| X2 | 1 | 204.16666667 | 1.69668 | 0.4168 | 204.16666667 | 1.69668 | 0.4168 |
| X1X2 | 1 | 169.00000000 | 1.40443 | 0.4462 | 169.00000000 | 1.40443 | 0.4462 |

| SOURCE | B VALUES | T FOR HO: B = 0 | PROB > \|T\| | STD ERR B | STD B VALUES |
|---|---|---|---|---|---|
| INTERCEPT | 77.50000000 | 9.99134 | 0.0635 | 7.75671752 | 0.0 |
| X1 | 0.0 | 0.0 | 1.0000 | 5.48482756 | 0.0 |
| X1SQ | −10.00000000 | −1.05263 | 0.4837 | 9.50000000 | −0.46120428 |
| X2 | 5.83333333 | 1.30257 | 0.4168 | 4.47834295 | 0.57071118 |
| X1X2 | 6.50000000 | 1.18509 | 0.4462 | 5.48482756 | 0.51923901 |

c. Test the null hypothesis $H_0$: $\beta_4 = 0$. Give the level of significance for your test.

d. Compare the computer output to the calculations of example 8.15.

8.20. An instructor believes that true-false tests are as effective as problem-type tests in judging a student's proficiency in mathematics. A test consisting of half true-false questions and half problems was given to 10 calculus students selected at random. The test score results were as follows:

| *Student* | 1 | 2 | 3 | 4 | 5 | 6 | 7 | 8 | 9 | 10 |
|---|---|---|---|---|---|---|---|---|---|---|
| *T-F* | 48 | 40 | 25 | 10 | 16 | 21 | 23 | 19 | 35 | 32 |
| *Problems* | 45 | 47 | 20 | 12 | 12 | 15 | 25 | 16 | 30 | 32 |

a. Calculate the coefficient of linear correlation for the two sets of test scores.

b. Conduct a test of the research hypothesis that there is a positive linear correlation between the two sets of scores. Give the level of significance for the test. Interpret your experimental findings.

*8.21. An experiment was conducted to investigate the amplitude of the shock wave recorded on sensors placed at different distances from an explosive charge. The charge is to be detonated underground, with 3 sensors placed at each of the three different distances from the charge, as illustrated in figure 8.5. The shock wave amplitudes are recorded and summarized

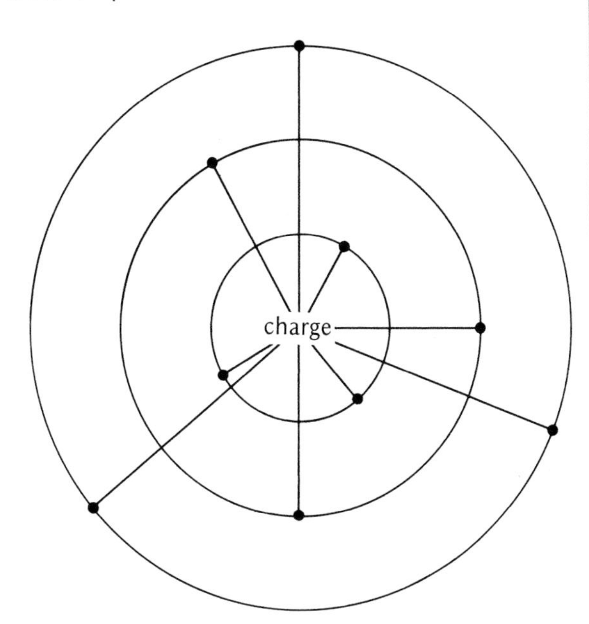


**Figure 8.5** Location of sensors from the charge for the shock wave experiment of exercise 8.21

according to the distance from the explosion. These data are given below.

| Distance, $x$ | 5 | 5 | 5 | 10 | 10 | 10 | 15 | 15 | 15 |
|---|---|---|---|---|---|---|---|---|---|
| Amplitude, $y$ | 8.6 | 8.2 | 8.1 | 5.8 | 6.2 | 6.1 | 5.2 | 4.8 | 4.7 |

a. Use these data and the matrix approach to fit the model

$$y = \beta_0 + \beta_1 x + \beta_2 x^2 + \epsilon$$

b. Compute and interpret the coefficient of determination.

8.22. Refer to exercise 8.21. Test the research hypothesis that $\beta_2 \neq 0$ (i.e., there is a significant quadratic effect for amplitude as a function of the distance from the charge). Give the level of significance for your test and interpret the results of your test.

8.23. Refer to exercise 8.15 (p. 217); a forester was interested in training an assistant to guess the timber volume of a standing tree. Having trained him, the forester calibrated his guessing against known timber volumes. Perhaps a better way to quantify the assistant's guess would be to base it on an objective reading, such as the basal area of the tree. If, indeed, volume is related to basal area, the assistant would have an objective way to estimate the timber volume of a tree. A random sample of 12 trees was obtained. For each tree included in the sample, the basal area $x$ was recorded along with the cubic-foot volume after the tree was felled. These data appear below.

| *Tree* | 1 | 2 | 3 | 4 | 5 | 6 | 7 | 8 | 9 | 10 | 11 | 12 |
|---|---|---|---|---|---|---|---|---|---|---|---|---|
| *Basal Area, x* | .3 | .5 | .4 | .9 | .7 | .2 | .6 | .5 | .8 | .4 | .8 | .6 |
| *Volume, y* | 6 | 9 | 7 | 19 | 15 | 5 | 12 | 9 | 20 | 9 | 18 | 13 |

a. Fit the linear model $y = \beta_0 + \beta_1 x + \epsilon$.
b. Compute and interpret the correlation coefficient between basal area and timber volume.

8.24. Refer to exercise 8.23.
a. Is there evidence to indicate a positive linear relationship between $x$ and $y$? Use $\alpha = .05$.
b. Predict the timber volume of a tree with a basal area of .75 using a 95% prediction interval.

8.25. An equal number of families from 8 different cities of various size were questioned on how much money they spend for food, clothing, and housing per year. The city sizes and average family responses are summarized below. (Both city size and expenditure are in 1000s.)

| *City Size* | 30 | 50 | 75 | 100 | 150 | 200 | 175 | 120 |
|---|---|---|---|---|---|---|---|---|
| *Expenditure* | 65 | 77 | 79 | 80 | 82 | 90 | 84 | 81 |

a. Plot the data.
b. Compute the coefficient of linear correlation.
c. Is there a significant positive linear relationship between the two variables? Use $\alpha = .05$.

*8.26. Refer to example 8.15 (p. 224).
a. Fit the models $y = \beta_0 + \beta_1 x_1 + \varepsilon$ and $x_2 = \alpha_0 + \alpha_1 x_1 + \varepsilon$
b. Use the residuals $z_y$ and $z_{x_2}$ from these models to fit the model

$$z_y = \beta_0 + \beta_1 z_{x_2} + \varepsilon$$

8.27. Refer to exercise 8.26.

a. Compute $r_{y1}$, $r_{y2}$, and $r_{12}$.

b. Use the results of part a to compute $r_{y1.2}$ and $r_{y2.1}$.

c. Note that $r_{y2.1}$ of part b should agree with $R$ of exercise 8.26, part b.

*8.28. Refer to exercise 8.17 (p. 217). Compute the coefficient of linear correlation between $y$ and $\hat{y}$. Compare your answer to $R$. (Hint: They should be the same except for rounding errors.)

8.29. Verify that SSE for the model $y = \beta_0 + \beta_1 x + \epsilon$ can be written as

$$\sum y^2 - \frac{\left(\sum y\right)^2}{n} - \frac{SS_{xy}^2}{SS_{xx}}$$

# References

Barr, A.J.; Goodnight, J.H.; Sall, J.P.; and Helwig, J.T. 1976. *A user's guide to SAS 76*. Raleigh, N.C.: SAS Institute, Inc.

Dixon, W.J. 1975. *BMDP, biomedical computer programs*. Berkeley: University of California Press.

Dixon, W.J., and Massey, F.J., Jr. 1969. *Introduction to statistical analysis*. 3d ed. New York: McGraw-Hill.

Draper, N.R., and Smith, H. 1966. *Applied regression analysis*. New York: Wiley.

Graybill, F.A. 1976. *Theory and application of the linear model*. N. Scituate, Mass.: Duxbury Press.

Harnett, D.L. 1970. *Introduction to statistical methods*. Reading, Mass.: Addison-Wesley.

Kirk, R.E. 1968. *Experimental design: Procedures for the behavioral sciences*. Belmont, Calif.: Brooks/Cole.

Mendenhall, W. 1968. *Introduction to linear models and the design and analysis of experiments*. Belmont, Calif.: Wadsworth.

Nie, N.H.; Hull, C.H.; Jenkins, J.G.; Steinbrenner, K.; and Bent, D.H. 1975. *Statistical package for the social sciences*. 2d ed. New York: McGraw-Hill.

Ostle, B. 1963. *Statistics in research*. 2d ed. Ames, Iowa: Iowa State University Press.

Pearson, E.S., and Hartley, H.O. 1966. *Biometrika tables for statisticians*. 3d ed. Vol. I. London: Cambridge University Press.

Service, J. 1972. *A user's guide to the statistical analysis system*. Raleigh, N.C.: Student Supply Stores, North Carolina State University.

Snedecor, G.W., and Cochran, W.G. 1967. *Statistical methods.* 6th ed. Ames, Iowa: Iowa State University Press.

Tanur, J.M.; Mosteller, F.; Kruskal, W.H.; Pieters, R.S.; and Rising, G.R. 1972. *Statistics: A guide to the unknown.* San Francisco: Holden-Day.

Winer, B.J. 1962. *Statistical principles in experimental design.* New York: McGraw-Hill.

# Elements of Experimental Design 9

## Introduction 9.1

The design of an experiment is the process of planning an experiment. Some of you may at some time in your professional career be required to design an experiment, to analyze data from someone else's designed experiment, or to do both. In this chapter we will present a brief preview of the design of an experiment. Information related to analyses for some standard experimental designs is presented in chapters 13 and 15.

In planning an experiment, typically we seek a design that will yield a specific amount of information at a minimum cost to the investigator. The statement of an objective is the first step in the design of an experiment. The second step is a statement of the amount of information required about the parameter(s) of interest. The third and fourth steps involve the selection of the experimental plan (called the *experimental design*) and the estimation or test procedure that gives the required information at minimum cost.

first step
second step
third step
fourth step

Suppose, for example, that the objective of an experiment is to estimate the mean monthly pay rate $\mu$ for clerical workers in moderate-sized (250,000 to 1 million people) metropolitan areas throughout the country. The statement of information about $\mu$ could be made by using a bound on the error of estimation; that is, we could require that our estimate be within $B$ units of $\mu$. Note that $B$ can be any value we choose. Hence we might want to estimate the mean pay rate for clerical personnel to within \$1.00. Here $B = 1.00$.

If the sample mean is used as a point estimate of $\mu$, then the bound on the error of estimation is $2\sigma/\sqrt{n}$. Finally, the sample size required to estimate $\mu$ to within $B$ units is found by solving the equation

$$\frac{2\sigma}{\sqrt{n}} = B$$

for $n$.

Since observations cost money, there is a price associated with executing this experimental plan and inference-making procedure. If there are alternative plans for achieving the same experimental objectives, we should consider the price associated with each one. Generally we would choose the design that produces the desired result at minimum cost. Alternatively, when an investigation must be conducted for a fixed cost (perhaps due to limited resources), we should determine the experimental design and inferential procedure that give the most information for the fixed cost constraint.

We begin our discussion of the design of experiments with a few necessary definitions.

---

**Definition 9.1** An *experimental unit* is the person or object on which a measurement is made.

---

Each person in a recent sample of 1000 motorists was asked his or her opinion concerning the proposed closing (for repairs) of a major access viaduct over the next two years. In this survey an individual motorist is an experimental unit; we obtain a measurement (1 or 0) depending on whether a person favors the closing or does not.

In monitoring the quality of radio transistors moving along a production line, an engineer has been instructed to inspect every twentieth transistor to determine whether the production is meeting specifications. An individual transistor would be the experimental unit in this sampling scheme.

---

**Definition 9.2** A factor level, or combination of factor levels, applied to an experimental unit is called a *treatment*.

---

In previous chapters we examined the effect of one or more quantitative variables on a dependent variable $y$. In doing so we experimented at various levels or combinations of levels for the different quantitative variables. For example, in examining the effect of weight loss $y$ of a compound when exposed to different drying times and different relative humidities (example 7.15, p. 165), the experimental unit was a measured amount of the compound. We then examined weight loss for 12 combinations of settings for the independent variables. Each of these factor-level combinations was then randomly assigned to an experimental unit to obtain weight loss. In this example the 12 factor-level combinations represent 12 different treatments that can be applied to experimental units.

factor-level combinations

It should be noted that we can also talk about levels and factor-level combinations of qualitative variables. For example, if three analysts are to run assays on a compound, the qualitative variable "analysts" has three levels, analyst 1, analyst 2, and analyst 3. Each of these levels would represent a treatment in an experiment to evaluate the effects of these three analysts on the assayed levels of a compound. Similarly, in comparing the gasoline mileage associated with four different ingredients (1, 2, 3, and 4) and two different car models (A and B), we would be examining the effects of factor-level combinations of the two qualitative variables "ingredient" and "car model" on the response "gasoline mileage."

To summarize, the design of an experiment consists of the steps outlined in the box.

**Steps in the Design of an Experiment**

1. Statement of the experimental objectives.
2. Statement of the amount of information required concerning the parameter(s) of interest.
3. Selection of an experimental design—which includes selection of the treatments to be studied and the plan for assigning treatments to experimental units.
4. Selection of an inference-making procedure and the sample size to achieve the objectives of the study.

Fortunately we will not have to consider alternative inference-making procedures for every experimental situation. In some cases the experimental design and the estimation or test procedure will be fixed. Hence the design of the experiment will consist of stating the objectives, specifying the amount of information required concerning any parameter(s) of interest, and determining the sample size necessary to achieve the desired amount of information. Such will be the case in section 9.2, where we will discuss briefly the sample size requirements for designing an experiment to estimate the parameters $\mu$ and $\mu_1 - \mu_2$ or to test an hypothesis about them. Then in sections 9.3 through 9.5 we will consider several extensions to more complicated plans for assigning treatments to experimental units.

# 9.2 Determining the Quantity of Information in the Sample Relevant to Parameter(s)

The experimental objective (step 1) in this section will be to estimate the parameters $\mu$ and $\mu_1 - \mu_2$ or to test an hypothesis about them. In order to state the amount of information required concerning these parameters (step 2) and the sample size necessary to achieve the objectives of the study (step 4), we must determine the quantity of information in an experiment related to a population parameter. In this section we will consider some ways to determine the quantity of information in an experiment relative to the parameter(s) of interest and how we can design an experiment to obtain this quantity of information.

Consider the general problem of estimating a parameter $\theta$, using a point estimate $\hat{\theta}$. If the sampling distribution of the point estimates is approximately normal, with mean $\theta$ and standard deviation $\sigma_{\hat{\theta}}$, then a rough measure of the information available for estimating $\theta$ is given by $2\sigma_{\hat{\theta}}$, the bound on the error of estimation. If we are concerned about obtaining a point estimate of the parameter $\theta$, we might specify that the bound on the error of estimation be equal to some value $B$. Since $\sigma_{\hat{\theta}}$ will, in general, depend on $n$, the sample size necessary to achieve this bound on the error can be found by solving the equation

solving for $n$

$$2\sigma_{\hat{\theta}} = B$$

for $n$. Two specific cases of interest are summarized in table 9.1.

**Table 9.1** Sample sizes required to obtain point estimates for $\mu$ and $\mu_1 - \mu_2$

| $\theta$ | $\hat{\theta}$ | $\sigma_{\hat{\theta}}$ | *Sample Size* |
|---|---|---|---|
| $\mu$ | $\bar{y}$ | $\frac{\sigma}{\sqrt{n}}$ | $n = \frac{4\sigma^2}{B^2}$ |
| $\mu_1 - \mu_2$ | $\bar{y}_1 - \bar{y}_2$ | $\sqrt{\frac{\sigma_1^2}{n} + \frac{\sigma_2^2}{n}}$ | $n = \frac{4(\sigma_1^2 + \sigma_2^2)}{B^2}$ |

If we are interested in estimating a population mean correct to within $B$ units by using the point estimate $\bar{y}$, the required sample size is found by solving the equation

$$\frac{2\sigma}{\sqrt{n}} = B$$

for $n$ (see table 9.1). Similarly, in estimating the difference between two population means $\mu_1 - \mu_2$, using $\bar{y}_1 - \bar{y}_2$, we can determine the sample size $n$ that must be drawn from each population to estimate $\mu_1 - \mu_2$ to within $B$ units by solving the equation

$$2\sqrt{\frac{\sigma_1^2}{n} + \frac{\sigma_2^2}{n}} = B$$

for $n$.

**EXAMPLE 9.1**

A production manager is interested in estimating the difference in mean breaking strengths for two different types of metal casings. Previous experimentation indicates that the range in breaking strengths for each type of casing is approximately 60 pounds per square inch. Determine the number of casings of each type that must be sampled to estimate $\mu_1 - \mu_2$ correct to within 5 pounds per square inch. Assume that we will use independent random samples of equal size.

**SOLUTION**

Since the range in breaking strengths for the two types of casings is the same, we can assume that the population variances, $\sigma_1^2$ and $\sigma_2^2$, are approximately equal. If we let $\sigma^2$ denote the common variance for the two populations, we can form a range estimate of $\sigma$ by using

$$\hat{\sigma} = \frac{\text{range}}{4} = \frac{60}{4} = 15$$

Substituting into the sample size formula, with $\sigma_1^2 = \sigma_2^2 = \sigma^2$ replaced by $\hat{\sigma}^2$, we find

$$n = \frac{4(\sigma_1^2 + \sigma_2^2)}{B^2}$$

$$\approx \frac{4(15^2 + 15^2)}{25} = \frac{4(450)}{25} = 72$$

Thus we must examine 72 metal casings of each type to estimate $\mu_1 - \mu_2$ with a bound on the error of estimation equal to 5 pounds.

If a sample of 72 observations per type represents too much of an expenditure, we could obtain a smaller sample size (and hence a lower cost) by increasing the desired bound on the error of estimation. Thus for $B = 10$ the required sample size is 18 (just one fourth of the original requirement).

The problem of determining the amount of information related to a parameter $\theta$ in interval estimation is a simple extension of the procedure just discussed for point estimation. We again let $\hat{\theta}$ be a point estimate of the parameter $\theta$, and we assume that the sampling distribution of the point estimates is approximately normal, with mean $\theta$ and standard deviation $\sigma_{\hat{\theta}}$. The confidence interval for $\theta$ with confidence coefficient $(1 - \alpha)$ is

$$\hat{\theta} \pm z_{\alpha/2}\,\sigma_{\hat{\theta}}$$

half width

One measure of the amount of information relevant to the parameter $\theta$ is the *half width* $z_{\alpha/2}\,\sigma_{\hat{\theta}}$ of the confidence interval. The sample size required to estimate a parameter $\theta$ by using a confidence interval with half width $B$ and confidence coefficient $(1 - \alpha)$ is found by solving

$$z_{\alpha/2}\,\sigma_{\hat{\theta}} = B$$

for $n$. Table 9.2 summarizes the results of interval estimation for the parameters $\mu$ and $\mu_1 - \mu_2$.

**Table 9.2** Sample sizes required to obtain an interval estimate for $\mu$ and $\mu_1 - \mu_2$

| $\theta$ | $\hat{\theta}$ | *Confidence Interval* | *Sample Size* |
|---|---|---|---|
| $\mu$ | $\bar{y}$ | $\bar{y} \pm z_{\alpha/2}\dfrac{\sigma}{\sqrt{n}}$ | $n = \dfrac{z^2_{\alpha/2}\,\sigma^2}{B^2}$ |
| $\mu_1 - \mu_2$ | $\bar{y}_1 - \bar{y}_2$ | $(\bar{y}_1 - \bar{y}_2) \pm z_{\alpha/2}\sqrt{\dfrac{\sigma_1^2}{n_1} + \dfrac{\sigma_2^2}{n_2}}$ | $n = \dfrac{z^2_{\alpha/2}(\sigma_1^2 + \sigma_2^2)}{B^2}$ |

**EXAMPLE 9.2**

A biologist would like to estimate the effect of an antibiotic on the growth of a particular bacterium by examining the mean amount of bacteria present per plate of culture when a fixed amount of the antibiotic is applied. Previous experimentation with the antibiotic on this type of bacterium indicates that the standard deviation of the amount of bacteria present is approximately 13 square centimeters. Use this information to determine the number of observations (cultures that must be developed and then tested) to estimate the mean amount of bacteria present, using a 99% confidence interval with a half width of 3 square centimeters.

**SOLUTION**

We use the sample size formula for $\mu$ from table 9.2, with $B = 3$, $\sigma^2 \approx 13^2 = 169$, and $1 - \alpha = .99$. The appropriate $z$-value for $\alpha = .01$ is 2.58. (To refresh your memory about the selection of $z$-values, refer back to section 3.4.) Thus

$$n = \frac{z_{\alpha/2}^2 \sigma^2}{B^2} \approx \frac{(2.58)^2(169)}{9} = 124.99 \approx 125$$

The biologist must grow and test 125 bacterial cultures to obtain the desired accuracy. Again note that we could decrease the sample size either by increasing the desired half width of the confidence interval $B$ or by decreasing the confidence coefficient $(1 - \alpha)$. Thus for a 90% confidence interval with the same half width, we would need

$$n = \frac{(1.645)^2(169)}{9} = 50.8 \approx 51$$

The quantity of information available for testing an hypothesis about a parameter $\theta$ is measured by the magnitudes of the type I error $\alpha$ and the type II error $\beta$ for a specified alternative hypothesis. Suppose that we are interested in testing the null hypothesis

$$H_0\colon \theta = \theta_0$$

against the one-sided alternative

$$H_a\colon \theta > \theta_0$$

In addition, assume that we wish the probability of a type I error to be $\alpha$ and the probability of a type II error to be $\beta$ or less when the actual value of $\theta$ lies a distance of $\Delta$ (Greek letter delta) or more above $\theta_0$. Then provided the distribution of $\hat{\theta}$ is approximately normal with mean $\theta_0$ and standard deviation $\sigma_{\hat{\theta}}$ under $H_0$, the required sample size is quite easy to compute. Table 9.3 gives the sample sizes for testing $\mu$ or $\mu_1 - \mu_2$.

$\Delta$

**Table 9.3** Sample sizes required for one-sided tests of $\mu$ and $\mu_1 - \mu_2$

| $\theta$ | $H_0$ | $\Delta$ | *Test Statistic* | *Sample Size* |
|---|---|---|---|---|
| $\mu$ | $\mu = \mu_0$ | $\lvert \mu - \mu_0 \rvert$ | $z = \dfrac{\bar{y} - \mu_0}{\sigma/\sqrt{n}}$ | $n = \dfrac{\sigma^2(z_\alpha + z_\beta)^2}{\Delta^2}$ |
| $\mu_1 - \mu_2$ | $\mu_1 - \mu_2 = D_0$ | $\lvert (\mu_1 - \mu_2) - D_0 \rvert$ | $z = \dfrac{(\bar{y}_1 - \bar{y}_2) - D_0}{\sqrt{(\sigma_1^2/n) + (\sigma_2^2/n)}}$ | $n = \dfrac{(\sigma_1^2 + \sigma_2^2)(z_\alpha + z_\beta)^2}{\Delta^2}$ |

The same sample size formula applies to the one-sided alternative $H_a\colon \theta < \theta_0$, with the exception that we would want the probability of a type II error to be of magnitude $\beta$ or less when the actual value of $\theta$ lies a distance of $\Delta$ or more below $\theta_0$.

**EXAMPLE 9.3**

A cereal packager is concerned that one of his machines has a mean fill per package of more than 16 ounces, the labeled net weight. While this is not bad from a public relations standpoint, it could cost the packager a great deal of money. Previous experience suggests that the standard deviation of the package fill weights is approximately .225. For

$H_0$: $\mu = 16$

$H_a$: $\mu > 16$

with $\alpha = .05$, determine the sample size required to make $\beta = .01$ or less if the actual mean is 16.1 ounces or more. By putting this restriction on $\beta$, the packager is saying that he wants a very small probability of falsely accepting $H_0$: $\mu = 16$ when in fact the actual mean is 16.1 ounces or more.

**SOLUTION**

From previous data the fill weights have a standard deviation approximately equal to .225. The appropriate $z$-values, $z_{.05}$ and $z_{.01}$, for $\alpha = .05$ and $\beta = .01$ are 1.645 and 2.33, respectively. Using $\Delta = 16.1 - 16 = .1$, the required sample size is

$$n = \frac{(.225)^2(1.645 + 2.33)^2}{(.1)^2} = 79.99 \approx 80$$

That is, the packager must obtain a random sample of $n = 80$ cartons to conduct this test under the conditions specified.

Suppose that after obtaining the sample, the computed value of

$$z = \frac{\bar{y} - 16}{\sigma_{\bar{y}}}$$

does not fall in the rejection region. What is our conclusion? In similar situations in previous chapters, our conclusions would have been that there was insufficient evidence to reject $H_0$. Now, however, knowing that $\beta \leq .01$ when $\mu \geq 16.1$, we would feel safe in our conclusion to accept $H_0$: $\mu = 16$. No further testing would be required.

With a slight modification of the sample size formula for the one-tailed tests, we can test

$H_0$: $\theta = \theta_0$

$H_a$: $\theta \neq \theta_0$

for a specified $\alpha$ and $\beta$ with $\Delta = |\theta - \theta_0|$. Formulas for approximate sample sizes when testing $\mu$ and $\mu_1 - \mu_2$ are presented in table 9.4 on the next page.

**EXAMPLE 9.4**

Consider the data of example 9.3; we were interested in testing $H_0$: $\mu = 16$. If the alternative is $H_a$: $\mu \neq 16$, determine the sample size required to test

**Table 9.4** Approximate sample sizes required for two-sided tests of $\mu$ and $\mu_1 - \mu_2$

| $\theta$ | $H_0$ | $\Delta$ | Test Statistic | Sample Size* |
|---|---|---|---|---|
| $\mu$ | $\mu = \mu_0$ | $\lvert\mu - \mu_0\rvert$ | $z = \dfrac{\bar{y} - \mu_0}{\sigma/\sqrt{n}}$ | $n \approx \dfrac{\sigma^2}{\Delta^2}(z_{\alpha/2} + z_\beta)^2$ |
| $\mu_1 - \mu_2$ | $\mu_1 - \mu_2 = D_0$ | $\lvert(\mu_1 - \mu_2) - D_0\rvert$ | $z = \dfrac{(\bar{y}_1 - \bar{y}_2) - D_0}{\sqrt{(\sigma_1^2/n) + (\sigma_2^2/n)}}$ | $n \approx \dfrac{(\sigma_1^2 + \sigma_2^2)}{\Delta^2}(z_{\alpha/2} + z_\beta)^2$ |

*The sample sizes obtained from these formulas will be conservative in the sense that they will be larger than we actually need for the desired value of $\beta$. However, the exact calculation of $n$ is more complicated and will be disregarded in favor of the approximations listed above.

$H_0$ with $\alpha = .05$ and $\beta \leq .01$ when the actual value of $\mu$ lies more than .1 units away from $\mu_0 = 16$.

**SOLUTION**

The sample results in example 9.3 showed $z_{.01} = 2.33$, and from previous work we know $z_{.025} = 1.96$. Using the formula for $n$ from table 9.4, an approximate sample size is

$$n \approx \frac{\sigma^2}{\Delta^2}(z_{\alpha/2} + z_\beta)^2 = \frac{(.225)^2}{(.1)^2}(1.96 + 2.33)^2 = 93.17$$

Thus, by rounding to the next highest integer, we need approximately $n = 94$ observations to conduct our statistical test. Note that the sample size obtained is only an approximation, but we do know that the computed value of $n$ tends to overestimate the required sample size.

We have discussed sample size requirements for specific inference-making procedures when the data were obtained from independent random samples. In the next section we will consider several different experimental designs. In practice, we would also have to consider accompanying inference-making procedures. However, a discussion of these procedures will be delayed until chapters 13 through 15.

# Exercises

9.1. Investigators would like to estimate the average annual taxable income of apartment dwellers in a city to within \$500. If we assume the annual incomes range from \$0 to \$20,000, determine the number of observations that should be included in the sample.

9.2. Refer to exercise 9.1. Determine the required sample size if the desired bound on the error of estimation is \$1000.

9.3. Previously we discussed an experiment to determine the effect on dairy cattle of a diet supplemented with liquid whey. While no differences were noted in milk production measures among cattle given a standard diet (7.5 kg of grain plus hay by choice) with water and those on the standard diet and liquid whey only, a considerable difference between the groups was noted in the amount of hay ingested. Suppose that prior to conducting this experiment, we wished to test the null hypothesis of no difference in mean hay consumption for the two diet groups of dairy cattle.
   a. For $\alpha = .05$, determine the approximate number of dairy cattle that should be included in each group if we want $\beta = .10$ for $|\mu_1 - \mu_2| > .5$. Previous experimentation has shown $\sigma$ to be approximately .8.
   b. Suppose we only wish to detect $\mu_1 - \mu_2 > 0$. Determine the sample size required per group for $\alpha = .05$ and $\beta = .10$ when $\mu_1 - \mu_2 > .5$.

$ 9.4. Union officials are concerned about reports of inferior wages paid to employees of a company under their jurisdiction. It is decided to obtain a random sample of wage sheets from the company to estimate the average hourly wage. If it is known that wages within the company have a range of $10, determine the sample size required to estimate the average hourly wage to within $0.50.

9.5. A study is conducted to compare the average number of years of service at the age of retirement for military personnel in 1965 versus 1975. A random sample of career records is to be obtained for each year. Previous information suggests that the common population standard deviation is approximately 4 years. Determine the number of observations to be collected from records for each year if we wish to estimate the mean difference in age using a 95% confidence interval with a half width of .5 years.

# 9.3 Completely Randomized Design, Randomized Blocks, and Latin Squares

completely randomized design

t treatments or populations

So far we have considered experimental designs for estimating a population mean or the difference between two population means. The *completely randomized design* extends these results to a comparison of $t (t \geq 2)$ population (treatment) means, $\mu_1, \mu_2, \ldots, \mu_t$. Here we assume that there are $t$ different populations from which we are to draw independent random samples of size

$n_1, n_2, \ldots, n_t$, respectively. Or, in the terminology of the design of experiments, we assume that there are $n_1 + n_2 + \cdots + n_t$ homogeneous experimental units. Then treatments are randomly allocated to the experimental units in such a way that $n_1$ units receive treatment 1, $n_2$ receive treatment 2, and so on. The objective of the experiment will be to make inferences about the corresponding treatment means. Although we will delay a discussion of the method for constructing a test concerning the $t$ treatment means until chapter 13, we will note here that one possible way to test these means would be to run certain pairwise tests using the $t$ test for comparing two means.

Consider the following example. A horticultural laboratory is interested in examining the leaves of apple trees to detect nutritional deficiencies using three different laboratory procedures. In particular, the laboratory would like to determine if there is a difference in mean assay readings for apple leaves utilizing three different laboratory procedures (A, B, C). The experimental units in this investigation are apple tree leaves and the treatments are the three levels of the qualitative variable "laboratory procedure." If a single analyst takes a random sample of nine leaves from the same tree, randomly assigns three leaves to each of the three procedures, and assays the leaves using the assigned treatment, we could use the three sample means to estimate the corresponding mean leaf nutritional deficiency for the three laboratory test procedures. We will show later in chapter 13 how to run a test of the hypothesis that all three treatment means are identical. The design used for this investigation is a completely randomized design with three observations for each treatment.

The completely randomized design has several advantages and disadvantages when used as an experimental design for comparing $t$ treatment means. These will become more apparent when we consider the analysis for a completely randomized design in chapter 13.

**Advantages and Disadvantages of a Completely Randomized Design**

Advantages

1. It is extremely easy to construct the design.
2. The design is easy to analyze even though the sample sizes might not be the same for each treatment (see chapter 15).
3. The design can be used for any number of treatments.

Disadvantages

1. Although the completely randomized design can be used for any number of treatments, it is best suited for situations where there are relatively few treatments.

2. The experimental units to which treatments are applied must be homogeneous, with no extraneous source of variability affecting them.

---

Let us now change the problem slightly and see how well the completely randomized design suits our needs. Suppose that, rather than relying upon one analyst, we use three analysts for the leaf assays. If we randomly assigned three apple leaves to each of the analysts, we might end up with a randomization scheme like the one listed in table 9.5.

**Table 9.5** Random assignment of the 9 leaves to the 3 analysts

| *Analyst* | | |
|---|---|---|
| *1* | *2* | *3* |
| A | B | C |
| A | B | C |
| A | B | C |

Even though we still have three observations for each treatment in this scheme, any differences that we may observe among the leaf determinations for the three laboratory procedures may be due entirely to differences among the analysts who assayed the leaves. For example, if we tested the hypothesis $H_0$: $\mu_A - \mu_B = 0$ against $H_a$: $\mu_A - \mu_B \neq 0$ and were lead to reject $H_0$, we would not be able to tell whether $\mu_A$ differs from $\mu_B$ because assays from Analyst 1 are different from those for Analyst 2 or because the properties of determinations by Procedure A differ markedly from those for Procedure B. This example illustrates a situation where the nine experimental units (tree leaves) are affected by an extraneous source of variability, analyst. In this case the units differ markedly and would not be a homogeneous set upon which we could base an evaluation of the effects of the three treatments.

The completely randomized design just described can be modified to gain additional information concerning the means $\mu_A$, $\mu_B$, and $\mu_C$. We can block out the undesirable variability among analysts by using the following experimental design. We restrict our randomization of treatments to experimental units to insure that each analyst performs a determination using each of the three procedures. The order of these determinations for each analyst is randomized. One such randomization is listed in table 9.6. Note that each analyst will assay three leaves, one leaf for each of the three procedures. Hence pairwise comparisons among the laboratory procedures that utilize the sample means will be free of any variability among analysts. For example, if we ran the test

$$H_0: \mu_A - \mu_B = 0$$

$$H_a: \mu_A - \mu_B \neq 0$$

**Table 9.6** A different assignment of leaves to analysts

| Analyst | | |
|---|---|---|
| *1* | *2* | *3* |
| A | B | A |
| C | A | B |
| B | C | C |

and rejected $H_0$, the difference between $\mu_A$ and $\mu_B$ would be due to a difference between the nutritional deficiencies detected by procedures A and B and not due to a difference among the analysts, since each analyst would have assayed one leaf for each of the three procedures.

randomized block design

blocks

This design, which represents an extension to the completely randomized design, is called a *randomized block design;* the analysts in our experiment are called *blocks*. By using this design, we have effectively filtered out any variability among the analysts, enabling us to make more precise comparisons among the *treatment* means, $\mu_A$, $\mu_B$, and $\mu_C$.

In general, we can use a randomized block design to compare $t$ different treatment means when an extraneous source of variability (blocks) is present. If there are $b$ different blocks, we would run each of the $t$ treatments in each block to filter out the block-to-block variability. In our example we had $t = 3$ treatment means (laboratory procedures) and $b = 3$ blocks (analysts).

The randomized block design has certain advantages and disadvantages which should be mentioned now, even though they will become more apparent when the analysis of the design is discussed in chapter 15.

**Advantages and Disadvantages of the Randomized Block Design**

Advantages

1. It is a useful design for comparing $t$ treatment means in the presence of a single extraneous source of variability.
2. The statistical analysis is simple (see chapter 15).
3. The design is easy to construct.
4. It can be used to accommodate any number of treatments in any number of blocks.

Disadvantages

1. Since the experimental units within a block must be homogeneous, the design is best suited for a relatively small number of treatments.
2. No second extraneous source of variability can affect the experimental units.
3. The effect of each treatment on the response must be approximately the same from block to block.

The apple leaf problem can be complicated further in the following way. Suppose that each leaf assay takes a long time and only one can be done by each analyst per day. If we used the randomized block design of table 9.6 and let the first row denote Day 1, the second row denote Day 2, and the third row denote Day 3, the design could be listed as shown in table 9.7.

**Table 9.7** A randomized block design for the leaf assay in the presence of a day effect

| | *Analyst* | | |
|---|---|---|---|
| *Day* | *1* | *2* | *3* |
| 1 | A | B | A |
| 2 | C | A | B |
| 3 | B | C | C |

Suppose now that we tested $H_0$: $\mu_A - \mu_B = 0$ against $H_a$: $\mu_A - \mu_B \neq 0$. Two Procedure A determinations were done on Day 1 and one on Day 2, while Procedure B was used on each of the three days. Thus if we reject $H_0$, we would not be certain whether $\mu_A$ differed from $\mu_B$ because of a difference in the laboratory procedures or because of a difference among the three days. Sometimes laboratory equipment must be calibrated daily and new chemical solutions must be prepared. Differences in determinations from day to day could be due to differences among the solutions or to differences in calibration accuracy.

This example illustrates a situation where the experimental units (leaves) are affected by a second extraneous source of variability, days. We can modify the randomized block design to filter out this second source of variability, the variability among days, in addition to filtering out the first source, variability among analysts. To do this we restrict our randomization to insure that each treatment appears in each row (day) and in each column (analyst). One such randomization is shown in table 9.8. Note that the test procedures have been assigned to analysts and to days so that each procedure is performed once a day and once by each analyst. Hence pairwise comparisons among treatment procedures that utilize the sample means are free of variability among days and analysts.

**Table 9.8** Assignment of leaves to analysts and days

| | *Analyst* | | |
|---|---|---|---|
| *Day* | *1* | *2* | *3* |
| 1 | A | B | C |
| 2 | B | C | A |
| 3 | C | A | B |

Latin square design

This experimental design is called a *Latin square design*. In general, a Latin square design can be used to compare $t$ treatment means in the presence of two extraneous sources of variability, which we block off into $t$ rows and $t$

columns. The $t$ treatments are then randomly assigned to the rows and columns so that each treatment appears in every row and every column of the design (see table 9.8).

The advantages and disadvantages of the Latin square design are listed in the box. As before, the pros and cons of this design will become more apparent when we consider the analysis of the design in chapter 15.

---

**Advantages and Disadvantages of the Latin Square Design**

Advantages

1. The design is particularly appropriate for comparing two treatment means in the presence of two sources of extraneous variation, each measured at $t$ levels.
2. The analysis is still quite simple (see chapter 15).

Disadvantages

1. While a Latin square can be constructed for any value of $t$, it is best suited for comparing $t$ treatments when $5 \leq t \leq 10$.
2. No additional extraneous source of variability can affect the experimental units.
3. The effect of each treatment on the response must be approximately the same across rows and columns.

---

We have delayed a discussion of the analyses for data generated from a completely randomized design, a randomized block design, and a Latin square design until later chapters. However, in the next section we will illustrate the analysis of a randomized block design for the special case where $t = 2$ treatments are to be compared in $b$ blocks. This particular randomized block design is often called a *paired difference experiment*.

paired difference experiment

# 9.4 Paired Difference Experiment: A Randomized Block Design with $t = 2$ Treatments

A randomized block design is frequently used to compare two population means ($t = 2$). We will illustrate the analysis of data generated from such an experiment by way of an example.

### EXAMPLE 9.5

Insurance adjusters are concerned about the high estimates they are receiving from Garage I in relation to Garage II. To verify their suspicions each of 15 cars recently involved in an accident was taken to both garages I and II for separate estimates of repair cost. Use the sample data shown in table 9.9 to test $H_0$: $\mu_1 - \mu_2 = 0$ against $H_a$: $\mu_1 - \mu_2 > 0$ (i.e., Garage I has higher estimates on the average than Garage II). Use $\alpha = .05$.

### SOLUTION

Initially we might be tempted to use the methods of section 5.4 (p. 114) to test the null and alternative hypotheses. But after taking a closer look at the data, we see that one of the assumptions of our usual $t$ test—that we obtain *independent* random samples from each population—has been violated. In fact, rather than having two samples with $n_1 = n_2 = 15$, there is really only one sample of $n = 15$ cars, but each car has been examined by both garages. From our discussion in section 9.3, we recognize this as a randomized block design with $b = 15$ blocks (cars) and $t = 2$ treatments (garages) per block.

The analysis of this experiment makes use of the $n = 15$ differences in repair estimates, which are recorded in column 4 of table 9.9. The sample

**Table 9.9** Repair estimates (in hundreds of dollars), example 9.5

| *Car* | *Garage I* | *Garage II* | *Difference, d* |
|---|---|---|---|
| 1 | 7.6 | 7.3 | .3 |
| 2 | 10.2 | 9.1 | 1.1 |
| 3 | 9.5 | 8.4 | 1.1 |
| 4 | 1.3 | 1.5 | −.2 |
| 5 | 3.0 | 2.7 | .3 |
| 6 | 6.3 | 5.8 | .5 |
| 7 | 5.3 | 4.9 | .4 |
| 8 | 6.2 | 5.3 | .9 |
| 9 | 2.2 | 2.0 | .2 |
| 10 | 4.8 | 4.2 | .6 |
| 11 | 11.3 | 11.0 | .3 |
| 12 | 12.1 | 11.0 | 1.1 |
| 13 | 6.9 | 6.1 | .8 |
| 14 | 7.6 | 6.7 | .9 |
| 15 | 8.4 | 7.5 | .9 |
| *Totals* | $\bar{y}_1 = 6.85$ | $\bar{y}_2 = 6.23$ | $\bar{d} = .61$ |

averages for the two garages and for the differences are shown at the bottom of the table. Note that (except for rounding errors in the calculations) $\bar{d} = \bar{y}_1 - \bar{y}_2$. In fact, the null hypothesis, $H_0$: $\mu_1 - \mu_2 = 0$, is equivalent to the hypothesis that $\mu_d$, the population mean difference, is zero. The test procedure is then similar to a one-sample test on $\mu$, where the $n$ differences represent the sample data. For this example the test procedure is as follows:

$\bar{d}$

$\mu_d$

$H_0$: $\mu_d = \mu_1 - \mu_2 = 0$.

$H_a$: $\mu_d > 0$.

T.S.: $t = \dfrac{\bar{d}}{s_d/\sqrt{n}}$.

R.R.: For df $= n - 1 = 14$, reject $H_0$ if $t > t_{.05}$.

Before computing $t$ we must first calculate $s_d$, the sample standard deviation of the differences. We can calculate $s_d$ by using our shortcut formula for a sample variance,

$s_d$

$$s_d^2 = \frac{1}{n-1}\left[\sum_{i=1}^{n} d_i^2 - \frac{\left(\sum_{i=1}^{n} d_i\right)^2}{n}\right]$$

For the data of table 9.9,

$$\sum_{i=1}^{15} d_i = .3 + 1.1 + 1.1 + \cdots + .9 = 9.2$$

$$\sum_{i=1}^{15} d_i^2 = (.3)^2 + (1.1)^2 + (1.1)^2 + \cdots + (.9)^2 = 7.82$$

Hence for $n = 15$ differences,

$$s_d^2 = \frac{1}{14}\left[7.82 - \frac{(9.2)^2}{15}\right] = .156$$

$$s_d = \sqrt{.156} = .394$$

Substituting into the test statistic $t$, we have

$$t = \frac{\bar{d} - 0}{s_d/\sqrt{n}} = \frac{.61}{.394/\sqrt{15}} = 6.00$$

For df $= n - 1 = 14$ and $\alpha = .05$, the critical value of $t$ is 1.761. Since the observed value of $t$ is greater than 1.761, we reject the null hypothesis $H_0$: $\mu_d = 0$ and conclude that the mean repair estimate for Garage I is greater than that for Garage II.

In the analysis of example 9.5, we used the sample differences rather than the individual repair estimates for each car at the two garages. If we had taken two independent random samples of 15 cars and obtained repair estimates for the first sample of 15 from Garage I and repair estimates for the second sample from Garage II, the difference in sample means may have been due to either variability in the two samples of damaged cars or variability in estimates for the two garages, or both. Hence in that situation any conclusion about $\mu_1$ and $\mu_2$ would be difficult to interpret. By using cars as blocks and comparing repair estimates for each car, we are able to filter out the car-to-car variability to obtain a meaningful comparison of $\mu_1$ and $\mu_2$.

*It should be noted that blocking will not always add information in an experiment. If we block when in fact there is no block-to-block variability, we cut the number of degrees of freedom in half.* Thus for our example the unpaired analysis comparing two population means (completely randomized design) would have had $n_1 + n_2 - 2 = 28$ degrees of freedom, while the paired difference experiment has only 14. Also, if there is a negative relationship between the experimental units of a block (such as when siblings are competing for nourishment), the randomized block design would sacrifice information relative to that gained over a completely randomized design. More information on the *efficiencies* of these block designs will be given in chapter 15.

---

**Statistical Test for a Paired Difference Experiment**

$H_0$: $\mu_d = D_0$ (specified).

$H_a$: 1. $\mu_d > D_0$.
2. $\mu_d < D_0$.
3. $\mu_d \neq D_0$.

T.S.: $t = \dfrac{\bar{d} - D_0}{s_d/\sqrt{n}}$.

R.R.: For a type I error $\alpha$ and df $= n - 1$,

1. reject $H_0$ if $t > t_\alpha$;
2. reject $H_0$ if $t < -t_\alpha$;
3. reject $H_0$ if $|t| > t_{\alpha/2}$.

Note: $n$ here denotes the number of differences.

---

A confidence interval for $\mu_d$ parallels the one-sample confidence interval on $\mu$.

---

**General Confidence Interval for $\mu_d$**

$$\bar{d} \pm t_{\alpha/2} \frac{s_d}{\sqrt{n}}$$

where $t$ is based on df $= n - 1$ and the confidence coefficient is $(1 - \alpha)$.

---

In sections 9.3 and 9.4 we have discussed briefly designs that are useful for comparing $t$ population (treatment) means in the presence of one or two sources of extraneous variability. These designs are frequently called block designs. In the next section we will consider a set of designs, called factorial experiments, that are useful for examining the effect of two or more independent variables (quantitative or qualitative) on a response.

# Exercises

9.6. An agricultural experimental station was interested in comparing the yields for two new varieties of corn. Because it was felt that there might be a great deal of variability in yield from one farm to another, each variety was randomly assigned to a different one-acre plot on each of 7 farms. The one-acre plots were planted; the corn was harvested at maturity. The results of the experiment (in bushels of corn) are listed below.

| *Farm* | 1 | 2 | 3 | 4 | 5 | 6 | 7 |
|---|---|---|---|---|---|---|---|
| *Variety A* | 48.2 | 44.6 | 49.7 | 40.5 | 54.6 | 47.1 | 51.4 |
| *Variety B* | 41.5 | 40.1 | 44.0 | 41.2 | 49.8 | 41.7 | 46.8 |

Use these data to test the null hypothesis that there is no difference in mean yields for the two varieties of corn. Use $\alpha = .05$.

9.7. Thirty sets of identical twins were asked to participate in a one-year study designed to measure certain social attitudes. One twin from each set was randomly assigned to live in the home of a minority family, while the other twin of the set stayed at home. After one year each person was asked to respond to a long questionnaire designed to detect and measure well-defined attitudes. Let sample 1 denote the combined questionnaire scores for those persons who lived at home and sample 2 denote the set of scores for those who lived with a family from a minority class. Test the null hypothesis

$H_0$: $\mu_1 - \mu_2 = 0$ (the population mean scores for those not exposed and those exposed to a minority environment are identical)

against the alternative

$H_a$: $\mu_1 - \mu_2 \neq 0$ (the population mean scores are different for the two environments)

Use $\alpha = .05$. The data are shown on the next page.

| Set of Twins | Home Environment, $y_1$ | Minority Environment, $y_2$ | Difference | Set of Twins | Home Environment, $y_1$ | Minority Environment, $y_2$ | Difference |
|---|---|---|---|---|---|---|---|
| 1 | 78 | 71 | 7 | 16 | 90 | 88 | 2 |
| 2 | 75 | 70 | 5 | 17 | 89 | 80 | 9 |
| 3 | 68 | 66 | 2 | 18 | 73 | 65 | 8 |
| 4 | 92 | 85 | 7 | 19 | 61 | 60 | 1 |
| 5 | 55 | 60 | −5 | 20 | 76 | 74 | 2 |
| 6 | 74 | 72 | 2 | 21 | 81 | 76 | 5 |
| 7 | 65 | 57 | 8 | 22 | 89 | 78 | 11 |
| 8 | 80 | 75 | 5 | 23 | 82 | 78 | 4 |
| 9 | 98 | 92 | 6 | 24 | 70 | 62 | 8 |
| 10 | 52 | 56 | −4 | 25 | 68 | 73 | −5 |
| 11 | 67 | 63 | 4 | 26 | 74 | 73 | 1 |
| 12 | 55 | 52 | 3 | 27 | 85 | 75 | 10 |
| 13 | 49 | 48 | 1 | 28 | 97 | 88 | 9 |
| 14 | 66 | 67 | −1 | 29 | 95 | 94 | 1 |
| 15 | 75 | 70 | 5 | 30 | 78 | 75 | 3 |
| | | | | | $\bar{y}_1 = 75.23$ | $\bar{y}_2 = 71.43$ | $\bar{d} = \bar{y}_1 - \bar{y}_2 = 3.8$ |

9.8. Suppose that we wish to estimate the difference between the mean monthly salaries of male and female sales representatives. Since there is a great deal of salary variability from company to company, it was decided to filter out the variability due to companies by making male-female comparisons within each company. One male and one female with the required background and work experience will be selected from each company. If the range of differences in salaries (between males and females) within a company is approximately \$300 per month, determine the number of companies that must be examined to estimate the difference in mean monthly salary for males and females. Use a 95% confidence interval with a half width of \$5.

## 9.5 Factorial Experiments

In sections 9.3 and 9.4 we examined the completely randomized design, the randomized block design, and the Latin square design as candidate designs to be used in situations where the experimenter is interested in comparing $t \geq 2$ population means. Suppose now that, rather than comparing $t$ population means, we wish to examine the effects of two or more independent variables on a response $y$. For example, suppose that we want to examine the effects of temperature $x_1$ and pressure $x_2$ on the bond strength $y$ of a new adhesive product. Two major problems arise. First, we must consider the number of levels and the actual settings of these levels for each independent variable. Second, having chosen the levels for each independent variable, we must choose the factor-level combinations (treatments) that will be applied to the experimental units.

To illustrate the importance of level selection for each of the independent variables, consider the simplified situation where an experimenter is interested in examining the effect of a single independent variable $x$ on a response $y$. The mean response $E(y)$ is the curved line shown in figure 9.1. If we choose to

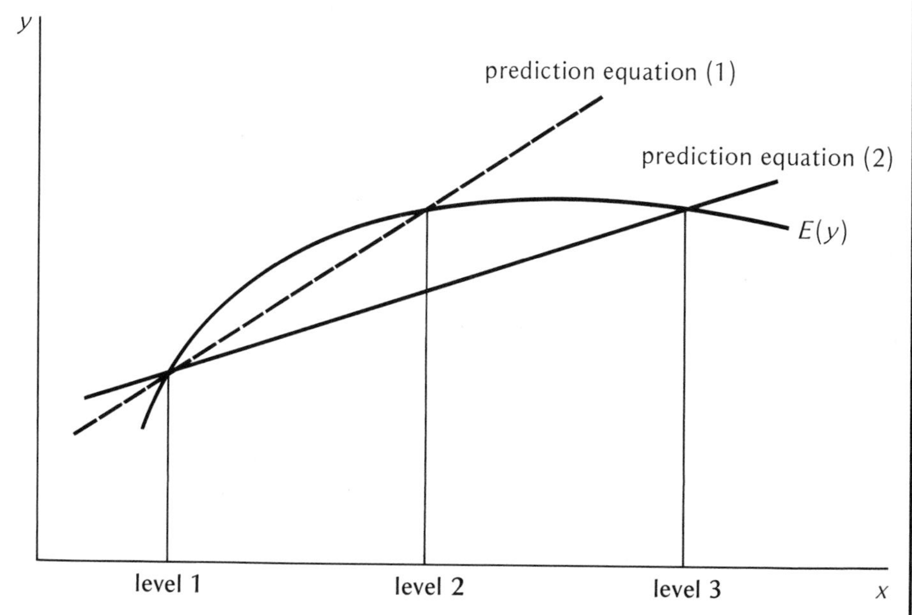


**Figure 9.1** Straight-line prediction for a curvilinear mean response $E(y)$

observe $y$ at only two settings of $x$, the best we can do is obtain a straight-line prediction equation through the two points. If we are to find a curvilinear prediction equation, we must design our experiment with more than two levels of $x$.

If the two design points chosen were levels 1 and 2 of $x$, the prediction equation would be equation (1) of figure 9.1; for levels of 1 and 3, the prediction equation would be equation (2). Obviously these two prediction equations would give vastly different predictions of $y$ for the same setting of $x$. Thus we see that the number of levels chosen is important and the actual settings of those levels are important in obtaining a prediction equation for $y$.

The ability to choose appropriate settings for the independent variables depends a great deal on the experimenter's knowledge of the physical situation under study. Then, assuming the experimenter has chosen the levels of each independent variable, he or she faces the task of deciding which factor-level combinations should be assigned to the experimental units. For purposes of illustration, suppose that an experimenter is interested in examining the effects of two independent variables, nitrogen and phosphorus, on the yield of a crop. For simplicity we will assume that two levels have been selected for the study of each factor: 40 and 60 pounds per plot for nitrogen, 10 and 20 pounds per plot for phosphorus. For this study the experimental units are small relatively homogeneous plots that have been partitioned from the acreage of a farm.

one-at-a-time approach

One approach suggested for examining the effects of two or more factors on a response is called the "one-at-a-time approach." To examine the effect of a single variable, an experimenter varies the levels of this variable while holding the levels of the other independent variables fixed. This process is continued until the effect of each variable on the response has been examined while holding the other independent variables constant. For our experiment the

**Table 9.10** Factor-level combinations for a one-at-a-time approach

| *Combination* | *Nitrogen* | *Phosphorus* |
|---|---|---|
| 1 | 60 | 10 |
| 2 | 40 | 10 |
| 3 | 40 | 20 |

factor-level combinations chosen might be as shown in table 9.10. These factor-level combinations are illustrated in figure 9.2.

From the graph in figure 9.2 we see that there is one difference that can be used to measure the effects of nitrogen and phosphorus separately. The

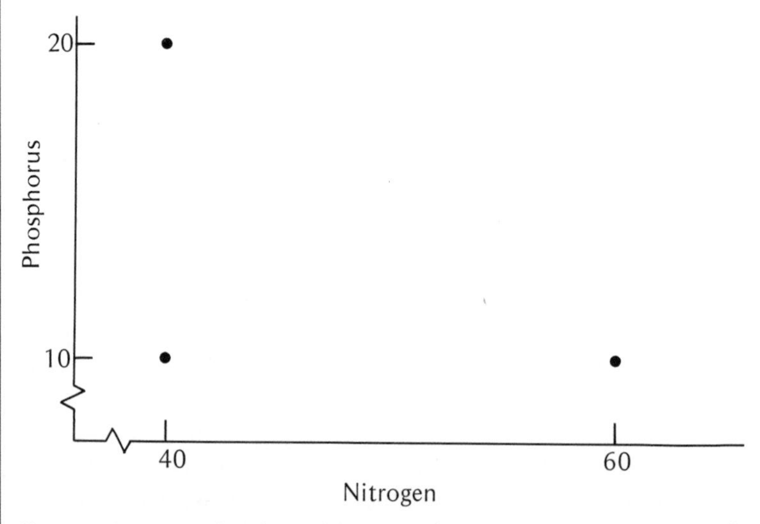


**Figure 9.2** Factor-level combinations for a one-at-a-time approach

difference in response for combinations 1 and 2 would estimate the effect of nitrogen; the difference in response for combinations 2 and 3 would estimate the effect of phosphorus.

Hypothetical yields corresponding to the three factor-level combinations of our experiment are given in table 9.11. Suppose the experimenter is interested

**Table 9.11** Yields for the three factor-level combinations

| *Observation (yield)* | *Nitrogen* | *Phosphorus* |
|---|---|---|
| 145 | 60 | 10 |
| 125 | 40 | 10 |
| 160 | 40 | 20 |
| ? | 60 | 20 |

in using the sample information to determine the factor-level combination that will give the maximum yield. From the table we see that crop yield increases when the nitrogen application is increased from 40 to 60 (holding phosphorus at 10). Yield also increases when the phosphorus setting is changed from 10 to 20 (at a fixed nitrogen setting of 40). Thus it might seem logical to predict that increasing both the nitrogen and phosphorus applications to the soil will result in a larger crop yield. The fallacy in this argument is that our prediction is based on the assumption that the effect of one factor is the same for both levels of the other factor.

We know from our investigation what happens to yield when the nitrogen application is increased from 40 to 60 for a phosphorus setting of 10. But will the yield also increase by approximately 20 units when the nitrogen application is changed from 40 to 60 at a setting of 20 for phosphorus?

To answer this question we could apply the factor-level combination of 60 nitrogen–20 phosphorus to another experimental plot and observe the crop yield. If the yield is 180, then the information obtained from the three factor-level combinations would be correct and would have been useful in predicting the factor-level combination that produces the greatest yield. However, suppose the yield obtained from the high settings of nitrogen and phosphorus turned out to be 110. If this happens, the two factors "nitrogen" and "phosphorus" are said to *interact*. That is, the effect of one factor on the response does not remain the same for different levels of the second factor, and the information obtained from the one-at-a-time approach would lead to a faulty prediction.

interaction

The two outcomes just discussed for the crop yield at the 60-20 setting is displayed in figure 9.3, along with the yields at the three initial design points. Figure 9.3(a) illustrates a situation with no interaction between the two factors.

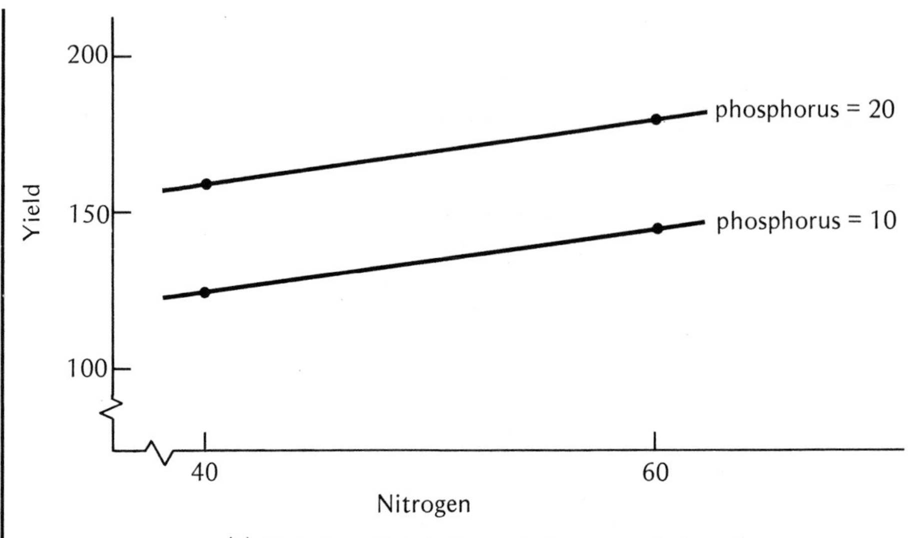

(a) No interaction between nitrogen and phosphorus

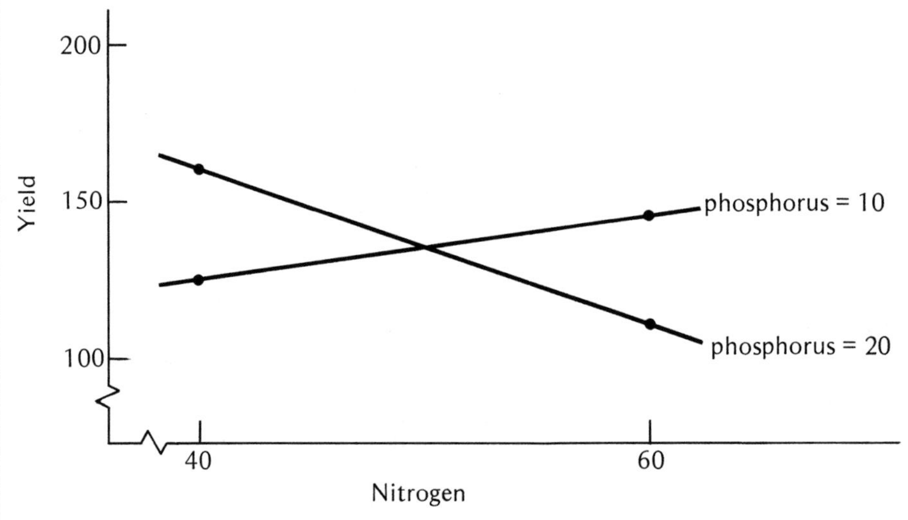


(b) An interaction between nitrogen and phosphorus

**Figure 9.3** Yields of the three design points and possible yield at a fourth design point

The effect of nitrogen on yield is the same for both levels of phosphorus. In contrast, figure 9.3(b) illustrates a case where the two factors nitrogen and phosphorus do interact.

We have seen that the one-at-a-time approach to investigating the effect of two factors on a response is suitable only for situations where the two factors do not interact. Although this was illustrated for the simple case where two factors were to be investigated at each of two levels, the inadequacies of a one-at-a-time approach are even more salient when trying to investigate the effects of more than two factors on a response.

Designs that are useful for examining the effects of two or more factors on a response $y$, whether or not interaction exists, are called *factorial experiments*. As before, the choice of the number of levels of each variable and the actual settings of these variables is important. But assuming we have made these selections with help from an investigator knowledgeable in the area being examined, we must decide at what factor-level combinations we will observe $y$.

factorial experiment

Classically, these designs have been referred to as "factorial-type experiments" since they deal with the choice of levels and the selection of factor-level combinations (treatments) rather than with how the treatments are assigned to experimental units. Unless otherwise specified, we will assume that treatments are assigned to experimental units at random, and we will refer to a factorial experiment as an experimental design.

**Definition 9.3**

A *factorial experiment* is an experiment in which the response $y$ is observed at all factor-level combinations of the independent variables.

Using our previous example, if we are interested in examining the effect of two levels of nitrogen $x_1$ at 40 and 60 pounds per plot and two levels of phosphorus $x_2$ at 10 and 20 pounds per plot on the yield of a crop, we must decide how to prepare plots to observe yield. A $2 \times 2$ factorial experiment for this example is shown in table 9.12. The four factor-level combinations are assigned at random to the experimental units.

**Table 9.12** $2 \times 2$ factorial experiment for crop yield

| *Factor-Level Combinations* | |
|---|---|
| $x_1$ | $x_2$ |
| 40 | 10 |
| 40 | 20 |
| 60 | 10 |
| 60 | 20 |

Similarly, if we wished to examine $x_1$ at the two levels 40 and 60 and $x_2$ at the three levels 10, 15, and 20, the $2 \times 3$ factorial experiment would have the factor-level combinations shown in table 9.13.

**Table 9.13** 2 × 3 factorial experiment for crop yield

| *Factor-Level Combinations* | |
|---|---|
| $x_1$ | $x_2$ |
| 40 | 10 |
| 40 | 15 |
| 40 | 20 |
| 60 | 10 |
| 60 | 15 |
| 60 | 20 |

### EXAMPLE 9.6

An auto manufacturer is interested in examining the effect of engine speed $x_1$, measured in revolutions per minute, and ground speed $x_2$, measured in miles per hour, on gasoline mileage. The investigators, in consultation with company mechanics and other personnel, decided to consider settings of $x_1$ at 800, 1000, and 1200 and settings of $x_2$ at 30, 50, and 70. Give the factor-level combinations to be used in a 3 × 3 factorial experiment.

### SOLUTION

Using the definition of factorial experiment, we would observe gasoline mileage at the following settings of $x_1$ and $x_2$:

| $x_1$ | 800 | 800 | 800 | 1000 | 1000 | 1000 | 1200 | 1200 | 1200 |
|---|---|---|---|---|---|---|---|---|---|
| $x_2$ | 30 | 50 | 70 | 30 | 50 | 70 | 30 | 50 | 70 |

The examples of factorial experiments presented in this section have concerned two independent variables. However, the procedure applies to any number of factors and levels per factor. Thus if we had four different factors $x_1$, $x_2$, $x_3$, and $x_4$ at 2, 3, 3, and 4 levels, respectively, we could formulate a 2 × 3 × 3 × 4 factorial experiment by considering all $2 \cdot 3 \cdot 3 \cdot 4 = 72$ factor-level combinations.

One final comparison should be made between the one-at-a-time approach and a factorial experiment. Not only do we get information concerning factor interactions using a factorial experiment, but also, when there are no interactions, we get the same amount of information about the effects of each individual factor using fewer observations. To illustrate this idea let us consider the 2 × 2 factorial experiment with nitrogen and phosphorus. If there is no interaction between the two factors, the data appear as shown in figure 9.3(a). For convenience, the data are reproduced in table 9.14, with the four sample combinations designated by the numbers 1 through 4. If a 2 × 2 factorial experiment is used and no interaction exists between the two factors, we can obtain two independent differences to examine the effects of each of the factors on the response. Thus from table 9.14, the differences between obser-

**Table 9.14** Factor-level combinations for a $2 \times 2$ factorial experiment

| *Combination* | *Yield* | *Nitrogen* | *Phosphorus* |
|---|---|---|---|
| 1 | 145 | 60 | 10 |
| 2 | 125 | 40 | 10 |
| 3 | 165 | 40 | 20 |
| 4 | 180 | 60 | 20 |

vations 1 and 4 and the difference between observations 2 and 3 would be used to measure the effect of phosphorus. Similarly, the difference between observations 4 and 3 and the difference between observations 2 and 1 would be used to measure the effect of the two levels of nitrogen on plot yield.

If we employed a one-at-a-time approach for the same experimental situation, it would take six observations (two observations at each of the three initial factor-level combinations shown in table 9.10) to obtain the same number of independent differences for examining the separate effects of nitrogen and phosphorus when no interaction is present.

## Summary 9.6

In this chapter we have presented a brief introduction to experimental design. Basically, the design of an experiment consists of a statement of the experimental objectives, a statement of the amount of information required in the experiment, and a selection of the experimental design and inference-making procedure to satisfy the experimenter's design requirements (at a minimum cost). The first two steps of the design of an experiment can usually be specified without too much difficulty. However, the third and fourth steps involve translating the experimenter's requirements into an appropriate experimental design and inference-making procedure.

We first considered inferences (estimations or tests of hypotheses) concerning $\mu$ or $\mu_1 - \mu_2$. Since the experimental design and inference-making procedures had already been established in chapters 4 and 5, the fourth step of the design of the experiment consisted of specifying the sample size required to satisfy the first two steps. These sample size requirements were illustrated in detail.

Next we considered several basic experimental designs. The completely randomized design, the randomized block design, and the Latin square design illustrate how we can block out undesirable background variability to obtain more precise comparisons among treatment means. In contrast, the factorial experiment is useful in investigating the effect of one or more factors on a response.

Not all designs can be classified either as a block design or as a factorial experiment. Some designs represent combinations of block designs with factorial experiments. Thus an experimenter may wish to examine the effects of two or more factors on a response while blocking out one or more extraneous sources of variability. Examples of such designs will be discussed in chapter 15.

Except for the paired difference experiment, an example of a randomized block design, inference-making procedures were not discussed in this chapter. These topics will be presented in later chapters when we begin to use these experimental designs.

You should keep in mind that this chapter has presented a very brief introduction to experimental design. Sequences of courses at the undergraduate and graduate level can be given to prepare a person to cope with the third step in the design of an experiment. But this is the professional statistician's problem. At this point you need only be familiar with the concepts of experimental design and the application of these concepts in the simplified situations discussed in this chapter.

# Exercises

9.9. An experimenter was interested in examining the bond strength of a new adhesive product prepared under 3 different temperature settings (280°F, 300°F, and 320°F) and 4 different pressure settings (100, 150, 200, and 250 psi). If a single fixed amount of adhesive is to be prepared and tested at each temperature–pressure setting combination, identify the design.

9.10. An oil company has been experimenting with a new gasoline additive. As part of the testing program, the company examined the effect on mileage (mpg) of 4 additive concentrations and 5 different octane levels for the gasoline. If one gasoline mixture is to be made and tested at each concentration-octane combination, identify the experimental design.

$ 9.11. A company executive was interested in comparing the cost per mile for all cars of a particular brand with V-8 engines (351 horsepower) and all cars of the same make with 6-cylinder engines (250 horsepower). Since the entire company fleet of cars for salespeople consisted of approximately 600 automobiles, it was decided to obtain a random sample of data from 16 cars, 8 from each engine type. To avoid geographic variability, a random sample of one car of each engine type was selected from each of 8 geographic areas throughout the country. Cost per mile was determined for each car sampled during the period of January 1976 to March 1976. These data appear below.

| *Geographic Area* | 1 | 2 | 3 | 4 | 5 | 6 | 7 | 8 |
|---|---|---|---|---|---|---|---|---|
| *351 hp, V-8* | .0837 | .0564 | .0703 | .0502 | .0638 | .0483 | .0746 | .0694 |
| *250 hp, 6-cylinder* | .0523 | .0371 | .0464 | .0481 | .0535 | .0335 | .0444 | .0528 |

a. Do the data suggest a difference in the mean cost per mile for the two engine types?
b. Give the level of significance for your test. Draw some conclusions.
c. Can you identify additional sources of variability that could be blocked?

# References

Barr, A.J.; Goodnight, J.H.; Sall, J.P.; and Helwig, J.T. 1976. *A user's guide to SAS 76*. Raleigh, N.C.: SAS Institute, Inc.

Cochran, W.G., and Cox, G.M. 1957. *Experimental design*. 2d ed. New York: Wiley.

Dixon, W.J. 1975. *BMDP, biomedical computer programs*. Berkeley: University of California Press.

Dixon, W.J., and Massey, F.J., Jr. 1969. *Introduction to statistical analysis*. 3d ed. New York: McGraw-Hill.

Draper, N.R., and Smith, H. 1966. *Applied regression analysis*. New York: Wiley.

Fisher, R.A. 1958. *Statistical methods for research workers*. 13th ed. New York: Hafner.

Graybill, F.A. 1976. *Theory and application of the linear model*. N. Scituate, Mass.: Duxbury Press.

Harnett, D.L. 1970. *Introduction to statistical methods*. Reading, Mass.: Addison-Wesley.

Kirk, R.E. 1968. *Experimental design: Procedures for the behavioral sciences*. Belmont, Calif.: Brooks/Cole.

Mendenhall, W. 1968. *Introduction to linear models and the design and analysis of experiments*. Belmont, Calif.: Wadsworth.

Nie, N.H.; Hull, C.H.; Jenkins, J.G.; Steinbrenner, K.; and Bent, D.H. 1975. *Statistical package for the social sciences*. 2d ed. New York: McGraw-Hill.

Ostle, B. 1963. *Statistics in research*. 2d ed. Ames, Iowa: Iowa State University Press.

Pearson, E.S., and Hartley, H.O. 1966. *Biometrika tables for statisticians*. 3d ed. Vol. I. London: Cambridge University Press.

Service, J. 1972. *A user's guide to the statistical analysis system*. Raleigh, N.C.: Student Supply Stores, North Carolina State University.

Snedecor, G.W., and Cochran, W.G. 1967. *Statistical methods*. 6th ed. Ames, Iowa: Iowa State University Press.

Tanur, J.M.; Mosteller, F.; Kruskal, W.H.; Pieters, R.S.; and Rising, G.R. 1972. *Statistics: A guide to the unknown*. San Francisco: Holden-Day.

Winer, B.J. 1962. *Statistical principles in experimental design*. New York: McGraw-Hill.

# 10 Count Data

## 10.1 Introduction

Up to this point we have been concerned primarily with sample data measured on a quantitative scale. However, we sometimes encounter situations where levels of the variable of interest are identified by name or rank only and we are interested in the number of observations occurring at each level of the variable. Data obtained from these types of variables are called *count data*. For example, an item coming off an assembly line may be classified into one of three quality classes: acceptable, second, or reject. Similarly, a traffic study might require a count and classification of the type of transportation used by commuters along a major access road into a city. A pollution study might be concerned with the number of different alga species identified in samples from a lake and the number of times each species is identified. A consumer protection agency might be interested in the results of a prescription fee survey to determine the markup or professional fee charged by pharmacists over and above the list price. We also might wish to classify styles of creative writing into one of several categories or rate teacher effectiveness as good, bad, or indifferent. Indeed, many experiments (especially in the social sciences) result in count or enumerative data of this sort.

count data

In this chapter we will examine specific inferences that can be made from experiments involving count data.

# 10.2 The Multinomial Experiment and Chi-square Goodness-of-fit Test

multinomial experiment

The examples in section 10.1 all exhibit, to a reasonable degree of approximation, the characteristics of a *multinomial experiment.*

---

**The Multinomial Experiment**

1. The experiment consists of $n$ identical trials.
2. Each trial results in one of $k$ outcomes.
3. The probability that a single trial will result in outcome $i$ is $p_i$, $i = 1, 2, \ldots, k$, and remains constant from trial to trial. (Note: $\Sigma_{i=1}^{k} p_i = 1$.)
4. The trials are independent.
5. We are interested in $n_i$, the number of trials resulting in outcome $i$. (Note: $\Sigma_{i=1}^{k} n_i = n$.)

---

multinomial distribution

The probability distribution for the number of observations resulting in each of the $k$ outcomes, called the *multinomial distribution,* is given by the formula

$$P(n_1, n_2, \ldots, n_k) = \frac{n!}{n_1!\, n_2! \cdots n_k!} p_1^{n_1} p_2^{n_2} \cdots p_k^{n_k}$$

$n!$, or $n$ factorial

(For a review of probability and the concept of a probability distribution, refer to sections 2.2 and 3.4.) The notation $n!$ (read "$n$ factorial") refers to the number computed as follows:

$$n! = n(n-1)(n-2)\cdots 1$$

For example, when $n = 4$,

$$4! = 4(3)(2)(1) = 24$$

Similarly, 2! and 5! are

$$2! = 2(1) = 2 \qquad 5! = 5(4)(3)(2)(1) = 120$$

The number 0! is always taken to be 1.

We can use the formula for the multinomial distribution to compute the probability of particular events.

**EXAMPLE 10.1**

Previous experience with the breeding of a particular herd of cattle suggests that the probability of obtaining 1 healthy calf from a mating is .83. Similarly, the probabilities of obtaining 0 or 2 healthy calves are, respectively, .15 and .02. If a farmer breeds 3 dams from the herd, find the probability of obtaining exactly 3 healthy calves.

**SOLUTION**

Assuming the 3 dams are chosen at random, this experiment can be viewed as a multinomial experiment with $n = 3$ trials and $k = 3$ outcomes. These outcomes are listed below with the corresponding probabilities.

| *Outcome* | *Number of Progeny* | *Probability, $p_i$* |
|---|---|---|
| 1 | 0 | .15 |
| 2 | 1 | .83 |
| 3 | 2 | .02 |

Note that outcomes 1, 2, and 3 refer to the events that a dam produces 0, 1, or 2 healthy calves, respectively. Similarly, $n_1$, $n_2$, and $n_3$ refer to the number of dams producing 0, 1, or 2 healthy progeny, respectively. To obtain exactly 3 healthy progeny, we must observe one of the following possible events.

$$A: \begin{cases} \text{1 dam gives birth to no healthy progeny:} & n_1 = 1 \\ \text{1 dam gives birth to 1 healthy progeny:} & n_2 = 1 \\ \text{1 dam gives birth to 2 healthy progeny:} & n_3 = 1 \end{cases}$$

$$B: \quad \text{3 dams give birth to 1 healthy progeny:} \begin{cases} n_1 = 0 \\ n_2 = 3 \\ n_3 = 0 \end{cases}$$

For event $A$ with $n = 3$ and $k = 3$,

$$P(n_1 = 1, n_2 = 1, n_3 = 1) = \frac{3!}{1!\,1!\,1!}(.15)^1\,(.83)^1\,(.02)^1 \approx .015$$

Similarly, for event $B$,

$$P(n_1 = 0, n_2 = 3, n_3 = 0) = \frac{3!}{0!\,3!\,0!}(.15)^0\,(.83)^3\,(.02)^0 = (.83)^3 \approx .572$$

Thus the probability of obtaining exactly 3 healthy progeny from 3 dams is the sum of the probabilities for events $A$ and $B$; namely, $.015 + .572 \approx .59$.

We turn now to a statistical test concerning the probabilities $p_1, p_2, \ldots, p_k$. We will hypothesize specific values for the $p$'s and then determine whether the sample data agree with the hypothesized values. One way to test such an hypothesis is to examine the observed number of trials resulting in each outcome and to compare this to the number we would *expect* to result in each outcome. For instance, in our previous example, we gave the probabilities associated with 0, 1, and 2 progeny as .15, .83, and .02. If we were to examine a sample of 100 mated dams, we would expect to observe 15 dams that produce no healthy progeny. Similarly, we would expect to observe 83 dams that produce one healthy calf and 2 dams that produce 2 healthy calves.

---

**Definition 10.1**

In a multinomial experiment where each trial can result in one of $k$ outcomes, the *expected number of outcomes* of type $i$ in $n$ trials is $np_i$, where $p_i$ is the probability that a single trial results in outcome $i$.

---

In 1900 Karl Pearson proposed the following test statistic to test the specified probabilities:

$\chi^2$

$$\chi^2 = \sum_{i=1}^{k} \left[ \frac{(n_i - E_i)^2}{E_i} \right]$$

where $n_i$ represents the number of trials resulting in outcome $i$ and $E_i$ represents the number of trials we would expect to result in outcome $i$ when the hypothesized probabilities represent the actual probabilities assigned to each outcome. (The symbol $\chi$ is the Greek letter chi.) Frequently we will refer to the probabilities $p_1, p_2, \ldots, p_k$ as *cell probabilities,* one cell corresponding to each of the $k$ outcomes. The observed numbers $n_1, n_2, \ldots, n_k$ corresponding to the $k$ outcomes will be called *observed cell counts,* and the expected numbers $E_1, E_2, \ldots, E_k$ will be referred to as *expected cell counts.*

cell probabilities

observed cell counts

expected cell counts

Suppose that we hypothesize values for the cell probabilities $p_1, p_2, \ldots, p_k$. We can then calculate the expected cell counts by using definition 10.1 to examine how well the observed data fit, or agree, with what we would expect to observe. Certainly if the hypothesized $p$-values are correct, the observed cell counts $n_i$ should not deviate greatly from the expected cell counts $E_i$, and the computed value of $\chi^2$ should be small. Similarly, when one or more of the hypothesized cell probabilities are incorrect, the observed and expected cell counts will differ substantially, making $\chi^2$ large.

chi-square distribution

The distribution of the quantity $\chi^2$ can be approximated by a *chi-square distribution* provided that the expected cell counts $E_i$ are fairly large. We will not give the mathematical formula for the chi-square probability distribution; instead, we will list its properties.

1. The chi-square distribution is a nonsymmetrical distribution (see figure 10.1).
2. There are many chi-square distributions. We obtain a particular one by specifying the degrees of freedom (df).

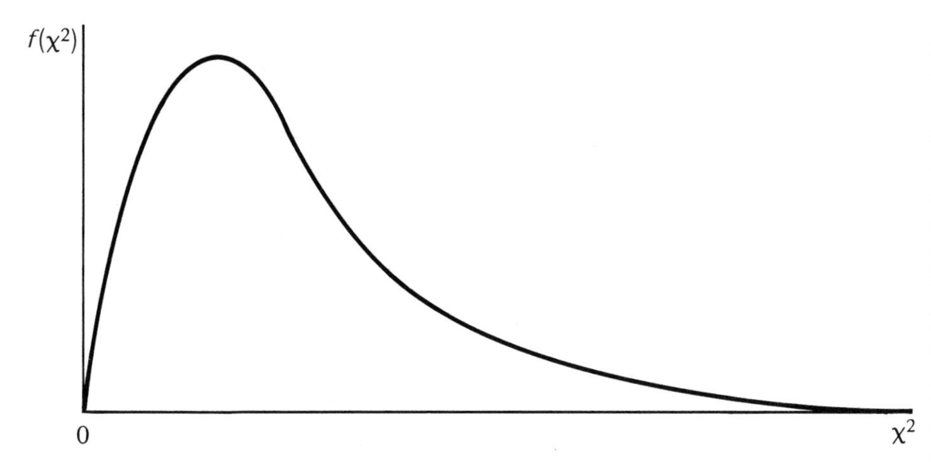


**Figure 10.1** Chi-square probability distribution for df = 4

The chi-square goodness-of-fit test based on $k$ specified cell probabilities will have $k - 1$ degrees of freedom. Upper-tail values of the test statistic

$$\chi^2 = \sum_{i=1}^{k} \left[ \frac{(n_i - E_i)^2}{E_i} \right]$$

can be found in table 3 of the appendix. Entries in the table are values of $\chi^2$ that have an area "a" to the right under the curve. The degrees of freedom are specified in the left-hand column of the table, and values of "a" are listed across the top of the table. Thus for df = 14, the value of chi-square with an area a = .10 to its right under the curve is 21.0642 (see figure 10.2).

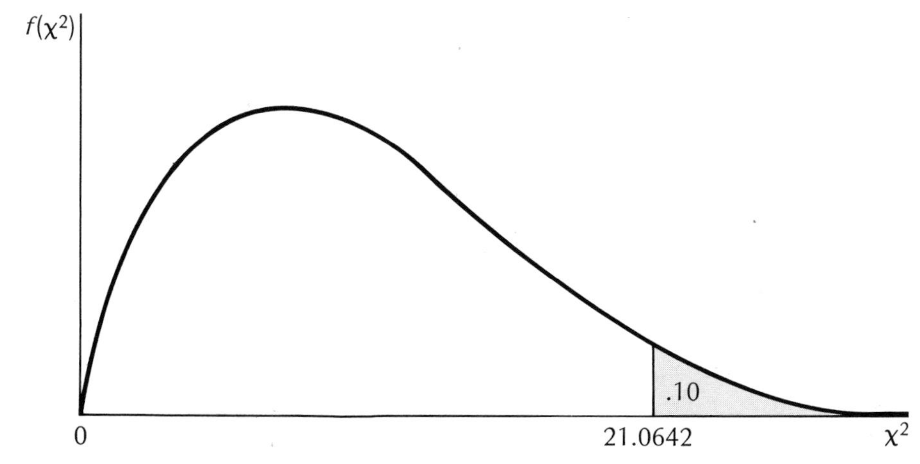


**Figure 10.2** Critical value of the chi-square distribution for a = .10 and df = 14

We can now summarize the chi-square goodness-of-fit test concerning $k$ specified cell probabilities.

**Chi-square Goodness-of-fit Test**

Null hypothesis: Each of $k$ cell probabilities is specified.

Alternative hypothesis: At least one of the cell probabilities differs from the hypothesized value.

Test statistic: $$\chi^2 = \sum_{i=1}^{k} \left[ \frac{(n_i - E_i)^2}{E_i} \right].$$

Rejection region: Reject $H_0$ if $\chi^2$ exceeds the tabulated critical value for $a = \alpha$ and df $= k - 1$.

Some researchers (see, for example, Siegel and Dixon and Massey) recommend that all the $E_i$'s should be 5 or more before performing this test. This requirement is perhaps too stringent. Cochran (1954) indicates that the approximation should be quite good if no $E_i$ is less than 1 and no more than 20% of the $E_i$'s are less than 5. We recommend applying Cochran's guidelines for determining whether $\chi^2$ can be approximated with a chi-square distribution. If some of the $E_i$'s are too small, we can combine categories, but care should be taken so that the combination of categories does not change the nature of the hypothesis to be tested.

**EXAMPLE 10.2**

A test drug is to be compared against a standard drug preparation useful in the maintenance of patients suffering from high blood pressure. Over many clinical trials at many different locations, patients suffering from comparable hypertension (as measured by a New York Heart Classification) have been administered the standard therapy. Responses to therapy for this large patient group were classified into one of four response categories. Table 10.1 lists the categories and percentages of patients

**Table 10.1** Results of clinical trials using the standard preparation, example 10.2

| *Category* | *Percentage* |
|---|---|
| marked decrease in BP | 50% |
| moderate decrease in BP | 25% |
| slight decrease in BP | 10% |
| stationary or slight increase in BP | 15% |

treated on the standard preparation who have been classified in each category.

A clinical trial is conducted with a random sample of 200 patients suffering from high blood pressure. All patients are required to be listed according to the same hypertensive categories of the New York Heart Classification as those studied under the standard preparation. Use the sample data in table 10.2 to test the hypothesis that the cell probabilities associated with the test preparation are identical to those for the standard. Use $\alpha = .05$.

**Table 10.2** Sample data for example 10.2

| *Category* | *Observed Cell Counts* |
|---|---|
| 1 | 120 |
| 2 | 60 |
| 3 | 10 |
| 4 | 10 |

**SOLUTION**

This experiment possesses the characteristics of a multinomial experiment, with $n = 200$ and $k = 4$ outcomes.

Outcome 1: A person's blood pressure will decrease markedly after treatment on the test drug.

Outcome 2: A person's blood pressure will decrease moderately after treatment on the test drug.

Outcome 3: A person's blood pressure will decrease slightly after treatment on the test drug.

Outcome 4: A person's blood pressure will remain stationary or increase slightly after treatment on the test drug.

The null and alternative hypotheses are then

$H_0$: $p_1 = .50$, $p_2 = .25$, $p_3 = .10$, $p_4 = .15$.

$H_a$: At least one of the cell probabilities is different from the hypothesized value.

Before computing the test statistic, we must determine the expected cell numbers. These data are given in table 10.3.

**Table 10.3** Observed and expected cell numbers for example 10.2

| *Category* | *Observed Cell Number, $n_i$* | *Expected Cell Number, $E_i$* |
|---|---|---|
| 1 | 120 | 200(.50) = 100 |
| 2 | 60 | 200(.25) = 50 |
| 3 | 10 | 200(.10) = 20 |
| 4 | 10 | 200(.15) = 30 |

Since all the expected cell numbers are large, we may calculate the chi-square statistic and compare it to a tabulated value of the chi-square distribution.

$$\chi^2 = \sum_{i=1}^{4} \left[ \frac{(n_i - E_i)^2}{E_i} \right]$$

$$= \frac{(120 - 100)^2}{100} + \frac{(60 - 50)^2}{50} + \frac{(10 - 20)^2}{20} + \frac{(10 - 30)^2}{30}$$

$$= 4 + 2 + 5 + 13.33 = 24.33$$

For the probability of a type I error set at $\alpha = .05$, we look up the value of the chi-square statistic for a $= .05$ and df $= k - 1 = 3$. The critical value from table 3, the appendix, is 7.81473.

R.R.: Reject $H_0$ if $\chi^2 > 7.81473$.

Conclusion: Since the computed value of $\chi^2$ is greater than 7.81473, we reject the null hypothesis and conclude that at least one of the cell probabilities differs from that specified under $H_0$. Practically, it appears that a much higher proportion of patients treated with the test preparation fall into the moderate and marked improvement categories.

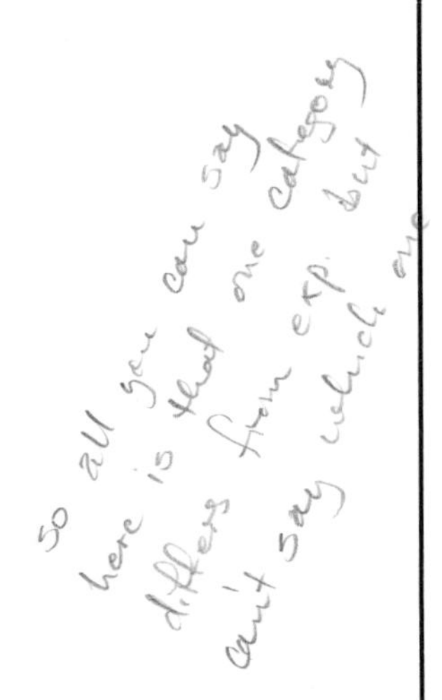

so all you can say
here is that one category
differs from exp. but
can't say which one


# Exercises

10.1. A work-study program was developed with a university and several industries in the surrounding community. Students were to work with industrial sociologists during a three-month internship. Equal numbers of students from the university were sent to a chemical, a textile, and a pharmaceutical industry. Students completing the program were classified according to the industry where they interned. Consider the following data as a random sample of the many students who could have completed the program. Test the null hypothesis that the probability that a finishing student interned in a pharmaceutical, chemical, or textile industry is $\frac{1}{3}$. Use $\alpha = .01$; $n_i$ is the number of students in group $i$ finishing the program.

| *Group* | $n_i$ |
|---|---|
| pharmaceutical | 20 |
| chemical | 13 |
| textile | 30 |

10.2. An experiment was conducted to determine if the proportion of mentally ill patients of each social class housed in a county facility agrees with the social class distribution of the county. The observed cell numbers for the 400 patients classified are given below.

| | |
|---|---|
| Lower: 215 | Upper-middle: 60 |
| Lower-middle: 100 | Upper: 25 |

Use these data to test the null hypothesis

$$p_1 = .25 \qquad p_2 = .48 \qquad p_3 = .20 \qquad p_4 = .07$$

which are the proportions of persons in the respective social class categories in the county. Use $\alpha = .05$.

10.3. In previous presidential elections in a given locality, 50% of the registered voters were Republicans, 40% were Democrats, and 10% were registered as independents. Prior to the upcoming election, a random sample of 200 registered voters showed that 90 were registered as Republicans, 80 as Democrats, and 30 as independents. Test the research hypothesis that the distribution of registered voters is different from previous election years. Use $\alpha = .05$.

+ 10.4. A local doctor suspects that there is a seasonal trend in the occurrence of the common cold. He estimates that 40% of the cases each year occur in the winter, 40% in the spring, 10% in the summer, and 10% in the fall. The following information was collected from a random sample of 1000 cases of the common cold over the past year:

| *Season* | *Frequency* |
|---|---|
| winter | 374 |
| spring | 292 |
| summer | 169 |
| fall | 165 |

Would you agree with the doctor's estimates, based on the sample information? Perform a statistical test of an hypothesis, using $\alpha = .05$.

10.5. Refer to exercise 10.4. What would the null hypothesis be if the doctor claimed that there are no differences in the number of cases over the seasons? Test the hypothesis that there is no seasonal trend in the occurrence of the common cold. Give the level of significance of your test.

10.6. Previous experimentation with a drug product developed for the relief of depression was conducted with normal adults with no signs of depression. We will assume a large data bank is available from studies conducted with normals and, for all practical purposes, the data bank can represent the population of responses for normals. Each of the adults participating in one of these studies was asked to rate the drug as ineffective, mildly effective, or effective. The percentages of respondents in these categories were 60%, 30%, and 10%, respectively. In a new study of depressed adults, a random sample of 85 adults responded as follows:

Ineffective: 30
Mildly effective: 35
Effective: 20

Is there evidence to indicate a different percentage distribution of responses for depressed adults than for normals? Give the level of significance for your test.

10.7. A random sample of 40 newspaper editors were interviewed to determine their opinions on the degree of future suppression of freedom of the press brought about by recent court decisions. The editors' opinions are summarized here.

| *Degree of Suppression* | *Frequency* |
|---|---|
| none | 8 |
| very little | 8 |
| moderate | 10 |
| severe | 14 |

Use these data to test the null hypothesis that each category is equally preferred. Use $\alpha = .05$.

# 10.3 The Binomial Distribution and the Normal Approximation to the Binomial

binomial experiment

The *binomial experiment* is a special case of the multinomial experiment, where each trial results in one of two outcomes, which we label, for lack of better terminology, either a "success" or a "failure." The binomial experiment is similar to a coin-tossing experiment where we observe either a head (success) or a tail (failure). The only difference is that we allow the probability of a success, denoted by $p$, to assume any value between 0 and 1. The probability of a failure, denoted by $q$, is then given by $q = 1 - p$.

Many experiments that scientists conduct are similar to a coin-tossing experiment. For example, a physician using a new drug preparation may classify patients as either improved or not improved. A political analyst interviewing potential voters may classify each as favoring the incumbent or not. A petroleum company must examine whether a proposed mileage ingredient does or does not increase gasoline mileage on a series of test runs for different makes of cars. Each of these situations involves the collection of dichotomous data, sometimes labeled as go-no-go or yes-no data.

The variable of interest in the binomial experiment is $y$, the number of successes in $n$ trials. The probability distribution for $y$ is given by the formula

probability distribution for $y$

$$P(y) = \frac{n!}{y!\,(n-y)!} p^y q^{n-y}$$

Note that this is a multinomial probability distribution (see p. 266) with

$$n_1 = y \qquad p_1 = p$$

$$n_2 = n - y \qquad p_2 = q$$

**EXAMPLE 10.3**

A drug company advertizes that a new product is effective in the treatment of a particular serious disease. However, the firm also states that an undesirable side effect occurs in 10% of the patients treated. If an attending physician has 4 unrelated patients for whom he could prescribe the drug, what is the probability that all 4 will experience the side effect?

**SOLUTION**

This example satisfies the characteristics of a binomial experiment, with $n = 4$ trials and $p$, the probability of observing a side effect, equal to .1. Substituting into the formula $P(y)$ with $y = 4, p = .1, q = 1 - p = .9$, and $n = 4$, we have

$$P(y = 4) = \frac{4!}{4!\,0!}(.1)^4(.9)^0 = (.1)^4 = .0001$$

Thus the probability that all 4 will experience the side effect is .0001.

**EXAMPLE 10.4**

Suppose the physician of example 10.3 treated all 4 patients and observed the indicated side effect in all 4. What might you conclude concerning the drug firm's claim?

**SOLUTION**

Since the probability of observing a side effect in all 4 is so small (.0001), assuming the side effect appears in only 10% of the patients treated as claimed by the drug firm, we would conclude that the company's claim is incorrect. Indeed, it appears that more than 10% of those treated experience the side effect.

Probabilities associated with values of $y$ can be computed for a binomial experiment for any values of $n$ or $p$, but as you might imagine, the task becomes more difficult when $n$ gets large. For example, suppose a sample of 1000 voters was polled to determine sentiment towards the consolidation of a city and county government. What would be the probability of observing 460 or fewer favoring consolidation if we assume that 50% of the entire population favor the change? Here we have a binomial experiment with $n = 1000$ and $p$, the probability of selecting a person favoring consolidation, equal to .5. To determine the probability of observing 460 or fewer favoring consolidation in the random sample of 1000 voters, we could compute $P(y)$ using the binomial formula for $y = 460, 459, \ldots, 0$. The derived probability would then be

$$P(y = 460) + P(y = 459) + \cdots + P(y = 0)$$

There would be 461 probabilities to calculate with each one being somewhat difficult due to the factorials. For example, the probability of observing 460 favoring consolidation is

$$P(y = 460) = \frac{1000!}{460!\,540!}(.5)^{460}(.5)^{540}$$

A similar calculation would be needed for all other values of $y$.

The normal distribution of chapter 3 can be used in many situations to approximate the binomial probability distribution, and areas under the normal curve can be used to *approximate* the actual binomial probabilities. The normal distribution that provides the best approximation to the binomial probability distribution has a mean and a standard deviation given by the formulas in the box.

**Normal Approximation to the Binomial Probability Distribution**

$$\mu = np \qquad \sigma = \sqrt{npq}$$

where $q = 1 - p$.

Note: This approximation can be used if

$$n \geq \frac{5}{\min(p, q)}$$

that is, if $n$ is greater than or equal to 5 divided by the minimum of $p$ and $q$.

---

### EXAMPLE 10.5

Use the normal approximation to the binomial to compute the probability of observing 460 or fewer in a sample of 1000 favoring consolidation if we assume that 50% of the entire population favor the change.

### SOLUTION

The normal distribution used to approximate the binomial distribution will have

$$\mu = np = 1000(.5) = 500$$

$$\sigma = \sqrt{npq} = \sqrt{1000(.5)(.5)} = 15.8$$

Note that

$$n \geq \frac{5}{.5} = 10$$

Hence we may use the normal approximation to the binomial. The desired probability is represented by the shaded area shown in figure 10.3.

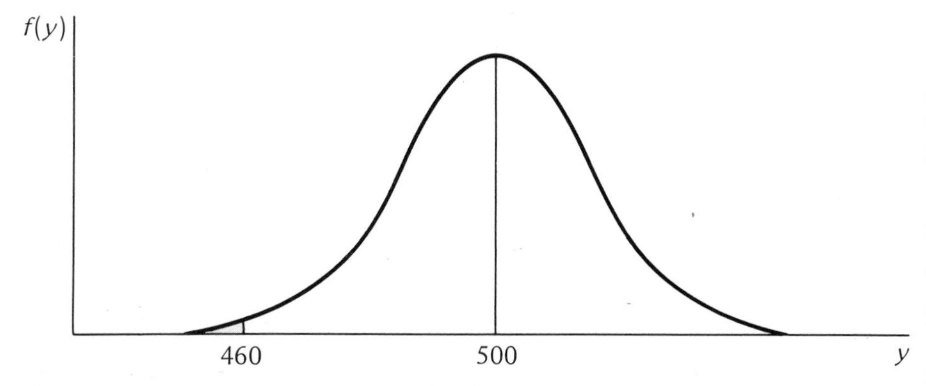


**Figure 10.3** Approximating normal distribution for the binomial distribution of example 10.5; $\mu = 500$ and $\sigma = 15.8$

We calculate the desired area by first computing

$$z = \frac{y - \mu}{\sigma} = \frac{460 - 500}{15.8} = -2.53$$

Referring to table 1 of the appendix, we find that the area under the normal curve between 460 and 500, that is, for $z = 2.53$, is .4943. Thus the probability of observing 460 or fewer favoring consolidation is approximately $.5 - .4943 = .0057$.

We have now considered calculating probabilities for the binomial distribution by using the formula $P(y)$ and by using the normal approximation to the binomial. In the next section we turn to inferences concerning the binomial parameter $p$.

# 10.4 Estimation and Tests of Hypotheses for $p$

The point estimate of the binomial parameter $p$ is one that we would choose intuitively. In a random sample of $n$ from a population in which the proportion of elements classified as successes is $p$, the best estimate of the parameter $p$ is the sample proportion of successes. Letting $y$ denote the number of successes in the $n$ sample trials, the sample proportion is

$\hat{p}$

$$\hat{p} = \frac{y}{n}$$

We observed in section 10.3 that $y$ possesses a mound-shaped probability distribution that can be approximated by using a normal curve when

$$n \geq \frac{5}{\min(p, q)}$$

In a similar way, the distribution of $\hat{p} = y/n$ can be approximated by a normal distribution with a mean and a standard deviation as given in the box.

**Mean and Standard Deviation of $\hat{p}$**

$$\mu_{\hat{p}} = p$$

$$\sigma_{\hat{p}} = \sqrt{\frac{pq}{n}}$$

The normal approximation to the distribution of $\hat{p}$ can be applied under the same condition as that for approximating $y$ by using a normal distribution. In fact, the approximation for both $y$ and $\hat{p}$ becomes more precise for large $n$. Henceforth in this text we will assume that $\hat{p}$ can be adequately approximated by using a normal distribution, and we will base all our inferences on results from our previous study of the normal distribution.

The reasoning applied in evaluating the goodness of $\hat{p} = y/n$ as a point estimate of $p$ is identical to that employed in chapter 4 when we used $\bar{y}$ as a point estimate of $\mu$. Approximately 95% of the point estimates $\hat{p}$ should lie within two standard deviations ($2\sigma_{\hat{p}}$) of their mean $p$ (see figure 10.4). Thus

$$2\sigma_{\hat{p}} = 2\sqrt{\frac{pq}{n}}$$

represents a bound on the error of estimation.

These results are summarized in the box.

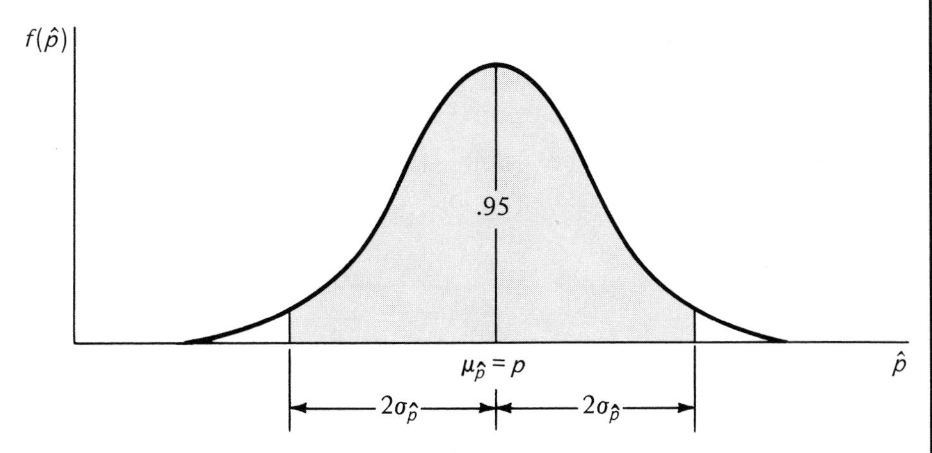


**Figure 10.4** Probability distribution of $\hat{p}$

---

**Point Estimation of the Binomial Parameter $p$**

Point estimate: $\hat{p} = \frac{y}{n}$.

Standard deviation of $p$: $\sigma_{\hat{p}} = \sqrt{\frac{pq}{n}}$.

Bound on the error of estimation: $2\sigma_{\hat{p}}$.

Note: Since $p$ is unknown, replace $p$ by $\hat{p}$ in $\sigma_{\hat{p}}$ to compute the bound on the error.

---

**EXAMPLE 10.6**

Legislators of a particular state were concerned that the enrollment (which affects budget allocations) at a particular university within the state system had been padded by allowing students to overenroll or to enroll for courses that required no academic work. To substantiate their initial findings, a random sample of 200 graduate students (from the 5000 currently enrolled) was interviewed. If 20 students stated they had been allowed to pad their enrollments the past quarter, use these data to determine a point estimate of the proportion $p$ of the entire student body with padded enrollments. Place a bound on the error of estimation.

**SOLUTION**

The point estimate of $p$ is

$$\hat{p} = \frac{y}{n} = \frac{20}{200} = .1$$

The corresponding bound on the error of estimation is

$$2\sigma_{\hat{p}} = 2\sqrt{\frac{(.1)(.9)}{200}} = 2(.021) = .042$$

Thus we are quite certain that .1 is within .042 of the actual proportion of padded enrollments in the entire graduate school.

An interval estimate of $p$ can be formulated using the methods of chapter 4. A general confidence interval for $p$ with a confidence coefficient of $(1 - \alpha)$ is given in the box.

**Confidence Interval for *p*, with Confidence Coefficient of $(1 - \alpha)$**

$$\hat{p} \pm z_{\alpha/2}\,\sigma_{\hat{p}}$$

where

$$\hat{p} = \frac{y}{n} \qquad \sigma_{\hat{p}} = \sqrt{\frac{pq}{n}}$$

Note: Since $p$ is unknown, replace $p$ by $\hat{p}$ in $\sigma_{\hat{p}}$.

**EXAMPLE 10.7**

Response to an advertising display was measured by counting the number of people who purchased the product out of the total number exposed to the display. If 330 purchased the product out of a total of 870 exposed, estimate the proportion of all persons exposed who will buy the product. Use a 90% confidence interval.

**SOLUTION**

For these data,

$$\hat{p} = \frac{y}{n} = \frac{330}{870} = .38$$

$$\sigma_{\hat{p}} = \sqrt{\frac{(.38)(.62)}{870}} = .016$$

The confidence coefficient for our example is .90. Recall from chapter 4 that we can obtain $z_{\alpha/2}$ by looking up the $z$-value in table 1 of the appendix corresponding to an area of $(1 - \alpha)/2$. For a confidence coefficient of .90, the $z$-value corresponding to an area of .45 is 1.645. Hence the 90% confidence interval on the proportion of persons who will purchase the product after exposure to this display is

$$.38 \pm 1.645(.016) \quad \text{or} \quad .38 \pm .026$$

A statistical test about a binomial parameter $p$ is very similar to the large-sample test concerning a population mean presented in chapter 4. These results are summarized in the box, with three different alternative hypotheses along with their corresponding rejection regions. Recall that only one alternative is chosen for a particular problem.

**Summary of a Statistical Test for $p$**

---

$H_0$: $p = p_0$ ($p_0$ is specified).

$H_a$: 1. $p > p_0$.
2. $p < p_0$.
3. $p \neq p_0$.

T.S.: $z = \dfrac{\hat{p} - p_0}{\sigma_{\hat{p}}}$.

R.R.: For a probability $\alpha$ of a type I error,

1. reject $H_0$ if $z > z_\alpha$;
2. reject $H_0$ if $z < -z_\alpha$;
3. reject $H_0$ if $|z| > z_{\alpha/2}$.

Note: Under $H_0$,

$$\sigma_{\hat{p}} = \sqrt{\frac{p_0 q_0}{n}}$$

---

**EXAMPLE 10.8**

Sports car owners in a town complain that their cars are judged differently from family style cars at the state vehicle inspection station. Previous records indicate that 30% of all passenger cars fail the inspection on the first time through. In a random sample of 150 sports cars, 60 failed the inspection on the first time through. Is there sufficient evidence to indicate that the percentage of first failures for sports cars is higher than the percentage for all passenger cars? Use $\alpha = .05$.

**SOLUTION**

The appropriate statistical test is as follows.

$H_0$: $p = .30$.

$H_a$: $p > .30$.

T.S.: $z = \dfrac{\hat{p} - p_0}{\sigma_{\hat{p}}}$.

R.R.: For $\alpha = .05$, we will reject $H_0$ if $z > 1.645$.

Using the sample data,

$$\hat{p} = \frac{60}{150} = .4 \quad \text{and} \quad \sigma_{\hat{p}} = \sqrt{\frac{(.3)(.7)}{150}} = .037$$

The test statistic is then

$$z = \frac{.4 - .3}{.037} = 2.7$$

Since the observed value of $z$ exceeds 1.645, we conclude that sports cars at the vehicle inspection station have a first-failure rate greater than .3. However, we must be careful not to attribute this difference to a difference in standards for sports cars and family style cars. Parallel testing of sports cars versus other cars would have to be conducted to eliminate other sources of variability that would perhaps account for the higher first-failure rate for sports cars.

# Exercises

10.8. The benign mucosal cyst is the most common lesion of a pair of sinuses in the upper jawbone. In a random sample of 800 males, 35 persons were observed to have a benign mucosal cyst.

a. Would it be appropriate to use a normal approximation in conducting a statistical test of the null hypothesis $H_0$: $p = .096$ (the highest incidence in previous studies among males)? Explain.

b. Conduct a statistical test of the research hypothesis $H_a$: $p < .096$. Use $\alpha = .05$.

10.9. National public opinion polls interview as few as 1500 persons in a random sampling of public sentiment towards one or more issues. These interviews are commonly done in person, because mail returns are poor and telephone interviews tend to reach older people, thus biasing the results. Suppose that a random sample of 1500 persons were surveyed to determine the proportion of the adult public in agreement with recent energy conservation proposals.

a. If 560 indicate they favor the policies set forth by the current administration, estimate $p$, the proportion of adults holding a "favor" opinion. Use a 95% confidence interval. What is the half width of the confidence interval?

b. How many persons must be surveyed to have a 95% confidence interval with a half width of .01? (Hint: The half width of the confidence interval is, in general,

$$z_{\alpha/2}\sqrt{\frac{pq}{n}}$$

with $p$ replaced by $\hat{p}$.)

10.10. A sample of 20 crayfish of all sizes was obtained from a large lake to estimate the proportion of crayfish that exhibit more than 9 (ppb) units of mercury. Of those sampled, 8 exceeded 9 units. Use these data to estimate $p$, the proportion of all crayfish in the lake with a mercury level greater than 9. Place a bound on the error of estimation.

e

*10.11. Simulate the binomial distribution for $n = 20$ and $p = .4$. Use a computer program. Do this by obtaining $y$, the number of successes in 20 trials when sampling from a binomial distribution with $p = .4$. Repeat this experiment 39 more times, for a total of 40 repetitions of the experiment.

a. Plot the sample data ($y$-values) in a relative frequency histogram.

b. Compute the sample mean and standard deviation. Compare your answers to the *actual* mean and standard deviation of $y$. (Hint: A Minitab program is given here for purposes of illustration.)

```
          BRANDOM 1 EXPERIMENT WITH N = 20, P = .4, PUT IN C1
40              :                   :           :
COMMANDS  BRANDOM 1 EXPERIMENT WITH N = 20, P = .4, PUT IN C1
          STOP
```

# Operating Characteristic Curves and Control Charts for the Binomial Parameter $p$ 10.5

OMIT

Two techniques are particularly appropriate for monitoring product quality as measured by $p$, the fraction of items that are defective. One technique, *control charts*, was discussed in chapter 5; they can be used to monitor $p$ as a measure of the ongoing quality of a manufacturing process. We are especially interested in detecting a shift in product quality. The second technique, called *lot acceptance sampling*, provides a means to screen (sample) ingoing raw materials or outgoing production from a plant where the product is shipped in

control charts

lot acceptance sampling

large quantities (often called "lots"). We begin by discussing lot acceptance sampling.

Not only are manufacturers interested in minimizing the amount or proportion of defective raw material to be used in the production process, but also they are interested in minimizing the proportion of defective finished products shipped from the plant. Thus they would like to sample, or screen, shipments of raw materials entering the plant and reject those shipments (lots) that contain too high a proportion of defectives. Similarly, they must screen the final product to make certain that a shipment does not contain too high a proportion of defectives.

The most obvious type of screen (sampling plan) to employ would be a careful inspection of each item from the lot. Unfortunately, this screen would be both costly and time-consuming. In addition, it would still be subject to errors in reporting brought about by human fatigue.

statistical sampling plan

Another type of screen is called a *statistical sampling plan*. Here we obtain a random sample of $n$ items from the lot. Each item of the sample is inspected, and if $y$, the number of defectives observed in the sample, is less than or equal to some predetermined number "a," we accept the lot. Thus a statistical sampling plan is designated by $n$, the sample size, and "a," the acceptance number. If the lot is accepted ($y \leq a$), we conclude that the proportion of defectives $p$ in the lot is small and acceptable. However, if $y > a$, we reject the lot and conclude $p$ is too large (above an acceptable level of defectives).

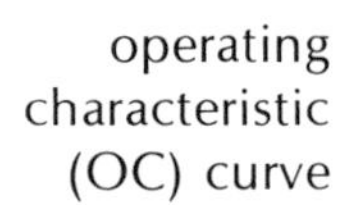


We can characterize the goodness of a particular sampling plan ($n$, a) by constructing an *operating characteristic* (*OC*) *curve*. The OC curve for a sampling plan is a graph displaying the probability of accepting a lot for various values of $p$, the proportion of defective items in the lot (see fig-

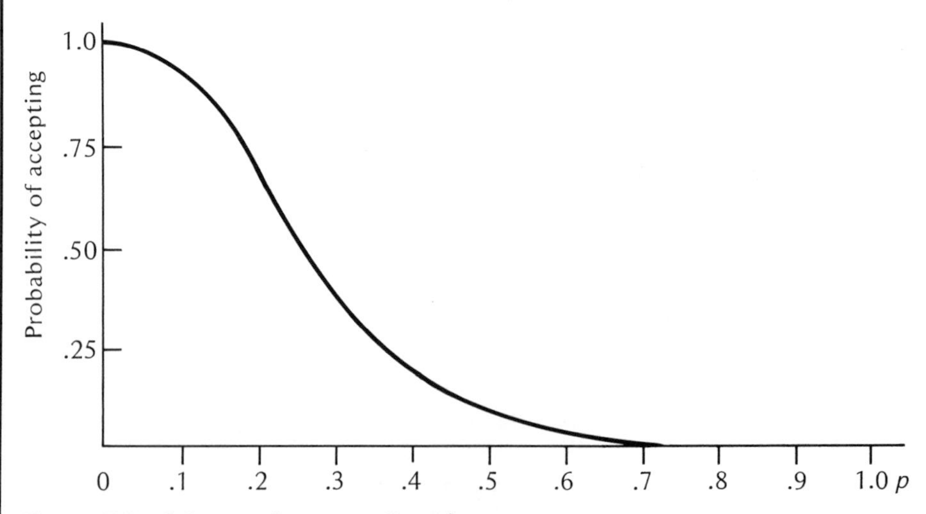


**Figure 10.5** OC curve for a sampling plan

ure 10.5). As you can see, the probability of accepting a lot decreases as the proportion of defectives within the lot increases.

We can construct an OC curve by computing the probability of accepting a lot, namely, $P(y \leq a)$, for several values of $p$. Consider the sampling plan $n = 4$ and $a = 0$. Here we sample 4 items and accept the lot if $y$, the number of defectives, is zero. Hence we must compute $P(y = 0)$ for $n = 4$ and for different values of $p$ to obtain the OC curve. We use the binomial probability distribution

$$P(y) = \frac{n!}{y!\,(n-y)!} p^y q^{n-y}$$

The results of the calculations are shown in table 10.4 for $p = .1, .2,$ and $.4$.

**Table 10.4** Calculation of $P(y = 0)$ for $n = 4$ and $p = .1, .2,$ and $.4$

| *Fraction of Defectives, p* | *Probability of Accepting, P(y = 0)* |
|---|---|
| .1 | $\frac{4!}{0!\,4!}(.1)^0(.9)^4 = .656$ |
| .2 | $\frac{4!}{0!\,4!}(.2)^0(.8)^4 = .410$ |
| .4 | $\frac{4!}{0!\,4!}(.4)^0(.6)^4 = .130$ |

Plotting these three points and connecting them, we have the OC curve shown in figure 10.6.

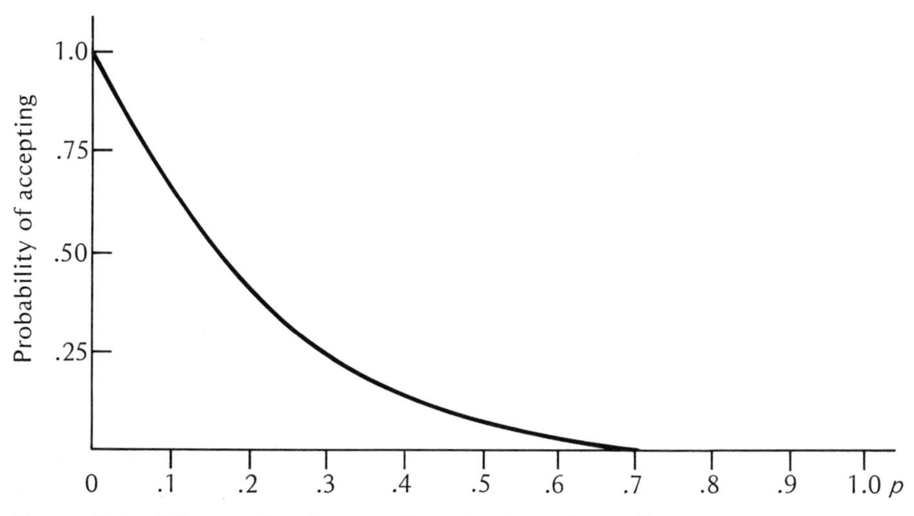


**Figure 10.6** OC curve for the sampling plan ($n = 4$, $a = 0$)

**EXAMPLE 10.9**

Construct an operating characteristic curve for the sampling plan ($n = 10$, a $= 1$).

**SOLUTION**

The probability of accepting the lot is given by $P(y \leq 1)$. Using the binomial probability distribution, we must calculate $P(y = 0) + P(y = 1)$ for $n = 10$ and various values of $p$. For $p = .1$ and $n = 10$,

$$P(y = 0) = \frac{10!}{0!\,10!}(.1)^0(.9)^{10} = .349$$

$$P(y = 1) = \frac{10!}{1!\,9!}(.1)^1(.9)^9 = .387$$

Hence the probability of accepting the lot is $.349 + .387 = .736$. Similarly, for $p = .2$ and $.4$, the probabilities of accepting the lot are found to be .376 and .046, respectively. Graphing our results we have the OC curve presented in figure 10.7.

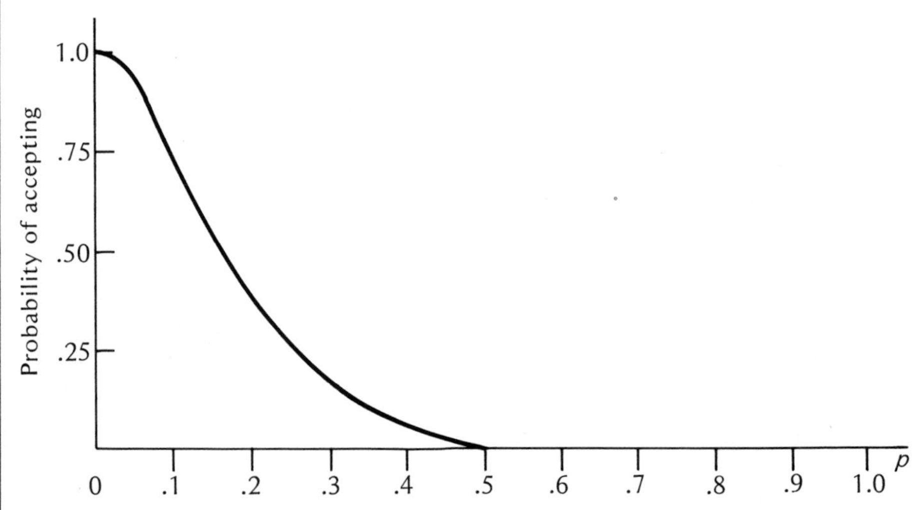


**Figure 10.7** OC curve for the sampling plan ($n = 10$, a $= 1$), example 10.9

Several comments should be made concerning statistical sampling plans. First, each plan ($n$, a) is unique; hence the inspector must choose a sampling plan that possesses characteristics suitable for his or her particular problem. In general, for a fixed value of "a," an increase in $n$ makes the graph of the probability of acceptance drop sharply as $p$ increases; increasing "a" for a fixed $n$ increases the probability of acceptance for values of $p$. Second, the OC curve for a particular sampling plan can be thought of as a plot of the probability of a type II error for the null hypothesis $H_0$: $p = 0$ for various actual values of $p$, when the rejection region is $y > a$.

control charts

The other method of monitoring product quality makes use of control charts for $p$. Recall from chapter 5 that a control chart typically consists of three lines. In a control chart for $p$, successive sample proportions $\hat{p}$ would be plotted and might appear as shown in figure 10.8. If one of the sample proportions falls

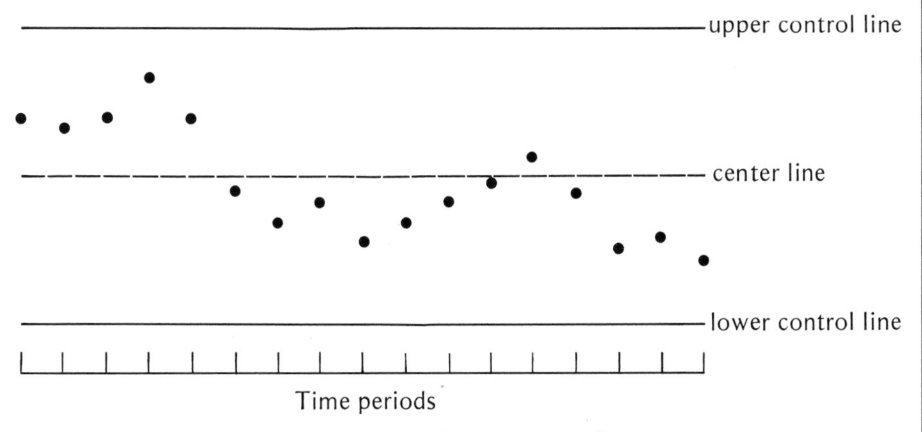


**Figure 10.8** Control chart for the sample proportion $\hat{p}$

outside either the upper or lower control line, the process is judged to be out of control; that is, the proportion of defectives $p$ in the production has shifted.

We can compute the three control lines in the following manner. The center line is designated by $\hat{p}_c$. If we obtain $k$ different random samples of $n$ observations each, $\hat{p}_c$ is the sample proportion of defectives for the entire set of $kn$ measurements, or, equivalently, $\hat{p}_c$ is the average of the sample proportions computed for the $k$ samples of $n$ measurements. The upper control line (UCL) and lower control line (LCL) are then

UCL

$$\text{UCL} = \hat{p}_c + 3\sqrt{\frac{\hat{p}_c\hat{q}_c}{n}}$$

LCL

$$\text{LCL} = \hat{p}_c - 3\sqrt{\frac{\hat{p}_c\hat{q}_c}{n}}$$

**EXAMPLE 10.10**

A pharmaceutical firm has been investigating the possibility of having hospital personnel supplied with small disposable vials that can be used to perform many of the standard laboratory analyses. For a particular analysis, such as blood sugar, the technician would insert a measured amount of fluid (perhaps blood) in an appropriate vial and observe its color when thoroughly mixed with the fluid already stored in the vial. By comparing the optical density of the combined fluid to a color-coded chart, the technician would have a reading on the blood sugar level of the patient. Quite obviously, the system must be tightly controlled to insure that the vials are correctly sealed with the proper amount of fluid prior to shipment to the hospital laboratories. The data in table 10.5 give the proportion of

**Table 10.5** 30 sample proportions each based on 50 observations, example 10.10

| *Sample* | *Sample Proportion Defective* | *Sample* | *Sample Proportion Defective* |
|---|---|---|---|
| 1 | .18 | 16 | .18 |
| 2 | .10 | 17 | .14 |
| 3 | .18 | 18 | .16 |
| 4 | .16 | 19 | .14 |
| 5 | .12 | 20 | .12 |
| 6 | .12 | 21 | .18 |
| 7 | .16 | 22 | .16 |
| 8 | .18 | 23 | .18 |
| 9 | .12 | 24 | .18 |
| 10 | .12 | 25 | .14 |
| 11 | .16 | 26 | .16 |
| 12 | .18 | 27 | .18 |
| 13 | .12 | 28 | .16 |
| 14 | .18 | 29 | .18 |
| 15 | .14 | 30 | .10 |

defectives in 30 different samples (taken from 30 different production hours) of 50 vials each. Use these data to construct the three control lines.

### SOLUTION

The sum of the 30 sample proportions is 4.58; hence

$$\hat{p}_c = \frac{4.58}{30} = .15 \qquad \text{and} \qquad \hat{q}_c = 1 - .15 = .85$$

Substituting into the formulas for the control limits, we have

$$\text{UCL} = .15 + 3\sqrt{\frac{(.15)(.85)}{50}} = .15 + 3(.05) = .30$$

$$\text{LCL} = .15 - 3\sqrt{\frac{(.15)(.85)}{50}} = .15 - 3(.05) = 0$$

We have discussed the binomial probability distribution and various count data problems which utilize the binomial distribution. In the next section we will consider another discrete random variable that is useful for analyzing count data.

# Exercises

10.12. Sketch the operating characteristic curve for the sampling plan ($n = 5$, a $= 0$).

10.13. Refer to exercise 10.12. Superimpose the OC curve for the sampling plan with $n = 10$ and a $= 0$ on that for ($n = 5$, a $= 0$). What is the effect of increasing the sample size $n$ while holding the acceptance number "a" constant?

10.14. Refer to the control limits for $p$ obtained in example 10.10. Suppose that you are now in charge of quality control for the vial production line. In the next 15 samples of 50 vials, you observe the following numbers of defectives:

| *Number* | 1 | 2 | 3 | 4 | 5 | 6 | 7 | 8 | 9 | 10 | 11 | 12 | 13 | 14 | 15 |
|---|---|---|---|---|---|---|---|---|---|---|---|---|---|---|---|
| *Defective* | 9 | 8 | 6 | 5 | 2 | 4 | 6 | 8 | 9 | 9 | 12 | 13 | 14 | 10 | 12 |

Determine whether the process has remained in control.

# The Poisson Distribution 10.6

In section 10.3 we indicated that the normal distribution provided a good approximation to the binomial probability distribution provided $n \geq 5/\min(p, q)$. This requirement was needed to ensure that the binomial distribution was reasonably symmetric. However, there are many instances when the binomial probability distribution is sufficiently skewed so as to render the normal approximation inappropriate. For example, in observing patients administered a new drug product in a properly conducted clinical trial, the number of persons experiencing a particular side effect might be quite small. Indeed, if $p$, the probability of observing a person with the side effect, is .001 and $n = 1000$, then $\min(p, q) = .001$. Hence the normal approximation to the binomial distribution is inappropriate for calculating probabilities in this circumstance.

1000 < 5/.001 = 5000

Since the binomial probabilities would still be difficult to calculate in situations where $p$ is small and $n$ is large, we seek another approximation. S. D. Poisson developed, in 1837, a discrete probability distribution, suitably called

Poisson distribution

the Poisson distribution, which provides a good approximation to the binomial when $p$ is small and $n$ is large but $np$ is less than 5. The probability of observing $y$ successes in the $n$ trials is given by the formula

$$P(y) = \frac{\mu^y e^{-\mu}}{y!}$$

$e = 2.71828$

where e is a constant, the number 2.71828, and $\mu$ is the average value of $y$. Table 7 of the appendix gives Poisson probabilities for various values of the parameter $\mu$.

**EXAMPLE 10.11**

Refer to the clinical trial alluded to at the beginning of this section, where $n = 1000$ patients were treated with a new drug. Compute the probability that none of the patients experiences a particular side effect (such as nausea). Assume $p = .001$.

**SOLUTION**

The mean of the binomial distribution is $\mu = np = 1000(.001) = 1$. Substituting into the Poisson probability distribution, with $\mu = 1$, we have

$$P(y = 0) = \frac{(1)^0 e^{-1}}{0!} = e^{-1} = \frac{1}{2.71828} = .367879$$

(Note also from table 7 of the appendix that the entry corresponding to $y = 0$ and $\mu = 1$ is .367879.)

**EXAMPLE 10.12**

Suppose after a clinical trial involving 1000 patients, none experienced nausea. Would it be reasonable to infer that less than .001 of the entire population would experience this side effect while taking the drug?

**SOLUTION**

Certainly not. We computed the probability of observing $y = 0$ in $n = 1000$ trials assuming $p = .001$ (i.e., assuming .1% of the population would experience nausea) to be .368. Since this probability is quite large, it would not be wise to infer that $p < .001$.

Although the Poisson distribution provides a useful approximation to the binomial under certain conditions, the application of the Poisson distribution is not limited to these situations. In particular, the Poisson distribution has been useful in finding the probability of $y$ occurrences of an event during a specified interval of time, volume, space, and so on, provided four assumptions are met.

assumptions for Poisson distribution

1. The interval can be subdivided such that the probability of an occurrence in a subinterval is small.
2. The probability of a single occurrence in a subinterval is constant across the subintervals.
3. The probability of more than one occurrence in a subinterval is zero.
4. The occurrence of an event in one subinterval does not affect or alter the probability of an occurrence in any other subinterval.

While these assumptions seem to be somewhat restrictive, many situations appear to satisfy these conditions. For example, the number of arrivals of customers at a checkout counter, parking lot toll booth, inspection station, or garage repair shop during a specified time interval (such as one minute) could be approximated with a Poisson probability distribution. Similarly, the number of clumps of algae of a particular species observed in a unit volume of lake water visible under a microscope could be approximated by a Poisson probability distribution.

Confronted with a set of measurements, we may now wish to check the assumption that the data follow a Poisson probability distribution. To do this we make use of the goodness-of-fit test of section 10.2, using the test statistic

tests using Poisson distribution

$$\chi^2 = \sum_{i=1}^{k} \left[ \frac{(n_i - E_i)^2}{E_i} \right]$$

1st hypothesis

There are two types of null hypotheses. The first hypothesis is that the data arise from a Poisson distribution with $\mu = \mu_0$; that is, we wish to test $H_0$: $\mu = \mu_0$ ($\mu_0$ is specified) against the alternative hypothesis $H_a$: $\mu \neq \mu_0$. The quantity $n_i$ denotes the number of observations in cell $i$ and $E_i$ is the expected number of observations in cell $i$ obtained from the probabilities for a Poisson distribution with mean $\mu_0$. The computed value of the test statistic is then compared to the tabulated chi-square value in table 3 of the appendix with $a = \alpha$ and $df = k - 1$.

The second null hypothesis we might be interested in is less specific. We test

2nd hypothesis

$H_0$: The observed cell counts all come from a common Poisson distribution with mean $\mu$ (unspecified)

The alternative is that not all cell counts come from a common Poisson distribution. The test statistic is

$$\chi^2 = \sum_{i=1}^{k} \left[ \frac{(n_i - E_i)^2}{E_i} \right]$$

where for all cells $E_i$ is the expected number of observations in cell $i$ obtained from the probabilities for a Poisson distribution with a mean estimated from the sample data. The rejection region is then located for a = $\alpha$ and df = $k - 2$. Note the difference in the degrees of freedom for the two null hypotheses. In the latter test we lose one degree of freedom because we must estimate the Poisson parameter $\mu$.

**EXAMPLE 10.13**

Environmental engineers often utilize information contained in the number of different alga species and the number of cell clumps per species to measure the health of a lake. Those lakes exhibiting only a few species but many cell clumps are classified as oligotrophic. In one such investigation a lake sample was analyzed under a microscope to determine the number of clumps of cells per microscopic field. These data are summarized below for 150 fields examined under a microscope. Here $y_i$ denotes the number of cell clumps per field and $n_i$ denotes the number of fields with $y_i$ cell clumps.

| $y_i$ | 0 | 1 | 2 | 3 | 4 | 5 | 6 | $\geq 7$ |
|---|---|---|---|---|---|---|---|---|
| $n_i$ | 6 | 23 | 29 | 31 | 27 | 13 | 8 | 13 |

Test the null hypothesis that the sample data were drawn from a Poisson probability distribution. Use $\alpha = .05$.

**SOLUTION**

Before we can compute the value of $\chi^2$, we must first estimate the Poisson parameter $\mu$ and then compute the expected cell counts. The Poisson mean $\mu$ is estimated by using the sample mean $\bar{y}$. For these data,

$$\bar{y} = \frac{\sum_{i=1}^{k} n_i y_i}{n} = \frac{486}{150} \approx 3.3$$

It should be noted that the sample mean was computed to be 3.3 by using all the sample data before the 13 largest values were collapsed into the final cell. This is why the sample mean computed here was rounded up to 3.3.

The Poisson probabilities for $y = 0, 1, \ldots, 7$ or more can be found in table 7 of the appendix with $\mu = 3.3$. These probabilities are shown below.

| $y_i$ | 0 | 1 | 2 | 3 | 4 | 5 | 6 | $\geq 7$ |
|---|---|---|---|---|---|---|---|---|
| $P(y_i)$ for $\mu = 3.3$ | .037 | .121 | .201 | .221 | .182 | .120 | .066 | .051 |

The expected cell count $E_i$ can be computed for any cell using the formula

$$E_i = n\,P(y_i)$$

Hence for our data (with $n = 150$), the expected cell counts are as shown below.

| $y_i$ | 0 | 1 | 2 | 3 | 4 | 5 | 6 | $\geq 7$ |
|---|---|---|---|---|---|---|---|---|
| $E_i$ | 5.55 | 18.15 | 30.15 | 33.15 | 27.30 | 18.00 | 9.90 | 7.65 |

Substituting these values into the test statistic, we have

$$\chi^2 = \sum_{i=1}^{8} \left[\frac{(n_i - E_i)^2}{E_i}\right]$$

$$= \frac{(6 - 5.55)^2}{5.55} + \frac{(23 - 18.15)^2}{18.15} + \cdots + \frac{(13 - 7.65)^2}{7.65} = 7.01$$

The tabulated value of chi-square for a = .05 and df = $k - 2 = 6$ is 12.59. Since the computed value of chi-square does not exceed 12.59, we have insufficient evidence to reject the null hypothesis that the data were collected from a Poisson distribution.

A word of caution is given here for situations where we are considering this test procedure. As we mentioned previously, when using a chi-square statistic, we should have all expected cell counts fairly large. In particular, we want all $E_i > 1$ and not more than 20% less than 5. If values of $y \geq 7$ had been considered individually, the $E_i$'s would not have satisfied the criteria for the use of $\chi^2$. That is why we combined all values of $y \geq 7$ into one category.

sample sizes & lumping

# Exercises

10.15. One portion of a study to determine the effectiveness of an exclusive bus lane was directed at examining the number of conflicts (driving situations that could result in an accident) at a major intersection during a specified period of time. A previous study prior to the installation of the exclusive bus lane indicated that the number of conflicts per 5 minutes during the 7–9 A.M. peak period could be adequately approximated by a Poisson distribution with $\mu = 2$. The following data were based on a sample of 10 days; $y_i$ denotes the number of conflicts and $n_i$ denotes the number of 5-minute periods during which $y$ was observed.

| $y_i$ | 0 | 1 | 2 | 3 | 4 | 5 | $\geq 6$ |
|---|---|---|---|---|---|---|---|
| $n_i$ | 90 | 230 | 240 | 130 | 68 | 30 | 12 |

Use these data to test the research hypothesis that the mean number of conflicts per 5 minutes differs from 2.

$ 10.16. The number of shutdowns per day caused by a breaking of the thread was noted for a nylon spinning process over a period of 10 days. Use the sample data below to determine if the number of shutdowns per day follows a Poisson distribution. Use $\alpha = .05$. In the listing of the data, $y_i$ denotes the number of shutdowns per day and $n_i$ denotes the number of days with $y_i$ shutdowns.

| $y_i$ | 0 | 1 | 2 | 3 | 4 | $\geq 5$ |
|---|---|---|---|---|---|---|
| $n_i$ | 20 | 28 | 15 | 8 | 7 | 12 |

## 10.7 $r \times c$ Contingency Tables: Chi-square Test of Independence

In all our calculations so far in this text, we have assumed that only one measurement is taken on each sampling unit. We might obtain the yield for an acre planted in wheat, a blood pressure reading on a patient who is being administered an anesthetic, or a measurement on the number of potential conflicts at a highway intersection during a one-hour period. However, research problems in the sciences frequently involve more than one variable. If measurements are taken on two (or more) variables for each sampling unit, we say that we have *bivariate* (or *multivariate*) *data*.

bivariate and multivariate data

As with univariate count data, where the data may be summarized in a table, we frequently arrange bivariate data in a two-way table. For example, in a study of the public approval of a proposed high-speed bus lane for commuters, the interviewers might also ask individuals information about their occupations. We could then classify each person by his or her opinion concerning the new lane (favor, do not favor, undecided) and his or her occupation (white-collar worker, blue-collar worker, laborer).

What is the objective of such a classification? In most studies either we wish to determine whether the two variables are related (dependent) or we wish to predict one variable based on knowledge of the other. This section deals with a test of independence for bivariate count data arranged in a two-way table. The two-way tables are sometimes called *contingency tables* because the alternative hypothesis in our test is that the two variables are dependent (i.e., there is a contingency between the two variables).

contingency tables

Consider the problem where we would like to determine whether the following two variables are dependent: employee classification (staff, faculty, administrator) at a university and an employee's opinion about whether or not the local chapter of the teachers' union should be the sole collective bargaining agent for employee benefits. A random sample of 200 employees is taken from employee records and each employee is classified according to both variables. Suppose the results of the survey appear as shown in table 10.6. Is there evidence to indicate that a person's opinion concerning collective bargaining depends on his or her employment status? That is, can we conclude that the two variables are dependent?

**Table 10.6** Classification of 200 employees by classification and opinion on collective bargaining

| | *Opinion on Collective Bargaining by Teachers' Union* | | | |
|---|---|---|---|---|
| *Employee Classification* | *Favor* | *Do Not Favor* | *Undecided* | *Totals* |
| staff | 30 | 15 | 15 | 60 |
| faculty | 40 | 50 | 10 | 100 |
| administrator | 10 | 25 | 5 | 40 |
| *Totals* | 80 | 90 | 30 | 200 |

To answer this question we must define the concept of *independence*.

independence

**Definition 10.2**

Two variables that have been categorized in a two-way table are *independent* if the probability that a measurement is classified into a given cell of the table is equal to the probability of being classified into that row times the probability of being classified into that column. This must be true for all cells of the table.

For example, suppose that the probability of selecting a person favoring the teachers' union in the university survey is $p_1$, the probability of selecting one who does not favor the union for collective bargaining is $p_2$, and the probability of selecting a person being undecided is $p_3$ (note: $p_1 + p_2 + p_3 = 1$). Similarly, suppose that the probabilities of selecting a staff member, a faculty member, or an administrator are, respectively, $p_A$, $p_B$, and $p_C$ (where $p_A + p_B + p_C = 1$). Then the two variables, employee classification and opinion concerning the teachers' union, are independent if the probability of classifying a person into a specific cell of the two-way table is obtained by multiplying the respective row and column probabilities. These ideas are illustrated in table 10.7.

**Table 10.7** Cell probabilities showing independence for the collective bargaining survey

| | *Opinion* | | |
|---|---|---|---|
| *Employee Classification* | *Favor,* $p_1$ | *Do Not Favor,* $p_2$ | *Undecided,* $p_3$ |
| staff, $p_A$ | $p_Ap_1$ | $p_Ap_2$ | $p_Ap_3$ |
| faculty, $p_B$ | $p_Bp_1$ | $p_Bp_2$ | $p_Bp_3$ |
| administrator, $p_C$ | $p_Cp_1$ | $p_Cp_2$ | $p_Cp_3$ |

A test of the independence of two variables arranged in a two-way table makes use of the test statistic

test statistic

$$\chi^2 = \sum_i \sum_j \left[\frac{(n_{ij} - E_{ij})^2}{E_{ij}}\right]$$

where $n_{ij}$ and $E_{ij}$ are, respectively, the observed and expected number of measurements falling in the cell for the $i$th row and the $j$th column.

**Definition 10.3**

The expected number of measurements $E_{ij}$ falling in the $i, j$ cell (cell of the $i$th row and $j$th column of the table) is taken to be

$$E_{ij} = \frac{(\text{row } i \text{ total})(\text{column } j \text{ total})}{n}$$

when the two variables are independent.

**EXAMPLE 10.14**

Compute the expected number of measurements falling into each cell of table 10.6.

**SOLUTION**

The expected number of measurements falling in the 1, 1 cell (1st row, 1st column) is

$$E_{11} = \frac{\text{(row 1 total)(column 1 total)}}{n} = \frac{(60)(80)}{200} = 24$$

Similarly, the expected number of measurements in the 3, 2 cell is

$$E_{32} = \frac{\text{(row 3 total)(column 2 total)}}{n} = \frac{(40)(90)}{200} = 18$$

These and the remaining cell counts appear in table 10.8. Note that the expected counts in a row sum to the same row total as the observed cell counts. The same applies for columns.

**Table 10.8** Expected cell counts for the collective bargaining survey

| *Employee Classification* | *Opinion* | | | |
|---|---|---|---|---|
| | *Favor* | *Do Not Favor* | *Undecided* | *Totals* |
| staff | 24 | 27 | 9 | 60 |
| faculty | 40 | 45 | 15 | 100 |
| administrator | 16 | 18 | 6 | 40 |
| *Totals* | 80 | 90 | 30 | 200 |

chi-square test of independence

We can now summarize the chi-square test of independence for data arranged in a two-way table.

**Chi-square Test of Independence**

Null hypothesis: The two variables are independent.

Alternative hypothesis: The two variables are dependent.

Test statistic: $\chi^2 = \sum_i \sum_j \left[\frac{(n_{ij} - E_{ij})^2}{E_{ij}}\right]$.

Rejection region: Reject $H_0$ if $\chi^2$ exceeds the tabulated value of chi-square (table 3 of the appendix) for $a = \alpha$ and $\text{df} = (r - 1)(c - 1)$, where

$r$ = number of rows in the table

$c$ = number of columns in the table

$r$

$c$

The guidelines that Cochran (1954) proposed for the $E_i$'s (see p. 270) are still in effect when we use the chi-square test of independence. While agreeing with Cochran, Conover (1971) goes even further by stating that when the $E_i$'s are all of about the same magnitude and both $r$ and $c$ are large, then even if the $E_i$'s are as small as 1, the approximation by a chi-square distribution will still be good. These guidelines give us a great deal of flexibility in applying the chi-square test without having to collapse some of the categories.

**EXAMPLE 10.15**

Conduct a chi-square test of independence for the teachers' union data in table 10.6. Use $\alpha = .05$.

**SOLUTION**

Using the observed cell counts of table 10.6 and the estimated expected cell counts of table 10.8, we can substitute these values into the test statistic.

$$\chi^2 = \sum_i \sum_j \left[ \frac{(n_{ij} - E_{ij})^2}{E_{ij}} \right]$$

$$= \frac{(30-24)^2}{24} + \frac{(15-27)^2}{27} + \frac{(15-9)^2}{9} + \frac{(40-40)^2}{40} + \frac{(50-45)^2}{45}$$

$$+ \frac{(10-15)^2}{15} + \frac{(10-16)^2}{16} + \frac{(25-18)^2}{18} + \frac{(5-6)^2}{6}$$

$$= 18.2$$

The critical value of $\chi^2$ for a = .05 and df = $(r-1)(c-1) = 2(2) = 4$ is 9.48773. Since the computed value, 18.2, exceeds 9.48773, we reject $H_0$ and conclude that the two variables are dependent. In particular, we say that the proportion of persons favoring the teachers' union as the collective bargaining agent varies depending on the employee status. From table 10.6 we see that a much higher proportion of the staff members favor the teachers' union as the collective bargaining agent than either the faculty or administrators.

# 10.8 Summary

In this chapter we have considered the analysis of count (enumerative) data. Although by no means have we studied all possible methods of analysis, we have examined some of the more important methods. We first considered the multinomial experiment which gives rise to count data. The chi-square goodness-of-fit test was then used to test that the cell probabilities of a multinomial experiment are equal to certain specified values.

Having discussed goodness-of-fit tests for data generated from a multinomial experiment, we considered two other discrete probability distributions. The binomial, which is a special case of the multinomial, and the Poisson distribution. For $n \geq 5/\min(p, q)$, the normal distribution can be used to approximate the binomial distribution. Similarly, for $n$ large, $p$ small, and $np < 5$, the Poisson is a useful approximation to the binomial. Specific test and estimation problems were discussed for the binomial and Poisson distributions.

Finally, we considered a test of independence for bivariate count data arranged in a two-way contingency table.

In the next chapter we turn to a discussion of a new class of statistical tests. These tests, which are not specifically related to parameters for the distribution of the sample measurements, are called nonparametric statistical tests.

# Exercises

10.17. A study was conducted to investigate whether there is any relationship between voting record and education. One hundred citizens selected at random were interviewed as to how often they vote and what level of formal education they had achieved. The results were as follows:

| | *How Often Do You Vote?* | | | |
|---|---|---|---|---|
| *Education* | *Never* | *Some Elections* | *All Elections* | *Totals* |
| less than h.s. level | 11 | 12 | 11 | 34 |
| high school | 7 | 15 | 13 | 35 |
| college | 2 | 20 | 14 | 36 |
| *Totals* | 20 | 47 | 38 | 105 |

Is there evidence to indicate a relationship between level of education and frequency of voting? Use $\alpha = .05$.

10.18. An extension of the traffic study for the implementation of a priority bus lane involved the sampling of public opinion concerning the bus lane during various phases of the study. Three different phases were to be studied. Phase 0 required the bus drivers to use the existing traffic lanes. In Phase 1, bus drivers made use of the exclusive bus lane with no preemption of the traffic signals at the intersection; Phase 2 allowed the bus drivers to extend the "green time" on a traffic signal to allow them to pass through before the light changed. Use the sample data below to determine whether the distribution of persons favoring, not favoring, or undecided changes from phase to phase. Use $\alpha = .05$.

| | *Opinion* | | | |
|---|---|---|---|---|
| *Phase* | *Favor* | *Do Not Favor* | *Undecided* | *Totals* |
| 0 | 80 | 90 | 30 | 200 |
| 1 | 60 | 112 | 28 | 200 |
| 2 | 50 | 125 | 25 | 200 |
| | 190 | 327 | 83 | 600 |

10.19. A random sample of 145 people of various occupations was taken to investigate the public opinion on police treatment. Each person was questioned as to whether he would expect the police to treat him as good, better, or worse than a common criminal. The following table summarizes the results:

| | *Expected Treatment* | | | |
|---|---|---|---|---|
| *Occupation* | *Better* | *As Good* | *Worse* | *Totals* |
| unemployed | 6 | 23 | 11 | 40 |
| blue-collar worker | 17 | 30 | 8 | 55 |
| white-collar worker | 16 | 28 | 6 | 50 |
| *Totals* | 39 | 81 | 25 | 145 |

Is there sufficient evidence to indicate that the expected treatment is independent of occupation? Use $\alpha = .10$.

*10.20. A sociological study was conducted to determine whether there is a relationship between the length of time blue-collar workers remain in their first job and the amount of their education. From union membership records, a random sample of persons was classified. The data are shown below.

a. Use the SPSS (1975) computer output that follows to identify the expected cell numbers.

b. Test the research hypothesis that the variable "length of time on first job" is related to the variable "amount of education."

| | *Years of Education* | | | |
|---|---|---|---|---|
| *Years on First Job* | *0–4.5* | *4.5–9* | *9–13.5* | *13.5* |
| 0–2.5 | 5 | 21 | 30 | 33 |
| 2.5–5 | 15 | 35 | 40 | 30 |
| 5–7.5 | 22 | 16 | 15 | 30 |
| 7.5 | 28 | 10 | 8 | 10 |

c. Give the level of significance for the test.

d. Draw your conclusions.

TABLE NO. 1 B (VAR 2) VS A (VAR 1)
USING THE COUNTS IN FRQUENCY (VAR 3)

| CELL FREQUENCY COUNTS | | | A | | (VAR 1) | | |
|---|---|---|---|---|---|---|---|
| | | | LT 4.5 | 4.5T 9.0 | 9. T13.5 | GT 13.5 | TOTAL |
| | | | 1.00 | 2.00 | 3.00 | 4.00 | |
| B | LT 2.5 | 1.00 | 5 | 21 | 30 | 33 | 89 |
| (VAR 2) | 2.5T 5.0 | 2.00 | 15 | 35 | 40 | 30 | 120 |
| | 5.0T 7.5 | 3.00 | 22 | 16 | 15 | 30 | 83 |
| | GT 7.5 | 4.00 | 28 | 10 | 8 | 10 | 56 |
| | TOTAL | | 70 | 82 | 93 | 103 | 348 |

********************************************************************************

STATISTICS BASED ON THE FREQUENCY TABLE

| STATISTIC | VALUE | D.F. | PROBABILITY | STATISTIC | VALUE | D.F. | PROBABILITY |
|---|---|---|---|---|---|---|---|
| CHISQUARE | 57.830 | 9 | 0.0 | | | | |
| MAX. LIKELIHOOD CHISQUARE | 55.606 | 9 | 0.0 | | | | |
| PHI | 0.408 | | | CRAMER'S V | 0.235 | | |
| CONTINGENCY COEFFICIENT C | 0.377 | | | | | | |
| MCNEMAR'S CHISQUARE | 35.663 | 6 | 0.0000 | | | | |

| STATISTIC | VALUE | ASYMPTOTIC ST. ERROR | VALUE /ASE | DEP. VAR. | STATISTIC | VALUE | ASYMPTOTIC ST. ERROR | VALUE /ASE | DEP. VAR. |
|---|---|---|---|---|---|---|---|---|---|
| PRODUCT MOMENT CORRELATION | −0.274 | 0.052 | −5.306 | | SPEARMAN'S CORRELATION COEF. | −0.252 | 0.052 | −4.836 | |
| GAMMA | −0.282 | 0.059 | −4.760 | | KENDALL'S TAU-B | −0.212 | 0.036 | −5.901 | |
| | | | | | STUART'S TAU-C | −0.209 | 0.045 | −4.641 | |
| SOMER'S D | −0.210 | 0.045 | −4.675 | 2 | SOMER'S D | −0.214 | 0.046 | −4.674 | 1 |
| LAMBDA-SYMMETRIC | 0.093 | 0.037 | 2.511 | | | | | | |
| LAMBDA-ASYMMETRIC | 0.070 | 0.043 | 1.624 | 2 | LAMBDA-ASYMMETRIC | 0.114 | 0.039 | 2.922 | 1 |
| LAMBDA-STAR-ASYMMETRIC | 0.081 | 0.044 | 1.869 | 2 | LAMBDA-STAR-ASYMMETRIC | 0.143 | 0.041 | 3.493 | 1 |
| TAU-ASYMMETRIC | 0.049 | 0.013 | 3.787 | 2 | TAU-ASYMMETRIC | 0.050 | 0.014 | 3.612 | 1 |
| UNCERTAINTY COEF.-SYMMETRIC | 0.029 | 0.0 | 0.0 | | | | | | |
| UNCERTAINTY COEF.-ASYMMETRIC | 1.835 | 0.623 | 2.945 | 2 | UNCERTAINTY COEF.-ASYMMETRIC | 1.870 | 0.556 | 3.362 | 1 |

********************************************************************************

PERCENTAGES OF THE TOTAL FREQUENCY

| | | LT 4.5 | 4.5T 9.0 | 9. T13.5 | GT 13.5 | TOTAL |
|---|---|---|---|---|---|---|
| | | 1.00 | 2.00 | 3.00 | 4.00 | |
| LT 2.5 | 1.00 | 1.44 | 6.03 | 8.62 | 9.48 | 25.57 |
| 2.5T 5.0 | 2.00 | 4.31 | 10.06 | 11.49 | 8.62 | 34.48 |
| 5.0T 7.5 | 3.00 | 6.32 | 4.60 | 4.31 | 8.62 | 23.85 |
| GT 7.5 | 4.00 | 8.05 | 2.87 | 2.30 | 2.87 | 16.09 |
| TOTAL | | 20.11 | 23.56 | 26.72 | 29.60 | 100.00 |

PERCENTAGES OF THE ROW TOTALS

| | | LT 4.5<br>1.00 | 4.5T 9.0<br>2.00 | 9. T13.5<br>3.00 | GT 13.5<br>4.00 | TOTAL |
|---|---|---|---|---|---|---|
| LT 2.5 | 1.00 | 5.62 | 23.60 | 33.71 | 37.08 | 100.00 |
| 2.5T 5.0 | 2.00 | 12.50 | 29.17 | 33.33 | 25.00 | 100.00 |
| 5.0T 7.5 | 3.00 | 26.51 | 19.28 | 18.07 | 36.14 | 100.00 |
| GT 7.5 | 4.00 | 50.00 | 17.86 | 14.29 | 17.86 | 100.00 |
| TOTAL | | 20.11 | 23.56 | 26.72 | 29.60 | 100.00 |

PERCENTAGES OF THE COLUMN TOTALS

| | | LT 4.5<br>1.00 | 4.5T 9.0<br>2.00 | 9. T13.5<br>3.00 | GT 13.5<br>4.00 | TOTAL |
|---|---|---|---|---|---|---|
| LT 2.5 | 1.00 | 7.14 | 25.61 | 32.26 | 32.04 | 25.57 |
| 2.5T 5.0 | 2.00 | 21.43 | 42.68 | 43.01 | 29.13 | 34.48 |
| 5.0T 7.5 | 3.00 | 31.43 | 19.51 | 16.13 | 29.13 | 23.85 |
| GT 7.5 | 4.00 | 40.00 | 12.20 | 8.60 | 9.71 | 16.09 |
| TOTAL | | 100.00 | 100.00 | 100.00 | 100.00 | 100.00 |

EXPECTED CELL VALUES

| | | LT 4.5<br>1.00 | 4.5T 9.0<br>2.00 | 9. T13.5<br>3.00 | GT 13.5<br>4.00 | TOTAL |
|---|---|---|---|---|---|---|
| LT 2.5 | 1.00 | 17.90 | 20.97 | 23.78 | 26.34 | 89.00 |
| 2.5T 5.0 | 2.00 | 24.14 | 28.28 | 32.07 | 35.52 | 120.00 |
| 5.0T 7.5 | 3.00 | 16.70 | 19.56 | 22.18 | 24.57 | 83.00 |
| GT 7.5 | 4.00 | 11.26 | 13.20 | 14.97 | 16.57 | 56.00 |
| TOTAL | | 70.00 | 82.00 | 93.00 | 103.00 | 348.00 |

DIFFERENCES = OBSERVED-EXPECTED

| | | LT 4.5<br>1.00 | 4.5T 9.0<br>2.00 | 9. T13.5<br>3.00 | GT 13.5<br>4.00 | TOTAL |
|---|---|---|---|---|---|---|
| LT 2.5 | 1.00 | −12.90 | 0.03 | 6.22 | 6.66 | 25.57 |
| 2.5T 5.0 | 2.00 | −9.14 | 6.72 | 7.93 | −5.52 | 34.48 |
| 5.0T 7.5 | 3.00 | 5.30 | −3.56 | −7.18 | 5.43 | 23.85 |
| GT 7.5 | 4.00 | 16.74 | −3.20 | −6.97 | −6.57 | 16.09 |
| TOTAL | | 100.00 | 100.00 | 100.00 | 100.00 | 100.00 |

$ 10.21. A lot acceptance sampling plan was constructed to monitor the proportion of $10\frac{1}{4}$-oz soup cans with improperly sealed labels. In a particular shipment, 20 boxes were sampled.

a. If the acceptance number is 1, that is, we accept the shipment if no more than 1 box is observed with a defective can label, determine the probability of acceptance if $p = .05$.

b. Construct an OC curve.

10.22. Consider a lot acceptance sampling plan with $n = 40$ and $a = 3$. Use the normal approximation to the binomial to compute the acceptance probability when $p = .2$.

10.23. Refer to exercise 10.22.

a. Compute the probability of acceptance using the probability that $y \leq 3.5$.

b. Discuss why the result in part a might be more precise than the result in exercise 10.22. (Hint: Draw a normal curve superimposed over a binomial distribution and label the horizontal axis as $y$.)

10.24. The personnel department of a large corporation was interested in determining the relationship between performance ratings of recently hired employees and their college grade point averages. To do this, a random sample of 90 records was obtained, examined, and classified in the following two-way table:

| | *College Grade Point Average* | | |
|---|---|---|---|
| *Performance Rating* | *A* | *B* | *C* |
| above average | 19 | 8 | 3 |
| average | 9 | 12 | 15 |
| below average | 6 | 5 | 13 |

Is there evidence to indicate a relationship between the two variables "performance rating" and "college grade point average"? Use $\alpha = .05$.

*10.25. A random sample of faculty members of a state university system were polled and classified by university and by which of the three collective bargaining agents (union 101, union 102, union 103) was preferred. The data appear below, and a computer output follows.

| | *Bargaining Agent* | | |
|---|---|---|---|
| *University* | *101* | *102* | *103* |
| 1 | 42 | 29 | 12 |
| 2 | 31 | 23 | 6 |
| 3 | 26 | 28 | 2 |
| 4 | 8 | 17 | 37 |

a. Identify the expected cell numbers.

b. Use the computer output to determine whether there is evidence to indicate a difference in the distributions of preference across the four state universities.

c. Give the level of significance for the test.
d. Draw your conclusions.

```
TYPE #ROWS, #COLS, (FREE FORMAT), TITLE (UP TO 20 CHAR.)
? 4,3,JOB 2
TYPE OBSERVATIONS - - ONE ROW PER LINE
? 42,29,12
? 31,23,6
? 26,28,2
? 8,17,37

OBSERVED VALUES      TITLE: JOB 2
 42.       29.       12.       83.
 31.       23.        6.       60.
 26.       28.        2.       56.
  8.       17.       37.       62.
107.       97.       57.      261.

EXPECTED VALUES
34.0       30.8       18.1
24.6       22.3       13.1
23.0       20.8       12.2
25.4       23.0       13.5

OBS. MINUS EXP.
   8.0       -1.8        -6.1
   6.4        0.7        -7.1
   3.0        7.2       -10.2
 -17.4       -6.0        23.5

CHI SQUARE BY CELL
 1.9       0.1        2.1
 1.7       0.0        3.9
 0.4       2.5        8.6
11.9       1.6       40.6

CHI SQUARE = 75.197    6 DEGREE(S) OF FREEDOM
```

+ 

10.26. An advertising firm selected to conduct a market awareness study for a brand of house paint obtained information from a national survey of 1500 randomly selected home owners. Each home owner selected for the survey was asked if he or she was familiar with a newly marketed line of interior latex paints. If 465 responded affirmatively, use the sample data to test the research hypothesis that the company has reached more than 30% of the home owners with recent advertising. Use $\alpha = .05$.

# References

Barr, A.J.; Goodnight, J.H.; Sall, J.P.; and Helwig, J.T. 1976. *A user's guide to SAS 76*. Raleigh, N.C.: SAS Institute, Inc.

Cochran, W.G. 1954. Some methods for strengthening the common $\chi^2$ test. *Biometrics* 10:417–451.

Conover, W.J. 1971. *Practical nonparametric statistics*. New York: Wiley.

Dixon, W.J. 1975. *BMDP, biomedical computer programs*. Berkeley: University of California Press.

Dixon, W.J., and Massey, F.J., Jr. 1969. *Introduction to statistical analysis*. 3d ed. New York: McGraw-Hill.

Hollander, M., and Wolfe, D.A. 1973. *Nonparametric statistical methods*. New York: Wiley.

Nie, N.H.; Hull, C.H.; Jenkins, J.G.; Steinbrenner, K.; and Bent, D.H. 1975. *Statistical package for the social sciences*. 2d ed. New York: McGraw-Hill.

Ostle, B. 1963. *Statistics in research*. 2d ed. Ames, Iowa: Iowa State University Press.

Pearson, E.S., and Hartley, H.O. 1966. *Biometrika tables for statisticians*. 3d ed. Vol. I. London: Cambridge University Press.

Service, J. 1972. *A user's guide to the statistical analysis system*. Raleigh, N.C.: Student Supply Stores, North Carolina State University.

Siegel, S. 1956. *Nonparametric statistics for the behavioral sciences*. New York: McGraw-Hill.

# 11 Nonparametric Methods

## 11.1 Introduction

ordinal data

Some studies yield data identified by rank only—*ordinal data*—because of the crudeness of the measuring equipment employed or the inability of the investigator to further quantify the measurements. For example, in comparing the damage done by cutworms to tomato plants sprayed with a new insecticide and others that were not sprayed, we might label the damage per plant as none, little, moderate, or heavy. Similarly, in classifying sides of beef into grades of meat, we might classify a particular side as prime, choice, good, or other. In the social sciences such variables as prestige, power, and alienation are measured by ordinal scales; in the behavioral sciences variables such as pain threshold level, emotional stability, and drug reaction might be measured on an ordinal scale.

nonparametric statistical tests

When the variable of interest is measured on an ordinal scale, the test procedures discussed thus far are inappropriate, and we must resort to *nonparametric statistical tests* to provide a means for analyzing these data. The word "nonparametric" evolves from the type of hypothesis usually tested when dealing with ordinal data. Nonparametric tests do not involve inferences about parameters from the original distribution of measurements. For example, instead of hypothesizing that two populations have the same mean (as in section 5.4), we could hypothesize that the two populations from which the samples were drawn are identical. Note that the practical implications of these two hypotheses are not equivalent because the latter hypothesis is less clearly defined. Two distributions of measurements could be different and still have the same mean.

In sections 11.2 through 11.4 we will discuss several nonparametric statistical tests for comparing two or more populations. Although they have been developed specifically for ordinal-level data, they are also appropriate for interval- or ratio-level data when one or more of the assumptions underlying a corresponding parametric statistical test have been violated. For example, in conducting a $t$ test comparing two population means, we made the assumptions that the two populations are normal and have a common variance $\sigma^2$. If either of these assumptions is violated, the usual $t$ test for comparing independent samples is inappropriate.

advantages of nonparametric procedures

The advantages of nonparametric procedures are listed below (see Hollander and Wolfe 1973).

1. Many of the usual assumptions regarding the population from which the data are obtained can be relaxed. One very important assumption, that of normality of the sampled populations, is not necessary for nonparametric procedures.
2. Nonparametric procedures can be used when parametric procedures relying on normality cannot.
3. Most nonparametric procedures are easy to apply and easy to understand.
4. Many people have felt that by ignoring certain sample information, nonparametric procedures are less efficient than their parametric counterparts. That is, by ignoring some of the sample information in the experiment and employing a nonparametric test in place of a parametric test, we might have less chance to correctly declare a result significant. Recent research has shown this to be false. Only when normality holds do the parametric procedures have a clear-cut advantage over the corresponding nonparametric tests.

In sections 11.5 and 11.6 we will consider the runs test, a test for randomness in a sequence, and several procedures that enable an experimenter to reach a preliminary conclusion in some situations with very little effort.

# The Sign Test 11.2

In section 9.4 we considered a paired $t$ test for comparing two population means. The *sign test* can be used for comparing two populations based on *paired data* measured on an ordinal scale, and it also provides an alternative to the paired $t$ test when one or more of the underlying assumptions are violated.

paired data

In particular, if we have a sample of pairs of observations, one observation from population 1 and another from population 2, the sign test is a procedure for testing that the two populations have identical probability distributions. Note that we are not running a test about any particular parameter.

The null and alternative hypotheses for a two-tailed sign test are

$H_0$: The two populations have identical probability distributions.
$H_a$: The two populations have different probability distributions.

The test statistic for the sign test makes use of $y$, the number of sample pairs for which the observation in population 1 is larger than the corresponding observation from population 2. Letting an observation from population 1 be denoted by $y_1$ and one from population 2 by $y_2$, we must examine the difference $y_1 - y_2$ for each pair. The sign test is based on the signs of these differences, and we are interested in the number of pairs for which we obtain a plus sign, that is, the number of pairs for which $y_1 - y_2 > 0$.

$y_1 - y_2$

If $H_0$ is true, the probability that $y_1 - y_2 > 0$ for any pair is $p = .5$. Thus a test of the null hypothesis that the distributions are identical is equivalent to a test concerning a binomial parameter $p$.

---

**Sign Test for Paired Data**

$H_0$: $p = .5$ (the distributions are identical).

$H_a$: 1. $p > .5$ (the probability distribution for population 1 is shifted to the right of the distribution for population 2).
2. $p < .5$ (the probability distribution for population 1 is shifted to the left of the distribution for population 2).
3. $p \neq .5$ (the two populations have different probability distributions).

We let $n$ denote the number of sample pairs where $y_1 - y_2 \neq 0$ (i.e., the number of pairs ignoring ties). The test statistic is

$$\text{T.S.: } z = \frac{y - .5n}{\sqrt{.25n}}.$$

R.R.: For a specified value of $\alpha$,

1. reject $H_0$ if $z > z_\alpha$;
2. reject $H_0$ if $z < -z_\alpha$;
3. reject $H_0$ if $|z| > z_{\alpha/2}$.

This test is valid provided $n \geq 10$.

---

**EXAMPLE 11.1**

Fifteen judges were asked to rate leaf samples from two different varieties of tobacco (I and II) on a scale of 1 to 5 points. Use the sample data in table 11.1 to test the research hypothesis that the distributions of ratings are different for the two varieties of tobacco. Use $\alpha = .05$.

**Table 11.1** Judges' ratings of leaf samples from two varieties of tobacco, example 11.1

| | *Tobacco Leaf* | | |
|---|---|---|---|
| *Judge* | *Variety I,* $y_1$ | *Variety II,* $y_2$ | *Sign of Difference* $y_1 - y_2$ |
| 1 | 1 | 2 | − |
| 2 | 4 | 3 | + |
| 3 | 4 | 3 | + |
| 4 | 2 | 1 | + |
| 5 | 4 | 3 | + |
| 6 | 5 | 4 | + |
| 7 | 5 | 3 | + |
| 8 | 4 | 2 | + |
| 9 | 5 | 3 | + |
| 10 | 3 | 1 | + |
| 11 | 4 | 4 | eliminate |
| 12 | 2 | 3 | − |
| 13 | 4 | 2 | + |
| 14 | 5 | 3 | + |
| 15 | 4 | 3 | + |

## SOLUTION

In applying the sign test to these data, we make use of the sign of the difference $y_1 - y_2$ for each judge. By letting $y$ denote the number of plus signs, our test procedure can be restated in terms of $p$, the probability of obtaining a plus sign for a given pair (judge).

$H_0$: $p = .5$ (the distributions of ratings are identical).

$H_a$: $p \neq .5$ (the distributions of ratings are different, and one is preferred to the other).

T.S.: $z = \dfrac{y - .5n}{\sqrt{.25n}}$

where $n$ is the number of pairs ignoring ties in the measurements. We must eliminate Judge 11 from the sign test since that judge was not able to discriminate between varieties I and II. Thus $n = 14$ and $y$, the number of plus signs, is 12. Substituting into the test statistic, we have

$$z = \frac{12 - .5(14)}{\sqrt{.25(14)}} = \frac{12 - 7}{\sqrt{3.5}} = \frac{5}{1.87} = 2.67$$

R.R.: For $\alpha = .05$, we will reject $H_0$ if $|z| > 1.96$.

Since the observed value of $z$ exceeds 1.96, we reject $H_0$ and conclude that the distributions of ratings are different for varieties I and II. Practically, it appears that Variety I has a distribution of scores that is shifted to the right (higher) than the distribution of scores for Variety II.

The sign test can also be applied as an alternative to a one-sample $t$ test for a population mean. Unlike the $t$ test, which requires that the sample measurements be drawn from a normal population, the sign test can be used to test the null hypothesis $\mu = \mu_0$ provided that the sample is drawn from a population with a symmetrical distribution. Applying the sign test, we subtract $\mu_0$ from each of the sample measurements. If the null hypothesis is true ($\mu = \mu_0$), the probability of observing a positive difference is equal to the probability of observing a negative difference, namely, .5. Letting $y$ denote the number of plus signs (positive differences), we proceed with the mechanics of the test as in the sign test for paired data.

---

**One-Sample Sign Test: An Alternative to a One-Sample $t$ Test**

$H_0$: $p = .5$ ($\mu = \mu_0$).

$H_a$: 1. $p > .5$ ($\mu > \mu_0$).
2. $p < .5$ ($\mu < \mu_0$).
3. $p \neq .5$ ($\mu \neq \mu_0$).

T.S.: $z = \dfrac{y - .5n}{\sqrt{.25n}}$.

R.R.: For a specified value of $\alpha$,
1. reject $H_0$ if $z > z_\alpha$;
2. reject $H_0$ if $z < -z_\alpha$;
3. reject $H_0$ if $|z| > z_{\alpha/2}$.

Note: Here $n$ denotes the number of positive and negative differences. This test is valid provided $n \geq 10$.

---

In summary, the sign test provides a useful nonparametric alternative to both the one-sample $t$ test for $\mu$ and the two-sample paired $t$ test. Because of its simplicity, an experimenter may choose to use a sign test even when the assumptions for the corresponding parametric test are fulfilled.

# Exercises

Ψ 11.1. Thirty sets of identical twins were asked to participate in a one-year study designed to measure certain social attitudes. One twin from each set was randomly assigned to live in the home of a minority family while the other twin of the set stayed at home. After one year, each person was asked to respond to a questionnaire designed to detect and measure well-defined attitudes. The sample data appear on the next page.

| *Set of Twins* | *Home Environment*, $y_1$ | *Minority Environment*, $y_2$ | $y_1 - y_2$ |
|---|---|---|---|
| 1 | 78 | 71 | 7 |
| 2 | 75 | 70 | 5 |
| 3 | 68 | 66 | 2 |
| 4 | 92 | 85 | 7 |
| 5 | 55 | 60 | −5 |
| 6 | 74 | 72 | 2 |
| 7 | 65 | 57 | 8 |
| 8 | 80 | 75 | 5 |
| 9 | 98 | 92 | 6 |
| 10 | 52 | 56 | −4 |
| 11 | 67 | 63 | 4 |
| 12 | 55 | 52 | 3 |
| 13 | 49 | 48 | 1 |
| 14 | 66 | 67 | −1 |
| 15 | 75 | 70 | 5 |
| 16 | 90 | 88 | 2 |
| 17 | 89 | 80 | 9 |
| 18 | 73 | 65 | 8 |
| 19 | 61 | 60 | 1 |
| 20 | 76 | 74 | 2 |
| 21 | 81 | 76 | 5 |
| 22 | 89 | 78 | 11 |
| 23 | 82 | 78 | 4 |
| 24 | 70 | 62 | 8 |
| 25 | 68 | 73 | −5 |
| 26 | 74 | 73 | 1 |
| 27 | 85 | 75 | 10 |
| 28 | 97 | 88 | 9 |
| 29 | 95 | 94 | 1 |
| 30 | 78 | 75 | 3 |

Run a sign test, using $\alpha = .05$, to compare the two populations of scores.

11.2. Two judges were asked to rate separately each of 22 inmates on his rehabilitative potential. These data appear below.

| *Inmate* | *Judge 1* | *Judge 2* | *Inmate* | *Judge 1* | *Judge 2* |
|---|---|---|---|---|---|
| 1 | 6 | 5 | 12 | 9 | 8 |
| 2 | 12 | 11 | 13 | 10 | 8 |
| 3 | 3 | 4 | 14 | 6 | 7 |
| 4 | 9 | 10 | 15 | 12 | 9 |
| 5 | 5 | 2 | 16 | 4 | 3 |
| 6 | 8 | 6 | 17 | 5 | 5 |
| 7 | 1 | 2 | 18 | 6 | 4 |
| 8 | 12 | 9 | 19 | 11 | 8 |
| 9 | 6 | 5 | 20 | 5 | 3 |
| 10 | 7 | 4 | 21 | 10 | 9 |
| 11 | 6 | 6 | 22 | 10 | 11 |

ψ

Use a sign test with $\alpha = .05$ to determine whether the distributions of scores are different for the two judges.

11.3. The effect of Benzedrine on the heart rate of dogs (in beats per minute) was examined in an experiment; 14 dogs were chosen for the study. Each dog was to serve as his own control, with half of the dogs assigned to receive Benzedrine during the first study period and the other half assigned to receive a placebo (saline solution). All dogs were examined to determine the heart rates after two hours on the medication. After two weeks in which no medication was given, the regimens for the dogs were switched for the second study period. The dogs on Benzedrine were given the placebo while the others received Benzedrine. Again heart rates were measured at the two-hour time period.

The sample data below are not arranged in the order in which they were taken but have been summarized by regimen.

| *Dog* | *Placebo* | *Benzedrine* | *Dog* | *Placebo* | *Benzedrine* |
|---|---|---|---|---|---|
| 1 | 250 | 258 | 8 | 296 | 305 |
| 2 | 271 | 285 | 9 | 301 | 319 |
| 3 | 243 | 245 | 10 | 298 | 308 |
| 4 | 252 | 250 | 11 | 310 | 320 |
| 5 | 266 | 268 | 12 | 286 | 293 |
| 6 | 272 | 278 | 13 | 306 | 305 |
| 7 | 293 | 280 | 14 | 309 | 313 |

Use these data to test the research hypothesis that the distribution of heart rates for the dogs when receiving Benzedrine is shifted to the right of that for the same animals when on the placebo. Use a one-tailed test, with $\alpha = .05$.

11.4. A structural engineer has been consulted to determine whether steel beams made of a new design will maintain a mean breaking strength of 100,000 pounds per square inch (psi), the same as that for steel beams using the standard design. To test this hypothesis a random sample of 16 steel beams of the new design is tested. The breaking strengths (in 1000 psi) are listed below.

| | | | | | | | |
|---|---|---|---|---|---|---|---|
| 104 | 92 | 108 | 85 | 82 | 105 | 96 | 98 |
| 96 | 98 | 90 | 94 | 93 | 97 | 89 | 103 |

a. Use the methods of section 5.2 to conduct a $t$ test of the null hypothesis $\mu = 100$. Use a two-tailed test, with $\alpha = .05$.

b. Instead of going through all the calculations of part a, use a sign test to examine the same null hypothesis.
c. Are your conclusions in parts a and b identical? When might they differ?

11.5. The pulp content of a paper manufacturing process must be carefully monitored so that the quality and texture of the final paper product is as uniform as possible. At one stage in the process, a manufacturer specifies that the pulp content must average 82%. Of course, he also requires the variability of the pulp contents to be small. Assuming the variability of the process is in control, a random sample of 12 paper pulp sheets is obtained to test the manufacturers claim. The data, in percentage of pulp content, are given below. Use a two-tailed sign test, with $\alpha = .05$.

$

| | | | | | |
|---|---|---|---|---|---|
| 84 | 85 | 86 | 85 | 84 | 82 |
| 80 | 83 | 84 | 83 | 81 | 85 |

11.6. Refer to exercise 11.5. Suppose the manufacturer was only interested in a one-tailed test to determine if the mean had drifted upwards. Give the level of significance of the test.

11.7. Refer to exercise 11.5. Is there any other test that might be appropriate to test the indicated null hypothesis?

# Wilcoxon's Signed-Rank Test 11.3

The Wilcoxon signed-rank test, which makes use of the sign and the magnitude of the rank of the differences between pairs of measurements, provides an alternative to the paired *t* test (of section 9.4), as did the sign test of the previous section. Utilizing the number of pairs of measurements with a nonzero difference, we rank the differences from lowest to highest, ignoring their signs. If two or more measurements have the same nonzero difference (ignoring sign), we assign each difference a rank equal to the average of the occupied ranks. The appropriate sign is then attached to the rank of each difference.

Before summarizing the Wilcoxon signed-rank test, we define the following notation:

notation

$n =$ the number of pairs of observations with a nonzero difference

$T_+ =$ the sum of the positive ranks

$T_- =$ the sum of the negative ranks

$T =$ the smaller of $T_+$ and $T_-$, ignoring their signs

$\mu_T$

$$\mu_T = \frac{n(n+1)}{4}$$

$\sigma_T$

$$\sigma_T = \sqrt{\frac{n(n+1)(2n+1)}{24}}$$

g groups

If we group all differences assigned the same rank together, and there are $g$ such groups, the variance of $T$ is

$$\sigma_T^2 = \frac{1}{24}\left[n(n+1)(2n+1) - \frac{1}{2}\sum_{j=1}^{g} t_j(t_j - 1)(t_j + 1)\right]$$

$t_j$

where $t_j$ is the number of tied ranks in the $j$th group. Note that if there are no tied ranks, $g = n$ and $t_j = 1$ for all groups. The formula then reduces to

$$\sigma_T^2 = \frac{n(n+1)(2n+1)}{24}$$

The large-sample Wilcoxon signed-rank test is presented in the box.

---

**Large-Sample Wilcoxon Signed-Rank Test**

$H_0$: The two distributions are identical.

$H_a$: The two distributions are different (for a two-tailed test).

T.S.: $z = \dfrac{T - \mu_T}{\sigma_T}$.

R.R.: For a specified $\alpha$, reject $H_0$ if

$|z| > z_{\alpha/2}$

Note: The test is valid provided $n > 50$.

---

one-tailed test

It should be noted that the large-sample Wilcoxon signed-rank test can also be used as a one-tailed test to detect specific directional differences between the two populations. Then the computed value of $z$ would be compared to $z_\alpha$ or $-z_\alpha$, depending on the alternative hypothesis.

For $n \leq 50$ we have the small-sample test procedure given in the box.

**Small-Sample Wilcoxon Signed-Rank Test**

$H_0$: The distributions are identical.

$H_a$: The distributions are different (for a two-tailed test).

T.S.: $T$.

R.R.: For a specified value of $\alpha$ (either .10, .05, .02, or .01) and sample size $n$, obtain the entry from table 9 of the appendix. If $T$ is less than or equal to the tabulated value, reject $H_0$.

Note: The test is valid provided $n \leq 50$.

### EXAMPLE 11.2

Two different brands of fertilizer (A and B) were compared on each of 10 different two-acre plots. Each plot was subdivided into one-acre subplots, with Brand A randomly assigned to one subplot and Brand B to the other. Fertilizers were then applied to subplots at the rate of 60 pounds per acre. The data, barley yields, in bushels per acre, are listed in table 11.2 by fertilizer and plot.

**Table 11.2** Barley yields (in bushels) by plot and by fertilizer, example 11.2

| | *Barley Yield* | | |
|---|---|---|---|
| *Plot* | *Fertilizer A,* $y_1$ | *Fertilizer B,* $y_2$ | *Difference,* $y_1 - y_2$ |
| 1 | 312 | 346 | −34 |
| 2 | 333 | 372 | −39 |
| 3 | 356 | 392 | −36 |
| 4 | 316 | 351 | −35 |
| 5 | 310 | 330 | −20 |
| 6 | 352 | 364 | −12 |
| 7 | 389 | 375 | 14 |
| 8 | 313 | 315 | −2 |
| 9 | 316 | 327 | −11 |
| 10 | 346 | 378 | −32 |

Use the Wilcoxon signed-rank test to test the hypothesis that the distributions of barley yields for the two brands of fertilizer are identical against the alternative that they are different. Use $\alpha = .05$.

### SOLUTION

We must first rank (from lowest to highest) the absolute values of the $n = 10$ differences. These ranks appear in column 2 of table 11.3. The appropriate sign is then attached to each rank (see column 3 in table

**Table 11.3** Rankings for the data of table 11.2

| *Plot* | *Rank of Difference* $\lvert y_1 - y_2 \rvert$ | *Rank with Appropriate Sign* |
|---|---|---|
| 1 | 7 | −7 |
| 2 | 10 | −10 |
| 3 | 9 | −9 |
| 4 | 8 | −8 |
| 5 | 5 | −5 |
| 6 | 3 | −3 |
| 7 | 4 | 4 |
| 8 | 1 | −1 |
| 9 | 2 | −2 |
| 10 | 6 | −6 |

11.3). The sum of the positive and negative ranks are, respectively,

$T_+ = 4$

$T_- = -7 + (-10) + \cdots + (-6) = -51$

Thus $T$, the smaller of $T_+$ and $T_-$, ignoring the sign, is 4.

For a two-tailed test with $n = 10$ and $\alpha = .05$, we see from table 9 that we will reject $H_0$ if $T$ is less than or equal to 8. Thus we reject $H_0$ and conclude that the distributions of barley yields for the two brands of fertilizers are different.

one-tailed test

It should be noted that we can modify the small-sample Wilcoxon signed-rank test for one-tailed tests. For the data in example 11.2, the experimenter might have been interested in the alternative that the distribution of barley yields for Brand B was shifted to the right of that for Brand A. A large value of $T_-$, or, equivalently, a small value of $T_+$, would provide evidence to support this hypothesis. Thus we would compare $T = T_+$ to the one-sided entry in table 9 for a specified value of $\alpha$. Similarly, if the alternative hypothesis was that the distribution for Brand A was shifted to the right of that for Brand B, we would expect $T_+$ to be large and $T_-$ small. Hence $T = |T_-|$ would be compared with the one-tailed entry in table 9 of the appendix for a specified value of $\alpha$.

As with the sign test for paired data, the Wilcoxon signed-rank test can be modified to provide yet another alternative to a one-sample $t$ test for a population mean. Although Wilcoxon's test is not as simple to apply as the sign test, it only requires that the distribution of sample measurements be selected from a symmetrical distribution. As might be expected, because Wilcoxon's test utilizes both the sign and the magnitude of the sample differences, it provides more information than the sign test for reaching a decision for both paired data and the one-sample test. For the null hypothesis $H_0$: $\mu = \mu_0$, subtract $\mu_0$ from each sample measurement. Treat these differences the same as differences for paired data and use Wilcoxon's signed-rank test.

# Wilcoxon's Rank Sum Test | 11.4

The Wilcoxon rank sum test (not to be confused with the Wilcoxon signed-rank test of section 11.3) provides a procedure for testing that two populations are identical. Since the two populations are assumed to be identical under the null hypothesis, independent random samples from the respective populations should be similar. One way to measure the similarity between the samples is to jointly rank (from lowest to highest) the measurements from the combined samples and examine the sum of the ranks for measurements in sample 1 (or, equivalently, sample 2). We let $T$ denote the sum of the ranks for sample 1. Intuitively, if $T$ is extremely small (or large), we would have evidence to reject the null hypothesis that the two populations are identical.

As with the Wilcoxon signed-rank test, if there are ties among the combined sample measurements, we assign each measurement a rank equal to the average of the occupied ranks. If we group all measurements assigned the same rank together and there are $g$ such groups, then when $n_1$ and $n_2$ are both larger than 10, $T$ will be approximately normally distributed with a mean and a variance given by

$$\mu_T = \frac{n_1(n_1 + n_2 + 1)}{2}$$

$\mu_T$

$$\sigma_T^2 = \frac{n_1 n_2}{12}\left[(n_1 + n_2 + 1) - \frac{\sum_{j=1}^{g} t_j(t_j^2 - 1)}{(n_1 + n_2)(n_1 + n_2 - 1)}\right]$$

$\sigma_T^2$

where $t_j$ denotes the number of tied ranks in the $j$th group. Note that when there are no tied ranks,

$$g = n_1 + n_2 \qquad \text{and} \qquad \sigma_T^2 = \frac{n_1 n_2(n_1 + n_2 + 1)}{12}$$

For situations where the sample sizes are too small to use the normal approximation, critical values of $T$ have been tabulated (see, for example, Hollander and Wolfe 1973).

---

**Wilcoxon's Rank Sum Test***

$H_0$: The two populations are identical.

$H_a$: 1. Population 1 is shifted to the right of population 2.
2. Population 1 is shifted to the left of population 2.

*This test is equivalent to the Mann-Whitney $U$ test (Conover 1971).

3. The two populations are different.

T.S.: $z = \dfrac{T - \mu_T}{\sigma_T}$

where $T$ denotes the sum of the ranks in sample 1.

R.R.: For a specified value of $\alpha$,

1. reject $H_0$ if $z > z_\alpha$;
2. reject $H_0$ if $z < -z_\alpha$;
3. reject $H_0$ if $|z| > z_{\alpha/2}$.

Note: This test is valid provided $n_1 > 10$ and $n_2 > 10$.

---

**EXAMPLE 11.3**

Environmental engineers were interested in determining whether a cleanup project on a nearby lake was effective. Prior to initiation of the project, 12 samples of water had been obtained at random from the lake and analyzed for the amount of dissolved oxygen (in ppm). Due to diurnal fluctuations in the dissolved oxygen, all measurements were obtained at the 2 P.M. peak period. The before and after data are presented in table 11.4. Use $\alpha = .05$ to test the following hypotheses:

**Table 11.4** Dissolved oxygen measurements (in ppm), example 11.3

| *Before Cleanup* | | *After Cleanup* | |
|---|---|---|---|
| 11.0 | 11.6 | 10.2 | 10.8 |
| 11.2 | 11.7 | 10.3 | 10.8 |
| 11.2 | 11.8 | 10.4 | 10.9 |
| 11.2 | 11.9 | 10.6 | 11.1 |
| 11.4 | 11.9 | 10.6 | 11.1 |
| 11.5 | 12.1 | 10.7 | 11.3 |

$H_0$: The distributions of measurements for before cleanup and six months after the cleanup project began are identical.

$H_a$: The distribution of dissolved oxygen measurements before the cleanup project is shifted to the right of the corresponding distribution of measurements for six months after initiating the cleanup project. (It should be noted that a cleanup project has been effective in one sense if the dissolved oxygen drops over a period of time.)

For convenience, the data have been arranged in ascending order in table 11.4.

**SOLUTION**

We must first jointly rank the combined sample of 24 observations by assigning the rank of 1 to the smallest observation, the rank of 2 to the next smallest, and so on. When two or more measurements are the same, we assign all of them a rank equal to the average of the ranks they occupy. The sample measurements and associated ranks (shown in parentheses) are listed in table 11.5.

**Table 11.5** Dissolved oxygen measurements and ranks, example 11.3

| *Before Cleanup* | | *After Cleanup* | |
|---|---|---|---|
| 11.0 | (10) | 10.2 | (1) |
| 11.2 | (14) | 10.3 | (2) |
| 11.2 | (14) | 10.4 | (3) |
| 11.2 | (14) | 10.6 | (4.5) |
| 11.4 | (17) | 10.6 | (4.5) |
| 11.5 | (18) | 10.7 | (6) |
| 11.6 | (19) | 10.8 | (7.5) |
| 11.7 | (20) | 10.8 | (7.5) |
| 11.8 | (21) | 10.9 | (9) |
| 11.9 | (22.5) | 11.1 | (11.5) |
| 11.9 | (22.5) | 11.1 | (11.5) |
| 12.1 | (24) | 11.3 | (16) |
| $T = 216$ | | | |

If we are trying to detect a shift to the left in the distribution after the cleanup, we would expect the sum of the ranks for the observations in sample 1 to be large. Thus we will reject $H_0$ for large values of $z = (T - \mu_T)/\sigma_T$.

Grouping the measurements with tied ranks, we have $g = 18$ groups. These groups are listed below with the corresponding values of $t_j$, the number of tied ranks in the group.

| *Rank(s)* | *Group* | $t_j$ |
|---|---|---|
| 1 | 1 | 1 |
| 2 | 2 | 1 |
| 3 | 3 | 1 |
| 4.5, 4.5 | 4 | 2 |
| 6 | 5 | 1 |
| 7.5, 7.5 | 6 | 2 |
| 9 | 7 | 1 |
| 10 | 8 | 1 |
| 11.5, 11.5 | 9 | 2 |
| 14, 14, 14 | 10 | 3 |
| 16 | 11 | 1 |
| 17 | 12 | 1 |
| 18 | 13 | 1 |
| 19 | 14 | 1 |
| 20 | 15 | 1 |
| 21 | 16 | 1 |
| 22.5, 22.5 | 17 | 2 |
| 24 | 18 | 1 |

For all groups with $t_j = 1$, there is no contribution for

$$\frac{\sum_{j=1}^{18} t_j(t_j^2 - 1)}{(n_1 + n_2)(n_1 + n_2 - 1)}$$

in $\sigma_T^2$ since $t_j^2 - 1 = 0$. Thus we will need only $t_j = 2, 3$.

Substituting our data in the formulas, we obtain

$$\mu_T = \frac{n_1(n_1 + n_2 + 1)}{2} = \frac{12(12 + 12 + 1)}{2} = 150$$

$$\sigma_T^2 = \frac{n_1 n_2}{12}\left[(n_1 + n_2 + 1) - \frac{\sum_{j=1}^{g} t_j(t_j^2 - 1)}{(n_1 + n_2)(n_1 + n_2 - 1)}\right]$$

$$= \frac{12(12)}{12}\left[25 - \frac{6 + 6 + 6 + 24 + 6}{24(23)}\right] = 12(25 - .0870) = 298.956$$

$$\sigma_T = 17.29$$

The computed value of $z$ is

$$z = \frac{T - \mu_T}{\sigma_T} = \frac{216 - 150}{17.29} = 3.82$$

Since this value exceeds 1.645, we reject $H_0$ and conclude that the distribution of before cleanup measurements is shifted to the right of the corresponding distribution of after cleanup measurements.

The concept of a rank sum test can be extended to a comparison of more than two populations. In particular, suppose that $n_1$ observations are drawn at random from population 1, $n_2$ from population 2, . . . , and $n_k$ from population $k$. We may wish to test the hypothesis that the $k$ samples were drawn from identical distributions. The following test procedure, sometimes called the Kruskal-Wallis test, is then appropriate.

---

**Extension of the Rank Sum Test for More Than Two Populations**

*H*

$H_0$: The $k$ distributions are identical.

$H_a$: Not all the distributions are the same.

$$\text{T.S.: } H = \frac{12}{n(n+1)} \sum_{i=1}^{k} \frac{T_i^2}{n_i} - 3(n + 1)$$

where $n_i$ is the number of observations from sample $i$ ($i = 1, 2, \ldots, k$), $n$ is the combined sample size, that is, $n = \Sigma_{i=1}^{k} n_i$, and $T_i$ denotes the sum of the ranks for the measurements in sample $i$ after the combined sample measurements have been ranked.

R.R.: For a specified value of $\alpha$, reject $H_0$ if $H$ exceeds the critical value of $\chi^2$ for a $= \alpha$ and df $= k - 1$.

Note: When there are ties in the ranks of two or more sample measurements, use

$$H' = \frac{H}{1 - \left[\sum_{j=1}^{g} (t_j^3 - t_j)/(n^3 - n)\right]}$$

$H'$

### EXAMPLE 11.4

Three random samples of clergymen were drawn, one containing 10 Methodist ministers, the second containing 10 Catholic priests, and the third containing 10 Pentecostal ministers. Each of the clergymen was then examined, using a test to measure his knowledge about causes of mental illness. These test scores are listed in table 11.6.

**Table 11.6** Mental illness knowledge scores for the clergymen, example 11.4

| *Methodist* | *Catholic* | *Pentecostal* |
|---|---|---|
| 32 | 32 | 28 |
| 30 | 32 | 21 |
| 30 | 26 | 15 |
| 29 | 26 | 15 |
| 26 | 22 | 14 |
| 23 | 20 | 14 |
| 20 | 19 | 14 |
| 19 | 16 | 11 |
| 18 | 14 | 9 |
| 12 | 14 | 8 |

Use the data to determine if the three groups of clergymen differ with respect to their knowledge about the causes of mental illness. Use $\alpha = .05$.

### SOLUTION

The research and null hypotheses for this example can be stated as follows:

$H_a$: At least one of the three groups of clergymen differs from the others with respect to knowledge about causes of mental illness.

$H_0$: There is no difference among the three groups with respect to knowledge about the causes of mental illness (i.e., the samples of scores were drawn from identical populations).

Before computing $H$ we must first jointly rank the 30 test scores from lowest to highest. From table 11.7 we see that 8 is the lowest test score, and this clergyman is assigned the rank of 1. Similarly, the scores 9, 11, and 12 receive the ranks 2, 3, and 4, respectively. Five clergymen have a

**Table 11.7** Test scores and ranks for the clergymen study

| *Methodist* | | *Catholic* | | *Pentecostal* | |
|---|---|---|---|---|---|
| 32 | (29) | 32 | (29) | 28 | (24) |
| 30 | (26.5) | 32 | (29) | 21 | (18) |
| 30 | (26.5) | 26 | (22) | 15 | (10.5) |
| 29 | (25) | 26 | (22) | 15 | (10.5) |
| 26 | (22) | 22 | (19) | 14 | (7) |
| 23 | (20) | 20 | (16.5) | 14 | (7) |
| 20 | (16.5) | 19 | (14.5) | 14 | (7) |
| 19 | (14.5) | 16 | (12) | 11 | (3) |
| 18 | (13) | 14 | (7) | 9 | (2) |
| 12 | (4) | 14 | (7) | 8 | (1) |
| $n_1 = 10,\ T_1 = 197$ | | $n_2 = 10,\ T_2 = 178$ | | $n_3 = 10,\ T_3 = 90$ | |

test score of 14, and since these 5 scores occupy the ranks 5, 6, 7, 8, and 9, we assign each one a rank of 7, the average of the occupied ranks. In a similar way we can assign the remaining ranks to test scores. Table 11.7 lists the 30 test scores and associated ranks (in parentheses).

Note from table 11.7 that the sums of the ranks for the three groups are 197, 178, and 90. Hence the computed value of $H$ is

$$H = \frac{12}{30(30+1)}\left[\frac{(197)^2}{10} + \frac{(178)^2}{10} + \frac{(90)^2}{10}\right] - 3(30+1)$$

$$= \frac{12}{930}(3880.9 + 3168.4 + 810) - 93 = 8.4$$

Since there are groups of tied ranks, we must use $H'$ rather than $H$ as the test statistic. To do this we form the g groups composed of identical ranks shown below.

| *Rank* | *Group* | $t_j$ |
|---|---|---|
| 1 | 1 | 1 |
| 2 | 2 | 1 |
| 3 | 3 | 1 |
| 4 | 4 | 1 |
| 7, 7, 7, 7, 7 | 5 | 5 |
| 10.5, 10.5 | 6 | 2 |
| 12 | 7 | 1 |
| 13 | 8 | 1 |
| 14.5, 14.5 | 9 | 2 |
| 16.5, 16.5 | 10 | 2 |
| 18 | 11 | 1 |
| 19 | 12 | 1 |
| 20 | 13 | 1 |
| 22, 22, 22 | 14 | 3 |
| 24 | 15 | 1 |
| 25 | 16 | 1 |
| 26.5, 26.5 | 17 | 2 |
| 29, 29, 29 | 18 | 3 |

From this information we calculate the quantity

$$\frac{\sum_{j=1}^{g}(t_j^3 - t_j)}{n^3 - n} = \frac{1}{30^3 - 30}[(5^3 - 5) + (2^3 - 2) + (2^3 - 2) + (2^3 - 2) + (3^3 - 3) + (2^3 - 2) + (3^3 - 3)]$$

$$= \frac{192}{26{,}970} = .0071$$

Substituting this value into the formula for $H'$, we have

$$H' = \frac{H}{1 - .0071} = \frac{8.4}{.9929} = 8.46$$

The critical value of chi-square with a = .05 and df = $k - 1 = 2$ can be found using table 3, the appendix. This value is 5.99. Since the observed value of $H'$ is greater than 5.99, we reject the null hypothesis and conclude that at least one of the clergy groups has more knowledge about the causes of mental illness than the other two groups.

# Exercises

11.8. Refer to exercise 11.1 (p. 310). Use the data given in exercise 11.1 to run the Wilcoxon signed-rank test. Use $\alpha = .05$. Which of the two procedures, the sign test or the Wilcoxon signed-rank test, would you think is more powerful; that is, which has a higher probability of declaring that the distributions are different when in fact they really are different?

11.9. Refer to exercise 11.2 (p. 311). Compare the results of a Wilcoxon signed-rank test to those of the sign test. Use $\alpha = .05$.

11.10. A single leaf was taken from each of 11 different tobacco plants. Each was divided in half; one half was chosen at random and treated with Preparation I and the other half received Preparation II. The object of the experiment was to compare the effects of the two preparations of mosaic virus on the number of lesions on the half leaves after a fixed period of time. These data are recorded on the next page.

For $\alpha = .05$, use Wilcoxon's signed-rank test to examine the research hypothesis that the distributions of lesions are different for the two populations.

11.11. Refer to exercise 11.10. Conduct a sign test and compare your results to those of exercise 11.10. Use $\alpha = .05$.

| Tobacco Plant | Number of Lesions on the Half Leaf | |
|---|---|---|
| | Preparation I | Preparation II |
| 1 | 18 | 14 |
| 2 | 20 | 15 |
| 3 | 9 | 6 |
| 4 | 14 | 12 |
| 5 | 38 | 32 |
| 6 | 26 | 30 |
| 7 | 15 | 9 |
| 8 | 10 | 2 |
| 9 | 25 | 18 |
| 10 | 7 | 3 |
| 11 | 13 | 6 |

11.12. An experiment was conducted to compare the weights of the combs of roosters fed two different vitamin-supplemented diets. Twenty eight healthy roosters were randomly divided into two groups, with one group receiving Diet I and the other receiving Diet II. After the study period the comb weight (in milligrams) was recorded for each rooster. These data are given below.

| *Diet I* | 73 | 130 | 115 | 144 | 127 | 126 | 112 | 76 | 68 | 101 | 126 | 49 | 110 | 123 |
|---|---|---|---|---|---|---|---|---|---|---|---|---|---|---|
| *Diet II* | 80 | 72 | 73 | 60 | 55 | 74 | 67 | 89 | 75 | 66 | 93 | 75 | 68 | 76 |

a. Use the Wilcoxon rank sum test to determine if there is a difference in the distributions of comb weights for the two groups. Use $\alpha = .05$.
b. Can you suggest other statistical procedures that might be appropriate for analyzing the same data? Which would you suggest?

11.13. Refer to exercise 11.12. Suppose the experimenter was interested in determining whether the comb weights for Diet I were selected from a distribution shifted above (to the right of) that for comb weights from Diet II. Run an appropriate Wilcoxon rank sum test, with $\alpha = .05$.

11.14. Refer to exercise 11.4 (p. 312).
a. Perform a Wilcoxon signed-rank test on these data, with $\alpha = .05$.
b. Compare the results you obtain with those for a sign test or a $t$ test.

11.15. Using the data of exercise 11.5 (p. 313) and a Wilcoxon signed-rank test, conduct a statistical test of the research hypothesis that the mean of the pulp content has increased above 82%. Use $\alpha = .05$. Check your results with those of exercise 11.6.

11.16. The yields (in pounds) of 5 different varieties (A, B, C, D, E) of four-year-old orange trees were to be compared in one orchard. A random sample of 7 trees of each variety was obtained from the orchard. The yields for these trees are presented below.

| A | B | C | D | E |
|---|---|---|---|---|
| 13 | 27 | 40 | 17 | 36 |
| 19 | 31 | 44 | 28 | 32 |
| 39 | 36 | 41 | 41 | 34 |
| 38 | 29 | 37 | 45 | 29 |
| 22 | 45 | 36 | 15 | 25 |
| 25 | 32 | 38 | 13 | 31 |
| 10 | 44 | 35 | 20 | 30 |

Conduct a test of the null hypothesis that the 5 varieties have the same yield distributions. Use $\alpha = .01$.

# The Runs Test: A Test for Randomness 11.5

In this section we will consider a procedure for testing whether events occur in a random order. For example, a state vehicle inspection station is operated by two inspectors who alternate inspecting cars. Of course, we would hope that both inspectors have been thoroughly trained and use the same criteria for passing or failing a car. To insure uniform treatment of vehicles during inspection, the state periodically examines the number of cars passed and failed during a one-hour period. Since inspectors alternate, we could identify each car by whether it passed or failed and by its inspector. If only every other car in the one-hour period is passed, intuitively we would conclude that the inspectors are not using the same criteria to judge cars.

This type of periodicity during the one-hour period is indicative of nonrandomness in the occurrence of successes (and failures) over time and can be detected by a *runs test*. The runs test studies a sequence of events, each of which can be classified as either a success (S) or a failure (F). Thus a one-hour inspection period may have yielded the following sequence of successes and failures:

runs test

S S F F S S S S F F F S S S S

Is there evidence to indicate nonrandomness in the sequence? To answer this question we must define a *run* of like elements in a sequence of elements.

**Definition 11.1**

A *run* is a subsequence of like elements with the first and last element of the run being preceded and followed, respectively, by an unlike element. Of course, when the first (last) element of the run is the first (last) element of the entire sequence, it cannot be preceded (followed) by an unlike symbol.

**EXAMPLE 11.5**

Determine the number of runs in the following sequence of successes and failures for car inspection data:

S S F F S S S S F F F S S S S

**SOLUTION**

Grouping the sequence of symbols into subsequences of like symbols that are surrounded by unlike symbols, we have

S S  F F  S S S S  F F F  S S S S

Thus there are 5 runs in this sequence of successes and failures.

Intuitively we would suspect nonrandomness in a sequence if we observed an extremely large number of runs, such as would be obtained from the sequence

S F S F S F S F S F S F S F S

or an extremely small number of runs, as in the sequence

S S S S S S S F F F F F F F F

$r$

$\ell$

In a sequence of symbols, we let $r$ denote the number of runs, $n_1$ the number of successes, and $n_2$ the number of failures. We can obtain the probability that $r$ is less than or equal to a specified value $\ell$ from table 10 in the appendix. For example, with $n_1 = 3$ successes and $n_2 = 9$ failures, the probability that $r \leq 2$ is .009. Similarly, for $n_1 = n_2 = 8$, the probability that $r \geq 8$ is

$$1 - P(r \leq 7) = 1 - .214 = .786$$

It should also be noted that table 10 provides only combinations of $n_1 \leq n_2$ where both are no more than 10. If we should need a combination such as (10, 4), we use the probability corresponding to the (4, 10) combination of $n_1$ and $n_2$.

**Small-Sample Runs Test**

$H_0$: The sequence is a random arrangement of successes and failures.

$H_a$: The sequence is not a random arrangement of successes and failures.

T.S.: $r = \ell$, the observed number of runs.

R.R.: For a two-tailed test with a given combination of $(n_1, n_2)$ and $\alpha$, determine

$$P(r \leq \ell) \qquad \text{and} \qquad P(r \geq \ell)$$

using table 10 of the appendix. If either of these probabilities is less than or equal to $\alpha/2$, we reject $H_0$.

For a one-tailed test where we are looking for an extremely large (or small) number of runs, we must place all the rejection region in one tail of the distribution of $r$.

Note: This test is valid when $n_1 \leq 10$ and $n_2 \leq 10$.

**EXAMPLE 11.6**

We determined previously that the sequence of successes and failures for the car inspection data had $r = 5$ runs. Is there sufficient evidence to indicate that the sequence is not random? Use $\alpha = .05$.

**SOLUTION**

The sequence is

S S F F S S S S F F F S S S S

with 10 successes, 5 failures, and $r = 5$. For these data let us assume we are interested in a one-tailed test for determining an extremely large number of runs (which would indicate that the two inspectors are not using the same criteria for judging cars). Thus we will locate all the rejection region in the upper tail of the distribution of $r$. From table 10 for the (5, 10) combination, we find $P(r \leq 4) = .029$; hence the probability of observing 5 or more runs is $1 - .029 = .971$. Since this is so high (and certainly not less than $\alpha = .05$), we have insufficient evidence to indicate a lack of randomness in the sequence.

large-sample runs test

The runs test can be modified for large-sample tests in the following way. When $n_1$ and $n_2$ are both more than 10, $r$ is approximately normally distributed, with

$$\mu_r = \frac{2n_1n_2}{n_1 + n_2} + 1$$

$$\sigma_r = \sqrt{\frac{2n_1n_2(2n_1n_2 - n_1 - n_2)}{(n_1 + n_2)^2(n_1 + n_2 - 1)}}$$

We can then employ a $z$ test with

$$z = \frac{r - \mu_r}{\sigma_r}$$

Critical values of $z$ are given in table 1 of the appendix for any specified value of $\alpha$.

# Exercises

$ 11.17. Periodically during the mass production of items such as flash cubes, appliances, wines, and so on, the producer should monitor the quality of the outgoing product. While control charts have a place in such situations, the use of a runs test will often suffice as a quick check on quality.

Suppose that every tenth bottle of champagne coming off the production line is checked carefully to determine if a proper seal has been achieved. Of the last 30 bottles examined, we have the following sequence of successes (proper seal) and failures (improper seal):

S S S S S F F S S S S S S S S S S F S S S S S S S S S S S S

a. Is there evidence to indicate that the sequence of successes and failures is nonrandom? Use $\alpha = .05$.

b. Refer to part a. List several kinds of nonrandom sequences and indicate whether or not they are desirable from the producer's viewpoint.

$ 11.18. Two machines (A and B) are used in the spinning of nylon yarn. Every so often the polymer breaks during the spinning process, and an operator must shut the machine down to rethread the polymer through the machine. The polymer breaks over the last week were monitored and a tally was made for each machine. The following sequence indicates the order of machine shutdowns as they occurred over the past week. Is there evidence to indicate a lack of randomness in the ordering of breaks? Use $\alpha = .05$.

B B A A A B B B B A B A B B B B A A A B A A A B B A A A A B B B A A A

# Quick Portable Statistics | 11.6

In addition to providing alternative analyses when underlying assumptions are violated, some nonparametric statistical techniques are so easy to remember and use that they can be quickly applied without a desk calculator, a computer, or a reference table for critical values. In short, they can be carried anywhere and applied in many situations to provide a quick preliminary conclusion. The techniques we have chosen for inclusion in this section do not represent an exhaustive list of such techniques. However, they are representative of some of the portable statistics available to you. For additional information about portable statistics, see the references at the end of this chapter.

Tukey-Duckworth two-sample test

The first technique we will consider is the Tukey-Duckworth two-sample test (Tukey 1959) to determine if two independent samples were drawn from identical populations. It can be used for sample sizes satisfying the following inequalities:

$$4 \leq n_1 \leq n_2 \leq 30 \qquad n_2 \leq \frac{4n_1}{3} + 3$$

It should be noted that the designation of population 1 and population 2 is completely arbitrary. Modifications of this procedure have been suggested by other authors since 1959 (see, for example, Neave 1966 and 1975). Refer to the references at the end of this chapter for additional material on the subject.

---

**Tukey-Duckworth Two-Sample Test**

$H_0$: The populations are identical.

$H_a$: The populations are different (a two-tailed test).

Test Procedure

1. Determine the largest and smallest measurement in each sample.
2. For the sample that contains the largest value in the combined samples, count all measurements that are larger than the largest measurement in the other sample.
3. For the other sample, count all measurements that are smaller than the smallest measurement of the first sample.
4. Let $C$ denote the sum of the two counts. For $\alpha = .05$, $.01$, or $.001$, reject $H_0$ if $C \geq 7$, 10, or 13, respectively.

$C$

---

**EXAMPLE 11.7**

Thirty different one-acre plots were randomly divided into two groups, with 15 plots per group. The plots in the first group were fertilized with Brand A fertilizer and those in the second group were fertilized with Brand B. Each of the 30 one-acre plots was then planted in corn. Yields (in bushels) are presented in table 11.8 for each of the plots. Use these data to determine if there is a difference in yields for the two brands of fertilizers. Use $\alpha = .05$.

**Table 11.8** Yields of corn (in bushels) for two different brands of fertilizer, example 11.7

| *Group 1 (Brand A)* | | *Group 2 (Brand B)* | |
|---|---|---|---|
| 96 | 89 | 98 | 92 |
| 92 | 94 | 94 | 89 |
| 98 | 80 | 92 | 95 |
| 82 | 97 | 84 | 92 |
| 86 | 84 | 99 | 96 |
| 87 | 85 | 96 | 101 |
| 93 | 83 | 98 | 103 |
| 81 | | 96 | |

**SOLUTION**

We can proceed immediately with the Tukey-Duckworth two-sample test since our sample sizes satisfy the criteria $4 \leq n_1 \leq n_2 \leq 30$ and $n_2 \leq (4n_1/3) + 3$. We must first determine the largest and smallest measurements for each sample. This is shown below.

| | *Group 1* | *Group 2* |
|---|---|---|
| *Largest* | 98 | 103 |
| *Smallest* | 80 | 84 |

Group 2 contains the largest measurement (103) for the combined samples. The number of measurements in Group 2 larger than 98, which is the largest measurement in Group 1, is 3. Similarly, the number of measurements in Group 1 less than 84, which is the smallest measurement in Group 2, is 4. Since $C = 3 + 4 = 7$, we reject the null hypothesis that the populations of corn yields corresponding to the two fertilizers are identical, at the $\alpha = .05$ level.

quadrant sum test for association

The second portable statistical test we present is the quadrant sum test for association (Omstead and Tukey 1947) to determine if two variables are correlated. This statistical test can be used when each of two variables is measured on 8 or more sampling units.

## Quadrant Sum Test for Association

$H_0$: The two variables are not correlated.

$H_a$: The two variables are correlated (a two-tailed test).

Test Procedure

1. Plot the data using a scatter diagram (see chapter 3).
2. Draw a median line parallel to each axis. The median line parallel to the vertical axis will designate the median (midpoint) value of the variable plotted along the horizontal axis. Similarly, the median line parallel to the horizontal axis will correspond to the median of the variable plotted along the vertical axis (see figure 11.1).
3. Beginning in the upper right-hand quadrant and moving counter-clockwise, label the quadrants +, −, +, −, respectively (see figure 11.1).

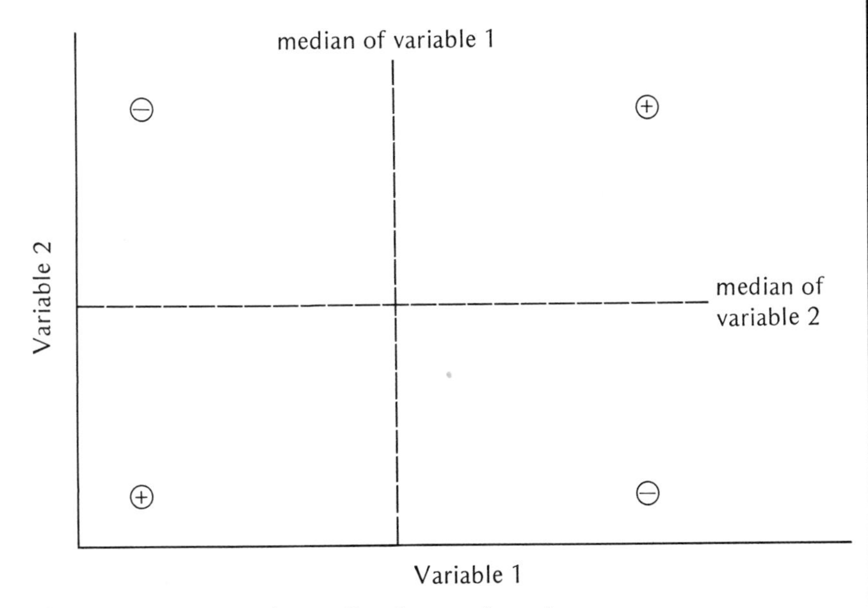


**Figure 11.1** Locating the median lines and quadrants for the quadrant sum test

4. Obtain the following four counts:
   (a) Beginning from the right-hand side of the scatter diagram and moving toward the left along the horizontal median line, count all observations (dots) that are on the same side of the horizontal median line. Stop counting when you encounter the first obser-

vation on the other side of the horizontal median line. Attach the sign of the quadrant to this count.

(b) Beginning from the top of the scatter diagram and moving down along the vertical median line, count all observations that are on the same side of the median line. Stop counting when you encounter the first observation on the other side of the vertical median line. Attach the sign of the quadrant to this count.

(c) Repeat moving from left to right.

(d) Repeat moving from bottom to top.

C

5. Let C denote the sum of the counts (with their appropriate signs) obtained in part 4.
6. For $\alpha = .10, .05$, or $.01$, reject $H_0$ if $|C| \geq 9, 11$, or $13$, respectively.
7. If C is positive, the two variables are positively correlated; if C is negative, the two variables are negatively correlated.

---

**EXAMPLE 11.8**

Data were obtained on the reduction in cholesterol count (in milligrams per 100 milliliters of blood serum) for a random sample of 15 male volunteers participating in a study involving a low-cholesterol diet. Each volunteer participated in the study for four weeks. A prestudy cholesterol reading was obtained prior to beginning the diet, and the reduction in the count was observed for the four-week period. In addition to cholesterol levels, ages of the volunteers were also recorded. These data are given in table 11.9. Use the quadrant sum test with $\alpha = .05$ to determine if

**Table 11.9** Age and cholesterol reduction data for example 11.8

| Age | *Reduction in Cholesterol* | Age | *Reduction in Cholesterol* |
|---|---|---|---|
| 45 | 30 | 31 | 40 |
| 43 | 52 | 26 | 17 |
| 46 | 45 | 22 | 28 |
| 49 | 38 | 58 | 44 |
| 50 | 62 | 60 | 61 |
| 37 | 55 | 52 | 58 |
| 34 | 25 | 27 | 45 |
| 30 | 30 | | |

there is a relationship between the age of a volunteer and the four-week cholesterol reduction.

**SOLUTION**

We must first construct a scatter diagram. This is shown in figure 11.2. The four quadrants are labeled +, −, +, − going counterclockwise from the

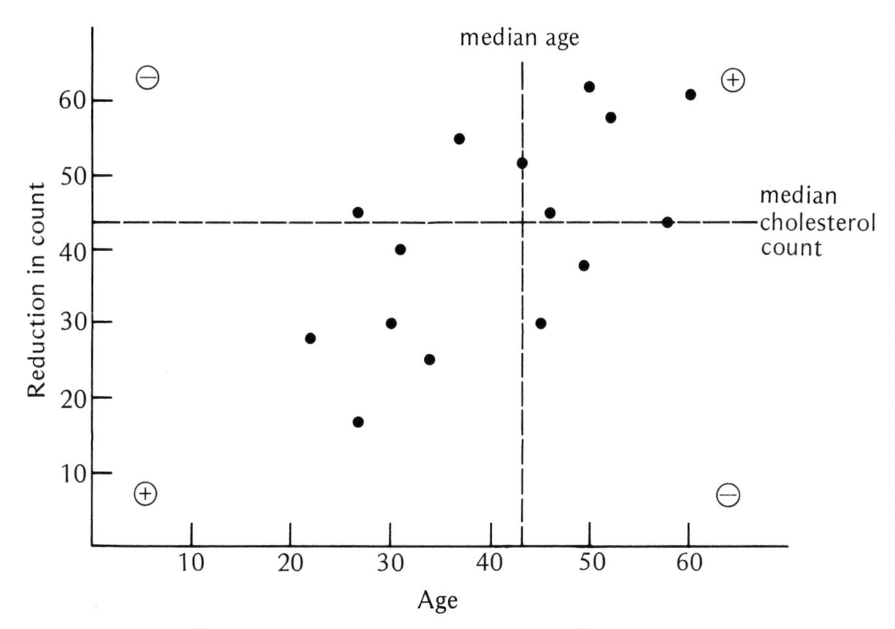


**Figure 11.2** Scatter diagram for the data of table 11.9

upper right. Using the age and cholesterol count data in the table, we find the median age to be 43 and the median cholesterol count to be 44. The median lines have been drawn on the scatter diagram by using dotted lines.

1. Beginning at the extreme right, we count observations until we must cross the horizontal median to count the next observation (dot). There are 4 observations (see figure 11.3, next page). We attach a plus sign to this count since the 4 observations were in the upper-right quadrant.
2. Beginning from the top, we count 3 observations before we must cross the vertical median line. Again we assign a plus sign to these 3 measurements.
3. Beginning from the extreme left, we count 2 observations before we must cross the horizontal median line. Since this observation was in the lower-left quadrant, it receives a plus sign.
4. Similarly, from the bottom we count 4 observations before we must cross the vertical median line. This count is assigned a plus sign because the measurements are in the lower-left quadrant.
5. The combined count (with appropriate sign) is

   $C = 4 + 3 + 2 + 4 = 13$

   Since $|C| > 11$, we reject the null hypothesis, at the $\alpha = .05$ level, that the variables are uncorrelated. The sign of $C$ indicates that there is a positive correlation between the age and cholesterol count reduction observed for persons treated with this diet.

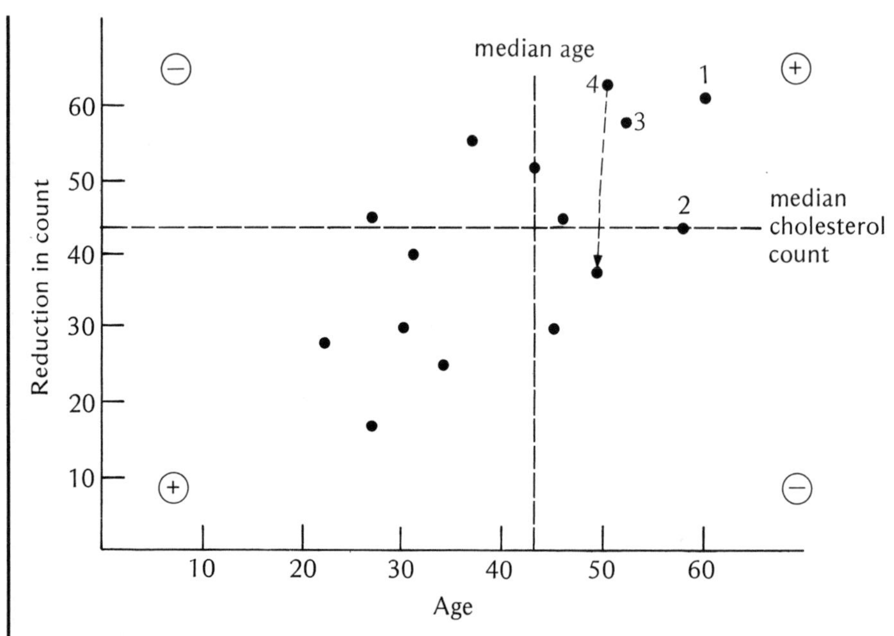


**Figure 11.3** Beginning the count of observations in example 11.8

approximating the least squares prediction equation

The last portable statistical technique we consider is not a statistical test; it is a shortcut method for approximating the least squares prediction equation for the model

$$y = \beta_0 + \beta_1 x + \epsilon$$

The procedure can be used provided we obtain measurements on each of the two variables ($x$ and $y$) for three or more sampling units.

---

**Procedure for Approximating the Least Squares Prediction Equation**

1. Plot the data in a scatter diagram (see chapter 7).
2. Using two lines parallel to the vertical axis, divide the data into three groups of *roughly* the same number of observations.
3. For the lower-end group, find the median point, that is, the point corresponding to the medians for $x$ and $y$ based on measurements in that group.
4. Repeat for the upper-end group.
5. Connect the two median points using a straight line. This line represents an approximation to the least squares prediction equation $\hat{y} = \hat{\beta}_0 + \hat{\beta}_1 x$.

---

**EXAMPLE 11.9**

Use the data of example 11.8 to approximate the least squares prediction equation $\widehat{y} = \widehat{\beta}_0 + \widehat{\beta}_1 x$.

**SOLUTION**

We begin with a scatter diagram of the data (see figure 11.4). The data are then divided into three approximately equal groups with two vertical

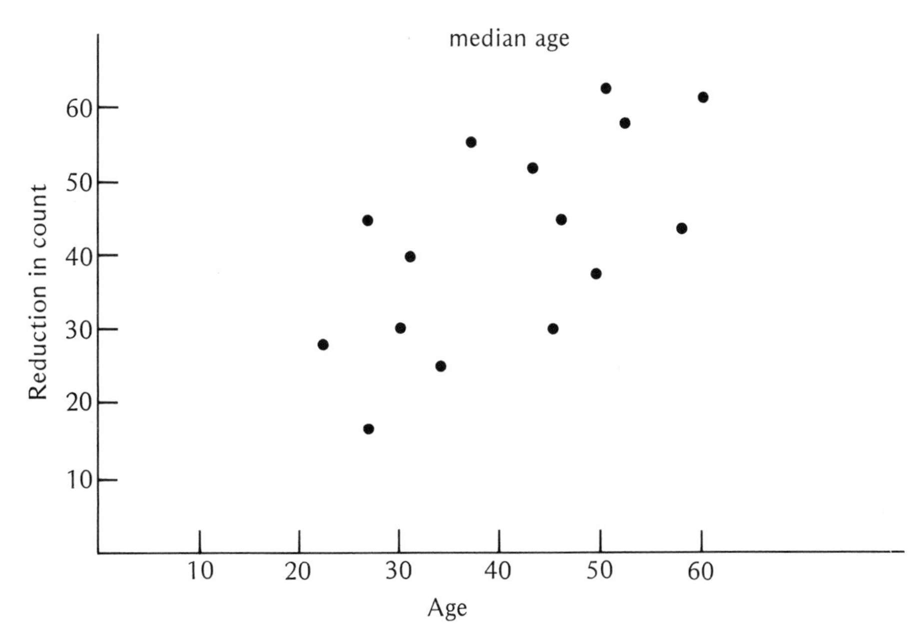


**Figure 11.4** Scatter diagram of the age and cholesterol count data, example 11.9

lines. These have been inserted onto the diagram in figure 11.5 on the next page. The median points for the lower-end group can be found by drawing a vertical line that evenly divides the $x$-values and a horizontal line that evenly divides the $y$-values. The intersection of these two lines is the median point for the lower-end group (see figure 11.5). Similarly, for the upper-end group, we find the median point indicated in figure 11.5. The straight line joining the median points for the two end groups is the line that approximates the least squares equation for these data.

# Summary 11.7

We have presented a few of the many nonparametric statistical procedures that are available to users of statistics. It is important to be aware of these

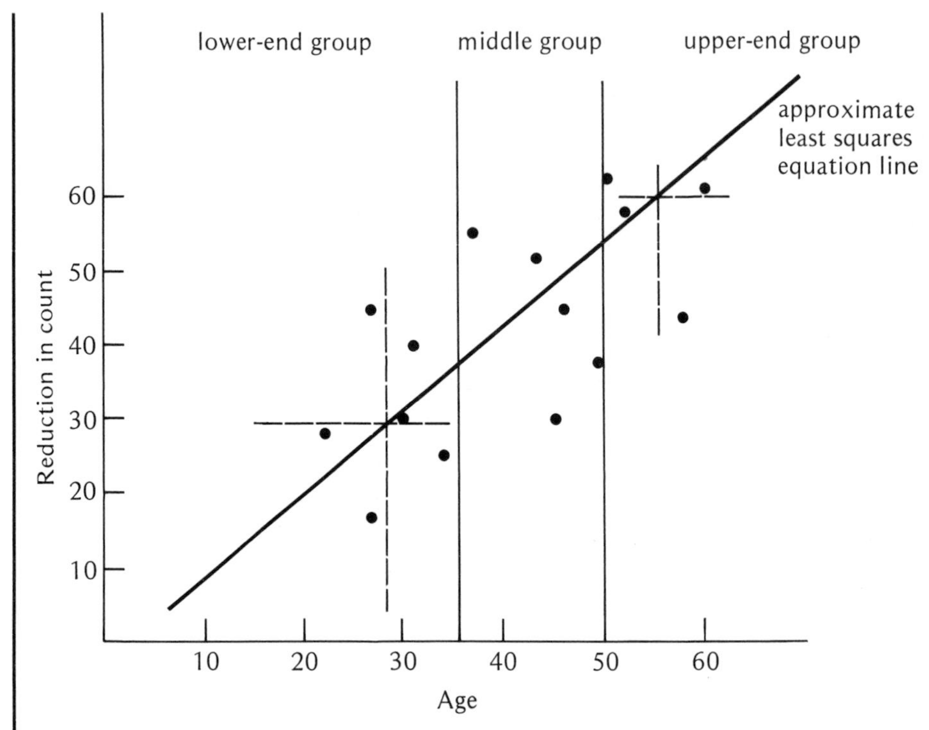


**Figure 11.5** Approximate least squares solution for example 11.9

techniques as potential alternatives to standard parametric procedures when one or more of the underlying assumptions are violated.

The sign test, perhaps the simplest of all nonparametric statistical tests, can be used to compare two populations based on either paired or unpaired data. In addition, we showed how the sign test could be used on a simple one-sample test of $H_0$: $\mu = \mu_0$.

The Wilcoxon signed-rank test was presented to provide an alternative to the two-sample paired $t$ test. Under the assumption that a sample of measurements is selected from a symmetrical distribution, the Wilcoxon signed-rank test can also be used as a one-sample test for $H_0$: $\mu = \mu_0$.

The Wilcoxon rank sum test can be used as an alternative to the two-sample unpaired $t$ test when one or more of the underlying assumptions for the parametric test are violated. The concept of a rank sum test was then extended to a comparison of more than two populations using the (Kruskal-Wallis) $H$ test.

The runs test presented in this chapter can be used as a nonparametric test for monitoring product quality. Items from a production line are classified as either a "success" or a "failure." The process is judged to be out of control if we observe either too many or too few runs of successes and failures.

Finally, we discussed several portable statistical procedures: the Tukey-Duckworth two-sample test, the quadrant sum test of association, and an approximation to the least squares prediction equation. Each of these techniques is simple to apply, easy to remember, and does not need the use of reference tables.

Although the nonparametric procedures of this text have been combined into this one chapter, we used this format to aid in the presentation of the material and to offer a convenient reference section in the text. It should be apparent where these tests (and estimation procedures) can be applied in previous sections. In later sections of the text, we will refer you to this chapter to indicate alternative tests whenever appropriate. Also, we did not intend to present a comprehensive list of suitable nonparametric tests in this chapter. Rather, we attempted to present an assortment of some of the more widely used tests that offer alternatives to the standard parametric tests of previous chapters.

We return in chapter 12 to additional parametric statistical procedures. In particular, chapter 12 is concerned with estimation and statistical tests for a population variance and the ratio of two population variances. In previous chapters we have been concerned with the population mean as the primary numerical descriptive measure of a population. As we will see, the variability of a population may be of more concern to the experimenter than the population mean.

# Exercises

11.19. An experiment was conducted to examine the relationship between the weight of plaque scraped from teeth and the DNA concentration of the plaque. To do this a random sample of 10 males (ages 18 to 20 years) was obtained. Each person was given a normal diet plus 15 two-milligram tablets of sucrose per day. No toothbrushing was allowed. At the end of four days, plaque samples were obtained from the 10 subjects. These data are shown below.

| *Person* | 1 | 2 | 3 | 4 | 5 | 6 | 7 | 8 | 9 | 10 |
|---|---|---|---|---|---|---|---|---|---|---|
| *Plaque Weight (mg)* | 42.7 | 52.3 | 24.6 | 33.4 | 41.8 | 36.7 | 27.0 | 47.3 | 31.4 | 33.9 |
| *DNA (μg)* | 260 | 303 | 175 | 214 | 226 | 246 | 181 | 251 | 154 | 247 |

Use the quadrant sum test to determine if there is a significant linear correlation between plaque weight and DNA concentration.

11.20. Use the data of exercise 11.19 to approximate the least squares prediction equation

$$\hat{y} = \hat{\beta}_0 + \hat{\beta}_1 x$$

11.21. Refer to exercise 11.12 (p. 324); an experiment was conducted to compare the comb weights of roosters fed on one of two diets. Use the Tukey-Duckworth test to compare the two distributions of weights. Does your answer here agree with the results you obtained in exercise 11.12?

e 11.22. Refer to exercise 7.18 (p. 168); the thermal pollution of automobiles was studied for 1971 and 1972 models. The data relating automobile weight to Btu (in 1000s) per vehicle mile are repeated in the table that is shown below.

a. Plot the data in a scatter diagram.

b. Use the method of section 11.6 to approximate the least squares prediction equation $\hat{y} = \hat{\beta}_0 + \hat{\beta}_1 x$.

| *Weight, x (in 1000 lb)* | *Btu per Vehicle Mile, y (in 1000s)* |
|---|---|
| 1.8 | 4 |
| 2.6 | 5.2 |
| 4.2 | 8.5 |
| 5.0 | 11.6 |
| 4.8 | 10.1 |
| 3.4 | 6.3 |

11.23. Refer to exercise 11.22.

a. Compare the approximate prediction equation of part b to the least squares prediction equation obtained in exercise 7.18.

b. Plot both equations (the approximate prediction equation and the least squares line) on your graph of part a of exercise 11.22.

*11.24. Examine the computer output that follows for the data of example 11.3 (p. 318).

a. Locate T.

b. Set up all parts of a test of the null hypothesis that the two populations are identical.

c. Give the level of significance for your test.

WILCOXON-MANN-WHITNEY U-TEST

FILES:

| OBSERVATIONS: | BEFORE CLEANUP | AFTER CLEANUP |
|---|---|---|
| | 11.00 | 10.20 |
| | 11.20 | 10.30 |
| | 11.20 | 10.40 |
| | 11.20 | 10.60 |
| | 11.40 | 10.60 |
| | 11.50 | 10.70 |
| | 11.60 | 10.80 |
| | 11.70 | 10.80 |
| | 11.80 | 10.90 |
| | 11.90 | 11.10 |
| | 11.90 | 11.10 |
| | 12.10 | 11.30 |

| | CONTROL | TEST CMPD |
|---|---|---|
| NO. OBSERVATIONS | 12.00 | 12.00 |
| MEDIANS | 11.55 | 10.75 |
| SUM OF RANKS | 216.00 | 84.00 |
| U-STATISTIC | 6.00 | 138.00 |

# References


Conover, W.J. 1971. *Practical nonparametric statistics.* New York: Wiley.

Hollander, M., and Wolfe, D.A. 1973. *Nonparametric statistical methods.* New York: Wiley.

Neave, H.R. 1966. A development of Tukey's quick test for location. *Journal of the American Statistical Association* 61: 897–1262.

Neave, H.R. 1975. A quick and simple technique for general slippage problems. *Journal of the American Statistical Association* 70: 721–726.

Omstead, P.S., and Tukey, J.W. 1947. A corner test for association. *Annals of Mathematical Statistics* 18: 495–513.

Siegel, S. 1956. *Nonparametric statistics for the behavioral sciences.* New York: McGraw-Hill.

Tukey, J.W. 1959. A quick compact two-sample test to Duckworth's specifications. *Technometrics* 1: 31–48.

Wilcoxon, F. 1964. Some rapid approximate statistical procedures. Pearl River, New York: Lederle Laboratories.

# 12 Inferences: Population Variance

## 12.1 Introduction

variability

When most people think of statistical inference, they think of inferences concerning population means. However, the population parameter that answers an experimenter's practical questions will vary from one situation to another, and sometimes the variability of a population is more important than its mean. For example, the producer of a drug product is certainly concerned with controlling the mean potency of pills, but he must also worry about the variation in potency from one pill to another. Excessive potency or an underdose could be very harmful to a patient. Hence the manufacturer would like to produce pills with the desired mean potency and with as little variation in potency (as measured by $\sigma$ or $\sigma^2$) as possible.

Inferential problems about a population variance are similar to those for a population mean. We can estimate or test hypotheses about a single population variance or compare two variances.

# Estimation and Tests for a Population Variance 12.2

The sample variance

$s^2$

$$s^2 = \frac{\sum (y - \bar{y})^2}{n - 1}$$

can be used for inferences concerning a population variance $\sigma^2$. For a random sample of $n$ measurements drawn from a normal population with mean $\mu$ and variance $\sigma^2$, $s^2$ provides an unbiased point estimate of $\sigma^2$. In addition, the quantity $(n - 1)s^2/\sigma^2$ follows a chi-square distribution with df $= n - 1$. We can use this information to form a confidence interval for $\sigma^2$.

point estimate for $\sigma^2$

---

**Confidence Interval for a Population Variance $\sigma^2$ with Confidence Coefficient of $(1 - \alpha)$**

$$\frac{(n-1)s^2}{\chi^2_U} < \sigma^2 < \frac{(n-1)s^2}{\chi^2_L}$$

where $\chi^2_U$ is the upper-tail value of chi-square for df $= n - 1$ with area $\alpha/2$ to its right, and $\chi^2_L$ is the lower-tail value with area $\alpha/2$ to its left (see figure 12.1). We can determine $\chi^2_U$ and $\chi^2_L$ for a specific value of df by obtaining the critical value in table 3 of the appendix corresponding to a $= \alpha/2$ and a $= 1 - \alpha/2$, respectively.

---

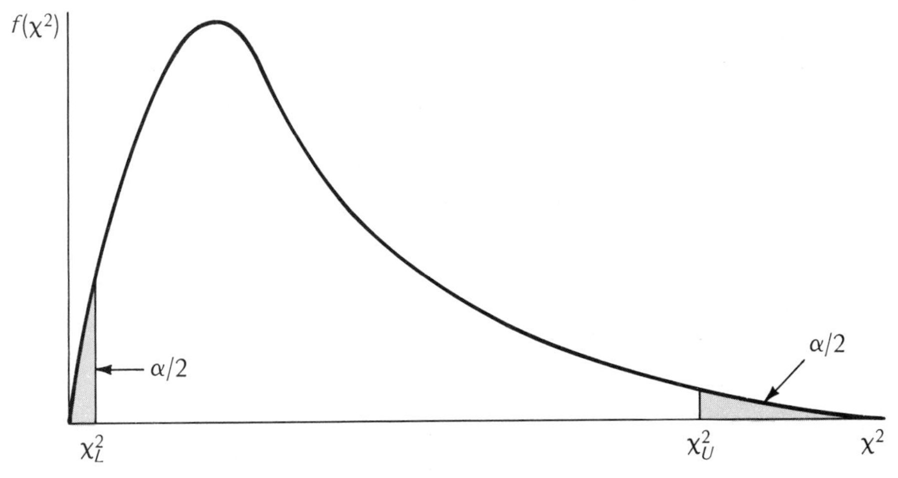


**Figure 12.1** Upper-tail and lower-tail values of chi-square

**EXAMPLE 12.1**

The variability in milk production for a 305-day lactation period was observed for a random sample of 15 holsteins. Use the milk yield data in table 12.1 to estimate $\sigma^2$, the population variance of milk yields, using a 95% confidence interval.

**Table 12.1** Milk production data (in 1000 pounds), example 12.1

| | | |
|---|---|---|
| 12.928 | 13.812 | 11.036 |
| 12.120 | 14.358 | 9.248 |
| 14.972 | 8.998 | 9.980 |
| 14.044 | 10.620 | 11.990 |
| 14.788 | 14.744 | 14.786 |

**SOLUTION**

For these data we find

$\sum y = 188.424 \qquad \sum y^2 = 2431.470$

Substituting into the shortcut formula for $s^2$ (chapter 2), we have

$$s^2 = \frac{1}{n-1}\left[\sum y^2 - \frac{\left(\sum y\right)^2}{n}\right] = \frac{1}{14}\left[2431.470 - \frac{(188.424)^2}{15}\right] = 4.612$$

The confidence coefficient for our example is $1 - \alpha = .95$. The upper-tail chi-square value can be obtained from table 3, the appendix, for $\text{df} = n - 1 = 14$ and $a = \alpha/2 = .025$. Similarly, the lower-tail chi-square value is obtained from table 3 with $a = 1 - \alpha/2 = .975$. Thus

$\chi_U^2 = 26.1190 \qquad \chi_L^2 = 5.62872$

The 95% confidence interval is then

$$\frac{14(4.612)}{26.1190} < \sigma^2 < \frac{14(4.612)}{5.62872}$$

or $2.472 < \sigma^2 < 11.471$.

In addition to estimating a population variance, we can construct a statistical test of the null hypothesis that $\sigma^2$ equals a specified value, $\sigma_0^2$. This test procedure is summarized in the box.

**Test of an Hypothesis Concerning a Population Variance $\sigma^2$**

$H_0$: $\sigma^2 = \sigma_0^2$ ($\sigma_0^2$ is specified).

$H_a$: 1. $\sigma^2 > \sigma_0^2$.
2. $\sigma^2 < \sigma_0^2$.
3. $\sigma^2 \neq \sigma_0^2$.

T.S.: $\chi^2 = \dfrac{(n-1)s^2}{\sigma_0^2}$.

R.R.: For a specified value of $\alpha$,

1. reject $H_0$ if $\chi^2$ is greater than $\chi_U^2$, the upper-tail value for a = $\alpha$ and df = $n - 1$;
2. reject $H_0$ if $\chi^2$ is less than $\chi_L^2$, the lower-tail value for a = $1 - \alpha$ and df = $n - 1$;
3. reject $H_0$ if $\chi^2$ is greater than $\chi_U^2$, based on a = $\alpha/2$ and df = $n - 1$, or less than $\chi_L^2$, based on a = $1 - \alpha/2$ and df = $n - 1$.

---

**EXAMPLE 12.2**

A manufacturer of a specific pesticide useful in the control of household bugs claims that his product retains most of its potency for a period of at least six months. More specifically, he claims that the drop in potency from zero to six months will vary in the interval from 0% to 8%. To test the manufacturer's claim, we obtained a random sample of 20 containers of pesticide from the manufacturer. Each can was tested for potency and then stored for a period of six months. After the storage period, each can was again tested for potency. The drop in potency was recorded for each can and the sample variance for the drops in potencies was computed to be $s^2 = 6.2$. Use these data to determine if there is sufficient evidence to indicate that the population of potency drops has more variability than that claimed by the manufacturer. Use $\alpha = .05$.

**SOLUTION**

The manufacturer has claimed that the population of potency reductions has a range of 8%. Dividing the range by 4, we obtain an approximate population standard deviation of $\sigma = 2\%$ (or $\sigma^2 = 4$).

The appropriate null and alternative hypotheses are

$H_0$: $\sigma^2 = 4$ (i.e., we assume the manufacturer's claim is correct).

$H_a$: $\sigma^2 > 4$ (i.e., there is more variability than claimed by the manufacturer).

Using the computed sample variance based on 20 observations, the test statistic and rejection region are as follows:

$$\text{T.S.: } \chi^2 = \frac{(n-1)s^2}{\sigma_0^2} = \frac{19(6.2)}{4} = 29.45.$$

R.R.: For $\alpha = .05$, we will reject $H_0$ if the computed value of chi-square is greater than 30.1435, obtained from table 3 of the appendix for a = .05 and df = 19.

Conclusion: Since the computed value of chi-square, 29.45, is less than the critical value, 30.1435, we have insufficient evidence to reject the manufacturer's claim.

# Exercises

e 12.1. In chapter 4, exercise 4.12 (p. 95), we discussed several alternatives to alleviating $SO_2$ pollution from smokestacks in operations that used coal as a major source of fuel or energy. One suggestion was to employ a scrubber that cleans the gases after the coal has been burned but before the gases are released into the atmosphere. In 50 samples of gases emitted from a stack equipped with a new scrubber, the sample mean and standard deviation $SO_2$ emission were .13 and .05 pounds per million Btu, respectively. Use these data to construct a 95% confidence interval on $\sigma^2$.

12.2. Refer to exercise 12.1. Suppose that testing on the leading commercial scrubber on a comparable stack has indicated an $SO_2$ emission variance of .50 pounds per million Btu. Is there sufficient evidence that the new scrubber (exercise 12.1) has a lower population variance? Use $\alpha = .05$.

12.3. As part of a detailed drivers training program, school officials are requiring teenagers to take a depth perception test. In one phase of this test, the student is asked to judge the distance between a parked vehicle and a pedestrian stationed a given distance from the student. The recorded distances in feet are listed below for 15 driver education students.

| | | | | | | | |
|---|---|---|---|---|---|---|---|
| 5 | 8 | 7 | 7 | 10 | 6 | 4 | 11 |
| 6 | 8 | 4 | 9 | 9 | 6 | 5 | |

Use these data to construct a 99% confidence on $\sigma^2$, the variance of the depth perception distances.

## 12.3 Estimation and Tests: Comparing Two Population Variances

We are frequently interested in making an inference about two population variances. For example, we may wish to compare the variability in temperatures for a newly designed filament for kitchen ovens to the temperature variability of the filament presently in use. First we hypothesize two populations of measurements that are normally distributed, one corresponding to temperatures run at a given power setting for Filament 1 and one for temperatures corresponding to Filament 2. We label these populations as 1 and 2, respectively. A temperature for either filament is obtained by bringing a single oven to a stable temperature for a fixed power setting. Independent random samples of $n_1$ and $n_2$ ovens are drawn from the two populations, and the temperature of each oven is observed at the fixed power setting. We are interested in comparing the variance of population 1, $\sigma_1^2$, to the variance for population 2, $\sigma_2^2$.

ratio $s_1^2/s_2^2$

Although we used the difference in sample means $\bar{y}_1 - \bar{y}_2$ to compare two population means in chapter 5, it is convenient to consider the *ratio* of the sample variances, $s_1^2/s_2^2$, for comparing two population variances. When independent random samples are drawn from two normal populations, the quantity

$$\frac{s_1^2}{\sigma_1^2} \cdot \frac{\sigma_2^2}{s_2^2}$$

F distribution

possesses a probability distribution in repeated sampling referred to as an *F distribution*. The formula for the probability distribution is omitted here, but we will specify its properties.

**Properties of the F Distribution**

1. Unlike $t$ or $z$, $F$ can assume only positive values.
2. The $F$ distribution, unlike the normal distribution or the $t$ distribution, is nonsymmetrical. (See figure 12.2.)
3. There are many $F$ distributions and each one has a different shape. We specify a particular one by designating the degrees of freedom associated with $s_1^2$ and $s_2^2$. We denote these quantities by $df_1$ and $df_2$, respectively.
4. Tail values for the $F$ distribution are tabulated and appear in tables 4 and 5 of the appendix.

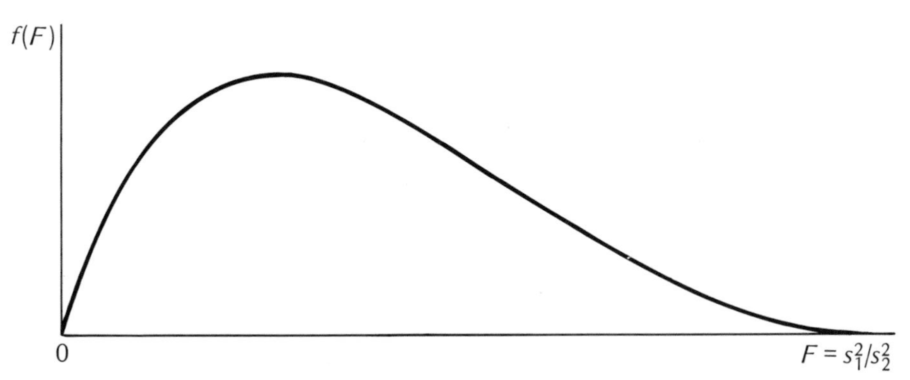


**Figure 12.2** Distribution of $s_1^2/s_2^2$, the $F$ distribution

Table 4 of the appendix records the upper-tail value of $F$ that has an area equal to .05 to its right (see figure 12.3). The degrees of freedom for $s_1^2$, designated by $df_1$, are indicated across the top of the table, while $df_2$, the degrees of freedom for $s_2^2$, appears in the first column to the left. Thus for

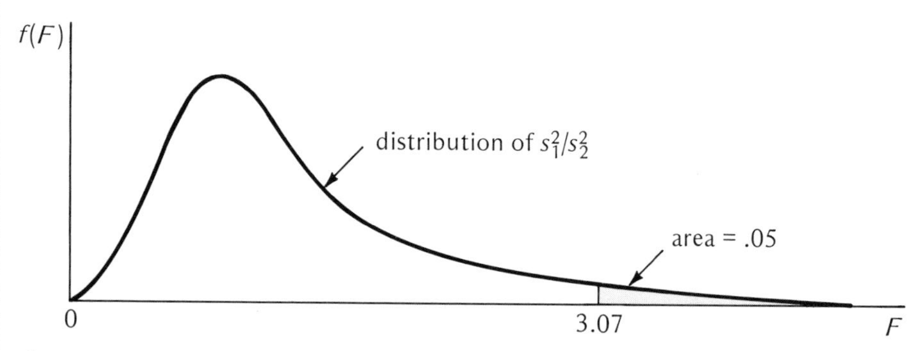


**Figure 12.3** Critical value for the $F$ distribution; $df_1 = 8$ and $df_2 = 10$

$df_1 = 8$ and $df_2 = 10$, the tabulated value is 3.07. Only 5% of the measurements from an $F$ distribution with $df_1 = 8$ and $df_2 = 10$ would exceed 3.07 in repeated sampling.

Table 5 of the appendix gives the upper-tail values for the $F$ distribution with an area of .01 to its right. Thus the .01 value of the $F$ distribution with $df_1 = 6$ and $df_2 = 19$ is 3.94.

We can now formulate a confidence interval for the ratio $\sigma_1^2/\sigma_2^2$.

---

**Confidence Interval for $\sigma_1^2/\sigma_2^2$ with Confidence Coefficient of $(1 - \alpha)$**

$$\frac{s_1^2}{s_2^2} F_L < \frac{\sigma_1^2}{\sigma_2^2} < \frac{s_1^2}{s_2^2} F_U$$

If $F_{df_1,df_2}$ represents the $\alpha/2$ upper-tail value of an $F$ distribution with $df_1$ and $df_2$ degrees of freedom and $F_{df_2,df_1}$ represents the $\alpha/2$ upper-tail value of an $F$ distribution with the degrees of freedom reversed, then

$$F_L = \frac{1}{F_{df_1,df_2}} \quad \text{and} \quad F_U = F_{df_2,df_1}$$

Note: $df_1 = n_1 - 1$ and $df_2 = n_2 - 1$.

---

It should be noted that although our estimation procedure for $\sigma_1^2/\sigma_2^2$ is appropriate for any confidence coefficient $(1 - \alpha)$, tables 4 and 5 limit us to 90% ($\alpha/2 = .05$) and 98% ($\alpha/2 = .01$) confidence intervals, respectively. For more detailed tables of the $F$ distribution, see *Biometrika Tables for Statisticians* (1966).

### EXAMPLE 12.3

The length of life of an electrical component was studied under two operating voltages, $V_1$ and $V_2$. Ten different components were randomly assigned to each of the two operating voltages. Use the data below to find a 90% confidence interval for $\sigma_1^2/\sigma_2^2$, the ratio of the variances in lengths of life for the two populations, populations 1 and 2, corresponding to the components studied under $V_1$ and $V_2$, respectively.

Voltage $V_1$: $n_1 = 10$ $s_1^2 = .51$

Voltage $V_2$: $n_2 = 10$ $s_2^2 = .20$

### SOLUTION

Before constructing our confidence interval, we must obtain $F_{\mathrm{df}_1,\mathrm{df}_2}$ and $F_{\mathrm{df}_2,\mathrm{df}_1}$. For $n_1 = n_2 = 10$, $\mathrm{df}_1 = \mathrm{df}_2 = 9$, and hence $F_{\mathrm{df}_1,\mathrm{df}_2}$ and $F_{\mathrm{df}_2,\mathrm{df}_1}$ are the same. For a 90% confidence interval (i.e., $1 - \alpha = .90$), we must look up the .05 $F$-value based on $\mathrm{df}_1 = 9$ and $\mathrm{df}_2 = 9$. This value is 3.18. The quantities $F_L$ and $F_U$ are

$$F_L = \frac{1}{3.18} \quad \text{and} \quad F_U = 3.18$$

Substituting into the confidence interval formula, we have

$$\frac{.51}{.20}\left(\frac{1}{3.18}\right) < \frac{\sigma_1^2}{\sigma_2^2} < \frac{.51}{.20}(3.18)$$

$$.80 < \frac{\sigma_1^2}{\sigma_2^2} < 8.11$$

We are 90% confident that the ratio of population variances corresponding to voltages $V_1$ and $V_2$ lies in the interval .80 to 8.11.

### EXAMPLE 12.4

Refer to example 12.3. Suppose one of the components on $V_1$ was damaged by the experimenter midway through the test period and had to be removed from the study. Then with $n_1 = 9$ and $n_2 = 10$, $\mathrm{df}_1 = 8$ and $\mathrm{df}_2 = 9$. Assuming $s_1^2$ and $s_2^2$ are as given in example 12.3, set up a 90% confidence interval for $\sigma_1^2/\sigma_2^2$.

### SOLUTION

The appropriate .05 $F$-values can be obtained from table 4 of the appendix.

$$F_{8,9} = 3.23 \quad \text{and} \quad F_L = \frac{1}{3.23}$$

$$F_{9,8} = 3.39 \quad \text{and} \quad F_U = 3.39$$

We then have the confidence interval

$$\frac{.51}{.20}\left(\frac{1}{3.23}\right) < \frac{\sigma_1^2}{\sigma_2^2} < \frac{.51}{.20}(3.39)$$

or $.79 < \sigma_1^2/\sigma_2^2 < 8.64$.

A statistical test of the null hypothesis $\sigma_1^2 = \sigma_2^2$ utilizes the test statistic $s_1^2/s_2^2$. When $H_0$ is true, $s_1^2/s_2^2$ follows an $F$ distribution with $\text{df}_1 = n_1 - 1$ and $\text{df}_2 = n_2 - 1$. If upper-tail and lower-tail values of $F$ were given in tables 4 and 5 of the appendix, we would have no difficulty in performing the test. Unfortunately, only upper-tail values of $F$ are given. To alleviate this situation, we are at liberty to identify either of the two populations as population 1. For a one-tailed alternative hypothesis, the populations are designated 1 and 2 so that $H_a$ is of the form $\sigma_1^2 > \sigma_2^2$. Then the rejection region is located in the upper-tail of the $F$ distribution. For a two-tailed alternative, we designate the population with the larger sample variance as population 1. By this convention, we again are concerned only with upper-tail rejection regions. The upper-tail $F$-value for a two-tailed test can then be obtained from table 4 or table 5 of the appendix.

We summarize the test procedure in the box.

**A Statistical Test Comparing Two Population Variances**

$H_0$: $\sigma_1^2 = \sigma_2^2$.

$H_a$: 1. $\sigma_1^2 > \sigma_2^2$.
2. $\sigma_1^2 \neq \sigma_2^2$.

T.S.: $F = \dfrac{s_1^2}{s_2^2}$.

R.R.: For a specified value of $\alpha$,

1. reject $H_0$ if $F$ exceeds the tabulated value of $F$ for $a = \alpha$, $\text{df}_1 = n_1 - 1$, and $\text{df}_2 = n_2 - 1$;
2. reject $H_0$ if $F$ exceeds the tabulated value of $F$ for $a = \alpha/2$, $\text{df}_1 = n_1 - 1$, and $\text{df}_2 = n_2 - 1$.

Note: Since tables 4 and 5 have upper-tail critical values for a = .05 and .01, respectively, we can run only a one-tailed test with $\alpha = .05$ or .01 or a two-tailed test with $\alpha = .10$ or .02.

**EXAMPLE 12.5**

Two consumer research groups are vying for a large government contract. Since subjective evaluations of consumer products will be made by judges during the study, government officials prefer to award the contract to a company that utilizes judges with consistent ratings (of course, other qualifications are also evaluated before awarding the contract). One measure of consistency is the variability of judges' scores on the same item.

Before issuing the contract, a test is conducted in which 25 judges from each company are asked to rate a single item. The sample variances are given on the next page.

Company A: $s_1^2 = .50$ Company B: $s_2^2 = .15$

Use these data to test the hypothesis that the variances of the judges' ratings are the same for the two populations. The alternative hypothesis is that the variances are different. Use $\alpha = .10$.

**SOLUTION**

We will use a two-tailed test with $\alpha = .10$. Recall that we are at liberty to designate which population is 1. Hence for a two-tailed alternative, Company A, the company with the larger sample variance, is designated as population 1. The test procedure is then as follows:

$H_0$: $\sigma_1^2 = \sigma_2^2$.

$H_a$: $\sigma_1^2 \neq \sigma_2^2$.

T.S.: $F = \dfrac{s_1^2}{s_2^2} = \dfrac{.50}{.15} = 3.33$.

Since there were 25 judges from each company, $df_1 = df_2 = 24$. We locate the critical value of $F$ for $a = \alpha/2 = .05$, $df_1 = 24$, and $df_2 = 24$.

R.R.: From table 4, we will reject $H_0$ if $F > 1.98$.

Conclusion: Since the observed value of $F$ is greater than 1.98, we reject $H_0$ and conclude that the two population variances are different. Practically, since $s_1^2$ exceeds $s_2^2$, we conclude that the ratings of Company A judges have a higher variability than those of Company B.

It should be noted that another use of the $F$ test presented here is to check one of the assumptions in running a two-sample $t$ test for comparing two population means. You will recall we made the assumption that the two populations were identical (i.e., $\sigma_1^2 = \sigma_2^2$). The test of $H_0$: $\sigma_1^2 = \sigma_2^2$ could be used to check this assumption prior to running the $t$ test. An extension of this procedure will be presented in the chapter on assumptions (chapter 21) for testing the equality of $k > 2$ population variances.

## Summary 12.4

In this chapter we discussed procedures for making inferences concerning a population variance and the ratio of two population variances. Estimation and statistical tests concerning $\sigma^2$ make use of the chi-square probability distribution with $df = n - 1$. Inferences concerning the ratio of two population variances utilize an $F$ distribution with $df_1 = n_1 - 1$ and $df_2 = n_2 - 1$.

The need for inferences concerning one or more population variances can be traced to our discussion of numerical descriptive measures of a population in chapter 2. To describe or make inferences about a population of measurements, we cannot always rely on the mean, a measure of central tendency. Many times in evaluating or comparing the performance of individuals on a psychological test, the consistency of manufactured products emerging from a production line, or the yields of a particular variety of corn, we gain important information by studying the population variance.

In the next chapter we return to a comparison of population means. In particular, we will consider a single test procedure, called the analysis of variance, for comparing two or more population means.

# Exercises

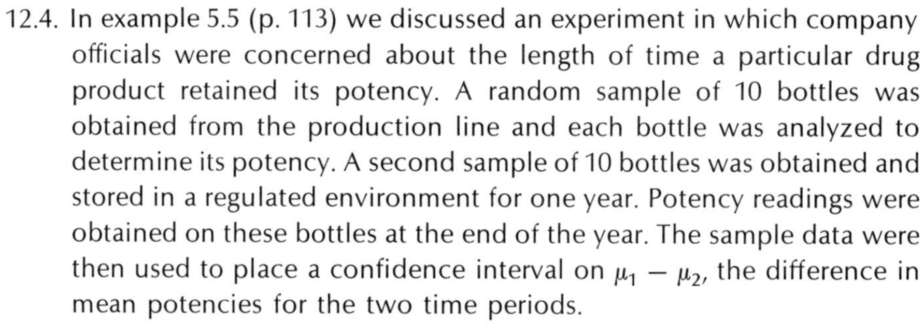

12.4. In example 5.5 (p. 113) we discussed an experiment in which company officials were concerned about the length of time a particular drug product retained its potency. A random sample of 10 bottles was obtained from the production line and each bottle was analyzed to determine its potency. A second sample of 10 bottles was obtained and stored in a regulated environment for one year. Potency readings were obtained on these bottles at the end of the year. The sample data were then used to place a confidence interval on $\mu_1 - \mu_2$, the difference in mean potencies for the two time periods.

Although we did not stress this at the time, in order to use $t$ in the confidence interval or in a statistical test, we do require that the samples be drawn from normal populations with possibly different means *but* with a common variance. Use the sample data summarized below to test the equality of the population variances. Use $\alpha = .02$. Sample 1 data are the readings taken immediately after production and sample 2 data are the readings taken one year after production.

Sample 1: $\sum y = 103.7$ $\quad \sum y^2 = 1076.31$

Sample 2: $\sum y = 98.3$ $\quad \sum y^2 = 966.81$

12.5. Refer to exercise 11.12 (p. 324); we were interested in comparing the weights of the combs of roosters fed one of two vitamin-supplemented diets. The Wilcoxon rank sum test was suggested as a test of the hypothesis that the two populations were identical. Would it have been appropriate to run a $t$ test comparing the two population means? Explain.

12.6. A consumer protection magazine was interested in comparing tires purchased from two different companies, each claiming their tires would last the same number of miles. A sample of 5 tires of each brand was obtained and tested under simulated road conditions. The number of miles before significant deterioration in tread was recorded for all tires. The data are given below (in 1000 miles).

| | | | | | |
|---|---|---|---|---|---|
| *Brand I* | 40.6 | 35.9 | 48.5 | 36.4 | 38.3 |
| *Brand II* | 40.9 | 40.2 | 42.5 | 39.1 | 42.6 |

a. Construct a 98% confidence interval for the ratio of the two population variances.
b. Use the sample data to conduct a test of the equality of the population variances. Use $\alpha = .10$.

12.7. A random sample of 20 patients, each of whom has suffered from depression, was selected from a mental hospital, and each patient was administered the Brief Psychiatric Rating Scale. The scale consists of a series of adjectives that the patient scores according to his or her mood. Extensive testing in the past has shown that ratings in certain mood adjectives tend to be similar and hence are grouped together as jointly measuring one or more components of one's mood. For example, a group consisting of certain adjectives seems to be measuring depression. Let us suppose that the mean and standard deviation of the 20 patients in the group are 13.2 and 4.6, respectively. Place a 99% confidence interval on $\sigma^2$, the variance of the population of patients' scores from which this sample was drawn.

12.8. Refer to exercise 12.7. Suppose that extensive testing in a large number of depressed patients throughout the century has indicated that the population standard deviation of scores for the depression adjectives is 5.9. Use the sample data of exercise 12.7 to test the research hypothesis that the standard deviation for all patients who might be treated for depression in this hospital is less than 5.9. Use $\alpha = .05$.

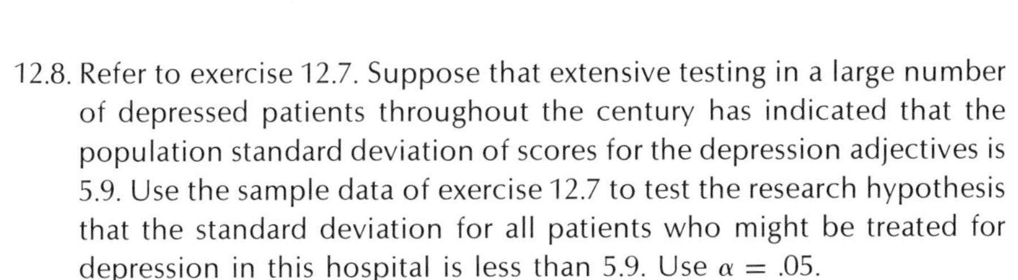

12.9. A pharmaceutical company manufactures a particular brand of antihistamine tablets. In the quality control division, certain tests are routinely performed to determine if the product being manufactured meets specific performance criteria prior to release of the product onto the market. In particular, the company requires that the potencies of the tablets lie in the range of 90% to 110% of the labeled drug amount.

a. If the company is manufacturing 25-milligram (mg) tablets, within what limits must tablet potencies lie?

b. A random sample of 30 tablets is obtained from a recent batch of antihistamine tablets. The data for the potencies of the tablets are given below. Translate the company's 90% to 110% specifications on the range of the product potency into a statistical test concerning the population variance for potencies. Use $\alpha = .05$.

| | | | | | |
|---|---|---|---|---|---|
| 24.1 | 27.2 | 26.7 | 23.6 | 26.4 | 25.2 |
| 25.8 | 27.3 | 23.2 | 26.9 | 27.1 | 26.7 |
| 22.7 | 26.9 | 24.8 | 24.0 | 23.4 | 25.0 |
| 24.5 | 26.1 | 25.9 | 25.4 | 22.9 | 24.9 |
| 26.4 | 25.4 | 23.3 | 23.0 | 24.3 | 23.8 |

$ 12.10. A study was conducted to compare the variabilities in strengths of one-inch-square sections of a synthetic fiber produced under two different procedures. A random sample of 9 squares from each process was obtained and tested. Use the data (psi) below to test the research hypothesis that the population variances corresponding to the two procedures are different. Use $\alpha = .10$.

| | | | | | | | | | |
|---|---|---|---|---|---|---|---|---|---|
| *Procedure 1* | 74 | 90 | 103 | 86 | 75 | 102 | 97 | 85 | 69 |
| *Procedure 2* | 59 | 66 | 73 | 68 | 70 | 71 | 82 | 69 | 74 |

# References


Dixon, W.J., and Massey, F.J., Jr. 1969. *Introduction to statistical analysis*. 3d ed. New York: McGraw-Hill.

Draper, N.R., and Smith, H. 1966. *Applied regression analysis*. New York: Wiley.

Harnett, D.L. 1970. *Introduction to statistical methods*. Reading, Mass.: Addison-Wesley.

Mendenhall, W. 1975. *Introduction to probability and statistics*. 4th ed. N. Scituate, Mass.: Duxbury Press.

Mendenhall, W.; Ott, L.; and Larson, R. 1974. *Statistics: A tool for the social sciences*. N. Scituate, Mass.: Duxbury Press.

Mendenhall, W., and Ott, L. 1976. *Understanding statistics*. 2d ed. N. Scituate, Mass: Duxbury Press.

Ostle, B. 1963. *Statistics in research*. 2d ed. Ames, Iowa: Iowa State University Press.

Pearson, E.S., and Hartley, H.O. 1966. *Biometrika tables for statisticians*. 3d ed. Vol. I. London: Cambridge University Press.

Snedecor, G.W., and Cochran, W.G. 1967. *Statistical methods*. 6th ed. Ames, Iowa: Iowa State University Press.

Tanur, J.M.; Mosteller, F.; Kruskal, W.H.; Pieters, R.S.; and Rising, G.R. 1972. *Statistics: A guide to the unknown*. San Francisco: Holden-Day.

# 13 Introduction to the Analysis of Variance: One-way Classification

## 13.1 Introduction

Methods for comparing two population means, based on independent random samples, were presented in chapter 5. Very often the two-sample problem is a simplification of what we encounter in practical situations. For example, suppose we wish to compare the mean hourly wage for nonunion farm laborers from three different ethnic groups (black, white, and Spanish-American) employed by a large produce company. Independent random samples of farm laborers would be selected from each of the three ethnic groups (populations). Then using the information from the three sample means, we would try to make an inference about the corresponding population mean hourly wages. Most likely, the sample means would differ, but this does not necessarily imply a difference among the population means for the three ethnic groups. How do you decide whether the differences among the sample means are large enough to imply that the corresponding population means are different? We will answer this question using a statistical testing procedure called an *analysis of variance*.

# The Logic Behind an Analysis of Variance 13.2

The reason we call the testing procedure an analysis of variance can be seen by using the example in section 13.1. Assume that we wish to compare the three ethnic mean hourly wages based on samples of 5 workers selected from each of the ethnic groups. Although a sample of size 5 from each of the populations seems pitifully small, we can illustrate the basic ideas.

Suppose the sample data (hourly wages, in dollars) appear as shown in table 13.1. Do these data present sufficient evidence to indicate differences

**Table 13.1** A comparison of three sample means (small amount of variation within samples)

| *Sample from Population* | | |
|---|---|---|
| *1* | *2* | *3* |
| 2.90 | 2.51 | 2.01 |
| 2.92 | 2.50 | 2.00 |
| 2.91 | 2.50 | 1.99 |
| 2.89 | 2.49 | 1.98 |
| 2.88 | 2.50 | 2.02 |
| $\bar{y}_1 = 2.90$ | $\bar{y}_2 = 2.50$ | $\bar{y}_3 = 2.00$ |

among the three population means? A brief visual inspection of the data indicates very little variation within a sample, while the variability among the sample means is much larger. Since the variability among the sample means is so large *in comparison to the within-sample variation,* we might conclude intuitively that the corresponding population means are different.

within-sample variation

Table 13.2 illustrates a situation where the sample means are the same as

**Table 13.2** A comparison of three sample means (large amount of within-sample variation)

| *Sample from Population* | | |
|---|---|---|
| *1* | *2* | *3* |
| 2.90 | 3.31 | 1.52 |
| 1.42 | .54 | 3.93 |
| 4.51 | 1.73 | 1.48 |
| 4.89 | 4.20 | 2.55 |
| .78 | 2.72 | .52 |
| $\bar{y}_1 = 2.90$ | $\bar{y}_2 = 2.50$ | $\bar{y}_3 = 2.00$ |

given in table 13.1 but the variability within a sample is much larger. In contrast to the data in table 13.1, the *between-sample variability* is small relative to the within-sample variability. We would be less likely to conclude that the corresponding population means differ based on these data.

between-sample variation

The variations in the two sets of data, tables 13.1 and 13.2, are shown graphically in figure 13.1. The strong evidence to indicate a difference in population means for the data of table 13.1 is apparent in figure 13.1(a). The

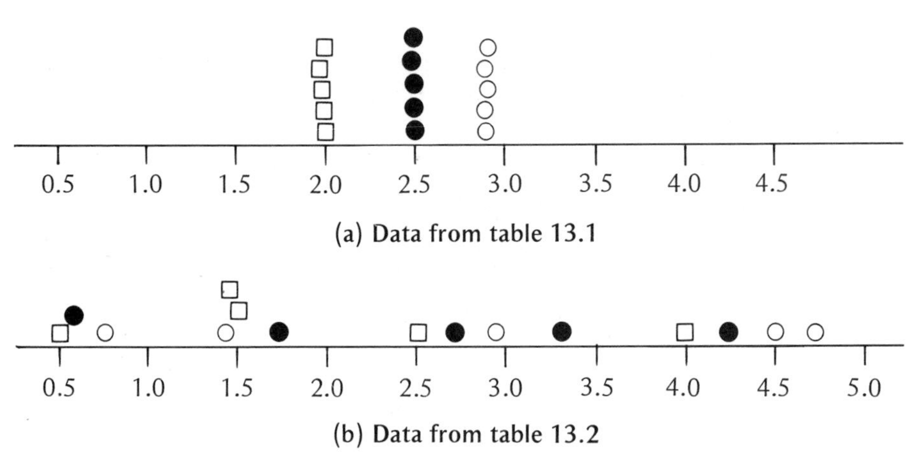


**Figure 13.1** Dot diagrams for the data of table 13.1 and table 13.2: ○, measurement from sample 1; ●, measurement from sample 2; □, measurement from sample 3

lack of evidence to indicate a difference in population means for the data of table 13.2 is indicated by the overlapping of data points for the samples in figure 13.1(b).

analysis of variance

The preceding discussion, with the aid of figure 13.1, should indicate what we mean by an *analysis of variance*. All differences in sample means are judged statistically significant (or not) by comparing them to the variation within samples. The details of the testing procedure will be presented in the next section.

## 13.3 A Test of an Hypothesis About $t > 2$ Population Means: An Analysis of Variance

In chapter 5, we presented a method for testing the equality of two population means. We hypothesized two normal populations (1 and 2) with means denoted by $\mu_1$ and $\mu_2$, respectively, and a common variance $\sigma^2$. To test the null hypothesis that $\mu_1 = \mu_2$, independent random samples of sizes $n_1$ and $n_2$ were drawn from the two populations. The sample data were then used to compute the value of the test statistic

$$t = \frac{\bar{y}_1 - \bar{y}_2}{s\sqrt{(1/n_1) + (1/n_2)}}$$

where

$$s^2 = \frac{(n_1 - 1)s_1^2 + (n_2 - 1)s_2^2}{(n_1 - 1) + (n_2 - 1)} = \frac{(n_1 - 1)s_1^2 + (n_2 - 1)s_2^2}{n_1 + n_2 - 2}$$

is a pooled estimate of the common population variance $\sigma^2$. The rejection region for a specified value of $\alpha$, the probability of a type I error, was then found using table 2, the appendix.

Suppose now that we wish to extend this method to test the equality of more than two population means. The test procedure described above applies only to two means and therefore is inappropriate. Hence we will employ a more general method of data analysis, the analysis of variance. We illustrate its use with the following example.

example

Students from 5 different campuses throughout the country were surveyed to determine their attitudes towards industrial pollution. Each student sampled was asked a specific number of questions and then given a total score for the interview. Suppose that 9 students are surveyed at each of the 5 campuses and we wish to compare the responses among the campuses. In particular, suppose we wish to examine the average student score for each of the 5 campuses.

We label the set of all test scores that could have been obtained from Campus I as population I, and we will assume that this population possesses a mean $\mu_1$. A random sample of $n_1 = 9$ measurements (scores) is obtained from this population to monitor student attitudes towards pollution. The set of all scores which could have been obtained from students on Campus II is labeled population II (which has a mean $\mu_2$). The data from a random sample of $n_2 = 9$ scores are obtained from this population. Similarly $\mu_3$, $\mu_4$, and $\mu_5$ represent the means of the populations for scores from campuses III, IV, and V, respectively. We also obtain random samples of 9 student scores from each of these populations.

From each of these 5 samples, we calculate a sample mean and variance. The sample results can then be summarized as shown in table 13.3, next page.

If we are interested in testing the equality of the population means (i.e., $\mu_1 = \mu_2 = \mu_3 = \mu_4 = \mu_5$), we might be tempted to run all possible pairwise comparisons of two population means. Hence if we assume that the 5 distributions are approximately normal with the same variance $\sigma^2$, we could run 10 $t$ tests comparing two means, as listed on the next page (see section 5.4).

multiple $t$ tests

**Table 13.3** Summary of the sample results for 5 populations

| | Population | | | | |
|---|---|---|---|---|---|
| | *I* | *II* | *III* | *IV* | *V* |
| *Sample Mean* | $\bar{y}_1$ | $\bar{y}_2$ | $\bar{y}_3$ | $\bar{y}_4$ | $\bar{y}_5$ |
| *Sample Variance* | $s_1^2$ | $s_2^2$ | $s_3^2$ | $s_4^2$ | $s_5^2$ |

Null hypotheses:

$\mu_1 = \mu_2$ $\quad \mu_1 = \mu_4$ $\quad \mu_2 = \mu_3$ $\quad \mu_2 = \mu_5$ $\quad \mu_3 = \mu_5$

$\mu_1 = \mu_3$ $\quad \mu_1 = \mu_5$ $\quad \mu_2 = \mu_4$ $\quad \mu_3 = \mu_4$ $\quad \mu_4 = \mu_5$

One obvious disadvantage to this test procedure is that it is tedious and time-consuming. But a more important and less apparent disadvantage of running multiple $t$ tests to compare means is that the probability of falsely rejecting at least one of the hypotheses increases as the number of $t$ tests increases. Thus, although we may have the probability of a type I error fixed at $\alpha = .05$ for each individual test, the probability of falsely rejecting *at least one* of those tests is larger than .05. In other words, the combined probability of a type I error for the set of 10 hypotheses would be larger than the value .05 set for each individual test. Indeed, it could be as large as .40.

What we need is a single test of the hypothesis "all five population means are equal," which will be less tedious than the individual $t$ tests and can be performed with a specified probability of a type I error (say, .05). This test is the analysis of variance.

First we assume that the five sets of measurements are normally distributed, with means given by $\mu_1$, $\mu_2$, $\mu_3$, $\mu_4$, and $\mu_5$ and with a common variance $\sigma^2$. Next we consider the quantity

$s_W^2$

$$s_W^2 = \frac{(n_1 - 1)s_1^2 + (n_2 - 1)s_2^2 + (n_3 - 1)s_3^2 + (n_4 - 1)s_4^2 + (n_5 - 1)s_5^2}{(n_1 - 1) + (n_2 - 1) + (n_3 - 1) + (n_4 - 1) + (n_5 - 1)}$$

$$= \frac{(n_1 - 1)s_1^2 + (n_2 - 1)s_2^2 + (n_3 - 1)s_3^2 + (n_4 - 1)s_4^2 + (n_5 - 1)s_5^2}{n_1 + n_2 + n_3 + n_4 + n_5 - 5}$$

Note that this quantity is merely an extension of

$$s^2 = \frac{(n_1 - 1)s_1^2 + (n_2 - 1)s_2^2}{n_1 + n_2 - 2}$$

which is used as an estimate of the common variance for two populations for a test of the hypothesis $\mu_1 = \mu_2$ (section 5.4). Thus $s_W^2$ represents a combined estimate of the common variance $\sigma^2$, and it measures the variability of the observations within the five populations. (The subscript $W$ refers to the within-population variability.)

Next we consider a quantity that measures the variability between or among the population means. If the null hypothesis $\mu_1 = \mu_2 = \mu_3 = \mu_4 = \mu_5$ is true, then the populations are identical, with mean $\mu$ and variance $\sigma^2$. Drawing single samples from the five populations is then equivalent to drawing five different samples from the same population. What kind of variation might be expected for these sample means? If the variation is too great, we would reject the hypothesis that $\mu_1 = \mu_2 = \mu_3 = \mu_4 = \mu_5$.

To discuss the variation from sample mean to sample mean, we need to know the distribution of the mean of a sample of 9 observations in repeated sampling. The distribution of sample means will have the same mean $\mu$ and variance $\sigma^2/9$. Since we have drawn 5 samples of 9 observations each, we can estimate the variance of the distribution of sample means, $\sigma^2/9$, using the formula

$$\begin{array}{l}\text{sample variance}\\ \text{(of the means)}\end{array} = \frac{\sum \bar{y}^2 - \left[\left(\sum \bar{y}\right)^2 / 5\right]}{5 - 1}$$

Note that we merely consider the $\bar{y}$'s as a sample of five observations and calculate the "sample variance." This quantity estimates $\sigma^2/9$ and hence $9 \times$ (sample variance of the means) estimates $\sigma^2$. We designate this quantity as $s_B^2$ (the subscript $B$ designates a measure of the variability among the sample means for the 5 populations).

$s_B^2$

Under the null hypothesis that all 5 population means are identical, we have two estimates of $\sigma^2$, namely, $s_W^2$ and $s_B^2$. Suppose the ratio

$$\frac{s_B^2}{s_W^2}$$

is used as the test statistic to test the hypothesis that $\mu_1 = \mu_2 = \mu_3 = \mu_4 = \mu_5$. What is the distribution of this quantity if we were to repeat the experiment over and over again, each time calculating $s_B^2$ and $s_W^2$?

For our example $s_B^2/s_W^2$ follows an $F$ distribution, with degrees of freedom that can be shown to be $\text{df}_1 = 4$ for $s_B^2$ and $\text{df}_2 = 40$ for $s_W^2$. The proof of these remarks is beyond the scope of this text. However, we make use of this result for testing the null hypothesis $\mu_1 = \mu_2 = \mu_3 = \mu_4 = \mu_5$.

The decision maker used to test equality of the population means is

$$F = \frac{s_B^2}{s_W^2}$$

test statistic

When the null hypothesis is true, both $s_B^2$ and $s_W^2$ estimate $\sigma^2$, and $F$ would be expected to assume a value near $F = 1$. When the hypothesis of equality is false, $s_B^2$ will tend to be larger than $s_W^2$ due to the differences among the population means. Hence we will reject the null hypothesis in the upper tail of the distribution of $F = s_B^2/s_W^2$. For $\alpha$, the probability of a type I error, equal to .05 or .01, we can locate the rejection region for this one-tailed test using tables 4 or 5 in the appendix, with $\text{df}_1 = 4$ and $\text{df}_2 = 40$. Thus for $\alpha = .05$ the critical value of $F = s_B^2/s_W^2$ is 2.61. (See figure 13.2.) If the calculated value of $F$ falls in

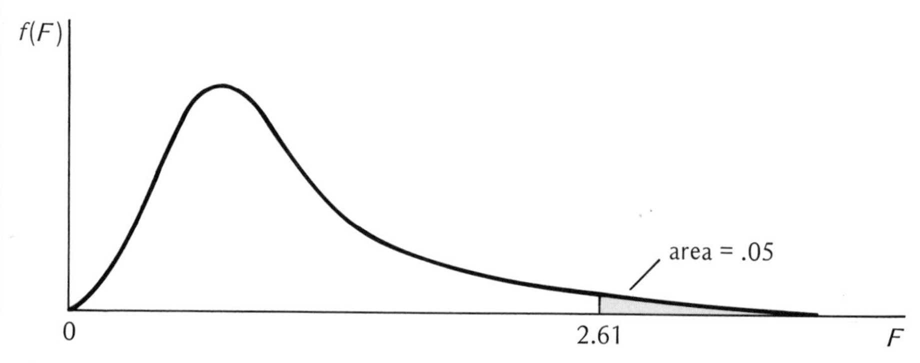


**Figure 13.2** Critical value of $F$ for $\alpha = .05$, $\text{df}_1 = 4$, and $\text{df}_2 = 40$

the rejection region, we conclude that not all five population means are identical.

This procedure can be generalized (and simplified) with only slight modifications in the formulas to test the equality of $t$ (where $t$ is an integer equal to or greater than 2) population means from normal populations with a common variance $\sigma^2$. Random samples of sizes $n_1, n_2, \ldots, n_t$ are drawn from the respective populations. We then compute the sample means and variances. The null hypothesis $\mu_1 = \mu_2 = \cdots = \mu_t$ is tested against the alternative that at least one of the population means is different from the others.

Before presenting the generalized test procedure, it is convenient to introduce the notation to be used in the shortcut computational formulas for $s_B^2$ and $s_W^2$.

Consider a set of sample data, where $y_{ij}$ denotes the $j$th observation obtained from population $i$. For a comparison among 5 populations with 4 observations per sample, the data could be organized in a two-way table, as shown in table 13.4. The arrangement of data in a table such as this is called a *one-way classification* because each observation is classified according to the population from which it was drawn. Using table 13.4, we can introduce notation that is helpful when performing an analysis of variance (AOV) for a one-way classification.

two-way table

one-way classification

AOV

**Table 13.4** Summary of sample data for a one-way classification

| *Sample* | *Data* | | | | *Totals* |
|---|---|---|---|---|---|
| 1 | $y_{11}$ | $y_{12}$ | $y_{13}$ | $y_{14}$ | $T_1$ |
| 2 | $y_{21}$ | $y_{22}$ | $y_{23}$ | $y_{24}$ | $T_2$ |
| 3 | $y_{31}$ | $y_{32}$ | $y_{33}$ | $y_{34}$ | $T_3$ |
| 4 | $y_{41}$ | $y_{42}$ | $y_{43}$ | $y_{44}$ | $T_4$ |
| 5 | $y_{51}$ | $y_{52}$ | $y_{53}$ | $y_{54}$ | $T_5$ |

**Notation Needed for the AOV of a One-way Classification**

$y_{ij}$: The $j$th sample observation selected from population $i$. For example, $y_{23}$ denotes the 3d sample observation drawn from population 2.

$n_i$: The number of sample observations selected from population $i$. In our data set, $n_1$, the number of observations obtained from population 1, is 4. Similarly, $n_2 = n_3 = n_4 = n_5 = 4$. It should be noted, however, that the sample sizes need not be the same. Thus we might have $n_1 = 12$, $n_2 = 3$, $n_3 = 6$, $n_4 = 10$, and so forth.

$n$: The total sample size; $n = \Sigma\, n_i$. For the data in table 13.4, $n = n_1 + n_2 + n_3 + n_4 + n_5 = 20$.

$T_i$: The sum (total) of the sample observations obtained from population $i$. These totals are displayed in the right-hand column of table 13.4.

$\bar{T}_i$: The average of the $n_i$ sample observations drawn from population $i$; $\bar{T}_i = T_i/n_i$.

$G$: The sum (grand total) of all sample observations drawn from the populations; $G = \Sigma\, T_i$.

$\bar{y}$: The average of all sample observations; $\bar{y} = G/n$.

With this notation it is possible to establish the following algebraic identities. (Although we will use these results in later calculations for $s_W^2$ and $s_B^2$, the proofs of these identities are beyond the scope of this text.) The variability of the $n$ sample measurements about their mean $\bar{y}$ can be measured using the sum of the squared deviations $(y_{ij} - \bar{y})^2$. This quantity,

$$\text{TSS} = \sum_i \sum_j (y_{ij} - \bar{y})^2$$

total sum of squares

is called the *total sum of squares* of the measurements about their mean. The double summation in TSS means that we must sum the squared deviations for all rows ($i$) and columns ($j$) of the one-way classification.

It is possible to partition the total sum of squares as follows:

$$\sum_i \sum_j (y_{ij} - \bar{y})^2 = \sum_i \sum_j (y_{ij} - \bar{T}_i)^2 + \sum_i n_i(\bar{T}_i - \bar{y})^2$$

The first quantity on the right-hand side of the equation measures the variability of an observation $y_{ij}$ about its sample mean $\bar{T}_i$. Thus

$$SSW = \sum_i \sum_j (y_{ij} - \bar{T}_i)^2$$

within-sample sum of squares

is a measure of the *within*-sample variability. SSW is referred to as the *within-sample sum of squares* and will be used to compute $s_W^2$.

between-sample sum of squares

The second expression in the total sum of squares equation measures the variability of the sample means $\bar{T}_i$ about the overall mean $\bar{y}$. This quantity, which measures the variability *between* (or among) the sample means, is referred to as the *sum of squares between samples* (SSB) and will be used to compute $s_B^2$.

$$SSB = \sum_i n_i(\bar{T}_i - \bar{y})^2$$

Although the formulas for TSS, SSW, and SSB are easily interpreted, they are not easy to use for calculations. Instead, we use the shortcut formulas shown in the box.

**Shortcut Sum of Squares Formulas for a One-way Classification**

$$TSS = \sum_i \sum_j y_{ij}^2 - \frac{G^2}{n}$$

$$SSB = \sum_i \frac{T_i^2}{n_i} - \frac{G^2}{n}$$

$$SSW = TSS - SSB$$

An analysis of variance for a one-way classification for $t$ populations has the following null and alternative hypotheses:

$H_0$: $\mu_1 = \mu_2 = \mu_3 = \cdots = \mu_t$ (i.e., the $t$ population means are equal).
$H_a$: At least one of the $t$ population means differs from the rest.

The quantities $s_B^2$ and $s_W^2$ can be computed using the shortcut formulas

$$s_B^2 = \frac{SSB}{t - 1} \qquad s_W^2 = \frac{SSW}{n - t}$$

where $t - 1$ and $n - t$ are the degrees of freedom for $s_B^2$ and $s_W^2$, respectively.

Historically, people have referred to a sum of squares divided by its degrees of freedom as a *mean square*. Hence $s_B^2$ is often called the *mean square between samples* and $s_W^2$ the *mean square within samples*.

mean square

The null hypothesis of equality of the $t$ population means is rejected if

$$F = \frac{s_B^2}{s_W^2}$$

exceeds the tabulated value of $F$ for $a = \alpha$, $df_1 = t - 1$, and $df_2 = n - t$.

After completing the $F$ test, the results of a study are then summarized in an *analysis of variance table*. The format of an analysis of variance table is shown in table 13.5. The analysis of variance table lists the sources of variability in the

AOV table

**Table 13.5** An example of an AOV table for a one-way classification

| Source | *Sum of Squares* | *Degrees of Freedom* | *Mean Square* | *F Test* |
|---|---|---|---|---|
| between samples | SSB | $t - 1$ | $s_B^2 = SSB/t - 1$ | $s_B^2/s_W^2$ |
| within samples | SSW | $n - t$ | $s_W^2 = SSW/n - t$ | |
| *Totals* | TSS | $n - 1$ | | |

first column. The second column lists the sums of squares associated with each source of variability. Since we showed that the total sum of squares (TSS) can be partitioned into two parts, then SSB and SSW must add up to TSS in the AOV table. The third column of the table gives the degrees of freedom associated with the sources of variability. Again we have a check, $(t - 1) + (n - t)$ must add to $n - 1$. Finally, the mean squares are found in the fourth column and the $F$ test for the equality of the $t$ population means is given in the fifth column.

**EXAMPLE 13.1**

A horticulturist was investigating the phosphorous content of tree leaves from 3 different varieties of apple trees (1, 2, and 3). Random samples of 5 leaves from each of the 3 varieties were analyzed for phosphorous content. The data are given in table 13.6. Use these data to test the hypothesis of equality of the mean phosphorous levels for the 3 varieties. Use $\alpha = .05$.

**Table 13.6** Phosphorous content of leaves from three different trees, example 13.1

| *Variety* | | | *Phosphorous Content* | | | *Totals* |
|---|---|---|---|---|---|---|
| 1 | .35 | .40 | .58 | .50 | .47 | 2.30 |
| 2 | .65 | .70 | .90 | .84 | .79 | 3.88 |
| 3 | .60 | .80 | .75 | .73 | .66 | 3.54 |
| *Total* | | | | | | 9.72 |

**SOLUTION**

The null and alternative hypotheses for this example are

$H_0$: $\mu_1 = \mu_2 = \mu_3$.

$H_a$: At least one of the population means differs from the rest.

The sample sizes are $n_1 = n_2 = n_3 = 5$, for which $n = 15$. From the sample data we see that the total (sum) for all observations on Variety 1 is $T_1 = 2.30$. Similarly, the totals for varieties 2 and 3 are $T_2 = 3.88$ and $T_3 = 3.54$. The sum of all sample measurements is then

$G = T_1 + T_2 + T_3 = 9.72$

Using the sample measurements, the total sum of squares is

$$\text{TSS} = \sum_i \sum_j y_{ij}^2 - \frac{G^2}{n} = (.35)^2 + (.40)^2 + \cdots + (.66)^2 - \frac{(9.72)^2}{15}$$

$$= 6.673 - 6.299 = .374$$

The sample totals can then be used to compute the sum of squares between samples, SSB.

$$\text{SSB} = \sum_i \frac{T_i^2}{n_i} - \frac{G^2}{n} = \frac{(2.30)^2 + (3.88)^2 + (3.54)^2}{5} - 6.299$$

$$= 6.575 - 6.299 = .276$$

Then the sum of squares within samples is

$\text{SSW} = \text{TSS} - \text{SSB} = .374 - .276 = .098$

The analysis of variance table for these data is shown in table 13.7.

**Table 13.7** Analysis of variance table for the data of example 13.1

| *Source* | *Sum of Squares* | *Degrees of Freedom* | *Mean Square* | *F Test* |
|---|---|---|---|---|
| between samples | .276 | 2 | .276/2 = .138 | .138/.008 = 17.25 |
| within samples | .098 | 12 | .098/12 = .008 | |
| *Totals* | .374 | 14 | | |

The critical value of $F = s_B^2/s_W^2$ is 3.89, which is obtained from table 4 of the appendix for a = .05, $df_1 = 2$, and $df_2 = 12$. Since the computed value of $F$, 17.25, exceeds 3.89, we reject the null hypothesis of equality of the mean phosphorous content for the three varieties. It appears from the data that the mean for Variety 1 is smaller than the means for varieties 2 and 3.

**EXAMPLE 13.2**

A clinical psychologist wished to compare three methods for reducing hostility levels in university students. A certain test (HLT) was used to measure the degree of hostility. A high score on this test indicated great hostility. Eleven students obtaining high and nearly equal scores were used in the experiment. Four were selected at random from among the 11 problem cases and treated with Method 1. Four of the remaining 7 students were selected at random and treated with Method 2. The remaining 3 students were treated with Method 3. All treatments were continued for a one-semester period. Each student was given the HLT test at the end of the semester, with the results shown in table 13.8. Use these

**Table 13.8** HLT test scores, example 13.2

| *Method* | | *Test Scores* | | | *Totals* |
|---|---|---|---|---|---|
| 1 | 80 | 92 | 87 | 83 | 342 |
| 2 | 70 | 81 | 78 | 74 | 303 |
| 3 | 63 | 76 | 70 | | 209 |
| *Total* | | | | | 854 |

data to perform an analysis of variance to determine if there are differences among mean scores for the three methods. Use $\alpha = .05$.

**SOLUTION**

The null and alternative hypotheses are

$H_0$: $\mu_1 = \mu_2 = \mu_3$.

$H_a$: At least one of the population means differs from the rest.

For $n_1 = 4$, $n_2 = 4$, and $n_3 = 3$, we have a total sample size of $n = 11$. The totals from table 13.8 are

$T_1 = 342 \qquad T_2 = 303 \qquad T_3 = 209$

$G = T_1 + T_2 + T_3 = 854$

Substituting into the computational formulas for TSS and SSB, we have

$$\text{TSS} = \sum_i \sum_j y_{ij}^2 - \frac{G^2}{n} = (80)^2 + (92)^2 + \cdots + (70)^2 - \frac{(854)^2}{11}$$

$$= 66{,}988 - 66{,}301.45 = 686.55$$

$$\text{SSB} = \sum_i \frac{T_i^2}{n_i} - \frac{G^2}{n} = \frac{(342)^2}{4} + \frac{(303)^2}{4} + \frac{(209)^2}{3} - 66{,}301.45$$

$$= 66{,}753.58 - 66{,}301.45 = 452.13$$

Then

$\text{SSW} = 686.55 - 452.13 = 234.42$

The analysis of variance table for these data is shown in table 13.9.

The critical value of $F$ is obtained from table 4 of the appendix for $a = .05$, $df_1 = 2$, and $df_2 = 8$; this value is 4.46. Since the computed value

**Table 13.9** AOV table for the data of example 13.2

| *Source* | *SS* | *df* | *MS* | *F* |
|---|---|---|---|---|
| between samples | 452.13 | 2 | 226.07 | 226.07/29.3 = 7.72 |
| within samples | 234.42 | 8 | 29.30 | |
| *Totals* | 686.55 | 10 | | |

of $F$, 7.72, exceeds the tabulated value, 4.46, we reject the null hypothesis of equality of the mean scores for the three groups.

# 13.4 The Model for Observations in a One-way Classification

We formulated a model (equation) to relate a response $y$ to a set of quantitative independent variables in chapter 6. In this section we will consider a model for the one-way classification. While the model at first may appear to be quite different from those of chapter 6, we will see later that it is very similar.

assumptions

We make the following assumptions concerning the sample measurements and the populations from which they were drawn:

1. The samples are independent random samples. Results from one sample in no way affect the measurements observed in another sample.
2. Each sample is selected from a normal population.
3. The mean and variance for population $i$ are, respectively, $\mu_i$ and $\sigma^2$ $(i = 1, 2, \ldots, t)$.
4. To summarize, we assume that the $t$ populations are normally distributed with different means but a common variance $\sigma^2$.

We can now formulate a model (equation) which encompasses the assumptions listed above. Recall that we previously let $y_{ij}$ denote the $j$th sample observation from population $i$.

model

$$y_{ij} = \mu + \alpha_i + \epsilon_{ij}$$

terms

This model states that $y_{ij}$, the $j$th sample measurement selected from population $i$, is the sum of three terms. The term $\mu$ denotes an overall mean which is an unknown constant. The term $\alpha_i$ denotes an effect due to population $i$; $\alpha_i$ is an unknown constant. The term $\epsilon_{ij}$ denotes a random error associated with the $j$th observation from population $i$. We assume that $\epsilon_{ij}$ is normally distributed, with a mean of 0 and a variance $\sigma^2$. In addition, the errors are independent; that is, the error associated with one observation in no way affects the error associated with another observation.

Since the $\epsilon$'s are normally distributed with mean 0, the mean or expected value of $y_{ij}$, denoted by $E(y_{ij})$, is

$$E(y_{ij}) = \mu + \alpha_i$$

That is, $y_{ij}$ has been selected from a population with mean $\mu + \alpha_i$. Since $\alpha_i$ may assume a positive, zero, or negative value, the mean for population $i$ can be greater than, equal to, or less than $\mu$, the overall mean. The variance for each of the $t$ populations can be shown to be $\sigma^2$. Finally, because the $\epsilon$'s are normally distributed, each of the $t$ populations is normal. A summary of the assumptions for a one-way classification is shown in table 13.10.

**Table 13.10** Summary of some of the assumptions for a one-way classification

| *Population* | *Population Mean* | *Population Variance* | *Sample Measurements* |
|---|---|---|---|
| 1 | $\mu + \alpha_1$ | $\sigma^2$ | $y_{11}, y_{12}, \ldots, y_{1n_1}$ |
| 2 | $\mu + \alpha_2$ | $\sigma^2$ | $y_{21}, y_{22}, \ldots, y_{2n_2}$ |
| $\vdots$ | $\vdots$ | $\vdots$ | $\vdots$ |
| $t$ | $\mu + \alpha_t$ | $\sigma^2$ | $y_{t1}, y_{t2}, \ldots, y_{tn_t}$ |

The null hypothesis for one analysis of variance is that $\mu_1 = \mu_2 = \cdots = \mu_t$. Using our model, this would be equivalent to the null hypothesis

$$H_0\colon \alpha_1 = \alpha_2 = \cdots = \alpha_t = 0$$

If $H_0$ is true, then all populations have the same unknown mean $\mu$. Indeed, many textbooks use this latter null hypothesis for the analysis of variance in a one-way classification. The corresponding alternative hypothesis is

$H_a$: At least one of the $\alpha_i$'s differs from zero

In this section we have presented a brief discussion of the model associated with the analysis of variance for a one-way classification. Although some authors bypass a discussion of the model, we feel it is a necessary part of an analysis of variance discussion. We utilized a model to relate a response $y$ to a set of quantitative independent variables in chapter 6. Now we have also introduced a model for our discussion of the analysis of variance. While the two types of models may appear to be quite different, we will show later (in chapters 16 and 17) that they are very similar.

You may be concerned with checking the validity of the underlying assumptions in an analysis of variance. In practice, you should always make at least a rough check before proceeding. At present we will bypass this step and emphasize the analysis of data to develop facility in working problems. A separate chapter (chapter 21) will be devoted to procedures for checking the underlying assumptions. However, the Kruskal Wallis test of section 11.4 (p. 320), also referred to as the Kruskal-Wallis one-way analysis of variance by ranks, can be used for ordinal data or for continuous data when the underlying assumption of normality of the treatment groups does not hold for the $F$ test. The null hypothesis for the Kruskal-Wallis test is that the $t$ populations are identical.

## 13.5 Summary

Methods for extending the results of chapter 5 were presented in this chapter to include a comparison among $t$ population means. An independent random sample is drawn from each of the $t$ populations. A measure of the within-sample variability is computed as $s_W^2 = \text{SSW}/(n - t)$. Similarly, a measure of the between-sample variability is obtained as $s_B^2 = \text{SSB}/(t - 1)$.

The decision to accept or reject the null hypothesis of equality of the $t$ population means depends on the computed value of $F = s_B^2/s_W^2$.

Under $H_0$, both $s_B^2$ and $s_W^2$ estimate $\sigma^2$, the variance common to all $t$ populations. Under the alternative hypothesis, $s_B^2$ estimates $\sigma^2 + \theta$, where $\theta$ is a positive quantity, while $s_W^2$ still estimates $\sigma^2$. Thus large values of $F$ indicate a rejection of $H_0$. Critical values for $F$ are obtained from tables 4 and 5 of the appendix for $\text{df}_1 = t - 1$ and $\text{df}_2 = n - t$. This test procedure, called an analysis of variance, is usually summarized in an analysis of variance (AOV) table.

You might be puzzled at this point with the following question: Suppose we reject $H_0$ and conclude that at least one of the means differs from the rest; Which ones differ from the others? While chapter 13 has not answered this question, chapter 14 attacks this very problem through procedures called multiple comparisons.

## Exercises

+  13.1. Some researchers have conjectured that the disease stem pitting in peach tree seedlings might be related to the presence or absence of nematodes in the soil. Hence weed and soil treatment using herbicides might be effective in promoting seedling growth. An experiment was conducted to compare peach tree seedling growth with soil and weeds treated with one of 3 herbicides:

A: control (no herbicide)
B: herbicide with Nemagone
C: herbicide without Nemagone

Of the 18 seedlings chosen for the study, 6 were randomly assigned to each treatment group. Soil and weeds in the growing areas for the 3 groups were treated with the appropriate herbicide. At the end of the study period, the height (in centimeters) was recorded for each seedling. Use the sample data below to run an analysis of variance for detecting differences among the seedling heights for the 3 groups. Use $\alpha = .05$. Draw your conclusions.

| | | | | | | |
|---|---|---|---|---|---|---|
| *Herbicide A* | 66 | 67 | 74 | 73 | 75 | 64 |
| *Herbicide B* | 85 | 84 | 76 | 82 | 79 | 86 |
| *Herbicide C* | 91 | 93 | 88 | 87 | 90 | 86 |

13.2. An experiment was conducted to compare the starch content of tomato plants grown in sandy soil supplemented by one of 3 different nutrients, A, B, or C. Eighteen tomato seedlings of one particular variety were selected for the study, with 6 assigned to each of the nutrient groups. All seedlings were planted in a sand culture and maintained under a controlled environment. Those seedlings assigned to Nutrient A served as the control group (receiving distilled water only). Plants assigned to Nutrient B were fed a weak concentration of Hoagland nutrient, while those assigned to Nutrient C received the Hoagland nutrient at full strength. The stem starch contents were determined 25 days after planting and are recorded below, in micrograms per milligram.

| | | | | | | |
|---|---|---|---|---|---|---|
| *Nutrient A* | 22 | 20 | 21 | 18 | 16 | 14 |
| *Nutrient B* | 12 | 14 | 15 | 10 | 9 | 6 |
| *Nutrient C* | 7 | 9 | 7 | 6 | 5 | 3 |

a. Run an analysis of variance to test for differences in starch content for the three nutrient groups. Use $\alpha = .05$.
b. Draw your conclusions.

13.3. The department of fruit crops at a university was interested in comparing 4 different preservatives to be used in freezing strawberries. The yield from a strawberry patch was prepared for freezing and randomly divided into 4 equal groups. Within each group the strawberries were treated with the appropriate preservative and packaged into 8 small plastic bags for freezing at 0°C. Those in Group I served as a control group, while those in groups II, III, and IV were assigned one of three newly developed preservatives. After all 32 bags of strawberries were prepared, they were stored at 0°C for a period of six months. At the end of this time, the contents of each bag were allowed to thaw and then rated on a scale of 1 to 10 points for discoloration. (Note that a low score indicates little discoloration.) These ratings are given below.

| | | | | | | | | |
|---|---|---|---|---|---|---|---|---|
| *Group I* | 10 | 8 | 7.5 | 8 | 9.5 | 9 | 7.5 | 7 |
| *Group II* | 6 | 7.5 | 8 | 7 | 6.5 | 6 | 5 | 5.5 |
| *Group III* | 3 | 5.5 | 4 | 4.5 | 3 | 3.5 | 4 | 4.5 |
| *Group IV* | 2 | 1 | 2.5 | 3 | 4 | 3.5 | 2 | 2 |

a. For these data we might be concerned with the normality of the data. To avoid any problems, refer to section 11.4 to run a Kruskal-Wallis one-way analysis of variance by ranks. Use $\alpha = .05$.
b. Run a one-way analysis of variance, with $\alpha = .05$. Compare your results with those of part a.

13.4. Refer to exercise 13.3. Suppose that some of the 32 bags were stored improperly and were not analyzable at the end of the six-month period. In particular, the sample sizes in the 4 groups were, respectively, 8, 6, 5, and 7. These data appear below.

| | | | | | | | | |
|---|---|---|---|---|---|---|---|---|
| *Group I* | 10 | 8 | 7.5 | 8 | 9.5 | 9 | 7.5 | 7 |
| *Group II* | 6 | 7.5 | 8 | 7 | 6.5 | 6 | | |
| *Group III* | 3 | 5.5 | 4 | 4.5 | 3 | | | |
| *Group IV* | 2 | 1 | 2.5 | 3 | 4 | 3.5 | 2 | |

The analysis of variance is less sensitive to deviations from normality and equality of variances among the groups when the sample sizes are equal. While the sample sizes are nearly equal for this example, refer to section 11.4 to run a Kruskal-Wallis test (one-way analysis of variance by ranks) instead of the usual analysis of variance. Use $\alpha = .05$.

+ 13.5. Although we often have well-planned experiments with equal numbers of observations per treatment, we still end up with unequal numbers at the end of a study. Suppose that in spite of allocating 6 plants to each of the nutrient groups of exercise 13.2, only 5 survived in Group B and 4 in Group C. The data for the heights of seedlings are given below.

| | | | | | | |
|---|---|---|---|---|---|---|
| *Nutrient A* | 22 | 20 | 21 | 18 | 16 | 14 |
| *Nutrient B* | 12 | 14 | 15 | 10 | 9 | |
| *Nutrient C* | 7 | 9 | 7 | 6 | | |

a. Write an appropriate model for this experimental situation. Define all terms.
b. Assuming that nutrients B and C did not cause the plants to die, perform an analysis of variance to compare the treatment means. Use $\alpha = .05$.

13.6. Salary disputes and their eventual resolutions often leave both employers and employees embittered by the entire ordeal. To assess employee reactions to a recently devised salary and fringe benefits plan, the personnel department obtained random samples of 15 employees from each of 3 divisions, manufacturing, marketing, and research. Each employee sampled was asked to respond (in confidence) to a series of questions. Several employees refused to cooperate, as reflected in the unequal sample sizes. The data are given below.

| | *Manufacturing* | *Marketing* | *Research* |
|---|---|---|---|
| *Sample Size* | 12 | 14 | 11 |
| *Sample Mean* | 25.2 | 32.6 | 28.1 |
| *Sample Variance* | 3.6 | 4.8 | 5.3 |

a. Write a model for this experimental situation.
b. Use the summary of the scored responses to compare the means for the 3 divisions (the higher a score, the higher the employee acceptance). Use $\alpha = .01$.

13.7. The yields of corn, in bushels per plot, were recorded for 4 different varieties of corn, A, B, C, and D. In a controlled greenhouse environment, each variety was randomly assigned to 8 of 32 plots available for the study. The yields are listed on the next page.

| | | | | | | | | |
|---|---|---|---|---|---|---|---|---|
| A | 2.5 | 3.6 | 2.8 | 2.7 | 3.1 | 3.4 | 2.9 | 3.5 |
| B | 3.6 | 3.9 | 4.1 | 4.3 | 2.9 | 3.5 | 3.8 | 3.7 |
| C | 4.3 | 4.4 | 4.5 | 4.1 | 3.5 | 3.4 | 3.2 | 4.6 |
| D | 2.8 | 2.9 | 3.1 | 2.4 | 3.2 | 2.5 | 3.6 | 2.7 |

a. Write an appropriate statistical model.
b. Perform an analysis of variance on these data and draw your conclusions. Use $\alpha = .05$.

13.8. Refer to exercise 13.7. Perform a Kruskal-Wallis analysis of variance by ranks (with $\alpha = .05$) and compare your results to those in exercise 13.7.

$ 13.9. Many corporations make use of the Wide Area Telephone System (WATS), where, for a fixed rent per month, the corporation can make as many long distance calls as it likes. Depending on the area of the country in which the corporation is located, it can rent a WATS line for certain geographic bands. For example, in Ohio these bands might include the following states:

Band I: Ohio

Band II: Indiana, Kentucky, Maryland, Michigan, North Carolina, Pennsylvania, Tennessee, Virginia, West Virginia, Washington, D.C.

Band III: 32 states shown in the map, plus Washington, D.C.

To monitor the use of the WATS lines, a corporation selected a random sample of 12 calls from each of the following areas in a given month. The length of the conversation (in minutes) was recorded for each call. (Band III excludes states in Band II and Ohio.)

| | | | | | | | | | | | | |
|---|---|---|---|---|---|---|---|---|---|---|---|---|
| *Ohio* | 2 | 3 | 5 | 8 | 4 | 6 | 18 | 19 | 9 | 6 | 7 | 5 |
| *Band II* | 6 | 8 | 10 | 15 | 19 | 21 | 10 | 12 | 13 | 2 | 5 | 7 |
| *Band III* | 12 | 14 | 13 | 20 | 25 | 30 | 5 | 6 | 21 | 22 | 28 | 11 |

Perform an analysis of variance to compare the mean lengths of calls for each of the 3 areas. Use $\alpha = .05$.

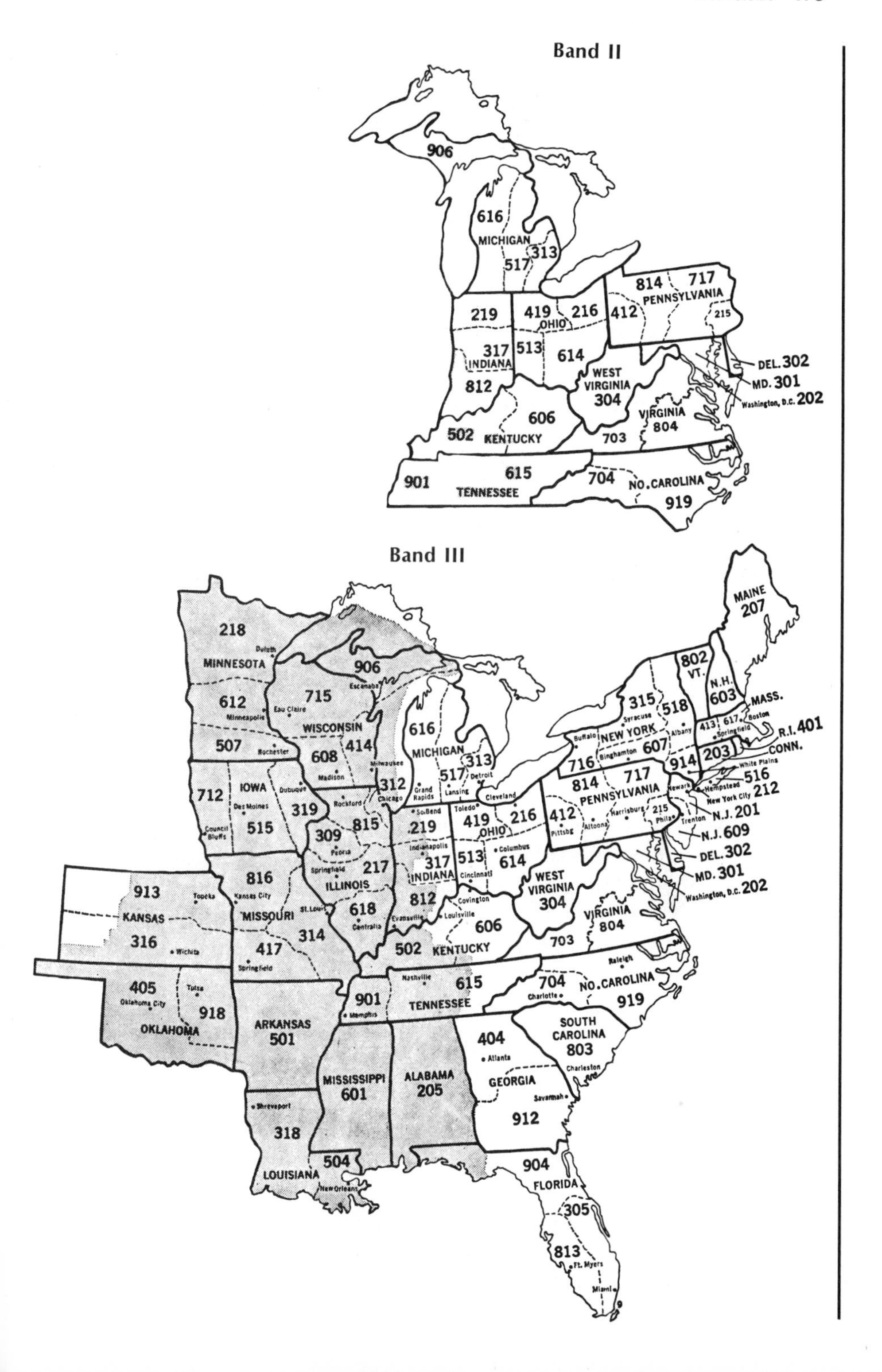
Band II
906
616
MICHIGAN
313
517
219
419
OHIO
216
412
814
717
PENNSYLVANIA
215
317
INDIANA
513
614
812
WEST VIRGINIA
304
DEL. 302
MD. 301
Washington, D.C. 202
VIRGINIA
804
703
606
502
KENTUCKY
901
615
TENNESSEE
704
NO. CAROLINA
919
Band III
218
MINNESOTA
Duluth
612
Minneapolis
507
Rochester
Eau Claire
715
WISCONSIN
906
Escanaba
414
608
Madison
Milwaukee
712
IOWA
Des Moines
515
Council Bluffs
Dubuque
319
Rockford
312
Chicago
815
309
Peoria
217
Springfield
ILLINOIS
618
Centralia
816
Kansas City
MISSOURI
St. Louis
314
417
Springfield
913
Topeka
KANSAS
316
Wichita
405
Oklahoma City
Tulsa
918
OKLAHOMA
ARKANSAS
501
MISSISSIPPI
601
ALABAMA
205
Shreveport
318
LOUISIANA
504
New Orleans
616
MICHIGAN
313
517
Detroit
Grand Rapids
Lansing
So. Bend
219
Indianapolis
317
INDIANA
812
Evansville
Toledo
419
OHIO
216
Cleveland
513
Columbus
614
Cincinnati
Covington
Louisville
606
502
KENTUCKY
901
Memphis
Nashville
615
TENNESSEE
WEST VIRGINIA
304
VIRGINIA
804
703
704
Charlotte
Raleigh
NO. CAROLINA
919
SOUTH CAROLINA
803
Charleston
404
Atlanta
GEORGIA
Savannah
912
904
FLORIDA
305
813
Ft. Myers
Miami
MAINE
207
802
VT.
N.H.
603
MASS.
413
617
Springfield
Boston
R.I. 401
203
CONN.
315
Syracuse
518
Albany
Buffalo
NEW YORK
716
607
Binghamton
914
White Plains
516
Hempstead
New York City 212
Newark
814
717
PENNSYLVANIA
Harrisburg
412
Pittsbg
Altoona
215
Phila
Trenton
N.J. 201
N.J. 609
DEL. 302
MD. 301
Washington, D.C. 202

$ 13.10. Refer to exercise 13.9. Suppose that rather than 12 calls from each area, we obtained a random sample of 15 calls. Use the additional 3 measurements recorded below for the total of 15 observations per area.

| | | | |
|---|---|---|---|
| *Ohio* | 10 | 11 | 4 |
| *Band II* | 8 | 28 | 31 |
| *Band III* | 29 | 50 | 120 |

Analyze the sample data to compare the mean number of call times for the 3 areas. Use $\alpha = .05$.

*13.11. Obtain a computer solution for the data in exercise 13.7. Compare your results to those obtained in exercise 13.7.

*13.12. Refer to the data of exercise 13.9. Obtain a computer solution and compare the results to your hand calculations of exercise 13.9.

*13.13. Obtain a computer solution for the data in exercise 13.10.

13.14. Using either data from a laboratory course or data that you can collect from another source, make a comparison of three or more methods, varieties, or plans. Use an analysis of variance for a one-way classification. Let $\alpha = .05$.

13.15. Refer to exercise 13.14.
a. Write an appropriate statistical model.
b. Perform a Kruskal-Wallis one-way analysis of variance by ranks and compare your results to those of exercise 13.14. Use $\alpha = .05$.

13.16. Refer to exercise 13.9. Some researchers would argue that durations of telephone calls may not be normally distributed. Perform a Kruskal-Wallis one-way analysis of variance by ranks and compare your results to those of exercise 13.9. Use $\alpha = .05$.

13.17. Examine the SAS computer output that follows for the analysis of variance performed in example 13.2.
a. Identify the sums of squares and degrees of freedom for methods, error, and total.
b. Give the mean squares and $F$ test for the equality of method means.
c. Give the level of significance for the test in part b.
d. What is the coefficient of determination?
e. Compare your results to those of example 13.2.

```
STATISTICAL ANALYSIS SYSTEM

DATA HLT;
INPUT Y 1-4 METHOD 6;
CARDS;


11 OBSERVATIONS IN DATA SET HLT     2 VARIABLES

PROC PRINT;


OBS        Y      METHOD

  1       80           1
  2       92           1
  3       87           1
  4       83           1
  5       70           2
  6       81           2
  7       78           2
  8       74           2
  9       63           3
 10       76           3
 11       70           3


PROC REGR;
CLASSES METHOD;
MODEL Y = METHOD;
TITLE 'EXERCISE 13.17';

***********************************************************************************************

PROC REGR : EXERCISE 13.17

DATA SET   : HLT      NUMBER OF VARIABLES = 1      NUMBER OF CLASSES = 1

VARIABLES  : Y

***********************************************************************************************

CLASSES      LEVELS      VALUES

METHOD          3        1 2 3

***********************************************************************************************

EFFECTS                  ASSIGNED

METHOD                   DUMMY001 - DUMMY002

***********************************************************************************************
```

EXERCISE 13.17

ANALYSIS OF VARIANCE TABLE, REGRESSION COEFFICIENTS, AND STATISTICS OF FIT FOR DEPENDENT VARIABLE Y

| SOURCE | DF | SUM OF SQUARES | MEAN SQUARE | F VALUE | PROB > F | R-SQUARE | C.V. |
|---|---|---|---|---|---|---|---|
| REGRESSION | 2 | 452.12878788 | 226.06439394 | 7.71496 | 0.0137 | 0.65855623 | 6.97243% |
| ERROR | 8 | 234.41666667 | 29.30208333 | | | STD DEV | Y MEAN |
| CORRECTED TOTAL | 10 | 686.54545455 | | | | 5.41313988 | 77.63636 |

| SOURCE | DF | SEQUENTIAL SS | F VALUE | PROB > F | PARTIAL SS | F VALUE | PROB > F |
|---|---|---|---|---|---|---|---|
| METHOD | 2 | 452.12878788 | 7.71496 | 0.0137 | 452.12878788 | 7.71496 | 0.0137 |

# References


Barr, A.J.; Goodnight, J.H.; Sall, J.P.; and Helwig, J.T. 1976. *A user's guide to SAS 76.* Raleigh, N.C.: SAS Institute, Inc.

Dixon, W.J. 1975. *BMDP, biomedical computer programs.* Berkeley: University of California Press.

Dixon, W.J., and Massey, F.J., Jr. 1969. *Introduction to statistical analysis.* 3d ed. New York: McGraw-Hill.

Harnett, D.L. 1970. *Introduction to statistical methods.* Reading, Mass.: Addison-Wesley.

Kirk, R.E. 1968. *Experimental design: Procedures for the behavioral sciences.* Belmont, Calif.: Brooks/Cole.

Mendenhall, W. 1968. *Introduction to linear models and the design and analysis of experiments.* Belmont, Calif.: Wadsworth.

Mendenhall, W.; Ott, L.; and Larson, R. 1974. *Statistics: A tool for the social sciences.* N. Scituate, Mass.: Duxbury Press.

Mendenhall, W., and Ott, L. 1976. *Understanding statistics.* 2d ed. N. Scituate, Mass.: Duxbury Press.

Nie, N.H.; Hull, C.H.; Jenkins, J.G.; Steinbrenner, K.; and Bent, D.H. 1975. *Statistical package for the social sciences,* 2d ed. New York: McGraw-Hill.

Ostle, B. 1963. *Statistics in research.* 2d ed. Ames, Iowa: Iowa State University Press.

Pearson, E.S., and Hartley, H.O. 1966. *Biometrika tables for statisticians.* 3d ed. Vol. I. London: Cambridge University Press.

Service, J. 1972. *A user's guide to the statistical analysis system*. Raleigh, N.C.: Student Supply Stores, North Carolina State University.

Snedecor, G.W., and Cochran, W.G. 1967. *Statistical methods*. 6th ed. Ames, Iowa: Iowa State University Press.

Steel, R.G.D., and Torrie, J.H. 1960. *Principles and procedures of statistics*. New York: McGraw-Hill.

Tanur, J.M.; Mosteller, F.; Kruskal, W.H.; Pieters, R.S.; and Rising, G.R. 1972. *Statistics: A guide to the unknown*. San Francisco: Holden-Day.

Winer, B.J. 1962. *Statistical principles in experimental design*. New York: McGraw-Hill.

# 14 Multiple Comparisons

## 14.1 Introduction

*F* test

In chapter 13 we introduced a procedure for testing the equality of $t$ population means. The test statistic $F = s_B^2/s_W^2$ was used to determine whether the between-sample variability was large relative to the within-sample variability. If the computed value of $F$ for the sample data exceeded the critical value obtained from the tables in the appendix, the null hypothesis $H_0$: $\mu_1 = \mu_2 = \cdots = \mu_t$ was rejected in favor of the alternative hypothesis.

$H_a$: At least one of the $t$ population means differs from the rest

multiple comparison procedures

While rejection of the null hypothesis does give us some information concerning the population means, we do not know which means differ from each other. For example, does $\mu_1$ differ from $\mu_2$ or $\mu_3$? Does $\mu_3$ differ from $\mu_4$, $\mu_5$, and $\mu_6$? *Multiple comparison procedures* have been developed to answer questions such as these. While many multiple comparison procedures have been proposed, we intend to focus on just a few of the more common methods. After studying these few procedures, you should be able to evaluate the results of most published material utilizing multiple comparisons or to suggest an appropriate multiple comparison procedure in an experimental situation.

# Notation and Definitions | 14.2

Before developing several different multiple comparison procedures, we need the following notation and definitions. Consider a one-way classification where we wish to make comparisons among the $t$ population means $\mu_1, \mu_2, \ldots, \mu_t$. These comparisons among $t$ population means can be written in the form

$$\ell = a_1\mu_1 + a_2\mu_2 + \cdots + a_t\mu_t = \sum_{i=1}^{t} a_i\mu_i$$

$\ell$

where the $a_i$'s are constants satisfying the property that $\Sigma\, a_i = 0$. For example, if we wanted to compare $\mu_1$ to $\mu_2$, we could write the linear form

$$\ell = \mu_1 - \mu_2$$

Note here that $a_1 = 1$, $a_2 = -1$, $a_3 = a_4 = \cdots = a_t = 0$, and $\Sigma_{i=1}^{t}\, a_i = 0$. Similarly, we could compare the mean for population 1 to the average of the means for populations 2 and 3. Then $\ell$ would be of the form

$$\ell = \mu_1 - \frac{(\mu_2 + \mu_3)}{2}$$

where $a_1 = 1$, $a_2 = a_3 = -\frac{1}{2}$, $a_4 = a_5 = \cdots = a_t = 0$, and $\Sigma_{i=1}^{t}\, a_i = 0$.

An estimate of the linear form $\ell$, designated by $\hat{\ell}$, is formed by replacing the $\mu_i$'s in $\ell$ with their corresponding sample means $\bar{y}_i$. The estimate $\hat{\ell}$ is called a *linear contrast*.

$\hat{\ell}$

linear contrast

---

**Definition 14.1**

$$\hat{\ell} = a_1\bar{y}_1 + a_2\bar{y}_2 + \cdots + a_t\bar{y}_t = \sum_{i=1}^{t} a_i\bar{y}_i$$

is called a *linear contrast* among the $t$ sample means and can be used to estimate $\ell = \Sigma_{i=1}^{t}\, a_i\mu_i$. The $a_i$'s are constants satisfying the constraint $\Sigma_{i=1}^{t}\, a_i = 0$.

---

The variance of the linear contrast $\hat{\ell}$ can be estimated as follows:

$$\hat{V}(\hat{\ell}) = s_W^2\left[\frac{a_1^2}{n_1} + \frac{a_2^2}{n_2} + \cdots + \frac{a_t^2}{n_t}\right] = s_W^2 \sum_{i=1}^{t} \frac{a_i^2}{n_i}$$

$\hat{V}(\hat{\ell})$

where $n_i$ is the number of sample observations selected from population $i$ and $s_W^2$ is the mean square within samples obtained from the analysis of variance table for the one-way classification. If all the sample sizes are the same (i.e., all $n_i = n$), then

$$\hat{V}(\hat{\ell}) = \frac{s_W^2}{n} \sum_{i=1}^{t} a_i^2$$

There are many different contrasts that can be formed among the $t$ sample means. If each of the sample means is based on the same number of observations (i.e., $n_i = n$), however, we have the following definition.

---

**Definition 14.2**

Two contrasts $\hat{\ell}_1$ and $\hat{\ell}_2$, where

$$\hat{\ell}_1 = \sum_{i=1}^{t} a_i\bar{y}_i \qquad \text{and} \qquad \hat{\ell}_2 = \sum_{i=1}^{t} b_i\bar{y}_i$$

orthogonal contrasts

are said to be *orthogonal* if

$$a_1b_1 + a_2b_2 + \cdots + a_tb_t = \sum_{i=1}^{t} a_ib_i = 0$$

Note: The sample sizes must be the same.

---

mutually orthogonal

A set of contrasts is said to be *mutually orthogonal* if all pairs of contrasts in the set are orthogonal.

**EXAMPLE 14.1**

Consider a one-way classification for comparing $t = 4$ population means. Are the following contrasts orthogonal?

$\hat{\ell}_1 = \bar{y}_1 - \bar{y}_2 \qquad \hat{\ell}_2 = \bar{y}_3 - \bar{y}_4$

**SOLUTION**

We can rewrite the contrasts in the following form:

$\ell_1 = \bar{y}_1 - \bar{y}_2 + 0(\bar{y}_3) + 0(\bar{y}_4)$

$\ell_2 = 0(\bar{y}_1) + 0(\bar{y}_2) + \bar{y}_3 - \bar{y}_4$

where we see that $a_1 = 1$, $a_2 = -1$, $a_3 = 0$, $a_4 = 0$, and $b_1 = 0$, $b_2 = 0$, $b_3 = 1$, $b_4 = -1$. It is then apparent that

$$\sum_{i=1}^{4} a_i b_i = a_1 b_1 + a_2 b_2 + a_3 b_3 + a_4 b_4 = 0$$

and hence the contrasts are orthogonal.

**EXAMPLE 14.2**

Refer to example 14.1. Are the contrasts given below orthogonal?

$\hat{\ell}_1 = \bar{y}_1 - \bar{y}_2$ and $\hat{\ell}_2 = \bar{y}_1 - \bar{y}_3$

**SOLUTION**

Rewriting the contrasts as

$\hat{\ell}_1 = \bar{y}_1 - \bar{y}_2 + 0(\bar{y}_3) + 0(\bar{y}_4)$

$\hat{\ell}_2 = \bar{y}_1 + 0(\bar{y}_2) - \bar{y}_3 + 0(\bar{y}_4)$

we have that

$$\sum_{i=1}^{4} a_i b_i = (1)(1) + (-1)(0) + (0)(-1) + (0)(0) = 1$$

which indicates that the two contrasts are not orthogonal.

The concept of orthogonality between linear contrasts is important in the study of multiple comparison procedures. Recall that prior to running an analysis of variance among the $t$ population means in a one-way classification, we assumed that:

1. the $t$ populations were normally distributed with a common variance $\sigma^2$ but different means (under $H_0$ we assume that the means are equal);
2. independent random samples were obtained from the $t$ populations.

If we assume that each of the sample means is based on the same number of observations, then it can be shown that $t - 1$ orthogonal contrasts can be formed using the $t$ sample means. These $t - 1$ contrasts form a set of mutually orthogonal contrasts. (An easy way to remember $t - 1$ is to refer to the number of degrees of freedom associated with the between-sample source of variability in the AOV table.) In addition, it can be shown that the sums of squares for the $t - 1$ contrasts will add up to the treatment sum of squares. We will discuss this partitioning of the sum of squares for treatments in chapter 17. Mutual orthogonality is desirable because it leads to independence of the $t - 1$ sums of squares associated with the orthogonal contrasts. As we will see later in the chapter, methods have also been developed that use nonorthogonal contrasts among the sample means. These methods will be particularly appropriate when the experimenter is making all pairwise comparisons among $t$ treatment means.

$t - 1$ contrasts

# Exercises

14.1. Consider the expressions

$$\hat{\ell}_1 = \bar{y}_1 + \bar{y}_2 - 2\bar{y}_3$$

$$\hat{\ell}_2 = \bar{y}_1 + \bar{y}_2 - 2\bar{y}_4$$

a. Are $\hat{\ell}_1$ and $\hat{\ell}_2$ linear contrasts?
b. Are $\hat{\ell}_1$ and $\hat{\ell}_2$ orthogonal?

14.2. Refer to exercise 14.1. Construct a set of 3 orthogonal contrasts for comparing 4 population means.

14.3. Write down a set of 4 orthogonal contrasts to be used in comparing 5 population means.

## 14.3 Which Error Rate Is Controlled?

Let us suppose that an experimenter wishes to compare $t$ population means using c independent (orthogonal) contrasts. Each comparison among the $t$ population means can be tested using a $t$ test of the following form:

*t* test

$H_0$: $\ell = 0$.

$H_a$: $\ell \neq 0$ (for a two-tailed test).

$$\text{T.S.: } t = \frac{\hat{\ell}}{\sqrt{\hat{V}(\hat{\ell})}} = \frac{\hat{\ell}}{\sqrt{s_W^2 \sum_{i=1}^{t} a_i^2/n_i}}.$$

The rejection region for the computed value of the test statistic can be located by using table 2 of the appendix with $a = \alpha/2$ and df $= n - t$.

$1 - (1 - \alpha)^c$ overall error rate

If each of the comparisons is tested with the same value of $\alpha$, and if we assume that $s_W^2$ has an infinite number of degrees of freedom (so the tests are independent), then when all the null hypotheses are true, the probability of falsely rejecting $H_0$ on at least one of the $t$ tests can be shown to be $1 - (1 - \alpha)^c$. This quantity is sometimes called an *overall error rate* for the c comparisons. We can see from table 14.1 that as c increases for a given value of $\alpha$, the probability of falsely rejecting $H_0$ on at least one of the $t$ tests becomes quite large. Hence if an experimenter wished to compare $t = 20$ population means by using $c = 10$ orthogonal contrasts, the probability of falsely rejecting

**Table 14.1** A comparison of the overall error rate for c independent contrasts among $t$ ($t > c$) sample means

| c, Number of Contrasts | $\alpha$, Probability of a Type I Error on an Individual Test | | |
|---|---|---|---|
| | *.10* | *.05* | *.01* |
| 1 | .100 | .050 | .010 |
| 2 | .190 | .097 | .020 |
| 3 | .271 | .143 | .030 |
| 4 | .344 | .186 | .039 |
| 5 | .410 | .226 | .049 |
| ⋮ | ⋮ | ⋮ | ⋮ |
| 10 | .651 | .401 | .096 |

$H_0$ on at least one of the $t$ tests could be as high as .401 when each individual test was performed with $\alpha = .05$.

The results of table 14.1 are disturbing and may lead us to question significant results when they appear. The problem can be alleviated somewhat by controlling the overall error rate rather than the error rate (type I error) for the individual $t$ test. Suppose, for example, that we wished the overall error rate for $c = 10$ orthogonal contrasts among $t = 20$ population means to be .10. What value of $\alpha$ must we use on the individual $t$ tests to achieve an overall error rate of .10? Assuming $s_W^2$ is based on a large number of degrees of freedom, this problem can be solved by determining the value of $\alpha$ for which

controlling overall error rate

$$1 - (1 - \alpha)^{10} = .10$$

The method of solution for this equation is not important now; we can see from table 14.1 that by using $\alpha = .01$ for all 10 tests, the overall error rate would be approximately .10.

While controlling the overall error rate for comparisons using orthogonal contrasts is fairly simple, it is difficult to obtain an expression equivalent to $1 - (1 - \alpha)^c$ for comparisons made with nonorthogonal contrasts. For example, suppose we wish to make all pairwise comparisons among $t = 4$ population means. Previous results indicate that we could make $t - 1 = 3$ orthogonal (independent) contrasts, but there are 6 possible pairwise comparisons among the population means (1 and 2, 1 and 3, 1 and 4, 2 and 3, 2 and 4, and 3 and 4). If each of these 6 comparisons is made using a $t$ test with $\alpha = .05$, what is the overall error rate? Pearson and Hartley (1942, 1943) and Harter (1957) attacked the problem of determining the probability of falsely rejecting $H_0$ on at least one of the $t$ tests for nonindependent contrasts. The solution is not easy, however, and is beyond the scope of this text. One alternative is to

error rate for nonorthogonal contrasts

redefine the overall error rate and determine a testing procedure that controls the overall error rate at a desired level. Indeed, a major difference among the multiple comparison procedures we will discuss in the following sections is the error rate that each procedure controls.

## 14.4 Fisher's Least Significant Difference

Recall that we are interested in determining which population means differ after we have rejected the hypothesis of equality of $t$ population means in an analysis of variance. R. A. Fisher (1949) developed a procedure for making pairwise comparisons among a set of $t$ population means that is particularly appropriate following a significant $F$ test in the analysis of variance. Fisher's procedure is summarized in the box.

**Fisher's Least Significant Difference Procedure**

LSD

1. Perform an analysis of variance to test $H_0\colon \mu_1 = \mu_2 = \cdots = \mu_t$ against the alternative hypothesis that at least one of the means differs from the rest.
2. If there is insufficient evidence to reject $H_0$ using $F = s_B^2/s_W^2$, we proceed no further.
3. If $H_0$ is rejected, define the least significant difference (LSD) to be the observed difference between two sample means necessary to declare the corresponding population means different.
4. For a specified value of $\alpha$, the least significant difference for comparing $\mu_i$ to $\mu_j$ is

   $$\text{LSD} = t_{\alpha/2}\sqrt{s_W^2\left(\frac{1}{n_i} + \frac{1}{n_j}\right)}$$

   where $n_i$ and $n_j$ are the respective sample sizes from population $i$ and $j$ and $t$ is the critical $t$-value (table 2 of the appendix) for $a = \alpha/2$ and df denoting the degrees of freedom for $s_W^2$. Note that for $n_i = n_j = n$,

   $$\text{LSD} = t_{\alpha/2}\sqrt{\frac{2s_W^2}{n}}$$

5. All pairs of sample means are then compared. If $|\bar{y}_i - \bar{y}_j| \geq \text{LSD}$, we declare the corresponding population means $\mu_i$ and $\mu_j$ different.
6. For each pairwise comparison of population means, the probability of a type I error is fixed at a specified value of $\alpha$.

Some researchers use Fisher's LSD procedure without performing an $F$ test for comparing the $t$ population means, but they modify the error rate for an individual comparison to be $\alpha/c$, where c is the number of comparisons they wish to make. This method is called *Fisher's unprotected LSD,* whereas the method that is preceded by an $F$ test is called Fisher's *protected LSD* (or simply Fisher's LSD). For the unprotected LSD the error rate is controlled on a per-comparison basis. For the protected LSD, simulation studies (Carmer and Swanson 1973) have shown that the error rate is controlled on an experiment-wise basis at a level approximately equal to the $\alpha$-value for the $F$ test.

unprotected LSD
protected LSD

**EXAMPLE 14.3**

Hydrochloric acid (HCL) is used in the preparation of certain dyes. Six different batches of HCL were used to produce a particular dye. Five measurements on the yield (in grams of dye) were obtained from each batch. A summary of the sample data is given in tables 14.2 and 14.3. Use Fisher's least significant difference procedure to make all pairwise comparisons among the 6 population (batch) mean yields. Use $\alpha = .05$.

**Table 14.2** Summary of the dye yields for example 14.3

| *Batch* | *Sample Mean* |
|---|---|
| 1 | 505 |
| 2 | 528 |
| 3 | 564 |
| 4 | 498 |
| 5 | 600 |
| 6 | 470 |

**Table 14.3** AOV table for the data of example 14.3

| *Source* | *df* | *SS* | *MS* | *F* |
|---|---|---|---|---|
| between batches | 5 | 56,360 | 11,272 | 4.60 |
| within batches | 24 | 58,824 | 2,451 | |
| *Total* | 29 | | | |

**SOLUTION**

We can solve this problem by following the five steps listed for the LSD procedure.

steps for LSD procedure

Step 1. We use the AOV table in table 14.3. The $F$ test of $H_0$: $\mu_1 = \mu_2 = \cdots = \mu_6$ is based on

$$F = \frac{s_B^2}{s_W^2} = 4.60$$

For $\alpha = .05$ with $df_1 = 5$ and $df_2 = 24$, we reject $H_0$ if $F$ exceeds 2.62 (see table 4 of the appendix).

Steps 2, 3. Since 4.60 is greater than 2.62, we reject $H_0$ and conclude that at least one of the population means differs from the rest.

Step 4. The least significant difference for comparing two means based on samples of size 5 is then

$$\text{LSD} = t_{\alpha/2}\sqrt{\frac{2\,s_W^2}{5}} = 2.064\sqrt{\frac{2(2451)}{5}} = 64.63$$

Note that the appropriate $t$-value (2.064) was obtained from table 2 with $a = \alpha/2 = .025$ and df $= 24$.

Step 5. When we have equal sample sizes, it is convenient to use the following procedure rather than making all pairwise comparisons among the sample means, because the same LSD is to be used for all comparisons.

a. We rank the sample means from lowest to highest.

| *Population* | 6 | 4 | 1 | 2 | 3 | 5 |
|---|---|---|---|---|---|---|
| *Sample Mean* | 470 | 498 | 505 | 528 | 564 | 600 |

b. We compute the sample difference

$\bar{y}_{\text{largest}} - \bar{y}_{\text{smallest}}$

If this difference is greater than the LSD, we declare the corresponding population means significantly different from each other. Next we compute the sample difference

$\bar{y}_{\text{2d largest}} - \bar{y}_{\text{smallest}}$

and compare the result to the LSD. We continue to make comparisons with $\bar{y}_{\text{smallest}}$:

$\bar{y}_{\text{3d largest}} - \bar{y}_{\text{smallest}}$

and so on, until we find either that all sample differences involving $\bar{y}_{\text{smallest}}$ exceed the LSD (and hence the corresponding population means are different) or that a sample difference involving $\bar{y}_{\text{smallest}}$ is less than the LSD. In the latter case we stop and make no further comparisons with $\bar{y}_{\text{smallest}}$. For our data, comparisons with $\bar{y}_{\text{smallest}}$, $\bar{y}_6$, give the following results:

| *Comparison* | *Conclusion* |
|---|---|
| $\bar{y}_{\text{largest}} - \bar{y}_{\text{smallest}} = \bar{y}_5 - \bar{y}_6 = 130$ | $>$ LSD; proceed |
| $\bar{y}_{\text{2d largest}} - \bar{y}_{\text{smallest}} = \bar{y}_3 - \bar{y}_6 = 94$ | $>$ LSD; proceed |
| $\bar{y}_{\text{3d largest}} - \bar{y}_{\text{smallest}} = \bar{y}_2 - \bar{y}_6 = 58$ | $<$ LSD; stop |

To summarize our results we make the following diagram:

summary diagram

population $\underline{6 \quad 4 \quad 1 \quad 2}$ 3 5

Those populations joined by the underline have means which are not significantly different from $\bar{y}_6$. Note that populations 3 and 5 have sample differences with population 6 that exceed the LSD and hence they are not underlined.

c. We now make similar comparisons with $\bar{y}_{\text{2d smallest}}$, $\bar{y}_4$ in this case, using the procedures of part b.

| *Comparison* | *Conclusion* |
|---|---|
| $\bar{y}_5 - \bar{y}_4 = 102$ | > LSD; proceed |
| $\bar{y}_3 - \bar{y}_4 = 66$ | > LSD; proceed |
| $\bar{y}_2 - \bar{y}_4 = 30$ | < LSD; stop |

population $\quad 6 \quad \underline{4 \quad 1 \quad 2} \quad 3 \quad 5$

d. Continue with $\bar{y}_{\text{3d smallest}}$, or $\bar{y}_1$ in our example.

| *Comparison* | *Conclusion* |
|---|---|
| $\bar{y}_5 - \bar{y}_1 = 95$ | > LSD; proceed |
| $\bar{y}_3 - \bar{y}_1 = 59$ | < LSD; stop |

population $\quad 6 \quad 4 \quad \underline{1 \quad 2 \quad 3} \quad 5$

e. Continue with $\bar{y}_{\text{4th smallest}}$, or $\bar{y}_2$ in our example.

| *Comparison* | *Conclusion* |
|---|---|
| $\bar{y}_5 - \bar{y}_2 = 72$ | > LSD; proceed |
| $\bar{y}_3 - \bar{y}_2 = 36$ | < LSD; stop |

population $\quad 6 \quad 4 \quad 1 \quad \underline{2 \quad 3} \quad 5$

f. Continue with $\bar{y}_{\text{5th smallest}}$, or $\bar{y}_3$ in our example.

| *Comparison* | *Conclusion* |
|---|---|
| $\bar{y}_5 - \bar{y}_3 = 36$ | < LSD; stop |

population $\quad 6 \quad 4 \quad 1 \quad 2 \quad \underline{3 \quad 5}$

g. We can summarize steps a through f as follows:

population $\quad \underline{6 \quad 4 \quad 1 \quad 2} \quad 3 \quad 5$

Those populations not underlined by a common line are declared to have means that are significantly different according to the least significant difference criterion. Note that we can eliminate the second and fourth lines from the top of part g since they are part of the first and third lines, respectively. The revised summary of significant and nonsignificant results is

```
population   6   4   1   2   3   5
             ---------------
                     -----------
                             -------
```

In conclusion, we have $\mu_6$, $\mu_4$, $\mu_1$, and $\mu_2$ significantly less than $\mu_5$. Also, $\mu_6$ and $\mu_4$ are significantly less than $\mu_3$.

While the LSD procedure described in example 14.3 may seem quite laborious, its application is quite simple. We first run an analysis of variance. If we reject the null hypothesis of equality of the population means, we compute the LSD for all pairs of sample means. When the sample sizes are the same, this difference is a single number for all pairs. We can then use the stepwise procedure described in steps 5a through 5g of example 14.3. We need not write all those steps down, only the summary lines. The final summary (as given in step 5g) gives a handy visual display of the pairwise comparisons using Fisher's LSD.

Several remarks should be made concerning the LSD method for pairwise comparisons.

First, there is the possibility that the overall $F$ test in our analysis of variance is significant but that no pairwise differences are significant using the LSD procedure. This apparent anomaly can occur because the null hypothesis $H_0$: $\mu_1 = \mu_2 = \cdots = \mu_t$ for the $F$ test is equivalent to the hypothesis that all possible comparisons (paired or otherwise) among the population means are zero. For a given set of data, the comparisons that are significant might not be of the form $\mu_i - \mu_j$, the form we are using in our paired comparisons.

Fisher confidence interval

Second, Fisher's LSD procedure can also be used to form a confidence interval for $\mu_i - \mu_j$. A $100(1 - \alpha)\%$ confidence interval has the form

$$(\bar{y}_i - \bar{y}_j) \pm \text{LSD}$$

LSD for equal sample sizes

Third, when all the sample sizes are the same, the LSD for all pairs is

$$t_{\alpha/2}\sqrt{\frac{2s_W^2}{n}}$$

# Tukey's *W* Procedure | 14.5

We are aware of the major drawback of a multiple comparison procedure with a controlled per-comparison error rate. Even when $\mu_1 = \mu_2 = \cdots = \mu_t$, unless $\alpha$, the per-comparison error rate (such as with Fisher's unprotected LSD) is quite small, there is a high probability of declaring at least one pair of means significantly different when running multiple comparisons. To avoid this, other multiple comparison procedures have been developed that control different error rates.

Tukey (1953) proposed a procedure that makes use of the *Studentized range distribution*. When more than two sample means are being compared, to test the largest and smallest sample means, we could use the test statistic

Studentized range distribution

$$\frac{\bar{y}_{\text{largest}} - \bar{y}_{\text{smallest}}}{s\sqrt{1/n}}$$

where $n$ is the number of observations in each sample and $s$ is a pooled estimate of the common population standard deviation $\sigma$. This test statistic is very similar to that for comparing two means (section 5.4), but it does not possess a $t$ distribution. One reason it does not is that we have waited to determine which two sample means (and hence population means) we would compare until we observed the largest and smallest sample means. This procedure is quite different from that of specifying $H_0: \mu_1 - \mu_2 = 0$, observing $\bar{y}_1$ and $\bar{y}_2$, and forming a $t$ statistic.

The quantity

$$\frac{\bar{y}_{\text{largest}} - \bar{y}_{\text{smallest}}}{s\sqrt{1/n}}$$

follows a Studentized range distribution. We will not discuss the properties of this distribution, but we will illustrate its use in Tukey's multiple comparison procedure.

---

**Tukey's *W* Procedure**

1. Rank the $t$ sample means.
2. Two population means are declared significantly different if the absolute value of their difference exceeds

$$W = q_\alpha(t, v)\sqrt{\frac{s_W^2}{n}}$$

W

$q_\alpha(t, v)$

experimentwise error rate

where $s_W^2$ is the mean square within samples based on v degrees of freedom, $q_\alpha(t, v)$ is the upper-tail critical value of the Studentized range (with a = $\alpha$) for comparing *t* different populations, and *n* is the number of observations in each sample. A discussion follows showing how to obtain values of $q_\alpha(t, v)$ from table 11 in the appendix.

3. The error rate that is controlled is an *experimentwise error rate*. Thus the probability of observing an experiment with one or more pairwise comparisons falsely declared significant is specified at $\alpha$.

---

We can obtain values of $q_\alpha(t, v)$ from table 11 in the appendix. Values of v are listed along the left-hand column of the table with values of *t* across the top row. Upper-tail values for the Studentized range are then presented for a = .05 and .01. For example, in comparing 10 population means based on 9 degrees of freedom for $s_W^2$, the .05 upper-tail critical value for the Studentized range is $q_{.05}(10, 9) = 5.74$.

**EXAMPLE 14.4**

Refer to the data of example 14.3 (p. 385). Use Tukey's *W* procedure with $\alpha = .05$ to make pairwise comparisons among the 6 population means.

**SOLUTION**

We can eliminate step 1 since the sample means were ranked in example 14.3. For

$t = 6$ (we are making pairwise comparisons among 6 means)

$v = 24$ ($s_W^2$ had 24 degrees of freedom in the AOV)

$\alpha = .05$ (we specified the experimentwise error rate at .05)

$n = 5$ (there were 5 sample observations selected from each population)

we find

$q_{.05}(6, 24) = 4.37$

The absolute value of each difference in sample means must then be compared to

$$W = q_\alpha(t, v)\sqrt{\frac{s_W^2}{n}} = 4.37\sqrt{\frac{2451}{5}} = 96.75$$

By substituting *W* for the LSD, we can use the same stepwise procedure for comparing sample means that we used in step 5 of the solution to example 14.3.

Having ranked the sample means from low to high, comparisons against $\bar{y}_{\text{smallest}}$, which is $\bar{y}_6$, yield

population $\underline{6 \quad 4 \quad 1 \quad 2 \quad 3}$ 5

Comparisons with $\bar{y}_{2d\ smallest}$ ($\bar{y}_4$) yield

population 6 4 1 2 3 5

Similarly, comparisons with $\bar{y}_1$, $\bar{y}_2$, and $\bar{y}_3$ yield

population 6 4 1 2 3 5

Combining our results we obtain

population 6 4 1 2 3 5

which simplifies to

population 6 4 1 2 3 5

All populations not underlined by a common line have population means that are significantly different from each other. That is, $\mu_6$ and $\mu_4$ are significantly less than $\mu_5$.

By examining the multiple comparison summaries using the least significant difference (example 14.3) and Tukey's *W* procedure (example 14.4), we see that Tukey's procedure is more conservative (declares fewer significant differences) than the LSD procedure. For example, applying Tukey's procedure to the data of table 14.2 shows that $\mu_3$ is no longer significantly larger than $\mu_6$ and $\mu_4$ (see p. 388). Similarly, $\mu_5$ is no longer significantly larger than $\mu_1$ and $\mu_2$. The explanation for this is that although both procedures have an experimentwise error rate, the per-comparison error rate of the protected LSD method has been shown to be larger than that for Tukey's *W* procedure.

restriction

One restriction placed on Tukey's procedure that was not required in applying Fisher's LSD is that all samples must be the same size, although an approximate multiple comparison procedure can be used if the sample sizes are nearly equal. In this approximation *n* is replaced by

$\tilde{n}$

$$\tilde{n} = \frac{t}{\frac{1}{n_1} + \frac{1}{n_2} + \cdots + \frac{1}{n_t}}$$

in the formula for $W$, where $t$ is the number of means to be compared and $n_i$ is the number of observations from population $i$ ($i = 1, 2, \ldots, t$). The rest of Tukey's procedure remains the same.

Tukey confidence interval

Tukey's procedure can also be used to construct confidence intervals for comparing two means. However, unlike the confidence intervals that can be formed from Fisher's LSD, Tukey's procedure enables us to construct *simultaneous confidence intervals* for all pairs of treatment differences. For a specified $\alpha$ level from which we compute $W$, the overall probability is $1 - \alpha$ that all differences $\mu_i - \mu_j$ will be included in an interval of the form

$$(\bar{y}_i - \bar{y}_j) \pm W$$

That is, the probability is $1 - \alpha$ that all the intervals $(\bar{y}_i - \bar{y}_j) \pm W$ include the corresponding population difference $\mu_i - \mu_j$.

# 14.6 Duncan's New Multiple Range Test

Duncan (1955) developed a multiple comparison procedure for obtaining all pairwise comparisons among $t$ sample means. Although his procedure makes use of the Studentized range, his error rate is neither on an experimentwise basis (as with Tukey's) nor on a per-comparison basis. When the sample means have been ranked from lowest to highest, the error rate is designated in the following way. In comparing two populations, the experimenter takes note of how many ordered steps lie between the sample means. For example, two sample means that are adjacent to each other when all the means have been ranked are said to be two steps apart. Similarly, two means that are separated by another sample mean are said to be three steps apart. In general, if two sample means are $r$ steps apart, Duncan defines the *protection level* as

protection level

$$(1 - \alpha)^{r-1}$$

The probability of falsely rejecting the equality of two population means when the sample means are $r$ steps apart is then taken to be

error rate

$$1 - (1 - \alpha)^{r-1}$$

For $\alpha = .05$ we illustrate the concept of a protection level in table 14.4.

Duncan's reasons for allowing the protection level to decrease for increasing

**Table 14.4** Duncan protection level using $\alpha = .05$ when the sample means are $r$ steps apart

| *Number of Steps Apart, r* | *Protection Level,* $(1 - .05)^{r-1}$ | *Probability of Falsely Rejecting* $H_0$, $1 - (1 - .05)^{r-1}$ |
|---|---|---|
| 2 | .950 | .050 |
| 3 | .903 | .097 |
| 4 | .857 | .143 |
| 5 | .815 | .185 |
| 6 | .774 | .226 |
| 7 | .735 | .265 |

values of $r$ have their basis in results presented in section 14.2. As we indicated there, it is possible to form $t - 1$ orthogonal contrasts for comparing $t$ treatment means. And using those contrasts, we can partition the treatment sum of squares into $t - 1$ single-degree-of-freedom sums of squares. (We will discuss this partitioning in detail in chapter 17.) If we assume the degrees of freedom for $s_W^2$ are quite large, then the $(t - 1)$ $F$ statistics are nearly independent. Then when each $F$ test is conducted at a preset $\alpha$-value, and we assume $\mu_1 = \mu_2 = \cdots = \mu_t$, the probability of rejecting $H_0$ for one or more contrasts is approximately

$$1 - (1 - \alpha)^{t-1}$$

Duncan argued that since experimenters have little or no reservations in performing these multiple $F$ tests for orthogonal contrasts even though the overall $\alpha$ level increases with $t$, it is reasonable to construct a multiple comparison test for which the protection level decreases with the number of sample means included in a comparison. Thus Duncan uses a $\alpha$-value equal to the quantity $1 - (1 - \alpha)^{r-1}$ when a pair of sample means are $r$ steps apart $(r = 2, \ldots, t)$.

Because the protection level decreases with increasing $r$, *Duncan's multiple range test is very powerful*. That is, there is a high probability of declaring a difference when there is actually a difference between the population means. This has been one of the reasons Duncan's procedure has been extremely popular among researchers.

We summarize Duncan's new multiple range test for pairwise comparisons of $t$ population means in the box.

**Duncan's New Multiple Range Test**

1. Rank the $t$ sample means.
2. Two population means are declared significantly different if the absolute value of their sample differences exceeds

$W_r$

$$W_r = q'_\alpha(r, v)\sqrt{\frac{s_W^2}{n}}$$

$q'_\alpha(r, v)$

where $n$ is the number of observations in each sample mean, $s_W^2$ is the mean square within samples obtained from the analysis of variance table, $v$ is the number of degrees of freedom for $s_W^2$, and $q'_\alpha(r, v)$ is the critical value of the Studentized range required for Duncan's procedure when the means being compared are $r$ steps apart. Values of $q'_\alpha(r, v)$ are given in table 12 of the appendix for $\alpha = .05$ or $.01$ and various combinations of $r$ and $v$.

---

We illustrate the use of Duncan's procedure with the data of example 14.3.

### EXAMPLE 14.5

Refer to the data of example 14.3 (p. 385). Run Duncan's multiple range test with $\alpha = .05$ to make all pairwise comparisons among the 6 population means.

### SOLUTION

Recall from example 14.3 that $n = 5$, $s_W^2 = 2451$, and $v = 24$. Using this information we can set up the following table for $r$, $q'_\alpha(r, v)$, and $W_r$.

| $r$ | 2 | 3 | 4 | 5 | 6 |
|---|---|---|---|---|---|
| $q'_\alpha(r, v)$ | 2.92 | 3.07 | 3.15 | 3.22 | 3.28 |
| $W_r$ | 64.7 | 68.0 | 69.7 | 71.3 | 72.6 |

For example, when two means are $r = 2$ steps apart, $q'_{.05}(2, 24)$ is 2.92. Then

$$W_2 = q'_{.05}(2, 24)\sqrt{\frac{s_W^2}{n}} = 2.92\sqrt{\frac{2451}{5}} = 64.7$$

Thus two sample means $r = 2$ steps apart will be declared significantly different if the absolute value of their difference exceeds 64.7. The remainder of the entries in the table for different values of $r$ were computed in a similar manner.

The sample means, ranked in order from lowest to highest, are

| *Population* | 6 | 4 | 1 | 2 | 3 | 5 |
|---|---|---|---|---|---|---|
| *Sample Mean* | 470 | 498 | 505 | 528 | 564 | 600 |

Beginning with the largest mean, $\bar{y}_5$, each sample mean is compared to the smallest mean, $\bar{y}_6$, using the appropriate value of $W_r$. For example, $\bar{y}_5$ and $\bar{y}_6$ are $r = 6$ steps apart and their difference must be compared to $W_6 = 72.6$. The comparisons with $\bar{y}_{\text{smallest}}$ are shown in the table.

| *Comparison* | *Conclusion* |
|---|---|
| $\bar{y}_5 - \bar{y}_6 = 130$ | $> 72.6$; proceed |
| $\bar{y}_3 - \bar{y}_6 = 94$ | $> 71.3$; proceed |
| $\bar{y}_2 - \bar{y}_6 = 58$ | $< 69.7$; stop |

Similarly, comparisons are made with $\bar{y}_4$, $\bar{y}_1$, $\bar{y}_2$, and $\bar{y}_3$ as follows:

| *Comparison* | *Conclusion* |
|---|---|
| $\bar{y}_5 - \bar{y}_4 = 102$ | $> 71.3$; proceed |
| $\bar{y}_3 - \bar{y}_4 = 66$ | $< 69.7$; stop |
| $\bar{y}_5 - \bar{y}_1 = 95$ | $> 69.7$; proceed |
| $\bar{y}_3 - \bar{y}_1 = 59$ | $< 68.0$; stop |
| $\bar{y}_5 - \bar{y}_2 = 72$ | $> 68.0$; proceed |
| $\bar{y}_3 - \bar{y}_2 = 36$ | $< 64.7$; stop |
| $\bar{y}_5 - \bar{y}_3 = 36$ | $< 64.7$; stop |

The results of these pairwise comparisons can be summarized as usual.

population 6 4 1 2 3 5

In conclusion, $\mu_6$, $\mu_4$, $\mu_1$, and $\mu_2$ are significantly less than $\mu_5$; $\mu_6$ is significantly less than $\mu_3$.

We can compare the results obtained by using Duncan's new multiple range test to those obtained for Fisher's LSD and Tukey's procedure. Judging from the summary lines for the three procedures (pp. 388 and 391), we see that Duncan's procedure for the data of table 14.2 results in conclusions more like the LSD than Tukey's. The only difference in the conclusions for the two is that in using the LSD the means for populations 3 and 4 were found to be different while for Duncan's they were not.

Unlike Tukey's procedure and Fisher's LSD, Duncan's new multiple range test cannot be used to form confidence intervals for the pairwise differences $\mu_i - \mu_j$. We can, however, adopt Duncan's procedure for unequal sample sizes if the $n_i$'s are nearly equal by replacing $n$ by

$$\tilde{n} = \frac{t}{\frac{1}{n_1} + \frac{1}{n_2} + \cdots + \frac{1}{n_t}}$$

in the formula for $W_r$. This same adjustment, which is also used for Tukey's procedure, was suggested by Bancroft (1968). Duncan (1957) and Kramer (1957) have also presented extensions for situations where the population variances are different and where the sample means are correlated.

# 14.7 Scheffé's $S$ Method

The three multiple comparison procedures discussed so far have been developed for pairwise comparisons among $t$ population means. A more general procedure, proposed by Scheffé (1953), can be used to make all possible comparisons among the $t$ population means. Although Scheffé's procedure can be applied to pairwise comparisons among the $t$ population means, it is more conservative (less sensitive) than any of the other three multiple comparison procedures for detecting significant differences among pairs of population means.

**Scheffé's $S$ Method for Multiple Comparisons**

1. Consider any linear comparison among the $t$ population means of the form

   $$\ell = a_1\mu_1 + a_2\mu_2 + \cdots + a_t\mu_t$$

   We wish to test the null hypothesis

   $$H_0\colon \ell = 0$$

   against the alternative

   $$H_a\colon \ell \neq 0$$

2. The test statistic is

   $$\hat{\ell} = a_1\bar{y}_1 + a_2\bar{y}_2 + \cdots + a_t\bar{y}_t$$

3. Let

   $S$

   $$S = \sqrt{\hat{V}(\hat{\ell})}\,\sqrt{(t-1)F_{\alpha,\mathrm{df}_1,\mathrm{df}_2}}$$

   where, from section 14.2,

   $$\hat{V}(\hat{\ell}) = s_W^2 \sum_i \frac{a_i^2}{n_i}$$

   $t$ is the total number of population means, and $F_{\alpha,\mathrm{df}_1,\mathrm{df}_2}$ is the upper-tail critical value of the $F$ distribution with $a = \alpha$, $\mathrm{df}_1 = t - 1$, and $\mathrm{df}_2$ the degrees of freedom for $s_W^2$.
4. For a specified value of $\alpha$, we reject $H_0$ if $|\hat{\ell}| > S$.
5. The error rate that is controlled is an *experimentwise error rate*. If we consider all imaginable contrasts, the probability of observing an experiment with one or more contrasts falsely declared significant is designated by $\alpha$.

**EXAMPLE 14.6**

Refer to the data of table 14.2 (p. 385). Suppose that three of the batches (6, 4, and 2) were prepared from one concentration of HCl and the other three batches (1, 3, and 5) from another concentration of HCl. Use the sample data and Scheffé's procedure to compare the mean dye yields for the two different concentrations of HCl. Let $\alpha = .05$.

**SOLUTION**

Assume the even-numbered batches are from Concentration I of HCl and the odd-numbered batches from Concentration II. A contrast of particular importance is

$$\hat{\ell} = \bar{y}_1 + \bar{y}_3 + \bar{y}_5 - \bar{y}_2 - \bar{y}_4 - \bar{y}_6$$

which compares the means for batches from Concentration II to those for Concentration I. In particular, we would like to test

$H_0$: $\ell = 0$

$H_a$: $\ell \neq 0$

The estimated value of $\ell$ is

$$\hat{\ell} = \bar{y}_1 + \bar{y}_3 + \bar{y}_5 - \bar{y}_2 - \bar{y}_4 - \bar{y}_6$$

$$= 505 + 564 + 600 - 528 - 498 - 470 = 173$$

To compute

$$S = \sqrt{\hat{V}(\hat{\ell})}\ \sqrt{(t-1)F_{\alpha,\mathrm{df}_1,\mathrm{df}_2}}$$

we must first calculate $\hat{V}(\hat{\ell})$. Using the formula

$$\hat{V}(\hat{\ell}) = s_W^2 \sum_i \frac{a_i^2}{n_i}$$

with all sample sizes equal to 5 and $s_W^2 = 2451$, we have

$$\hat{V}(\hat{\ell}) = 2451[\tfrac{1}{5} + \tfrac{1}{5} + \tfrac{1}{5} + \tfrac{1}{5} + \tfrac{1}{5} + \tfrac{1}{5}] = 2941.2$$

From table 4 for $\alpha = .05$, $\mathrm{df}_1 = t - 1 = 5$, and $\mathrm{df}_2 = 24$ (the degrees of freedom for $s_W^2$),

$$F_{.05,5,24} = 2.62$$

The computed value of $S$ is then

$$S = \sqrt{2941.2}\ \sqrt{5(2.62)} = (54.23)(3.62) = 196.31$$

Since the absolute value of $\hat{\ell}$, 173, does not exceed $S = 196.31$, we have insufficient evidence to indicate that the means for batches from Concentration II differ from those for Concentration I.

Scheffé confidence interval

Scheffé's method can also be used for constructing a simultaneous confidence interval for all possible (not necessarily pairwise) contrasts using the $t$ treatment means. In particular, there is a probability equal to $1 - \alpha$ that all possible comparisons of the form $\ell = \Sigma a_i \mu_i$, where $\Sigma a_i = 0$, will be encompassed by intervals of the form

$$\hat{\ell} - S < \ell < \hat{\ell} + S$$

## 14.8 The *k* Ratio Rule for Making Pairwise Comparisons Among Treatment Means

In this chapter we have spent considerable time discussing the merits of different multiple comparison procedures. Fisher's LSD procedure can be used to make all pairwise comparisons among $t$ population means while the controlled error rate for each comparison is at a specified level $\alpha$. We found that even when $\mu_1 = \mu_2 = \cdots = \mu_t$, unless the per-comparison error rate $\alpha$ is quite small, the probability of falsely declaring at least one pair of means significantly different is quite large for Fisher's LSD.

To avoid this difficulty, Tukey proposed a multiple comparison procedure utilizing an experimentwise error rate. Numerous other authors have offered solutions to this same problem, as, for example, Duncan's multiple range test and Scheffé's procedure, but none of these has adequately answered the question of which error rate to control.

The dilemma (see Duncan [1975]) of multiple comparisons can be illustrated with the following simplistic example. Suppose we wish to make pairwise comparisons among $t$ ($t$ very large) population means. In addition, suppose the null and alternative hypotheses for each of the pairwise comparisons is of the form

$$H_0\colon \mu_i - \mu_j = 0$$

$$H_a\colon \mu_i - \mu_j \neq 0$$

and that for each test the LSD, based on $\alpha = .05$, is used to determine whether two population means are significantly different.

*Case I.* Suppose that only 5% of all the differences in population means were declared significant. Intuitively, since we set $\alpha = .05$, we would be inclined to think we erred by declaring 5% of the differences significant when in fact all the population means are identical. To guard against declaring too many differences significant, we would certainly want to increase the magnitude of the absolute difference (LSD) in sample means required for significance. This is precisely the approach taken by Tukey (and others, such as Scheffé), who required that the absolute difference in sample means exceed $W$ (where $W > \text{LSD}$) before declaring significance. Certainly the use of Tukey's procedure would be preferred to Fisher's LSD, to protect against a situation as described in case I.

*Case II*. Suppose, however, that only 5% of the population differences were declared to be nonsignificant. Then, intuitively, we might think we had erred on the opposite side and declared 5% of the population differences nonsignificant when in fact all $t$ population means are different. To protect against such an occurrence, we would decrease the magnitude of the absolute difference in sample means required to declare significance. Clearly in situations such as this, Fisher's LSD procedure would be preferable to Tukey's $W$ procedure.

Waller and Duncan (1969) proposed a multiple comparison procedure that uses the sample data to help in determining whether we need to use a conservative rule (much like Tukey's) or a nonconservative rule (much like Fisher's). The procedure makes use of the computed value of $F = s_B^2/s_W^2$, the test statistic for the null hypothesis $H_0$: $\mu_1 = \mu_2 = \cdots = \mu_t$. If the computed value of $F$ is small, then the sample data tend to indicate that the population means are rather homogeneous. For this situation, to protect against case I, we would require a large absolute difference in sample means to declare significance. In contrast, if the computed value of $F$ is large, the sample data would tend to indicate or confirm that the population means are heterogeneous. For this situation, to protect against case II, we would require a small absolute difference in sample means to declare significance.

While a complete explanation of this procedure is well beyond the scope of this text, adequate tables have been developed to enable us to apply it to multiple comparisons resulting from the analysis of variance for a completely randomized design (chapter 13) and the other standard experimental designs to be discussed in chapter 15. There are two restrictions that are placed on the Waller-Duncan procedure. First, we require all sample sizes to be the same, and, second, the procedure should not be used when a priori we would expect certain of the population means to differ more than others.

Fisher's LSD required us to choose the comparisonwise error rate $\alpha$. Now we must specify the *error weight ratio k*, which designates the seriousness of a type I error relative to a type II error. While we cannot go into a detailed discussion of the rationale for choosing $k$, some guidelines can be given. Whereas typical values of $\alpha$ for Fisher's LSD may be .10, .05, or .01, corresponding values of $k$, the error weight ratio, are 50, 100, or 500.

*k*, error weight ratio

The Waller-Duncan test procedure is summarized in the box.

**Waller-Duncan $k$ Ratio Procedure**

1. Choose $k$, the error weight ratio.
2. Perform an analysis of variance to obtain the computed value of $F$ for $H_0$: $\mu_1 = \mu_2 = \cdots = \mu_t$.
3. Compute

$$\text{LSD} = t_c\sqrt{s_W^2\left(\frac{2}{n}\right)}$$

$t_c$

where $t_c$ is obtained from tables 13 or 14 in the appendix and is based on $k$, $df_1$, $df_2$, and $F$; $s_W^2$ is the mean square error from the AOV table; and $n$ is the number of observations selected from each population. (Note that this procedure requires that we have the same number of observations from each population.)

4. Follow a stepwise procedure similar to that for Fisher's LSD to declare $\mu_i - \mu_j \neq 0$ if $|\bar{y}_i - \bar{y}_j| > \text{LSD}$.

Before we proceed with an example, let us consider the format of tables 13 and 14. Because of the extensiveness of the tables, only two values of $k$ (100 and 500) have been included. (Tables for $k = 50$ are presented in Waller and Duncan [1972].) The table entry is the $t_c$-value corresponding to specific values of $k$, $df_1$ (df for MST), $df_2$ (df for MSE), and $F = \text{MST}/\text{MSE}$. Thus for $k = 100$, $df_1 = 6$, $df_2 = 8$, and $F = 3.0$, the $t$-value for Waller-Duncan's least significant difference is 2.61.

Because of the complexity of the table, not all values of $df_1$, $df_2$, and $F$ can be given for a fixed error weight ratio. Fortunately, it is possible to interpolate to obtain the appropriate value of $t_c$.

interpolation

Interpolation will often be required for values of $F$. To obtain the appropriate value of $t_c$ when the computed value of $F$ falls between two tabulated $F$-values, interpolate linearly using one of two quantities

$$a = \frac{1}{\sqrt{F}} \quad \text{or} \quad b = \sqrt{\frac{F}{F-1}}$$

Table 14.5 indicates under what situations we use "a" and under what conditions we interpolate using $b$.

**EXAMPLE 14.7**

Refer to example 14.3 (p. 385). We were interested in comparisons among 6 batches used to produce a particular dye. The computed value of $F$ was 4.60, based on $df_1 = 5$ and $df_2 = 24$. Find the value of $t_c$ to be used in a $k$ ratio procedure for pairwise comparisons among the 6 batch means. Use an error weight ratio of 100.

**Table 14.5** Situations in which to use a or $b$ for interpolation; the Waller-Duncan procedure

| $F$ | $df_1$ | $df_2$ $\leq 100$ | $df_2$ $> 100$ |
|---|---|---|---|
| $\leq 2.4$ | $\leq 60$ | a | a |
| | $> 60$ | a | $b$ |
| | | $df_2$ $\leq 20$ | $df_2$ $> 20$ |
| $> 2.4$ | $\leq 20$ | a | $b$ |
| | $> 20$ | $b$ | $b$ |

**SOLUTION**

From table 13 we note that there are no $t_c$-values listed for $F = 4.60$; we must interpolate between the values listed for $F = 4.0$ and $F = 6.0$. Note also that there are no $t_c$-values for $\text{df}_1 = 5$. However, very little difference is observed whether we use $\text{df}_1 = 4$ or 6. For $k = 100$, $\text{df}_1 = 4$, $\text{df}_2 = 24$, and $F = 4.0$, the table entry is 2.18. For the same settings except $\text{df}_1 = 6$, $t_c = 2.19$. There is no change in $t_c$ for $k = 100$, $\text{df}_2 = 24$, and $F = 6.0$ when $\text{df}_1 = 4$ or 6. In both cases $t_c = 2.06$.

Because $\text{df}_1$ has very little effect on $t_c$ in the range of 4 to 6 for this problem, we will work with $\text{df}_1 = 4$. To obtain that $t_c$-value corresponding to 4.60, we see from table 14.5 that we must interpolate using

$$b = \sqrt{\frac{F}{F - 1}}$$

From table 13 of the appendix, for $F = 4.0$, then $b = 1.155$, and for $F = 6.0$, then $b = 1.095$. The distance between 1.155 and 1.095 is .060. Computing $b$ for $F = 4.60$, we have

$$b = \sqrt{\frac{4.60}{3.60}} = 1.13$$

We now interpolate linearly with respect to $b$. First we draw a graph with the horizontal axis labeled with $b$ and the vertical axis with $t_c$. We plot the values of $t_c$ corresponding to $b = 1.155$ and $b = 1.095$. Then we connect these two points with a straight line. We use the straight line to determine the interpolated value of $t_c$ corresponding to $b = 1.13$. (See figure 14.1.) As can be seen from figure 14.1, the interpolated value of $t_c$ corresponding to $b = 1.13$ is $t_c = 2.13$.

**EXAMPLE 14.8**

Refer to example 14.3 and perform all pairwise comparisons between population means using the $k$ ratio procedure with $k = 100$.

**SOLUTION**

In example 14.7 we found the appropriate value of $t_c$ to be 2.13. Recalling that $s_W^2 = 2451$ and that there were 5 observations per batch, the least

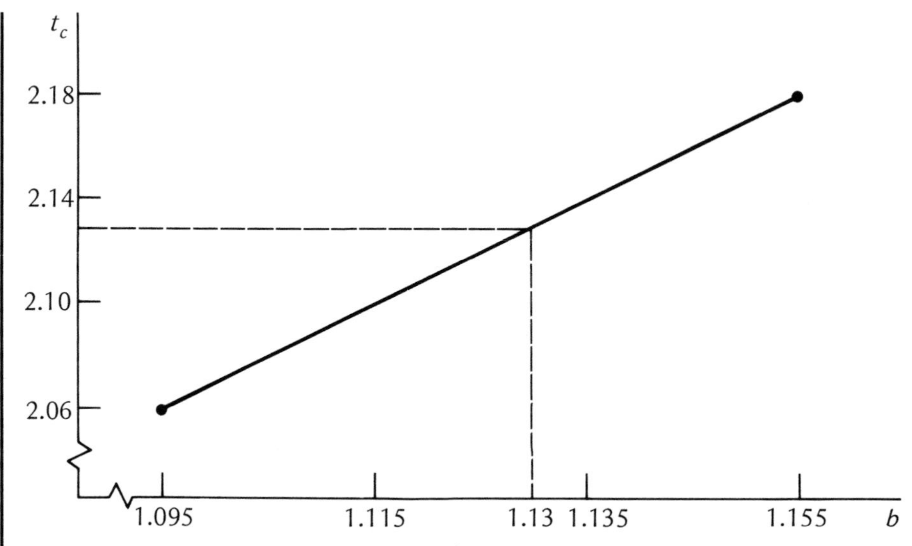


**Figure 14.1** Interpolating linearly with respect to $b$ to find $t_c$; example 14.7.

significant difference using the Waller-Duncan approach is

$$\text{LSD} = t_c\sqrt{s_W^2\left(\frac{2}{n}\right)} = 2.13\sqrt{2451\left(\frac{2}{5}\right)} = 66.69$$

The stepwise procedure of Fisher's LSD can be employed to summarize our sample findings. The lines summarizing our results are seen to bbe

population 6 4 1 2 3 5

(6 4 1 2 underlined; 4 1 2 underlined; 3 5 underlined)

It is interesting to compare our results in example 14.8 with those of previous multiple comparison procedures for the same set of data. We see that our results are identical to Duncan's multiple range test results (p. 395) and very similar to Fisher's LSD results (p. 388). The reason that the $k$ ratio results are similar to the results for the less conservative multiple comparison procedures is due to the fact that the value of $F = s_B^2/s_W^2$ is fairly large, indicating that the population means are more heterogeneous. This situation requires a non-conservative rule. If the computed value of $F$ was small, we would have found the results of the $k$ ratio procedure similar to one of the more conservative rules.

# Summary | 14.9

In this chapter we have tried to attack the problem of determining significant differences among $t$ population means for a one-way classification. Three of the procedures described are particularly appropriate for running all pairwise comparisons among the population means. All three are easy to apply. The major difference between these methods is the error rate that is controlled.

Fisher's least significant difference procedure sets the error rate for each pairwise comparison. Recent studies of the LSD have shown that the error rate is controlled at a level approximately equal to the $\alpha$ level of the overall $F$ test for treatments. Fisher's unprotected LSD controls the comparisonwise error rate. Tukey's $W$ procedure controls the error rate for all pairwise comparisons on an experimentwise basis.

In contrast, Duncan's new multiple range test has an error rate that stems from the concept of a protection level. When the $t$ means are arranged in order of magnitude, the comparison of two means $r$ steps apart has a protection level of $(1 - \alpha)^{r-1}$, where $\alpha$ is specified. The error rate is then $1 - (1 - \alpha)^{r-1}$. Since the error rate changes with $r$, we have neither an experimentwise or comparisonwise error rate.

The error rate for Scheffé's $S$ method is an experimentwise error rate that applies to all possible contrasts among the $t$ population means. Since it is more general than the other three procedures presented, it is not recommended for conducting pairwise comparisons.

The last multiple comparison procedure we presented was the Waller-Duncan $k$ ratio test. Since critical differences depend on the computed value of the $F$ test on treatments, it can be either a conservative or a nonconservative rule, depending on the weight of the sample evidence for rejection of the null hypothesis.

We have presented some of the more popular multiple comparison procedures in use today. There are, of course, others. If, for example, we wish to make comparisons with a standard (control) preparation, Dunnett (1955) has developed an appropriate multiple range test. The purpose of this chapter was not to present an exhaustive discussion of multiple comparison procedures but to familiarize you with some of the more common ones. For a more comprehensive discussion of multiple comparison procedures, see Chew (1976).

In the next chapter we return to a discussion of the analysis of variance for some standard experimental designs. As you will see, the multiple comparison procedures presented here can be extended to these more complicated designs.

# Exercises

*14.4. Refer to the data of example 13.1 (p. 363). Recall a horticulturist was investigating the phosphorous content of the leaves from 3 different varieties of apple trees.

a. Perform an analysis of variance.
b. Use Duncan's multiple range test procedure to run all pairwise comparisons. Use $\alpha = .05$.
c. Compare your conclusions in part b to those in the SAS computer output shown here.

```
                                   DATA VARIETY

DUNCAN'S 5 PERCENT LEVEL NEW MULTIPLE RANGE TEST FOR VARIABLE Y

NO.                  MEAN VARIETY

  1      +      0.77600      2
         |
  2      +      0.70800      3
  3      +      0.46000      1
```

14.5. An experiment was conducted to compare the effectiveness of 5 different weight-reducing agents. A random sample of 150 males was randomly divided into 5 equal groups, with Preparation A assigned to the first group, B to the second group, and so on. Each person in the experiment was given a prestudy physical and told how many pounds he was overweight. A comparison of the mean number of pounds overweight for the groups showed no significant differences. The study program was then begun, with each group taking the prescribed preparation for a fixed period of time. At the end of the study period, weight losses were recorded. The data are given below.

| | | | | | | | | | | |
|---|---|---|---|---|---|---|---|---|---|---|
| A | 12.4 | 10.7 | 11.9 | 11.0 | 12.4 | 12.3 | 13.0 | 12.5 | 11.2 | 13.1 |
| B | 9.1 | 11.5 | 11.3 | 9.7 | 13.2 | 10.7 | 10.6 | 11.3 | 11.1 | 11.7 |
| C | 8.5 | 11.6 | 10.2 | 10.9 | 9.0 | 9.6 | 9.9 | 11.3 | 10.5 | 11.2 |
| D | 8.7 | 9.3 | 8.2 | 8.3 | 9.0 | 9.4 | 9.2 | 12.2 | 8.5 | 9.9 |
| E | 12.7 | 13.2 | 11.8 | 11.9 | 12.2 | 11.2 | 13.7 | 11.8 | 11.5 | 11.7 |

Run an analysis of variance to determine if there are any significant differences among the weight losses for the 5 diet preparations. Use $\alpha = .05$.

14.6. Refer to exercise 14.5. Run all pairwise comparisons of population means using the following procedures.

a. Fisher's LSD, $\alpha = .05$

b. Tukey's $W$, $\alpha = .05$

c. Duncan's multiple range test, $\alpha = .05$

d. Compare the conclusions for the three procedures.

*14.7. Use a computer program to run an analysis of variance for the data of example 14.5 (p. 394).

14.8. Refer to exercise 14.5. Examine the widths of 95% confidence intervals for pairwise comparisons of diets A, B, . . . , E, using Fisher's LSD, Tukey's W, and Scheffé's $S$ method.

14.9. Refer to exercise 14.5. Suppose that Preparation D was a placebo. Use Scheffé's procedure to make the following comparisons. Set $\alpha = .05$.

a. $\mu_D - \frac{1}{4}(\mu_A + \mu_B + \mu_C + \mu_E)$

b. $\mu_A - \mu_E$

c. $\mu_A - \frac{1}{2}(\mu_B + \mu_E)$

d. $\mu_A - \frac{1}{3}(\mu_B + \mu_C + \mu_E)$

14.10. Refer to exercise 14.9. Give the corresponding Scheffé 95% confidence intervals for these four comparisons.

14.11. Refer to exercise 13.1 (p. 368).

a. Use Fisher's LSD procedure with $\alpha = .05$ to declare significant differences.

b. Use Tukey's $W$ method with $\alpha = .05$ to make all pairwise comparisons of population means for the data of exercise 13.1. Compare your conclusions to those of part a.

14.12. Refer to exercise 13.1.

a. Use the Waller-Duncan method, with $k = 100$.

b. Compare your results for parts a and b of exercise 14.11 and part a above.

14.13. Refer to exercise 13.2 (p. 369). Use Duncan's multiple range procedure to declare significant differences among the population means. Set $\alpha = .05$.

14.14. Refer to exercise 14.5. Determine the appropriate value of $t_c$ for the Waller-Duncan $k$ ratio procedure when $k = 100$ and $k = 500$.

14.15. Make all pairwise comparisons among the 5 diet preparations of exercise 14.5 using the Waller-Duncan method. Use an error weight ratio of $k = 100$. Compare your results to those of exercise 14.6.

14.16. Use Duncan's multiple range procedure to compare the 4 population means of exercise 13.7 (p. 371).

14.17. Refer to exercise 13.4 (p. 370). Perform all pairwise comparisons of the 4 population means. Be sure to identify the procedure you use and the error rate controlled. Use $\alpha = .05$.

14.18. Refer to exercise 14.17. Set up confidence intervals for all pairwise comparisons of treatment means in exercise 13.4, using Tukey's $W$ procedure. Interpret your findings.

14.19. Refer to the data of exercise 13.10 (p. 374). Make all pairwise comparisons of the three areas using the Waller-Duncan procedure with $k = 100$.

# References


Bancroft, T.A. 1968. *Topics in intermediate statistical methods*. Vol. I. Ames, Iowa: Iowa State University Press.

Boardman, T.J., and Moffitt, D.R. 1971. Graphical Monte Carlo type I error rates for multiple comparison procedures. *Biometrics* 27: 738–744.

Carmer, S.G., and Swanson, M.R. 1973. An evaluation of ten pairwise multiple comparison procedures by Monte Carlo methods. *Journal of the American Statistical Association* 68: 66–74.

Chew, V. 1976. *Comparing treatment means—A compendium*. Technical report no. 105. Agriculture Research Service of the United States Department of Agriculture and the Department of Statistics, University of Florida.

Cornell, J.A. 1971. A review of multiple comparison procedures for comparing a set of $k$ population means. *Proceedings of the Soil and Crop Science Society of Florida* 31: 92–97.

Duncan, D.B. 1955. Multiple range and multiple $F$ tests. *Biometrics* 11: 1–42.

———. 1957. Multiple range tests for correlated and heteroscedastic means. *Biometrics* 13: 164–176.

———. 1975. $T$ tests and intervals for comparisons suggested by the data. *Biometrics* 31: 339–359.

Dunnett, C.W. 1955. A multiple comparison procedure for comparing several treatments with a control. *Journal of the American Statistical Association* 50: 1096–1121.

Fisher, R.A. 1949. *The design of experiments*. Edinburgh: Oliver and Boyd Ltd.

Harter, H.L. 1957. Error rates and sample sizes for range tests in multiple comparisons. *Biometrics* 13: 511–536.

Kirk, R.E. 1968. *Experimental design: Procedures for the behavioral sciences*. Belmont, Calif.: Brooks/Cole.

Kramer, C.Y. 1957. Extension of multiple range tests to group correlated adjusted means. *Biometrics* 13: 13–18.

Pearson, E.S., and Hartley, H.O. 1942. The probability integral of the range in samples of *n* observations from a normal population. *Biometrics* 32: 301–310.

———. 1943. Tables of the probability integral of the studentized range. *Biometrics* 33: 89–99.

Scheffé, H. 1953. A method for judging all contrasts in an analysis of variance. *Biometrika* 40: 87–104.

———. 1959. *The analysis of variance*. New York: Wiley

Tukey, J.W. 1953. The problem of multiple comparisons. Mimeographed. Princeton, N.J.: Princeton University.

Waller, R.A., and Duncan, D.B. 1969. A Bayes rule for the symmetric multiple comparison problem. *Journal of the American Statistical Association* 64: 1484–1503.

———. 1972. Corrigenda. *Journal of the American Statistical Association* 67: 253–255.

# 15 Analysis of Variance for Some Standard Experimental Designs

## 15.1 Introduction

In chapter 9 we presented the four steps in the design of an experiment and briefly discussed several experimental designs: completely randomized designs, randomized block designs, Latin square designs, and factorial experiments. Recall that for a given study, we must define the experimental objectives, determine the amount of information required about the parameters of interest, and then choose an experimental design and an inferential procedure to achieve the specified objectives. In this chapter we present a widely used inferential technique, the *analysis of variance*. We considered the analysis of variance for a completely randomized design (one-way classification) in chapter 13. The analysis of variance technique will be extended in this chapter to a randomized block design, a Latin square design, and a $k$-way classification.

## 15.2 The Analysis of Variance for a Randomized Block Design

As stated in chapter 9, a randomized block design can be used to compare $t$ population (treatment) means when an additional source of variability (blocks) is present. The advantage of the randomized block design over the completely randomized design is that we are able to filter out the block-to-block variability by making comparisons among the treatments within blocks. For example, in analyzing the susceptibility of root stocks of 10 different plant varieties to a certain larva, an experimenter chose 5 different soil beds on which to make his comparisons. All 10 different plant varieties were randomly assigned within each of the 5 soil beds. By making damage comparisons among varieties within beds, the experimenter eliminated any bed-to-bed variability.

The definition of a randomized block design is given in the box.

randomized block design

**Definition 15.1**

A *randomized block design* is an experimental design for comparing $t$ treatments in $b$ blocks. Treatments are randomly assigned to experimental units within a block, with each treatment appearing exactly once in every block.

Consider the data for a randomized block design as arranged in table 15.1. Using table 15.1 we can introduce notation that is helpful in performing an analysis of variance. This notation is presented in the box.

**Table 15.1** A randomized block design

| | *Block* | | | | | |
|---|---|---|---|---|---|---|
| *Treatment* | 1 | 2 | ... | $b$ | *Totals* | *Mean* |
| 1 | $y_{11}$ | $y_{12}$ | ... | $y_{1b}$ | $T_1$ | $\bar{T}_1$ |
| 2 | $y_{21}$ | $y_{22}$ | ... | $y_{2b}$ | $T_2$ | $\bar{T}_2$ |
| ⋮ | ⋮ | ⋮ | | ⋮ | ⋮ | ⋮ |
| $t$ | $y_{t1}$ | $y_{t2}$ | ... | $y_{tb}$ | $T_t$ | $\bar{T}_t$ |
| *Totals* | $B_1$ | $B_2$ | ... | $B_b$ | $G$ | |
| *Mean* | $\bar{B}_1$ | $\bar{B}_2$ | ... | $\bar{B}_b$ | | $\bar{y}$ |

**Notation Needed for the AOV of a Randomized Block Design**

$y_{ij}$: The observation for treatment $i$ in block $j$.

$t$: The number of treatments.

$b$: The number of blocks.

$n$: The total number of sample measurements; $n = bt$.

$T_i$: The total for all observations receiving treatment $i$.

$B_j$: The total for all observations in block $j$.

$G$: The total for all sample observations.
$\bar{T}_i$: The sample mean for treatment $i$; $\bar{T}_i = T_i/b$.
$\bar{B}_j$: The sample mean for block $j$; $\bar{B}_j = B_j/t$.
$\bar{y}$: The overall sample mean; $\bar{y} = G/n$.

---

total sum of squares

The *total sum of squares* of the measurements about their mean $\bar{y}$ is defined as before:

$$\text{TSS} = \sum_i \sum_j (y_{ij} - \bar{y})^2$$

partitioning TSS

It is possible to partition the total sum of squares into three separate sources of variability: one due to the variability among treatments, one due to the variability among blocks, and one due to the variability among the $y_{ij}$'s which is not accounted for by either treatments or blocks. We call this final source of variability "error." The partition of TSS can be shown to take the following form:

$$\sum_i \sum_j (y_{ij} - \bar{y})^2 = b \sum_i (\bar{T}_i - \bar{y})^2 + t \sum_j (\bar{B}_j - \bar{y})^2 + \sum_i \sum_j (y_{ij} - \bar{T}_i - \bar{B}_j + \bar{y})^2$$

The first quantity on the right-hand side of the equation measures the variability of the treatment means $\bar{T}_i$ from the overall mean. Thus

$$\text{SST} = b \sum_i (\bar{T}_i - \bar{y})^2$$

between-treatment sum of squares

called the *between-treatment sum of squares,* is a measure of the between-treatment variability. Similarly, the second quantity,

$$\text{SSB} = t \sum_j (\bar{B}_j - \bar{y})^2$$

between-block sum of squares

measures the variability between the block means $\bar{B}_j$ and the overall mean. It is called the *between-block sum of squares*. The third source of variability, referred to as the *sum of squares for error* (SSE), represents the variability in the $y_{ij}$'s not accounted for by the block and treatment sources.

Although the sum of squares formulas just discussed are instructive, they are not convenient to use in calculations. The shortcut formulas given in the box are more convenient tools for calculations.

**Shortcut Sums of Squares Formulas for a Randomized Block Design**

$$\text{TSS} = \sum_i \sum_j y_{ij}^2 - \frac{G^2}{n}$$

$$\text{SST} = \sum_i \frac{T_i^2}{b} - \frac{G^2}{n}$$

$$\text{SSB} = \sum_j \frac{B_j^2}{t} - \frac{G^2}{n}$$

$$\text{SSE} = \text{TSS} - \text{SST} - \text{SSB}$$

The model for an observation in a randomized block design can be written in the form

model

$$y_{ij} = \mu + \alpha_i + \beta_j + \epsilon_{ij}$$

where the terms of the model are defined as follows:

$\mu$: An overall mean, which is an unknown constant.

$\alpha_i$: An effect due to treatment $i$; $\alpha_i$ is an unknown constant.

$\beta_j$: An effect due to block $j$; $\beta_j$ is an unknown constant.

$\epsilon_{ij}$: A random error associated with the response on treatment $i$, block $j$. We assume that the $\epsilon_{ij}$'s are normally distributed with mean 0 and unknown variance $\sigma^2$. In addition, the errors are assumed to be independent. (Technically speaking, we need not assume the $\epsilon_{ij}$'s are normally distributed at this point, but we must make this assumption prior to running an analysis of variance.)

The above assumptions for our model can be shown to imply that $y_{ij}$, the response on treatment $i$ in block $j$, is normally distributed with mean $\mu + \alpha_i + \beta_j$ and variance $\sigma^2$. A table of population means (expected values) for the data of table 15.1 is shown in table 15.2.

Several comments should be made concerning the table of expected values. First, two observations that receive the same treatment (appear in the same row of table 15.2) have population means that differ by block effects only. For

**Table 15.2** Expected values for the $y_{ij}$'s in a randomized block design

| | Block | | | |
|---|---|---|---|---|
| Treatment | 1 | 2 | ... | b |
| 1 | $E(y_{11}) = \mu + \alpha_1 + \beta_1$ | $E(y_{12}) = \mu + \alpha_1 + \beta_2$ | ... | $E(y_{1b}) = \mu + \alpha_1 + \beta_b$ |
| 2 | $E(y_{21}) = \mu + \alpha_2 + \beta_1$ | $E(y_{22}) = \mu + \alpha_2 + \beta_2$ | ... | $E(y_{2b}) = \mu + \alpha_2 + \beta_b$ |
| $\vdots$ | $\vdots$ | $\vdots$ | | $\vdots$ |
| $t$ | $E(y_{t1}) = \mu + \alpha_t + \beta_1$ | $E(y_{t2}) = \mu + \alpha_t + \beta_2$ | ... | $E(y_{tb}) = \mu + \alpha_t + \beta_b$ |

example, the expected values associated with $y_{11}$ and $y_{12}$ (two observations receiving treatment 1) are, respectively, $\mu + \alpha_1 + \beta_1$ and $\mu + \alpha_1 + \beta_2$. Thus the difference in their means is $\beta_1 - \beta_2$, which accounts for the fact that $y_{11}$ appeared in block 1 and $y_{12}$ in block 2. Second, two observations appearing in the same block (in the same column of table 15.2) have means that differ by a treatment effect only. For example, $y_{11}$ and $y_{21}$ both appear in block 1. The difference in their means is, from table 15.2,

$$(\mu + \alpha_1 + \beta_1) - (\mu + \alpha_2 + \beta_1) = \alpha_1 - \alpha_2$$

which accounts for the fact that the observations received different treatments. Finally, when two observations receive a different treatment *and* appear in different blocks, their expected values differ by effects due to treatments and to blocks. Thus $y_{11}$ and $y_{22}$ have expectations that differ by

$$(\mu + \alpha_1 + \beta_1) - (\mu + \alpha_2 + \beta_2) = (\alpha_1 - \alpha_2) + (\beta_1 - \beta_2)$$

filtering

Using the information we have learned concerning the model for a randomized block design, we can illustrate the concept of *filtering* and show how the randomized block design filters out the variability due to blocks. Consider a randomized block design with $t = 3$ treatments (I, II, and III) laid out in $b = 3$ blocks as shown in table 15.3.

**Table 15.3** Randomized block design with $t = 3$ treatments and $b = 3$ blocks

| Block | Treatment | | |
|---|---|---|---|
| 1 | I | II | III |
| 2 | I | III | II |
| 3 | III | I | II |

The model for this randomized block design is

$$y_{ij} = \mu + \alpha_i + \beta_j + \epsilon_{ij} \qquad (i = 1, 2, 3;\ j = 1, 2, 3)$$

Suppose we wish to estimate the difference in mean response for treatments II and I, namely, $\alpha_2 - \alpha_1$. The difference in sample means, $\bar{y}_{II} - \bar{y}_I$, would represent a point estimate of $\alpha_2 - \alpha_1$. By substituting into our model, we have

$$\bar{y}_I = \sum_{j=1}^{3} \frac{y_{ij}}{3}$$

$$= \tfrac{1}{3}[(\mu + \alpha_1 + \beta_1 + \epsilon_{11}) + (\mu + \alpha_1 + \beta_2 + \epsilon_{12}) + (\mu + \alpha_1 + \beta_3 + \epsilon_{13})]$$

$$= \mu + \alpha_1 + \bar{\beta} + \bar{\epsilon}_1$$

where $\bar{\beta}$ represents the mean of the three block effects $\beta_1$, $\beta_2$, and $\beta_3$, and $\bar{\epsilon}_1$ represents the mean of the three random errors $\epsilon_{11}$, $\epsilon_{12}$, and $\epsilon_{13}$. Similarly, it is easy to show that

$$\bar{y}_{II} = \mu + \alpha_2 + \bar{\beta} + \bar{\epsilon}_2$$

and hence

$$\bar{y}_{II} - \bar{y}_I = (\alpha_2 - \alpha_1) + (\bar{\epsilon}_2 - \bar{\epsilon}_1)$$

Note how the block effects cancel, leaving the quantity $(\bar{\epsilon}_2 - \bar{\epsilon}_1)$ as the error of estimation when using $\bar{y}_{II} - \bar{y}_I$ to estimate $\alpha_2 - \alpha_1$.

If a completely randomized design had been employed instead of a randomized block design, treatments would have been assigned to experimental units at random and it is quite unlikely that each treatment would have appeared in each block. When the same treatment appears more than once in a block and we calculate an estimate of $\alpha_2 - \alpha_1$ using $\bar{y}_{II} - \bar{y}_I$, all block effects would not cancel out as they did previously. Then the error of estimation would include not only $\bar{\epsilon}_2 - \bar{\epsilon}_1$ but also the block effects that do not cancel. That is,

$$\bar{y}_{II} - \bar{y}_I = \alpha_2 - \alpha_1 + (\bar{\epsilon}_2 - \bar{\epsilon}_1) + (\text{block effects that do not cancel})$$

Hence the randomized block design filters out variability due to blocks by decreasing the error of estimation for a comparison of treatment means.

Having indicated the shortcut formulas for calculating sums of squares, and having specified the model associated with a randomized block design, we can now formulate the analysis of variance. The null hypothesis of no difference among the treatment means is equivalent to testing

statistical test

$$H_0\colon \alpha_1 = \alpha_2 = \cdots = \alpha_t = 0$$

As we observed in table 15.2, anytime we compare the mean response of two treatments (say $i$ and $i'$) in the same block, the difference in their mean response is $\alpha_i - \alpha_{i'}$. Thus under $H_0$ we are assuming that treatments have the same mean response within a block.

The alternative hypothesis is

$H_a$: At least one $\alpha_i$ is different from zero (i.e., at least one of the treatment means differs from the rest)

The test statistic is

$$F = \frac{\text{MST}}{\text{MSE}}$$

where MST and MSE are mean squares computed from the appropriate sums of squares in the AOV table of table 15.4.

**Table 15.4** Analysis of variance table for a randomized block design

| *Source* | *SS* | *df* | *MS* | *F* |
|---|---|---|---|---|
| treatments | SST | $t - 1$ | MST = SST/$(t - 1)$ | MST/MSE |
| blocks | SSB | $b - 1$ | MSB = SSB/$(b - 1)$ | MSB/MSE |
| error | SSE | $(b - 1)(t - 1)$ | MSE = SSE/$(b - 1)(t - 1)$ | |
| *Totals* | TSS | $bt - 1$ | | |

unbiased estimates

When $H_0$: $\alpha_1 = \alpha_2 = \cdots = \alpha_t = 0$ is true, both MST and MSE are *unbiased estimates* of $\sigma^2$, the variance associated with the observations in our model. That is, when $H_0$ is true, both MST and MSE have a mean value in repeated sampling equal to $\sigma^2$, and we would expect $F$ = MST/MSE to be near 1. When $H_a$ is true, the mean of MSE, called the *expected mean square for error*, is still

expected mean square for error

$$E(\text{MSE}) = \sigma^2$$

However, MST is no longer unbiased for $\sigma^2$. In fact, the expected mean square for treatments can be shown to be

$$E(\text{MST}) = \sigma^2 + \theta_T$$

where $\theta_T$ is a positive function of the $\alpha_i$'s. Because MST will tend to overestimate $\sigma^2$ under $H_a$, the ratio $F$ = MST/MSE will be greater than 1, and we will reject $H_0$ in the upper-tail of the distribution of $F$.

For a specified probability of a type I error, the $F$ test for $H_0$: $\alpha_1 = \alpha_2 = \cdots = \alpha_t = 0$ will reject $H_0$ if the computed value of $F$ exceeds the critical value of $F$ for a $= \alpha$, $df_1 = t - 1$, and $df_2 = (b - 1)(t - 1)$. Note that $df_1$ and $df_2$ correspond to the degrees of freedom for MST and MSE, respectively, in the AOV table.

test for effects of blocks

We may also be interested in testing whether it was advantageous to block. That is, is there an effect due to blocks that we have effectively filtered out? Recall that the expected values for two observations from different blocks receiving the same treatment differed by an effect due to blocks only (see table 15.2). The hypothesis of no effect due to blocks can be written in the form

$$H_0: \beta_1 = \beta_2 = \cdots = \beta_b = 0$$

The alternative hypothesis and test statistic are then

$H_a$: At least one of the $\beta$'s differs from zero.

T.S.: $F = \dfrac{\text{MSB}}{\text{MSE}}$.

Under $H_0$, both MSB and MSE are unbiased estimates of $\sigma^2$ [i.e., $E(\text{MSB}) = E(\text{MSE}) = \sigma^2$]. But under $H_a$, the expected mean squares are

$$E(\text{MSB}) = \sigma^2 + \theta_B \qquad \text{and} \qquad E(\text{MSE}) = \sigma^2$$

where $\theta_B$ is a positive function of the $\beta$'s. Since MSB will tend to overestimate $\sigma^2$ under $H_a$, we will reject $H_0$ for large values of $F$. For a specified value of $\alpha$, we can locate the rejection region in the $F$ table of the appendix for a $= \alpha$, $df_1 = b - 1$, and $df_2 = (b - 1)(t - 1)$.

**EXAMPLE 15.1**

An experiment was conducted to compare the effect of 3 different insecticides on a particular variety of string beans. Four different plots were prepared, with each plot subdivided into 3 rows. A suitable distance was maintained between the rows within a plot. Each row was planted with 100 seeds and then maintained under the insecticide assigned to the row. The insecticides were randomly assigned to the rows within a plot so that each insecticide appeared in one row in all 4 plots. The response of interest was the number of seedlings that emerged per row. These data are given in table 15.5.

**Table 15.5** Number of seedlings by insecticide and plot, example 15.1

| | *Plot* | | | |
|---|---|---|---|---|
| *Insecticide* | *1* | *2* | *3* | *4* |
| 1 | 56 | 49 | 65 | 60 |
| 2 | 84 | 78 | 94 | 93 |
| 3 | 80 | 72 | 83 | 85 |

a. Set up an appropriate statistical model for this experimental situation.
b. Run an analysis of variance to compare the three insecticides.
c. Summarize your results in an AOV table. Use $\alpha = .05$.

## SOLUTION

We recognize this experimental design as a randomized block design with $b = 4$ blocks (plots) and $t = 3$ treatments (insecticides) per block. The appropriate statistical model is

$$y_{ij} = \mu + \alpha_i + \beta_j + \epsilon_{ij} \qquad (i = 1, 2, 3;\ j = 1, 2, \ldots, 4)$$

From the sample data, the treatment and block totals can be shown to be

| Insecticides: | | Plots: | |
|---|---|---|---|
| | $T_1 = 230$ | | $B_1 = 220$ |
| | $T_2 = 349$ | | $B_2 = 199$ |
| | $T_3 = 320$ | | $B_3 = 242$ |
| | $G = 899$ | | $B_4 = 238$ |
| | | | $G = 899$ |

Substituting into the corresponding shortcut formulas for the sums of squares, we have

$$\text{SST} = \sum_i \frac{T_i^2}{b} - \frac{G^2}{n} = \frac{(230)^2 + (349)^2 + (320)^2}{4} - \frac{(899)^2}{12}$$

$$= 69{,}275.25 - 67{,}350.08 = 1925.17$$

$$\text{SSB} = \sum_j \frac{B_j^2}{t} - \frac{G^2}{n} = \frac{(220)^2 + (199)^2 + (242)^2 + (238)^2}{3} - \frac{(899)^2}{12}$$

$$= 67{,}736.33 - 67{,}350.08 = 386.25$$

Then

$$\text{TSS} = \sum_i \sum_j y_{ij}^2 - \frac{G^2}{n} = (56)^2 + (49)^2 + \cdots + (85)^2 - 67{,}350.08$$

$$= 2334.92$$

By subtraction

$$\text{SSE} = \text{TSS} - \text{SST} - \text{SSB} = 2334.92 - 1925.17 - 386.25 = 23.50$$

The analysis of variance table in table 15.6 summarizes our results. Note that the mean square for a source in the AOV table is computed by dividing the sum of squares for that source by its degrees of freedom.

**Table 15.6** AOV table for the data of example 15.1

| *Source* | *SS* | *df* | *MS* | *F* |
|---|---|---|---|---|
| treatments | 1925.17 | 2 | 962.59 | $962.59/3.92 = 245.56$ |
| blocks | 386.25 | 3 | 128.75 | $128.75/3.92 = 32.84$ |
| error | 23.50 | 6 | 3.92 | |
| *Totals* | 2334.92 | 11 | | |

The $F$ test for treatments, namely

$H_0$: $\alpha_1 = \alpha_2 = \alpha_3 = 0$ (no differences among treatment means)

makes use of the $F$ statistic MST/MSE. Since the computed value of $F$, 245.56, is greater than the tabulated $F$-value, 5.14, based on $df_1 = 2$, $df_2 = 6$, and a = .05, we reject $H_0$ and conclude that there are differences among the three treatment (insecticide) means.

The test for blocks

$H_0$: $\beta_1 = \beta_2 = \cdots = \beta_4 = 0$ (no differences among plot means)

utilizes the computed value of $F$ = MSB/MSE as the test statistic. Since the computed value of $F$, 32.84, is greater than the tabulated value, 4.76, based on $df_1 = 3$, $df_2 = 6$, and a = .05, we reject the null hypothesis and conclude that there are differences among the plot means.

We have discussed randomized block designs briefly in chapter 9 and in more detail here. But we might still ask whether blocking has increased our precision for comparing treatment means in a given experiment. Let $MSE_{RB}$ and $MSE_{CR}$ denote the mean square errors for a randomized block design and a completely randomized design, respectively. One measure of precision for the two designs is the variance of $\bar{y}_i$, the $i$th treatment mean ($i = 1, 2, \ldots, t$). For a randomized block design, the estimated variance for $\bar{y}_i$ is $MSE_{RB}/b$. A similar expression for a completely randomized design is $MSE_{CR}/r$, where $r$ is the number of observations (replications) of each treatment required to satisfy the relationship

$$\frac{MSE_{CR}}{r} = \frac{MSE_{RB}}{b} \quad \text{or} \quad \frac{MSE_{CR}}{MSE_{RB}} = \frac{r}{b}$$

relative efficiency

The quantity $r/b$ is called the *relative efficiency* of the randomized block design. The larger the value of $MSE_{CR}$ to $MSE_{RB}$, the larger $r$ must be to obtain the same precision for a treatment mean as obtained with the randomized block design.

Although we never perform an analysis of variance for both the completely randomized design and the randomized block design every time we employ a randomized block design, we can use the mean squares from the analysis of variance for the randomized block design to obtain the relative efficiency by using the formula

$$\frac{MSE_{CR}}{MSE_{RB}} = \frac{(b-1)MSB + b(t-1)MSE}{(bt-1)MSE}$$

**EXAMPLE 15.2**

Refer to example 15.1 and compute the relative efficiency of the randomized block design.

**SOLUTION**

From the AOV table (p. 416), MSB = 128.75 and MSE = 3.92. Hence the relative efficiency of this randomized block design relative to a completely randomized design is

$$\frac{\text{MSE}_{\text{CR}}}{\text{MSE}_{\text{RB}}} = \frac{3(128.75) + 4(2)(3.92)}{11(3.92)} = 9.68$$

That is, approximately 10 times as many observations of each treatment would be required in a completely randomized design to obtain the same precision for treatment comparisons as with this randomized block design.

# Exercises

15.1. An experiment was conducted to compare 4 different mixtures of the components oxidizer, binder, and fuel used in the manufacturing of rocket propellant. The 4 mixtures under test, corresponding to settings of the mixture proportions for oxide, are shown below.

| *Mixture* | *Oxidizer* | *Binder* | *Fuel* |
|---|---|---|---|
| 1 | 0.4 | 0.4 | 0.2 |
| 2 | 0.4 | 0.2 | 0.4 |
| 3 | 0.6 | 0.2 | 0.2 |
| 4 | 0.5 | 0.3 | 0.2 |

To compare the 4 mixtures, 5 different samples of propellant were prepared from each mixture and readied for testing. Each of 5 investigators was randomly assigned one sample of each of the 4 mixtures and asked to measure the propellant thrust. These data are summarized below.

| | *Investigator* | | | | |
|---|---|---|---|---|---|
| *Mixture* | *1* | *2* | *3* | *4* | *5* |
| 1 | 2340 | 2355 | 2362 | 2350 | 2348 |
| 2 | 2658 | 2650 | 2665 | 2640 | 2653 |
| 3 | 2449 | 2458 | 2432 | 2437 | 2445 |
| 4 | 2403 | 2410 | 2418 | 2397 | 2405 |

a. Identify the blocks and treatments for this experimental design.
b. Indicate the method of randomization.

15.2. Refer to exercise 15.1.

a. Perform an analysis of variance. Use $\alpha = .05$.

b. What conclusions can you draw concerning the best mixture from the 4 tested? (Note: The higher the response value, the better the rocket propellant thrust.)

c. Compute the relative efficiency of the randomized block design.

15.3. A study was undertaken to compare the starting salaries of bachelor's degree candidates at a particular university for the academic years 1972–73 and 1973–74. Since there could be a great deal of variability in salaries from one discipline to another, the investigator blocked on curricula within the university. Then a random sample of one bachelor's candidate for 1972–73 and one for 1973–74 was obtained from the most recent lists of graduates to ascertain the starting salary. It should be noted that only those students who had accepted a job were considered in this study. Use the following sample data of starting monthly salaries (in $100s) to conduct an analysis of variance, testing separately for an effect due to treatments and blocks. Use $\alpha = .05$ for both tests.

| *Curriculum* | *1973–74* | *1972–73* | *Totals* |
|---|---|---|---|
| accounting | 9.0 | 8.5 | 17.5 |
| agricultural sciences | 7.6 | 7.3 | 14.9 |
| biological sciences | 7.0 | 6.8 | 13.8 |
| business (general) | 7.8 | 7.6 | 15.4 |
| chemistry | 8.4 | 8.2 | 16.6 |
| computer science | 8.9 | 8.5 | 17.4 |
| engineering—civil | 9.3 | 8.9 | 18.2 |
| engineering—chemical | 10.0 | 9.6 | 19.6 |
| humanities | 6.5 | 6.6 | 13.1 |
| mathematics | 7.7 | 8.1 | 15.8 |
| social sciences | 6.9 | 6.8 | 13.7 |
| *Totals* | 89.1 | 86.9 | 176.0 |

*15.4. Examine the BMDP computer output shown here for the data of example 15.1.

a. Identify the sums of squares and degrees of freedom for plots, insecticides, and error.

b. Give the $F$ test and $p$-values for the null hypothesis of no difference in insecticide means. Draw a conclusion.

c. Give the $F$ test and $p$-value for the null hypothesis of no difference in plot means. Draw a conclusion.

d. Compare your results to those of example 15.1.

| VARIABLE NO. NAME | BEFORE TRANSFORMATION MINIMUM LIMIT | MAXIMUM LIMIT | MISSING CODE | CATEGORY CODE | CATEGORY NAME | INTERVAL RANGE GREATER THAN | LESS THAN OR EQUAL TO |
|---|---|---|---|---|---|---|---|
| 1 GROUP | | | | 1.00000 | 1.0000 | | |
| | | | | 2.00000 | 2.0000 | | |
| | | | | 3.00000 | 3.0000 | | |
| 2 LEVEL | | | | 1.00000 | 1.0000 | | |
| | | | | 2.00000 | 2.0000 | | |
| | | | | 3.00000 | 3.0000 | | |
| | | | | 4.00000 | 4.0000 | | |

NUMBER OF CASES READ .............................. 12

GROUP STRUCTURE

| GROUP | LEVEL | COUNT |
|---|---|---|
| 1.0000 | 1.0000 | 1. |
| 1.0000 | 2.0000 | 1. |
| 1.0000 | 3.0000 | 1. |
| 1.0000 | 4.0000 | 1. |
| 2.0000 | 1.0000 | 1. |
| 2.0000 | 2.0000 | 1. |
| 2.0000 | 3.0000 | 1. |
| 2.0000 | 4.0000 | 1. |
| 3.0000 | 1.0000 | 1. |
| 3.0000 | 2.0000 | 1. |
| 3.0000 | 3.0000 | 1. |
| 3.0000 | 4.0000 | 1. |

CELL MEANS FOR 1-ST DEPENDENT VARIABLE

| | | | | | | | | | | |
|---|---|---|---|---|---|---|---|---|---|---|
| GROUP = | 1.0000 | 1.0000 | 1.0000 | 1.0000 | 2.0000 | 2.0000 | 2.0000 | 2.0000 | 3.0000 | 3.0000 |
| LEVEL = | 1.0000 | 2.0000 | 3.0000 | 4.0000 | 1.0000 | 2.0000 | 3.0000 | 4.0000 | 1.0000 | 2.0000 |
| RESPONSE | 56.00000 | 49.00000 | 65.00000 | 60.00000 | 84.00000 | 78.00000 | 94.00000 | 93.00000 | 80.00000 | 72.00000 |
| COUNT | 1 | 1 | 1 | 1 | 1 | 1 | 1 | 1 | 1 | 1 |
| COUNT | 12 | | | | | | | | | |

| | | | MARGINAL |
|---|---|---|---|
| GROUP = | 3.0000 | 3.0000 | |
| LEVEL = | 3.0000 | 4.0000 | |
| RESPONSE | 83.00000 | 85.00000 | 74.91666 |
| COUNT | 1 | 1 | 12 |

STANDARD DEVIATIONS FOR 1-ST DEPENDENT VARIABLE

| GROUP = | 1.0000 | 1.0000 | 1.0000 | 1.0000 | 2.0000 | 2.0000 | 2.0000 | 2.0000 | 3.0000 | 3.0000 |
|---|---|---|---|---|---|---|---|---|---|---|
| LEVEL = | 1.0000 | 2.0000 | 3.0000 | 4.0000 | 1.0000 | 2.0000 | 3.0000 | 4.0000 | 1.0000 | 2.0000 |
| RESPONSE | 0.0 | 0.0 | 0.0 | 0.0 | 0.0 | 0.0 | 0.0 | 0.0 | 0.0 | 0.0 |

| GROUP = | 3.0000 | 3.0000 |
|---|---|---|
| LEVEL = | 3.0000 | 4.0000 |
| RESPONSE | 0.0 | 0.0 |

ANALYSIS OF VARIANCE FOR 1-ST DEPENDENT VARIABLE

| SOURCE | SUM OF SQUARES | DEGREES OF FREEDOM | MEAN SQUARE | F | PROB. F EXCEEDED |
|---|---|---|---|---|---|
| MEAN | 67350.06250 | 1 | 67350.06250 | 17195.78991 | 0.000 |
| G | 1925.16382 | 2 | 962.58179 | 246.39508 | 0.000 |
| L | 386.24829 | 3 | 128.74942 | 32.87225 | 0.000 |
| GL | 23.49994 | 6 | 3.91666 | | |

## 15.3 The Analysis of Variance for a Latin Square Design

The Latin square design can be used to compare $t$ treatment means when there are two additional sources of variability (usually referred to as rows and columns) that we wish to filter out. For example, in chapter 9 we saw that an investigator could obtain precise comparisons of three different laboratory procedures (A, B, C) for detecting nutritional deficiencies of apple tree leaves by filtering out both analyst-to-analyst variability and day-to-day variability from the assay measurements. Similarly, in examining the wholesale price of prime, choice, and good sides of beef, we could obtain precise comparisons of the average selling prices by blocking on days and on beef auction locations throughout the country.

The definition of a Latin square design is given in the box.

**Definition 15.2**

A $t \times t$ *Latin square design* contains $t$ rows and $t$ columns. The $t$ treatments are randomly assigned to experimental units within the rows and columns so that each treatment appears in every row and in every column.

A typical randomization scheme for a $4 \times 4$ Latin square comparing the treatments I, II, III, and IV is shown in table 15.7. Note that each treatment appears in all 4 rows and all 4 columns.

**Table 15.7** A $4 \times 4$ Latin square design

| | *Column* | | | |
|---|---|---|---|---|
| *Row* | *1* | *2* | *3* | *4* |
| 1 | I | II | III | IV |
| 2 | II | III | IV | I |
| 3 | III | IV | I | II |
| 4 | IV | I | II | III |

The notation for a Latin square design is only slightly more complicated than that for a randomized block design.

**Notation Needed for the AOV of a $t \times t$ Latin Square Design**

$y_{ijk}$: The response on treatment $i$ in row $j$ and column $k$.

$t$: The number of treatments; also the number of rows and the number of columns.

$n$: The total number of sample measurements; $n = t^2$.

$T_i$: The total for all observations receiving treatment $i$.

$\bar{T}_i$: The sample mean for treatment $i$; $\bar{T}_i = T_i/t$.

$R_j$: The total for all observations in row $j$.

$\bar{R}_j$: The sample mean for row $j$; $\bar{R}_j = R_j/t$.

$C_k$: The total for all observations in column $k$.

$\bar{C}_k$: The sample mean for column $k$; $\bar{C}_k = C_k/t$.

$G$: The total for all sample measurements.

$\bar{y}$: The overall sample mean; $\bar{y} = G/n$.

partitioning TSS

With this notation we can show a partition of the total sum of squares into four components. The first three components measure variability among the treatments, rows, and columns, respectively. The other source is due to random error.

$$\sum_i \sum_j (y_{ijk} - \bar{y})^2 = t \sum_i (\bar{T}_i - \bar{y})^2 + t \sum_j (\bar{R}_j - \bar{y})^2 + t \sum_k (\bar{C}_k - \bar{y})^2 + \sum_{\substack{\text{over} \\ \text{all } y\text{'s}}} (y_{ijk} - \bar{T}_i - \bar{R}_j - \bar{C}_k + 2\bar{y})^2$$

Note that the total sum of squares is obtained by summing over only two of the three subscripts. Even though observations are identified by treatment, row, and column, by summing over treatments ($i$) and rows ($j$), we have also summed over columns ($k$).

The algebraic verification of the partitioning of TSS into the four components is unimportant here and beyond the scope of this text. We will concentrate instead on the interpretation of the partitioning.

The first quantity on the right-hand side of the equation for TSS measures the variability of the treatment means $\bar{T}_i$ about the overall mean $\bar{y}$. As before, we call this source of variability the sum of squares between treatments:

$$\text{SST} = t \sum_i (\bar{T}_i - \bar{y})^2$$

Similarly, the second and third terms of the equation measure, respectively, the variability between rows and the variability between columns. These are designated by

$$\text{SSR} = t \sum_j (\bar{R}_j - \bar{y})^2$$

$$\text{SSC} = t \sum_k (\bar{C}_k - \bar{y})^2$$

The final source of variability, designated as the sum of squares for error (SSE), represents all additional variability in the measurements not accounted for by rows, columns, or treatments.

The simplified computational formulas useful in the AOV of a Latin square are given in the box.

---

**Shortcut Sums of Squares Formulas for a Latin Square Design**

$$\text{TSS} = \sum_i \sum_j y_{ijk}^2 - \frac{G^2}{n}$$

$$\text{SST} = \sum_i \frac{T_i^2}{t} - \frac{G^2}{n}$$

$$\text{SSR} = \sum_j \frac{R_j^2}{t} - \frac{G^2}{n}$$

$$SSC = \sum_k \frac{C_k^2}{t} - \frac{G^2}{n}$$

$$SSE = TSS - SST - SSR - SSC$$

---

model

The model for a response in a Latin square design is the same as that for a randomized block design, with the addition of one more term to account for the second blocking variable. Thus

$$y_{ijk} = \mu + \alpha_i + \beta_j + \gamma_k + \epsilon_{ijk}$$

where the terms are defined as follows:

$y_{ijk}$: The response on treatment $i$ in row $j$ and column $k$.
$\mu$: An overall mean; $\mu$ is a constant.
$\alpha_i$: An effect due to treatment $i$; $\alpha_i$ is a constant.
$\beta_j$: An effect due to row $j$; $\beta_j$ is a constant.
$\gamma_k$: An effect due to column $k$; $\gamma_k$ is a constant.
$\epsilon_{ijk}$: A random error associated with the response for treatment $i$ in row $j$ and column $k$. We assume the $\epsilon_{ijk}$'s are normally distributed with mean 0 and unknown variance $\sigma^2$. As before, the $\epsilon$'s are assumed to be independent.

These assumptions for this model imply that $y_{ijk}$, the response for treatment $i$ in row $j$ and column $k$, is normally distributed with mean

$$E(y_{ijk}) = \mu + \alpha_i + \beta_j + \gamma_k$$

and variance $\sigma^2$.

filtering

We can use the model to illustrate how a Latin square design filters out extraneous variability due to rows and columns. For purposes of illustration we will consider the Latin square design shown in table 15.7. If we wish to estimate $\alpha_3 - \alpha_1$, the difference in mean response for treatments III and I, using the sample difference $\bar{y}_{III} - \bar{y}_I$, we can substitute into our model to obtain expressions for $\bar{y}_{III}$ and $\bar{y}_I$. If $y_{ijk}$ denotes the observation in treatment $i$ in row $j$ and column $k$, we have from table 15.7

$$\begin{aligned}\bar{y}_I &= \tfrac{1}{4}(y_{111} + y_{142} + y_{133} + y_{124}) \\ &= \mu + \alpha_1 + \tfrac{1}{4}(\beta_1 + \beta_2 + \beta_3 + \beta_4) + \tfrac{1}{4}(\gamma_1 + \gamma_2 + \gamma_3 + \gamma_4) + \bar{\epsilon}_1.\end{aligned}$$

where $\bar{\epsilon}_1$ is the mean of the random errors for the 4 observations on treatment 1. Similarly,

$$\bar{y}_{\text{III}} = \tfrac{1}{4}(y_{331} + y_{322} + y_{313} + y_{344})$$
$$= \mu + \alpha_3 + \tfrac{1}{4}(\beta_1 + \beta_2 + \beta_3 + \beta_4) + \tfrac{1}{4}(\gamma_1 + \gamma_2 + \gamma_3 + \gamma_4) + \bar{\epsilon}_3$$

Then the sample difference is

$$\bar{y}_{\text{III}} - \bar{y}_{\text{I}} = \alpha_3 - \alpha_1 + (\bar{\epsilon}_3 - \bar{\epsilon}_1)$$

and the error of estimation for $\alpha_3 - \alpha_1$ is $\bar{\epsilon}_3 - \bar{\epsilon}_1$.

If a randomized block design had been used with blocks representing rows, treatments would be randomized within the rows only. It is quite possible for the same treatment to appear more than once in the same column. Then the sample difference would be

$$\bar{y}_{\text{III}} - \bar{y}_{\text{I}} = \alpha_3 - \alpha_1 + (\bar{\epsilon}_3 - \bar{\epsilon}_1) + (\text{column effects that do not cancel})$$

Thus the error of estimation would be inflated by the column effects that do not cancel out. Following the same reasoning, if a completely randomized design were used when a Latin square design was appropriate, the error of estimation would be inflated by both row and column effects that do not cancel out.

statistical tests

We can test specific hypotheses concerning the parameters in our model. In particular, we may wish to test the hypothesis of no difference among the $t$ treatment means. This hypothesis can be stated in the form

$H_0$: $\alpha_1 = \alpha_2 = \cdots = \alpha_t = 0$ (i.e., the $t$ treatment means are identical)

The alternative hypothesis would be

$H_a$: At least one of the $\alpha_i$'s differs from the rest (i.e., at least one treatment mean is different from the others)

The test statistic for our test would be

$$F = \frac{\text{MST}}{\text{MSE}}$$

For our model,

$$E(\text{MSE}) = \sigma^2 \qquad \text{and} \qquad E(\text{MST}) = \sigma^2 + \theta_T$$

Since it can be shown that $\theta_T$ is zero under $H_0$ and is a positive function of the $\alpha$'s under $H_a$, we will reject $H_0$ in the upper-tail of the $F$ distribution. The appropriate degrees of freedom are obtained from the AOV table shown in table 15.8.

**Table 15.8** AOV table for a $t \times t$ Latin square design

| *Source* | *SS* | *df* | *MS* | *F* |
|---|---|---|---|---|
| treatments | SST | $t - 1$ | MST = SST/$(t - 1)$ | MST/MSE |
| rows | SSR | $t - 1$ | MSR = SSR/$(t - 1)$ | MSR/MSE |
| columns | SSC | $t - 1$ | MSC = SSC/$(t - 1)$ | MSC/MSE |
| error | SSE | $t^2 - 3t + 2$ | MSE = SSE/$(t^2 - 3t + 2)$ | |
| *Totals* | TSS | $t^2 - 1$ | | |

SSE
df for SSE

We should note that we can compute SSE and the degrees of freedom for SSE by subtraction. Thus knowing the degrees of freedom for treatments, rows, and columns, we can subtract this sum from $t^2 - 1$ to obtain the degrees of freedom for error.

test for rows

Tests similar to that for treatments can be formulated for rows and columns. The test for rows is as follows:

$H_0$: $\beta_1 = \beta_2 = \cdots = \beta_t = 0$ (i.e., no effect due to rows).

$H_a$: At least one of the $\beta$'s differs from zero.

$$\text{T.S.: } F = \frac{\text{MSR}}{\text{MSE}}.$$

test for columns

The test for columns is as follows:

$H_0$: $\gamma_1 = \gamma_2 = \cdots = \gamma_t = 0$ (i.e., no effect due to columns).

$H_a$: At least one of the $\gamma$'s differs from zero.

$$\text{T.S.: } F = \frac{\text{MSC}}{\text{MSE}}.$$

**EXAMPLE 15.3**

A traffic engineer wished to compare the total unused green time (the total amount of time the signal is green but no vehicles are progressing through the intersection) for 4 different signal-control sequencing devices at 4 different intersections of a city. It was assumed that the intersections were far enough apart that they, in effect, acted independently, regardless of the signal sequencing device employed.

In addition to comparing the devices at the 4 different intersections, the engineer wished to compare the devices at different time periods during the day (A.M. peak traffic, A.M. lull, P.M. lull, P.M. peak). The $4 \times 4$ Latin square design shown in table 15.9 was chosen for the experimental design, where Roman numerals are used to denote the four signal-control devices under test.

**Table 15.9** 4 × 4 Latin square assignment for the traffic delay experiment of example 15.3

| | *Time Period* | | | |
|---|---|---|---|---|
| *Intersection* | A.M. *Peak* | A.M. *Lull* | P.M. *Lull* | P.M. *Peak* |
| 1 | IV | II | III | I |
| 2 | II | III | I | IV |
| 3 | III | I | IV | II |
| 4 | I | IV | II | III |

The signal-control devices were then assigned to experimental units (intersection and traffic peak combinations) according to the following scheme: On the appointed test day, Device IV was set up at Intersection 1, Device II at Intersection 2, Device III at Intersection 3, and Device I at Intersection 4. Readings on traffic delay were then obtained for the 7–8 A.M. peak rush hour at each intersection. Following the A.M. peak, the devices were reassigned to the appropriate intersection for the A.M. lull (10–11 A.M.) according to the assignment scheme of table 15.9. Delay readings were then obtained. In a similar way, the remainder of the data were obtained for the P.M. lull and the P.M. peak rush hour. Use the sample data in table 15.10 to perform an analysis of variance. Make all appropriate tests using $\alpha = .05$.

**Table 15.10** Sample data for the traffic delay study (unused green time, in minutes) of example 15.3

| | *Time Period* | | | | |
|---|---|---|---|---|---|
| *Intersection* | A.M. *Peak* | A.M. *Lull* | P.M. *Peak* | P.M. *Lull* | *Totals* |
| 1 | IV 15.5 | II 33.9 | III 13.2 | I 29.1 | 91.7 |
| 2 | II 16.3 | III 26.6 | I 19.4 | IV 22.8 | 85.1 |
| 3 | III 10.8 | I 31.1 | IV 17.1 | II 30.3 | 89.3 |
| 4 | I 14.7 | IV 34.0 | II 19.7 | III 21.6 | 90.0 |
| *Totals* | 57.3 | 125.6 | 69.4 | 103.8 | 356.1 |

**SOLUTION**

Using table 15.10, we find the treatment totals to be

$T_{\text{I}} = 94.3 \qquad T_{\text{II}} = 100.2 \qquad T_{\text{III}} = 72.2 \qquad T_{\text{IV}} = 89.4$

Thus

$$\text{SST} = \sum_i \frac{T_i^2}{t} - \frac{G^2}{n} = \frac{(94.3)^2 + (100.2)^2 + (72.2)^2 + (89.4)^2}{4} - \frac{(356.1)^2}{16}$$

$$= \frac{32{,}137.73}{4} - 7925.45 = 108.98$$

Similarly,

$$\text{SSR} = \sum_j \frac{R_j^2}{t} - \frac{G^2}{n} = \frac{(91.7)^2 + \cdots + (90.0)^2}{4} - 7925.45$$

$$= 7931.35 - 7925.45 = 5.9$$

$$\text{SSC} = \sum_k \frac{C_k^2}{t} - \frac{G^2}{n} = \frac{(57.3)^2 + \cdots + (103.8)^2}{4} - 7925.45$$

$$= 8662.36 - 7925.45 = 736.91$$

$$\text{TSS} = \sum_i \sum_j y_{ijk}^2 - \frac{G^2}{n} = (15.5)^2 + (33.9)^2 + \cdots + (21.6)^2 - 7925.45$$

$$= 8801.05 - 7925.45 = 875.6$$

The sum of squares for error can be found by subtraction:

$$\text{SSE} = \text{TSS} - \text{SST} - \text{SSR} - \text{SSC} = 875.6 - 108.98 - 5.9 - 736.91 = 23.81$$

The results of these calculations and $F$ tests for treatments, rows, and columns can be summarized in an analysis of variance table, as shown in table 15.11.

**Table 15.11** AOV table for the data of example 15.3

| *Source* | *SS* | *df* | *MS* | *F* |
|---|---|---|---|---|
| treatments (devices) | 108.98 | 3 | 36.33 | 9.15 |
| rows (intersections) | 5.90 | 3 | 1.97 | 0.50 |
| columns (time periods) | 736.91 | 3 | 245.64 | 61.87 |
| error | 23.81 | 6 | 3.97 | |
| *Totals* | 875.6 | 15 | | |

The $F$-values for a $= .05$, $\text{df}_1 = 3$, and $\text{df}_2 = 6$ is 4.76. Since the computed value of $F$ for treatments and for columns exceeds 4.76, we conclude that there are significant differences among the devices and among the time periods.

relative efficiency

As with the randomized block design, we can compare the efficiency of the Latin square design to the completely randomized design. Let $\text{MSE}_{\text{LS}}$ and $\text{MSE}_{\text{CR}}$ denote the mean square errors, respectively, for a Latin square design and a completely randomized design. The relative efficiency is

$$\frac{\text{MSE}_{\text{CR}}}{\text{MSE}_{\text{LS}}} = \frac{\text{MSR} + \text{MSC} + (t-1)\text{MSE}}{(t+1)\text{MSE}}$$

### EXAMPLE 15.4

Refer to the data of example 15.3. Compute the efficiency of the Latin square design relative to a completely randomized design.

### SOLUTION

For these data, $t = 4$, MSR = 1.97, MSC = 245.64, and MSE = 3.97. Substituting into the formula for relative efficiency, we have

$$\frac{\text{MSE}_{\text{CR}}}{\text{MSE}_{\text{LS}}} = \frac{1.97 + 245.64 + 3(3.97)}{5(3.97)} = 13.07$$

That is, it would take approximately 13 times as many observations in using a completely randomized design to gather the same amount of information on the treatment means when using the Latin square design.

# Exercises

15.5. An experiment was planned to compare 2 different fertilizer placements (broadcast band) and 2 different rates of fertilizer flow on watermelon yields. Recent research has shown that broadcast application (scattering over the outer area) of fertilizer is superior to bands of fertilizer applied near the seed for watermelon yields. For this experiment the investigators wished to compare 2 nitrogen-phosphorous-potassium (broadcast and band) fertilizers applied at a rate of 160-70-135 lb/acre and 2 brands of micronutrients (A and B). These 4 combinations were to be studied in a Latin square field plot.

The treatments were randomly assigned according to a Latin square design conducted over a large farm plot, which was divided into rows and columns. A watermelon plant dry weight was obtained for each row-column combination 30 days after the emergence of the plants. These data are shown below.

| | *Column* | | | |
|---|---|---|---|---|
| *Row* | *1* | *2* | *3* | *4* |
| 1 | I 1.75 | III 1.43 | IV 1.28 | II 1.66 |
| 2 | II 1.70 | I 1.78 | III 1.40 | IV 1.31 |
| 3 | IV 1.35 | II 1.73 | I 1.69 | III 1.41 |
| 4 | III 1.45 | IV 1.36 | II 1.65 | I 1.73 |

Treatment I: Broadcast, A  Treatment III: Band, A
Treatment II: Broadcast, B  Treatment IV: Band, B

a. Write an appropriate statistical model for this experiment.
b. Use the data to run an analysis of variance. Use $\alpha = .05$ for all tests. Draw your conclusions.

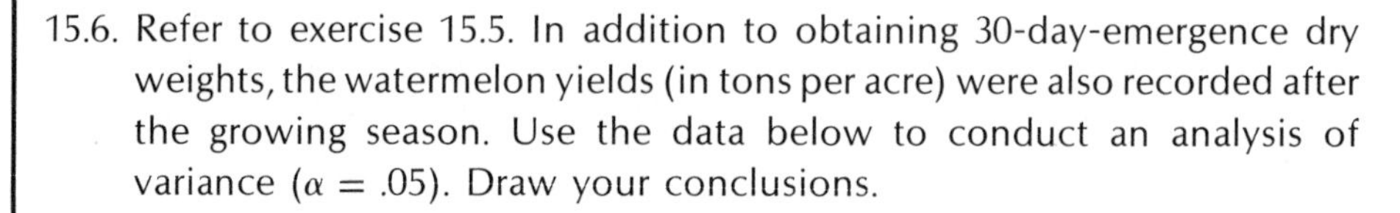

15.6. Refer to exercise 15.5. In addition to obtaining 30-day-emergence dry weights, the watermelon yields (in tons per acre) were also recorded after the growing season. Use the data below to conduct an analysis of variance ($\alpha = .05$). Draw your conclusions.

| | Column | | | |
|---|---|---|---|---|
| *Row* | *1* | *2* | *3* | *4* |
| 1 | 9.5 | 6.8 | 4.9 | 7.1 |
| 2 | 7.9 | 9.1 | 6.6 | 5.3 |
| 3 | 5.6 | 7.6 | 8.7 | 6.7 |
| 4 | 7.1 | 5.4 | 6.9 | 8.8 |

*15.7. Refer to the data of example 15.3.
a. Perform an analysis of variance using a computer program available to you.
b. Compare your results to those of example 15.3.

15.8. Refer to table 15.9 (p. 427). Derive the error of estimation for estimating $\alpha_4 - \alpha_2$ using $\bar{y}_{IV} - \bar{y}_{II}$.

## 15.4 The Analysis of Variance for a *k*-way Classification and the Factorial Experiment

differences among designs

In chapter 9 we defined a factorial experiment to be an experiment where the response $y$ is observed at all factor-level combinations of the independent variables. The major difference among the randomized block, Latin square, and factorial designs lies in their design objectives. With the randomized block design, an experimenter wishes to block out (or filter) variability due to a nuisance variable so that precise comparisons may be made among the treatments. Similarly, a Latin square design filters out variability due to two nuisance variables in order to obtain precise comparisons among the treatments of interest. In contrast, the factorial experiment is used to investigate the effect(s) of $k$ factors (independent variables) on the response.

The differences in design objectives lead to different methods of randomization of treatments to experimental units. In the randomized block design, treatments are applied at random to experimental units within a block, with the stipulation that each treatment appear one time in a block. Treatments for a Latin square can be assigned at random across columns within a row, subject to the restriction that each treatment appear in every row and every column. Treatments for a factorial experiment correspond to factor-level combinations. Treatments are then assigned at random to the experimental units, with the only restriction being that each factor-level combination appear the same number of times.

As mentioned in chapter 9, not all designs can be classified as either a block design or a factorial experiment. Sometimes the objectives of a study are such that we wish to investigate the effects of certain factors on a response while blocking out certain other extraneous sources of variability. Such experiments may require an experimental design that is a combination of a block design and a factorial experiment. For example, suppose we wish to examine the effects of 2 factors (each measured at 3 levels) on a response $y$. If we can perform 9 factor-level combinations per day, we could run one complete replication of the $3 \times 3$ factorial experiment on a given day. Then if additional observations are required, we could run complete replications of the $3 \times 3$ factorial experiment on successive days. If a total of $b$ complete runs are made (9 factor-level combinations on each of $b$ days), we can think of this experimental design as a randomized block design with $b$ blocks (days) and $t = 9$ treatments (the 9 factor-level combinations).

This particular design satisfies two major design objectives. We can investigate the effect of the two factors on the response and block out background variability due to blocks. Similarly, other factorial experiments could be combined with block designs to obtain more precise evaluations of the effects of the factors on the response of interest.

A factorial experiment can also be called a *k-way classification,* which is defined in the box.

---

**Definition 15.3**

An experiment produces a *k-way classification* of the data if all factor-level combinations of the $k$ independent variables are observed at least once.

---

You should note that except for a one-way classification ($k = 1$), a $k$-way classification does not define a specific experimental design. Rather, the term refers to how the data may be arranged in a table.

Most of the $k$-way classifications we will consider will have an equal number of observations at each factor-level combination. A randomized block design is then a 2-way classification ($k = 2$) of treatments and blocks. Similarly, a $2 \times 3$ factorial is a 2-way classification and a $2 \times 4 \times 3$ factorial is a 3-way classification of data. In fact, any factorial experiment may be referred to as a $k$-way classification. But a Latin square design is not a 3-way classification of rows, columns, and treatments.

One of the simplest examples of a 2-way classification is the $2 \times 2$ factorial experiment (two variables, each at two levels). For example, we may wish to examine the effect of two different strengths of an antiserum and two different concentrations of a buffer on the potency of a particular solution. To do this, 4 different samples of the same solution are randomly assigned one each to the 4 antiserum-buffer combinations. The data for such an experiment could be listed as in table 15.12.

**Table 15.12** Data for a $2 \times 2$ factorial experiment

| *Factor A, Antiserum* | *Factor B, Buffer* | | | |
|---|---|---|---|---|
| | *Concentration 1* | *Concentration 2* | *Totals* | *Mean* |
| level 1 | $y_{11}$ | $y_{12}$ | $A_1$ | $\bar{A}_1$ |
| level 2 | $y_{21}$ | $y_{22}$ | $A_2$ | $\bar{A}_2$ |
| *Totals* | $B_1$ | $B_2$ | $G$ | |
| *Mean* | $\bar{B}_1$ | $\bar{B}_2$ | | $\bar{y}$ |

model

The model for this experimental situation is

$$y_{ij} = \mu + \alpha_i + \beta_j + \epsilon_{ij}$$

with the usual assumptions for $\alpha_i$, $\beta_j$, and $\epsilon_{ij}$. The observation $y_{ij}$ in row $i$ and column $j$ is normally distributed with a mean of $\mu + \alpha_i + \beta_j$ and an unknown variance $\sigma^2$. The table of expected values is given in table 15.13.

**Table 15.13** Expected values for the observations in table 15.12

| *Factor A, Antiserum* | *Factor B, Buffer* | |
|---|---|---|
| | *Concentration 1* | *Concentration 2* |
| level 1 | $\mu + \alpha_1 + \beta_1$ | $\mu + \alpha_1 + \beta_2$ |
| level 2 | $\mu + \alpha_2 + \beta_1$ | $\mu + \alpha_2 + \beta_2$ |

additive model

Entries can be formed by adding the corresponding row and column effects to the overall mean. Hence the model is sometimes referred to as an *additive model.*

Note the additivity property of the expected values associated with a $2 \times 2$

factorial experiment. This same phenomenon was noted for the randomized block design (also a 2-way classification). For example, the difference in mean response for levels 1 and 2 is the same value, $\alpha_1 - \alpha_2$, no matter what buffer we are considering. Thus a test for no differences among the two levels of antiserum would be of the form $H_0$: $\alpha_1 - \alpha_2 = 0$. Similarly, the difference between buffer concentrations 1 and 2 is $\beta_1 - \beta_2$ for either level of antiserum, and a test of no difference between the buffer means is $H_0$: $\beta_1 - \beta_2 = 0$.

Consider now a 2-way classification with $r$ observations (sometimes called $r$ replications) per cell. We have the model

$$y_{ijk} = \mu + \alpha_i + \beta_j + \gamma_{ij} + \epsilon_{ijk}$$

where $y_{ijk}$ is the response obtained for the $k$th observation in row $i$ and column $j$ of the 2-way table and $\gamma_{ij}$ is an effect due to row $i$ and column $j$. The expected values for a $2 \times 2$ factorial experiment with $r$ observations per cell are presented in table 15.14.

**Table 15.14** Expected values for a $2 \times 2$ factorial experiment, with replications

| Factor A, Antiserum | Factor B, Buffer | |
|---|---|---|
| | *Concentration 1* | *Concentration 2* |
| level 1 | $\mu + \alpha_1 + \beta_1 + \gamma_{11}$ | $\mu + \alpha_1 + \beta_2 + \gamma_{12}$ |
| level 2 | $\mu + \alpha_2 + \beta_1 + \gamma_{21}$ | $\mu + \alpha_2 + \beta_2 + \gamma_{22}$ |

As can be seen from table 15.14, the difference in mean response for levels 1 and 2 on buffer Concentration 1 is now

$$(\alpha_1 - \alpha_2) + (\gamma_{11} - \gamma_{21})$$

but for Concentration 2 this difference is

$$(\alpha_1 - \alpha_2) + (\gamma_{12} - \gamma_{22})$$

Since the difference in mean response for levels 1 and 2 is *not* the same for different buffer concentrations, the model is no longer additive, and we say that the two factors antiserum level and buffer concentration *interact*.

interaction

**Definition 15.4**

Two factors $A$ and $B$ are said to *interact* if the difference in mean responses for two levels of one factor is not constant across levels of the second factor.

For example, when $r = 1$ observation is obtained at each factor-level combination of $A$ and $B$, the difference in mean response for levels 1 and 2 of factor $A$ in the additive model (p. 432) is $\alpha_1 - \alpha_2$, no matter what level of $B$ is chosen. Now, for $r > 1$, using the model

$$y_{ijk} = \mu + \alpha_i + \beta_j + \gamma_{ij} + \epsilon_{ijk}$$

the difference in mean response for levels 1 and 2 of factor $A$ changes for levels of factor $B$. Hence factors $A$ and $B$ interact and we cannot discuss one without discussing the other.

In measuring the octane rating of gasoline, interaction can occur when two components of the blend are combined to form a gasoline mixture. The octane properties of the blended mixture may be quite different than would be expected by examining each component of the mixture. Interaction in this situation could have a positive or negative effect on the performance of the blend, in which case the components are said to potentiate or antagonize one another.

We can amplify the notion of an interaction with the diagrams shown in figure 15.1. As we see from figure 15.1(a), when no interaction is present, the difference in the mean response between levels 1 and 2 of factor $B$ (as indicated by the braces) is the same for both levels of factor $A$. However, for the two illustrations in figures 15.1(b) and (c), we see that the difference between the levels of factor $B$ changes from level 1 to level 2 of factor $A$. For these cases we have an interaction between the two factors.

It should be noted that an interaction is not restricted to 2 factors. In a 3-way classification of factors $A$, $B$, and $C$, we might have an interaction between factors $A$ and $B$, $A$ and $C$, and $B$ and $C$, and the 2-factor interactions would have interpretations that follow immediately from definition 15.4. Thus the presence of an $AC$ interaction indicates that the difference in mean response for levels of one factor (either $A$ or $C$) varies across levels of the other factors. A 3-way interaction between factors $A$, $B$, and $C$ might indicate that the difference in mean response for levels of $C$ changes across combinations of levels for factors $A$ and $B$.

In general, the analysis of variance table for a $k$-way classification depends on whether or not we have $r > 1$ replications per cell. Before presenting these tables, we need the notation defined in the box.

**Notation Needed for the AOV of a $k$-way Classification, with $r > 1$ Replications Per Cell**

$n_A$: The number of observations at each level of factor $A$.

$A_i$: The sum of the $n_A$ observations receiving the $i$th level of factor $A$ ($i = 1, 2, \ldots, a$).

$B_j$: The sum of the $n_B$ observations receiving the $j$th level of factor $B$ ($j = 1, 2, \ldots, b$).

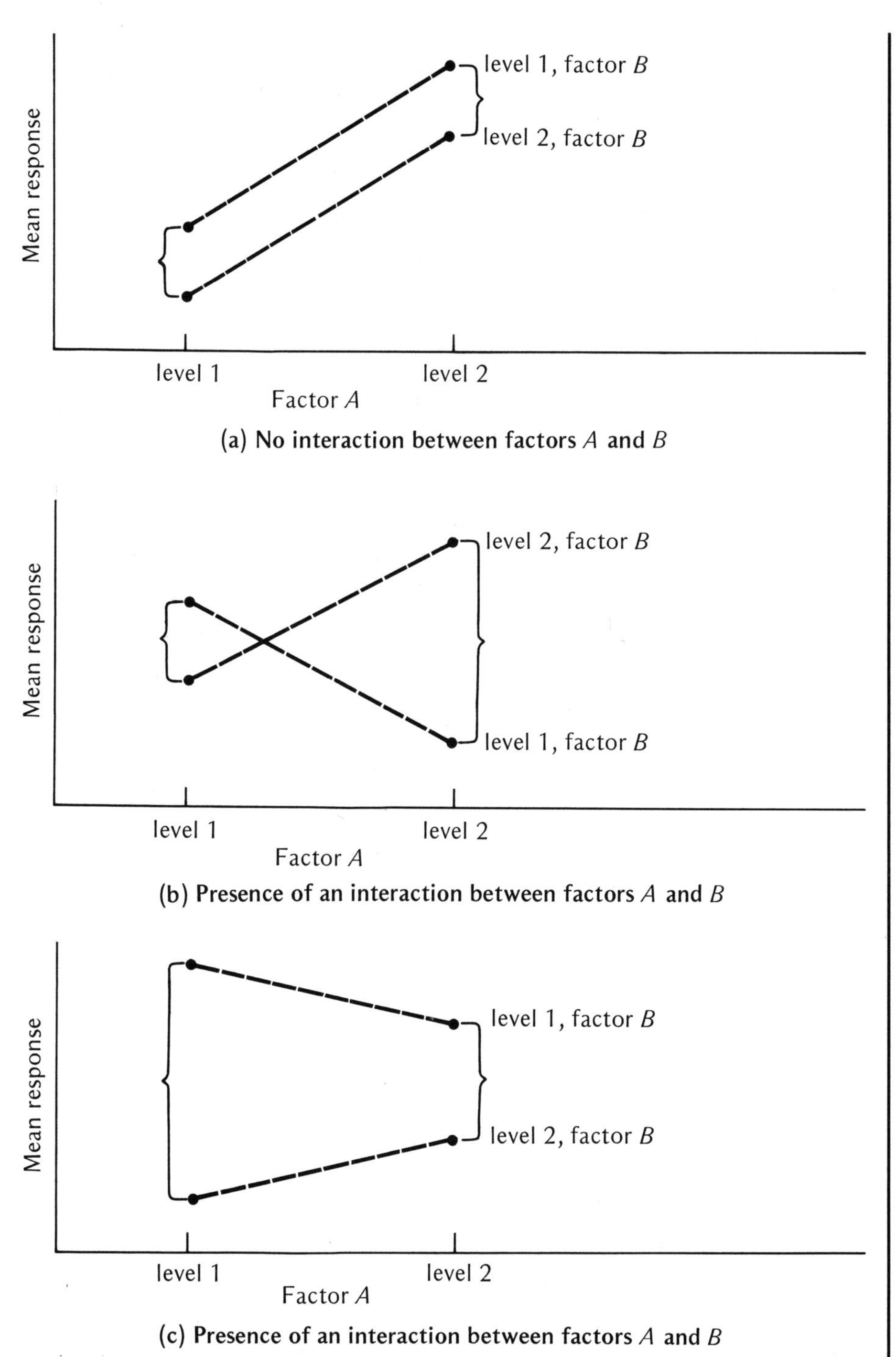


**Figure 15.1** Illustrations of the absence and presence of interaction in a $2 \times 2$ factorial experiment.

$C_k$: The sum of the $n_C$ observations receiving the $k$th level of factor $C$ ($k = 1, 2, \ldots, c$).

⋮

$n_{AB}$: The number of observations at each combination of levels of factors $A$ and $B$.

$(AB)_{ij}$: The sum of the $n_{AB}$ observations receiving the $i$th level of $A$ and $j$th level of $B$.

$(AC)_{ik}$: The sum of the $n_{AC}$ observations receiving the $i$th level of $A$ and $k$th level of $C$.

$(BC)_{jk}$: The sum of the $n_{BC}$ observations receiving the $j$th level of $B$ and $k$th level of $C$.

⋮

$n_{ABC}$: The number of observations at each combination of levels of the three factors $A$, $B$, and $C$.

$(ABC)_{ijk}$: The sum of the $n_{ABC}$ observations receiving the $i$th level of $A$, $j$th level of $B$, and $k$th level of $C$.

⋮

---

partitioning sums of squares

The appropriate AOV formulas for a $k$-way classification with $r$ observations per cell can be subdivided into sums of squares for main effects (variability between levels of a single factor), 2-way interactions, 3-way interactions, and so on.

The sums of squares for main effects are

main effects

$$\text{SSA} = \sum_i \frac{A_i^2}{n_A} - \frac{G^2}{n}$$

$$\text{SSB} = \sum_j \frac{B_j^2}{n_B} - \frac{G^2}{n}$$

$$\text{SSC} = \sum_k \frac{C_k^2}{n_C} - \frac{G^2}{n}$$

and so on.

The sums of squares for 2-way interactions are

2-way interactions

$$\text{SSAB} = \sum_{i,j} \frac{(AB)_{ij}^2}{n_{AB}} - \text{SSA} - \text{SSB} - \frac{G^2}{n}$$

$$SSAC = \sum_{i,k} \frac{(AC)_{ik}^2}{n_{AC}} - SSA - SSC - \frac{G^2}{n}$$

$$SSBC = \sum_{j,k} \frac{(BC)_{jk}^2}{n_{BC}} - SSB - SSC - \frac{G^2}{n}$$

and so on.

3-way interactions

The sums of squares for 3-way interactions are

$$SSABC = \sum_{i,j,k} \frac{(ABC)_{ijk}^2}{n_{ABC}} - SSAB - SSAC - SSBC - SSA - SSB - SSC - \frac{G^2}{n}$$

and so on.

You will note that all the main effects formulas are identical except for interchanging of the appropriate letters. Similar substitutions apply to the interaction sums of squares. And, in general, if we wished to compute the sums of squares associated with the *ABC* . . . interaction, we would use the formula

$$SSABC\ldots = \sum_{i,j,k\ldots} \frac{(ABC\ldots)^2}{n_{ABC\ldots}} - \begin{bmatrix}\text{sums of squares associated} \\ \text{with } \textit{all} \text{ subcombinations} \\ \text{of the factors } A, B, C, \ldots\end{bmatrix} - \frac{G^2}{n}$$

We illustrate the AOV table for a 3-way classification (i.e., $k = 3$), with "a" levels of $A$, $b$ levels of $B$, c levels of C, and $r = 1$ observation per cell. The AOV table is given in table 15.15.

**Table 15.15** AOV for a 3-way classification without replication

| *Source* | *SS* | *df* | *MS* |
|---|---|---|---|
| Main Effects | | | |
| *A* | SSA | a − 1 | MSA = SSA/(a − 1) |
| *B* | SSB | $b$ − 1 | MSB = SSB/($b$ − 1) |
| *C* | SSC | c − 1 | MSC = SSC/(c − 1) |
| Interactions | | | |
| *AB* | SSAB | (a − 1)($b$ − 1) | MSAB = SSAB/(a − 1)($b$ − 1) |
| *AC* | SSAC | (a − 1)(c − 1) | MSAC = SSAC/(a − 1)(c − 1) |
| *BC* | SSBC | ($b$ − 1)(c − 1) | MSBC = SSBC/($b$ − 1)(c − 1) |
| *ABC* | SSABC | (a − 1)($b$ − 1)(c − 1) | MSABC = SSABC/(a − 1)($b$ − 1)(c − 1) |
| *Totals* | TSS | abc − 1 | |

The total sum of squares formula is as before:

$$TSS = \sum y^2 - \frac{G^2}{n}$$

You will note, however, that there is no source of variability designated as error since there are no more degrees of freedom. This is true for any $k$-way classification with $r = 1$ observation per cell.

*F* tests

To construct $F$ tests for the sources of variability listed in the AOV table, we must do one of two things:

1. Assume one or more of the higher-order interactions are negligible. The affected sums of squares are then combined to form an error sum of squares. (Note that this is why we illustrated the $2 \times 2$ factorial [with $r = 1$] using an additive model.)
2. Replicate the experiment to generate additional degrees of freedom for error.

### EXAMPLE 15.5

An experiment was conducted to determine the effect of 4 different pesticides on the yield of fruit from 3 different varieties ($B_1$, $B_2$, $B_3$) of a citrus tree. Four trees from each variety were randomly selected from an orchard. The 4 pesticides were then randomly assigned to trees of a particular variety and applications were made according to recommended levels. Yields of fruit, in bushels per tree, were obtained after the test period. These data appear in table 15.16. Set up an analysis of variance

**Table 15.16** Data for example 15.5; yield of fruit (in bushels per.tree)

| | *Pesticide, A* | | | | |
|---|---|---|---|---|---|
| *Variety, B* | *1* | *2* | *3* | *4* | *Totals* |
| 1 | 29 | 50 | 43 | 53 | 175 |
| 2 | 41 | 58 | 42 | 73 | 214 |
| 3 | 66 | 85 | 69 | 85 | 305 |
| *Totals* | 136 | 193 | 154 | 211 | 694 |

table, computing all the sums of squares and the mean squares. Use $\alpha = .05$ for all $F$ tests.

### SOLUTION

The experiment just described is a $3 \times 4$ factorial with one observation per cell. Factor $A$, pesticides, is investigated at $a = 4$ levels and factor $B$, varieties, at $b = 3$ levels. The sources of variability and the corresponding degrees of freedom are shown below.

| *Source* | *df* |
|---|---|
| pesticide, *A* | 3 |
| variety, *B* | 2 |
| *AB* | 6 |
| *Total* | 11 |

Using the AOV formulas, we have, from table 15.16,

$$n = ab = 12 \qquad n_A = 3 \qquad n_B = 4$$

$$\frac{G^2}{n} = \frac{(694)^2}{12} = 40{,}136.33$$

$$\text{SSA} = \sum_i \frac{A_i^2}{n_A} - \frac{G^2}{n}$$

$$= \frac{(136)^2 + (193)^2 + (154)^2 + (211)^2}{3} - 40{,}136.33$$

$$= 41{,}327.33 - 40{,}136.33 = 1191$$

$$\text{SSB} = \sum_j \frac{B_j^2}{n_B} - \frac{G^2}{n} = \frac{(175)^2 + (214)^2 + (305)^2}{4} - 40{,}136.33$$

$$= 42{,}361.5 - 40{,}136.33 = 2225.17$$

$$\text{TSS} = \sum y^2 - \frac{G^2}{n} = (29)^2 + (50)^2 + \cdots + (85)^2 - 40{,}136.33$$

$$= 43{,}704 - 40{,}136.33 = 3567.67$$

The formula for the $AB$ interaction is

$$\text{SSAB} = \sum_i \sum_j \frac{(AB)_{ij}^2}{n_{AB}} - \text{SSA} - \text{SSB} - \frac{G^2}{n}$$

For $n_{AB} = 1$ observation at each $AB$ combination, we have

$$\sum_i \sum_j \frac{(AB)_{ij}^2}{n_{AB}} = (29)^2 + (50)^2 + \cdots + (85)^2 = 43{,}704$$

Then

$$\text{SSAB} = 43{,}704 - 1191 - 2225.17 - 40{,}136.33 = 151.50$$

This same result could be obtained by subtraction as

$$\text{SSAB} = \text{TSS} - \text{SSA} - \text{SSB} = 3567.67 - 1191 - 2225.17 = 151.50$$

The analysis of variance table is then as shown in table 15.17. As can be

**Table 15.17** AOV table for the fruit yield experiment of example 15.5

| *Source* | *SS* | *df* | *MS* |
|---|---|---|---|
| pesticide, $A$ | 1191.00 | 3 | 397 |
| varieties, $B$ | 2225.17 | 2 | 1112.59 |
| $AB$ | 151.50 | 6 | 25.25 |
| *Totals* | 3567.67 | 11 | |

seen from this AOV table, we have no degrees of freedom remaining for error.

If we are willing to assume that the $AB$ interaction is negligible, we can designate the $AB$ interaction source as error and use SSAB as SSE. For our data, if we are willing to assume that the difference in mean yield for the two pesticides remains constant for all varieties, the AOV table would be as shown in table 15.18.

**Table 15.18** AOV table for the fruit yield experiment of example 15.5, with the $AB$ interaction designated as error

| *Source* | *SS* | *df* | *MS* | *F* |
|---|---|---|---|---|
| $A$ | 1191.00 | 3 | 397.00 | $397/25.25 = 15.72$ |
| $B$ | 2225.17 | 2 | 1112.59 | $1112.59/25.25 = 44.06$ |
| error | 151.50 | 6 | 25.25 | |
| *Totals* | 3567.67 | 11 | | |

To test for no difference among pesticides, with $\alpha = .05$, we use the test statistic

$$F = \frac{\text{MSA}}{\text{MSE}} = \frac{397.00}{25.25} = 15.72$$

Since the computed value of $F$ exceeds 4.76, the tabulated value for $a = .05$, $df_1 = 3$, and $df_2 = 6$, we reject $H_0$ and conclude that there are differences among the mean yields for the 4 pesticides.

Similarly, to test for no difference among varieties, we use

$$F = \frac{\text{MSB}}{\text{MSE}} = 44.06$$

Based on $\alpha = .05$, the computed value of $F$ exceeds the tabulated value of 5.14 for $a = .05$, $df_1 = 2$, and $df_2 = 6$, and we conclude that there are differences among the mean yields for varieties.

### EXAMPLE 15.6

Refer to example 15.5. Suppose that prior to beginning the study, the experimenter realized that he would not be willing to assume that the $AB$ interaction was negligible in order to perform $F$ tests in an analysis of variance. Thus he decided to use 24 citrus trees (8 of each variety) and randomly assigned 2 trees to each factor-level combination. Use the yield data of table 15.19 to conduct an analysis of variance. Use $\alpha = .05$ for all $F$ tests.

### SOLUTION

Before constructing the AOV table, we must compute the sums of squares for the main effects of factors $A$ and $B$, the $AB$ interaction, the error, and the total. We use the data of table 15.19.

$$\frac{G^2}{n} = \frac{(1475)^2}{24} = 90{,}651.04$$

**Table 15.19** Data for the 3 × 4 factorial experiment of fruit yield, example 15.6, with $r = 2$ observations per cell

| | Pesticide, A | | | | |
|---|---|---|---|---|---|
| *Variety, B* | *1* | *2* | *3* | *4* | *Totals* |
| 1 | 49<br>39 | 50<br>55 | 43<br>38 | 53<br>48 | 375 |
| 2 | 55<br>41 | 67<br>58 | 53<br>42 | 85<br>73 | 474 |
| 3 | 66<br>68 | 85<br>92 | 69<br>62 | 85<br>99 | 626 |
| *Totals* | 318 | 407 | 307 | 443 | 1475 |

$$\text{SSA} = \sum_i \frac{A_i^2}{n_A} - \frac{G^2}{n} = \frac{(318)^2 + (407)^2 + (307)^2 + (443)^2}{6} - 90{,}651.04$$

$$= 92{,}878.5 - 90{,}651.04 = 2227.46$$

$$\text{SSB} = \sum_j \frac{B_j^2}{n_B} - \frac{G^2}{n} = \frac{(375)^2 + (474)^2 + (626)^2}{8} - 90{,}651.04$$

$$= 94{,}647.13 - 90{,}651.04 = 3996.09$$

To compute the $AB$ interaction, we must calculate the totals $(AB)_{ij}$ for all cells of the table. These are given in table 15.20.

**Table 15.20** $(AB)_{ij}$ totals for the data of table 15.19

| | Pesticide, A | | | | |
|---|---|---|---|---|---|
| *Variety, B* | *1* | *2* | *3* | *4* | *Totals* |
| 1 | 88 | 105 | 81 | 101 | 375 |
| 2 | 96 | 125 | 95 | 158 | 474 |
| 3 | 134 | 177 | 131 | 184 | 626 |
| *Totals* | 318 | 407 | 307 | 443 | 1475 |

Using the data of table 15.20, we compute SSAB.

$$\text{SSAB} = \sum_{i,j} \frac{(AB)_{ij}^2}{n_{AB}} - \text{SSA} - \text{SSB} - \frac{G^2}{n}$$

$$= \frac{(88)^2 + (105)^2 + \cdots + (184)^2}{2} - 2227.46 - 3996.09 - 90{,}651.04$$

$$= 97{,}331.5 - 2227.46 - 3996.09 - 90{,}651.04 = 456.91$$

From table 15.19, the total sum of squares is computed as

$$\text{TSS} = \sum y^2 - \frac{G^2}{n} = (49)^2 + (39)^2 + (50)^2 + \cdots + (85)^2 + (99)^2 - 90{,}651.04$$

$$= 97{,}839 - 90{,}651.04 = 7187.96$$

The sum of squares and degrees of freedom for error can be obtained by subtraction:

$$\text{SSE} = \text{TSS} - \text{SSA} - \text{SSB} - \text{SSAB}$$

$$= 7187.96 - 2227.46 - 3996.09 - 456.91 = 507.50$$

$$\text{df} = 23 - 3 - 2 - 6 = 12$$

The analysis of variance table for this $3 \times 4$ factorial experiment with 2 observations per cell is given in table 15.21.

**Table 15.21** AOV table for the fruit yield experiment of example 15.6

| *Source* | *SS* | *df* | *MS* | *F* |
|---|---|---|---|---|
| *A* | 2227.46 | 3 | 742.49 | 742.49/42.29 = 17.56 |
| *B* | 3996.09 | 2 | 1998.05 | 1998.05/42.29 = 47.25 |
| *AB* | 456.91 | 6 | 76.15 | 76.15/42.29 = 1.80 |
| error | 507.50 | 12 | 42.29 | |
| *Totals* | 7187.96 | 23 | | |

The first test of significance would be to test for no interaction between factors *A* and *B*. The *F* statistic is

$$F = \frac{\text{MSAB}}{\text{MSE}} = \frac{76.15}{42.29} = 1.80$$

The computed value of *F* does not exceed the tabulated value of 3.00 for a = .05, $\text{df}_1 = 6$, and $\text{df}_2 = 12$. Hence we have insufficient evidence to indicate an interaction between *A* and *B*. We then proceed as in the AOV table and test separately for differences among pesticides and among varieties. Comparing the computed *F*-values to the appropriate critical value from the tables, we find a significant effect due to both factors.

It should be noted that if we had concluded that there was a significant interaction between pesticides and varieties in example 15.6, it would not have been necessary to test for main effects for *A* or *B*. In such a situation it would not be important to know whether there is a significant variability among the mean response for different varieties because the mean response is affected by the level of pesticide used. Hence we recommend conducting an *F* test on main effects for a factor only if that factor has not appeared in a significant test for interaction.

# The Estimation of Treatment Differences and Multiple Comparisons 15.5

We have emphasized the analysis of variance associated with a randomized block design, a Latin square design, and $k$-way classifications. There are times, however, when we might be more interested in estimating the difference in mean response for two treatments (different levels of the same factor or different combinations of levels). For example, an environmental engineer might be more interested in estimating the difference in the mean dissolved oxygen content for a lake before and after rehabilitative work than in testing to see whether there is a difference. Thus he is asking the question, "What is the difference in mean dissolved oxygen content?" instead of the question, "Is there a difference between the mean content before and after the cleanup project?"

The same formula can be used to estimate the difference in treatment means for a randomized block design, a Latin square design, and $k$-way classifications. Let $\bar{T}_i$ denote the mean response for treatment $i$, $\bar{T}_{i'}$ denote the mean response for treatment $i'$, and $n_t$ denote the number of observations in each treatment. A $100(1 - \alpha)\%$ confidence interval on $\mu_i - \mu_{i'}$, the difference in mean response for the two treatments, is defined as shown in the box.

**100(1 − α)% Confidence Interval for the Difference in Treatment Means**

$$(\bar{T}_i - \bar{T}_{i'}) \pm t_{a/2}\, s\sqrt{\frac{2}{n_t}}$$

where $s$ is the square root of MSE in the AOV table and $t_{\alpha/2}$ can be obtained from table 2 in the appendix for $a = \alpha/2$ and the degrees of freedom for MSE.

**EXAMPLE 15.7**

A company was interested in comparing 3 different display panels for use by air traffic controllers. Each display panel was to be examined under 5 different simulated emergency conditions. Thirty highly trained air traffic controllers with similar work experience were enlisted for the study. A random assignment of controllers to display panel–emergency conditions was made, with two controllers assigned to each factor-level combination. The time (in seconds) required to stabilize the emergency situation was recorded for each controller at a panel-emergency condition. These data appear in table 15.22.

**Table 15.22** Display panel data for example 15.7 (time, in seconds)

| | *Emergency Condition, A* | | | | | |
|---|---|---|---|---|---|---|
| *Display Panel, B* | *1* | *2* | *3* | *4* | *5* | *Totals* |
| 1 | 18<br>16 | 31<br>35 | 22<br>27 | 39<br>36 | 15<br>12 | 251 |
| 2 | 13<br>15 | 33<br>30 | 24<br>21 | 35<br>38 | 10<br>16 | 235 |
| 3 | 24<br>28 | 42<br>46 | 40<br>37 | 52<br>57 | 28<br>24 | 378 |
| *Totals* | 114 | 217 | 171 | 257 | 105 | 864 |

Set up an analysis of variance table listing the sources and the degrees of freedom for this experimental design. Assume that there is no interaction between the variables "display panels" and "emergency conditions." If SSE = 150, obtain a 95% confidence interval on the difference in mean response (reaction time) for display panels 3 and 2. Note that it would not make sense to estimate the difference in mean response for panels 3 and 2 if there was an interaction between display panels and emergency conditions. Instead, we might estimate the difference in mean response for two display panel-emergency condition combinations.

## SOLUTION

The experimental design is a $3 \times 5$ factorial with 2 observations per cell. The degrees of freedom (assuming no *AB* interaction) are shown below.

| *Source* | *df* |
|---|---|
| emergency condition, *A* | 4 |
| display panel, *B* | 2 |
| error | 23 |
| *Total* | 29 |

Now assuming SSE = 150, the mean square error is MSE = 150/23 = 6.52. From table 15.22, the averages for panels 3 and 2 are

$$\bar{T}_3 = \frac{378}{10} = 37.8 \qquad \bar{T}_2 = \frac{235}{10} = 23.5$$

The $t$-value for a = .025 and df = 23 is 2.069. Hence the 95% confidence interval for the difference in mean reaction times recorded for persons using panels 3 and 2 is

$$(\bar{T}_3 - \bar{T}_2) \pm t_{a/2}\, s\sqrt{\frac{2}{n_t}}$$

$$(37.8 - 23.5) \pm 2.069(2.55)\sqrt{\frac{2}{10}}$$

$$14.3 \pm 2.36$$

or 11.94 to 16.66.

Assume now that we have chosen analysis of variance rather than estimation as the inferential technique for answering our practical questions. We proceed to compute the appropriate sums of squares and perform $F$ tests on the sources of variability. Having rejected an hypothesis of the form "no difference in mean response for levels of factor $A$," do we stop and draw no further conclusion? No, we do not, because the $F$ test is only a preliminary test to determine if there are any differences among the treatment means. We must then try to determine which levels of the factor differ from the rest.

As mentioned in chapter 14, we could perform multiple $t$ tests among the factor-level means, but the overall error rate could be quite large. Presumably we would proceed with one of the multiple comparison procedures, such as Tukey's, Duncan's, or Scheffé's, with a controlled error rate. All these procedures can be performed following an analysis of variance for a randomized block design, a Latin square design, or a $k$-way classification. The degrees of freedom for MSE are obtained from the AOV table. (For details refer to chapter 14.)

using multiple comparison procedures

**EXAMPLE 15.8**

Refer to example 15.7 and the data in table 15.22. Perform an analysis of variance that includes a test for interaction, with $\alpha = .05$. If there is not a significant interaction between display panels and emergency conditions, test for main effects. Use Tukey's $W$ procedure to locate significant differences among display panels.

**SOLUTION**

Substituting into the appropriate sums of squares formulas and using the data from table 15.22, we have

$$\frac{G^2}{n} = \frac{(864)^2}{30} = 24{,}883.2$$

$$\text{SSA} = \sum_i \frac{A_i^2}{n_A} - \frac{G^2}{n} = \frac{(114)^2 + (217)^2 + \cdots + (105)^2}{6} - 24{,}883.2$$

$$= 27{,}733.33 - 24{,}883.2 = 2850.13$$

$$\text{SSB} = \sum_j \frac{B_j^2}{n_B} - \frac{G^2}{n} = \frac{(251)^2 + (235)^2 + (378)^2}{10} - 24{,}883.2$$

$$= 26{,}111 - 24{,}883.2 = 1227.8$$

Combining the two observations in each cell, we obtain a table of $(AB)_{ij}$ totals, as shown in table 15.23.

Substituting into the formula for SSAB, we have

**Table 15.23** $(AB)_{ij}$ totals for the data of table 15.22

| | *Emergency Condition, A* | | | | | |
|---|---|---|---|---|---|---|
| *Display Panel, B* | *1* | *2* | *3* | *4* | *5* | *Totals* |
| 1 | 34 | 66 | 49 | 75 | 27 | 251 |
| 2 | 28 | 63 | 45 | 73 | 26 | 235 |
| 3 | 52 | 88 | 77 | 109 | 52 | 378 |
| *Totals* | 114 | 217 | 171 | 257 | 105 | 864 |

$$\text{SSAB} = \sum_{i,j} \frac{(AB)_{ij}^2}{n_{AB}} - \text{SSA} - \text{SSB} - \frac{G^2}{n}$$

$$= \frac{(34)^2 + (66)^2 + \cdots + (52)^2}{2} - 2850.13 - 1227.8 - 24{,}883.2$$

$$= 29{,}006 - 2850.13 - 1227.8 - 24{,}883.2 = 44.87$$

The total sum of squares can be computed using the original observations in table 15.22.

$$\text{TSS} = \sum y^2 - \frac{G^2}{n}$$

$$= (18)^2 + (16)^2 + (31)^2 + \cdots + (28)^2 + (24)^2 - 24{,}883.2$$

$$= 29{,}112 - 24{,}883.2 = 4228.8$$

By subtraction we find SSE to be

$$\text{SSE} = \text{TSS} - \text{SSA} - \text{SSB} - \text{SSAB}$$

$$= 4228.8 - 2850.13 - 1227.8 - 44.87 = 106$$

We can summarize our computations in an AOV table, table 15.24.

**Table 15.24** AOV table for the data of table 15.22

| *Source* | *SS* | *df* | *MS* | *F* |
|---|---|---|---|---|
| emergency conditions, *A* | 2850.13 | 4 | 712.53 | 712.53/7.07 = 100.78 |
| display panel, *B* | 1227.8 | 2 | 613.9 | 613.9/7.07 = 86.83 |
| *AB* | 44.87 | 8 | 5.61 | 5.61/7.07 = .79 |
| error | 106 | 15 | 7.07 | |
| *Totals* | 4228.8 | 29 | | |

The first test of significance should be a test for interaction between factors *A* and *B*. Since the computed value of *F*, .79, is less than the critical value of *F*, 2.64, for a = .05, $df_1 = 8$, and $df_2 = 15$, we have insufficient evidence to indicate an interaction between emergency conditions and display panels. That is, we are unable to contradict the statement that the difference in mean reaction time for controllers on two different display panels is constant for all 5 emergency conditions.

We then proceed to test separately for main effects due to factors $A$ and $B$. The critical $F$-values for these tests are (from table 4 in the appendix) 3.06 and 3.68, respectively. Hence both main effects are significant, and we conclude that there are differences among the mean reaction times for the 5 emergency conditions and among the mean reaction times for the 3 display panels. Practically, it would be important to follow up the AOV $F$ tests to determine which display panels and which emergency conditions are significantly different from the others. For purposes of illustration, we will examine differences among the display panels.

For Tukey's $W$ procedure we must compute (see p. 389)

$$W = q_\alpha(t, v)\sqrt{\frac{s_W^2}{n}}$$

where $s_W^2$ is MSE from the AOV table, based on $v = 15$ degrees of freedom, and $q_\alpha(t, v)$ is the upper-tail critical value of the Studentized range (with $a = \alpha$) for comparing $t$ different population means. The value of $q_\alpha(t, v)$ from table 11 in the appendix for comparing the 3 display panel means is

$$q_{.05}(3, 15) = 3.67$$

For $n = 10$ observations per mean,

$$W = q_\alpha(t, v)\sqrt{\frac{s_W^2}{n}} = 3.67\sqrt{\frac{7.07}{10}} = 3.09$$

The display panel means are, from table 15.22,

$$\bar{B}_1 = \frac{251}{10} = 25.1 \qquad \bar{B}_2 = \frac{235}{10} = 23.5 \qquad \bar{B}_3 = \frac{378}{10} = 37.8$$

We first rank the sample means from lowest to highest:

| *Display Panel* | 2 | 1 | 3 |
|---|---|---|---|
| *Means* | 23.5 | 25.1 | 37.8 |

For two means that differ (in absolute value) by more than $W = 3.09$, we declare them to be significantly different from each other. The results of our multiple comparison procedure are summarized below.

display panel $\underline{2 \quad 1}$ $\quad 3$

Thus display panels 1 and 2 both have a mean reaction time significantly lower than display panel 3, but we are unable to detect any difference between panels 1 and 2.

## Summary 15.6

In this chapter we extended our discussion of the analysis of variance to a randomized block design, a Latin square design, and a $k$-way classification. For each situation discussed, we presented an appropriate model, the notation needed for the computational formulas of the sums of squares, and an analysis of variance table that summarized our calculations. Appropriate tests of significance were also discussed.

We then considered estimation of the difference between two treatment means using a confidence interval. And, finally, we completed our examination of the analysis of variance for randomized block designs, Latin squares, and $k$-way classifications by employing the multiple comparison procedures of chapter 14 to detect specific treatment differences following a significant $F$ test.

# Exercises

15.9. Write the model and complete an analysis of variance table for a $3 \times 5$ factorial experiment.

a. How many degrees of freedom do you have for the error term when the two-way interaction is included?

b. How many degrees of freedom do you have for the error term if you assume there is no two-way interaction?

*15.10. Examine the analysis of variance for the data of example 15.6 shown in the BMDP and SAS computer outputs that follow.

a. Construct an analysis of variance table from the computer output.

b. Give the $p$-value and conclusion for the $F$ test on the variety-by-treatment interaction. Is it appropriate to proceed with tests for main effects?

c. Draw conclusions from the $p$-values listed for varieties and pesticides.

d. Compare your results from the two sets of output to those of example 15.6.

| VARIABLE NO. NAME | BEFORE TRANSFORMATION MINIMUM LIMIT | MAXIMUM LIMIT | MISSING CODE | CATEGORY CODE | CATEGORY NAME | INTERVAL RANGE GREATER THAN | LESS THAN OR EQUAL TO |
|---|---|---|---|---|---|---|---|
| 1 GROUP | | | | 1.00000 | 1.0000 | | |
| | | | | 2.00000 | 2.0000 | | |
| | | | | 3.00000 | 3.0000 | | |
| 2 LEVEL | | | | 1.00000 | 1.0000 | | |
| | | | | 2.00000 | 2.0000 | | |
| | | | | 3.00000 | 3.0000 | | |
| | | | | 4.00000 | 4.0000 | | |

NUMBER OF CASES READ . . . . . . . . . . . . . . . . . . . . . . . . . . . . 24

GROUP STRUCTURE

| GROUP | LEVEL | COUNT |
|---|---|---|
| 1.0000 | 1.0000 | 2. |
| 1.0000 | 2.0000 | 2. |
| 1.0000 | 3.0000 | 2. |
| 1.0000 | 4.0000 | 2. |
| 2.0000 | 1.0000 | 2. |
| 2.0000 | 2.0000 | 2. |
| 2.0000 | 3.0000 | 2. |
| 2.0000 | 4.0000 | 2. |
| 3.0000 | 1.0000 | 2. |
| 3.0000 | 2.0000 | 2. |
| 3.0000 | 3.0000 | 2. |
| 3.0000 | 4.0000 | 2. |

CELL MEANS FOR 1-ST DEPENDENT VARIABLE

| | | | | | | | | | | |
|---|---|---|---|---|---|---|---|---|---|---|
| GROUP = | 1.0000 | 1.0000 | 1.0000 | 1.0000 | 2.0000 | 2.0000 | 2.0000 | 2.0000 | 3.0000 | 3.0000 |
| LEVEL = | 1.0000 | 2.0000 | 3.0000 | 4.0000 | 1.0000 | 2.0000 | 3.0000 | 4.0000 | 1.0000 | 2.0000 |
| RESPONSE | 44.00000 | 52.50000 | 40.50000 | 50.50000 | 48.00000 | 62.50000 | 47.50000 | 79.00000 | 67.00000 | 88.50000 |
| COUNT | 2 | 2 | 2 | 2 | 2 | 2 | 2 | 2 | 2 | 2 |
| COUNT | 24 | | | | | | | | | |

| | | | MARGINAL |
|---|---|---|---|
| GROUP = | 3.0000 | 3.0000 | |
| LEVEL = | 3.0000 | 4.0000 | |
| RESPONSE | 65.50000 | 92.00000 | 61.45833 |
| COUNT | 2 | 2 | 24 |

STANDARD DEVIATIONS FOR 1-ST DEPENDENT VARIABLE

| | | | | | | | | | | |
|---|---|---|---|---|---|---|---|---|---|---|
| GROUP = | 1.0000 | 1.0000 | 1.0000 | 1.0000 | 2.0000 | 2.0000 | 2.0000 | 2.0000 | 3.0000 | 3.0000 |
| LEVEL = | 1.0000 | 2.0000 | 3.0000 | 4.0000 | 1.0000 | 2.0000 | 3.0000 | 4.0000 | 1.0000 | 2.0000 |
| RESPONSE | 7.07107 | 3.53553 | 3.53553 | 3.53553 | 9.89950 | 6.36396 | 7.77817 | 8.48528 | 1.41421 | 4.94975 |

| | | |
|---|---|---|
| GROUP = | 3.0000 | 3.0000 |
| LEVEL = | 3.0000 | 4.0000 |
| RESPONSE | 4.94975 | 9.89950 |

ANALYSIS OF VARIANCE FOR 1-ST DEPENDENT VARIABLE

| SOURCE | SUM OF SQUARES | DEGREES OF FREEDOM | MEAN SQUARE | F | PROB. F EXCEEDED |
|---|---|---|---|---|---|
| MEAN | 90650.93750 | 1 | 90650.93750 | 2143.45068 | 0.000 |
| G | 3996.07349 | 2 | 1998.03662 | 47.24377 | 0.000 |
| L | 2227.44238 | 3 | 742.48071 | 17.55603 | 0.000 |
| GL | 456.91479 | 6 | 76.15247 | 1.80063 | 0.182 |
| ERROR | 507.50464 | 12 | 42.29205 | | |

```
STATISTICAL ANALYSIS SYSTEM



DATA FRUIT;
INPUT Y 1-4 VARIETY 6-8 PEST 10-12;
CARDS;


  24 OBSERVATIONS IN DATA SET FRUIT     3 VARIABLES


PROC PRINT;


OBS       Y       VARIETY      PEST

  1      49             1         1
  2      39             1         1
  3      55             2         1
  4      41             2         1
  5      66             3         1
  6      68             3         1
  7      50             1         2
  8      55             1         2
  9      67             2         2
 10      58             2         2
 11      85             3         2
 12      92             3         2
 13      43             1         3
 14      38             1         3
 15      53             2         3
 16      42             2         3
 17      69             3         3
 18      62             3         3
 19      53             1         4
 20      48             1         4
 21      85             2         4
 22      73             2         4
 23      85             3         4
 24      99             3         4


PROC REGR;
CLASSES VARIETY PEST;
MODEL Y = VARIETY PEST VARIETY*PEST;
TITLE: 'EXERCISE 15.10';
```

EXERCISE 15.10

ANALYSIS OF VARIANCE TABLE, REGRESSION COEFFICIENTS, AND STATISTICS OF FIT FOR DEPENDENT VARIABLE Y

| SOURCE | DF | SUM OF SQUARES | MEAN SQUARE | F VALUE | PROB > F | R-SQUARE | C.V. |
|---|---|---|---|---|---|---|---|
| REGRESSION | 11 | 6680.45833333 | 607.31439394 | 14.36014 | 0.0001 | 0.92939581 | 10.58149% |
| ERROR | 12 | 507.50000000 | 42.29166667 | | | STD DEV | Y MEAN |
| CORRECTED TOTAL | 23 | 7187.95833333 | | | | 6.50320434 | 61.45833 |

| SOURCE | DF | SEQUENTIAL SS | F VALUE | PROB > F | PARTIAL SS | F VALUE | PROB > F |
|---|---|---|---|---|---|---|---|
| VARIETY | 2 | 3996.08333333 | 47.24433 | 0.0001 | 3996.08333333 | 47.24433 | 0.0001 |
| PEST | 3 | 2227.45833333 | 17.55632 | 0.0002 | 2227.45833333 | 17.55632 | 0.0002 |
| VARIETY*PEST | 6 | 456.91666667 | 1.80066 | 0.1815 | 456.91666667 | 1.80066 | 0.1815 |

15.11. A study was conducted to compare the yield of soybeans in a factorial experiment consisting of 4 manganese rates (from $MnSO_4$) and 4 copper rates (from $CuSO_4\ 5H_2O$). A large plot was subdivided into 16 separate subplots, to which the 16 factor-level combinations were applied. Soybeans were then planted over the entire plot in rows 3 feet apart. The sample data appear below (in kilograms/hectare).

| | *Mn* | | | |
|---|---|---|---|---|
| *Cu* | *20* | *50* | *80* | *110* |
| 1 | 1558 | 2003 | 2490 | 2830 |
| 3 | 1590 | 2020 | 2620 | 2860 |
| 5 | 1550 | 2010 | 2490 | 2750 |
| 7 | 1328 | 1760 | 2280 | 2630 |

Treating this 4 × 4 factorial as a two-way classification, write an appropriate statistical model.

15.12. Refer to exercise 15.11.

a. Perform an analysis of variance. Can you test for an interaction? Use $\alpha = .05$.

b. Assuming there is no interaction between the two variables Cu and Mn, write a model expressing the soybean yield in terms of two quantitative variables (Cu and Mn) using the methods of chapter 6. (Note: We will show the relationship between this model and the one of part a in chapter 16.)

15.13. Write the model and complete an analysis of variance table for a $4 \times 3 \times 6$ factorial experiment. (Assume the 3-way interaction is nonsignificant.)

15.14. An experiment was set up to compare the effect of different soil pH and calcium additives on the increase in trunk diameters for orange trees. Annual applications of elemental sulfur, gypsum, soda ash, and other ingredients were applied to provide pH levels of 4, 5, 6, and 7. Three levels of a calcium supplement (100, 200, and 300 pounds per acre) were also applied. All factor-level combinations of these two variables were used in the experiment. At the end of a two-year period, 3 diameters were examined at each factor-level combination. These data appear below.

| | *Calcium* | | |
|---|---|---|---|
| *pH* | *100* | *200* | *300* |
| | 5.2 | 7.4 | 6.3 |
| 4.0 | 5.9 | 7.0 | 6.7 |
| | 6.3 | 7.6 | 6.1 |
| | 7.1 | 7.4 | 7.3 |
| 5.0 | 7.4 | 7.3 | 7.5 |
| | 7.5 | 7.1 | 7.2 |
| | 7.6 | 7.6 | 7.2 |
| 6.0 | 7.2 | 7.5 | 7.3 |
| | 7.4 | 7.8 | 7.0 |
| | 7.2 | 7.4 | 6.8 |
| 7.0 | 7.5 | 7.0 | 6.6 |
| | 7.2 | 6.9 | 6.4 |

a. Write an appropriate statistical model.

b. Treat the data as a 2-way classification and perform an analysis of variance. Use $\alpha = .05$.

15.15. Refer to exercise 15.5 (p. 429). Use Fisher's LSD to determine significant differences among the 4 treatments (broadcast and band methods of fertilizer applications).

15.16. Refer to exercise 15.11. Use Tukey's $W$ procedure to determine differences among the 4 manganese rates. Use $\alpha = .05$.

15.17. Refer to exercise 15.14. Use Duncan's multiple range test (for $\alpha = .05$) to declare significant differences.

15.18. An experiment was conducted to compare the average oral body temperature for persons taking one of 9 different medications often prescribed for a specific disorder. To do this, each of 3 investigators with prior clinical experience of this nature was to obtain a random sample of patients from his or her practice who satisfy the study entrance criteria. Then the investigator was to randomly allocate the medications one to each person. Each patient in the study was given the assigned medication at 6:00 A.M. of the assigned study day. Temperatures were taken at hourly intervals beginning at 8:00 A.M. and continuing for 10 hours. During this time the patients were not allowed to do any physical activity and had to lie in bed. To eliminate the variability of temperature readings within a day, the average of the hourly determinations was the response of interest. These data are given below.

a. Write an appropriate statistical model and identify the parameters of the model.

b. Perform an analysis of variance to test for a difference in mean temperatures for medications and then for investigators. Summarize your results in an AOV table. Use $\alpha = .05$. (Hint: To simplify some of your calculations, subtract a constant, such as 97, from each of the sample measurements.)

| | Medication | | | | | | | | |
|---|---|---|---|---|---|---|---|---|---|
| *Investigator* | *A* | *B* | *C* | *D* | *E* | *F* | *G* | *H* | *I* |
| | 97.8 | 98.1 | 98.0 | 97.3 | 97.9 | 97.9 | 97.1 | 98.0 | 97.8 |
| | 97.2 | 98.1 | 97.8 | 97.3 | 97.8 | 97.9 | 97.6 | 97.8 | 98.0 |
| 1 | 97.6 | 98.0 | 98.1 | 97.5 | 97.8 | 97.8 | 97.3 | 98.0 | 97.7 |
| | 97.2 | 97.7 | 97.8 | 97.5 | 97.7 | 97.8 | 97.7 | 97.9 | 97.9 |
| | 97.6 | 97.7 | 97.9 | 97.6 | 97.8 | 97.6 | 97.5 | 98.0 | 97.8 |
| | 97.6 | 97.8 | 97.9 | 97.5 | 97.8 | 98.0 | 97.6 | 97.9 | 98.0 |
| | 97.4 | 97.7 | 98.1 | 97.4 | 97.8 | 97.7 | 97.5 | 98.0 | 97.6 |
| 2 | 97.3 | 97.6 | 97.8 | 97.5 | 97.7 | 97.8 | 97.6 | 97.9 | 98.0 |
| | 97.5 | 97.7 | 97.8 | 97.6 | 97.7 | 97.9 | 97.5 | 97.9 | 97.9 |
| | 97.5 | 97.7 | 97.6 | 97.7 | 97.8 | 97.8 | 97.3 | 97.8 | 97.9 |
| | 97.5 | 97.6 | 98.0 | 97.9 | 97.7 | 97.9 | 97.4 | 97.8 | 98.0 |
| | 97.9 | 97.7 | 97.8 | 97.8 | 97.8 | 98.0 | 97.8 | 97.8 | 98.1 |
| 3 | 97.6 | 97.9 | 98.1 | 97.8 | 97.9 | 97.7 | 97.4 | 98.0 | 97.9 |
| | 97.6 | 97.9 | 97.7 | 97.8 | 98.0 | 97.9 | 97.6 | 97.9 | 98.1 |
| | 97.7 | 97.8 | 98.7 | 97.6 | 98.1 | 97.9 | 97.6 | 97.8 | 97.9 |

*15.19. Refer to exercise 15.18. Use a computer program to run an analysis of variance. Compare your results to those of exercise 15.18, where we coded the measurements for ease of calculations.

15.20. A physician was interested in examining the relationship between work performed by an individual in an exercise tolerance test and the excess weight (as determined by standard weight-height tables) they carried. To do this, a random sample of 28 healthy adult females, ranging in age from 25 to 40, was selected from the community clinic during routine visits for physical examinations. The selection process was restricted so that 7 persons were selected from each of the following weight classifications.

normal
1-10% overweight
11-20% overweight
more than 20% overweight

As part of the physical examination, each person was required to exercise on a bicycle ergometer until the onset of fatigue. The time to fatigue (in minutes) was recorded for each person. These data are given below.

| *Classification* | *Fatigue Time* |
|---|---|
| normal | 25, 28, 19, 27, 23, 30, 35 |
| 1-10% | 24, 26, 18, 16, 14, 12, 17 |
| 11-20% | 15, 18, 17, 25, 12, 10, 23 |
| more than 20% | 10, 9, 18, 14, 6, 4, 15 |

a. Identify the experimental design and write an appropriate statistical model.
b. Perform an analysis of variance. Use $\alpha = .05$.

15.21. Refer to exercise 15.20.
a. How would you design an experiment to investigate the effects of age, sex, and excess weight on fatigue time?
b. Suppose the physician wanted to investigate the relationship among the quantitative variables "percentage overweight," "age," and "fatigue time." Write a possible model.

15.22. An experiment was conducted to compare the heat loss for 5 different designs for commercial thermal panes. To do this, a sample of 5 panes of each design was obtained. The panes were then randomly assigned to the 5 exterior temperature settings (in °F) listed below so that each design appeared in each temperature setting. The interior temperature of the test was controlled at 68° Fahrenheit. Use the sample data (p. 455) to compare the heat loss associated with the 5 pane designs.

| | Pane Design | | | | |
|---|---|---|---|---|---|
| Exterior Temperature Setting | A | B | C | D | E |
| 80 | 8.4 | 8.6 | 9.2 | 9.1 | 10.3 |
| 60 | 8.4 | 8.7 | 9.3 | 9.4 | 10.7 |
| 40 | 8.9 | 9.1 | 9.7 | 9.9 | 10.9 |
| 20 | 10.4 | 10.7 | 10.6 | 10.5 | 11.3 |
| 0 | 10.8 | 11.2 | 11.1 | 11.3 | 11.6 |

a. Identify the experimental design and write an appropriate statistical model.
b. Run an analysis of variance. Use $\alpha = .05$.

15.23. Refer to exercise 15.22. Use Tukey's $W$ procedure to compare the treatment means. Set $\alpha = .05$.

15.24. Refer to exercise 15.14. Suppose that rather than running 3 observations in each cell of the $4 \times 3$ factorial experiment at the same orange grove, a separate factorial experiment with one observation per cell is run at each of 3 different orange groves, as shown below.

| Grove 1 | Grove 2 | Grove 3 |
|---|---|---|
| $4 \times 3$ factorial | $4 \times 3$ factorial | $4 \times 3$ factorial |

Assume the first, second, and third observations at each factor-level combination of exercise 15.14 represent observations from groves 1, 2, and 3, respectively.
a. Identify the experimental design.
b. Write an appropriate linear statistical model, assuming there is no 3-way interaction.
c. Perform an analysis of variance and give the level of significance for each test.

*15.25. An experiment was conducted to examine the effects of different levels of reinforcement and different levels of isolation on children's ability to recall. A single analyst was to work with a random sample of 36 children selected from a relatively homogeneous group of fourth-grade students. Two levels of reinforcement (none and verbal) and 3 levels of isolation (20, 40, and 60 minutes) were to be used. Students were randomly assigned to the 6 treatment groups, with a total of 6 students being assigned to each group.

Each student was to spend a 30-minute session with the analyst. During this time the student was to memorize a specific passage, with reinforcement provided as dictated by the group to which the student was assigned. Following the 30-minute session, the student was isolated for the time specified for his or her group and then tested for recall of the memorized passage. These data appear below.

| Level of Reinforcement | Time of Isolation (minutes) | | | | | |
|---|---|---|---|---|---|---|
| | 20 | | 40 | | 60 | |
| none | 26 | 19 | 30 | 36 | 6 | 10 |
| | 23 | 18 | 25 | 28 | 11 | 14 |
| | 28 | 25 | 27 | 24 | 17 | 19 |
| verbal | 15 | 16 | 24 | 26 | 31 | 38 |
| | 24 | 22 | 29 | 27 | 29 | 34 |
| | 25 | 21 | 23 | 21 | 35 | 30 |

Use the computer output shown here to draw your conclusions.

EXAMPLE OF TWO-WAY ANOVA WITH INTERACTION

FILE NONAME

*************************************** ANALYSIS OF VARIANCE ***************************************

VAR03
VAR01
VAR02

******************************************************************************************

| SOURCE OF VARIATION | SUM OF SQUARES | DF | MEAN SQUARE | F | SIGNIF OF F |
|---|---|---|---|---|---|
| MAIN EFFECTS | 352.223 | 3 | 117.408 | 7.441 | 0.001 |
| VAR01 | 196.000 | 1 | 196.000 | 12.423 | 0.001 |
| VAR02 | 156.223 | 2 | 78.112 | 4.951 | 0.014 |
| 2-WAY INTERACTIONS | 1058.665 | 2 | 529.333 | 33.549 | 0.001 |
| VAR01 VAR02 | 1058.665 | 2 | 529.333 | 33.549 | 0.001 |
| EXPLAINED | 1410.888 | 5 | 282.177 | 17.885 | 0.001 |
| RESIDUAL | 473.330 | 30 | 15.778 | | |
| TOTAL | 1884.218 | 35 | 53.835 | | |

36 CASES WERE PROCESSED.
0 CASES (0.0 PCT) WERE MISSING.

# References

Barr, A.J.; Goodnight, J.H.; Sall, J.P.; and Helwig, J.T. 1976. *A user's guide to SAS 76*. Raleigh, N.C.: SAS Institute, Inc.

Cochran, W.G., and Cox, G.M. 1957. *Experimental design*. 2d ed. New York: Wiley.

Davies, O.L. 1956. *The design and analysis of industrial experiments*. New York: Hafner.

Dixon, W.J. 1975. *BMDP, Biomedical computer programs*. Berkeley: University of California Press.

Dixon, W.J., and Massey, F.J., Jr. 1969. *Introduction to statistical analysis*. 3d ed. New York: McGraw-Hill.

Draper, N.R., and Smith, H. 1966. *Applied regression analysis*. New York: Wiley.

Fisher, R.A. 1949. *The design of experiments*. Edinburgh: Oliver and Boyd Ltd.

Graybill, F.A. 1976. *Theory and application of the linear model*. N. Scituate, Mass.: Duxbury Press.

Harnett, D.L. 1970. *Introduction to statistical methods*. Reading, Mass.: Addison-Wesley.

Kirk, R.E. 1968. *Experimental design: Procedures for the behavioral sciences*. Belmont, Calif.: Brooks/Cole.

Mendenhall, W. 1968. *Introduction to linear models and the design and analysis of experiments*. Belmont, Calif.: Wadsworth.

Nie, N.H.; Hull, C.H.; Jenkins, J.G.; Steinbrenner, K.; and Bent, D.H. 1975. *Statistical package for the social sciences*. 2d ed. New York: McGraw-Hill.

Ostle, B. 1963. *Statistics in research*. 2d ed. Ames, Iowa: Iowa State University Press.

Pearson, E.S., and Hartley, H.O. 1966. *Biometrika tables for statisticians*. 3d ed. Vol. I. London: Cambridge University Press.

Service, J. 1972. *A user's guide to the statistical analysis system*. Raleigh, N.C.: Student Supply Stores, North Carolina State University.

Snedecor, G.W., and Cochran, W.G. 1967. *Statistical methods*. 6th ed. Ames, Iowa: Iowa State University Press.

Sokal, R.R., and Rohlf, F.J. 1969. *Biometry*. San Francisco: W.H. Freeman.

Steel, R.G.D., and Torrie, J.H. 1960. *Principles and procedures of statistics*. New York: McGraw-Hill.

Winer, B.J. 1962. *Statistical principles in experimental design*. New York: McGraw-Hill.

# 16 Models Relating a Response to a Set of Independent Variables

## 16.1 Introduction

In chapters 13 and 15 we discussed the analysis of variance for a completely randomized design, a randomized block design, a Latin square design, and a $k$-way classification. An important part of the discussion of an analysis of variance for a particular design was the presentation of an appropriate model. By understanding the terms included in the model, we were able to formulate statistical tests in terms of parameters in the model.

alternative model for qualitative variables

In this chapter we will consider an alternative model for relating a response $y$ to a set of *qualitative independent variables,* which will be consistent with our discussion of the general linear model for quantitative variables of chapter 6.

While it may seem confusing at first to consider a second procedure for relating a response to a set of qualitative variables (such as in a completely randomized design, randomized block design, or Latin square design), we are doing this for several important reasons. First, many of the standard methods textbooks use the models as discussed in chapters 13 and 15. Hence it is important to be familiar with this approach. Second, while these models are easily understood, they appear to be quite different from the standard models relating a response to a set of quantitative variables (the general linear model of chapter 6). The method presented in this chapter for writing models for qualitative independent variables will be very similar to the models of chapter 6. We will see that all models can be written using a common format. Finally, the procedure to be discussed in this chapter can be used to incorporate both qualitative and quantitative variables in a single model and is readily adaptable to a computer solution.

The emphasis in this chapter will be on an understanding of the new procedure for writing models. We will stress the similarity between our work here and the material of chapter 6. We will also give a relationship between the method of least squares and the analysis of variance. In chapter 17 we will pursue the general problem of relating parameters of one model to those of another.

# Review of Models for Quantitative Independent Variables 16.2

In section 6.3 we considered a general linear model for relating a response $y$ to a set of $k$ quantitative independent variables ($k \geq 1$). This model was of the form

$$y = \beta_0 + \beta_1 x_1 + \beta_2 x_2 + \cdots + \beta_k x_k + \epsilon$$

where $x_1, x_2, \ldots, x_k$ are quantitative independent variables measured without error; $\beta_0, \beta_1, \beta_2, \ldots, \beta_k$ are unknown parameters; and $\epsilon$ is a random error term with an expected value of zero. All polynomial models in a set of quantitative variables can then be written as a special case of the general linear model.

Each term of the general linear model can be classified by its exponent. Hence we defined a term of the form $x_i^p$ to be a $p$th degree term and a term of the form $x_i^p x_j^q$ to be a $(p + q)$th degree term.

In the next section we will use the format of the general linear model to relate a response to a set of qualitative independent variables.

# Models for a Set of Qualitative Independent Variables 16.3

This new approach to formulating a model for qualitative independent variables can be introduced best by way of an example. Suppose we wish to compare the average number of lightning discharges per minute for a storm, as

example

measured from two different tracking posts located 30 miles apart. If we let $y$ denote the number of discharges recorded on an oscilloscope during a one-minute period, we could write the following two models:

for tracking post 1: $y = \mu_1 + \epsilon$

for tracking post 2: $y = \mu_2 + \epsilon$

Thus we assume that observations at tracking post 1 randomly "bob" about a population mean $\mu_1$. Similarly, at tracking post 2, observations differ from a population mean $\mu_2$ by a random amount $\epsilon$. These two models are not new and could have been used to describe observations when comparing two population means in chapter 5. What is new is that we can combine these two models into a single model of the form

$$y = \beta_0 + \beta_1 x_1 + \epsilon$$

dummy variable

where $\beta_0$ and $\beta_1$ are unknown parameters, $\epsilon$ is a random error term, and $x_1$ is a *dummy variable* with the following interpretation. We let

$x_1 = 1$ if an observation is obtained from tracking post 2

$x_1 = 0$ if an observation is obtained from tracking post 1

For observations obtained from tracking post 1, we substitute $x_1 = 0$ into our model to obtain

$$y = \beta_0 + \beta_1(0) + \epsilon = \beta_0 + \epsilon$$

Hence $\beta_0 = \mu_1$, the population mean for observations from tracking post 1. Similarly, by substituting $x_1 = 1$ into our model, the equation for observations from tracking post 2 is

$$y = \beta_0 + \beta_1(1) + \epsilon = \beta_0 + \beta_1 + \epsilon$$

Since $\beta_0 = \mu_1$ and $\beta_0 + \beta_1$ must equal $\mu_2$, we have $\beta_1 = \mu_2 - \mu_1$, the difference in means between observations from tracking posts 2 and 1.

completely randomized design

Dummy variables can be used to extend our results to a completely randomized design comparing $t$ treatment means. For example, in comparing $t = 4$ treatments, we could write the single model

$$y = \beta_0 + \beta_1 x_1 + \beta_2 x_2 + \beta_3 x_3 + \epsilon$$

where

$x_1 = 1$ if treatment 2 $\quad x_1 = 0$ otherwise

$x_2 = 1$ if treatment 3 $\quad x_2 = 0$ otherwise

$x_3 = 1$ if treatment 4 $\quad x_3 = 0$ otherwise

To interpret the $\beta$'s in this equation, it is convenient to construct a table of expected values. Since $\epsilon$ has expectation zero, the expected value of $y$ is

$$E(y) = \beta_0 + \beta_1 x_1 + \beta_2 x_2 + \beta_3 x_3$$

Observations on treatment 1 (with $x_1 = x_2 = x_3 = 0$) would have an expected value equal to $\beta_0$. Similarly, for the other treatments, we substitute values for $x_1$, $x_2$, and $x_3$ to obtain the results given in table 16.1.

**Table 16.1** Expected values for the completely randomized design, with $t = 4$

| Treatment | | | |
|---|---|---|---|
| 1 | 2 | 3 | 4 |
| $E(y) = \beta_0$ | $E(y) = \beta_0 + \beta_1$ | $E(y) = \beta_0 + \beta_2$ | $E(y) = \beta_0 + \beta_3$ |

If we identify the mean of treatment 1 as $\mu_1$, the mean of treatment 2 as $\mu_2$, and so on, then from table 16.1 it is clear that

$\beta_0 = \mu_1 \qquad \beta_2 = \mu_3 - \mu_1$

$\beta_1 = \mu_2 - \mu_1 \qquad \beta_3 = \mu_4 - \mu_1$

Any comparison among the treatment means can be phrased in terms of the $\beta$'s. For example, the comparison $\mu_4 - \mu_3$ could be written as $\beta_3 - \beta_2$. Likewise, $\mu_3 - \mu_2$ could be written as $\beta_2 - \beta_1$.

**EXAMPLE 16.1**

Use dummy variables to write a model for a completely randomized design with $t$ treatments. Identify the $\beta$'s.

**SOLUTION**

The model could be written in the form

$$y = \beta_0 + \beta_1 x_1 + \beta_2 x_2 + \cdots + \beta_{t-1} x_{t-1} + \epsilon$$

where

$x_1 = 1$ if treatment 2 $\quad x_1 = 0$ otherwise

$x_2 = 1$ if treatment 3 $\quad x_2 = 0$ otherwise

$\vdots$

$x_{t-1} = 1$ if treatment $t$ $\quad x_{t-1} = 0$ otherwise

The table of expected values would be

| *Treatment* | | | |
|---|---|---|---|
| 1 | 2 | ... | $t$ |
| $E(y) = \beta_0$ | $E(y) = \beta_0 + \beta_1$ | ... | $E(y) = \beta_0 + \beta_{t-1}$ |

from which we obtain

$$\beta_0 = \mu_1$$

$$\beta_1 = \mu_2 - \mu_1$$

$$\vdots$$

$$\beta_{t-1} = \mu_t - \mu_1$$

The procedure just described for a completely randomized design can be applied to any experimental situation where a response $y$ is related to one or more qualitative independent variables. In the completely randomized design, we have a response related to the variable "treatments," and for $t$ levels of the treatments, we enter $(t - 1)$ $\beta$'s into our model, using dummy variables.

randomized block design

For a randomized block design, we have a response related to two qualitative independent variables, blocks and treatments. Applying our previous results, if there are $b$ blocks and $t$ treatments, we will enter $(b - 1)$ $\beta$'s for blocks and $(t - 1)$ $\beta$'s for treatments into our model.

**EXAMPLE 16.2**

Write a model for a randomized block design to compare $t = 4$ treatments in $b = 3$ blocks. The design is shown in table 16.2.

| *Blocks* | *Treatments* | | | |
|---|---|---|---|---|
| 1 | I | III | II | IV |
| 2 | II | IV | III | I |
| 3 | IV | I | II | III |

**Table 16.2** Design for $b = 3$ blocks and $t = 4$ treatments, example 16.2

**SOLUTION**

The model can be written in the form

$$y = \beta_0 + \overbrace{\beta_1 x_1 + \beta_2 x_2}^{\text{blocks}} + \overbrace{\beta_3 x_3 + \beta_4 x_4 + \beta_5 x_5}^{\text{treatments}} + \epsilon$$

where

$x_1 = 1$ if block 2 $\quad x_1 = 0$ otherwise

$x_2 = 1$ if block 3 $\quad x_2 = 0$ otherwise

$x_3 = 1$ if treatment II $\quad x_3 = 0$ otherwise

$x_4 = 1$ if treatment III $\quad x_4 = 0$ otherwise

$x_5 = 1$ if treatment IV $\quad x_5 = 0$ otherwise

We can easily interpret the $\beta$'s using a table of expected values. To obtain the expected value of an observation for a given block-treatment combination, we substitute appropriate values for $x_1, x_2, \ldots, x_5$ in the formula

$$E(y) = \beta_0 + \beta_1 x_1 + \beta_2 x_2 + \cdots + \beta_5 x_5$$

For example, the expected value for an observation in block 2 on treatment II would have $x_1 = 1$, $x_2 = 0$, $x_3 = 1$, $x_4 = 0$, and $x_5 = 0$, giving

$$E(y) = \beta_0 + \beta_1(1) + \beta_2(0) + \beta_3(1) + \beta_4(0) + \beta_5(0) = \beta_0 + \beta_1 + \beta_3$$

The table of expected values is given in table 16.3.

**Table 16.3** Table of expected values for a randomized block design, with $b = 3$ and $t = 4$; example 16.2

| | *Treatment* | | | |
|---|---|---|---|---|
| *Block* | *I* | *II* | *III* | *IV* |
| 1 | $\beta_0$ | $\beta_0 + \beta_3$ | $\beta_0 + \beta_4$ | $\beta_0 + \beta_5$ |
| 2 | $\beta_0 + \beta_1$ | $\beta_0 + \beta_1 + \beta_3$ | $\beta_0 + \beta_1 + \beta_4$ | $\beta_0 + \beta_1 + \beta_5$ |
| 3 | $\beta_0 + \beta_2$ | $\beta_0 + \beta_2 + \beta_3$ | $\beta_0 + \beta_2 + \beta_4$ | $\beta_0 + \beta_2 + \beta_5$ |

The mean response for block 1 and treatment I is $\beta_0$. If we consider any treatment and compare blocks 2 and 1, the difference in the mean response is $\beta_1$. For example, when using treatment II, the difference in the mean response for blocks 2 and 1 is, from table 16.3,

$$(\beta_0 + \beta_1 + \beta_3) - (\beta_0 + \beta_3) = \beta_1$$

Note that this is true for any treatment. Hence $\beta_1 = \mu_2 - \mu_1$, the difference in the mean response for blocks 2 and 1. Similarly, in comparing blocks 3 and 1 for a given treatment, the difference is $\beta_2 = \mu_3 - \mu_1$.

In the same way, we can compare two treatments for a given block. For example it follows immediately that $\beta_3 = \mu_{II} - \mu_I$, the difference in mean response for treatments II and I. Similarly, in block 3 the difference in mean response for treatments II and I is

$$(\beta_0 + \beta_2 + \beta_3) - (\beta_0 + \beta_2) = \beta_3$$

In the same fashion, we can show that

$\beta_4 = \mu_{III} - \mu_I$

$\beta_5 = \mu_{IV} - \mu_I$

Latin square design

These results can easily be extended to a Latin square design.

**EXAMPLE 16.3**

Nylon is spun on a series of machines. When a break in the nylon thread occurs during the spinning process, the machine operator must stop the machine and rethread the nylon prior to continuing. Investigators would like to compare the output of 3 different spinning machines (I, II, III) using 3 different operators (A, B, C). To control day-to-day variability in addition to operator variability, it was decided to use a $3 \times 3$ Latin square design, as shown in table 16.4. Write an appropriate model for this Latin square design, using dummy variables. Interpret the $\beta$'s for the model.

**Table 16.4** Latin square design for the spinning machines experiment, example 16.3

| | *Day* | | |
|---|---|---|---|
| *Operator* | *1* | *2* | *3* |
| A | I | II | III |
| B | II | III | I |
| C | III | I | II |

**SOLUTION**

The model for this Latin square design must relate a response to three qualitative independent variables, rows (operators), columns (days), and treatments (machines). For each qualitative variable, we will include two $\beta$'s.

$$y = \beta_0 + \overbrace{\beta_1 x_1 + \beta_2 x_2}^{\text{operators}} + \overbrace{\beta_3 x_3 + \beta_4 x_4}^{\text{days}} + \overbrace{\beta_5 x_5 + \beta_6 x_6}^{\text{machines}} + \epsilon$$

where

$x_1 = 1$ if Operator B $\quad x_1 = 0$ otherwise

$x_2 = 1$ if Operator C $\quad x_2 = 0$ otherwise

$x_3 = 1$ if Day 2 $\quad x_3 = 0$ otherwise

$x_4 = 1$ if Day 3 $\quad x_4 = 0$ otherwise

$x_5 = 1$ if Machine II $\quad x_5 = 0$ otherwise

$x_6 = 1$ if Machine III $\quad x_6 = 0$ otherwise

For this model, with $E(\epsilon) = 0$, we can write the expected value of $y$ as

$$E(y) = \beta_0 + \beta_1 x_1 + \beta_2 x_2 + \beta_3 x_3 + \beta_4 x_4 + \beta_5 x_5 + \beta_6 x_6$$

By substituting appropriate values for the dummy variables $x_1, x_2, \ldots, x_6$, we obtain the table of expected values, table 16.5.

**Table 16.5** Expected values for the 3 × 3 Latin square design of the spinning machines experiment, example 16.3

| | Day | | |
|---|---|---|---|
| *Operator* | *1* | *2* | *3* |
| A | I $\beta_0$ | II $\beta_0 + \beta_3 + \beta_5$ | III $\beta_0 + \beta_4 + \beta_6$ |
| B | II $\beta_0 + \beta_1 + \beta_5$ | III $\beta_0 + \beta_1 + \beta_3 + \beta_6$ | I $\beta_0 + \beta_1 + \beta_4$ |
| C | III $\beta_0 + \beta_2 + \beta_6$ | I $\beta_0 + \beta_2 + \beta_3$ | II $\beta_0 + \beta_2 + \beta_4 + \beta_5$ |

Although it is readily apparent that $\beta_0$ represents the mean response for Machine I run by Operator A during Day 1, it is more difficult to identify the other $\beta$'s. Since each machine appears in each row (operator) and each column (day), the $\beta$'s due to rows and columns will be eliminated when comparing two machines. If we average the three expected values for observations on Machine I and the three expected values from table 16.5 for Machine II, we have

$$\text{average for Machine I} = \tfrac{1}{3}[\beta_0 + (\beta_0 + \beta_2 + \beta_3) + (\beta_0 + \beta_1 + \beta_4)]$$
$$= \beta_0 + \tfrac{1}{3}(\beta_1 + \beta_2 + \beta_3 + \beta_4)$$

$$\text{average for Machine II} = \tfrac{1}{3}[(\beta_0 + \beta_1 + \beta_5) + (\beta_0 + \beta_3 + \beta_5) + (\beta_0 + \beta_2 + \beta_4 + \beta_5)]$$
$$= \beta_0 + \beta_5 + \tfrac{1}{3}(\beta_1 + \beta_2 + \beta_3 + \beta_4)$$

The difference in these averages represents $\mu_{II} - \mu_I$. Subtracting, we have

$$\beta_5 = \mu_{II} - \mu_I$$

Similarly, by comparing Machine III to Machine I, we can show that

$$\beta_6 = \mu_{III} - \mu_I$$

The same reasoning can be used to obtain

$$\beta_1 = \mu_B - \mu_A$$

$$\beta_2 = \mu_C - \mu_A$$

$$\beta_3 = \mu_2 - \mu_1$$

$$\beta_4 = \mu_3 - \mu_1$$

Now that the $\beta$'s have been identified, these interpretations are exactly what we might have imagined, having examined the completely randomized design and a randomized block design.

The models that we have written in this section have had only *main effects terms*—terms involving only one $x$—for each of the qualitative independent variables. But it is also possible to write models containing *interaction terms*—terms involving the product of $x$'s between two or more variables.

main effects terms

interaction terms

Consider a 2 × 3 factorial experiment in which an experimenter would like to compare 3 different diet preparations (A, B, C) under 2 different diet plans.

factorial experiment

A fixed number of overweight persons would be assigned to each of the 6 factor-level combinations. Since it is possible that the difference in mean response (weight loss) for 2 different diet preparations is not the same for each diet plan, we must include interaction terms as well as main effects in our model. An appropriate model is given by

model

$$y = \beta_0 + \overbrace{\beta_1 x_1 + \beta_2 x_2 + \beta_3 x_3}^{\text{main effects}} + \overbrace{\beta_4 x_1 x_2 + \beta_5 x_1 x_3}^{\text{interaction}} + \epsilon$$

where

$x_1 = 1$ if Plan 2 $\quad x_1 = 0$ otherwise

$x_2 = 1$ if Diet B $\quad x_2 = 0$ otherwise

$x_3 = 1$ if Diet C $\quad x_3 = 0$ otherwise

The main effects terms in our model are

$\beta_1 x_1$ for diet plans

$\beta_2 x_2$ and $\beta_3 x_3$ for diet preparations

Interaction terms are formed from cross products of the $x$'s involved in main effects. Thus from the product of $x_1$ and $x_2$, we have the interaction term $\beta_4 x_1 x_2$. Similarly, we have $\beta_5 x_1 x_3$ from the product of $x_1$ and $x_3$. It might also appear that we should have an interaction term involving the product of $x_2$ with $x_3$. However, *interaction terms are always formed by products of x-values from different variables*. Intuitively this makes sense, since an interaction measures how two (or more) variables react when combined. Hence no term involving levels of the same variable could contribute towards measuring this effect.

with interaction

To interpret the $\beta$'s associated with our model, we again form a table of expected values by substituting appropriate combinations for $x_1$, $x_2$, and $x_3$ into the general formula

$$E(y) = \beta_0 + \beta_1 x_1 + \beta_2 x_2 + \beta_3 x_3 + \beta_4 x_1 x_2 + \beta_5 x_1 x_3$$

For example, the expected response on Plan 2 and Diet B can be found by substituting $x_1 = 1$, $x_2 = 1$, and $x_3 = 0$ into $E(y)$.

$$\begin{aligned} E(y) &= \beta_0 + \beta_1(1) + \beta_2(1) + \beta_3(0) + \beta_4(1)(1) + \beta_5(1)(0) \\ &= \beta_0 + \beta_1 + \beta_2 + \beta_4 \end{aligned}$$

The table of expected values is given in table 16.6.

**Table 16.6** Expected values for a 2 × 3 factorial experiment

| | *Diet Preparation* | | |
|---|---|---|---|
| *Diet Plan* | *A* | *B* | *C* |
| 1 | $\beta_0$ | $\beta_0 + \beta_2$ | $\beta_0 + \beta_3$ |
| 2 | $\beta_0 + \beta_1$ | $\beta_0 + \beta_1 + \beta_2 + \beta_4$ | $\beta_0 + \beta_1 + \beta_3 + \beta_5$ |

From table 16.6 we have the following interpretations for main effects:

$\beta_0$: mean response for Plan 1, Diet A

$\beta_1$: difference in mean response for plans 2 and 1 on Diet A

$\beta_2$: difference in mean response for diets B and A on Plan 1

$\beta_3$: difference in mean response for diets C and A on Plan 1

The interaction $\beta$'s are slightly more complicated to interpret since they measure the failure of diet preparations to have the same effect across different diet plans. Using the first two columns of table 16.6, we can find $\beta_4$ as the sum of the expected values in the diagonal left-to-right direction minus the sum of the expected values in the diagonal right-to-left direction.

| | *A* | *B* |
|---|---|---|
| 1 | $\beta_0$ | $\beta_0 + \beta_2$ |
| 2 | $\beta_0 + \beta_1$ | $\beta_0 + \beta_1 + \beta_2 + \beta_4$ |

$$[\beta_0 + (\beta_0 + \beta_1 + \beta_2 + \beta_4)] - [(\beta_0 + \beta_2) + (\beta_0 + \beta_1)] = \beta_4$$

Similarly, taking the first and third columns of table 16.6 and subtracting the right-to-left sum from the left-to-right sum, we have

$$[\beta_0 + (\beta_0 + \beta_1 + \beta_3 + \beta_5)] - [(\beta_0 + \beta_3) + (\beta_0 + \beta_1)] = \beta_5$$

no interaction

We should note that if there were no interaction between the variables "diet plan" and "diet preparation," the parameters $\beta_4$ and $\beta_5$ would both equal zero. Using table 16.6, with $\beta_4 = \beta_5 = 0$, we have the following interpretations for the parameters in the reduced model:

$$y = \beta_0 + \beta_1 x_1 + \beta_2 x_2 + \beta_3 x_3 + \epsilon$$

$\beta_0$: mean response for Diet A, Plan 1

$\beta_1 = \mu_2 - \mu_1$: the difference in mean response for plans 2 and 1

$\beta_2 = \mu_B - \mu_A$: the difference in mean response for diets B and A

$\beta_3 = \mu_C - \mu_A$: the difference in mean response for diets C and A

Note that in the absence of interaction, the interpretation of a main effect term for one variable does not depend on the level of another variable. For example, with interaction present,

$\beta_1 = \mu_2 - \mu_1$ on Diet A (i.e., the difference in mean response for plans 2 and 1 while on Diet A)

without interaction,

$\beta_1 = \mu_2 - \mu_1$ (i.e., the difference in mean response for plans 2 and 1 for any diet preparation)

**EXAMPLE 16.4**

Refer to the previous experiment, which involved the investigation of diet plans and diet preparations. Write a complete model (indicate main effects and interactions) for an experiment to compare 4 diet preparations using 3 diet plans.

**SOLUTION**

The model must relate a response to two qualitative independent variables, diet plans and diet preparations. The variable "diet plan" will have $3 - 1 = 2$ main effects terms; the variable "diet preparation" will have $4 - 1 = 3$ main effects terms, and there will be $2(3) = 6$ interaction terms formed from products of $x$-values, one from each of the two variables. The model can be written as follows:

$$y = \beta_0 + \overbrace{\beta_1 x_1 + \beta_2 x_2}^{\text{diet plans}} + \overbrace{\beta_3 x_3 + \beta_4 x_4 + \beta_5 x_5}^{\text{diet preparations}}$$

$$+ \overbrace{\beta_6 x_1 x_3 + \beta_7 x_1 x_4 + \beta_8 x_1 x_5 + \beta_9 x_2 x_3 + \beta_{10} x_2 x_4 + \beta_{11} x_2 x_5}^{\text{interaction terms}} + \epsilon$$

where

$x_1 = 1$ if Plan 2 $\quad x_1 = 0$ otherwise

$x_2 = 1$ if Plan 3 $\quad x_2 = 0$ otherwise

$x_3 = 1$ if Diet B $\quad x_3 = 0$ otherwise

$x_4 = 1$ if Diet C $\quad x_4 = 0$ otherwise

$x_5 = 1$ if Diet D $\quad x_5 = 0$ otherwise

# Exercises

16.1. Using dummy variables, write a statistical model for a randomized block design with $b = 3$ blocks and $t = 5$ treatments. Identify the parameters in the model.

16.2. Using dummy variables, write a statistical model for a $3 \times 3$ Latin square design. Identify the parameters in your model.

16.3. An experiment is conducted to compare the effect of 3 different fertilizers (A, B, C) dispensed at 3 different rates (30, 50, and 80 pounds per plot) on the yield of lima beans. Write an appropriate statistical model. Interpret the parameters of the general linear model in terms of three separate models corresponding to the 3 fertilizers.

16.4. In exercise 15.1 (p. 418) we considered an experiment to investigate the propellant thrust of 4 different mixtures of the components oxidizer, binder, and fuel. Five different samples of propellant were prepared from each mixture, with one sample of each mixture randomly assigned to each of 5 investigators. Using dummy variables, write a general linear model for this experimental design. Treat the different mixtures as levels of a qualitative variable. Identify the parameters in the model.

16.5. Refer to example 15.3 (p. 426). Write a general linear model for the experimental design indicated. Interpret the parameters in the model.

# 16.4 Testing a Set of $\beta$'s in the General Linear Model

The model we formulated in section 16.3 related a response $y$ to a set of qualitative independent variables through the use of dummy variables. The format we used was chosen to conform to the general linear model discussed in chapter 6. In chapter 6 we were concerned with relating a response to a set of quantitative variables, and the general linear model took the form

general linear model

$$y = \beta_0 + \beta_1 x_1 + \beta_2 x_2 + \cdots + \beta_k x_k + \epsilon$$

Recall that the independent variables $x_1, x_2, \ldots, x_k$ could be chosen to make the model a polynomial of a specified degree.

In this chapter we have extended the general linear model to relate a response to a set of $k$ independent variables (qualitative or quantitative or both). Hence for qualitative variables, the $x$'s in the general linear model would represent dummy variables or the product of dummy variables associated with the qualitative variables. The $x$'s associated with quantitative independent variables would have a quantitative interpretation (such as initial weight, amount of water added, stirring time, etc.). For example, the model in section 16.3 that related weight loss to 3 diet preparations and 2 diet plans was

$$y = \beta_0 + \beta_1 x_1 + \beta_2 x_2 + \beta_3 x_3 + \beta_4 x_1 x_2 + \beta_5 x_1 x_3 + \epsilon$$

which is equivalent to the general linear model with $k = 5$, where

$$x_1 = x_1 \qquad x_4 = x_1 x_2$$

$$x_2 = x_2 \qquad x_5 = x_1 x_3$$

$$x_3 = x_3$$

The ability to write both qualitative and quantitative independent variables in the framework of a general linear model will enable us to develop one procedure for simultaneously testing the significance of one or more $\beta$'s of the model. For example, a model might be proposed for relating the sales volume $y$ to advertising expenditure $x_1$, with the form

$$y = \beta_0 + \beta_1 x_1 + \beta_2 x_1^2 + \epsilon$$

In chapter 8 we considered the problem of testing the significance of a single $\beta$ in the general linear model. Thus a test of $H_0$: $\beta_2 = 0$ would be a test of the null hypothesis that $y$ is linearly related to $x_1$. However, an experimenter might wish to test

tests for set of $\beta$'s, each equal to zero

$$H_0\text{: } \beta_1 = \beta_2 = 0$$

which hypothesizes that $y$ is *not related* in a linear or quadratic way to the independent variable $x_1$. Similarly, for the $2 \times 3$ factorial experiment of section 16.3, we might wish to test $H_0$: $\beta_4 = \beta_5 = 0$; that is, we wish to test the hypothesis of no interaction between the variables "diet plan" and "diet preparation."

The procedure for simultaneously testing that a set of $\beta$'s is equal to zero in the general linear model will have the following null and alternative hypotheses:

$H_0$: $\beta_{g+1} = \beta_{g+2} = \cdots = \beta_k = 0 \qquad (k > g)$.
$H_a$: At least one of the $\beta$'s is nonzero.

To formulate a test statistic, we must specify two models, model 1—often referred to as the *complete model*—and model 2—referred to as the *reduced model*.

complete model
reduced model

$$\text{Model 1: } y = \beta_0 + \beta_1 x_1 + \beta_2 x_2 + \cdots + \beta_g x_g + \beta_{g+1} x_{g+1} + \cdots + \beta_k x_k + \epsilon \qquad (k > g)$$

$$\text{Model 2: } y = \beta_0 + \beta_1 x_1 + \beta_2 x_2 + \cdots + \beta_g x_g + \epsilon$$

You will note that model 1 represents the general linear model and model 2 is a general linear model under the assumption that $H_0$ is true. Thus model 1 contains $k$ independent variables, model 2 contains $g$ independent variables, and we are testing that a set of $(k - g)$ $\beta$'s is equal to zero. The $x$'s in the two models represent either quantitative independent variables or dummy variables associated with qualitative independent variables.

Using the method of least squares (see section 7.4), we fit both models separately, and for each model we calculate the sum of squares for error. [A computational formula for SSE is given in section 8.2 (p. 193).] Letting $SSE_1$ and $SSE_2$ denote the sum of squares for error for models 1 and 2, respectively, we can examine the difference $SSE_2 - SSE_1$. For $n$ sample observations, the degrees of freedom for a sum of squares for error can be computed as $n -$ (the number of parameters in the model). Hence $SSE_1$ and $SSE_2$ have, respectively, $n - (k + 1)$ and $n - (g + 1)$ degrees of freedom. When $H_0$ is true, $SSE_2$ and $SSE_1$ will be of approximately the same magnitude, although $SSE_1$ will be less than $SSE_2$ owing to the fact that the complete model has more terms in it. When $H_a$ is true and at least one of the $\beta$'s under test is different from zero, $SSE_2$ will be much larger than $SSE_1$. Since $SSE_2 - SSE_1$ will be greater than zero under either $H_0$ or $H_a$, we call the difference $SSE_2 - SSE_1$ the *drop in the sum of squares for error attributable to the variables* $x_{g+1}, x_{g+2}, \ldots, x_k$. If the sum of squares drop is large, this implies that the sum of squares for error has been greatly reduced by including the variables $x_{g+1}, x_{g+2}, \ldots, x_k$ in the model, and intuitively we would reject $H_0$.

$SSE_1 - SSE_2$

df

$SS_{\text{drop}}$

How large must $SSE_2 - SSE_1$ be in order to reject the hypothesis that $x_{g+1}, x_{g+2}, \ldots, x_k$ are unrelated to the response $y$? When $H_0$ is true, the quantity $SS_{\text{drop}} = SSE_2 - SSE_1$ divided by the number of parameters under test in $H_0$ $(k - g)$ provides an unbiased estimate of $\sigma^2$, the variance associated with an observation in the general linear model. We designate this estimate as the mean square drop:

$MS_{drop}$

$$MS_{drop} = \frac{SSE_2 - SSE_1}{k - g}$$

The mean square error for model 1,

$$MSE_1 = \frac{SSE_1}{n - (k + 1)}$$

also provides an unbiased estimate of $\sigma^2$. It can be shown that when $H_0$ is true, the ratio

*F* statistic

$$\frac{MS_{drop}}{MSE_1}$$

follows an $F$ distribution, with $k - g$ and $n - (k + 1)$ degrees of freedom, respectively.

When $H_a$ is true, $MSE_1$ is still an unbiased estimate of $\sigma^2$, while $MS_{drop}$ is an unbiased estimate of $\sigma^2$ + (a positive function of $\beta_{g+1}, \beta_{g+2}, \ldots, \beta_k$). Thus large values of $F = MS_{drop}/MSE_1$ will indicate rejection of the null hypothesis.

**EXAMPLE 16.5**

In example 7.15 (p. 165) of chapter 7 we considered a situation where a chemist was interested in determining the weight loss $y$ of a compound as a function of the length of time the compound was exposed to the air and the relative humidity of the environment during exposure. A $3 \times 4$ factorial experiment was performed using 3 relative humidities and 4 exposure times, with weight loss (in pounds) recorded for each factor-level combination. We obtained a least squares fit to a first-order model, relating the response to the two variables.

Now let us consider a model of the form

$$y = \beta_0 + \beta_1 x_1 + \beta_2 x_2 + \beta_3 x_2^2 + \beta_4 x_1 x_2 + \beta_5 x_1 x_2^2 + \epsilon$$

where

$x_1$ = coded exposure time

$x_2$ = coded relative humidity

Use the data, which are repeated here in table 16.7, to fit complete and reduced models for testing

$H_0$: $\beta_4 = \beta_5 = 0$

(i.e., that there is no interaction between the exposure time and relative humidity). Use $\alpha = .05$.

**Table 16.7** Data for the weight loss experiment of example 16.5

| *Exposure Time (in hours)* | *Relative Humidity* .20 | .30 | .40 |
|---|---|---|---|
| 4 | 4.3 | 4.0 | 2.0 |
| 5 | 5.5 | 5.2 | 4.0 |
| 6 | 6.8 | 6.6 | 5.7 |
| 7 | 8.0 | 7.5 | 6.5 |

**SOLUTION**

In chapter 7 we found it convenient computationally to define $x_1$ and $x_2$ as follows:

$$x_1 = (\text{exposure time} - \text{mean exposure time})$$
$$= (\text{exposure time} - 5.5)$$
$$x_2 = (\text{relative humidity} - \text{mean relative humidity})$$
$$= (\text{relative humidity} - .30)$$

Using the coded variables, the **Y** and **X** matrices for the complete model are

$$\mathbf{Y} = \begin{bmatrix} 4.3 \\ 5.5 \\ 6.8 \\ 8.0 \\ 4.0 \\ 5.2 \\ 6.6 \\ 7.5 \\ 2.0 \\ 4.0 \\ 5.7 \\ 6.5 \end{bmatrix} \quad \mathbf{X} = \begin{bmatrix} 1 & -1.5 & -.1 & .01 & .15 & -.015 \\ 1 & -.5 & -.1 & .01 & .05 & -.005 \\ 1 & .5 & -.1 & .01 & -.05 & .005 \\ 1 & 1.5 & -.1 & .01 & -.15 & .015 \\ 1 & -1.5 & 0 & 0 & 0 & 0 \\ 1 & -.5 & 0 & 0 & 0 & 0 \\ 1 & .5 & 0 & 0 & 0 & 0 \\ 1 & 1.5 & 0 & 0 & 0 & 0 \\ 1 & -1.5 & .1 & .01 & -.15 & -.015 \\ 1 & -.5 & .1 & .01 & -.05 & -.005 \\ 1 & .5 & .1 & .01 & .05 & .005 \\ 1 & 1.5 & .1 & .01 & .15 & .015 \end{bmatrix}$$

From the Statistical Analysis System (SAS) (see the computer program that follows), we find that

$$(\mathbf{X'X})^{-1} = \begin{bmatrix} .25 & 0 & 0 & -25 & 0 & 0 \\ 0 & .20 & 0 & 0 & 0 & -20 \\ 0 & 0 & 12.50 & 0 & 0 & 0 \\ -25.00 & 0 & 0 & 3750 & 0 & 0 \\ 0 & 0 & 0 & 0 & 10 & 0 \\ 0 & -20.00 & 0 & 0 & 0 & 3000 \end{bmatrix} \quad \hat{\boldsymbol{\beta}} = \begin{bmatrix} 5.825 \\ 1.190 \\ -8.000 \\ -47.500 \\ 1.400 \\ 19.000 \end{bmatrix}$$

The sum of squares for error for the complete model is $SSE_1 = .427$, based on df $= n - (k + 1) = 12 - 6 = 6$.

For the reduced model (model 2),

$$y = \beta_0 + \beta_1 x_1 + \beta_2 x_2 + \beta_3 x_2^2 + \epsilon$$

$$(\mathbf{X}'\mathbf{X})^{-1} = \begin{bmatrix} .25 & 0 & 0 & -25 \\ 0 & .067 & 0 & 0 \\ 0 & 0 & 12.5 & 0 \\ -25.00 & 0 & 0 & 3750 \end{bmatrix} \qquad \hat{\boldsymbol{\beta}} = \begin{bmatrix} 5.825 \\ 1.317 \\ -8.000 \\ -47.500 \end{bmatrix}$$

and $SSE_2 = .743$, based on 8 degrees of freedom.

Combining the results from the complete and reduced models, we find

$$SS_{drop} = SSE_2 - SSE_1 = .743 - .427 = .316$$

$$MS_{drop} = \frac{SS_{drop}}{k - g} = \frac{.316}{2} = .158$$

$$MSE_1 = \frac{SSE_1}{6} = \frac{.427}{6} = .071$$

For $H_0$: $\beta_4 = \beta_5 = 0$,

$$F = \frac{MS_{drop}}{MSE_1} = \frac{.158}{.071} = 2.23$$

The critical value of $F$ for $\alpha = .05$, $df_1 = 2$, and $df_2 = 6$ is 5.14. Since the observed value of $F$ does not exceed the critical value, we have insufficient evidence to reject $H_0$. Practically, we would say we are unable to detect a significant interaction between time of exposure and relative humidity.

The computer output for the solution to both the complete and reduced models follows.

```
STATISTICAL ANALYSIS SYSTEM

DATA WT;
INPUT Y 1-4  X1 6-10 X2 11-15;
X2SQ = X2**2;
X1X2 = X1*X2;
X1X2SQ = X1*X2SQ;
CARDS;


 12 OBSERVATIONS IN DATA SET WT    6 VARIABLES
PROC PRINT;
```

| OBS | Y | X1 | X2 | X2SQ | X1X2 | X1X2SQ |
|---|---|---|---|---|---|---|
| 1 | 4.3 | −1.5 | −.1 | .01 | .15 | −.015 |
| 2 | 5.5 | −0.5 | −.1 | .01 | .05 | −.005 |
| 3 | 6.8 | 0.5 | −.1 | .01 | −.05 | .005 |
| 4 | 8.0 | 1.5 | −.1 | .01 | −.15 | .015 |

| OBS | Y | X1 | X2 | X2SQ | X1X2 | X1X2SQ |
|---|---|---|---|---|---|---|
| 5 | 4.0 | −1.5 | .0 | .00 | .00 | .000 |
| 6 | 5.2 | −0.5 | .0 | .00 | .00 | .000 |
| 7 | 6.6 | 0.5 | .0 | .00 | .00 | .000 |
| 8 | 7.5 | 1.5 | .0 | .00 | .00 | .000 |
| 9 | 2.0 | −1.5 | .1 | .01 | −.15 | −.015 |
| 10 | 4.0 | −0.5 | .1 | .01 | −.05 | −.005 |
| 11 | 5.7 | 0.5 | .1 | .01 | .05 | .005 |
| 12 | 6.5 | 1.5 | .1 | .01 | .15 | .015 |

```
PROC REGR;
MODEL Y = X1 X2 X2SQ X1X2 X1X2SQ / X I P CLM;
MODEL Y = X1 X2 X2SQ / X I P CLM;
TITLE 'EXAMPLE 16.5';
```

```
************************************************************************************************

MODEL         : EXAMPLE 16.5

DEPENDENTS    : Y

INDEPENDENTS  : X1 X2 X2SQ X1X2 X1X2SQ

************************************************************************************************
```

EXAMPLE 16.5

THE X'X MATRIX

| | INTERCPT | X1 | X2 | X2SQ | X1X2 | X1X2SQ |
|---|---|---|---|---|---|---|
| INTERCPT | 12.00000000 | 0.0 | 0.0 | 0.08000000 | 0.0 | 0.0 |
| X1 | 0.0 | 15.00000000 | 0.0 | 0.0 | 0.00000000 | 0.10000000 |
| X2 | 0.0 | 0.0 | 0.08000000 | 0.00000000 | 0.0 | 0.0 |
| X2SQ | 0.08000000 | 0.0 | 0.00000000 | 0.00080000 | 0.0 | 0.0 |
| X1X2 | 0.0 | 0.00000000 | 0.0 | 0.0 | 0.10000000 | 0.00000000 |
| X1X2SQ | 0.0 | 0.10000000 | 0.0 | 0.0 | 0.00000000 | 0.00100000 |

EXAMPLE 16.5

THE X′X INVERSE MATRIX, RANK = 6

| | INTERCPT | X1 | X2 | X2SQ | X1X2 | X1X2SQ |
|---|---|---|---|---|---|---|
| INTERCPT | 0.25000000 | 0.0 | 0.00000000 | −25.00000000 | 0.0 | 0.0 |
| X1 | 0.0 | 0.20000000 | 0.0 | 0.0 | 0.00000000 | −20.00000000 |
| X2 | 0.00000000 | 0.0 | 12.50000000 | −0.00000000 | 0.0 | 0.0 |
| X2SQ | −25.00000000 | 0.0 | −0.00000000 | 3750.00000000 | 0.0 | 0.0 |
| X1X2 | 0.0 | 0.00000000 | 0.0 | 0.0 | 10.00000000 | −0.00000000 |
| X1X2SQ | 0.0 | −20.00000000 | 0.0 | 0.0 | −0.00000000 | 3000.00000000 |

EXAMPLE 16.5

ANALYSIS OF VARIANCE TABLE, REGRESSION COEFFICIENTS, AND STATISTICS OF FIT FOR DEPENDENT VARIABLE Y

| SOURCE | DF | SUM OF SQUARES | MEAN SQUARE | F VALUE | PROB > F | R-SQUARE | C.V. |
|---|---|---|---|---|---|---|---|
| REGRESSION | 5 | 32.04216667 | 6.40843333 | 90.04824 | 0.0002 | 0.98684906 | 4.84304% |
| ERROR | 6 | 0.42700000 | 0.07116667 | | | STD DEV | Y MEAN |
| CORRECTED TOTAL | 11 | 32.46916667 | | | | 0.26677081 | 5.50833 |

| SOURCE | DF | SEQUENTIAL SS | F VALUE | PROB > F | PARTIAL SS | F VALUE | PROB > F |
|---|---|---|---|---|---|---|---|
| X1 | 1 | 26.00416667 | 365.39813 | 0.0001 | 7.08050000 | 99.49180 | 0.0001 |
| X2 | 1 | 5.12000000 | 71.94379 | 0.0001 | 5.12000000 | 71.94379 | 0.0001 |
| X2SQ | 1 | 0.60166667 | 8.45433 | 0.0271 | 0.60166667 | 8.45433 | 0.0271 |
| X1X2 | 1 | 0.19600000 | 2.75410 | 0.1481 | 0.19600000 | 2.75410 | 0.1481 |
| X1X2SQ | 1 | 0.12033333 | 1.69087 | 0.2412 | 0.12033333 | 1.69087 | 0.2412 |

| SOURCE | B VALUES | T FOR HO:B = 0 | PROB > \|T\| | STD ERR B | STD B VALUES |
|---|---|---|---|---|---|
| INTERCEPT | 5.82500000 | 43.67044 | 0.0001 | 0.13338541 | 0.0 |
| X1 | 1.19000000 | 9.97456 | 0.0001 | 0.11930353 | 0.80882957 |
| X2 | −8.00000000 | −8.48197 | 0.0001 | 0.94317725 | −0.39709956 |
| X2SQ | −47.50000000 | −2.90763 | 0.0271 | 16.33630925 | −0.13612641 |
| X1X2 | 1.40000000 | 1.65955 | 0.1481 | 0.84360338 | 0.07769489 |
| X1X2SQ | 19.00000000 | 1.30033 | 0.2412 | 14.61163920 | 0.10544307 |

| OBS NUMBER | OBSERVED VALUE | PREDICTED VALUE | RESIDUAL | LOWER 95% CL FOR MEAN | UPPER 95% CL FOR MEAN |
|---|---|---|---|---|---|
| 1 | 4.30000000 | 4.29000000 | 0.01000000 | 3.74385775 | 4.83614225 |
| 2 | 5.50000000 | 5.53000000 | −0.03000000 | 5.17246597 | 5.88753403 |
| 3 | 6.80000000 | 6.77000000 | 0.03000000 | 6.41246597 | 7.12753403 |
| 4 | 8.00000000 | 8.01000000 | −0.01000000 | 7.46385775 | 8.55614225 |
| 5 | 4.00000000 | 4.04000000 | −0.04000000 | 3.49385775 | 4.58614225 |
| 6 | 5.20000000 | 5.23000000 | −0.03000000 | 4.87246597 | 5.58753403 |
| 7 | 6.60000000 | 6.42000000 | 0.18000000 | 6.06246597 | 6.77753403 |
| 8 | 7.50000000 | 7.61000000 | −0.11000000 | 7.06385775 | 8.15614225 |
| 9 | 2.00000000 | 2.27000000 | −0.27000000 | 1.72385775 | 2.81614225 |
| 10 | 4.00000000 | 3.79000000 | 0.21000000 | 3.43246697 | 4.14753403 |
| 11 | 5.70000000 | 5.31000000 | 0.39000000 | 4.95246597 | 5.66753403 |
| 12 | 6.50000000 | 6.83000000 | −0.33000000 | 6.28385775 | 7.37614225 |

| | | |
|---|---|---|
| SUM OF RESIDUALS | = | 0.00000000 |
| SUM OF SQUARED RESIDUALS | = | 0.42700000 |
| SUM OF SQUARED RESIDUALS − ERROR SS | = | 0.00000000 |
| FIRST ORDER AUTOCORRELATION OF RESIDUALS | = | −0.26838076 |
| DURBIN-WATSON D | = | 2.20796253 |

****************************************************************************************************

MODEL : EXAMPLE 16.5

DEPENDENTS : Y

INDEPENDENTS : X1 X2 X2SQ

****************************************************************************************************

EXAMPLE 16.5

THE X′X MATRIX

| | INTERCPT | X1 | X2 | X2SQ |
|---|---|---|---|---|
| INTERCPT | 12.00000000 | 0.0 | 0.0 | 0.08000000 |
| X1 | 0.0 | 15.00000000 | 0.0 | 0.0 |
| X2 | 0.0 | 0.0 | 0.08000000 | 0.00000000 |
| X2SQ | 0.08000000 | 0.0 | 0.00000000 | 0.00080000 |

EXAMPLE 16.5

THE X'X INVERSE MATRIX, RANK = 4

| | INTERCPT | X1 | X2 | X2SQ |
|---|---|---|---|---|
| INTERCPT | 0.25000000 | 0.0 | 0.00000000 | −25.00000000 |
| X1 | 0.0 | 0.06666667 | 0.0 | 0.0 |
| X2 | 0.00000000 | 0.0 | 12.50000000 | −0.00000000 |
| X2SQ | −25.00000000 | 0.0 | −0.00000000 | 3750.00000000 |

EXAMPLE 16.5

ANALYSIS OF VARIANCE TABLE, REGRESSION COEFFICIENTS, AND STATISTICS OF FIT FOR DEPENDENT VARIABLE Y

| SOURCE | DF | SUM OF SQUARES | MEAN SQUARE | F VALUE | PROB > F | R-SQUARE | C.V. |
|---|---|---|---|---|---|---|---|
| REGRESSION | 3 | 31.72583333 | 10.57527778 | 113.81465 | 0.0001 | 0.97710649 | 5.53384% |
| ERROR | 8 | 0.74333333 | 0.09291667 | | | STD DEV | Y MEAN |
| CORRECTED TOTAL | 11 | 32.46916667 | | | | 0.30482235 | 5.50833 |

| SOURCE | DF | SEQUENTIAL SS | F VALUE | PROB > F | PARTIAL SS | F VALUE | PROB > F |
|---|---|---|---|---|---|---|---|
| X1 | 1 | 26.00416667 | 279.86547 | 0.0001 | 26.00416667 | 279.86547 | 0.0001 |
| X2 | 1 | 5.12000000 | 55.10314 | 0.0001 | 5.12000000 | 55.10314 | 0.0001 |
| X2SQ | 1 | 0.60166667 | 6.47534 | 0.0345 | 0.60166667 | 6.47534 | 0.0345 |

| SOURCE | B VALUES | T FOR HO:B = 0 | PROB > \|T\| | STD ERR B | STD B VALUES |
|---|---|---|---|---|---|
| INTERCEPT | 5.82500000 | 38.21898 | 0.0001 | 0.15241118 | 0.0 |
| X1 | 1.31666667 | 16.72918 | 0.0001 | 0.07870479 | 0.89492347 |
| X2 | −8.00000000 | −7.42315 | 0.0001 | 1.07770976 | −0.39709956 |
| X2SQ | −47.50000000 | −2.54467 | 0.0345 | 18.66648065 | −0.13612641 |

| OBS NUMBER | OBSERVED VALUE | PREDICTED VALUE | RESIDUAL | LOWER 95% CL FOR MEAN | UPPER 95% CL FOR MEAN |
|---|---|---|---|---|---|
| 1 | 4.30000000 | 4.17500000 | 0.12500000 | 3.73042925 | 4.61957075 |
| 2 | 5.50000000 | 5.49166667 | 0.00833333 | 5.12867617 | 5.85465716 |
| 3 | 6.80000000 | 6.80333333 | −0.00833333 | 6.44534284 | 7.17132383 |
| 4 | 8.00000000 | 8.12500000 | −0.12500000 | 7.68042925 | 8.56957075 |
| 5 | 4.00000000 | 3.85000000 | 0.15000000 | 3.40542925 | 4.29457075 |
| 6 | 5.20000000 | 5.1666667 | 0.03333333 | 4.80367617 | 5.52965716 |
| 7 | 6.60000000 | 6.48333333 | 0.11666667 | 6.12034284 | 6.84832383 |

| OBS NUMBER | OBSERVED VALUE | PREDICTED VALUE | RESIDUAL | LOWER 95% CL FOR MEAN | UPPER 95% CL FOR MEAN |
|---|---|---|---|---|---|
| 8 | 7.50000000 | 7.80000000 | −0.30000000 | 7.35542925 | 8.24457075 |
| 9 | 2.00000000 | 2.57500000 | −0.57500000 | 2.13042925 | 3.01957075 |
| 10 | 4.00000000 | 3.89166667 | 0.10833333 | 3.52867617 | 4.25465716 |
| 11 | 5.70000000 | 5.20833333 | 0.49166667 | 4.84534284 | 5.57132383 |
| 12 | 6.50000000 | 6.52500000 | −0.02500000 | 6.08042925 | 6.96957075 |

SUM OF RESIDUALS = 0.00000000

SUM OF SQUARED RESIDUALS = 0.74333333

SUM OF SQUARED RESIDUALS − ERROR SS = 0.00000000

FIRST ORDER AUTOCORRELATION OF RESIDUALS = 0.14735819

DURBIN-WATSON D = 1.68665919

### EXAMPLE 16.6

In example 15.5 (p. 438) of chapter 15 we considered an experiment to investigate the effects of 4 different pesticides on the yield of fruit from 3 different varieties of citrus trees. Write a model that contains main effects for both qualitative variables but no interaction. Use the data, reproduced in table 16.8, to test for no difference among mean yields for pesticides by fitting complete and reduced models. Use $\alpha = .05$.

**Table 16.8** Data for the fruit yield experiment of example 16.6

| | *Pesticide* | | | |
|---|---|---|---|---|
| *Variety* | 1 | 2 | 3 | 4 |
| 1 | 29 | 50 | 43 | 53 |
| 2 | 41 | 58 | 42 | 73 |
| 3 | 66 | 85 | 69 | 85 |

### SOLUTION

The complete model is

$$y = \beta_0 + \overbrace{\beta_1 x_1 + \beta_2 x_2}^{\text{varieties}} + \overbrace{\beta_3 x_3 + \beta_4 x_4 + \beta_5 x_5}^{\text{pesticides}} + \epsilon$$

where $x_1, x_2, \ldots, x_5$ are dummy variables defined in the usual way. For example,

$x_1 = 1$ if Variety 2 $\quad x_1 = 0$ otherwise

Similarly,

$x_3 = 1$ if Pesticide 2 $\quad x_3 = 0$ otherwise

We use a computer to fit this model (the computer solution follows).

We obtain

$$\hat{\boldsymbol{\beta}} = (\mathbf{X}'\mathbf{X})^{-1}\mathbf{X}'\mathbf{Y} = \begin{bmatrix} 31.25 \\ 9.75 \\ 32.50 \\ 19.00 \\ 6.00 \\ 25.00 \end{bmatrix}$$

and a sum of squares for error, $SSE_1 = 151.50$.

The null hypothesis for a test of no difference among mean yields for pesticides is $H_0$: $\beta_3 = \beta_4 = \beta_5 = 0$. The corresponding reduced model is

$$y = \beta_0 + \beta_1 x_1 + \beta_2 x_2 + \epsilon$$

for which we obtain

$$\hat{\boldsymbol{\beta}} = \begin{bmatrix} 43.75 \\ 9.75 \\ 32.50 \end{bmatrix}$$

and $SSE_2 = 1342.50$.

```
STATISTICAL ANALYSIS SYSTEM

DATA CITRUS;
INPUT Y 1-2 X1 4 X2 6 X3 8 X4 10 X5 12;
CARDS;


  12 OBSERVATIONS IN DATA SET CITRUS     6 VARIABLES
PROC PRINT;


OBS        Y       X1        X2        X3        X4        X5

  1       29        0         0         0         0         0
  2       41        1         0         0         0         0
  3       66        0         1         0         0         0
  4       50        0         0         1         0         0
  5       58        1         0         1         0         0
  6       85        0         1         1         0         0
  7       43        0         0         0         1         0
  8       42        1         0         0         1         0
  9       69        0         1         0         1         0
 10       53        0         0         0         0         1
 11       73        1         0         0         0         1
 12       85        0         1         0         0         1


PROC REGR;
MODEL Y = X1 X2 X3 X4 X5 / X I P CLM;
MODEL Y = X1 X2 / X I;
TITLE 'EXAMPLE 16.6';
```

****************************************************************************************************

MODEL : EXAMPLE 16.6

DEPENDENTS : Y

INDEPENDENTS : X1 X2 X3 X4 X5

****************************************************************************************************

EXAMPLE 16.6

THE X'X MATRIX

| | INTERCPT | X1 | X2 | X3 | X4 | X5 |
|---|---|---|---|---|---|---|
| INTERCPT | 12.00000000 | 4.00000000 | 4.00000000 | 3.00000000 | 3.00000000 | 3.00000000 |
| X1 | 4.00000000 | 4.00000000 | 0.0 | 1.00000000 | 1.00000000 | 1.00000000 |
| X2 | 4.00000000 | 0.0 | 4.00000000 | 1.00000000 | 1.00000000 | 1.00000000 |
| X3 | 3.00000000 | 1.00000000 | 1.00000000 | 3.00000000 | 0.0 | 0.0 |
| X4 | 3.00000000 | 1.00000000 | 1.00000000 | 0.0 | 3.00000000 | 0.0 |
| X5 | 3.00000000 | 1.00000000 | 1.00000000 | 0.0 | 0.0 | 3.00000000 |

EXAMPLE 16.6

THE X'X INVERSE MATRIX, RANK = 6

| | INTERCPT | X1 | X2 | X3 | X4 | X5 |
|---|---|---|---|---|---|---|
| INTERCPT | 0.50000000 | −0.25000000 | −0.25000000 | −0.33333333 | −0.33333333 | −0.33333333 |
| X1 | −0.25000000 | 0.50000000 | 0.25000000 | 0.0 | 0.0 | 0.0 |
| X2 | −0.25000000 | 0.25000000 | 0.50000000 | 0.0 | 0.0 | 0.0 |
| X3 | −0.33333333 | 0.0 | 0.0 | 0.66666667 | 0.33333333 | 0.33333333 |
| X4 | −0.33333333 | 0.0 | 0.0 | 0.33333333 | 0.66666667 | 0.33333333 |
| X5 | −0.33333333 | 0.0 | 0.0 | 0.33333333 | 0.33333333 | 0.66666667 |

EXAMPLE 16.6

ANALYSIS OF VARIANCE TABLE, REGRESSION COEFFICIENTS, AND STATISTICS OF FIT FOR DEPENDENT VARIABLE Y

| SOURCE | DF | SUM OF SQUARES | MEAN SQUARE | F VALUE | PROB > F | R-SQUARE | C.V. |
|---|---|---|---|---|---|---|---|
| REGRESSION | 5 | 3416.16666667 | 683.23333333 | 27.05875 | 0.0010 | 0.95753527 | 8.68865% |
| ERROR | 6 | 151.50000000 | 25.25000000 | | | STD DEV | Y MEAN |
| CORRECTED TOTAL | 11 | 3567.66666667 | | | | 5.02493781 | 57.83333 |

| SOURCE | DF | SEQUENTIAL SS | F VALUE | PROB > F | PARTIAL SS | F VALUE | PROB > F |
|---|---|---|---|---|---|---|---|
| X1 | 1 | 112.66666667 | 4.46205 | 0.0791 | 190.12500000 | 7.52970 | 0.0336 |
| X2 | 1 | 2112.50000000 | 83.66337 | 0.0001 | 2112.50000000 | 83.66337 | 0.0001 |
| X3 | 1 | 169.00000000 | 6.69307 | 0.0414 | 541.50000000 | 21.44554 | 0.0036 |
| X4 | 1 | 84.50000000 | 3.34653 | 0.1171 | 54.00000000 | 2.13861 | 0.1940 |
| X5 | 1 | 937.50000000 | 37.12871 | 0.0009 | 937.50000000 | 37.12871 | 0.0009 |

| SOURCE | B VALUES | T FOR HO:B = 0 | PROB > \|T\| | STD ERR B | STD B VALUES |
|---|---|---|---|---|---|
| INTERCEPT | 31.25000000 | 8.79497 | 0.0001 | 3.55316760 | 0.0 |
| X1 | 9.75000000 | 2.74403 | 0.0336 | 3.55316760 | 0.26656115 |
| X2 | 32.50000000 | 9.14677 | 0.0001 | 3.55316760 | 0.88853715 |
| X3 | 19.00000000 | 4.63093 | 0.0036 | 4.10284454 | 0.47714758 |
| X4 | 6.00000000 | 1.46240 | 0.1940 | 4.10284454 | 0.15067818 |
| X5 | 25.00000000 | 6.09333 | 0.0009 | 4.10284454 | 0.62782576 |

| OBS NUMBER | OBSERVED VALUE | PREDICTED VALUE | RESIDUAL | LOWER 95% CL FOR MEAN | UPPER 95% CL FOR MEAN |
|---|---|---|---|---|---|
| 1 | 29.00000000 | 31.25000000 | −2.25000000 | 22.55570966 | 39.94429034 |
| 2 | 41.00000000 | 41.00000000 | 0.00000000 | 32.30570966 | 49.69429034 |
| 3 | 66.00000000 | 63.75000000 | 2.25000000 | 55.05570966 | 72.44429034 |
| 4 | 50.00000000 | 50.25000000 | −0.25000000 | 41.55570966 | 58.94429034 |
| 5 | 58.00000000 | 60.00000000 | −2.00000000 | 51.30570966 | 68.69429034 |
| 6 | 85.00000000 | 82.75000000 | 2.25000000 | 74.05570966 | 91.44429034 |
| 7 | 43.00000000 | 37.25000000 | 5.75000000 | 28.55570966 | 45.94429034 |
| 8 | 42.00000000 | 47.00000000 | −5.00000000 | 38.30570966 | 55.69429034 |
| 9 | 69.00000000 | 69.75000000 | −0.75000000 | 61.05570966 | 78.44429034 |
| 10 | 53.00000000 | 56.25000000 | −3.25000000 | 47.55570966 | 64.94429034 |
| 11 | 73.00000000 | 66.00000000 | 7.00000000 | 57.30570966 | 74.69429034 |
| 12 | 85.00000000 | 88.75000000 | −3.75000000 | 80.05570966 | 97.44429034 |

SUM OF RESIDUALS = 0.00000000

SUM OF SQUARED RESIDUALS = 151.50000000

SUM OF SQUARED RESIDUALS − ERROR SS = 0.00000000

FIRST ORDER AUTOCORRELATION OF RESIDUALS = −0.44540224

DURBIN-WATSON D = 2.70792079

**************************************************************************************************

MODEL : EXAMPLE 16.6

DEPENDENTS : Y

INDEPENDENTS : X1 X2

**************************************************************************************************

EXAMPLE 16.6

THE X'X MATRIX

| | INTERCPT | X1 | X2 |
|---|---|---|---|
| INTERCPT | 12.00000000 | 4.00000000 | 4.00000000 |
| X1 | 4.00000000 | 4.00000000 | 0.0 |
| X2 | 4.00000000 | 0.0 | 4.00000000 |

EXAMPLE 16.6

THE X'X INVERSE MATRIX, RANK = 3

| | INTERCPT | X1 | X2 |
|---|---|---|---|
| INTERCPT | 0.25000000 | −0.25000000 | −0.25000000 |
| X1 | −0.25000000 | 0.50000000 | 0.25000000 |
| X2 | −0.25000000 | 0.25000000 | 0.50000000 |

EXAMPLE 16.6

ANALYSIS OF VARIANCE TABLE, REGRESSION COEFFICIENTS, AND STATISTICS OF FIT FOR DEPENDENT VARIABLE Y

| SOURCE | DF | SUM OF SQUARES | MEAN SQUARE | F VALUE | PROB > F | R-SQUARE | C.V. |
|---|---|---|---|---|---|---|---|
| REGRESSION | 2 | 2225.16666667 | 1112.58333333 | 7.45866 | 0.0124 | 0.62370363 | 21.11824% |
| ERROR | 9 | 1342.50000000 | 149.16666667 | | | STD DEV | Y MEAN |
| CORRECTED TOTAL | 11 | 3567.66666667 | | | | 12.21338064 | 57.83333 |

| SOURCE | DF | SEQUENTIAL SS | F VALUE | PROB > F | PARTIAL SS | F VALUE | PROB > F |
|---|---|---|---|---|---|---|---|
| X1 | 1 | 112.66666667 | 0.75531 | 0.4074 | 190.12500000 | 1.27458 | 0.2881 |
| X2 | 1 | 2112.50000000 | 14.16201 | 0.0045 | 2112.50000000 | 14.16201 | 0.0045 |

| SOURCE | B VALUES | T FOR HO:B = 0 | PROB > \|T\| | STD ERR B | STD B VALUES |
|---|---|---|---|---|---|
| INTERCEPT | 43.75000000 | 7.16427 | 0.0001 | 6.10669032 | 0.0 |
| X1 | 9.75000000 | 1.12897 | 0.2881 | 8.63616427 | 0.26656115 |
| X2 | 32.50000000 | 3.76324 | 0.0045 | 8.63616427 | 0.88853715 |

The drop in the sum of squares due to pesticides (the variables $x_3$, $x_4$, and $x_5$) is

$$SS_{drop} = SSE_2 - SSE_1 = 1342.50 - 151.50 = 1191.00$$

$$MS_{drop} = \frac{SS_{drop}}{3} = 397.00$$

With $MSE_1 = SSE_1/6 = 151.50/6 = 25.25$, the test statistic for $H_0$: $\beta_3 = \beta_4 = \beta_5 = 0$ is

$$F = \frac{MS_{drop}}{MSE_1} = \frac{397.00}{25.25} = 15.72$$

For $\alpha = .05$, we will reject the null hypothesis if the observed value of $F$ exceeds the value found in table 4 of the appendix for $df_1 = 3$ and $df_2 = 6$, which is 4.76. Since 15.72 exceeds 4.76, we conclude that at least one of the pesticides gives a mean yield different from the rest.

# 16.5 A Relationship Between Least Squares and the Analysis of Variance

Many people who have studied statistics often do not see that there is a relationship between the method of least squares, which is used in fitting complete and reduced models, and the analysis of variance. Instead they see least squares and the analysis of variance as separate methods with no linkage. The experiment of example 16.6 can be used to show a relationship between least squares and the analysis of variance of chapters 13 and 15. In particular, the hypothesis of no difference among the pesticide means, $H_0$: $\beta_3 = \beta_4 = \beta_5 = 0$, is the same as that specified for an analysis of variance $F$

test on pesticides. In addition the sum of squares drop ($SS_{drop}$) obtained by fitting models 1 and 2 is identical to the sum of squares due to pesticides of example 15.5 (p. 438).

$$\text{Model 1:} \quad y = \beta_0 + \overbrace{\beta_1 x_1 + \beta_2 x_2}^{\text{varieties}} + \overbrace{\beta_3 x_3 + \beta_4 x_4 + \beta_5 x_5}^{\text{pesticides}} + \epsilon$$

$$SSE_1 = 151.50$$

$$\text{Model 2:} \quad y = \beta_0 + \beta_1 x_1 + \beta_2 x_2 + \epsilon$$

$$SSE_2 = 1342.50$$

$$SS_{drop} \text{ (due to pesticides)} = SSE_2 - SSE_1 = 1191.00$$

In the same way, the sum of squares drop obtained for testing no difference among the variety means ($H_0$: $\beta_1 = \beta_2 = 0$) is identical to the sum of squares due to varieties in the analysis of variance of example 15.5. Here

$$\text{Model 1:} \quad y = \beta_0 + \overbrace{\beta_1 x_1 + \beta_2 x_2}^{\text{varieties}} + \overbrace{\beta_3 x_3 + \beta_4 x_4 + \beta_5 x_5}^{\text{pesticides}} + \epsilon$$

$$SSE_1 = 151.50$$

$$\text{Model 2:} \quad y = \beta_0 + \beta_3 x_3 + \beta_4 x_4 + \beta_5 x_5 + \epsilon$$

$$SSE_2 = 2376.67$$

Although we have not fit model 2, it can be shown that the sum of squares for error is as given here. Also,

$$SS_{drop} \text{ (due to varieties)} = SSE_2 - SSE_1 = 2225.17$$

Note that $SSE_1$, the sum of squares for error for the complete model, is SSE from the AOV table of example 15.5.

The method of obtaining sums of squares ($SS_{drop}$) due to various sources of variability by fitting complete and reduced models can be used for *any* experimental design. All we need to do is to specify an appropriate general linear model. Then by hypothesizing that various parameters in the model are zero, we can obtain $SS_{drop}$ and run an appropriate $F$ test. The $F$ test is identical to that used in a standard analysis of variance. *In fact, as we have developed this chapter, we see that an analysis of variance consists of testing hypotheses concerning parameters in the general linear model.*

Why don't we fit complete and reduced models every time we conduct an analysis of variance? While this would certainly be possible, it is not practical. The reason is that many designs are *balanced* designs.

balanced design

**Definition 16.1**

A *balanced design* has each level of one independent variable appearing the same number of times with each level of another independent variable, and this is true for all pairs of independent variables.

For many balanced designs it is possible to obtain shortcut computational formulas for calculating the sum of squares drop for a particular source of variability. For example, by definition 16.1, it is clear that the factorial experiment, Latin square, randomized block, and completely randomized designs are all balanced designs. As indicated in chapters 13 and 15, we have shortcut formulas for calculating the sum of squares associated with the various sources of variability in an AOV table. These formulas are simplified expressions for calculating $SS_{\text{drop}}$ by fitting a complete and reduced model.

At this point you might be concerned with identifying an unbalanced design. An example of an unbalanced design would be any balanced design with one or more missing observations. Suppose an experimenter was interested in comparing the drop in potency for 3 different concentrations of a drug product stored at 3 different temperatures. Assume further that 2 bottles were to be stored for six weeks at each factor-level combination. Since final potency determinations must be made after the six-week period in order to compute the drop in potency, it is possible that one or more bottles would be broken. If such an accident did occur, we would have an unequal number of observations at the different factor-level combinations, making the design *unbalanced*. The shortcut formulas developed in chapter 15 would no longer be appropriate for computing sums of squares for sources in the AOV table. While the general procedure for fitting complete and reduced models can be used, we will delay any further discussion of analyzing unbalanced designs until chapter 19.

# Exercises

*16.6. Refer to example 16.5 (p. 472). Under the assumption that there is no interaction between the variables "exposure time" and "relative humidity," start with the complete model

$$y = \beta_0 + \beta_1 x_1 + \beta_2 x_2 + \beta_3 x_2^2 + \epsilon$$

Use an appropriate reduced model to test the hypothesis of no linear or quadratic effect due to $x_2$. Use $\alpha = .05$.

*16.7. Refer to exercise 16.6.

a. Fit a complete and reduced model to test for no linear effect due to $x_1$. Use $\alpha = .05$.

b. How else might you test this same hypothesis without fitting complete and reduced models?

*16.8. Refer to example 16.6 (p. 479). We considered an experiment to investigate the effects of 4 different pesticides on the yields of 3 different varieties of citrus trees. Use complete and reduced models to test the hypothesis of no differences among mean yields for the 3 varieties. Use $\alpha = .05$. Compare your results with those obtained in example 15.5 (p. 439) by using the treatment formulas.

*16.9. Refer to example 15.3 (p. 426). We compared the total unused green time using 4 different signal sequencing devices at 4 locations at 4 different time periods. With the aid of a computer, fit complete and reduced models to obtain the sums of squares for sources in the AOV table for this $4 \times 4$ Latin square design.

*16.10. While we will not indicate a specific exercise here, the procedure for fitting complete and reduced models can be used to obtain the sums of squares for sources of variability in *any* of the analysis of variance exercises or examples of chapter 15. You may wish to choose additional problems (under the guidance of your professor) to obtain additional practice in computer solutions to AOV problems by fitting complete and reduced models.

# Slopes of Two Regression Lines: Qualitative and Quantitative Independent Variables 16.6

We have shown that the general linear model can be used to relate a response $y$ to a set of quantitative or qualitative independent variables, but we have not yet illustrated how we could include one or more quantitative and qualitative independent variables in the same model. The best way to illustrate this is by way of an example.

### EXAMPLE 16.7

An investigator is interested in comparing the responses of rats to different doses of two drug products (A and B). The study calls for a sample of 60 rats of a particular strain to be randomly allocated into two equal groups. The first group of rats are to receive Drug A, with 10 rats randomly assigned to each of 3 doses (5, 10, and 20 mg). Similarly, the 30 rats in Group 2 are to receive Drug B, with 10 rats randomly assigned to the 5-, 10-, and 20-mg doses. In the study each rat received its assigned dose, and after a 30-minute observation period, it was scored for signs of anxiety on a 0-to-30-point scale. If a rat's anxiety score is assumed to be a linear function of the dosage of the drug, write a model relating a rat's score to the two independent variables "drug product" and "drug dose." Interpret the $\beta$'s.

### SOLUTION

For this experimental situation, we have one qualitative variable (drug product) and one quantitative variable (drug dose). Letting $x$ denote the drug dose, we have the model

$$y = \beta_0 + \beta_1 x_1 + \beta_2 x_2 + \beta_3 x_1 x_2 + \epsilon$$

where

$x_1$ = drug dose

$x_2 = 1$ if Product B $\quad x_2 = 0$ otherwise

The expected value for $y$ in our model is

$$E(y) = \beta_0 + \beta_1 x_1 + \beta_2 x_2 + \beta_3 x_1 x_2$$

Substituting $x_2 = 0$ and $x_2 = 1$, respectively, for drug products A and B, we obtain the expected rat anxiety score for a given dose:

Product A: $E(y) = \beta_0 + \beta_1 x_1$

Product B: $E(y) = \beta_0 + \beta_1 x_1 + \beta_2 + \beta_3 x_1 = (\beta_0 + \beta_2) + (\beta_1 + \beta_3) x_1$

linear regression lines

These two expected values represent linear regression lines. The parameters in the model can be interpreted in terms of the slopes and intercepts associated with these regression lines. In particular,

y-intercept

slope

$\beta_0$: $y$-intercept for Product A regression line

$\beta_1$: slope of Product A regression line

$\beta_2$: difference in the $y$-intercepts of the regression lines for products B and A

$\beta_3$: difference in the slopes of the regression lines for products B and A

intersecting lines

parallel lines

Figure 16.1(a) indicates a situation where $\beta_3 \neq 0$ (i.e., there is an interaction between the two variables "drug product" and "drug dose"). Thus the regression lines are not parallel. Figure 16.1(b) indicates a case where $\beta_3 = 0$ (no interaction), which results in parallel regression lines.

Indeed, many other experimental situations are possible, depending on the signs and magnitudes of the parameters $\beta_0$, $\beta_1$, $\beta_2$, and $\beta_3$.

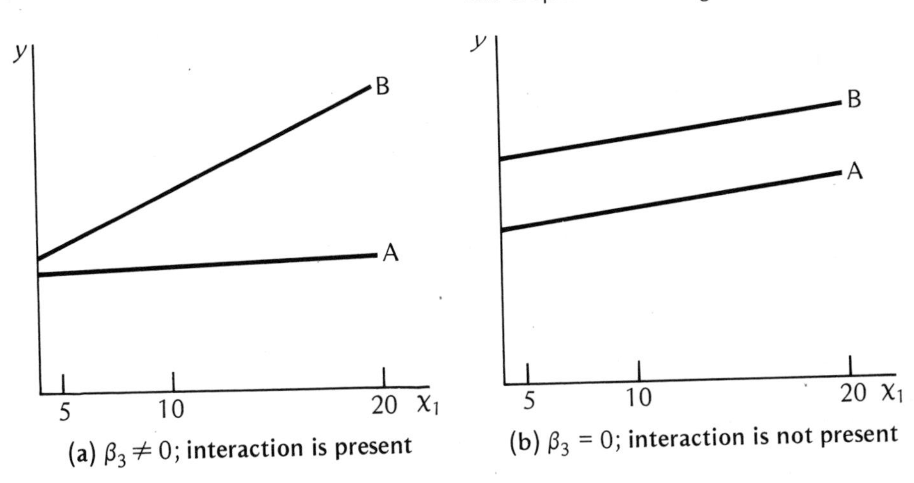


(a) $\beta_3 \neq 0$; interaction is present

(b) $\beta_3 = 0$; interaction is not present

**Figure 16.1** Comparing two regression lines

**EXAMPLE 16.8**

Sample data for the experiment discussed in example 16.7 are listed in table 16.9. The response of interest is an anxiety score obtained from

**Table 16.9** Rat anxiety scores, example 16.8

| | *Drug Dose (mg)* | | |
|---|---|---|---|
| *Drug Product* | 5 | 10 | 20 |
| A | 15 16 | 18 16 | 20 17 |
| | 16 15 | 17 15 | 19 18 |
| | 18 16 | 18 19 | 21 21 |
| | 13 17 | 19 18 | 18 20 |
| | 19 15 | 20 16 | 19 17 |
| | av = 16 | av = 17.6 | av = 19.0 |
| B | 16 15 | 19 18 | 24 23 |
| | 17 15 | 21 20 | 25 24 |
| | 18 18 | 22 21 | 23 22 |
| | 17 17 | 23 22 | 25 26 |
| | 15 16 | 20 19 | 25 24 |
| | av = 16.4 | av = 20.5 | av = 24.1 |

trained investigators. Use these data to fit the general linear model

$$y = \beta_0 + \beta_1 x_1 + \beta_2 x_2 + \beta_3 x_1 x_2 + \epsilon$$

Of particular interest to the experimenter is a comparison between the slopes of the regression lines. A difference in slopes would indicate that the drug products have different effects on the anxiety of the rats. Conduct a statistical test of the equality of the two slopes. Use $\alpha = .05$.

## SOLUTION

Using the complete model

$$y = \beta_0 + \beta_1 x_1 + \beta_2 x_2 + \beta_3 x_1 x_2 + \epsilon$$

we obtain a least squares fit of

$$\hat{\boldsymbol{\beta}} = \begin{bmatrix} 15.30 \\ .19 \\ -.70 \\ .30 \end{bmatrix}$$

with $SSE_1 = 133.63$ (see the computer output that follows).

The reduced model corresponding to the null hypothesis $H_0$: $\beta_3 = 0$ (i.e., the slopes are the same) is

$$y = \beta_0 + \beta_1 x_1 + \beta_2 x_2 + \epsilon$$

for which we obtain

$$\hat{\boldsymbol{\beta}} = \begin{bmatrix} 13.55 \\ .34 \\ 2.80 \end{bmatrix}$$

and $SSE_2 = 186.13$. The reduction in the sum of squares for error attributed to $x_1 x_2$ is

$$SS_{drop} = SSE_2 - SSE_1 = 186.13 - 133.63 = 52.50$$

Using $MSE_1 = SSE_1/56 = 133.63/56 = 2.39$ and $MS_{drop} = 52.50$ (since we are testing only one $\beta$),

$$F = \frac{MS_{drop}}{MSE_1} = \frac{52.50}{2.39} = 22.00$$

Since the observed value of $F$ exceeds 4.00, the table value for $df_1 = 1$, $df_2 = 56$ (actually, 60), and $a = .05$, we reject $H_0$ and conclude that the slopes for the two groups are different. (Also see exercise 16.26.)

```
STATISTICAL ANALYSIS SYSTEM

DATA ANXIETY;
INPUT Y 1-4 X1 6-9 X2 11;
X1X2 = X1* X2;
CARDS;


   60 OBSERVATIONS IN DATA SET ANXIETY     4 VARIABLES


PROC PRINT;
```

| OBS | Y | X1 | X2 | X1X2 |
|---|---|---|---|---|
| 1 | 15 | 5 | 0 | 0 |
| 2 | 16 | 5 | 0 | 0 |
| 3 | 18 | 5 | 0 | 0 |
| 4 | 13 | 5 | 0 | 0 |
| 5 | 19 | 5 | 0 | 0 |
| 6 | 16 | 5 | 0 | 0 |
| 7 | 15 | 5 | 0 | 0 |
| 8 | 16 | 5 | 0 | 0 |
| 9 | 17 | 5 | 0 | 0 |
| 10 | 15 | 5 | 0 | 0 |
| 11 | 16 | 5 | 1 | 5 |
| 12 | 17 | 5 | 1 | 5 |
| 13 | 18 | 5 | 1 | 5 |
| 14 | 17 | 5 | 1 | 5 |
| 15 | 15 | 5 | 1 | 5 |
| 16 | 15 | 5 | 1 | 5 |
| 17 | 15 | 5 | 1 | 5 |
| 18 | 18 | 5 | 1 | 5 |
| 19 | 17 | 5 | 1 | 5 |
| 20 | 16 | 5 | 1 | 5 |
| 21 | 18 | 10 | 0 | 0 |
| 22 | 17 | 10 | 0 | 0 |
| 23 | 18 | 10 | 0 | 0 |
| 24 | 19 | 10 | 0 | 0 |
| 25 | 20 | 10 | 0 | 0 |
| 26 | 16 | 10 | 0 | 0 |
| 27 | 15 | 10 | 0 | 0 |
| 28 | 19 | 10 | 0 | 0 |
| 29 | 18 | 10 | 0 | 0 |
| 30 | 16 | 10 | 0 | 0 |
| 31 | 19 | 10 | 1 | 10 |
| 32 | 21 | 10 | 1 | 10 |
| 33 | 22 | 10 | 1 | 10 |
| 34 | 23 | 10 | 1 | 10 |
| 35 | 20 | 10 | 1 | 10 |
| 36 | 18 | 10 | 1 | 10 |
| 37 | 20 | 10 | 1 | 10 |
| 38 | 21 | 10 | 1 | 10 |
| 39 | 22 | 10 | 1 | 10 |
| 40 | 19 | 10 | 1 | 10 |
| 41 | 20 | 20 | 0 | 0 |
| 42 | 19 | 20 | 0 | 0 |
| 43 | 21 | 20 | 0 | 0 |
| 44 | 18 | 20 | 0 | 0 |
| 45 | 19 | 20 | 0 | 0 |
| 46 | 17 | 20 | 0 | 0 |
| 47 | 18 | 20 | 0 | 0 |
| 48 | 21 | 20 | 0 | 0 |
| 49 | 20 | 20 | 0 | 0 |
| 50 | 17 | 20 | 0 | 0 |
| 51 | 24 | 20 | 1 | 20 |
| 52 | 25 | 20 | 1 | 20 |

| OBS | Y | X1 | X2 | X1X2 |
|---|---|---|---|---|
| 53 | 23 | 20 | 1 | 20 |
| 54 | 25 | 20 | 1 | 20 |
| 55 | 25 | 20 | 1 | 20 |
| 56 | 23 | 20 | 1 | 20 |
| 57 | 24 | 20 | 1 | 20 |
| 58 | 22 | 20 | 1 | 20 |
| 59 | 26 | 20 | 1 | 20 |
| 60 | 24 | 20 | 1 | 20 |

```
PROC REGR;
MODEL Y = X1 X2 X1X2;
MODEL Y=X1 X2;
TITLE 'EXAMPLE 16.8';
```

************************************************************************************************************

PROC REGR : EXAMPLE 16.8

DATA SET : ANXIETY NUMBER OF VARIABLES = 4 NUMBER OF CLASSES = 0

VARIABLES : Y X1 X2 X1X2

************************************************************************************************************

EXAMPLE 16.8

ANALYSIS OF VARIANCE TABLE, REGRESSION COEFFICIENTS, AND STATISTICS OF FIT FOR DEPENDENT VARIABLE Y

| SOURCE | DF | SUM OF SQUARES | MEAN SQUARE | F VALUE | PROB > F | R-SQUARE | C.V. |
|---|---|---|---|---|---|---|---|
| REGRESSION | 3 | 442.10476190 | 147.36825397 | 61.75792 | 0.0001 | 0.76789850 | 8.15884% |
| ERROR | 56 | 133.62857143 | 2.38622449 | | | STD DEV | Y MEAN |
| CORRECTED TOTAL | 59 | 575.73333333 | | | | 1.54474091 | 18.93333 |

| SOURCE | DF | SEQUENTIAL SS | F VALUE | PROB > F | PARTIAL SS | F VALUE | PROB > F |
|---|---|---|---|---|---|---|---|
| X1 | 1 | 272.00476190 | 113.98959 | 0.0001 | 42.75238095 | 17.91633 | 0.0001 |
| X2 | 1 | 117.60000000 | 49.28287 | 0.0001 | 1.63333333 | 0.68448 | 0.4116 |
| X1X2 | 1 | 52.50000000 | 22.00128 | 0.0001 | 52.50000000 | 22.00128 | 0.0001 |

| SOURCE | B VALUES | T FOR HO:B = 0 | PROB > \|T\| | STD ERR B | STD B VALUES |
|---|---|---|---|---|---|
| INTERCEPT | 15.30000000 | 25.57350 | 0.0001 | 0.59827558 | 0.0 |
| X1 | 0.19142857 | 4.23277 | 0.0001 | 0.04522538 | 0.38537582 |
| X2 | −0.70000000 | −0.82734 | 0.4116 | 0.84608944 | −0.11298817 |
| X1X2 | 0.30000000 | 4.69055 | 0.0001 | 0.06395835 | 0.70819085 |

EXAMPLE 16.8

ANALYSIS OF VARIANCE TABLE, REGRESSION COEFFICIENTS, AND STATISTICS OF FIT FOR DEPENDENT VARIABLE Y

| SOURCE | DF | SUM OF SQUARES | MEAN SQUARE | F VALUE | PROB > F | R-SQUARE | C.V. |
|---|---|---|---|---|---|---|---|
| REGRESSION | 2 | 389.60476190 | 194.80238095 | 59.65627 | 0.0001 | 0.67671045 | 9.54425% |
| ERROR | 57 | 186.12857143 | 3.26541353 | | | STD DEV | Y MEAN |
| CORRECTED TOTAL | 59 | 575.73333333 | | | | 1.80704553 | 18.93333 |

| SOURCE | DF | SEQUENTIAL SS | F VALUE | PROB > F | PARTIAL SS | F VALUE | PROB > F |
|---|---|---|---|---|---|---|---|
| X1 | 1 | 272.00476190 | 83.29872 | 0.0001 | 272.00476190 | 83.29872 | 0.0001 |
| X2 | 1 | 117.60000000 | 36.01382 | 0.0001 | 117.60000000 | 36.01382 | 0.0001 |

| SOURCE | B VALUES | T FOR HO:B = 0 | PROB > \|T\| | STD ERR B | STD B VALUES |
|---|---|---|---|---|---|
| INTERCEPT | 13.55000000 | 24.76649 | 0.0001 | 0.54711020 | 0.0 |
| X1 | 0.34142857 | 9.12681 | 0.0001 | 0.03740940 | 0.68734941 |
| X2 | 2.80000000 | 6.00115 | 0.0001 | 0.46657715 | 0.45195269 |

# 16.7 Testing for Lack of Fit of a Polynomial Model

In most of the models that we have written for experimental situations, we have assumed that either our model is correct or that we can test the significance of individual parameters or sets of parameters in the model. However, in most experimental situations we do not know the actual relationship between the response $y$ and a set of $k$ independent variables. Consequently we try to formulate a reasonable general linear model.

Suppose the assumed model (the model we fit) is of a lower order than the actual model relating a response $y$ to a set of quantitative independent variables. Although the testing procedures discussed so far are not appropriate, it is possible to formulate a test of the inadequacy of the fitted polynomial model.

By way of illustration, suppose we are interested in investigating the effect of two physical characteristics ($x_1$ and $x_2$) on the hardness of steel clamps. To do so, we observe $y$ at the 8 factor-level combinations of a $2 \times 4$ factorial experiment. The model proposed is

$$y = \beta_0 + \beta_1 x_1 + \beta_2 x_2 + \beta_3 x_2^2 + \beta_4 x_1 x_2 + \beta_5 x_1 x_2^2 + \epsilon$$

lack of fit

It is quite possible that the assumed model is incorrect, and although we can perform tests for individual $\beta$'s or sets of $\beta$'s included in the model or compute $R^2$, the coefficient of determination, we have not discussed means for determining the inadequacy (*lack of fit*) of the model due to higher-degree terms that should be included in the statistical model.

When there is more than one observation per factor-level combination of the independent variables, we can conduct a test for lack of fit of the fitted model by partitioning SSE in the analysis of variance table into two parts, one due to "pure experimental error" and the other due to "lack of fit." If there are $n_i$ observations at the $i$th factor-level combination of the independent variables, then no matter what the form of the assumed general linear model, the quantity

$$\sum_{j=1}^{n_i} (y_{ij} - \bar{y}_i)^2$$

pure experimental error

provides a measure of what we will call *pure experimental error.* This sum of squares has $n_i - 1$ degrees of freedom.

Similarly, for each of the other factor-level combinations, we can compute a sum of squares due to pure experimental error. If there are $s$ different factor-level combinations, the pooled sum of squares

$SSP_{exp}$

$$SSP_{exp} = \sum_{i=1}^{s} \sum_{j=1}^{n_i} (y_{ij} - \bar{y}_i)^2$$

$SS_{Lack}$

called the *sum of squares for pure experimental error,* has $\Sigma_{i=1}^{s} (n_i - 1)$ degrees of freedom. With $SS_{Lack}$ representing the remaining portion of SSE, we have

$$\underset{\text{due to pure experimental error}}{\text{SSE} = SSP_{exp}} + \underset{\text{due to lack of fit}}{SS_{Lack}}$$

If SSE is based on $n - (k + 1)$ degrees of freedom in the general linear model, then $SS_{Lack}$ will have df $= [n - (k + 1)] - \Sigma_{i=1}^{s} (n_i - 1)$.

Under the null hypothesis that our model is correct, we can form independent estimates of $\sigma^2$, the model error variance. In particular, by dividing $SSP_{exp}$ and $SS_{Lack}$ by their respective degrees of freedom, we have two unbiased estimates of $\sigma^2$, $MSP_{exp}$ and $MS_{Lack}$.

The test for lack of fit is summarized in the box.

**Test for Lack of Fit**

$H_0$: The assumed model is correct.

$H_a$: There are additional polynomial terms that should be included in the model.

T.S.: $F = \dfrac{MS_{Lack}}{MSP_{exp}}$.

R.R: For a specified value of $\alpha$, reject $H_0$ (the adequacy of the model) if the computed value of $F$ exceeds the table value for $df_1$ = degrees of freedom for lack of fit and $df_2 = \Sigma_{i=1}^{s} (n_i - 1)$.

**EXAMPLE 16.9**

The amount of heat loss for 3 brands of thermal panes (A, B, C) was compared in a laboratory test room under controlled conditions. The data for the 3 × 3 factorial experiment are given in table 16.10. Random

**Table 16.10** Heat loss data for the experiment of example 16.9

| | *Outdoor Temperature (°F)* | | |
|---|---|---|---|
| *Brand of Thermal Pane* | *20* | *40* | *60* |
| | 86 | 56 | 38 |
| A | 80 | 51 | 39 |
| | 72 | 54 | 30 |
| | 91 | 58 | 35 |
| B | 93 | 53 | 32 |
| | 87 | 50 | 28 |
| | 78 | 49 | 26 |
| C | 71 | 51 | 38 |
| | 70 | 44 | 34 |

assignment of the order of testing was maintained, with triplicate determinations being made at each of the factor-level combinations. For each trial, the indoor temperature was controlled at 68°F and 50% relative humidity.

a. Fit the model

$$y = \beta_0 + \beta_1 x_1 + \beta_2 x_2 + \beta_3 x_3 + \beta_4 x_1 x_3 + \beta_5 x_2 x_3 + \epsilon$$

$x_1 = 1$ if Pane B $\qquad x_1 = 0$ otherwise

$x_2 = 1$ if Pane C $\qquad x_2 = 0$ otherwise

$x_3 =$ outdoor temperature

b. Test for lack of fit of the assumed model.

## SOLUTION

From the SAS computer output for the fitted model (the computer output follows), we have SSE $= 558.33$. To partition SSE into the sum of squares for pure experimental error and the sum of squares for lack of fit, we must first compute

$$\sum_{j=1}^{n_i} (y_{ij} - \bar{y}_i)^2$$

for each cell. Using the shortcut formula

$$\sum_{j=1}^{n_i} y_{ij}^2 - \frac{\left(\sum_{j=1}^{n_i} y_{ij}\right)^2}{n_i}$$

or a hand calculator, we find the contribution to pure experimental error for each cell to be

| | 20 | 40 | 60 |
|---|---|---|---|
| A | 98.67 | 12.67 | 48.67 |
| B | 18.67 | 32.67 | 24.67 |
| C | 38.00 | 26.00 | 74.67 |

Summing these results, we have

$$\text{SSP}_{\text{exp}} = \sum_{i=1}^{9} \sum_{j=1}^{3} (y_{ij} - \bar{y}_i)^2 = 374.69$$

By subtraction,

$$\text{SS}_{\text{Lack}} = \text{SSE} - \text{SSP}_{\text{exp}} = 558.33 - 374.69 = 183.64$$

The sum of squares due to pure experimental error has 18 degrees of freedom ($n_i = 3$ for all cells). Using df $= 21$ for SSE in the analysis of variance table, then $\text{SS}_{\text{Lack}}$ has 3 degrees of freedom. We find that

$$\text{MSP}_{\text{exp}} = 20.82 \qquad \text{and} \qquad \text{MS}_{\text{Lack}} = 61.21$$

The $F$ test is

$$F = \frac{MS_{\text{Lack}}}{MSP_{\text{exp}}} = \frac{61.21}{20.82} = 2.94$$

The table value of $F$, based on $df_1 = 3$, $df_2 = 18$, and $a = .05$, is 3.16. Since the computed value of $F$ does not exceed the table value, we have insufficient evidence to indicate a lack of fit for the model.

```
STATISTICAL ANALYSIS SYSTEM

DATA HEAT;
INPUT Y 1-2 X1 4 X2 6 X3 8-9;
X1X3=X1*X3;
X2X3=X2*X3;


CARDS;

  27 OBSERVATIONS IN DATA SET HEAT     9 VARIABLES
 PROC PRINT;
```

| OBS | Y | X1 | X2 | X3 | X1X3 | X2X3 |
|---|---|---|---|---|---|---|
| 1 | 86 | 0 | 0 | 20 | 0 | 0 |
| 2 | 80 | 0 | 0 | 20 | 0 | 0 |
| 3 | 72 | 0 | 0 | 20 | 0 | 0 |
| 4 | 91 | 1 | 0 | 20 | 20 | 0 |
| 5 | 93 | 1 | 0 | 20 | 20 | 0 |
| 6 | 87 | 1 | 0 | 20 | 20 | 0 |
| 7 | 78 | 0 | 1 | 20 | 0 | 20 |
| 8 | 71 | 0 | 1 | 20 | 0 | 20 |
| 9 | 70 | 0 | 1 | 20 | 0 | 20 |
| 10 | 56 | 0 | 0 | 40 | 0 | 0 |
| 11 | 51 | 0 | 0 | 40 | 0 | 0 |
| 12 | 54 | 0 | 0 | 40 | 0 | 0 |
| 13 | 58 | 1 | 0 | 40 | 40 | 0 |
| 14 | 53 | 1 | 0 | 40 | 40 | 0 |
| 15 | 50 | 1 | 0 | 40 | 40 | 0 |
| 16 | 49 | 0 | 1 | 40 | 0 | 40 |
| 17 | 51 | 0 | 1 | 40 | 0 | 40 |
| 18 | 44 | 0 | 1 | 40 | 0 | 40 |
| 19 | 38 | 0 | 0 | 60 | 0 | 0 |
| 20 | 39 | 0 | 0 | 60 | 0 | 0 |
| 21 | 30 | 0 | 0 | 60 | 0 | 0 |
| 22 | 35 | 1 | 0 | 60 | 60 | 0 |
| 23 | 32 | 1 | 0 | 60 | 60 | 0 |
| 24 | 28 | 1 | 0 | 60 | 60 | 0 |
| 25 | 26 | 0 | 1 | 60 | 0 | 60 |
| 26 | 38 | 0 | 1 | 60 | 0 | 60 |
| 27 | 34 | 0 | 1 | 60 | 0 | 60 |

```
PROC REGR;
MODEL Y= X1 X2 X3 X1X3 X2X3;
TITLE 'EXAMPLE 16.9';
```

```
**************************************************************************************************************

PROC REGR      :  EXAMPLE 16.9

DATA SET       :  HEAT     NUMBER OF VARIABLES = 9   NUMBER OF CLASSES = 0

VARIABLES      :  Y X1 X2 X3 X1X3 X2X3

**************************************************************************************************************
```

EXAMPLE 16.9

ANALYSIS OF VARIANCE TABLE, REGRESSION COEFFICIENTS, AND STATISTICS OF FIT FOR DEPENDENT VARIABLE Y

| SOURCE | DF | SUM OF SQUARES | MEAN SQUARE | F VALUE | PROB > F | R-SQUARE | C.V. |
|---|---|---|---|---|---|---|---|
| REGRESSION | 5 | 10715.66666667 | 2143.13333333 | 80.60740 | 0.0001 | 0.95047602 | 9.31859% |
| ERROR | 21 | 558.33333333 | 26.58730159 | | | | |
| | | | | | | STD DEV | Y MEAN |
| CORRECTED TOTAL | 26 | 11274.00000000 | | | | | |
| | | | | | | 5.15628758 | 55.33333 |

| SOURCE | DF | SEQUENTIAL SS | F VALUE | PROB > F | PARTIAL SS | F VALUE | PROB > F |
|---|---|---|---|---|---|---|---|
| X1 | 1 | 140.16666667 | 5.27194 | 0.0321 | 193.14285714 | 7.26448 | 0.0136 |
| X2 | 1 | 112.50000000 | 4.23134 | 0.0523 | 44.64285714 | 1.67910 | 0.2091 |
| X3 | 1 | 10176.88888889 | 382.77254 | 0.0001 | 2860.16666667 | 107.57642 | 0.0001 |
| X1X3 | 1 | 277.77777778 | 10.44776 | 0.0040 | 168.75000000 | 6.34701 | 0.0199 |
| X2X3 | 1 | 8.33333333 | 0.31343 | 0.5815 | 8.33333333 | 0.31343 | 0.5815 |

| SOURCE | B VALUES | T FOR HO:B = 0 | PROB > \|T\| | STD ERR B | STD B VALUES |
|---|---|---|---|---|---|
| INTERCEPT | 99.88888889 | 21.96607 | 0.0001 | 4.54741821 | 0.0 |
| X1 | 17.33333333 | 2.69527 | 0.0136 | 6.43102050 | 0.39986989 |
| X2 | −8.33333333 | −1.29580 | 0.2091 | 6.43102050 | −0.19224514 |
| X3 | −1.09166667 | −10.37191 | 0.0001 | 0.10525228 | −0.87240327 |
| X1X3 | −0.37500000 | −2.51933 | 0.0199 | 0.14884920 | −0.38688588 |
| X2X3 | 0.08333333 | 0.55985 | 0.5815 | 0.14884920 | 0.08597464 |

# Lincat: A Computer Procedure for Multidimensional Contingency Tables 16.8

In this section we will present an example of a procedure that can be used to compare $s$ different multinomial populations, where each population has $r$ possible responses. The procedure represents an extension of our study of two-dimensional contingency tables (chapter 10). We present the material in this chapter to take advantage of what we have learned about writing models for experimental situations. The example we will discuss makes use of methodology developed by Grizzle, Starmer, and Koch (1969), with the format explained in more detail by Forthofer, Starmer, and Grizzle (1971).

The subject of analyses for multidimensional contingency tables has been widely studied, and we can only hope to scratch the surface in this section. However, by following the example, you can see the utility of the approach and then refer to the references for more details of additional applications.

Assume there are $s$ different multinomial populations, each with $r$ categories of response. Let $\pi_{ij}$ (Greek letter pi) denote the cell probability for the $i$th row and $j$th column. Then the cell probabilities for the $s$ populations are as listed in table 16.11. Note that for any row we require

$\pi_{ij}$

$$\sum_{j=1}^{r} \pi_{ij} = 1$$

**Table 16.11** Cell probabilities for the $s$ populations

| | Category | | | | |
|---|---|---|---|---|---|
| *Population* | *1* | *2* | *3* | ... | *r* |
| 1 | $\pi_{11}$ | $\pi_{12}$ | $\pi_{13}$ | ... | $\pi_{1r}$ |
| 2 | $\pi_{21}$ | $\pi_{22}$ | $\pi_{23}$ | ... | $\pi_{2r}$ |
| 3 | $\pi_{31}$ | $\pi_{32}$ | $\pi_{33}$ | ... | $\pi_{3r}$ |
| ⋮ | ⋮ | ⋮ | ⋮ | | ⋮ |
| $s$ | $\pi_{s1}$ | $\pi_{s2}$ | $\pi_{s3}$ | ... | $\pi_{sr}$ |

The procedure we will consider for analyzing categorical data is similar to the general linear model approach of chapters 7 and 8. In particular, we construct $u$ different functions of the multinomial population probabilities by fitting the model

model

$$\mathbf{A}\boldsymbol{\pi} = \mathbf{X}\boldsymbol{\beta}$$

(Note that this model is similar to the general linear model $\mathbf{Y} = \mathbf{X}\boldsymbol{\beta}$.) The matrix **A** in our model is a matrix of constants that defines the $u$ different functions of the $\pi$'s and is used to specify the response. The matrix $\boldsymbol{\pi}$ is given by

$$\boldsymbol{\pi} = \begin{bmatrix} \pi_{11} \\ \pi_{12} \\ \vdots \\ \pi_{sr} \end{bmatrix}$$

The design matrix **X** is similar to those used in the general linear model, and $\boldsymbol{\beta}$ is a matrix of $v$ unknown parameters.

$$\boldsymbol{\beta} = \begin{bmatrix} \beta_1 \\ \beta_2 \\ \vdots \\ \beta_v \end{bmatrix} \qquad v \leq u$$

To bring this procedure to life, consider the following example. Suppose that a multiclinic investigation was conducted to compare 4 different treatments (nasal sprays) for their effectiveness in relieving the symptoms of ragweed allergy. Each investigator was to obtain a random sample of patients who responded positively to a skin test for ragweed allergy and who volunteered to participate in the study. A double blind procedure was used, in which each physician was supplied with a random code to assign one of the 4 sprays to patients but neither the investigator nor the patient knew which medication he was to receive. On the appointed day, volunteers were asked to report to the physician's office, where he or she was examined for allergic symptoms due to ragweed. Then each person was shown how to administer to himself one dose of the assigned nasal spray. After a four-hour period, each patient was rated as to whether he or she had improved while on the drug (yes or no).

The data appear in table 16.12. An entry is the number of yes or no responses for a treatment-investigator combination.

Note that this example differs from the two-dimensional contingency tables of section 10.7. In particular, we have $s = 24$ (6 investigators times 4 treatments) different multinomial populations with each population having $r = 2$ possible categories (yes or no).

The $\boldsymbol{\pi}$ matrix of unknown probabilities associated with our data is shown

**Table 16.12** Yes and no responses for the multiclinic investigation

| *Investigator* | *Treatment* | *Response* No | *Response* Yes | *Sample Size* |
|---|---|---|---|---|
| 1 | A | 1 | 2 | 3 |
| | B | 0 | 1 | 1 |
| | C | 2 | 1 | 3 |
| | D | 1 | 2 | 3 |
| 2 | A | 2 | 3 | 5 |
| | B | 1 | 2 | 3 |
| | C | 0 | 3 | 3 |
| | D | 0 | 4 | 4 |
| 3 | A | 10 | 11 | 21 |
| | B | 13 | 5 | 18 |
| | C | 12 | 7 | 19 |
| | D | 7 | 11 | 18 |
| 4 | A | 1 | 0 | 1 |
| | B | 3 | 0 | 3 |
| | C | 0 | 1 | 1 |
| | D | 2 | 0 | 2 |
| 5 | A | 2 | 4 | 6 |
| | B | 4 | 2 | 6 |
| | C | 7 | 6 | 13 |
| | D | 4 | 4 | 8 |
| 6 | A | 15 | 4 | 19 |
| | B | 14 | 4 | 18 |
| | C | 13 | 5 | 18 |
| | D | 14 | 8 | 22 |
| | | | | 218 |

below. Several of the individual $\pi$'s have been identified to show you how the matrix has been constructed using a single subscript.

**$\pi$ matrix**

$$\boldsymbol{\pi} = \begin{bmatrix} \pi_1 \\ \pi_2 \\ \pi_3 \\ \vdots \\ \pi_8 \\ \pi_9 \\ \vdots \\ \pi_{48} \end{bmatrix} \begin{array}{l} \text{probability of a no response on Treat. A, Invest. 1} \\ \text{probability of a yes response on Treat. A, Invest. 1} \\ \text{probability of a no response on Treat. B, Invest. 1} \\ \\ \text{probability of a yes response on Treat. D, Invest. 1} \\ \text{probability of a no response on Treat. A, Invest. 2} \\ \\ \text{probability of a yes response on Treat. D, Invest. 6} \end{array}$$

tests

There are a number of different hypotheses that we might wish to pose concerning the cell probabilities for the 24 multinomial populations.

For example, we might hypothesize that the proportion of no responses is linearly related to a treatment and an investigator effect. To do this, we define the following parameters:

$\mu$: overall mean effect
$\alpha_1$: effect due to Investigator 1
$\alpha_2$: effect due to Investigator 2
$\vdots$
$\alpha_5$: effect due to Investigator 5
$\beta_1$: effect due to Treatment A
$\beta_2$: effect due to Treatment B
$\beta_3$: effect due to Treatment C

Note that we have not defined an effect for Treatment D or for Investigator 6 since these effects can be stated in terms of the previous effects. In particular, we take

Investigator 6 effect: $\alpha_6 = -\alpha_1 - \alpha_2 - \alpha_3 - \alpha_4 - \alpha_5$
Treatment D effect: $\beta_4 = -\beta_1 - \beta_2 - \beta_3$

We now define the following matrix model relating the theoretical cell probabilities (the $\pi$'s) to a linear function of the $\alpha$'s, $\beta$'s, and $\mu$. This is given by our general model $\mathbf{A}\boldsymbol{\pi} = \mathbf{X}\boldsymbol{\beta}$, where $\mathbf{A}$ is a $24 \times 48$ matrix of the form

**A** matrix

$$\mathbf{A} = \begin{bmatrix} 1\ 0 & & & & & \\ & 1\ 0 & & & & \\ & & 1\ 0 & & & \\ & & & \cdot & & \\ & & & & \cdot & \\ & & & & & \cdot \\ & & & & & & 1\ 0 \end{bmatrix}$$

In general, we can write the matrix **A** as

**A***

$$\mathbf{A} = \begin{bmatrix} \mathbf{A}^* & & & & \\ & \mathbf{A}^* & & & \\ & & \cdot & & \\ & & & \cdot & \\ & & & & \cdot \\ & & & & & \mathbf{A}^* \end{bmatrix}$$

where, for our case, $\mathbf{A}^* = [1\ 0]$.

**X** matrix

The **X** matrix is the design matrix corresponding to the $\boldsymbol{\pi}$ matrix. For our example,

$$
\mathbf{X} = \begin{array}{c} \\ \end{array}
\begin{array}{ccccccccc}
\mu & \alpha_1 & \alpha_2 & \alpha_3 & \alpha_4 & \alpha_5 & \beta_1 & \beta_2 & \beta_3
\end{array}
$$

$$
\mathbf{X} = \begin{bmatrix}
1 & 1 & 0 & 0 & 0 & 0 & 1 & 0 & 0 \\
1 & 1 & 0 & 0 & 0 & 0 & 0 & 1 & 0 \\
1 & 1 & 0 & 0 & 0 & 0 & 0 & 0 & 1 \\
1 & 1 & 0 & 0 & 0 & 0 & -1 & -1 & -1 \\
1 & 0 & 1 & 0 & 0 & 0 & 1 & 0 & 0 \\
1 & 0 & 1 & 0 & 0 & 0 & 0 & 1 & 0 \\
1 & 0 & 1 & 0 & 0 & 0 & 0 & 0 & 1 \\
1 & 0 & 1 & 0 & 0 & 0 & -1 & -1 & -1 \\
1 & 0 & 0 & 1 & 0 & 0 & 1 & 0 & 0 \\
1 & 0 & 0 & 1 & 0 & 0 & 0 & 1 & 0 \\
1 & 0 & 0 & 1 & 0 & 0 & 0 & 0 & 1 \\
1 & 0 & 0 & 1 & 0 & 0 & -1 & -1 & -1 \\
1 & 0 & 0 & 0 & 1 & 0 & 1 & 0 & 0 \\
1 & 0 & 0 & 0 & 1 & 0 & 0 & 1 & 0 \\
1 & 0 & 0 & 0 & 1 & 0 & 0 & 0 & 1 \\
1 & 0 & 0 & 0 & 1 & 0 & -1 & -1 & -1 \\
1 & 0 & 0 & 0 & 0 & 1 & 1 & 0 & 0 \\
1 & 0 & 0 & 0 & 0 & 1 & 0 & 1 & 0 \\
1 & 0 & 0 & 0 & 0 & 1 & 0 & 0 & 1 \\
1 & 0 & 0 & 0 & 0 & 1 & -1 & -1 & -1 \\
1 & -1 & -1 & -1 & -1 & -1 & 1 & 0 & 0 \\
1 & -1 & -1 & -1 & -1 & -1 & 0 & 1 & 0 \\
1 & -1 & -1 & -1 & -1 & -1 & 0 & 0 & 1 \\
1 & -1 & -1 & -1 & -1 & -1 & -1 & -1 & -1
\end{bmatrix}
$$

For our data, the matrix of parameters is

$\boldsymbol{\beta}$ matrix

$$
\boldsymbol{\beta} = \begin{bmatrix}
\mu \\ \alpha_1 \\ \alpha_2 \\ \alpha_3 \\ \alpha_4 \\ \alpha_5 \\ \beta_1 \\ \beta_2 \\ \beta_3
\end{bmatrix}
$$

With these three matrices, we see that the model

$$\mathbf{A}\boldsymbol{\pi} = \mathbf{X}\boldsymbol{\beta}$$

indicates that the probability of a no response for a particular multinomial population is linearly related to the treatment and investigator effect for that cell. For example, from the first row of both sides of the equation, we have $\pi_1$, the probability of a no response on Investigator 1 and Treatment A, equal to $\mu + \alpha_1 + \beta_1$.

test for fit

Having specified a model, we can perform a number of different tests similar to those indicated by an analysis of variance table for the general linear model. The first test we consider is a test of the adequacy of the linear model $\mathbf{A}\boldsymbol{\pi} = \mathbf{X}\boldsymbol{\beta}$. Some people refer to this as a test for the treatment-by-investigator interaction. Since this test will be performed automatically by the computer program we will use, we will not go into any details here. Then provided there is no significant interaction, we proceed to examine the main effects due to treatments (sprays) and investigators.

Unlike the analysis of variance tests for the general linear model, our tests will utilize chi-square rather than $F$ statistics. The degrees of freedom for the chi-square test will be the same as the numerator degrees of freedom in a comparable $F$ test. The details of the calculations are left to the computer solution (Lincat).

As with the general linear model, we wish to test that sets of parameters equal zero. Main effects null hypotheses can be stated in matrix notation as

$$H_0\text{: } \mathbf{C}\boldsymbol{\beta} = \mathbf{0}$$

investigator effects

For testing that there is no effect due to investigators ($\alpha_1 = \alpha_2 = \cdots = \alpha_5 = 0$), we can use the $\mathbf{C}$ matrix

$$\mathbf{C} = \begin{bmatrix} 0 & 1 & 0 & 0 & 0 & 0 & 0 & 0 & 0 \\ 0 & 0 & 1 & 0 & 0 & 0 & 0 & 0 & 0 \\ 0 & 0 & 0 & 1 & 0 & 0 & 0 & 0 & 0 \\ 0 & 0 & 0 & 0 & 1 & 0 & 0 & 0 & 0 \\ 0 & 0 & 0 & 0 & 0 & 1 & 0 & 0 & 0 \end{bmatrix}$$

treatment effects

Similarly, for testing $H_0$: $\beta_1 = \beta_2 = \beta_3 = 0$ (no treatment effects), we use

$$\mathbf{C} = \begin{bmatrix} 0 & 0 & 0 & 0 & 0 & 0 & 1 & 0 & 0 \\ 0 & 0 & 0 & 0 & 0 & 0 & 0 & 1 & 0 \\ 0 & 0 & 0 & 0 & 0 & 0 & 0 & 0 & 1 \end{bmatrix}$$

No **C** matrix need be specified for the interaction test; it is performed automatically.

### EXAMPLE 16.10

Use the previous data to fit a model relating the probability of a no response to a linear function of a treatment and an investigator effect. Test for interaction and main effects. Use $\alpha = .05$ for each test.

### SOLUTION

The computer program we will use is described in detail in Forthofer, Starmer, and Grizzle (1971) and is available from the Biostatistics Department, University of North Carolina. Another program, Gencat, is an improved version of Lincat and is available from the Department of Biostatistics, University of Michigan.

We will identify the computer deck (without the appropriate control cards; these are dictated by the computer center you use) necessary to run this job. A copy of the ouput follows.

*Parameter card.*

| *Columns* | *Description* |
|---|---|
| 1–5 | number of categories of response (enter $r$) |
| 6–10 | number of multinomial populations (enter $s$) |
| 11–15 | 1 |
| 16–20 | 1 |
| 21–25 | 0 |
| 26–30 | number of **C** matrices |
| 31–35 | 0 |
| 36–40 | 0 |

For our data, the parameter card is

2 24 1 1 0 2 0 0

Note that we are not showing the entire data card here but only the numbers punched on the card.

*Data cards.* The data cards are punched by populations, beginning with a new card for each population. The number of responses falling into the first category of a population is recorded anywhere in the first 10 columns of the card, with the decimal punched. The first 10 columns are called a *10-column field.* The number of responses falling into the second category of a population is recorded in the next 10-column field (columns 11–20), and so on. For cells with no responses, we enter $1/r$ (in our case, 0.5) rather than a zero. Our data cards are presented here.

```
1.0    2.0
0.5    1.0
2.0    1.0
1.0    2.0
2.0    3.0
1.0    2.0
```

```
 0.5       3.0
 0.5       4.0
10.0      11.0
13.0       5.0
12.0       7.0
 7.0      11.0
 1.0       0.5
 3.0       0.5
 0.5       1.0
 2.0       0.5
 2.0       4.0
 4.0       2.0
 7.0       6.0
 4.0       4.0
15.0       4.0
14.0       4.0
13.0       5.0
14.0       8.0
```

**A*** *matrix*. The **A*** matrix is entered by rows in 10-column fields with the decimals punched.

```
1.0       0.0
```

**X** *matrix*. The first card for the **X** matrix contains the number of parameters in our model entered on the right side (right-justified) of the first 5-column field (columns 1–5). For example, if the **X** matrix contains 6 parameters, the number 6 would be entered in column 5. For an **X** matrix containing 12 parameters, the number 12 would be entered in columns 4 and 5. Then with the second card, we enter the **X** matrix *by columns* in fields of 5, with the decimal punched. Each column starts a new card. These cards are shown for our example.

```
    9
1.0  1.0  1.0  1.0  1.0  1.0  1.0  1.0  1.0  1.0  1.0  1.0  1.0  1.0  1.0  1.0
1.0  1.0  1.0  1.0  1.0  1.0  1.0  1.0
1.0  1.0  1.0  1.0  0.0  0.0  0.0  0.0  0.0  0.0  0.0  0.0  0.0  0.0  0.0  0.0
0.0  0.0  0.0  0.0  -1.0  -1.0  -1.0  -1.0
0.0  0.0  0.0  0.0  1.0  1.0  1.0  1.0  0.0  0.0  0.0  0.0  0.0  0.0  0.0  0.0
0.0  0.0  0.0  0.0  -1.0  -1.0  -1.0  -1.0
0.0  0.0  0.0  0.0  0.0  0.0  0.0  0.0  1.0  1.0  1.0  1.0  0.0  0.0  0.0  0.0
0.0  0.0  0.0  0.0  -1.0  -1.0  -1.0  -1.0
0.0  0.0  0.0  0.0  0.0  0.0  0.0  0.0  0.0  0.0  0.0  0.0  1.0  1.0  1.0  1.0
0.0  0.0  0.0  0.0  -1.0  -1.0  -1.0  -1.0
0.0  0.0  0.0  0.0  0.0  0.0  0.0  0.0  0.0  0.0  0.0  0.0  0.0  0.0  0.0  0.0
```

```
1.0  1.0  1.0  1.0  -1.0  -1.0  -1.0  -1.0
1.0  0.0  0.0  -1.0  1.0  0.0  0.0  -1.0  1.0  0.0  0.0  -1.0  1.0  0.0  0.0  -1.0
1.0  0.0  0.0  -1.0  1.0  0.0  0.0  -1.0
0.0  1.0  0.0  -1.0  0.0  1.0  0.0  -1.0  0.0  1.0  0.0  -1.0  0.0  1.0  0.0  -1.0
0.0  1.0  0.0  -1.0  0.0  1.0  0.0  -1.0
0.0  0.0  1.0  -1.0  0.0  0.0  1.0  -1.0  0.0  0.0  1.0  -1.0  0.0  0.0  1.0  -1.0
0.0  0.0  1.0  -1.0  0.0  0.0  1.0  -1.0
```

**C** *matrices*. For each **C** matrix needed for a specific null hypothesis, we first enter a card giving the number of rows of the **C** matrix in a 5-column field, right-justified. Succeeding cards contain the **C** matrix entered *by rows* in 5-column fields, with the decimal punched. Each row starts a new card.

```
    5
0.0  1.0  0.0  0.0  0.0  0.0  0.0  0.0  0.0
0.0  0.0  1.0  0.0  0.0  0.0  0.0  0.0  0.0
0.0  0.0  0.0  1.0  0.0  0.0  0.0  0.0  0.0
0.0  0.0  0.0  0.0  1.0  0.0  0.0  0.0  0.0
0.0  0.0  0.0  0.0  0.0  1.0  0.0  0.0  0.0
    3
0.0  0.0  0.0  0.0  0.0  0.0  1.0  0.0  0.0
0.0  0.0  0.0  0.0  0.0  0.0  0.0  1.0  0.0
0.0  0.0  0.0  0.0  0.0  0.0  0.0  0.0  1.0
```

A copy of the output follows. Comments have been made within the output to explain the parts of the output we are interested in.

```
                        GENERALIZED CHI-SQUARE ANALYSIS

R =       2   R IS THE NUMBER OF CATEGORIES OF RESPONSE

S =      24   S IS THE NUMBER OF POPULATIONS

U =       1   U IS THE RANK OF THE A MATRIX

MM =      1   MM = 1 IF LEAST SQUARES ANALYSIS IS USED AND ZERO OTHERWISE

ML =      0   ML = 0 IF TESTING A LINEAR HYPOTHESIS AND ML = 1 IF TESTING A LOGARITHMIC HYPOTHESIS

NC =      2   NC IS THE NUMBER OF SETS OF CONTRASTS TO BE TESTED BY LEAST SQUARES ANALYSIS

IK =      0   IK = 1 IF K IS THE IDENTITY MATRIX AND ZERO OTHERWISE

ISW =     0   ISW = 1 IF YOU WISH TO REANALYZE THE DATA ENTERED IN THE PRECEDING PROBLEM AND ZERO OTHERWISE
```

FREQUENCY TABLE (S × R)

| | |
|---|---|
| 1. | 2. |
| 1. | 1. |
| 2. | 1. |
| 1. | 2. |
| 2. | 3. |
| 1. | 2. |
| 1. | 3. |
| 1. | 4. |
| 10. | 11. |
| 13. | 5. |
| 12. | 7. |
| 7. | 11. |
| 1. | 1. |
| 3. | 1. |
| 1. | 1. |
| 2. | 1. |
| 2. | 4. |
| 4. | 2. |
| 7. | 6. |
| 4. | 4. |
| 15. | 4. |
| 14. | 4. |
| 13. | 5. |
| 14. | 8. |

This table gives the number of responses falling into the no and yes categories of each population. (Note: When a zero response was observed for a particular cell, we entered 0.5, which the computer then rounded to 1. for this frequency table.)

PROBABILITY TABLE (S × R)

| | |
|---|---|
| 0.33333 | 0.66667 |
| 0.33333 | 0.66667 |
| 0.66667 | 0.33333 |
| 0.33333 | 0.66667 |
| 0.40000 | 0.60000 |
| 0.33333 | 0.66667 |
| 0.14286 | 0.85714 |
| 0.11111 | 0.88889 |
| 0.47619 | 0.52381 |
| 0.72222 | 0.27778 |
| 0.63158 | 0.36842 |
| 0.38889 | 0.61111 |
| 0.66667 | 0.33333 |
| 0.85714 | 0.14286 |
| 0.33333 | 0.66667 |
| 0.80000 | 0.20000 |
| 0.33333 | 0.66667 |
| 0.66667 | 0.33333 |
| 0.53846 | 0.46154 |
| 0.50000 | 0.50000 |
| 0.78947 | 0.21053 |
| 0.77778 | 0.22222 |
| 0.72222 | 0.27778 |
| 0.63636 | 0.36364 |

This table gives the sample proportions of no and yes for each population, based on the frequency table entered into the computer (see p. 505).

CONTRAST MATRIX(U × RS) This is the **A*** matrix.

1.00000 0.0

F(P) LINEAR MODEL

| | | | | | | | |
|---|---|---|---|---|---|---|---|
| 0.33333D 00 | 0.33333D 00 | 0.66667D 00 | 0.33333D 00 | 0.40000D 00 | 0.33333D 00 | 0.14286D 00 | 0.11111D 00 |
| 0.47619D 00 | 0.72222D 00 | 0.63158D 00 | 0.38889D 00 | 0.66667D 00 | 0.85714D 00 | 0.33333D 00 | 0.80000D 00 |
| 0.33333D 00 | 0.66667D 00 | 0.53846D 00 | 0.50000D 00 | 0.78947D 00 | 0.77778D 00 | 0.72222D 00 | 0.63636D 00 |

VARIANCE COVARIANCE MATRIX

| | | | | | | | |
|---|---|---|---|---|---|---|---|
| 0.74074D-01 | 0.0 | 0.0 | 0.0 | 0.0 | 0.0 | 0.0 | 0.0 |
| 0.0 | 0.0 | 0.0 | 0.0 | 0.0 | 0.0 | 0.0 | 0.0 |
| 0.0 | 0.0 | 0.0 | 0.0 | 0.0 | 0.0 | 0.0 | 0.0 |
| 0.0 | 0.14815D 00 | 0.0 | 0.0 | 0.0 | 0.0 | 0.0 | 0.0 |
| 0.0 | 0.0 | 0.0 | 0.0 | 0.0 | 0.0 | 0.0 | 0.0 |
| 0.0 | 0.0 | 0.0 | 0.0 | 0.0 | 0.0 | 0.0 | 0.0 |
| 0.0 | 0.0 | 0.74074D-01 | 0.0 | 0.0 | 0.0 | 0.0 | 0.0 |
| 0.0 | 0.0 | 0.0 | 0.0 | 0.0 | 0.0 | 0.0 | 0.0 |
| 0.0 | 0.0 | 0.0 | 0.0 | 0.0 | 0.0 | 0.0 | 0.0 |
| 0.0 | 0.0 | 0.0 | 0.74074D-01 | 0.0 | 0.0 | 0.0 | 0.0 |
| 0.0 | 0.0 | 0.0 | 0.0 | 0.0 | 0.0 | 0.0 | 0.0 |
| 0.0 | 0.0 | 0.0 | 0.0 | 0.0 | 0.0 | 0.0 | 0.0 |
| 0.0 | 0.0 | 0.0 | 0.0 | 0.48000D-01 | 0.0 | 0.0 | 0.0 |
| 0.0 | 0.0 | 0.0 | 0.0 | 0.0 | 0.0 | 0.0 | 0.0 |
| 0.0 | 0.0 | 0.0 | 0.0 | 0.0 | 0.0 | 0.0 | 0.0 |
| 0.0 | 0.0 | 0.0 | 0.0 | 0.0 | 0.74074D-01 | 0.0 | 0.0 |
| 0.0 | 0.0 | 0.0 | 0.0 | 0.0 | 0.0 | 0.0 | 0.0 |
| 0.0 | 0.0 | 0.0 | 0.0 | 0.0 | 0.0 | 0.0 | 0.0 |
| 0.0 | 0.0 | 0.0 | 0.0 | 0.0 | 0.0 | 0.34985D-01 | 0.0 |
| 0.0 | 0.0 | 0.0 | 0.0 | 0.0 | 0.0 | 0.0 | 0.0 |
| 0.0 | 0.0 | 0.0 | 0.0 | 0.0 | 0.0 | 0.0 | 0.0 |
| 0.0 | 0.0 | 0.0 | 0.0 | 0.0 | 0.0 | 0.0 | 0.21948D-01 |
| 0.0 | 0.0 | 0.0 | 0.0 | 0.0 | 0.0 | 0.0 | 0.0 |
| 0.0 | 0.0 | 0.0 | 0.0 | 0.0 | 0.0 | 0.0 | 0.0 |
| 0.0 | 0.0 | 0.0 | 0.0 | 0.0 | 0.0 | 0.0 | 0.0 |
| 0.11878D-01 | 0.0 | 0.0 | 0.0 | 0.0 | 0.0 | 0.0 | 0.0 |
| 0.0 | 0.0 | 0.0 | 0.0 | 0.0 | 0.0 | 0.0 | 0.0 |
| 0.0 | 0.0 | 0.0 | 0.0 | 0.0 | 0.0 | 0.0 | 0.0 |
| 0.0 | 0.11145D-01 | 0.0 | 0.0 | 0.0 | 0.0 | 0.0 | 0.0 |
| 0.0 | 0.0 | 0.0 | 0.0 | 0.0 | 0.0 | 0.0 | 0.0 |
| 0.0 | 0.0 | 0.0 | 0.0 | 0.0 | 0.0 | 0.0 | 0.0 |
| 0.0 | 0.0 | 0.12247D-01 | 0.0 | 0.0 | 0.0 | 0.0 | 0.0 |
| 0.0 | 0.0 | 0.0 | 0.0 | 0.0 | 0.0 | 0.0 | 0.0 |
| 0.0 | 0.0 | 0.0 | 0.0 | 0.0 | 0.0 | 0.0 | 0.0 |
| 0.0 | 0.0 | 0.0 | 0.13203D-01 | 0.0 | 0.0 | 0.0 | 0.0 |
| 0.0 | 0.0 | 0.0 | 0.0 | 0.0 | 0.0 | 0.0 | 0.0 |
| 0.0 | 0.0 | 0.0 | 0.0 | 0.0 | 0.0 | 0.0 | 0.0 |
| 0.0 | 0.0 | 0.0 | 0.0 | 0.14815D 00 | 0.0 | 0.0 | 0.0 |
| 0.0 | 0.0 | 0.0 | 0.0 | 0.0 | 0.0 | 0.0 | 0.0 |
| 0.0 | 0.0 | 0.0 | 0.0 | 0.0 | 0.0 | 0.0 | 0.0 |
| 0.0 | 0.0 | 0.0 | 0.0 | 0.0 | 0.34985D-01 | 0.0 | 0.0 |
| 0.0 | 0.0 | 0.0 | 0.0 | 0.0 | 0.0 | 0.0 | 0.0 |
| 0.0 | 0.0 | 0.0 | 0.0 | 0.0 | 0.0 | 0.0 | 0.0 |
| 0.0 | 0.0 | 0.0 | 0.0 | 0.0 | 0.0 | 0.14815D 00 | 0.0 |
| 0.0 | 0.0 | 0.0 | 0.0 | 0.0 | 0.0 | 0.0 | 0.0 |
| 0.0 | 0.0 | 0.0 | 0.0 | 0.0 | 0.0 | 0.0 | 0.0 |
| 0.0 | 0.0 | 0.0 | 0.0 | 0.0 | 0.0 | 0.0 | 0.64000D-01 |
| 0.0 | 0.0 | 0.0 | 0.0 | 0.0 | 0.0 | 0.0 | 0.0 |
| 0.0 | 0.0 | 0.0 | 0.0 | 0.0 | 0.0 | 0.0 | 0.0 |

| | | | | | | | |
|---|---|---|---|---|---|---|---|
| 0.0 | 0.0 | 0.0 | 0.0 | 0.0 | 0.0 | 0.0 | 0.0 |
| 0.37037D-01 | 0.0 | 0.0 | 0.0 | 0.0 | 0.0 | 0.0 | 0.0 |
| 0.0 | 0.0 | 0.0 | 0.0 | 0.0 | 0.0 | 0.0 | 0.0 |
| 0.0 | 0.0 | 0.0 | 0.0 | 0.0 | 0.0 | 0.0 | 0.0 |
| 0.0 | 0.37037D-01 | 0.0 | 0.0 | 0.0 | 0.0 | 0.0 | 0.0 |
| 0.0 | 0.0 | 0.0 | 0.0 | 0.0 | 0.0 | 0.0 | 0.0 |
| 0.0 | 0.0 | 0.0 | 0.0 | 0.0 | 0.0 | 0.0 | 0.0 |
| 0.0 | 0.0 | 0.19117D-01 | 0.0 | 0.0 | 0.0 | 0.0 | 0.0 |
| 0.0 | 0.0 | 0.0 | 0.0 | 0.0 | 0.0 | 0.0 | 0.0 |
| 0.0 | 0.0 | 0.0 | 0.0 | 0.0 | 0.0 | 0.0 | 0.0 |
| 0.0 | 0.0 | 0.0 | 0.31250D-01 | 0.0 | 0.0 | 0.0 | 0.0 |
| 0.0 | 0.0 | 0.0 | 0.0 | 0.0 | 0.0 | 0.0 | 0.0 |
| 0.0 | 0.0 | 0.0 | 0.0 | 0.0 | 0.0 | 0.0 | 0.0 |
| 0.0 | 0.0 | 0.0 | 0.0 | 0.87476D-02 | 0.0 | 0.0 | 0.0 |
| 0.0 | 0.0 | 0.0 | 0.0 | 0.0 | 0.0 | 0.0 | 0.0 |
| 0.0 | 0.0 | 0.0 | 0.0 | 0.0 | 0.0 | 0.0 | 0.0 |
| 0.0 | 0.0 | 0.0 | 0.0 | 0.0 | 0.96022D-02 | 0.0 | 0.0 |
| 0.0 | 0.0 | 0.0 | 0.0 | 0.0 | 0.0 | 0.0 | 0.0 |
| 0.0 | 0.0 | 0.0 | 0.0 | 0.0 | 0.0 | 0.0 | 0.0 |
| 0.0 | 0.0 | 0.0 | 0.0 | 0.0 | 0.0 | 0.11145D-01 | 0.0 |
| 0.0 | 0.0 | 0.0 | 0 0 | 0.0 | 0.0 | 0.0 | 0.0 |
| 0.0 | 0.0 | 0.0 | 0.0 | 0.0 | 0.0 | 0.0 | 0.0 |
| 0.0 | 0.0 | 0.0 | 0.0 | 0.0 | 0.0 | 0.0 | 0.10518D-01 |

MINIMUM MODIFIED CHI-SQUARE ESTIMATES OF CELL PROBABILITIES

THESE ESTIMATES ARE VALID ONLY IF F(P) = 0

| | |
|---|---|
| 0.00000 | 1.00000 |
| 0.00000 | 1.00000 |
| 0.00000 | 1.00000 |
| 0.00000 | 1.00000 |
| 0.00000 | 1.00000 |
| 0.00000 | 1.00000 |
| 0.00000 | 1.00000 |
| 0.00000 | 1.00000 |
| 0.00000 | 1.00000 |
| 0.00000 | 1.00000 |
| 0.00000 | 1.00000 |
| 0.00000 | 1.00000 |
| 0.00000 | 1.00000 |
| 0.00000 | 1.00000 |
| 0.00000 | 1.00000 |
| 0.00000 | 1.00000 |
| 0.00000 | 1.00000 |
| 0.00000 | 1.00000 |
| 0.00000 | 1.00000 |
| 0.00000 | 1.00000 |
| 0.00000 | 1.00000 |
| 0.00000 | 1.00000 |
| 0.00000 | 1.00000 |
| 0.00000 | 1.00000 |

CHI-SQUARE = 418.1127 DF = 24 P = 0.0

DESIGN MATRIX — This is our **X** matrix.

| | | | | | | | | |
|---|---|---|---|---|---|---|---|---|
| 1. | 1. | 0. | 0. | 0. | 0. | 1. | 0. | 0. |
| 1. | 1. | 0. | 0. | 0. | 0. | 0. | 1. | 0. |
| 1. | 1. | 0. | 0. | 0. | 0. | 0. | 0. | 1. |
| 1. | 1. | 0. | 0. | 0. | 0. | -1. | -1. | -1. |
| 1. | 0. | 1. | 0. | 0. | 0. | 1. | 0. | 0. |
| 1. | 0. | 1. | 0. | 0. | 0. | 0. | 1. | 0. |
| 1. | 0. | 1. | 0. | 0. | 0. | 0. | 0. | 1. |
| 1. | 0. | 1. | 0. | 0. | 0. | -1. | -1. | -1. |
| 1. | 0. | 0. | 1. | 0. | 0. | 1. | 0. | 0. |
| 1. | 0. | 0. | 1. | 0. | 0. | 0. | 1. | 0. |
| 1. | 0. | 0. | 1. | 0. | 0. | 0. | 0. | 1. |
| 1. | 0. | 0. | 1. | 0. | 0. | -1. | -1. | -1. |
| 1. | 0. | 0. | 0. | 1. | 0. | 1. | 0. | 0. |
| 1. | 0. | 0. | 0. | 1. | 0. | 0. | 1. | 0. |
| 1. | 0. | 0. | 0. | 1. | 0. | 0. | 0. | 1. |
| 1. | 0. | 0. | 0. | 1. | 0. | -1. | -1. | -1. |
| 1. | 0. | 0. | 0. | 0. | 1. | 1. | 0. | 0. |
| 1. | 0. | 0. | 0. | 0. | 1. | 0. | 1. | 0. |
| 1. | 0. | 0. | 0. | 0. | 1. | 0. | 0. | 1. |
| 1. | 0. | 0. | 0. | 0. | 1. | -1. | -1. | -1. |
| 1. | -1. | -1. | -1. | -1. | -1. | 1. | 0. | 0. |
| 1. | -1. | -1. | -1. | -1. | -1. | 0. | 1. | 0. |
| 1. | -1. | -1. | -1. | -1. | -1. | 0. | 0. | 1. |
| 1. | -1. | -1. | -1. | -1. | -1. | -1. | -1. | -1. |

ESTIMATED MODE 3 PARAMETERS — These are least squares estimates of the model parameters.

0.53613D 00 -0.92943D-01 -0.30425D 00 0.20820D-01 0.19908D 00 -0.20553D-01 -0.14319D-01 0.10228D 00
0.14669D-01

VARIANCE COVARIANCE MATRIX OF THE ESTIMATED MODEL PARAMETERS

| | | | | | | | | |
|---|---|---|---|---|---|---|---|---|
| 0.16851D-02 | 0.18434D-02 | -0.12320D-03 | -0.11849D-02 | 0.12164D-02 | -0.47443D-03 | 0.10812D-03 | 0.78970D-05 | -0.96186D-05 |
| 0.18434D-02 | 0.15851D-01 | -0.33372D-02 | -0.23548D-02 | -0.48790D-02 | -0.30238D-02 | -0.24452D-03 | 0.40121D-03 | -0.12907D-03 |
| -0.12320D-03 | -0.33372D-02 | 0.79601D-02 | -0.40146D-03 | -0.28617D-02 | -0.10533D-02 | 0.10747D-03 | 0.46891D-03 | -0.52128D-04 |
| -0.11849D-02 | -0.23548D-02 | -0.40146D-03 | 0.37066D-02 | -0.16988D-02 | -0.31286D-04 | -0.12321D-03 | -0.87075D-04 | 0.24444D-04 |
| 0.12164D-02 | -0.48790D-02 | -0.28617D-02 | -0.16988D-02 | 0.13577D-01 | -0.25169D-02 | 0.39683D-03 | -0.94708D-03 | 0.51481D-03 |
| -0.47443D-03 | -0.30238D-02 | -0.10533D-02 | -0.31286D-04 | -0.25169D-02 | 0.65797D-02 | 0.97526D-04 | 0.20367D-03 | -0.47156D-03 |
| 0.10812D-03 | -0.24452D-03 | 0.10747D-03 | -0.12321D-03 | 0.39683D-03 | 0.97526D-04 | 0.28285D-02 | -0.95478D-03 | -0.93921D-03 |
| 0.78970D-05 | 0.40121D-03 | 0.46891D-03 | -0.87075D-04 | -0.94708D-03 | 0.20367D-03 | -0.95478D-03 | 0.28637D-02 | -0.97081D-03 |
| -0.96186D-05 | -0.12907D-03 | -0.52128D-04 | 0.24444D-04 | 0.51481D-03 | -0.47156D-03 | -0.93921D-03 | -0.97081D-03 | 0.28066D-02 |

```
CHI-SQUARE DUE TO ERROR =     7.0677 DF =     15     P = 0.9557
```

This is our chi-square test for interaction. The *p*-value is the probability of $x^2$ being greater than the observed value.

```
                    F(P)             LINEAR MODEL
 0.33333D 00  0.33333D 00  0.66667D 00  0.33333D 00  0.40000D 00  0.33333D 00  0.14286D 00  0.11111D 00
 0.47619D 00  0.72222D 00  0.63158D 00  0.38889D 00  0.66667D 00  0.85714D 00  0.33333D 00  0.80000D 00
 0.33333D 00  0.66667D 00  0.53846D 00  0.50000D 00  0.78947D 00  0.77778D 00  0.72222D 00  0.63636D 00

                    F(P) PREDICTED FROM FITTED MODEL
 0.42886D 00  0.54546D 00  0.45785D 00  0.34055D 00  0.21755D 00  0.33415D 00  0.24654D 00  0.12924D 00
 0.54263D 00  0.65922D 00  0.57161D 00  0.45432D 00  0.72089D 00  0.83749D 00  0.74988D 00  0.63258D 00
 0.50125D 00  0.61785D 00  0.53024D 00  0.41294D 00  0.71965D 00  0.83625D 00  0.74864D 00  0.63134D 00

                       F(P) - F(P) PREDICTED = RESIDUAL
-0.95531D-01 -0.21213D 00  0.20882D 00 -0.72208D-02  0.18245D 00 -0.81888D-03 -0.10368D 00 -0.18134D-01
-0.66436D-01  0.62998D-01  0.59965D-01 -0.65428D-01 -0.54225D-01  0.19654D-01 -0.41655D 00  0.16742D 00
-0.16792D 00  0.48816D-01  0.82207D-02  0.87056D-01  0.69823D-01 -0.58471D-01 -0.26416D-01  0.50226D-02

       C MATRIX
0.    1.    0.    0.    0.    0.    0.    0.    0.
0.    0.    1.    0.    0.    0.    0.    0.    0.
0.    0.    0.    1.    0.    0.    0.    0.    0.
0.    0.    0.    0.    1.    0.    0.    0.    0.
0.    0.    0.    0.    0.    1.    0.    0.    0.
```

The **C** matrix for investigators.

```
                        ESTIMATED MODEL CONTRASTS
-0.92943D-01   -0.30425D 00   0.20820D-01   0.19908D 00   -0.20553D-01

          STANDARD DEVIATIONS OF THE ESTIMATED MODEL CONTRASTS
 0.12590D 00    0.89219D-01   0.60882D-01   0.11652D 00    0.81115D-01


CHI-SQUARE =     25.0614 DF =     5     P = 0.0001
```

This is our chi-square test for investigators.

```
                        C MATRIX
0.    0.    0.    0.    0.    0.    1.    0.    0.
0.    0.    0.    0.    0.    0.    0.    1.    0.
0.    0.    0.    0.    0.    0.    0.    0.    1.
```

The **C** matrix for treatments.

```
          ESTIMATED MODEL CONTRASTS
-0.14319D-01       0.10228D 00       0.14669D-01

STANDARD DEVIATIONS OF THE ESTIMATED MODEL CONTRASTS
 0.53184D-01       0.53514D-01       0.52978D-01


CHI-SQUARE =     5.7173 DF =     3     P = 0.1262
```

This is our chi-square test for treatments.

To summarize the results of this analysis, the test for interaction between investigators and treatments (sprays) was nonsignificant ($p = .9557$). The tests for main effects showed a highly significant effect ($p = .0001$) due to differences among investigators, but the treatment effect only achieved a level of significance of .1262. Thus, although there were significant differences in the proportions of no responses for different investigators, there did not appear to be differences in the proportions of no responses for the different sprays.

# Summary 16.9

In this chapter we have extended the general linear model of chapter 6 to include both quantitative and qualitative independent variables. The procedure for including the main effects terms for a qualitative variable at $t$ levels is to assign $(t - 1)$ dummy variables, each of which is coded 0 or 1. Interaction terms are then formed from products of the $x$-values between main effects terms for different variables. An interaction term cannot be formed from the product of $x$-values for levels of the same independent variable.

The next step, after writing an appropriate model, is to define the parameters ($\beta$'s) which have been included in the model. If the model contains only qualitative variables, the $\beta$'s represent means or differences among means. For models containing only quantitative variables, the $\beta$'s define a curve (or surface) in terms of the intercept, slope, and so on. When both quantitative and qualitative independent variables are included in the same model, we obtain a separate curve (or surface) for each combination of levels of the qualitative variables. The $\beta$'s then distinguish between the different curves (or surfaces).

Having written a model and interpreted the $\beta$'s associated with the model, we may wish to make certain inferences using the sample data. In this chapter we presented a procedure for testing that each term in a specified set of $\beta$'s in the general linear model is equal to zero. The technique consists of fitting a complete model and a reduced model to obtain SSE for both. The drop in the sum of squares for error associated with the $\beta$'s under test can be used to form an $F$ test.

The fitting of complete and reduced models is a general procedure for obtaining the sum of squares associated with a source of variability in an analysis of variance table. Fortunately, for balanced designs these sums of squares drops can be computed with shortcut formulas (see chapters 13 and 15).

Finally, we discussed a procedure for analyzing multidimensional contingency tables. Although this material should logically have followed our discussion of contingency tables in chapter 10, we needed a discussion of linear models prior to presenting this methodology. One particular example was used to illustrate the analysis. Further reading on this topic can be found in the contingency table references at the end of the chapter.

# Exercises

16.11. An experimenter would like to compare the potencies of 3 different drug products. To do this, 12 test tubes were inoculated with a culture of the virus under study and incubated for 2 days at 35°C. Four dosage levels (.2, .4, .8, and 1.6 $\mu$g per tube) were to be used from each of the 3 drug products, with only one dose–drug product combination for each of the 12 test tube cultures. One means of comparing the drug products would be to examine their slopes (with respect to dose).

a. Write a general linear model relating the response $y$ to the independent variables "dose" and "drug product." Make the expected response a linear function of log dose ($x_1$). Identify the parameters in the model.

b. It would seem reasonable to assume that the three separate response lines have a common intercept $\beta_0$, since this would correspond to a zero dosage level of any of the drug products. Change the model of part a to reflect this change.

*16.12. Refer to exercise 16.11.

a. Use the data below to make a comparison among the three slopes. Fit a complete and a reduced model for your test. Use $\alpha = .05$.

| | Drug Product | | |
|---|---|---|---|
| *Dose* | *A* | *B* | *C* |
| .2 | 2.0 | 1.8 | 1.3 |
| .4 | 4.3 | 4.1 | 2.0 |
| .8 | 6.5 | 4.9 | 2.8 |
| 1.6 | 8.9 | 5.7 | 3.4 |

b. Is there evidence to indicate that the slopes are equal?

c. Suggest how you might test the null hypothesis that the intercepts are all equal to zero.

16.13. Use the data below to fit a model. Plot the data and suggest a polynomial model.

| $y$ | 7 | 8 | 6 | 12 | 15 | 13 | 7 | 10 | 11 | 14 | 16 | 17 |
|---|---|---|---|---|---|---|---|---|---|---|---|---|
| $x$ | 10 | 10 | 10 | 15 | 15 | 15 | 20 | 20 | 20 | 25 | 25 | 25 |

*16.14. Refer to the data of exercise 16.13.

a. Fit the model $y = \beta_0 + \beta_1 x + \beta_2 x^2 + \beta_3 x^3 + \epsilon$.

b. Test for lack of fit, using $\alpha = .05$.

*16.15. Refer to exercise 16.13. Suppose that the 3d, 5th, 6th, and 10th observations are missing.

a. Fit a cubic model.

b. Compare your results to those in exercise 16.14.

*16.16. An experiment was conducted to examine the weather resistance of a new commercial paint as a function of two independent variables, temperature $x_1$ and exposure time $x_2$. The sample data are listed below.

| $y$ | 120 | 101 | 110 | 105 | 92 | 130 |
|---|---|---|---|---|---|---|
| $x_1$ (°C) | −10 | −10 | 0 | 0 | 10 | 10 |
| $x_2$ (months) | 1 | 3 | 2 | 2 | 1 | 3 |

a. Assuming observations were obtained in a random order for the factor-level combinations, identify (or characterize) the experimental design.

b. Fit the model

$$y = \beta_0 + \beta_1 x_1 + \beta_2 x_2 + \beta_3 x_1 x_2 + \epsilon$$

*16.17. Refer to exercise 16.16.

a. Could we fit the following model?

$$y = \beta_0 + \beta_1 x_1 + \beta_2 x_1^2 + \beta_3 x_2 + \beta_4 x_2^2 + \beta_5 x_1 x_2 + \beta_6 x_1 x_2^2 + \beta_7 x_1^2 x_2 + \beta_8 x_1^2 x_2^2 + \epsilon$$

b. Test for lack of fit of the model in exercise 16.16, using $\alpha = .05$.

*16.18. Refer to example 16.10. If we wish to make pairwise comparisons among the 4 treatments, we can do so by identifying different **C** matrices for a null hypothesis of the form $\mathbf{C}\boldsymbol{\beta} = \mathbf{0}$. For example, to test for no significant difference between treatments A and B, we would use the **C** matrix.

$$\mathbf{C} = [0 \quad 0 \quad 0 \quad 0 \quad 0 \quad 0 \quad 1 \quad -1 \quad 0]$$

Identify the **C** matrices for the following comparisons. (Hint: For comparisons with Treatment D, recall that $\beta_4 = -\beta_1 - \beta_2 - \beta_3$.)

a. A versus C

b. A versus D

c. B versus D

d. C versus D

*16.19. Refer to exercise 16.18. Run the Lincat procedure to make the 5 pairwise treatment comparisons of the nasal sprays:

A versus B, A versus C, A versus D, B versus D, C versus D

Note that the number of **C** matrices to be used will change the parameter card of example 16.10.

*16.20. Refer to the data of exercise 7.27 (p. 183).

a. Can we test for lack of fit for the following model?

$$y = \beta_0 + \beta_1 x_1 + \beta_2 x_1^2 + \beta_3 x_2 + \beta_4 x_2^2 + \beta_5 x_1 x_2 + \beta_6 x_1 x_2^2 + \beta_7 x_1^2 x_2 + \beta_8 x_1^2 x_2^2 + \epsilon$$

b. Write the complete model for the sample data. Note that if there were replication at one or more design points, the number of degrees of freedom for $SS_{\text{Lack}}$ would be identical to the difference between the number of parameters in the complete model and the number of parameters in the model of part a.

*16.21. The solubility of a solution was examined for 6 different temperature settings.

| *y, Solubility by Weight* | *x, Temperature (in °C)* |
|---|---|
| 43, 45, 42 | 0 |
| 32, 33, 37 | 25 |
| 21, 28, 29 | 50 |
| 15, 14, 9 | 75 |
| 12, 10, 8 | 100 |
| 7, 6, 2 | 125 |

a. Plot the data, and fit as appropriate.
b. Test for lack of fit if possible. Use $\alpha = .05$.

*16.22. Refer to exercise 16.21. Suppose we are missing observations 5, 8, and 14.

a. Fit the model $y = \beta_0 + \beta_1 x + \beta_2 x^2 + \epsilon$.
b. Test for lack of fit, using $\alpha = .05$.

*16.23. Refer to the data of exercise 7.26 (p. 183).

a. Test for lack of fit of the model
$$y = \beta_0 + \beta_1 x_1 + \beta_2 x_1^2 + \beta_3 x_2 + \beta_4 x_1 x_2 + \beta_5 x_1^2 x_2 + \epsilon$$

b. Write the complete model for this experimental situation.

16.24. Refer to the data of exercise 7.23 (p. 176). Test for lack of fit of a quadratic model.

*16.25. In a random sample of 500 motorists on a major toll road, each driver was classified according to the following categories.

1. Transportation status: number of passengers (including the driver)

   1
   2
   3
   4 or more

2. Destination area of the city for the driver: NE, SE, NW, SW.
3. Whether or not the driver planned to use the proposed public transportation system (yes or no).

| *Yes* | | | | | *No* | | | | |
|---|---|---|---|---|---|---|---|---|---|
| | *Destination Area* | | | | | *Destination Area* | | | |
| *No. of Passengers* | *NE* | *SE* | *NW* | *SW* | *No. of Passengers* | *NE* | *SE* | *NW* | *SW* |
| 1 | 7 | 10 | 15 | 22 | 1 | 31 | 27 | 15 | 7 |
| 2 | 4 | 8 | 13 | 14 | 2 | 18 | 16 | 9 | 4 |
| 3 | 13 | 10 | 21 | 29 | 3 | 12 | 13 | 4 | 3 |
| 4 or more | 28 | 17 | 32 | 35 | 4 or more | 15 | 18 | 16 | 12 |

These data, the number of motorists in each classification, are summarized in two 2-way tables. Analyze these data using Lincat or some other multidimensional contingency table analysis program. If you use Lincat, follow the format of example 16.10. Draw your conclusions.

*16.26. Refer to example 16.8 (p. 489). Suggest another way to test for parallelism. Use the computer output given in example 16.8.

# References

Barr, A.J.; Goodnight, J.H.; Sall, J.P.; and Helwig, J.T. 1976. *A user's guide to SAS 76*. Raleigh, N.C.: SAS Institute, Inc.

Bishop, Y.M.M.; Fineberg, S.E.; and Holland, P.W. 1975. *Discrete multivariate analysis: Theory and practice*. Cambridge, Mass.: MIT Press.

Cochran, W.G., and Cox, G.M. 1957. *Experimental design*. 2d ed. New York: Wiley.

Dixon, W.J. 1975. *BMDP, biomedical computer programs*. Berkeley: University of California Press.

Dixon, W.J., and Massey, F.J., Jr. 1969. *Introduction to statistical analysis*. 3d ed. New York: McGraw-Hill.

Draper, N.R., and Smith, H. 1966. *Applied regression analysis*. New York: Wiley.

Forthofer, R.N.; Starmer, C.F.; and Grizzle, J.F. 1971. A program for the analysis of categorical data by linear models. *Journal of Biomedical Systems* 2: 3–48.

Goodman, L.A. 1971. The analysis of multidimensional contingency tables: Stepwise procedures and direct estimation methods for building models for multiple classifications. *Technometrics* 13: 33–61.

Graybill, F.A. 1976. *Theory and application of the linear model*. N. Scituate, Mass.: Duxbury Press.

Grizzle, J.E.; Starmer, C.F.; and Koch, G.G. 1969. Analysis of categorical data by linear models. *Biometrics* 25: 489–504.

Haberman, S.J. 1973. *C-tab analysis of multidimensional contingency tables by log-linear models*. Ann Arbor, Mich.: National Educational Resources, Inc.

Harnett, D.L. 1970. *Introduction to statistical methods*. Reading, Mass.: Addison-Wesley.

Kirk, R.E. 1968. *Experimental design: Procedures for the behavioral sciences*. Belmont, Calif.: Brooks/Cole.

Ku, H.H.; Varner, R.N.; and Kullback, S. 1971. Analysis of multidimensional contingency tables. *Journal of the American Statistical Association* 66: 55–64.

Landis, J.R.; Stanish, W.M.; and Koch, G.G. 1976. *A computer program for the generalized chi-square analysis of categorical data using weighted least squares to compute Wald statistics* (*Gencat*). Biostatistics technical report no. 8. Ann Arbor: Department of Biostatistics, University of Michigan.

Mendenhall, W. 1968. *Introduction to linear models and the design and analysis of experiments*. Belmont, Calif.: Wadsworth.

Nie, N.H.; Hull, C.H.; Jenkins, J.G.; Steinbrenner, K.; and Bent, D.H. 1975. *Statistical package for the social sciences*. 2d ed. New York: McGraw-Hill.

Ostle, B. 1963. *Statistics in research*. 2d ed. Ames, Iowa: Iowa State University Press.

Pearson, E.S., and Hartley, H.O. 1966. *Biometrika tables for statisticians*. 3d ed. Vol. I. London: Cambridge University Press.

Service, J. 1972. *A user's guide to the statistical analysis system*. Raleigh, N.C.: Student Supply Stores, North Carolina State University.

Snedecor, G.W., and Cochran, W.G. 1967. *Statistical methods*. 6th ed. Ames, Iowa: Iowa State University Press.

Sokal, R.R., and Rohlf, F.J. 1969. *Biometry*. San Francisco: W.H. Freeman.

Steel, R.G.D., and Torrie, J.H. 1960. *Principles and procedures of statistics.* New York: McGraw-Hill.

Wilcoxon, F., and Wilcox, R.A. 1964. *Some rapid approximate statistical procedures.* Pearl River, New York: Lederle Laboratories.

Winer, B.J. 1962. *Statistical principles in experimental design.* New York: McGraw-Hill.

Zahn, D.A. 1974. *Documentation for Contab—A computer program to aid in the analysis of multidimensional contingency tables using log-linear models.* FSU statistical report M292. Tallahassee: Department of Statistics, Florida State University.

# 17 Systems of Coding

## 17.1 Introduction

In chapters 13 and 15 we discussed the analysis of variance for a completely randomized design, randomized block design, Latin square design, and a $k$-way classification. For each analysis we posed a model and developed $F$ tests for making inferences about parameters in the model. Similarly, in chapter 16 we posed models for experimental situations using the format of the general linear model. To avoid confusion and the probability of conflicting conclusions based on different models for the same experimental situation, we must be able to express the parameters of one model in terms of parameters of a second model.

relationship between parameters of two models

For example, in chapter 15 we wrote the model for a randomized block design with $t = 2$ treatments in each of two blocks in the form

randomized block design

$$y_{ij} = \mu + \alpha_i + \gamma_j + \epsilon_{ij} \qquad (i = 1, 2;\ j = 1, 2)$$

where $\mu$ was an overall mean, $\alpha_i$ was an effect due to treatment $i$, and $\gamma_j$ was an effect due to block $j$. The terms $\mu$, $\alpha_i$, and $\gamma_j$ were assumed to be unknown constants. (Note: In chapter 15 we denoted the block effect by $\beta_j$. Here, to avoid confusion with the $\beta$'s in a general linear model, we will use the symbol $\gamma_j$.)

For the same experimental situation, we could write a general linear model as

general linear model

$$y = \beta_0 + \beta_1 x_1 + \beta_2 x_2 + \epsilon$$

where

$x_1 = 1$ if treatment 2 $\quad x_1 = 0$ otherwise

$x_2 = 1$ if block 2 $\quad x_2 = 0$ otherwise

What is the relationship between the parameters of the two models? The table for expected values, table 17.1, can be useful in interpreting the relationship between the two models. From table 17.1 we see immediately that $\beta_0$,

**Table 17.1** Expected values for two models for a randomized block design

| (a) | | | (b) | | |
|---|---|---|---|---|---|
| | *Treatments* | | | *Treatments* | |
| *Blocks* | 1 | 2 | *Blocks* | 1 | 2 |
| 1 | $\mu + \alpha_1 + \gamma_1$ | $\mu + \alpha_2 + \gamma_1$ | 1 | $\beta_0$ | $\beta_0 + \beta_1$ |
| 2 | $\mu + \alpha_1 + \gamma_2$ | $\mu + \alpha_2 + \gamma_2$ | 2 | $\beta_0 + \beta_2$ | $\beta_0 + \beta_1 + \beta_2$ |
| $E(y_{ij}) = \mu + \alpha_i + \gamma_j$ | | | $E(y) = \beta_0 + \beta_1 x_1 + \beta_2 x_2$ | | |

the mean response for treatment 1 and block 1, is equal to

$$\beta_0 = \mu + \alpha_1 + \gamma_1$$

Comparing the mean response for treatments 2 and 1 in block 1, we have

(a): $\mu + \alpha_2 + \gamma_1 - (\mu + \alpha_1 + \gamma_1) = \alpha_2 - \alpha_1$

(b): $\beta_0 + \beta_1 - \beta_0 = \beta_1$

For block 2 we also get $\alpha_2 - \alpha_1$ and $\beta_1$, respectively, as the difference in mean response for treatments 2 and 1. Since both models are explaining the same experimental situation, we have

$$\beta_1 = \alpha_2 - \alpha_1$$

A comparison of the mean response for blocks 2 and 1 for treatment 1 yields

(a): $\mu + \alpha_1 + \gamma_2 - (\mu + \alpha_1 + \gamma_1) = \gamma_2 - \gamma_1$

(b): $\beta_0 + \beta_2 - \beta_0 = \beta_2$

We obtain the same result when comparing blocks 2 and 1 for treatment 2. Thus

$$\beta_2 = \gamma_2 - \gamma_1$$

It is important to be able to relate parameters of one model to those of another for the same experimental situation, because not all people write models in the same way. To draw the same inferences, we must be certain we are testing (or estimating) the same parameter(s). For example, in the randomized block design just discussed, a test of no difference in treatment means can be stated in terms of the $\alpha$'s or $\beta$'s. In terms of the $\alpha$'s:

$$H_0\colon \alpha_2 - \alpha_1 = 0$$

but since we assume $\Sigma\,\alpha_i = 0$, this would imply $\alpha_1 = \alpha_2 = 0$. In terms of the $\beta$'s:

$$H_0\colon \beta_1 = 0$$

Both of these hypotheses are designed to test the same thing and we should not become confused if one experimenter phrases the null hypothesis in the framework of the $\alpha$'s and another experimenter uses the $\beta$'s of the general linear model.

In this chapter we will be concerned with the problem of relating the parameters associated with one system of coding to those of another system. This was illustrated for a simplified randomized block design in this section. In the remainder of this chapter, we will provide a procedure that defines the parameters of a general linear model in terms of parameters of any other model that might be written to describe the same experimental situation.

## 17.2 Relationship Between Two Systems of Coding

The general linear model was written in matrix notation when discussing quantitative independent variables in chapter 7. Since we utilize the same format whether we are concerned with qualitative or quantitative independent variables, the matrix notation is still the same.

For $n$ observations and the general linear model

$$y = \beta_0 + \beta_1 x_1 + \beta_2 x_2 + \cdots + \beta_k x_k + \epsilon$$

we have the matrix notation

$$\mathbf{Y} = \mathbf{X}\boldsymbol{\beta} + \boldsymbol{\epsilon}$$

matrix notation, linear model

where the matrices are defined as follows:

**Y:** An $n \times 1$ matrix containing the observations $y_1, y_2, \ldots, y_n$.

**X:** An $n \times (k + 1)$ matrix of settings for the $k$ independent variables, which is augmented by a column of 1s.

$\boldsymbol{\beta}$**:** A $(k + 1) \times 1$ matrix containing the unknown parameters $\beta_0, \beta_1, \ldots, \beta_k$ in the general linear model.

$\boldsymbol{\epsilon}$**:** An $n \times 1$ matrix of random errors associated with the $n$ observations.

Since the expected value of each random error is zero, we have

$$E(\mathbf{Y}) = \mathbf{X}\boldsymbol{\beta}$$

representing the expected values for all observations in our model.

A second system of coding, which employs different independent variables $x_0^*, x_1^*, x_2^*, \ldots, x_m^*$ to explain the *same* experimental situation, could be written by using the model

second system of coding

$$y = \beta_0^* + \beta_1^* x_1^* + \beta_2^* x_2^* + \cdots + \beta_m^* x_m^* + \epsilon \qquad m \geq k$$

The matrix notation would then be

matrix notation

$$\mathbf{Y} = \mathbf{X}^*\boldsymbol{\beta}^* + \boldsymbol{\epsilon}$$

where **Y** and $\boldsymbol{\epsilon}$ are defined as before and

**X*:** An $n \times (m + 1)$ matrix of settings for the $m$ new independent variables. This matrix is augmented also by a column of 1s.

$\boldsymbol{\beta}^*$**:** An $(m + 1) \times 1$ matrix of unknown parameters for the new model.

The expected value of the matrix **Y** is

$$E(\mathbf{Y}) = \mathbf{X}^*\boldsymbol{\beta}^*$$

What is the relationship between the two systems of coding? In particular, if someone uses a second system of coding different from the general linear model with which we are familiar, can we interpret the $\beta$'s of the general linear model in terms of those new parameters, the $\beta^*$'s?

We can proceed from the fact that

$$\mathbf{X}\boldsymbol{\beta} = \mathbf{X}^*\boldsymbol{\beta}^*$$

[since $E(\mathbf{Y})$ must be the same for two models describing the same experimental situation] to obtain the equation shown in the box. This equation gives the relationship between parameters of one system of coding and those of the general linear model.

**Relationship Between Parameters**

$$\boldsymbol{\beta} = (\mathbf{X}'\mathbf{X})^{-1}\mathbf{X}'\mathbf{X}^*\boldsymbol{\beta}^*$$

### EXAMPLE 17.1

Consider the randomized block design discussed in section 17.1, where $t = 2$ treatments appeared in $b = 2$ blocks. Using the two models (two systems of coding),

$$y_{ij} = \mu + \alpha_i + \gamma_j + \epsilon_{ij}$$

$$y = \beta_0 + \beta_1 x_1 + \beta_2 x_2 + \epsilon$$

obtain the relationship between the $\beta$'s of the general linear model and the parameters of the first model by way of the matrix result just described. Verify the answer with that obtained in section 17.1.

### SOLUTION

At first glance it might not appear that the first equation can be put into the framework

$$y = \beta_0^* + \beta_1^* x_1^* + \beta_2^* x_2^* + \cdots + \beta_m^* x_m^* + \epsilon$$

However, by making the following definitions,

$$\beta_0^* = \mu$$

$$\beta_1^* = \alpha_1 \quad \begin{cases} x_1^* = 1 \text{ if treatment 1} \\ x_1^* = 0 \text{ otherwise} \end{cases}$$

$$\beta_2^* = \alpha_2 \quad \begin{cases} x_2^* = 1 \text{ if treatment 2} \\ x_2^* = 0 \text{ otherwise} \end{cases}$$

$$\beta_3^* = \gamma_1 \quad \begin{cases} x_3^* = 1 \text{ if block 1} \\ x_3^* = 0 \text{ otherwise} \end{cases}$$

$$\beta_4^* = \gamma_2 \quad \begin{cases} x_4^* = 1 \text{ if block 2} \\ x_4^* = 0 \text{ otherwise} \end{cases}$$

we see that

$$y_{ij} = \mu + \alpha_i + \gamma_j + \epsilon_{ij}$$

is equivalent to a $\beta^*$ model with $m = 4$. The **Y** matrix for this design is

$$\mathbf{Y} = \begin{bmatrix} y_{11} \\ y_{12} \\ y_{21} \\ y_{22} \end{bmatrix}$$

The $\boldsymbol{\beta}^*$ and $\boldsymbol{\beta}$ matrices are

$$\boldsymbol{\beta}^* = \begin{bmatrix} \mu \\ \alpha_1 \\ \alpha_2 \\ \gamma_1 \\ \gamma_2 \end{bmatrix} \quad \text{and} \quad \boldsymbol{\beta} = \begin{bmatrix} \beta_0 \\ \beta_1 \\ \beta_2 \end{bmatrix}$$

The $\mathbf{X}^*$ and $\mathbf{X}$ matrices can be shown to be

$$\begin{array}{c} \\ \mathbf{X}^* = \end{array} \begin{array}{c} \begin{array}{ccccc} \mu & \alpha_1 & \alpha_2 & \gamma_1 & \gamma_2 \end{array} \\ \begin{bmatrix} 1 & 1 & 0 & 1 & 0 \\ 1 & 0 & 1 & 1 & 0 \\ 1 & 1 & 0 & 0 & 1 \\ 1 & 0 & 1 & 0 & 1 \end{bmatrix} \end{array} \qquad \begin{array}{c} \\ \mathbf{X} = \end{array} \begin{array}{c} \begin{array}{ccc} \beta_0 & \beta_1 & \beta_2 \end{array} \\ \begin{bmatrix} 1 & 0 & 0 \\ 1 & 1 & 0 \\ 1 & 0 & 1 \\ 1 & 1 & 1 \end{bmatrix} \end{array}$$

To assure yourself of the correctness of these matrices, check several of the expected values. For the first model

$$E(\mathbf{Y}) = \mathbf{X}^*\boldsymbol{\beta}^* = \begin{bmatrix} 1 & 1 & 0 & 1 & 0 \\ 1 & 0 & 1 & 1 & 0 \\ 1 & 1 & 0 & 0 & 1 \\ 1 & 0 & 1 & 0 & 1 \end{bmatrix} \begin{bmatrix} \mu \\ \alpha_1 \\ \alpha_2 \\ \gamma_1 \\ \gamma_2 \end{bmatrix} = \begin{bmatrix} \mu + \alpha_1 + \gamma_1 \\ \mu + \alpha_2 + \gamma_1 \\ \mu + \alpha_1 + \gamma_2 \\ \mu + \alpha_2 + \gamma_2 \end{bmatrix}$$

Note that these are precisely the expected values listed for $y_{11}, \ldots, y_{22}$ in table 17.1.

Similarly, by examining

$$E(\mathbf{Y}) = \mathbf{X}\boldsymbol{\beta}$$

we can verify the correctness of the matrices **X** and $\boldsymbol{\beta}$.

The formula for relating the matrix $\boldsymbol{\beta}$ to the matrix $\boldsymbol{\beta}^*$ requires the calculation of $(\mathbf{X}'\mathbf{X})^{-1}\mathbf{X}'\mathbf{X}^*$. Either by using a computer or by using the methods of chapter 7, it can be shown that

$$(\mathbf{X}'\mathbf{X}) = \begin{bmatrix} 4 & 2 & 2 \\ 2 & 2 & 1 \\ 2 & 1 & 2 \end{bmatrix} \qquad (\mathbf{X}'\mathbf{X})^{-1} = \begin{bmatrix} \frac{3}{4} & -\frac{1}{2} & -\frac{1}{2} \\ -\frac{1}{2} & 1 & 0 \\ -\frac{1}{2} & 0 & 1 \end{bmatrix}$$

$$(\mathbf{X}'\mathbf{X})^{-1}\mathbf{X}' = \begin{bmatrix} \frac{3}{4} & \frac{1}{4} & \frac{1}{4} & -\frac{1}{4} \\ -\frac{1}{2} & \frac{1}{2} & -\frac{1}{2} & \frac{1}{2} \\ -\frac{1}{2} & -\frac{1}{2} & \frac{1}{2} & \frac{1}{2} \end{bmatrix}$$

$$(\mathbf{X'X})^{-1}\mathbf{X'X^*} = \begin{bmatrix} 1 & 1 & 0 & 1 & 0 \\ 0 & -1 & 1 & 0 & 0 \\ 0 & 0 & 0 & -1 & 1 \end{bmatrix}$$

Substituting into the formula for $\beta$ in terms of $\beta^*$, we have

$$\boldsymbol{\beta} = (\mathbf{X'X})^{-1}\mathbf{X'X^*}\boldsymbol{\beta}^*$$

$$\begin{bmatrix} \beta_0 \\ \beta_1 \\ \beta_2 \end{bmatrix} = \begin{bmatrix} 1 & 1 & 0 & 1 & 0 \\ 0 & -1 & 1 & 0 & 0 \\ 0 & 0 & 0 & -1 & 1 \end{bmatrix} \begin{bmatrix} \mu \\ \alpha_1 \\ \alpha_2 \\ \gamma_1 \\ \gamma_2 \end{bmatrix}$$

Multiplying the two matrices on the right side of the equation, we have

$$\begin{bmatrix} \beta_0 \\ \beta_1 \\ \beta_2 \end{bmatrix} = \begin{bmatrix} \mu + \alpha_1 + \gamma_1 \\ \alpha_2 - \alpha_1 \\ \gamma_2 - \gamma_1 \end{bmatrix}$$

Note that these results are identical to those obtained in section 17.1.

While the matrix approach to relating two systems of coding may seem needlessly burdensome for the simple randomized block design, it is a procedure that can be used for any problem, complex or simple. All we need do is set up the **X** and **X*** matrices. Then either by hand or with a computer, we calculate $(\mathbf{X'X})^{-1}\mathbf{X'X^*}$. The relationship between the $\beta$'s and $\beta^*$'s then follows immediately.

# Exercises

17.1. Consider a completely randomized design with $t = 3$ treatments and the two equivalent models

$$y_{ij} = \mu + \alpha_i + \epsilon_{ij}$$

$$y = \beta_0 + \beta_1 x_1 + \beta_2 x_2 + \epsilon$$

$x_1 = 1$ if treatment 2 $\quad x_1 = 0$ otherwise

$x_2 = 1$ if treatment 3 $\quad x_2 = 0$ otherwise

a. Construct a table of expected values for the two models. Use this table to express the parameters of the second model in terms of the parameters in the first model.

b. Using the matrix expression

$$\boldsymbol{\beta} = (\mathbf{X}'\mathbf{X})^{-1}\mathbf{X}'\mathbf{X}^*\boldsymbol{\beta}^*$$

verify your results of part a.

17.2. Consider a randomized block design with $t = 4$ treatments in $b = 3$ blocks. Two equivalent models for this experimental situation are

$$y_{ij} = \mu + \alpha_i + \gamma_j + \epsilon_{ij}$$

$$y = \beta_0 + \overbrace{\beta_1 x_1 + \beta_2 x_2 + \beta_3 x_3}^{\text{treatments}} + \overbrace{\beta_4 x_4 + \beta_5 x_5}^{\text{blocks}} + \epsilon$$

a. Use a table of expected values to relate the parameters of the second model to those of the first model.

*b. Using a computer program, obtain the matrix $(\mathbf{X}'\mathbf{X})^{-1}\mathbf{X}'\mathbf{X}^*$ and verify your results of part a for the matrix expression

$$\boldsymbol{\beta} = (\mathbf{X}'\mathbf{X})^{-1}\mathbf{X}'\mathbf{X}^*\boldsymbol{\beta}^*$$

17.3. Consider the following two models for a $3 \times 3$ Latin square design.

$$y = \beta_0 + \overbrace{\beta_1 x_1 + \beta_2 x_2}^{\text{treatments}} + \overbrace{\beta_3 x_3 + \beta_4 x_4}^{\text{rows}} + \overbrace{\beta_5 x_5 + \beta_6 x_6}^{\text{columns}} + \epsilon$$

$$y_{ijk} = \mu + \alpha_i + \gamma_j + \delta_k + \epsilon_{ijk}$$

Relate the parameters in the first model to those in the second model.

# Orthogonal Linear Contrasts for Least Squares and the Analysis of Variance 17.3

In chapter 14 we alluded to the fact that it is possible to partition the sums of squares associated with each main effect of a $k$-way classification into single-degree-of-freedom sums of squares by using orthogonal linear contrasts among the observations. For example, if factor $A$ of a $k$-way classification has "a" levels, there are a − 1 degrees of freedom associated with SSA in the AOV table. It can be shown that we can partition SSA into a − 1 sums of squares, each with one degree of freedom, by using a − 1 orthogonal linear contrasts among the sample observations.

In chapter 14 (p. 379) we also gave a definition for a linear contrast among $k$ sample means. We can now do the same for a contrast among a set of $n$ observations (since an observation is a sample mean based on one observation).

**Definition 17.1**

A *linear contrast* of $n$ observations can be written in the form

$$\ell = a_1y_1 + a_2y_2 + \cdots + a_ny_n = \sum_{i=1}^{n} a_iy_i$$

where the $a_i$'s are constants satisfying the property $\Sigma_{i=1}^{n} a_i = 0$.

Note: We used the symbol $\hat{\ell}$ to denote a linear contrast in chapter 14. In this chapter we will use simply the letter $\ell$.

**Definition 17.2**

Two linear contrasts $\ell_1$ and $\ell_2$, where

$$\ell_1 = \sum_{i=1}^{n} a_iy_i \qquad \ell_2 = \sum_{i=1}^{n} b_iy_i$$

orthogonal

are *orthogonal* if

$$a_1b_1 + a_2b_2 + \cdots + a_nb_n = \sum_{i=1}^{n} a_ib_i = 0$$

$k = 1$ model

Consider first a one-way classification ($k = 1$) with $t = 4$ treatments and 3 observations per treatment. A general linear model for the experiment is

$$y = \beta_0 + \beta_1x_1 + \beta_2x_2 + \beta_3x_3 + \epsilon$$

where

$x_1 = 1$ if treatment 2 $\quad x_1 = 0$ otherwise

$x_2 = 1$ if treatment 3 $\quad x_2 = 0$ otherwise

$x_3 = 1$ if treatment 4 $\quad x_3 = 0$ otherwise

A test of the null hypothesis $\beta_1 = \beta_2 = \beta_3 = 0$ (no difference among means) makes use of SST, the sum of squares for treatments, based on 3 degrees of freedom. The **Y** and **X** matrices for this general linear model are

$$\mathbf{Y} = \begin{bmatrix} y_1 \\ y_2 \\ y_3 \\ \hline y_4 \\ y_5 \\ y_6 \\ \hline y_7 \\ y_8 \\ y_9 \\ \hline y_{10} \\ y_{11} \\ y_{12} \end{bmatrix} \begin{matrix} \\ \text{treatment 1} \\ \\ \\ \text{treatment 2} \\ \\ \\ \text{treatment 3} \\ \\ \\ \text{treatment 4} \\ \\ \end{matrix} \qquad \mathbf{X} = \begin{array}{c} \begin{matrix} & x_1 & x_2 & x_3 \end{matrix} \\ \left[\begin{array}{cccc} 1 & 0 & 0 & 0 \\ 1 & 0 & 0 & 0 \\ 1 & 0 & 0 & 0 \\ \hline 1 & 1 & 0 & 0 \\ 1 & 1 & 0 & 0 \\ 1 & 1 & 0 & 0 \\ \hline 1 & 0 & 1 & 0 \\ 1 & 0 & 1 & 0 \\ 1 & 0 & 1 & 0 \\ \hline 1 & 0 & 0 & 1 \\ 1 & 0 & 0 & 1 \\ 1 & 0 & 0 & 1 \end{array}\right] \end{array}$$

We can obtain SST either by fitting complete and reduced models to obtain $SS_{\text{drop}} = \text{SST}$ or by using the computational formula (of chapter 13)

$$\text{SST} = \sum_i \frac{T_i^2}{3} - \frac{G^2}{12}$$

coding by orthogonal linear contrasts

Using orthogonal linear contrasts among the 12 observations, it is possible to partition SST into three sums of squares, each with one degree of freedom. The single-degree-of-freedom sums of squares can be used to test individual hypotheses among the 4 different treatments. We will introduce the orthogonal linear contrasts as part of a new model (system of coding) for $y$.

Let

$$y = \beta_0^* + \beta_1^* x_1^* + \beta_2^* x_2^* + \beta_3^* x_3^* + \epsilon$$

The sample of 12 observations on $y$ can be written in matrix notation as

$$\mathbf{Y} = \mathbf{X}^* \boldsymbol{\beta}^* + \boldsymbol{\epsilon}$$

where

$$\mathbf{Y} = \begin{bmatrix} y_1 \\ y_2 \\ y_3 \\ \hline y_4 \\ y_5 \\ y_6 \\ \hline y_7 \\ y_8 \\ y_9 \\ \hline y_{10} \\ y_{11} \\ y_{12} \end{bmatrix} \begin{matrix} \\ \text{treatment 1} \\ \\ \\ \text{treatment 2} \\ \\ \\ \text{treatment 3} \\ \\ \\ \text{treatment 4} \\ \\ \end{matrix} \qquad \mathbf{X}^* = \begin{matrix} & x_1^* & x_2^* & x_3^* \\ \end{matrix} \begin{bmatrix} 1 & 1 & 1 & 1 \\ 1 & 1 & 1 & 1 \\ 1 & 1 & 1 & 1 \\ \hline 1 & -1 & 1 & 1 \\ 1 & -1 & 1 & 1 \\ 1 & -1 & 1 & 1 \\ \hline 1 & 0 & -2 & 1 \\ 1 & 0 & -2 & 1 \\ 1 & 0 & -2 & 1 \\ \hline 1 & 0 & 0 & -3 \\ 1 & 0 & 0 & -3 \\ 1 & 0 & 0 & -3 \end{bmatrix} \qquad \boldsymbol{\beta}^* = \begin{bmatrix} \beta_0^* \\ \beta_1^* \\ \beta_2^* \\ \beta_3^* \end{bmatrix}$$

and $\boldsymbol{\epsilon}$ is a matrix of random errors associated with the 12 observations.

The $x_1^*$, $x_2^*$, and $x_3^*$ columns of the $\mathbf{X}^*$ matrix give the coefficients of a set of three orthogonal linear contrasts for the $y$-values. The coefficients in the $x_1^*$ column refer to the linear contrast

$$\ell_1 = y_1 + y_2 + y_3 - y_4 - y_5 - y_6$$

Similarly, from the $x_2^*$ and $x_3^*$ columns of the $\mathbf{X}^*$ matrix, we have, respectively, the linear contrasts

$$\ell_2 = y_1 + y_2 + y_3 + y_4 + y_5 + y_6 - 2y_7 - 2y_8 - 2y_9$$

$$\ell_3 = y_1 + y_2 + y_3 + y_4 + y_5 + y_6 + y_7 + y_8 + y_9 - 3y_{10} - 3y_{11} - 3y_{12}$$

In addition, it can be seen that $\Sigma_{i=1}^{12} a_i b_i$ for any pair of contrasts equals zero. For example, matching coefficients for the $y$-values in $\ell_1$ and $\ell_2$, we have

$$\begin{aligned} \sum_{i=1}^{12} a_i b_i &= 1(1) + 1(1) + 1(1) - 1(1) - 1(1) - 1(1) \\ &\quad + 0(-2) + 0(-2) + 0(-2) \\ &= 0 \end{aligned}$$

Because the $x_1^*$, $x_2^*$, and $x_3^*$ columns represent the coefficients associated with orthogonal linear contrasts, the $\mathbf{X^{*\prime}X^*}$ matrix will be a diagonal matrix, with diagonal elements, in general, of the form

$$\mathbf{X^{*\prime}X^*} = \begin{bmatrix} n & & & & & \\ & \sum a_i^2 & & & & \\ & & \sum b_i^2 & & & \\ & & & \sum c_i^2 & & \\ & & & & \cdot & \\ & & & & & \cdot \\ & & & & & & \cdot \end{bmatrix}$$

where $\Sigma\, a_i^2$, $\Sigma\, b_i^2$, $\Sigma\, c_i^2$, and so on, represent the sum of squares of the coefficients of the orthogonal linear contrasts. Or, as we have seen, they represent the sum of squares of the entries in the columns of the $\mathbf{X}^*$ matrix. For our example with $n = 12$,

$$\mathbf{X^{*\prime}X^*} = \begin{bmatrix} 12 & & & \\ & 6 & & \\ & & 18 & \\ & & & 36 \end{bmatrix}.$$

The $\mathbf{X^{*\prime}Y}$ matrix also takes a very simple form. In general,

$$\mathbf{X^{*\prime}Y} = \begin{bmatrix} \sum y \\ \ell_1 \\ \ell_2 \\ \ell_3 \\ \vdots \end{bmatrix}$$

where the $\ell_i$ are the orthogonal linear contrasts of the measurements.

With a diagonal matrix for $\mathbf{X^{*\prime}X^*}$ and a simplified $\mathbf{X^{*\prime}Y}$ matrix, the least squares estimate of the $\beta^*$ parameters in matrix notation is

$$\widehat{\boldsymbol{\beta}}^* = (\mathbf{X}^{*\prime}\mathbf{X}^*)^{-1}\mathbf{X}^{*\prime}\mathbf{Y}$$

$$= \begin{bmatrix} \bar{y} \\ \dfrac{\ell_1}{\sum a_i^2} \\ \dfrac{\ell_2}{\sum b_i^2} \\ \dfrac{\ell_3}{\sum c_i^2} \\ \vdots \end{bmatrix} \begin{matrix} \widehat{\beta}_0^* \\ \widehat{\beta}_1^* \\ \widehat{\beta}_2^* \\ \widehat{\beta}_3^* \\ \vdots \end{matrix}$$

What interpretation can we give to $\widehat{\boldsymbol{\beta}}^*$? Since we know the form of $\ell_1, \ell_2, \ell_3, \ldots,$ we can see that the estimates have the following interpretations:

$\widehat{\beta}_0^*$: An estimate of the overall mean.

$\widehat{\beta}_1^* = \dfrac{\ell_1}{\sum a_i^2}$: A comparison of the means for treatments 1 and 2.

$\widehat{\beta}_2^* = \dfrac{\ell_2}{\sum b_i^2}$: A comparison of the means for treatments 1 and 2 versus treatment 3.

$\widehat{\beta}_3^* = \dfrac{\ell_3}{\sum c_i^2}$: A comparison of the means for treatments 1, 2, and 3 versus treatment 4.

tests

A test of $H_0$: $\beta_1^* = 0$ would be a test of the hypothesis that the difference in the mean response for treatments 1 and 2, $\mu_1 - \mu_2$, is zero. Similarly, for $H_0$: $\beta_2^* = 0$ we would be testing that the contrast $\mu_1 + \mu_2 - 2\mu_3 = 0$; and $H_0$: $\beta_3^* = 0$ would correspond to the hypothesis $\mu_1 + \mu_2 + \mu_3 - 3\mu_4 = 0$.

The sum of squares due to treatments, SST, which will have 3 degrees of freedom for our example, can now be represented as the sum

$$\text{SST} = \frac{\ell_1^2}{\sum a_i^2} + \frac{\ell_2^2}{\sum b_i^2} + \frac{\ell_3^2}{\sum c_i^2}$$

and each term has one degree of freedom. In general, SST could be written as the sum of $(t - 1)$ single-degree-of-freedom terms. Partitioning the sum of squares for treatments in a $k$-way classification into single-degree-of-freedom sums of squares, we have

$$\text{SST} = \frac{\ell_1^2}{\sum a_i^2} + \frac{\ell_2^2}{\sum b_i^2} + \frac{\ell_3^2}{\sum c_i^2} + \cdots$$

Each of the individual sums of squares can be used to test an hypothesis concerning a $\beta^*$ in the new model. In particular, for our example a test of $H_0$: $\beta_1^* = 0$ (no difference in the means for treatments 1 and 2) would use the statistic

test statistics

$$F = \frac{\ell_1^2/\sum a_i^2}{\text{MSE}}$$

The computed value of $F$ would then be compared to a table $F$-value for $\text{df}_1 = 1$ and $\text{df}_2 =$ degrees of freedom of MSE. Similarly, the tests for $H_0$: $\beta_2^* = 0$ and $H_0$: $\beta_3^* = 0$ would use the statistics

$$F = \frac{\ell_2^2/\sum b_i^2}{\text{MSE}} \quad \text{and} \quad F = \frac{\ell_3^2/\sum c_i^2}{\text{MSE}}$$

respectively, and would be compared to the same critical $F$-value.

To summarize, we have considered a way to partition the sum of squares due to treatments in a completely randomized design into $(t - 1)$ single-degree-of-freedom sums of squares of the form $\ell_1^2/\Sigma a_i^2, \ell_2^2/\Sigma b_i^2, \ldots,$ where the $\ell_i$'s $(i = 1, 2, \ldots, t - 1)$ represent orthogonal linear contrasts among the $n$ sample observations. We suggested using the set of coefficients for the $\ell_i$'s shown in table 17.2. Examples 17.2 through 17.5 of this section show how these contrasts are to be used for partitioning the sums of squares for main effects and interaction terms in an analysis of variance.

**Table 17.2** Coefficients for the $t - 1$ orthogonal linear contrasts

| *Observations on Treatments* | *Contrasts* | | | | |
|---|---|---|---|---|---|
| | *1* | *2* | *3* | $\cdots$ | $t - 1$ |
| 1 | 1 | 1 | 1 | $\cdots$ | 1 |
| 2 | −1 | 1 | 1 | $\cdots$ | 1 |
| 3 | 0 | −2 | 1 | $\cdots$ | 1 |
| 4 | 0 | 0 | −3 | $\cdots$ | 1 |
| 5 | 0 | 0 | 0 | $\cdots$ | 1 |
| $\vdots$ | $\vdots$ | $\vdots$ | $\vdots$ | | $\vdots$ |
| $t - 1$ | 0 | 0 | 0 | $\cdots$ | 1 |
| $t$ | 0 | 0 | 0 | $\cdots$ | $-(t - 1)$ |

The procedure of partitioning the sum of squares for treatments can be viewed as a new method of coding, since the coefficients of the orthogonal linear contrasts appear as columns of the $\mathbf{X}^*$ matrix. Because of the properties associated with orthogonal linear contrasts, the least square estimates for the $\beta^*$ parameters have a simplified form. In particular,

$$\widehat{\beta}_0^* = \bar{y}$$

$$\widehat{\beta}_1^* = \frac{\ell_1}{\sum a_i^2}$$

$$\widehat{\beta}_2^* = \frac{\ell_2}{\sum b_i^2}$$

An $F$ test can be made for each single-degree-of-freedom sum of squares in the partitioning of SST. Each test will relate to a $\beta^*$ in the new model and will involve a contrast among the treatment means.

A word of caution is given here about performing these multiple $F$ tests. From our work in chapter 14, we know that if MSE has a large number of degrees of freedom, the separate $F$ tests will be approximately independent. Then if each test is conducted at a preset $\alpha$-value, and if $\mu_1 = \mu_2 = \cdots = \mu_t$, the probability of rejecting $H_0$ on at least one test is $1 - (1 - \alpha)^{t-1}$. (Refer to table 14.1, p. 383, to examine the overall error rate for various values of $\alpha$ and $t$.) If many of these single-degree-of-freedom $F$ tests are to be performed, we should preset $\alpha$ at a fairly small value for an individual test. Or, better still, rather than presetting $\alpha$, we could examine the level of significance for each test. Then we would tend to believe only those results that are highly significant.

relationship to original parameters

The only thing that we have not considered in this section is relating the $\beta^*$ parameters of the new model to the original $\beta$'s of the general linear model. While we might be able to do this without going through a matrix solution, it is certainly possible to compute $(\mathbf{X}'\mathbf{X})^{-1}\,\mathbf{X}'\mathbf{X}^*$ to obtain the relationship

$$\boldsymbol{\beta} = (\mathbf{X}'\mathbf{X})^{-1}\mathbf{X}'\mathbf{X}^*\boldsymbol{\beta}^*$$

Most times when we use the orthogonal linear contrasts, we would not need to relate back to the general linear model. However, should the necessity arise, we can do so by using the above formula.

### EXAMPLE 17.2

In example 13.1 (p. 363) we referred to an experiment conducted by a horticulturist for investigating the phosphorous content of tree leaves from 3 different varieties of apple trees. Use the data, which have been reproduced here in table 17.3, to partition the sum of squares for treatments (varieties) into single-degree-of-freedom sums of squares. Identify the orthogonal linear contrasts used in the partitioning and run an appropriate $F$ test for each. Use $\alpha = .05$.

**Table 17.3** Phosphorous content data for tree leaves, example 17.2

| | *Variety* | | |
|---|---|---|---|
| | *1* | *2* | *3* |
| | .35 | .65 | .60 |
| | .40 | .70 | .80 |
| | .58 | .90 | .75 |
| | .50 | .84 | .73 |
| | .47 | .79 | .66 |
| *Totals* | 2.30 | 3.88 | 3.54 |

**SOLUTION**

Since this is a completely randomized design ($t = 3$) with an equal number of observations per treatment (5), we can partition the sum of squares due to varieties into two single-degree-of-freedom sums of squares. The **Y** matrix for these data and the **X*** matrix, which identifies the coefficients of the orthogonal linear contrasts, are

$$\mathbf{Y} = \begin{bmatrix} .35 \\ .40 \\ .58 \\ .50 \\ .47 \\ \hline .65 \\ .70 \\ .90 \\ .84 \\ .79 \\ \hline .60 \\ .80 \\ .75 \\ .73 \\ .66 \end{bmatrix} \begin{matrix} \\ \\ \text{Variety 1} \\ \\ \\ \\ \\ \text{Variety 2} \\ \\ \\ \\ \\ \text{Variety 3} \\ \\ \\ \end{matrix} \qquad \mathbf{X}^* = \begin{bmatrix} 1 & 1 & 1 \\ 1 & 1 & 1 \\ 1 & 1 & 1 \\ 1 & 1 & 1 \\ 1 & 1 & 1 \\ \hline 1 & -1 & 1 \\ 1 & -1 & 1 \\ 1 & -1 & 1 \\ 1 & -1 & 1 \\ 1 & -1 & 1 \\ \hline 1 & 0 & -2 \\ 1 & 0 & -2 \\ 1 & 0 & -2 \\ 1 & 0 & -2 \\ 1 & 0 & -2 \end{bmatrix} \begin{matrix} \\ x_1^* \; x_2^* \end{matrix}$$

Note that the $x_1^*$ and $x_2^*$ columns of the **X*** matrix use the coefficients for the first two contrasts in table 17.2. The two orthogonal linear contrasts are

$$\ell_1 = .35 + .40 + .58 + .50 + .47 - (.65 + .70 + .90 + .84 + .79)$$

or, from table 17.3,

$\ell_1 = 2.30 - 3.88 = -1.58$

and

$$\ell_2 = .35 + .40 + .58 + .50 + .47 + .65 + .70 + .90 + .84 + .79 - 2(.60 + .80 + .75 + .73 + .66)$$

or, from table 17.3,

$\ell_2 = 2.30 + 3.88 - 2(3.54) = -.9$

Note that in computing the values of the contrasts, we need only form contrasts of the totals for each variety.

The quantities $\Sigma a_i^2$ and $\Sigma b_i^2$ can be computed directly from the $x_1^*$ and $x_2^*$ columns of $\mathbf{X}^*$:

for $\ell_1$: $\sum a_i^2 = 10$

for $\ell_2$: $\sum b_i^2 = 30$

Thus we have

$$\text{SST} = \frac{\ell_1^2}{\sum a_i^2} + \frac{\ell_2^2}{\sum b_i^2}$$

$$= \frac{(-1.58)^2}{10} + \frac{(-.9)^2}{30} = .2496 + .0270 = .2766$$

which is identical to the result obtained in example 13.1 by using the shortcut formula

$$\text{SST} = \frac{\sum T_i^2}{n_i} - \frac{G^2}{n}$$

Actually, in example 13.1 we obtained SST = .276. The difference in the two results can be attributed to the fact that only three decimal places were carried in the calculations for example 13.1.

For our example here we could compute TSS and SSE = TSS − SST to obtain the analysis of variance table, shown in table 17.4.

As can be seen from the AOV table, we can compute an overall $F$ test to test the hypothesis of no difference among the 3 variety means. Or the

**Table 17.4** AOV table for the data of example 17.2

| *Source* | *SS* | *df* | *MS* | *F* |
|---|---|---|---|---|
| treatments | .2766 | 2 | .1383 | .1383/.0082 = 16.87 |
| $\ell_1$ | .2496 | 1 | .2496 | .2496/.0082 = 30.44 |
| $\ell_2$ | .0270 | 1 | .0270 | .0270/.0082 = 3.29 |
| error | .0978 | 12 | .0082 | |
| *Totals* | .3744 | 14 | | |

separate $F$ tests

$$F = \frac{.2496}{.0082} = 30.44$$

$$F = \frac{.0270}{.0082} = 3.29$$

can be used to test the hypotheses

For $\ell_1$: No difference in means for varieties 1 and 2 ($H_0$: $\beta_1^* = 0$).

For $\ell_2$: No difference in means for varieties 1 and 2 versus Variety 3 ($H_0$: $\beta_2^* = 0$).

The critical $F$-value for $df_1 = 1$, $df_2 = 12$, and $a = .05$ is 4.75. Thus from a test for $\ell_1$, we conclude that varieties 1 and 2 are significantly different from each other. The test for $\ell_2$ shows that, when combined, varieties 1 and 2 are not significantly different from Variety 3.

### EXAMPLE 17.3

In example 15.5 (p. 438) we considered an experiment that was conducted to determine the effect of 4 different pesticides on the yield of fruit from 3 different varieties of citrus trees. Four trees from each variety were randomly selected from an orchard. The 4 pesticides were then assigned at random to the trees, with each pesticide assigned to one tree of each variety. The data for example 15.5 have been reproduced here in table 17.5.

**Table 17.5** Yield of fruit (in bushels per tree) for a $3 \times 4$ factorial experiment, example 17.3

| | *Pesticide, A* | | | | |
|---|---|---|---|---|---|
| *Variety, B* | 1 | 2 | 3 | 4 | *Totals* |
| 1 | 29 | 50 | 43 | 53 | 175 |
| 2 | 41 | 58 | 42 | 73 | 214 |
| 3 | 66 | 85 | 69 | 85 | 305 |
| *Totals* | 136 | 193 | 154 | 211 | 694 |

For these data the analysis of variance table would have the sources and degrees of freedom listed below:

| *Source* | *df* |
|---|---|
| *A* | 3 |
| *B* | 2 |
| *AB* | 6 |
| *Total* | 11 |

If there were no replication of the experiment and we were willing to assume that there is no interaction between pesticides and varieties, the

*AB* interaction could be used as the error term for testing main effects.

a. Use these data to set up the $\mathbf{X}^*$ matrix containing the coefficients of the orthogonal linear contrasts for the main effects of factors from the model

$$y = \beta_0^* + \overbrace{\beta_1^* x_1^* + \beta_2^* x_2^* + \beta_3^* x_3^*}^{\text{pesticides}} + \overbrace{\beta_4^* x_4^* + \beta_5^* x_5^*}^{\text{varieties}} + \epsilon$$

b. Show the partitioning of the sums of squares for pesticides and varieties.

## SOLUTION

a. We can use the coefficients given in table 17.2 to form columns of the $\mathbf{X}^*$ matrix. The $\mathbf{Y}$ matrix for these data will establish the order of the $\mathbf{X}^*$ matrix.

| $\mathbf{Y}$ | | $\mathbf{X}^*$ | $x_1^*$ | $x_2^*$ | $x_3^*$ | $x_4^*$ | $x_5^*$ |
|---|---|---|---|---|---|---|---|
| 29 | | 1 | 1 | 1 | 1 | 1 | 1 |
| 41 | Pesticide 1 | 1 | 1 | 1 | 1 | −1 | 1 |
| 66 | | 1 | 1 | 1 | 1 | 0 | −2 |
| 50 | | 1 | −1 | 1 | 1 | 1 | 1 |
| 58 | Pesticide 2 | 1 | −1 | 1 | 1 | −1 | 1 |
| 85 | | 1 | −1 | 1 | 1 | 0 | −2 |
| 43 | | 1 | 0 | −2 | 1 | 1 | 1 |
| 42 | Pesticide 3 | 1 | 0 | −2 | 1 | −1 | 1 |
| 69 | | 1 | 0 | −2 | 1 | 0 | −2 |
| 53 | | 1 | 0 | 0 | −3 | 1 | 1 |
| 73 | Pesticide 4 | 1 | 0 | 0 | −3 | −1 | 1 |
| 85 | | 1 | 0 | 0 | −3 | 0 | −2 |

Note that the coefficients of $\ell_1$ given in the $x_1^*$ column compare pesticides 1 and 2. Similarly, $\ell_2$ and $\ell_3$ represent comparisons of pesticides 1 and 2 versus 3 and pesticides 1, 2, and 3 versus 4, respectively. The $x_4^*$ column lists the coefficients of $\ell_4$ and, as can be seen from the $\mathbf{X}^*$ matrix, $\ell_4$ compares varieties 1 and 2. Similarly, column $x_5^*$ ($\ell_5$) is a comparison of varieties 1 and 2 versus 3.

Because the data conform to a *k*-way classification, we are able to formulate contrasts separately by using table 17.2 for the main effects of factors *A* and *B* while maintaining the property that not only are the contrasts of the same factor orthogonal to each other, but they are also orthogonal to contrasts of other factors. This can be seen by comparing coefficients of pairs of contrasts in the $\mathbf{X}^*$ matrix. Not only are the contrasts $\ell_1$, $\ell_2$, and $\ell_3$ orthogonal to each other, but they are also orthogonal to the contrasts $\ell_4$ and $\ell_5$. Similarly, $\ell_4$ and $\ell_5$ are orthogonal to each other.

mutual orthogonality

This property of *mutual orthogonality* of the contrasts leads to a diagonal $\mathbf{X}^{*\prime}\mathbf{X}^*$ matrix and a least squares solution

$$\hat{\boldsymbol{\beta}}^* = \begin{bmatrix} \bar{y} \\ \dfrac{\ell_1}{\sum a_i^2} \\ \dfrac{\ell_2}{\sum b_i^2} \\ \dfrac{\ell_3}{\sum c_i^2} \\ \dfrac{\ell_4}{\sum d_i^2} \\ \dfrac{\ell_5}{\sum e_i^2} \end{bmatrix}$$

here $\Sigma\, a_i^2$, $\Sigma\, b_i^2, \ldots, \Sigma\, e_i^2$ represent the sums of squares of the coefficients for contrasts $\ell_1, \ell_2, \ldots, \ell_5$, respectively.

b. The partitioning of the main effects sum of squares for pesticides and varieties is given by

$$\text{pesticides: SSA} = \frac{\ell_1^2}{\sum a_i^2} + \frac{\ell_2^2}{\sum b_i^2} + \frac{\ell_3^2}{\sum c_i^2}$$

$$\text{varieties: SSB} = \frac{\ell_4^2}{\sum d_i^2} + \frac{\ell_5^2}{\sum e_i^2}$$

**EXAMPLE 17.4**

Compute the AOV table for the data of example 17.3, using the partitioned sums of squares. From previous calculations in chapter 15, we have TSS = 3567.67.

**SOLUTION**

As was shown in example 17.2, we need only form contrasts of the totals for each main effect to compute values of $\ell_1, \ell_2, \ldots, \ell_5$. Using the pesticide totals from table 17.5, we have

$$\ell_1 = 136 - 193 = -57$$

$$\ell_2 = 136 + 193 - 2(154) = 21$$

$$\ell_3 = 136 + 193 + 154 - 3(211) = -150$$

The quantities $\Sigma\, a_i^2$, $\Sigma\, b_i^2$, and $\Sigma\, c_i^2$ are computed by determining the sum of squares of elements in the $x_1^*$, $x_2^*$, and $x_3^*$ columns of the $\mathbf{X}^*$ matrix.

Hence

$$\sum a_i^2 = 6 \qquad \sum b_i^2 = 18 \qquad \sum c_i^2 = 36$$

The main effect sum of squares for pesticides is

$$\text{SSA} = \frac{\ell_1^2}{\sum a_i^2} + \frac{\ell_2^2}{\sum b_i^2} + \frac{\ell_3^2}{\sum c_i^2}$$

$$= \frac{(-57)^2}{6} + \frac{(21)^2}{18} + \frac{(-150)^2}{36}$$

$$= 541.5 + 24.5 + 625 = 1191$$

You will note that this is identical to the value obtained in example 15.5 (p. 439) by using the shortcut formula for the sum of squares due to pesticides.

Similarly, using the variety totals in table 17.5, we have

$$\ell_4 = 175 - 214 = -39$$

$$\ell_5 = 175 + 214 - 2(305) = -221$$

Using the $x_4^*$ and $x_5^*$ columns of $\mathbf{X}^*$, we have

$$\sum d_i^2 = 8 \qquad \sum e_i^2 = 24$$

The main effect sum of squares for varieties is

$$\text{SSB} = \frac{\ell_4^2}{\sum d_i^2} + \frac{\ell_5^2}{\sum e_i^2}$$

$$= \frac{(-39)^2}{8} + \frac{(-221)^2}{24}$$

$$= 190.13 + 2035.04 = 2225.17$$

**Table 17.6** AOV table for the data of examples 17.3 and 17.4

| *Source* | *SS* | | *df* | | *MS* | |
|---|---|---|---|---|---|---|
| pesticides, *A* | 1191 | | 3 | | 397 | |
| $\ell_1$ | | 541.5 | | 1 | | 541.5 |
| $\ell_2$ | | 24.5 | | 1 | | 24.5 |
| $\ell_3$ | | 625 | | 1 | | 625 |
| varieties, *B* | 2225.17 | | 2 | | 1112.59 | |
| $\ell_4$ | | 190.13 | | 1 | | 190.13 |
| $\ell_5$ | | 2035.04 | | 1 | | 2035.04 |
| error | 151.5 | | 6 | | 25.25 | |
| *Totals* | 3567.67 | | 11 | | | |

This is identical to the value obtained in example 15.5 by using the shortcut formula.

The sum of squares for error (assuming no interaction between pesticides and varieties) is

$$SSE = TSS - SSA - SSB = 3567.67 - 1191 - 2225.17 = 151.5$$

Combining our results we have the analysis of variance table shown in table 17.6.

Although $F$ tests have not been indicated here, we could proceed to test any of the single-degree-of-freedom contrasts against error and compare the computed $F$-value against a critical value with $df_1 = 1$ and $df_2 = 6$.

partitioning interaction SS

Although we have not considered the partitioning of an interaction sum of squares into single-degree-of-freedom sums of squares using orthogonal contrasts, the procedure is straightforward. If, in a $k$-way classification, factor A has $(a - 1)$ degrees of freedom and factor $B$ has $(b - 1)$ degrees of freedom, then the $AB$ interaction can be partitioned into $(a - 1)(b - 1)$ sums of squares, each with one degree of freedom. The coefficients of each of the $(a - 1)(b - 1)$ orthogonal contrasts can be obtained from the corresponding coefficients of the two orthogonal contrasts, one from factor $A$, one from factor $B$.

### EXAMPLE 17.5

Consider the experiment of examples 17.3 and 17.4. We compared the effects of 4 different pesticides on the yields of 3 varieties of citrus trees. If the experimenter is not willing to assume the absence of an $AB$ interaction, he would have to obtain more than one observation for each factor-level combination in order to obtain sufficient degrees of freedom for main-effects tests and for a test of the interaction between factors $A$ and $B$. The data for this $3 \times 4$ factorial experiment with 2 observations (fruit yields, in bushels per tree) per factor-level combination have been taken from example 15.6 (p. 440) and are reproduced here in table 17.7.

**Table 17.7** Yield of fruit (in bushels per tree) for a $3 \times 4$ factorial experiment, example 17.5

| | *Pesticide, A* | | | | |
|---|---|---|---|---|---|
| *Variety, B* | *1* | *2* | *3* | *4* | *Totals* |
| 1 | 49<br>39 | 50<br>55 | 43<br>38 | 53<br>48 | 375 |
| 2 | 55<br>41 | 67<br>58 | 53<br>42 | 85<br>73 | 474 |
| 3 | 66<br>68 | 85<br>92 | 69<br>62 | 85<br>99 | 626 |
| *Totals* | 318 | 407 | 307 | 443 | 1475 |

a. Construct the $\mathbf{X}^*$ matrix containing the coefficients of the orthogonal linear contrasts for the model

$$y = \beta_0^* + \overbrace{\beta_1^* x_1^* + \beta_2^* x_2^* + \beta_3^* x_3^*}^{\text{pesticides}} + \overbrace{\beta_4^* x_4^* + \beta_5^* x_5^*}^{\text{varieties}}$$

$$+ \overbrace{\beta_6^* x_1^* x_4^* + \beta_7^* x_1^* x_5^* + \beta_8^* x_2^* x_4^* + \beta_9^* x_2^* x_5^* + \beta_{10}^* x_3^* x_4^* + \beta_{11}^* x_3^* x_5^*}^{\text{interactions}} + \epsilon$$

b. Show the partitioning of the sums of squares for main effects and interaction.

## SOLUTION

a. The coefficients for main effects can be obtained from table 17.2. The coefficients for the interaction orthogonal linear contrasts are formed by multiplying corresponding coefficients from two main effects contrasts, one from $A$ and one from $B$. For example, each coefficient in the

| $\mathbf{Y}$ | | Pesticide, $A$ | | | Variety, $B$ | | $AB$ interaction | | | | | |
|---|---|---|---|---|---|---|---|---|---|---|---|---|
| | $\mathbf{X}^*$ | $x_1^*$ | $x_2^*$ | $x_3^*$ | $x_4^*$ | $x_5^*$ | $x_1^*x_4^*$ | $x_1^*x_5^*$ | $x_2^*x_4^*$ | $x_2^*x_5^*$ | $x_3^*x_4^*$ | $x_3^*x_5^*$ |
| 49 | 1 | 1 | 1 | 1 | 1 | 1 | 1 | 1 | 1 | 1 | 1 | 1 |
| 39 | 1 | 1 | 1 | 1 | 1 | 1 | 1 | 1 | 1 | 1 | 1 | 1 |
| 55 | 1 | 1 | 1 | 1 | −1 | 1 | −1 | 1 | −1 | 1 | −1 | 1 |
| 41 | 1 | 1 | 1 | 1 | −1 | 1 | −1 | 1 | −1 | 1 | −1 | 1 |
| 66 | 1 | 1 | 1 | 1 | 0 | −2 | 0 | −2 | 0 | −2 | 0 | −2 |
| 68 | 1 | 1 | 1 | 1 | 0 | −2 | 0 | −2 | 0 | −2 | 0 | −2 |
| 50 | 1 | −1 | 1 | 1 | 1 | 1 | −1 | −1 | 1 | 1 | 1 | 1 |
| 55 | 1 | −1 | 1 | 1 | 1 | 1 | −1 | −1 | 1 | 1 | 1 | 1 |
| 67 | 1 | −1 | 1 | 1 | −1 | 1 | 1 | −1 | −1 | 1 | −1 | 1 |
| 58 | 1 | −1 | 1 | 1 | −1 | 1 | 1 | −1 | −1 | 1 | −1 | 1 |
| 85 | 1 | −1 | 1 | 1 | 0 | −2 | 0 | 2 | 0 | −2 | 0 | −2 |
| 92 | 1 | −1 | 1 | 1 | 0 | −2 | 0 | 2 | 0 | −2 | 0 | −2 |
| 43 | 1 | 0 | −2 | 1 | 1 | 1 | 0 | 0 | −2 | −2 | 1 | 1 |
| 38 | 1 | 0 | −2 | 1 | 1 | 1 | 0 | 0 | −2 | −2 | 1 | 1 |
| 53 | 1 | 0 | −2 | 1 | −1 | 1 | 0 | 0 | 2 | −2 | −1 | 1 |
| 42 | 1 | 0 | −2 | 1 | −1 | 1 | 0 | 0 | 2 | −2 | −1 | 1 |
| 69 | 1 | 0 | −2 | 1 | 0 | −2 | 0 | 0 | 0 | 4 | 0 | −2 |
| 62 | 1 | 0 | −2 | 1 | 0 | −2 | 0 | 0 | 0 | 4 | 0 | −2 |
| 53 | 1 | 0 | 0 | −3 | 1 | 1 | 0 | 0 | 0 | 0 | −3 | −3 |
| 48 | 1 | 0 | 0 | −3 | 1 | 1 | 0 | 0 | 0 | 0 | −3 | −3 |
| 85 | 1 | 0 | 0 | −3 | −1 | 1 | 0 | 0 | 0 | 0 | 3 | −3 |
| 73 | 1 | 0 | 0 | −3 | −1 | 1 | 0 | 0 | 0 | 0 | 3 | −3 |
| 85 | 1 | 0 | 0 | −3 | 0 | −2 | 0 | 0 | 0 | 0 | 0 | 6 |
| 99 | 1 | 0 | 0 | −3 | 0 | −2 | 0 | 0 | 0 | 0 | 0 | 6 |

$x_1^* x_4^*$ column is obtained by multiplying corresponding coefficients from the $x_1^*$ and $x_4^*$ columns. Letting $\ell_1 = \Sigma\, a_i y_i$ and $\ell_4 = \Sigma\, d_i y_i$, the contrast corresponding to the $x_1^* x_4^*$ columns if $\ell_6 = \Sigma\, (a_i d_i) y_i$. In a similar way, we obtain the coefficients for the other interaction columns.

The **Y** and **X*** matrices can be written as shown on page 542.

b. The sums of squares for main effects could be computed using the data of table 17.7, much as we did in example 17.4. Thus

$$\text{SSA} = \frac{\ell_1^2}{\sum a_i^2} + \frac{\ell_2^2}{\sum b_i^2} + \frac{\ell_3^2}{\sum c_i^2}$$

$$\text{SSB} = \frac{\ell_4^2}{\sum d_i^2} + \frac{\ell_5^2}{\sum e_i^2}$$

If we let $\ell_6, \ell_7, \ldots, \ell_{11}$ designate the orthogonal linear contrasts formed for the *AB* interaction, then we have

$$\text{SSAB} = \frac{\ell_6^2}{\sum (a_i d_i)^2} + \frac{\ell_7^2}{\sum (a_i e_i)^2} + \cdots + \frac{\ell_{11}^2}{\sum (c_i e_i)^2}$$

It should be noted that a denominator of a term on the right-hand side of the equation represents the sums of squares of the coefficients for the appropriate interaction column of the **X*** matrix. For example, contrast $\ell_6$ refers to the $x_1^* x_4^*$ term in our model. The sum of squares of the coefficients in the $x_1^* x_4^*$ column is

$$\sum (a_i d_i)^2 = 1^2 + 1^2 + (-1)^2 + (-1)^2 + 0^2 + 0^2 + (-1)^2 + (-1)^2 + 1^2 + 1^2 + 0^2 + 0^2 + \cdots + 0^2 = 8$$

The AOV table would reflect the partitioning of the *A*, *B*, and *AB* sums of squares into single-degree-of-freedom components, and *F* tests could be run on individual components. However, it would not be advisable to run all single-degree-of-freedom *F* tests owing to the overall $\alpha$ level. If each *F* test specified a particular type I error, the probability of falsely rejecting at least one null hypothesis increases with the number of tests conducted.

# Exercises

17.4. In example 15.1 (p. 415) we considered an experiment to compare the effects of 3 different insecticides on a particular variety of string beans. Four different plots were prepared, with each plot subdivided into 3 rows. Each row was planted with 100 seeds and maintained under the assigned insecticide. The response of interest was the number of seedlings that emerged per row. These data appear in the table that is shown below.

| *Insecticide* | *Plot* 1 | 2 | 3 | 4 |
|---|---|---|---|---|
| 1 | 56 | 49 | 65 | 60 |
| 2 | 84 | 78 | 94 | 93 |
| 3 | 80 | 72 | 83 | 85 |

a. Construct the $\mathbf{X}^*$ matrix containing the coefficients of the orthogonal linear contrasts for the model

$$y = \beta_0^* + \overbrace{\beta_1^* x_1^* + \beta_2^* x_2^*}^{\text{insecticides}} + \overbrace{\beta_3^* x_3^* + \beta_4^* x_4^* + \beta_5^* x_5^*}^{\text{plots}} + \epsilon$$

b. Compute the AOV table for these data, using the partitioned sums of squares. Run appropriate $F$ tests, using $\alpha = .05$.

17.5. Refer to exercise 17.4. State two other ways that you could obtain the sum of squares for insecticides.

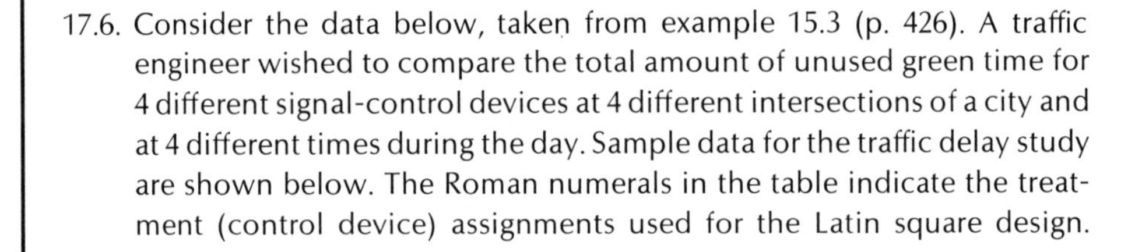

17.6. Consider the data below, taken from example 15.3 (p. 426). A traffic engineer wished to compare the total amount of unused green time for 4 different signal-control devices at 4 different intersections of a city and at 4 different times during the day. Sample data for the traffic delay study are shown below. The Roman numerals in the table indicate the treatment (control device) assignments used for the Latin square design.

| *Intersection* | *Time Period* A.M. *Peak* | | A.M. *Lull* | | P.M. *Peak* | | P.M. *Lull* | | *Totals* |
|---|---|---|---|---|---|---|---|---|---|
| 1 | IV | 15.5 | II | 33.9 | III | 13.2 | I | 29.1 | 91.7 |
| 2 | II | 16.3 | III | 26.6 | I | 19.4 | IV | 22.8 | 85.1 |
| 3 | III | 10.8 | I | 31.1 | IV | 17.1 | II | 30.3 | 89.3 |
| 4 | I | 14.7 | IV | 34.0 | II | 19.7 | III | 21.6 | 90.0 |
| *Totals* | | 57.3 | | 125.6 | | 69.4 | | 103.8 | 356.1 |

For the **Y** matrix

$$\mathbf{Y} = \begin{bmatrix} 15.5 \\ 16.3 \\ 10.8 \\ 14.7 \\ \vdots \\ 30.3 \\ 21.6 \end{bmatrix}$$

construct the **X*** matrix containing the coefficients of the orthogonal contrasts for the model

$$y = \beta_0^* + \overbrace{\beta_1^* x_1^* + \beta_2^* x_2^* + \beta_3^* x_3^*}^{\text{time periods}} + \overbrace{\beta_4^* x_4^* + \beta_5^* x_5^* + \beta_6^* x_6^*}^{\text{intersections}}$$

$$+ \overbrace{\beta_7^* x_7^* + \beta_8^* x_8^* + \beta_9^* x_9^*}^{\text{control devices}} + \epsilon$$

17.7. Refer to exercise 17.6. Compute the AOV table for these data, using the partitioned sums of squares. Compare your results to those obtained in example 15.3.

17.8. Refer to exercise 17.6. State how you would relate the $\beta$'s in a general linear model (with 0, 1 coding) to each $\beta^*$ of the model in exercise 17.6. Set up the appropriate **X** matrix.

17.9. The efficiencies of 4 different machines used for filling sustained-release capsules were tested by a pharmaceutical firm under 4 different modes of operation. Random assignments of modes to machines were made, with 2 observations obtained at each machine-mode combination. The sample efficiencies are recorded below.

| | *Mode of Operation* | | | |
|---|---|---|---|---|
| *Machine* | *1* | *2* | *3* | *4* |
| A | 8, 12 | 55, 62 | 70, 63 | 72, 68 |
| B | 28, 33 | 53, 47 | 58, 64 | 60, 57 |
| C | 15, 19 | 33, 31 | 35, 30 | 32, 27 |
| D | 33, 26 | 43, 46 | 49, 54 | 47, 45 |

Construct the **X*** matrix of the coefficients of the orthogonal linear contrasts for the model

$$y = \beta_0^* + \overbrace{\beta_1^* x_1^* + \beta_2^* x_2^* + \beta_3^* x_3^*}^{\text{machines}} + \overbrace{\beta_4^* x_4^* + \beta_5^* x_5^* + \beta_6^* x_6^*}^{\text{modes}}$$
$$+ \beta_7^* x_1^* x_4^* + \beta_8^* x_1^* x_5^* + \beta_9^* x_1^* x_6^* + \beta_{10}^* x_2^* x_4^* + \beta_{11}^* x_2^* x_5^* + \beta_{12}^* x_2^* x_6^*$$
$$+ \beta_{13}^* x_3^* x_4^* + \beta_{14}^* x_3^* x_5^* + \beta_{15}^* x_3^* x_6^* + \epsilon$$

with the **Y** matrix as shown here.

$$\mathbf{Y} = \begin{bmatrix} 8 \\ 12 \\ 28 \\ 33 \\ 15 \\ 19 \\ \vdots \\ 32 \\ 27 \\ 47 \\ 45 \end{bmatrix}$$

17.10. Refer to exercise 17.9. Construct an AOV table for the sample data. Partition all main effects and interactions into single-degree-of-freedom sums of squares.

17.11. Refer to exercise 17.9. Compute the sums of squares for main effects and interactions, using the shortcut formulas of chapter 15. Compare your results to those of exercise 17.10.

17.12. Refer to exercise 17.9. How would you relate the parameters ($\beta$'s) of a general linear model to those ($\beta^*$'s) in exercise 17.9? Set up the corresponding **X** matrix of the general linear model.

# 17.4 Orthogonal Polynomials

In the previous section we discussed how we could partition the sums of squares for sources of variability of a $k$-way classification into single-degree-of-freedom sums of squares by using orthogonal contrasts among the sample observations. The factors of the $k$-way classification were qualitative variables, such as pesticides, varieties, and so forth. We will show now that we can partition the sum of squares due to a quantitative variable $x$ in a polynomial model, using a system of coding called the method of *orthogonal polynomials*.

Let $x$ be a quantitative independent variable measured at $p$ levels, which are one unit apart. If the independent variable of interest assumes equally spaced values that are not one unit apart, we can code the values to obtain a spacing of one unit by dividing each value by the common interval width. For example, suppose we wish to examine the effect of the levels 10, 20, 30, and 40 pounds of fertilizer per plot on the yield $y$ of corn. The independent variable "fertilizer" assumes the $p = 4$ equally spaced values 10, 20, 30, and 40. To obtain 4 equally spaced values one unit apart, we divide each value by 10, the common interval width. Then we can use the independent variable $x$ that assumes the values 1, 2, 3, and 4. We will also assume that there are an equal number of observations at each of the $p$ levels of the independent variable.

$x$ for $p$ levels, one unit apart

The general linear model of chapter 6 can be written in the form

$$y = \beta_0 + \beta_1 x + \beta_2 x^2 + \cdots + \beta_{p-1} x^{p-1} + \epsilon$$

Using the methods of section 16.4, we can fit complete and reduced models to obtain a test of

$$H_0\colon \beta_1 = \beta_2 = \cdots = \beta_{p-1} = 0$$

that is, a test of no polynomial effect, up to degree $p - 1$, due to the quantitative independent variable $x$. Since we are hypothesizing that $(p - 1)$ $\beta$'s are simultaneously zero, the $SS_{\text{drop}}$ due to the variable $x$ would have $p - 1$ degrees of freedom. In some cases it would be desirable to partition the $SS_{\text{drop}}$ into $p - 1$ single-degree-of-freedom sums of squares corresponding to $p - 1$ orthogonal contrasts among the sample observations. Although the coefficients of table 17.2 could be used to form a set of contrasts for levels of $x$, the particular contrasts obtained from the method of orthogonal polynomials measure specific polynomial effects due to the independent variable $x$.

The method of orthogonal polynomials consists of defining a new model (system of coding)

model

$$y = \beta_0^* + \beta_1^* x_1^* + \beta_2^* x_2^* + \cdots + \beta_{p-1}^* x_{p-1}^* + \epsilon$$

where $x_1^*, x_2^*, \ldots, x_{p-1}^*$ are called *orthogonal polynomials*. Although the independent variables in our new model are called orthogonal polynomials, we retain the $x^*$ notation to indicate we are dealing with an alternative system of coding, which is consistent with our use of $x^*$ in sections 17.2 and 17.3. The first

orthogonal polynomials

three orthogonal polynomials have the following form:

$$x_1^* = \lambda_1(x - \bar{x})$$

$$x_2^* = \lambda_2\left[(x - \bar{x})^2 - \frac{p^2 - 1}{12}\right]$$

$$x_3^* = \lambda_3\left[(x - \bar{x})^3 - (x - \bar{x})\frac{3p^2 - 7}{20}\right]$$

The $\lambda$'s are constants related to $p$, the number of levels of $x$. Although we will not list the form of additional polynomials here, you can refer to Pearson and Hartley (1966) in the references for a more complete listing. Note that the first orthogonal polynomial represents a linear effect due to $x$; the second, a quadratic effect; the third, a combined cubic and linear effect.

Fortunately, tables have been constructed showing $x_i^*$ values for various values of $p$. In table 17.8 we show the values $x_1^*$ and $x_2^*$ for $p = 3$, taken from

**Table 17.8** Values of $x_1^*$, $x_2^*$ for $p = 3$

| *Level of x* | $x_1^*$ | $x_2^*$ |
|---|---|---|
| 1 | −1 | 1 |
| 2 | 0 | −2 |
| 3 | 1 | 1 |
| $\lambda_i$ | 1 | 3 |

table 15 in the appendix. Thus for the low level of the independent variable $x$, $x_1^*$ and $x_2^*$ assume the values of −1 and 1, respectively. These could be computed from the formulas

$$x_1^* = \lambda_1(x - \bar{x})$$

$$x_2^* = \lambda_2\left[(x - \bar{x})^2 - \frac{p^2 - 1}{12}\right]$$

with $p = 3$ from table 17.8, $\lambda_1 = 1$, and $\lambda_2 = 3$.

Similarly, table 15 of the appendix gives values for the orthogonal polynomials $x_1^*$, $x_2^*$, and $x_3^*$ for $p = 4, 5, \ldots, 10$. Note that for a given value of $p$, the values of the orthogonal polynomials form orthogonal contrasts among the observations.

It should be noted that table 15 of the appendix is similar to table 17.2, which gives the coefficients for orthogonal contrasts in a $k$-way classification. If we let $\ell_1, \ell_2, \ldots, \ell_{p-1}$ represent the orthogonal linear contrasts corresponding to the orthogonal polynomials $x_1^*, x_2^*, \ldots, x_{p-1}^*$, the $\mathbf{X}^{*\prime}\mathbf{X}^*$ and $\mathbf{X}^{*\prime}\mathbf{Y}$ matrices have the simplified form

$$\mathbf{X}^{*\prime}\mathbf{X}^{*} = \begin{bmatrix} n & & & \\ & \sum a_i^2 & & \\ & & \sum b_i^2 & \\ & & & \ddots \end{bmatrix} \qquad \mathbf{X}^{*\prime}\mathbf{Y} = \begin{bmatrix} \sum y_i \\ \ell_1 \\ \ell_2 \\ \vdots \\ \ell_{p-1} \end{bmatrix}$$

The least squares estimates can be computed as

$$\widehat{\boldsymbol{\beta}}^{*} = (\mathbf{X}^{*\prime}\mathbf{X}^{*})^{-1}\,\mathbf{X}^{*\prime}\mathbf{Y} = \begin{bmatrix} \bar{y} \\ \dfrac{\ell_1}{\sum a_i^2} \\ \dfrac{\ell_2}{\sum b_i^2} \\ \vdots \end{bmatrix}$$

Note that this matrix has the same form as it had with orthogonal coding in a $k$-way classification, except that $\ell_1, \ell_2, \ldots, \ell_{p-1}$ (and hence the parameters $\widehat{\beta}^{*}$) have different interpretations now. Because of the form of the orthogonal polynomials, they have the following interpretations:

$\widehat{\beta}_0^{*} = \bar{y}$: An estimate of the expected value of $y$.

$\widehat{\beta}_1^{*} = \dfrac{\ell_1}{\sum a_i^2}$: A measure of the linear trend due to $x$.

$\widehat{\beta}_2^{*} = \dfrac{\ell_2}{\sum b_i^2}$: A measure of the quadratic effect due to $x$.

and so on.

The sum of squares drop (due to the independent variable $x$ in the general linear model) can now be partitioned into $p - 1$ sums of squares, each with one degree of freedom. For example, with $x$ measured at $p = 3$ equally spaced levels one unit apart, we have 2 degrees of freedom for $SS_{\text{drop}}$, with

$$SS_{\text{drop}} = \frac{\ell_1^2}{\sum a_i^2} + \frac{\ell_2^2}{\sum b_i^2}$$

Similarly for $p = 4$,

$$SS_{drop} = \frac{\ell_1^2}{\sum_i a_i^2} + \frac{\ell_2^2}{\sum_i b_i^2} + \frac{\ell_3^2}{\sum_i c_i^2}$$

tests

Tests concerning individual parameters ($\beta^*$) in the model can be performed using all $F$ ratios of the form

$$F = \frac{\ell_1^2/\sum_i a_i^2}{\text{MSE}} \qquad F = \frac{\ell_2^2/\sum_i b_i^2}{\text{MSE}} \qquad \cdots$$

with each $F$ statistic to be compared against the table value for $a = \alpha$, $df_1 = 1$, and $df_2$ = degrees of freedom for MSE.

**EXAMPLE 17.6**

An investigator was interested in determining the effect of a sulfur-coated urea fertilizer (SCU), with a dissolution rate of 10, on the yield of winter ryegrass. The SCU was applied to plots at a rate of either 100, 200, or 300 pounds per hectare to a random sample of 9 equal-sized plots. Three observations were obtained at each of the application rates. Use the sample data of table 17.9 to set up the $\mathbf{X}^*$ matrix for the model

$$y = \beta_0^* + \beta_1^* x_1^* + \beta_2^* x_2^* + \epsilon$$

where $x_1^*$ and $x_2^*$ are the first- and second-order orthogonal polynomials in the independent variable $x$, the amount of SCU applied to each plot. Indicate an appropriate $F$ test for detecting a quadratic effect due to the application rate of SCU.

**Table 17.9** Yield of ryegrass for different application rates of SCU

| *Application Rate (pounds per hectare)* | | |
|---|---|---|
| *100* | *200* | *300* |
| 145 | 200 | 190 |
| 160 | 175 | 170 |
| 135 | 160 | 205 |

**SOLUTION**

First it would be good to plot the data for this experimental situation. The scatter diagram is shown in figure 17.1. Although there is a great deal of plot-to-plot variability in yields for a given application rate, it appears that there is some curvature over the region of experimentation.

Recall that before using orthogonal polynomials, we require that the independent variable be measured at $p$ equally spaced levels one unit apart. While the $p = 3$ levels of the variable application rate are equally

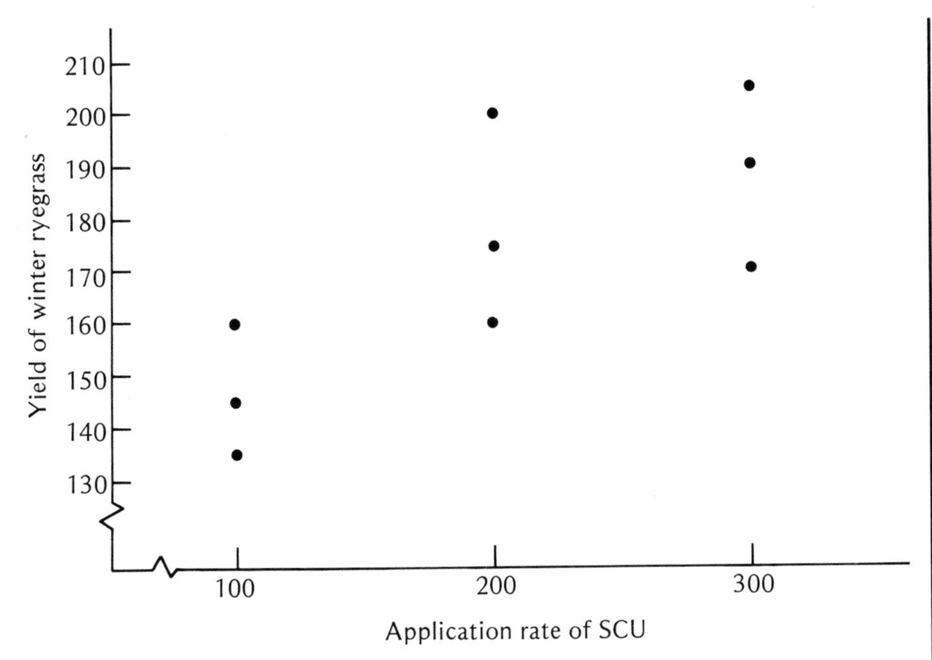


**Figure 17.1** Scatter diagram of the data in table 17.9

spaced, they are not one unit apart. However, by dividing each rate by 100, the common interval width, we obtain the values 1, 2, and 3 for our independent variable $x$.

Using table 15 of the appendix for $p = 3$, we can form the following **Y** and **X*** matrices:

$$\mathbf{Y} = \begin{bmatrix} 145 \\ 160 \\ 135 \\ \hline 200 \\ 175 \\ 160 \\ \hline 190 \\ 170 \\ 205 \end{bmatrix} \qquad \mathbf{X}^* = \begin{bmatrix} 1 & -1 & 1 \\ 1 & -1 & 1 \\ 1 & -1 & 1 \\ \hline 1 & 0 & -2 \\ 1 & 0 & -2 \\ 1 & 0 & -2 \\ \hline 1 & 1 & 1 \\ 1 & 1 & 1 \\ 1 & 1 & 1 \end{bmatrix}$$

A portion of the analysis of variance table for the experiment is given in table 17.10. We compute the total sum of squares with the usual shortcut formula

$$\text{TSS} = \sum y^2 - \frac{G^2}{n}$$

and SSE by subtraction.

The null hypothesis for testing the existence of a quadratic effect is

$H_0$: $\beta_2^* = 0$ (there is no quadratic effect between application rates and ryegrass yields)

**Table 17.10** Partial AOV table for the data of example 17.6

| *Source* | *SS* | *df* |
|---|---|---|
| $x_1^*$ | $\ell_1^2/\Sigma\, a_i^2$ | 1 |
| $x_2^*$ | $\ell_2^2/\Sigma\, b_i^2$ | 1 |
| error | by subtraction | 6 |
| *Totals* | TSS | 8 |

The test statistic, with $df_1 = 1$ and $df_2 = 6$, is

$$F = \frac{\ell_2^2/\sum b_i^2}{\text{MSE}}$$

From the **Y** matrix and the $x_2^*$ column of the $\mathbf{X}^*$ matrix,

$$\ell_2 = 145 + 160 + 135 - 2(200) - 2(175) - 2(160) + 190 + 170 + 205 = -65$$

$$\sum b_i^2 = 1^2 + 1^2 + 1^2 + (-2)^2 + (-2)^2 + (-2)^2 + 1^2 + 1^2 + 1^2 = 18$$

factorial experiment

In this section the use of orthogonal polynomials has been examined and illustrated for an experimental situation where an equal number of observations are obtained at each of $p$ equally spaced units of a quantitative independent variable. Similarly, if we run a factorial experiment, with factor $A$ measured at (a − 1) equally spaced levels, and so on, and an equal number of observations at each factor-level combination, we can employ orthogonal polynomials for each of the factors to obtain a partitioning of SSA, SSB, and so forth. In addition, by forming products of two columns of the $\mathbf{X}^*$ matrix, one from factor $A$ and one from factor $B$, we can partition the $AB$ interaction sum of squares into single-degree-of-freedom sums of squares. This method of partitioning an interaction sum of squares after establishing the coefficients for main effects (factor $A$, factor $B$, etc.) is identical to that employed in orthogonal coding for a $k$-way classification.

# 17.5 Summary

In this chapter we have been concerned with the relationships between systems of coding. In particular, we first considered the relationship between parameters of the general linear model and any other system of coding (model) for the same experimental situation. If the expected value of $y$ in matrix notation for the general linear model is given by

$$E(\mathbf{Y}) = \boldsymbol{X\beta}$$

and $\mathbf{X}^*\boldsymbol{\beta}^*$ represents the expected value of $y$ for another system of coding, then the relationship between the $\boldsymbol{\beta}$ and $\boldsymbol{\beta}^*$ matrices is

$$\boldsymbol{\beta} = (\mathbf{X}'\mathbf{X})^{-1}\mathbf{X}'\mathbf{X}^*\boldsymbol{\beta}^*$$

Orthogonal linear contrasts for partitioning each main effect of a $k$-way classification into single-degree-of-freedom sums of squares can be viewed as an alternative system of coding to that of the general linear model for the $k$-way classification. Coefficients for each of the orthogonal contrasts form a column of the $\mathbf{X}^*$ matrix. The $\widehat{\beta}^*$ estimates have a simplified form owing to the orthogonality of the columns of $\mathbf{X}^*$, and we can easily write down the partitioning of a main effect or an interaction sum of squares as a sum of single-degree-of-freedom sums of squares involving orthogonal linear contrasts. $F$ tests are available for each $\beta^*$ in the model.

Similarly, for a factorial experiment involving one or more quantitative independent variables, each with equally spaced levels and the same number of observations at all factor-level combinations, it is possible to use a coding system involving orthogonal polynomials. The coefficients associated with the orthogonal polynomials form the coefficients for the orthogonal linear contrasts among the sample observations. The coefficients for a particular polynomial appear as a column of the $\mathbf{X}^*$ matrix. Again because of the simplified form of the $\mathbf{X}^*$ matrix, the $\widehat{\beta}^*$ estimates have a simplified form, and we can readily partition the sum of squares due to an independent variable into single-degree-of-freedom sums of squares involving linear effects, quadratic effects, and so on. Tests of significance for individual parameters ($\beta^*$) of the model are of the form

$$F = \frac{\ell^2/\sum a_i^2}{\text{MSE}}$$

and are based on $df_1 = 1$ and $df_2 =$ degrees of freedom for MSE.

# Exercises

17.13. Quarterly sales information (in $1000s) was recorded for a small corporation over a two-year period. These data follow.

$

| $x$, Time Period | 1 | 2 | 3 | 4 | 5 | 6 | 7 | 8 |
|---|---|---|---|---|---|---|---|---|
| $y$, Sales Volume | 10 | 50 | 150 | 190 | 200 | 180 | 170 | 140 |

a. Plot the sample data using a scatter diagram.
b. Based on the scatter diagram, write an appropriate polynomial model relating $y$ to $x$.
c. Code the independent variable $x$ to fit the model

$$y = \beta_0^* + \beta_1^* x_1^* + \beta_2^* x_2^* + \epsilon$$

where $x_1^*$ and $x_2^*$ are the first- and second-order orthogonal polynomials.

17.14. Refer to exercise 17.13.
a. Relate the parameters of the model in part b to the parameters $\beta^*$ of part c by writing out the form of the orthogonal polynomials and equating coefficients of $x$ and $x^2$.
b. Fit the model of part c in exercise 17.13.
c. Find the $\widehat{\beta}$ estimates for $y = \beta_0 + \beta_1 x + \beta_2 x^2 + \epsilon$ by using the $\widehat{\beta}^*$ estimates of part b.

17.15. Refer to exercise 17.9 (p. 545), where the efficiencies of 4 different machines for filling sustained-release capsules were tested under 4 different modes of operation. Now suppose that the 4 modes of operation were the machine speeds 120, 180, 240, and 300 (in capsules per minute). The sample data efficiencies are given below.

| | Machine Speed | | | |
|---|---|---|---|---|
| Machine | 120 | 180 | 240 | 300 |
| A | 8, 12 | 55, 62 | 70, 63 | 72, 68 |
| B | 28, 33 | 53, 47 | 58, 64 | 60, 57 |
| C | 15, 19 | 33, 31 | 35, 30 | 32, 27 |
| D | 33, 26 | 43, 46 | 49, 54 | 47, 45 |

Construct the **X**$^*$ matrix containing the orthogonal linear contrasts for the qualitative variable "machine" and the orthogonal polynomials for the quantitative variable "machine speed." Use the model

$$\begin{aligned} y = {} & \beta_0^* + \beta_1^* x_1^* + \beta_2^* x_2^* + \beta_3^* x_3^* + \beta_4^* x_4^* + \beta_5^* x_5^* + \beta_6^* x_6^* \\ & + \beta_7^* x_1^* x_4^* + \beta_8^* x_1^* x_5^* + \beta_9^* x_1^* x_6^* + \beta_{10}^* x_2^* x_4^* + \beta_{11}^* x_2^* x_5^* + \beta_{12}^* x_2^* x_6^* \\ & + \beta_{13}^* x_3^* x_4^* + \beta_{14}^* x_3^* x_5^* + \beta_{15}^* x_3^* x_6^* + \epsilon \end{aligned}$$

and the **Y** matrix

$$\mathbf{Y} = \begin{bmatrix} 8 \\ 12 \\ 28 \\ 33 \\ \vdots \\ 32 \\ 27 \\ 47 \\ 45 \end{bmatrix}$$

17.16. Refer to exercise 17.15.

a. Run an analysis of variance, showing the partitioning of the sums of squares for main effects and interactions. Make appropriate tests, using $\alpha = .05$.

b. Relate the parameters of the model in exercise 17.15 to those in an equivalent model for exercise 17.9.

17.17. Refer to exercise 17.4 (p. 543); we used orthogonal linear contrasts to fit the model

$$y = \beta_0^* + \overbrace{\beta_1^* x_1^* + \beta_2^* x_2^*}^{\text{insecticides}} + \overbrace{\beta_3^* x_3^* + \beta_4^* x_4^* + \beta_5^* x_5^*}^{\text{plots}} + \epsilon$$

Relate the parameters of that model to those in the model

$$y = \beta_0 + \overbrace{\beta_1 x_1 + \beta_2 x_2}^{\text{insecticides}} + \overbrace{\beta_3 x_3 + \beta_4 x_4 + \beta_5 x_5}^{\text{plots}} + \epsilon$$

where $x_1, x_2, \ldots, x_5$ are the usual dummy variables in the general linear model.

17.18. Refer to the data in table 17.9 (p. 550).

a. Using orthogonal polynomials, fit the model

$$y = \beta_0^* + \beta_1^* x_1^* + \beta_2^* x_2^* + \epsilon$$

b. Complete an AOV table and draw your conclusions, using $\alpha = .05$.

*17.19. Refer to exercise 17.18.

a. Use a computer program to fit the model

$$y = \beta_0 + \beta_1 x_1 + \beta_2 x_1^2 + \epsilon$$

where $x_1$ is the application rate given in table 17.9.

b. Test separately for a linear and quadratic effect due to $x_1$.
c. Compare your conclusions to those obtained in part b of exercise 17.18.

17.20. Refer to exercise 15.14 (p. 452). Treating the data as a 2-way classification with $r = 3$ observations per cell, partition sums of squares for main effects and interactions into single-degree-of-freedom contrasts. Show the AOV table with this partitioning.

17.21. Refer to exercise 17.20.
a. Instead of using the orthogonal coding appropriate for a $k$-way classification, fit a second-order model relating the response $y$ to the independent variables pH and calcium by using orthogonal polynomials.
b. Predict the response when pH $= 4.5$ and calcium $= 250$. (Hint: Recall the formulas for orthogonal polynomials.)

17.22. Refer to exercise 17.21. Test the model for lack of fit. Use $\alpha = .05$. Examine the residuals $(y - \hat{y})$ and attempt to discuss the nature of the lack of fit.

17.23. Refer to the data of exercise 16.13 (p. 514).
a. Fit a cubic model using orthogonal polynomials.
b. Predict the response when $x = 17$, using a 95% prediction interval.
c. Relate the parameters in part a to the parameters of the model fit in exercise 16.14.

17.24. Use the data of exercise 16.21 (p. 516) to fit a quadratic model by using orthogonal polynomials. How would you test for lack of fit?

17.25. Use orthogonal polynomials to partition the sums of squares for main effects in the data of example 16.5 (table 16.7, p. 473). Set up the analysis of variance table.

# References

Barr, A.J.; Goodnight, J.H.; Sall, J.P.; and Helwig, J.T. 1976. *A user's guide to SAS 76*. Raleigh, N.C.: SAS Institute, Inc.

Brownlee, K.A. 1965. *Statistical theory and methodology in science and engineering*. 2d ed. New York: Wiley.

Cochran, W.G., and Cox, G.M. 1957. *Experimental design.* 2d ed. New York: Wiley.

Davies, O.L. 1956. *The design and analysis of experiments.* New York: Hafner.

Dixon W.J. 1975. *BMDP, biomedical computer programs.* Berkeley: University of California Press.

Dixon, W.J., and Massey, F.J., Jr. 1969. *Introduction to statistical analysis.* 3d ed. New York: McGraw-Hill.

Draper, N.R., and Smith, H. 1966. *Applied regression analysis.* New York: Wiley.

Fisher, R.A. 1949. *The design of experiments.* Edinburgh: Oliver and Boyd.

Graybill, F.A. 1976. *Theory and application of the linear model.* N. Scituate, Mass.: Duxbury Press.

Harnett, D.L. 1970. *Introduction to statistical methods.* Reading, Mass.: Addison-Wesley.

Kirk, R.E. 1968. *Experimental design: Procedures for the behavioral sciences.* Belmont, Calif.: Brooks/Cole.

Mendenhall, W. 1968. *Introduction to linear models and the design and analysis of experiments.* Belmont, Calif.: Wadsworth.

Nie, N.H.; Hull, C.H.; Jenkins, J.G.; Steinbrenner, K.; and Bent, D.H. 1975. *Statistical package for the social sciences.* 2d ed. New York: McGraw-Hill.

Ostle, B. 1963. *Statistics in research.* 2d ed. Ames, Iowa: Iowa State University Press.

Pearson, E.S., and Hartley, H.O. 1966. *Biometrika tables for statisticians.* 3d ed. Vol. I. London: Cambridge University Press.

Service, J. 1972. *A user's guide to the statistical analysis system.* Raleigh, N.C. Student Supply Stores, North Carolina State University.

Snedecor, G.W., and Cochran, W.G. 1967. *Statistical methods.* 6th ed. Ames, Iowa: Iowa State University Press.

Sokal, R.R., and Rohlf, F.J. 1969. *Biometry.* San Francisco: W.H. Freeman.

Steel, R.G.D., and Torrie, J.H. 1960. *Principles and procedures of statistics.* New York: McGraw-Hill.

Tanur, J.M.; Mosteller, F.; Kruskal, W.H.; Pieters, R.S.; and Rising, G.R. 1972. *Statistics: A guide to the unknown.* San Francisco: Holden-Day.

Winer, B.J. 1962. *Statistical principles in experimental design.* New York: McGraw-Hill.

# 18 Analysis of Variance for Some Fixed, Random, and Mixed Effects Models

## 18.1 Introduction

In previous chapters we have been able to write the model for a response $y$ in terms of $k$ independent variables, using the general linear model

general linear model

$$y_i = \beta_0 + \beta_1 x_{i1} + \beta_2 x_{i2} + \cdots + \beta_k x_{ik} + \epsilon_i$$

Initially we assumed $x_1, x_2, \ldots, x_k$ to be independent variables measured without error; $\beta_0, \beta_1, \ldots, \beta_k$ to be unknown parameters; and the random error $\epsilon_i$ associated with observation $i$ to have $E(\epsilon_i) = 0$. We then expanded these assumptions to include the following:

1. $\epsilon_1, \epsilon_2, \ldots, \epsilon_n$ are independent of each other.
2. For a given setting of the independent variables $x_1, x_2, \ldots, x_k$, the variance of $\epsilon_i$ is $\sigma^2$.

Thus although we had many terms in the model, there was only one source of random variation.

---

**Definition 18.1**

A model that can be written in the form of a general linear model with $k > 0$ independent variables and one random component is called a *fixed effects model.*

---

All models discussed so far in this text relating a response $y$ to one or more independent variables have fallen into the category of fixed effects models. Inferences for fixed effects models are stated in terms of one or more parameters in the general linear model.

Sometimes, however, we must account for more than one source of random variation in an experimental situation, and the model must be expanded to accommodate these additional *components of variation*. For example, the blocks in a randomized block design might represent a random sample of $b$ blocks taken from a population of all possible blocks. Then the effects due to blocks are considered to be random effects rather than fixed effects. Appropriate changes would be made in the model to reflect this difference in interpretation.

components of variation

Before we give examples of situations that might warrant the inclusion of more than one random source of variability in a model, we define two different types of models.

---

**Definition 18.2**

A model that can be written in the form of a general linear model with $k = 0$ independent variables and more than one random component is called a *random effects model*.

---

**Definition 18.3**

A model that can be written in the form of a general linear model with $k > 0$ independent variables and more than one random component is called a *mixed effects* (or simply *mixed*) *model*.

---

In this chapter we will consider various random effects and mixed models. For each model we will indicate the appropriate analysis of variance and also relate the analysis discussed to those that we would obtain with the corresponding fixed effects model.

# 18.2 A One-way Classification with Random Treatments: A Random Effects Model

The best way to illustrate the difference between the fixed and random effects models for a one-way classification is by way of an example. Suppose we want to compare readings made on the intensities of the electrostatic discharges of

example

lightning at 3 different tracking stations within a 20-mile radius of the central computing facilities of a university. If these 3 tracking stations are the only feasible tracking stations for such an operation and inferences are to be about these stations only, then we could write the fixed effects model in the form of a general linear model (see section 16.3):

fixed effects model

$$y = \beta_0 + \beta_1 x_1 + \beta_2 x_2 + \epsilon$$

where

$x_1 = 1$ if Tracking Station 2 $\quad x_1 = 0$ otherwise

$x_2 = 1$ if Tracking Station 3 $\quad x_2 = 0$ otherwise

Also, $\beta_0$, $\beta_1$, and $\beta_2$ are unknown parameters representing mean intensities or differences in mean intensities. Equivalently, using the results of section 13.4, we could write the fixed effects model as

$$y_{ij} = \mu + \alpha_i + \epsilon_{ij}$$

where $y_{ij}$ is the $j$th observation at Tracking Station $i$ ($i = 1, 2, 3$), $\mu$ is an overall mean, and $\alpha_i$ is a fixed effect due to Tracking Station $i$. For both of these models, $\epsilon$ is assumed to be normally distributed, with mean 0 and variance $\sigma^2$.

Suppose, however, that rather than being concerned about only these 3 tracking stations, we consider these stations as a random sample of 3 taken from the many possible locations for tracking stations. Inferences would now relate not just to what happened at the sampled locations but also to what might happen at other possible locations for tracking stations. A model that can account for this difference in interpretation is the random effects model

random effects model

$$y_{ij} = \mu + \alpha_i + \epsilon_{ij}$$

While the model looks the same as the previous fixed effects model, some of the assumptions are different.

assumptions

1. $\mu$ is still an overall mean, which is an unknown constant.
2. $\alpha_i$ is a random effect due to the $i$th tracking station. We assume that $\alpha_i$ is normally distributed, with mean 0 and variance $\sigma_\alpha^2$.
3. The $\alpha_i$'s are independent.
4. As before, $\epsilon_{ij}$ is normally distributed, with mean 0 and variance $\sigma^2$.
5. The $\epsilon_{ij}$'s are independent.
6. The random components $\alpha_i$ and $\epsilon_{ij}$ are independent.

The difference between the fixed effects model and the random effects model can be illustrated by supposing we were to repeat the experiment. For the fixed effects model, we would use the same 3 tracking stations, and hence it would make sense to make inferences about the mean intensities or differences in mean intensities at these 3 locations. However, for the random effects model, we would take another random sample of 3 tracking stations (i.e., take another sample of 3 $\alpha$'s). Now rather than concentrating on the effect of a particular group of 3 $\alpha$'s from one experiment, we should examine the variability of the population of all possible $\alpha$-values. This will be illustrated using the analysis of variance table given in table 18.1.

The analysis of variance table is the same for a fixed or random effects model, with the exception that the *expected mean squares* (EMS) columns are different. You will recall that this column was not used in our tables in chapters 13 and 15, since all mean squares except MSE had an expectation under the alternative hypothesis equal to $\sigma^2$ plus a positive constant $\theta$, which depended on the parameters under test. In general, with $t$ treatments (tracking stations) and $r$ observations per treatment, the AOV table would appear as shown in table 18.1. For the fixed effects model, $\theta_T$ is a positive function of the

EMS

**Table 18.1** An AOV table for a one-way classification: Fixed or random model

AOV table

| | | | | *EMS* | |
|---|---|---|---|---|---|
| *Source* | *SS* | *df* | *MS* | *Fixed Effects* | *Random Effects* |
| treatments | SST | $t - 1$ | MST | $\sigma^2 + \theta_T$ | $\sigma^2 + r\sigma_\alpha^2$ |
| error | SSE | $t(r - 1)$ | MSE | $\sigma^2$ | $\sigma^2$ |
| *Totals* | TSS | $rt - 1$ | | | |

constants $\alpha_i$, while $\sigma_\alpha^2$ represents the variance of the population of $\alpha_i$-values for the random effects model.

Referring to our example, a test for the equality of the mean intensities at the 3 tracking stations in the fixed effects model is (from chapter 13)

test for means

$H_0$: $\alpha_1 = \alpha_2 = \alpha_3 = 0$ (i.e., the 3 means are identical).

$H_a$: At least one $\alpha$ is different from zero.

T.S.: $F = \text{MST}/\text{MSE}$, based on $\text{df}_1 = t - 1$ and $\text{df}_2 = t(r - 1)$.

A test concerning the variability for the population of $\alpha$-values in the random effects model makes use of the same test statistic. The null hypothesis and alternative hypothesis are

test for $\sigma_\alpha^2$

$H_0$: $\sigma_\alpha^2 = 0$

$H_a$: $\sigma_\alpha^2 > 0$

Since we assumed that the $\alpha$'s sampled were selected from a normal population with mean 0 and variance $\sigma_\alpha^2$, the null hypothesis states that the $\alpha$'s were drawn from a normal population with mean 0 and variance 0; that is, all $\alpha$-values in the population are equal to zero.

Thus, although the forms of the null hypotheses are different for the two models, the meanings attached to them are very similar. For the fixed effects model, we are assuming that the sampled $\alpha$'s (which are the only $\alpha$'s) are identically zero, whereas in the random effects model, the null hypothesis leads us to assume that the sampled $\alpha$'s, as well as all other $\alpha$'s in the population, are zero.

The alternative hypotheses are also similar. In the fixed effects model, we are assuming that at least one of the $\alpha$'s is different from the rest; that is, there is some variability among the set of $\alpha$'s. For the random effects model, the alternative hypothesis is that $\sigma_\alpha^2 > 0$; that is, not all $\alpha$-values are the same in the population.

### EXAMPLE 18.1

Consider the problem we have used to illustrate a one-way classification with random treatments. Two graduate students working for a professor in electrical engineering have been funded to record lightning discharge intensities (intensities of the electric field) at 3 tracking stations. Because of the high frequency of thunderstorms in the summer months (in Florida, storms occur on 80 or more days per year), the graduate students were to choose a point at random on a map of the 20-mile-radius region and assemble their tracking equipment (provided they could get permission of the property owners). Each day during the hours from 8 A.M. to 5 P.M., they were to monitor their instruments until the maximum intensity had been recorded for 5 separate storms. The process was then repeated separately at the two other locations chosen at random. The sample data (in volts per meter) appear in table 18.2.

**Table 18.2** Lightning discharge intensities (in volts per meter), example 18.1

| *Tracking Station* | *Intensities* | | | | | *Totals* |
|---|---|---|---|---|---|---|
| 1 | 20 | 1050 | 3200 | 5600 | 50 | 9,920 |
| 2 | 4300 | 70 | 2560 | 3650 | 80 | 10,660 |
| 3 | 100 | 7700 | 8500 | 2960 | 3340 | 22,600 |
| *Total* | | | | | | 43,180 |

a. Write an appropriate statistical model, defining all terms.
b. Perform an analysis of variance and interpret your results. Use $\alpha = .05$.

### SOLUTION

A model appropriate for this one-way classification, with tracking stations chosen at random, is

$$y_{ij} = \mu + \alpha_i + \epsilon_{ij} \qquad (i = 1, 2, 3;\ j = 1, 2, \ldots, 5)$$

where the terms in the model were defined on page 560 of this section.

The computational formulas for sums of squares are identical to those for a fixed effects model.

$$\frac{G^2}{n} = \frac{(43{,}180)^2}{15} = 124{,}300{,}826.7$$

$$\text{SST} = \frac{\sum T_i^2}{5} - \frac{G^2}{n}$$

$$= 144{,}560{,}400 - 124{,}300{,}826.7 = 20{,}259{,}573.3$$

$$\text{TSS} = \sum \sum y_{ij}^2 - \frac{G^2}{n}$$

$$= 232{,}550{,}000 - 124{,}300{,}826.7 = 108{,}249{,}173.3$$

By subtraction,

$$\text{SSE} = \text{TSS} - \text{SST} = 87{,}989{,}600$$

We can use these calculations to construct an AOV table, as shown in table 18.3.

**Table 18.3** AOV table for the data of example 18.1

| *Source* | *SS* | *df* | *MS* | *EMS* | *F* |
|---|---|---|---|---|---|
| tracking stations | 20,259,573.3 | 2 | 10,129,786.65 | $\sigma^2 + 5\sigma_\alpha^2$ | 1.38 |
| error | 87,989,600 | 12 | 7,332,466.67 | $\sigma^2$ | |
| *Totals* | 108,249,173.3 | 14 | | | |

The $F$ test for $H_0$: $\sigma_\alpha^2 = 0$ is based on $\text{df}_1 = 2$ and $\text{df}_2 = 12$ degrees of freedom. Since the computed value of $F$, 1.38, does not exceed 3.89, the value in table 4 for $a = .05$, $\text{df}_1 = 2$, and $\text{df}_2 = 12$, we have insufficient evidence to indicate that there is a significant random component due to variability in intensities from tracking station to tracking station. Rather, as an electrical engineer postulated, it is probably best to work with a single tracking station, since most of the variability in intensities is related to the distance of the tracking station from the point of discharge, and we have no control of this source.

# Exercises

18.1. A pharmaceutical company would like to examine the potency of a liquid medication mixed in large vats. To do this, a random sample of 5 vats from a month's production was obtained, and 4 separate samples were selected from each vat.

a. Write a random effects model for this experimental situation. Identify all terms in the model.

b. Run an analysis of variance for the sample data given below. Use $\alpha = .05$.

| *Vat 1* | *Vat 2* | *Vat 3* | *Vat 4* | *Vat 5* |
|---|---|---|---|---|
| 3.2 | 2.6 | 3.4 | 4.2 | 1.8 |
| 3.8 | 2.9 | 3.9 | 4.4 | 2.3 |
| 3.5 | 2.8 | 3.3 | 4.3 | 1.9 |
| 3.0 | 2.0 | 3.1 | 4.2 | 2.1 |

18.2. Suppose the pharmaceutical company of exercise 18.1 wishes to estimate $\mu$, the expected potency for a measurement made on a vat selected at random. Although we have not discussed this topic in this section, as might be expected, we can estimate $\mu$ using $\bar{y}$, the mean of the sample data. However, it is not so obvious, but nonetheless true, that the variance of the estimator $\bar{y}$ is

$$\frac{\sigma^2}{20} + \frac{\sigma_\alpha^2}{5}$$

where $\sigma_\alpha^2$ is the variance of the random vat effect $\alpha_i$. Since we do not know $\sigma^2$ or $\sigma_\alpha^2$, we can form estimates by using the MS and EMS columns of the AOV table shown below.

| *Source* | *MS* | *EMS* |
|---|---|---|
| vats | MSA | $\sigma^2 + 4\sigma_\alpha^2$ |
| error | MSE | $\sigma^2$ |

From the abbreviated table, we see that MSA has expected mean square of $\sigma^2 + 4\sigma_\alpha^2$, and hence MSA/20 provides an estimate of the variance of $\bar{y}$, $(\sigma^2/20) + (\sigma_\alpha^2/5)$. In general, for a random effects model in a one-way classification with the same number of observations per treatment, $\bar{y}$ provides an estimate of $\mu$ and MSA/$n$ provides an estimate for the variance of the distribution of sample means.

a. Using the sample data of exercise 18.1, form a point estimate of $\mu$, the average potency for a measurement made on a vat selected at random.

b. Place an approximate bound on the error of estimation. Hint:

$$\text{bound} = 2\sqrt{\frac{\text{MSA}}{n}}$$

# A Two-way Classification with Both Factors Random: A Random Effects Model 18.3

The ideas presented for a random effects model in a one-way classification can be extended to any of the block and factorial experimental designs covered in chapters 13 and 15. Although we will not have time to cover all such situations, we will consider a two-variable factorial experiment in a two-way classification where both factors are random.

Consider an experiment to examine the effects of different analysts and subjects in chemical analyses for the DNA content of plaque. Three female subjects (ages 18–20) were chosen for the study. Each subject was allowed to maintain her usual diet, supplemented with 30 mg (15 tablets) of sucrose per day. No toothbrushing or mouthwashing was allowed during the study. At the end of the week, plaque was scraped from the entire dentition of each subject and divided into 3 samples. Each of 3 analysts chosen at random was then given an unmarked sample of plaque from each of the subjects and asked to perform an analysis for the DNA content (in micrograms). The two-way classification of sample data could then be organized as shown in table 18.4.

**Table 18.4** DNA concentrations for samples of plaque

| | *Subject* | | | |
|---|---|---|---|---|
| *Analyst* | *1* | *2* | *3* | *Totals* |
| 1 | | | | |
| 2 | | | | |
| 3 | | | | |
| *Totals* | | | | |

randomized block design

This experimental design is recognized as a randomized block design, with subjects representing blocks and analysts being the treatments. The experimental units are samples of plaque scraped from the dentition of subjects. If we assume that the 3 subjects represent a random sample from a large population of possible subjects, and, similarly, that the 3 analysts represent a random sample from a large population of possible analysts, we can write the following random effects model relating DNA concentration to the two factors "analysts" and "subjects":

random effects model

$$y_{ij} = \mu + \alpha_i + \beta_j + \epsilon_{ij}$$

We assume the following:

assumptions

1. $\mu$ is an overall unknown concentration mean.
2. $\alpha_i$ is a random effect due to the $i$th analyst. $\alpha_i$ is normally distributed, with mean 0 and variance $\sigma_\alpha^2$.
3. The $\alpha_i$'s are independent.
4. $\beta_j$ is a random effect due to the $j$th subject. $\beta_j$ is a normally distributed random variable, with mean 0 and variance $\sigma_\beta^2$.
5. The $\beta_j$'s are independent.
6. The $\alpha_i$'s and the $\beta_j$'s are independent.

Again note the difference between assuming that the treatments and blocks are random rather than fixed effects. If, for example, the 3 analysts chosen for the study were the only analysts of interest, we would be concerned with differences in mean DNA concentrations for these specific analysts. Now, however, treating the effect due to an analyst as a random variable, our inference will be about the population of analysts' effects. Since the mean of this normal population is assumed to be 0, we want to determine whether or not the variance $\sigma_\alpha^2$ is greater than zero.

Once again the analysis of variance tables are the same for fixed or random effects models in a two-way classification, with the exception that the expected mean squares columns are different. The AOV table for a general two-way classification of "a" levels of factor $A$ and $b$ levels of factor $B$ and no replication is shown in table 18.5. Note the difference in the expected mean squares columns.

AOV table

**Table 18.5** AOV table for a two-way classification with no replication

| | | | | *EMS* | |
|---|---|---|---|---|---|
| *Source* | *SS* | *df* | *MS* | *Fixed Effects* | *Random Effects* |
| *A* | SSA | a − 1 | MSA | $\sigma^2 + \theta_A$ | $\sigma^2 + b\sigma_\alpha^2$ |
| *B* | SSB | $b - 1$ | MSB | $\sigma^2 + \theta_B$ | $\sigma^2 + a\sigma_\beta^2$ |
| error | SSE | $(a - 1)(b - 1)$ | MSE | $\sigma^2$ | $\sigma^2$ |
| *Totals* | TSS | $ab - 1$ | | | |

The computation of sums of squares and mean squares would proceed exactly as shown in chapter 15, using the appropriate shortcut formulas. The difference in test procedures is illustrated in table 18.6 for factor $A$. Similar results would also apply to factor $B$.

Rather than proceed with an example at this point, we will discuss a random effects model for a factorial experiment arranged in a two-way classification, with $r > 1$ observations at each factor-level combination. Then we will illustrate the test procedure.

**Table 18.6** Difference in test procedures for factor $A$

| *Fixed Effects Model* | *Random Effects Model* |
|---|---|
| $H_0$: $\alpha_1 = \alpha_2 = \cdots = \alpha_a = 0$. | $H_0$: $\sigma_\alpha^2 = 0$. |
| $H_a$: At least one of the $\alpha$'s differs from the rest. | $H_a$: $\sigma_\alpha^2 > 0$. |
| T.S.: $F = \frac{\text{MSA}}{\text{MSE}}$. | T.S.: $F = \frac{\text{MSA}}{\text{MSE}}$. |
| R.R.: Based on $df_1 = a - 1$, $df_2 = (a - 1)(b - 1)$. | R.R.: Same. |

tests

In chapter 15 we considered the fixed effects model for a $2 \times 2$ factorial experiment with $r > 1$ observations per cell. The random effects model for an $a \times b$ factorial experiment would be of the same form as the corresponding fixed effects experiment, but with different assumptions.

$a \times b$ factorial, $r > 1$

$$y_{ijk} = \mu + \alpha_i + \beta_j + \gamma_{ij} + \epsilon_{ijk}$$

model

where $y_{ijk}$ is the response for the $k$th observation at the $i$th level of factor $A$ and the $j$th level of factor $B$; $\mu$, $\alpha_i$, $\beta_j$, and $\epsilon_{ijk}$ are defined as before for the random effects model without replication. In addition, we assume the following:

1. $\gamma_{ij}$ is a random effect due to the $i$th level of factor $A$ and the $j$th level of factor $B$. $\gamma_{ij}$ is normally distributed, with mean 0 and variance $\sigma_\gamma^2$.
2. The $\gamma_{ij}$'s are independent.
3. The $\alpha_i$'s, $\beta_j$'s, and $\gamma_{ij}$'s are independent.

assumptions

The appropriate AOV tables for fixed and random effects models are shown in table 18.7.

**Table 18.7** AOV table for a two-way classification with $r$ observations per cell

AOV table

| | | | | *EMS* | |
|---|---|---|---|---|---|
| *Source* | *SS* | *df* | *MS* | *Fixed Effects* | *Random Effects* |
| $A$ | SSA | $a - 1$ | MSA | $\sigma^2 + \theta_A$ | $\sigma^2 + r\sigma_\gamma^2 + br\sigma_\alpha^2$ |
| $B$ | SSB | $b - 1$ | MSB | $\sigma^2 + \theta_B$ | $\sigma^2 + r\sigma_\gamma^2 + ar\sigma_\beta^2$ |
| $AB$ | SSAB | $(a - 1)(b - 1)$ | MSAB | $\sigma^2 + \theta_{AB}$ | $\sigma^2 + r\sigma_\gamma^2$ |
| error | SSE | $ab(r - 1)$ | MSE | $\sigma^2$ | $\sigma^2$ |
| *Totals* | TSS | $abr - 1$ | | | |

The appropriate tests using the $AB$ interaction sum of squares are illustrated in table 18.8 for the two models.

Now unlike the one-way classification and the two-way classification without replication, the test statistics for main effects are different for the fixed

tests, interaction

**Table 18.8** A comparison of appropriate interaction tests for fixed and random effects models

| *Fixed Effects Model* | *Random Effects Model* |
|---|---|
| $H_0$: $\gamma_{11} = \gamma_{12} = \cdots = \gamma_{ab} = 0$. | $H_0$: $\sigma_\gamma^2 = 0$. |
| $H_a$: At least one $\gamma_{ij}$ differs from the rest. | $H_a$: $\sigma_\gamma^2 > 0$. |
| T.S.: $F = \dfrac{\text{MSAB}}{\text{MSE}}$. | T.S.: $F = \dfrac{\text{MSAB}}{\text{MSE}}$. |
| R.R.: Based on $df_1 = (a - 1)(b - 1)$, $df_2 = ab(r - 1)$. | R.R.: Same. |

and random effects models. In addition, for the random effects model, the tests for $\sigma_\alpha^2$ and $\sigma_\beta^2$ can proceed even when the test on the *AB* interaction ($\sigma_\gamma^2$) is significant. For the fixed effects model following a nonsignificant test on the *AB* interaction, we can test for main effects due to factors *A* and *B* by using

$$F = \frac{\text{MSA}}{\text{MSE}} \quad \text{and} \quad F = \frac{\text{MSB}}{\text{MSE}}$$

respectively. As we see from the expected mean squares column of table 18.7, no matter what the result of the test $H_0$: $\sigma_\gamma^2 = 0$, we can form an *F* test for the components $\sigma_\alpha^2$ and $\sigma_\beta^2$ using the test procedures shown in table 18.9. Note that

tests, main effects

**Table 18.9** Tests for a two-way classification with replication: Random effects model

| *Factor A* | *Factor B* |
|---|---|
| $H_0$: $\sigma_\alpha^2 = 0$. | $H_0$: $\sigma_\beta^2 = 0$. |
| $H_a$: $\sigma_\alpha^2 > 0$. | $H_a$: $\sigma_\beta^2 > 0$. |
| T.S.: $F = \dfrac{\text{MSA}}{\text{MSAB}}$. | T.S.: $F = \dfrac{\text{MSB}}{\text{MSAB}}$. |
| R.R.: Based on $df_1 = (a - 1)$, $df_2 = (a - 1)(b - 1)$. | R.R.: Based on $df_1 = (b - 1)$, $df_2 = (a - 1)(b - 1)$. |

the test statistics differ from those used in the fixed effects case, where the denominator of all *F* statistics is MSE.

**EXAMPLE 18.2**

Consider the experimental situation described for the data of table 18.4, where we were concerned with the effects of analysts and subjects on the DNA concentration of plaque in 18-to-20-year-old females. Suppose now that the plaque collection from each girl was subdivided into 6 samples and each analyst (unknown to him or her) made a DNA concentration determination on 2 samples for each subject. These data are recorded in table 18.10 (in units of 10 $\mu$g).

**Table 18.10** DNA concentration data, example 18.2

| Analyst, Factor B | Subject, Factor A 1 | 2 | 3 | Totals |
|---|---|---|---|---|
| 1 | 13.2<br>12.3<br>25.5 | 10.6<br>9.8<br>20.4 | 8.5<br>8.9<br>17.4 | 63.3 |
| 2 | 12.5<br>12.9<br>25.4 | 9.6<br>10.7<br>20.3 | 7.9<br>8.4<br>16.3 | 62.0 |
| 3 | 13.0<br>12.4<br>25.4 | 9.9<br>10.3<br>20.2 | 8.3<br>8.6<br>16.9 | 62.5 |
| *Totals* | 76.3 | 60.9 | 50.6 | 187.8 |

Perform an analysis of variance for this experiment. Conduct all tests with $\alpha = .05$, and draw your conclusions.

**SOLUTION**

Using the shortcut formulas of chapter 15, we obtain the sums of squares as follows:

$$\text{TSS} = \sum y_{ijk}^2 - \frac{G^2}{n} = 2017.38 - \frac{(187.8)^2}{18}$$

$$= 2017.38 - 1959.38 = 58$$

$$\text{SSA} = \frac{\sum A_i^2}{6} - \frac{G^2}{n} = 2015.14 - 1959.38 = 55.76$$

$$\text{SSB} = \frac{\sum B_j^2}{6} - \frac{G^2}{n} = 1959.52 - 1959.38 = .14$$

$$\text{SSAB} = \frac{\sum (AB)_{ij}^2}{2} - \text{SSA} - \text{SSB} - \frac{G^2}{n}$$

$$= 2015.46 - 55.76 - .14 - 1959.38 = .18$$

$$\text{SSE} = \text{TSS} - \text{SSA} - \text{SSB} - \text{SSAB} = 1.92$$

Our results are summarized in an analysis of variance table, table 18.11.

We can proceed with appropriate statistical tests, using the results presented in the AOV table. For the $AB$ interaction, we have:

$H_0$: $\sigma_\gamma^2 = 0$.

$H_a$: $\sigma_\gamma^2 > 0$.

**Table 18.11** AOV table for the data of example 18.2

| *Source* | *SS* | *df* | *MS* | *EMS* |
|---|---|---|---|---|
| *A* | 55.76 | 2 | 27.88 | $\sigma^2 + 2\sigma_\gamma^2 + 6\sigma_\alpha^2$ |
| *B* | .14 | 2 | .07 | $\sigma^2 + 2\sigma_\gamma^2 + 6\sigma_\beta^2$ |
| *AB* | .18 | 4 | .05 | $\sigma^2 + 2\sigma_\gamma^2$ |
| error | 1.92 | 9 | .21 | $\sigma^2$ |
| *Totals* | 58.00 | 17 | | |

T.S.: $F = \dfrac{\text{MSAB}}{\text{MSE}} = \dfrac{.05}{.21} = .24.$

R.R.: For $\alpha = .05$, we will reject $H_0$ if $F$ exceeds 3.63, the critical value for a = .05, $df_1 = 4$, and $df_2 = 9$.

Conclusion: There is insufficient evidence to reject $H_0$. There does not appear to be a significant variability in DNA concentrations due to combinations of analysts and subjects.

For factor *B* we have:

$H_0$: $\sigma_\beta^2 = 0$.

$H_a$: $\sigma_\beta^2 > 0$.

T.S.: $F = \dfrac{\text{MSB}}{\text{MSAB}} = \dfrac{.07}{.05} = 1.4.$

R.R.: For $\alpha = .05$, we will reject $H_0$ if $F$ exceeds 6.94, the critical value for a = .05, $df_1 = 2$, and $df_2 = 4$.

Conclusion: There is insufficient evidence to indicate a significant variability in DNA determinations from analyst to analyst.

Note that if $\sigma_\gamma^2 = 0$, both MSAB and MSE are unbiased estimates of $\sigma^2$, and MSB has expectation equal to $\sigma^2 + 6\sigma_\beta^2$. It can be argued that when MSAB is small relative to MSE, it would be wise to pool the sum of squares for *AB* and for error to form a combined estimate of $\sigma^2$. We can illustrate this procedure for our data. The pooled mean square based on 13 degrees of freedom is

$MS_{\text{pooled}}$

$$MS_{\text{pooled}} = \frac{\text{SSAB} + \text{SSE}}{4 + 9} = \frac{2.10}{13} = .16$$

and can be used to test for main effects. The $F$ test for $H_0$: $\sigma_\beta^2 = 0$ would be

$$F = \frac{\text{MSB}}{MS_{\text{pooled}}} = \frac{.07}{.16} = .44$$

Comparing the computed value of $F$ to the critical value 3.81 for a = .05, $df_1 = 2$, and $df_2 = 13$, we see that there is insufficient evidence to reject $H_0$.

Since this is the same conclusion we reached by using $F = \text{MSB}/\text{MSAB}$, we recommend pooling only in cases where MSAB is considerably less than MSE.

The test for factor $A$ follows.

$H_0$: $\sigma_\alpha^2 = 0$.

$H_a$: $\sigma_\alpha^2 > 0$.

T.S.: $F = \dfrac{\text{MSA}}{\text{MSE}} = \dfrac{27.88}{.21} = 132.76.$

R.R.: For $\alpha = .05$, we will reject $H_0$ if $F$ exceeds 4.26, the critical value based on $a = .05$, $df_1 = 2$, and $df_2 = 9$.

Conclusion: Since the observed value of $F$ is much larger than 4.26, we reject $H_0$ and conclude that there is a significant variability in DNA concentrations in plaque collections from females between the ages of 18 to 20.

In this section we have compared a random effects model to a fixed effects model for the completely randomized design and for the $a \times b$ factorial experiment with $r$ observations per cell. This study has been in no way exhaustive, but it has shown that there are alternatives to a fixed effects model. A more detailed study of the random effects model would certainly include factorial experiments with more than 2 factors and the nested sampling experiment. For the latter design, levels of factor $B$ are nested (rather than cross-classified) within levels of factor $A$. For example, in considering the potency of a chemical, we could sample different manufacturing plants, batches of chemicals within a plant, and determinations within a batch. Note that the factor "batches" is not cross-classified with the factor "plants" since, for example, Batch 1 for Plant 1 is different from Batch 1 for Plant 2.

nested sampling experiment

In section 18.4 we will consider extending the results of this section to include a mixed model for an $a \times b$ factorial experiment.

# Exercises

18.3. Refer to example 18.2 (p. 568). Suppose that only one observation was made by an analyst on a plaque sample from each subject. Taking the first observation for each factor-level combination, we have the following sample data.

| | Subject | | | |
|---|---|---|---|---|
| *Analyst* | *1* | *2* | *3* | *Totals* |
| 1 | 13.2 | 10.6 | 8.5 | 32.3 |
| 2 | 12.5 | 9.6 | 7.9 | 30.0 |
| 3 | 13.0 | 9.9 | 8.3 | 31.2 |
| *Totals* | 38.7 | 30.1 | 24.7 | 93.5 |

a. Write an appropriate linear statistical model identifying all terms in the model.

b. Write down the expected mean squares.

18.4. Refer to exercise 18.3. Perform an analysis of variance. Use $\alpha = .05$ for all tests.

$ 18.5. Officials of a marketing research corporation were interested in studying the effect of a new promotional campaign for an improved brand of D-cell batteries. The study was conducted in a random sample of 4 standard metropolitan statistical areas (SMSAs), which had outlet stores for a random sample of 3 chain stores (selected from a large list of grocery, drug, and department stores). Sales volumes (in dollars) were recorded for a random sample of two weeks following the promotional campaign in the designated areas. These data are shown below.

| | SMSA | | | |
|---|---|---|---|---|
| *Chain Store* | *1* | *2* | *3* | *4* |
| 1 | 98 | 149 | 79 | 340 |
| | 112 | 126 | 61 | 302 |
| 2 | 87 | 96 | 119 | 125 |
| | 75 | 138 | 104 | 133 |
| 3 | 140 | 159 | 169 | 460 |
| | 190 | 185 | 150 | 420 |

a. Write an appropriate linear statistical model. List the assumptions and identify terms.

b. Perform an analysis of variance, showing expected mean squares. Use $\alpha = .05$.

# A Two-way Classification, One Factor Fixed and One Random: A Mixed Effects Model 18.4

In section 18.3 we compared the analysis of variance tables for fixed and random effects models for a two-way classification resulting from an a $\times$ $b$ factorial experiment. Suppose, however, that we have a mixed effects model, where one factor ($A$) is fixed and the other is random. For example, in section 18.3 we considered an experiment to examine the effects of different subjects and different analysts on the DNA content of plaque. If the 3 subjects were selected at random and if the 3 analysts chosen were the only analysts of interest, we would have a mixed model with fixed analysts and random subjects.

mixed model

The model for the a $\times$ $b$ factorial experiment is the same as that given in section 18.3 except that there are different assumptions.

$$y_{ijk} = \mu + \alpha_i + \beta_j + \gamma_{ij} + \epsilon_{ij}$$

where we use the following assumptions:

assumptions

1. $\mu$ is an overall unknown mean.
2. $\alpha_i$ is fixed effect corresponding to the $i$th level of factor $A$.
3. $\beta_j$ is a random effect due to the $j$th level of factor $B$. $\beta_j$ is normally distributed, with mean 0 and variance $\sigma_\beta^2$.
4. $\gamma_{ij}$ is a random effect due to the $i$th level of $A$ and $j$th level of $B$. $\gamma_{ij}$ is normally distributed, with mean 0 and variance $\sigma_\gamma^2$.
5. The $\beta_j$'s and $\gamma_{ij}$'s are all independent.

Using these assumptions, the analysis of variance table (incorporating table 18.7) for either a fixed, random, or mixed model in a two-way classification with replication is as shown in table 18.12.

The expected mean squares column of table 18.12 can be helpful in determining appropriate tests of significance. The test for $\sigma_\gamma^2$ is the same in the mixed model as in the random effects model.

test for $\sigma_\gamma^2$

$H_0$: $\sigma_\gamma^2 = 0$.

$H_a$: $\sigma_\gamma^2 > 0$.

T.S.: $F = \dfrac{\text{MSAB}}{\text{MSE}}$.

R.R.: Based on $df_1 = (a - 1)(b - 1)$ and $df_2 = ab(r - 1)$.

AOV table

**Table 18.12** AOV for a two-way classification with $r$ observations per cell

| | | | | *EMS* | | |
|---|---|---|---|---|---|---|
| *Source* | *SS* | *df* | *MS* | *Fixed Effects* | *Random Effects* | *Mixed* |
| $A$ | SSA | $a - 1$ | MSA | $\sigma^2 + \theta_A$ | $\sigma^2 + r\sigma_\gamma^2 + br\sigma_\alpha^2$ | $\sigma^2 + r\sigma_\gamma^2 + \theta_A$ |
| $B$ | SSB | $b - 1$ | MSB | $\sigma^2 + \theta_B$ | $\sigma^2 + r\sigma_\gamma^2 + ar\sigma_\beta^2$ | $\sigma^2 + ar\sigma_\beta^2$ |
| $AB$ | SSAB | $(a - 1)(b - 1)$ | MSAB | $\sigma^2 + \theta_{AB}$ | $\sigma^2 + r\sigma_\gamma^2$ | $\sigma^2 + r\sigma_\gamma^2$ |
| error | SSE | $ab(r - 1)$ | MSE | $\sigma^2$ | $\sigma^2$ | $\sigma^2$ |
| *Totals* | TSS | $abr - 1$ | | | | |

No matter what the results of our test for $\sigma_\gamma^2$, we can proceed to use the following tests for factors $A$ and $B$, which follow from entries in the expected mean squares column of table 18.12. For factor $A$ we have

test, factor $A$

$H_0$: $\alpha_1 = \alpha_2 = \cdots = \alpha_a = 0$.

$H_a$: At least one of the $\alpha$'s differs from the rest.

T.S.: $F = \dfrac{\text{MSA}}{\text{MSAB}}$.

R.R.: Based on $df_1 = (a - 1)$ and $df_2 = (a - 1)(b - 1)$.

For factor $B$ we have

test, factor $B$

$H_0$: $\sigma_\beta^2 = 0$.

$H_a$: $\sigma_\beta^2 > 0$.

T.S.: $F = \dfrac{\text{MSB}}{\text{MSE}}$.

R.R.: Based on $df_1 = (b - 1)$ and $df_2 = ab(r - 1)$.

The analysis of variance procedure outlined for a mixed effects model in a two-way classification can be illustrated for a randomized block design, where blocks are assumed to be random.

### EXAMPLE 18.3

A corporation is interested in comparing two different sun screens ($s_1$ and $s_2$) for protecting the skin of persons who may wish to avoid burning or additional tanning while exposed to the sun. A random sample of females (ages 20–25) agreed to participate in the study. For each person two 1″ × 1″ squares were marked off on either side of the back, under the shoulder but above the small of the back. Sun screen $s_1$ was then randomly assigned to the two squares on one side of the back, with $s_2$ assigned to the other two squares. A reading based on the color of skin in a square was made prior to the application of a fixed amount of the assigned sun screen, and then again after application and exposure to the sun for a two-hour period. The data recorded in table 18.13 are differences (post-

**Table 18.13** Data (differences) for the sun screen experiment, example 18.3

| Sun Screen, A | Persons, B 1 | 2 | 3 | 4 | 5 | 6 | 7 | 8 | 9 | 10 | Totals |
|---|---|---|---|---|---|---|---|---|---|---|---|
| | 8.2 | 3.6 | 10.7 | 3.9 | 12.9 | 5.5 | 9.1 | 13.7 | 8.1 | 2.5 | |
| | 7.6 | 3.5 | 10.3 | 4.4 | 12.1 | 5.9 | 9.7 | 13.2 | 8.7 | 2.8 | 156.4 |
| | 15.8 | 7.1 | 21.0 | 8.3 | 25.0 | 11.4 | 18.8 | 26.9 | 16.8 | 5.3 | |
| | 6.1 | 4.3 | 9.6 | 2.3 | 12.4 | 4.8 | 8.3 | 12.9 | 8.0 | 2.1 | |
| 2 | 6.8 | 4.7 | 9.2 | 2.5 | 12.8 | 4.0 | 8.6 | 13.6 | 7.5 | 2.5 | 143.0 |
| | 12.9 | 9.0 | 18.8 | 4.8 | 25.2 | 8.8 | 16.9 | 26.5 | 15.5 | 4.6 | |
| *Totals* | 28.7 | 16.1 | 39.8 | 13.1 | 50.2 | 20.2 | 35.7 | 53.4 | 32.3 | 9.9 | 299.4 |

exposure minus preexposure) for the persons in the study. A high response indicates more burning.

Use these data to run an analysis of variance. Set $\alpha = .05$.

**SOLUTION**

We can compute the sums of squares for the sources of variability in the AOV table using the usual shortcut formulas.

$$\text{TSS} = \sum y_{ijk}^2 - \frac{G^2}{n}$$

$$= 2771.6 - \frac{(299.4)^2}{40} = 2771.6 - 2241.01 = 530.59$$

$$\text{SSA} = \frac{\sum A_i^2}{20} - \frac{G^2}{n} = 2245.50 - 2241.01 = 4.49$$

$$\text{SSB} = \frac{\sum B_j^2}{4} - \frac{G^2}{n} = 2758.50 - 2241.01 = 517.49$$

$$\text{SSAB} = \frac{\sum (AB)_{ij}^2}{2} - \text{SSA} - \text{SSB} - \frac{G^2}{n}$$

$$= 2768.96 - 4.49 - 517.49 - 2241.01 = 5.97$$

Then by subtraction, SSE = TSS − SSA − SSB − SSAB = 2.64.

Substituting a = 2, $b = 10$, and $r = 2$ into an AOV table similar to that shown in table 18.12, we have the results shown in table 18.14.

A test for the random component $\gamma_{ij}$ is as follows:

$H_0$: $\sigma_\gamma^2 = 0$.

$H_a$: $\sigma_\gamma^2 > 0$.

$$\text{T.S.: } F = \frac{\text{MSAB}}{\text{MSE}} = \frac{.66}{.13} = 5.08.$$

**Table 18.14** AOV table for the data of example 18.3

| *Source* | *SS* | *df* | *MS* | *EMS* *Mixed Model* |
|---|---|---|---|---|
| *A* | 4.49 | 1 | 4.49 | $\sigma^2 + 2\sigma_\gamma^2 + \theta_A$ |
| *B* | 517.49 | 9 | 57.50 | $\sigma^2 + 4\sigma_\beta^2$ |
| *AB* | 5.97 | 9 | .66 | $\sigma^2 + 2\sigma_\gamma^2$ |
| error | 2.64 | 20 | .13 | $\sigma^2$ |
| *Totals* | 530.59 | 39 | | |

R.R.: For $\alpha = .05$, we will reject $H_0$ if the computed value of $F$ exceeds 2.39, the value in table 4 for a = .05, $df_1 = 9$, and $df_2 = 20$.

Conclusion: Since 5.08 exceeds 2.39, we reject $H_0$ and conclude that $\sigma_\gamma^2 > 0$; that is, there is a significant source of random variation due to the combination of the *i*th level of *A* (sun screens) and the *j*th level of *B* (persons). We would infer from this that one sun screen may not necessarily be better than the other for all persons.

Even with this significant $F$ test for $\sigma_\gamma^2$, we can proceed to test for effects due to factors *A* and *B* separately. For factor *B* we have

$H_0$: $\sigma_\beta^2 = 0$.

$H_a$: $\sigma_\beta^2 > 0$.

T.S.: $F = \dfrac{\text{MSB}}{\text{MSE}} = \dfrac{57.50}{.13} = 442.31.$

R.R.: For $\alpha = .05$, we will reject $H_0$ if $F$ exceeds 2.39, the value in table 4 for a = .05, $df_1 = 9$, and $df_2 = 20$.

Conclusion: Since 442.31 exceeds 2.39, we reject $H_0$ and conclude that $\sigma_\beta^2 > 0$. Thus there is a significant source of random variation due to variability from person to person.

For factor *A* we have

$H_0$: $\alpha_1 = \alpha_2 = 0$.

$H_a$: $\alpha_1 \neq \alpha_2$.

T.S.: $F = \dfrac{\text{MSA}}{\text{MSAB}} = \dfrac{4.49}{.66} = 6.80.$

R.R.: For $\alpha = .05$, we will reject $H_0$ if $F$ exceeds 5.12, the value in table 4 for a = .05, $df_1 = 1$, and $df_2 = 9$.

Conclusion: Since $6.80 > 5.12$, we reject $H_0$ and conclude that the mean response (post minus pre) differs for the two sun screens. Since $\bar{y}_{s1} = 156.4/20 = 7.82$ and $\bar{y}_{s2} = 143/20 = 7.15$, we would conclude that $s_2$ offers more protection on the average than $s_1$. However, as noted previously, there are significant sources of variability due to persons and the combination of persons with sun screens.

This discussion of mixed models, which has been illustrated for the a $\times$ $b$ factorial with $r$ observations per cell, provides only a brief introduction to the study of mixed models. Indeed, we could spend one or more quarters of study at the graduate level covering topics appropriate for mixed models. For more advanced work, we could examine factorial experiments with three or more factors (some random, others fixed). In addition, when examining the effect of two factors (both fixed effects) on a response while blocking on a third factor (which is random), this *split-plot design* becomes an important alternative to a factorial experiment that is laid off in a randomized block design. The difference between this split-plot design and the factorial experiment set off in a randomized block design lies in the method of applying treatments (factor-level combinations) to experimental units. For each block, levels of factor 1 are randomly assigned to experimental units. Then levels of the second factor are randomly assigned to subunits within each level of factor 1. This randomization is quite different from the randomization used in a factorial experiment that is laid off in a randomized block design. For a discussion of this topic, see Steel and Torrie (1960) or Snedecor and Cochran (1967). Finally, a more extended discussion of mixed models could include the nested sampling experiment, mentioned in section 18.3, as applied to mixed models.

split-plot design

# Exercises

18.6. Prior to conducting a clinical trial that involves a subjective evaluation of a patient's progress, the participating physicians are asked to agree on certain criteria for reaching an evaluation. To examine the consistency in their evaluations before the initiation of a particular clinical trial, a pilot study was conducted on 4 patients who had been treated with a drug that was to be included in the trial. Each of the 5 physicians who were to participate in the study was asked to evaluate (on a 0-to-10-point scale) the degree of cure after a two-week treatment period. Since the clinical evaluations of a patient's cure was to be based on the results of a bacterial culture analysis, each physician analysed 2 cultures from each patient. This feature was unknown to the physicians, who were merely told they would be analyzing 8 separate bacterial cultures. The evaluations based on these cultures are recorded on the next page.

a. Treating physicians as fixed and patients as random, write an appropriate linear statistical model. Identify all terms in the model.

b. Show the expected mean squares column in the AOV table.

| | *Patient* | | | |
|---|---|---|---|---|
| *Physician* | *1* | *2* | *3* | *4* |
| 1 | 7.2<br>9.6 | 4.2<br>3.5 | 9.5<br>9.3 | 5.4<br>3.9 |
| 2 | 8.5<br>9.6 | 2.9<br>3.3 | 8.8<br>9.2 | 6.3<br>6.0 |
| 3 | 9.1<br>8.6 | 1.8<br>2.4 | 7.6<br>7.1 | 6.1<br>5.6 |
| 4 | 8.2<br>9.0 | 3.6<br>4.4 | 7.3<br>7.0 | 5.0<br>5.4 |
| 5 | 7.8<br>8.0 | 3.7<br>3.9 | 9.2<br>8.3 | 6.5<br>6.9 |

18.7. Refer to exercise 18.6. Perform an analysis of variance. Draw your conclusions, using $\alpha = .05$.

# 18.5 Rules for Obtaining Expected Mean Squares

We indicated in chapter 15 that an experiment produces a $k$-way classification of the data if all factor-level combinations of the $k$ independent variables are observed the same number of times. Thus all the experimental designs discussed in chapter 15 (randomized block, Latin square, and factorial) give rise to $k$-way classifications.

We will see in this section that for any $k$-way classification of data, with $r$ observations per factor-level combination, it is possible to write expected mean squares for all main effects and interactions for fixed, random, or mixed models using some rather simple rules. *The importance of these rules is that, having written down the expected mean squares for an unfamiliar experimental design, we often can construct appropriate F tests.* The assumptions for the fixed and random effects models will be the same as we have used in describing fixed, random, and mixed models in previous sections.

classifying interactions

Two rules for classifying interactions as fixed or random effects are needed before we can proceed with the rules for obtaining expected mean squares.

**Rules for the Classification of Interactions**

1. If a fixed effect interacts with another fixed effect, the resulting interaction term is a fixed effect.
2. If a random effect interacts with another effect (fixed or random), the resulting interaction term is a random component.

**EXAMPLE 18.4**

Consider a $3 \times 6$ factorial with 2 observations per factor-level combination. Classify the $AB$ interaction as fixed or random for the following situations:

a. $A$ and $B$ are both fixed effects.
b. $A$ is fixed and $B$ is random.
c. $A$ and $B$ are both random.

**SOLUTION**

We apply the rules for classifying interactions.

a. $AB$ is a fixed effect since $A$ (fixed) interacts with $B$ (fixed).
b. $AB$ is a random component since $A$ (fixed) interacts with $B$ (random).
c. $AB$ is random since $A$ (random) interacts with $B$ (random).

**EXAMPLE 18.5**

Consider a factorial experiment in the factors $A$, $B$, and $C$. Classify the $AB$, $AC$, $BC$, and $ABC$ interactions as fixed or random when $A$ and $B$ are fixed effects and $C$ is random.

**SOLUTION**

We apply the classification rules.

$AB$ is fixed; $A$ (fixed) interacts with $B$ (fixed).

$AC$ is random; $A$ (fixed) interacts with $C$ (random).

$BC$ is random; $B$ (fixed) interacts with $C$ (random).

$ABC$ is random; $A$ (fixed) interacts with $BC$ (random).

Since we now know how to classify interaction terms as fixed or random, we can state the rules for obtaining expected mean squares in a $k$-way classification.

**Rules for Obtaining the EMS in a $k$-way Classification**

1. MSE has expectation $\sigma^2$.
2. The expected mean square for any other source $S$ ($S$ could represent either a main effect or an interaction) is the sum of the following terms:

(a) $\sigma^2$;
(b) a separate variance term for the interaction of $S$ with any other random component; the coefficient of such an interaction is $r$ times the product of the number of levels in all factors not appearing in the resulting interaction terms;
(c) $\theta_S$, if $S$ is a fixed effect;
(d) a variance term for $S$, if $S$ is a random effect; the coefficient of $\sigma_S^2$ is $r$ times the product of the number of levels in all factors not appearing in the source $S$.

---

### EXAMPLE 18.6

Give the expected mean squares for a $3 \times 5 \times 2$ factorial, with $r = 4$ observations per cell, if $A$ and $B$ are fixed and $C$ is random.

### SOLUTION

For this example $a = 3$, $b = 5$, $c = 2$, and $r = 4$. The sources of variability are

| | |
|---|---|
| $A$ | fixed |
| $B$ | fixed |
| $C$ | random |
| $AB$ | fixed |
| $AC$ | random |
| $BC$ | random |
| $ABC$ | random |
| error | |
| total | |

We work from the bottom up and apply the rules:
$E(\text{MSE}) = \sigma^2$

Since $ABC$ is a random component, and since it does not interact with any other random component, its expected mean square is

$\sigma^2 + 4\sigma_{ABC}^2$

(Note: $r = 4$.)
Since $BC$ is a random component, and since it does not interact with any other random component, its expected mean square can be written as

$E(\text{MSBC}) = \sigma^2 + ra\sigma_{BC}^2 = \sigma^2 + 12\sigma_{BC}^2$

$AC$ is a random component. But since it does not interact with any other random component, we have

$E(\text{MSAC}) = \sigma^2 + rb\sigma_{AC}^2 = \sigma^2 + 20\sigma_{AC}^2$

$AB$ is a fixed effect. It does interact with the random component $C$.

Hence

$E(\text{MSAB}) = \sigma^2 + 4\sigma^2_{ABC} + \theta_{AB}$

$C$ is a random component that does not interact with any other random component. Hence

$E(\text{MSC}) = \sigma^2 + rab\sigma^2_C = \sigma^2 + 60\sigma^2_C$

$B$ is a fixed effect, and it interacts with the random effects $C$ and $AC$. Hence

$$E(\text{MSB}) = \sigma^2 + r\sigma^2_{ABC} + ra\sigma^2_{BC} + \theta_B$$
$$= \sigma^2 + 4\sigma^2_{ABC} + 12\sigma^2_{BC} + \theta_B$$

$A$ is a fixed effect that interacts with the random effects $C$ and $BC$. Hence

$$E(\text{MSA}) = \sigma^2 + r\sigma^2_{ABC} + rb\sigma^2_{AC} + \theta_A$$
$$= \sigma^2 + 4\sigma^2_{ABC} + 20\sigma^2_{AC} + \theta_A$$

We can summarize these results in an analysis of variance table, table 18.15.

**Table 18.15** AOV table for the terms of example 18.6

| *Source* | *SS* | *df* | *MS* | *EMS* |
|---|---|---|---|---|
| $A$ | SSA | 2 | MSA | $\sigma^2 + 4\sigma^2_{ABC} + 20\sigma^2_{AC} + \theta_A$ |
| $B$ | SSB | 4 | MSB | $\sigma^2 + 4\sigma^2_{ABC} + 12\sigma^2_{BC} + \theta_B$ |
| $C$ | SSC | 1 | MSC | $\sigma^2 + 60\sigma^2_C$ |
| $AB$ | SSAB | 8 | MSAB | $\sigma^2 + 4\sigma^2_{ABC} + \theta_{AB}$ |
| $AC$ | SSAC | 2 | MSAC | $\sigma^2 + 20\sigma^2_{AC}$ |
| $BC$ | SSBC | 4 | MSBC | $\sigma^2 + 12\sigma^2_{BC}$ |
| $ABC$ | SSABC | 8 | MSABC | $\sigma^2 + 4\sigma^2_{ABC}$ |
| error | SSE | 90 | MSE | $\sigma^2$ |
| *Totals* | TSS | 119 | | |

Valid $F$ tests can be formed for all sources except the main effects for factors $A$ and $B$. Thus, for example, the test statistic for $H_0$: $\sigma^2_{AC} = 0$ is $F = \text{MSAC}/\text{MSE}$. However, to construct an exact test for the main effects for factor $A$, for example, either $\sigma^2_{ABC}$ or $\sigma^2_{AC}$ must be equal to zero. Suppose the test for $\sigma^2_{AC}$ turned out to be nonsignificant (i.e., there was insufficient evidence to indicate $\sigma^2_{AC} > 0$). Then an appropriate test statistic for the null hypothesis $H_0$: $\theta_A = 0$ (i.e., $\alpha_1 = \alpha_2 = \alpha_3 = 0$) would be $F = \text{MSA}/\text{MSABC}$, based on $\text{df}_1 = 2$ and $\text{df}_2 = 8$.

It is not always possible to obtain valid $F$ tests for all sources of variability in an AOV table for a $k$-way classification. This problem occurs for both random and mixed effects models (as illustrated in example 18.6). We can always obtain valid tests for all sources of variability for fixed effects models. Some researchers have developed approximate $F$ tests that can be constructed for sources of variability in random or mixed models where no valid $F$ test is available [see, for example, Satterthwaite (1946) and Welch (1956)]. We will not cover this topic in this text since these tests can become quite involved and are open to some controversy.

estimation and prediction

For some random and mixed effects models, the objectives of the researcher might include estimation of the variances for random effects and prediction of the average value of $y$. We briefly discussed estimation of the expected value of $y$ and how we use the expected mean squares to obtain the variance of our estimates for a random effects model. Estimation of the average value of $y$ for a mixed model is more difficult and beyond the scope of this text.

estimates of variance components

Another use of expected mean squares is in the estimation of the variances associated with random effects in the model. For example, in a random effects model for a one-way classification of $t$ treatments and $r$ observations per treatment, the expected mean squares for treatments and error are $\sigma^2 + r\sigma_\alpha^2$ and $\sigma^2$, respectively. As before, MSE is our estimate of $\sigma^2$. Similarly, since MST estimates $\sigma^2 + r\sigma_\alpha^2$, by substituting MSE for $\sigma^2$, we can equate MST to the expected mean square for treatments and solve for $\sigma_\alpha^2$. The solution, $(\text{MST} - \text{MSE})/r$, is an unbiased estimate of the variance of the treatments' source of variability in a one-way classification. This procedure of equating mean squares to expected mean squares can be used for obtaining estimates of variance components (variances of random effects) in random and mixed effects models for balanced designs. The problem of variance component estimation for unbalanced designs is a difficult one and beyond the scope of this text. If you are interested in this topic, we refer you to two references for additional reading on the subject of variance component estimation: Mendenhall (1968), chapter 12 and Searle (1971), chapters 9 through 11.

# 18.6 Summary

Fixed, random, and mixed effects models are easily distinguished if we think in terms of the general linear model. The fixed effects model relates a response to $k \geq 1$ independent variables and one random component, while a random effects model is a general linear model with $k = 0$ and more than one random component. The mixed model, a combination of the fixed and random effects models, relates a response to $k \geq 1$ independent variables and more than one random component.

The application of random effects models to experimental situations was illustrated for the completely randomized design and for the $a \times b$ factorial experiment. Similarities were noted between tests of significance in an analysis of variance for a random effects model and for the corresponding fixed effects model. Inferences resulting from an analysis of variance for a mixed model were illustrated using the $a \times b$ factorial experiment.

Unfortunately, in an introductory course, only a limited amount of time can be devoted to a discussion of random and mixed effects models. To expand our discussion in the text, the results of section 18.5 are useful in developing the expected mean squares for sources of variability in the analysis of variance table for balanced designs. Using these expectations we can then attempt to construct appropriate test statistics for evaluating the significance of any of the fixed or random effects in the model.

The hardest part in any of these problems involving random or mixed models arises from trying to estimate $E(y)$, with an appropriate bound on error, for a random effects model and the average value of $y$ at some level or combination of levels for fixed effects in a mixed model. In exercise 18.2 we illustrated how to obtain an estimate of $E(y)$ for a random effects model and how to construct an approximate bound on the error of estimation. The problem becomes even more complicated for mixed models and hence is discussed in more advanced studies.

# Exercises

18.8. Consider a factorial experiment with "a" levels of factor $A$, $b$ levels of factor $B$, c levels of factor $C$, and $r$ observations per factor-level combination.

a. Write down the expected mean squares for all sources of variability in an analysis of variance table, treating $A$, $B$, and $C$ as fixed effects.

b. Note appropriate tests for all sources.

18.9. Refer to exercise 18.8.

a. Repeat part a, but now consider factors $A$, $B$, and $C$ as random effects.

b. Write down the test statistics for appropriate $F$ tests for sources of variability.

18.10. Refer to exercise 18.8. Suppose now that factor $A$ is fixed and factors $B$ and $C$ are random.

a. Write down expected mean squares for all sources in an AOV table.

b. Indicate the test statistics for those sources of variability that can be tested using an exact $F$ test.

18.11. The civil engineering department at a university was awarded a large grant to study the campus traffic problems and to recommend alternative solutions. One small phase of the study involved obtaining daily counts on the number of cars crossing, but not making use of, the campus facilities. To do this, a team of volunteers was stationed at each entrance to simultaneously monitor the license number and the time of entrance or exit for all cars passing through the checkpoint. By comparing lists for all checkpoints and allowing a reasonable time for cars to traverse the campus, the teams were able to determine the number of cars crossing but not using the campus facilities during the 8:00 A.M. to 5:00 P.M. time period. A random sample of six weeks throughout the academic year were used, with two midweek days selected for study in the weeks sampled. The traffic volume data appear below.

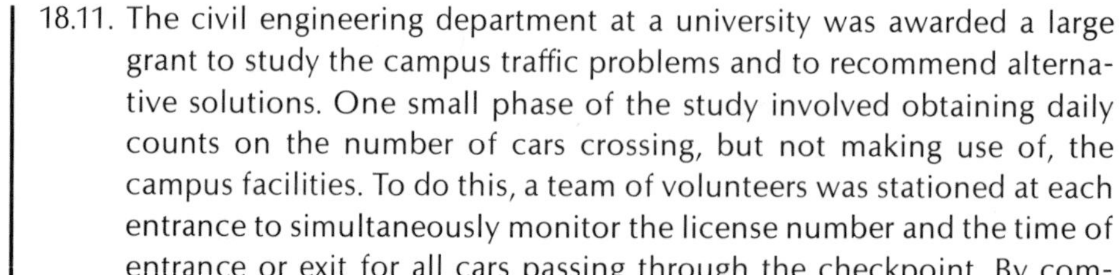

| *Week 1* | *Week 2* | *Week 3* | *Week 4* | *Week 5* | *Week 6* |
|---|---|---|---|---|---|
| 680 | 438 | 539 | 264 | 693 | 530 |
| 618 | 520 | 600 | 198 | 646 | 575 |

a. Write an appropriate linear statistical model. Identify all terms in the model.

b. Perform an analysis of variance, indicating expected mean squares. Use $\alpha = .05$.

18.12. Refer to exercise 18.11. Estimate the average number of cars crossing but not using the campus facilities for a midweek day of a randomly selected week. (Hint: Refer to exercise 18.2.) Place an approximate bound on the error of estimation.

18.13. Refer to exercise 18.6 (p. 577). Suppose the 5 physicians chosen for the pilot study were considered to be a random sample from many possible physicians.

a. Write an appropriate model. Indicate how the assumptions for this model differ from those of part a in exercise 18.6.

b. Compare the AOV table and conclusions which are drawn here to those of exercise 18.7.

c. Which model and analysis seem to be more appropriate?

d. Might you also consider a fixed effects model? Why or why not?

18.14. Refer to exercise 15.1 (p. 418). Suppose that we consider the 5 investigators as a random sample from a population of all possible investigators for the rocket propellant experiment.

a. Write an appropriate linear statistical model, identifying all terms and listing your assumptions.
b. Perform an analysis of variance. Include an expected mean squares column in the analysis of variance table.
c. Indicate the differences in the hypothesis under test and differences in the conclusions drawn.

18.15. Obtain the expected mean squares for the experiment described in example 15.3 (p. 426) if we assume the intersections were selected at random from the many possible locations in a large city.

18.16. Obtain expected mean squares for the experiment described in exercise 15.5 (p. 429) if we assume both rows and columns are random effects.

18.17. Refer to the data of exercise 15.18 (p. 453).
a. Give the expected mean squares under the assumption that investigators were selected at random from a large group of similarly qualified persons throughout the country.
b. Indicate how your analysis and conclusions would change from those in exercise 15.18 under the assumption of part a.

18.18. Refer to the data of example 18.2 (p. 568).
a. Using the expected mean squares, give formulas for estimates of $\sigma^2$, $\sigma^2_\gamma$, $\sigma^2_\beta$, and $\sigma^2_\alpha$.
b. Using the data of example 18.2 and the formulas of part a, find estimates for all the variance components.

18.19. Refer to exercise 18.5 (p. 572). Use the expected mean squares to obtain estimates of all the variance components.

# References

Barr, A.J.; Goodnight, J.H.; Sall, J.P.; and Helwig, J.T. 1976. *A user's guide to SAS 76*. Raleigh, N.C.; SAS Institute, Inc.

Cochran, W.G., and Cox, G.M. 1957. *Experimental design*. 2d ed. New York: Wiley.

Dixon, W.J. 1975. *BMDP, biomedical computer programs*. Berkeley: University of California Press.

Dixon, W.J., and Massey, F.J., Jr. 1969. *Introduction to statistical analysis.* 3d ed. New York: McGraw-Hill.

Draper, N.R., and Smith, H. 1966. *Applied regression analysis.* New York: Wiley.

Graybill, F.A. 1976. *Theory and application of the linear model.* N. Scituate, Mass.: Duxbury Press.

Harnett, D.L. 1970. *Introduction to statistical methods.* Reading, Mass.: Addison-Wesley.

Hurlburt, R.J., and Spiegel, D.K. 1976. Dependence of *F* ratios sharing a common denominator mean square. *American Statistician* 30: 74–78.

Kirk, R.E. 1968. *Experimental design: Procedures for the behavioral sciences.* Belmont, Calif.: Brooks/Cole.

Mendenhall, W. 1968. *Introduction to linear models and the design and analysis of experiments.* Belmont, Calif.: Wadsworth.

Nie, N.H.; Hull, C.H.; Jenkins, J.G.; Steinbrenner, K.; and Bent, D.H. 1975. *Statistical package for the social sciences.* 2d ed. New York: McGraw-Hill.

Ostle, B. 1963. *Statistics in research.* 2d ed. Ames, Iowa: Iowa State University Press.

Pearson, E.S., and Hartley, H.O. 1966. *Biometrika tables for statisticians.* 3d ed. Vol. I. London: Cambridge University Press.

Satterthwaite, F.E. 1946. An approximate distribution of estimates of variance components. *Biometric Bulletin* 2: 110–114.

Searle, S.R. 1971. *Linear models.* New York: Wiley.

Service, J. 1972. *A user's guide to the statistical analysis system.* Raleigh, N.C.: Student Supply Stores, North Carolina State University.

Snedecor, G.W., and Cochran, W.G. 1967. *Statistical methods.* 6th ed. Ames, Iowa: Iowa State University Press.

Sokal, R.R., and Rohlf, F.J. 1969. *Biometry.* San Francisco: W.H. Freeman.

Steel, R.G.D., and Torrie, J.H. 1960. *Principles and procedures of statistics.* New York: McGraw-Hill.

Tanur, J.M.; Mosteller, F.; Kruskal, W.H.; Pieters, R.S.; and Rising, G.R. 1972. *Statistics: A guide to the unknown.* San Francisco: Holden-Day.

Welch, B.L. 1956. On linear combinations of several variances. *Journal of the American Statistical Association* 51: 132–148.

Winer, B.J. 1962. *Statistical principles in experimental design.* New York: McGraw-Hill.

# Analysis of Variance for Some Unbalanced Designs 19

## Introduction 19.1

We examined the analysis of variance for balanced designs in chapters 13 and 15, where we used appropriate shortcut formulas (and corresponding computer solutions) to construct AOV tables and set up hypothesis tests. In chapter 16 we considered another way of performing an analysis of variance. We found that the null hypothesis under test in an analysis of variance can be expressed in terms of one or more $\beta$'s in the general linear model. We also saw that the sum of squares associated with a source of variability in the analysis of variance table can be found as the drop in the sum of squares for error obtained from fitting reduced and complete models. While we did not advocate the use of complete and reduced models for obtaining the sums of squares for sources of variability in balanced designs, we did indicate that the procedure was completely general and could be used for any experimental design. In particular, in this chapter we will make use of complete and reduced models for obtaining the sums of squares in the analyses for unbalanced designs, where shortcut formulas are no longer readily available and easy to apply.

You might ask why an experimenter would run a study using an unbalanced design, especially since unbalanced designs seem to be more difficult to analyze. In point of fact, most studies do begin by using a balanced design, but for any one of many different reasons, the experimenter is unable to obtain the same number of observations per cell as dictated by the balanced design being employed. Consider a study of 3 different weight reducing agents where 5 different clinics (blocks) are employed and patients are to be randomly assigned to the 3 treatment groups according to a randomized block design. Even if the experimenter plans to have 5 overweight persons assigned to each treatment at each clinic, the final count will almost certainly show an imbalance of persons assigned to each treatment group. Almost every clinic could be expected to have a few people who would not complete the study. Some people might move from the community, others might drop out due to a lack of efficacy, and so on. In addition, the experimenter might find that it is impossible to locate 15 overweight people at each clinic who are willing to participate in the study. Since a balanced design at the end of a study is a rarity in practice, we must learn how to analyze data arising from unbalanced designs.

## 19.2 A Randomized Block Design with One or More Missing Observations

unbalanced design

Any time the number of observations is not the same for all factor-level combinations, we call the design *unbalanced*. Thus a randomized block design or a Latin square design with one or more missing observations is an unbalanced design. We will begin our examination by considering a simple case, a randomized block design with one missing observation.

value of missing observation

The analysis of variance for a randomized block design with one missing observation can be performed rather easily by using the shortcut formulas for a balanced design, after we have estimated the value of the missing observation. The formula for the missing observation $M$ is given by

$$M = \frac{tT + bB - G}{(t-1)(b-1)}$$

where $t$ is the number of treatments, $b$ is the number of blocks, $T$ is the sum of all the observations on the treatment assigned to the missing observation, $B$ is the sum of all measurements in the block with the missing observation, and $G$ is the sum of all the measurements.

We illustrate the analysis of variance for this design with an example.

### EXAMPLE 19.1

An experiment was conducted to determine the nutritional value of diets for cows that are supplemented by whey. Five dairies were involved in the study. Each cow in a sample of 4 cows from a dairy was randomly assigned to one of the 4 treatment groups, so that a total of 5 cows were in each treatment group.

Treatment 1: water only

Treatment 2: whey plus 30.2 liters of water/day

Treatment 3: whey plus 15.1 liters of water/day

Treatment 4: whey only

In addition to the liquid portion of the diet listed for each treatment group, each cow was fed 7.5 kg of grain per day and hay as desired.

One response of interest was the amount of whey consumed per day. These data (in kilograms per animal) are listed in table 19.1. Unfortunately, as can be seen from the data, the cow on Diet 4 from Dairy 2 was dropped from the study and no replacement was made. The cow developed an infection (unrelated to the treatment) and was dropped from the study for safety reasons.

**Table 19.1** Consumption of whey for cows, example 19.1

| | *Treatment* | | | |
|---|---|---|---|---|
| *Dairy* | *1* | *2* | *3* | *4* |
| 1 | 15.4 | 9.6 | 9.5 | 8.4 |
| 2 | 14.8 | 9.3 | 9.4 | — |
| 3 | 15.9 | 9.8 | 9.7 | 9.3 |
| 4 | 15.5 | 9.4 | 9.2 | 8.1 |
| 5 | 14.7 | 9.2 | 9.0 | 7.9 |

Estimate the missing value and then perform an analysis of variance. Use $\alpha = .01$.

### SOLUTION

For this randomized block design with $b = 5$ and $t = 4$, the quantities $T$, $B$, and $G$ are defined as follows:

$T$ = sum of all observations on treatment 4

$= 8.4 + 9.3 + 8.1 + 7.9 = 33.7$

$B$ = sum of all observations in block 2

$= 14.8 + 9.3 + 9.4 = 33.5$

$G$ = sum of all measurements

$= 15.4 + 9.6 + \cdots + 7.9 = 204.1$

The estimate of the missing value is

$$M = \frac{tT + bB - G}{(t-1)(b-1)} = \frac{4(33.7) + 5(33.5) - 204.1}{3(4)}$$

$$= \frac{98.2}{12} = 8.2$$

Having estimated the missing value, we can compute sums of squares for our analysis of variance by using the shortcut formulas of chapter 15. The treatment and block totals are given by

| | | |
|---|---|---|
| | $T_1 = 76.3$ | $B_1 = 42.9$ |
| | $T_2 = 47.3$ | $B_2 = 41.7$ |
| | $T_3 = 46.8$ | $B_3 = 44.7$ |
| | $T_4 = 41.9$ | $B_4 = 42.2$ |
| | | $B_5 = 40.8$ |
| Totals | 212.3 | 212.3 |

Note that the new totals for treatment 4 and for block 2 incorporate the estimated missing observation. Similarly, the sum of all measurements includes the estimated missing value.

$$\text{SST} = \frac{(76.3)^2 + \cdots + (41.9)^2}{5} - \frac{(212.3)^2}{20} = 2400.97 - 2253.56 = 147.41$$

$$\text{SSB} = \frac{(42.9)^2 + \cdots + (40.8)^2}{4} - 2253.56 = 2.16$$

$$\text{TSS} = \sum_i \sum_j y_{ij}^2 - \frac{G^2}{20} = (15.4)^2 + (9.6)^2 + \cdots + (7.9)^2 - 2253.56$$

$$= 2403.69 - 2253.56 = 150.13$$

By subtraction, SSE = .56.

The only difference in the analysis variance table for unbalanced and balanced randomized block designs is that since $n$ refers to the number of actual observations, the error for an unbalanced design loses one degree of freedom for each missing observation when compared to the corresponding balanced design. The AOV table for our example is shown in table 19.2.

**Table 19.2** AOV table for the data of example 19.1

| *Source* | *SS* | *df* | *MS* | *F* |
|---|---|---|---|---|
| treatments | 147.41 | 3 | 49.14 | 982.80 |
| blocks | 2.16 | 4 | .54 | 10.80 |
| error | .56 | 11 | .05 | |
| *Totals* | 150.13 | 18 | | |

The $F$ tests for treatments and blocks are both significant, using $\alpha = .01$ (the critical values of $F$ are 6.22 and 5.67, respectively). As can be seen from the data, those cows on treatment 1 (water only) consumed much more hay than cows on any of the diets supplemented with whey.

comparisons among treatment means

Having seen an analysis of variance, we may wish to make certain comparisons among the treatment means. Unfortunately, the multiple comparison procedures of Tukey and Duncan presented in chapter 14 are inappropriate because they depend on equal sample sizes among treatment groups. And Scheffé's procedure is too conservative for our purposes here. So we will resort to running pairwise comparisons using Fisher's least significant difference. The least significant difference between the treatment with a missing observation and any other treatment mean is

LSD

$$\text{LSD} = t_{\alpha/2}\sqrt{\text{MSE}\left(\frac{2}{b} + \frac{t}{b(b-1)(t-1)}\right)}$$

For any pair of treatments with no missing value, the least significant difference is as before; namely,

$$\text{LSD} = t_{\alpha/2}\sqrt{\frac{2\text{MSE}}{b}}$$

fitting complete and reduced models

The formulas for estimating missing observations in a randomized block design become more complicated with more missing data, as do the formulas for least significant differences. Because of this, we will consider fitting complete and reduced models to analyze unbalanced designs. We will illustrate the procedure first by examining an unbalanced randomized block design.

Because it would require more data input for a computer solution using the general linear model format presented in chapters 16 and 17, we will represent the complete and reduced models for testing treatments as follows:

models

complete model: $y_{ij} = \mu + \beta_j + \alpha_i + \epsilon_{ij}$
(model 1)

reduced model: $y_{ij} = \mu + \beta_j + \epsilon_{ij}$
(model 2)

where $\beta_j$ is the $j$th block effect and $\alpha_i$ is the $i$th treatment effect.

By fitting model 1 (using SAS or any other computer routine), we obtain $\text{SSE}_1$. Similarly, a fit of model 2 yields $\text{SSE}_2$. The difference in the two sums of squares for error, $\text{SSE}_2 - \text{SSE}_1$, gives the drop in the sum of squares due to treatments. Since this is an unbalanced design, the block effects do not cancel out when comparing treatment means as they do in a balanced randomized block design (see chapter 15). The difference in the sums of squares, $SSE_2 - SSE_1$, has been adjusted for any effects due to blocks caused by the imbalance in the design. This difference is called the sum of squares due to treatments *adjusted for blocks*.

$SST_{adj}$

$$SSE_2 - SSE_1 = SST_{adj}$$

The sum of squares due to blocks *unadjusted for any treatment differences* is obtained by subtraction:

$$SSB = TSS - SST_{adj} - SSE$$

where SSE and TSS are sums of squares from the complete model. (Note: We could also obtain SSB, the uncorrected sum of squares for blocks, using the shortcut formula of section 15.2.)

The analysis of variance table for testing the effect of treatments is shown in table 19.3. In the table $n$ is the number of actual observations.

AOV table, treatments

**Table 19.3** AOV table for testing the effects of treatments, unbalanced randomized block design

| *Source* | *SS* | *df* | *MS* | *F* |
|---|---|---|---|---|
| blocks | SSB | $b - 1$ | — | — |
| $\text{treatments}_{adj}$ | $SST_{adj}$ | $t - 1$ | $MST_{adj}$ | $MST_{adj}/MSE$ |
| error | SSE | by subtraction | MSE | |
| *Totals* | TSS | $n - 1$ | | |

The corresponding sum of squares for testing the effect of blocks has the same complete model (model 1) as before, and

$$y_{ij} = \mu + \alpha_i + \epsilon_{ij}$$

$SSB_{adj}$

is the reduced model (model 2). The sum of squares drop, $SSE_2 - SSE_1$, is the sum of squares due to blocks after adjusting for the effects of treatments. By subtraction, we obtain

$$SST = TSS - SSB_{adj} - SSE$$

The AOV table is shown in table 19.4.

Note that SST and $SST_{adj}$ are not the same quantity in an unbalanced design; they will be the same only for a balanced design. Similarly, SSB and $SSB_{adj}$ are

AOV table, blocks

**Table 19.4** AOV table for testing effects of blocks, unbalanced randomized block design

| *Source* | *SS* | *df* | *MS* | *F* |
|---|---|---|---|---|
| $\text{blocks}_{\text{adj}}$ | $\text{SSB}_{\text{adj}}$ | $b - 1$ | $\text{MSB}_{\text{adj}}$ | $\text{MSB}_{\text{adj}}/\text{MSE}$ |
| treatments | SST | $t - 1$ | — | — |
| error | SSE | by subtraction | MSE | — |
| *Totals* | TSS | $n - 1$ | | |

different quantities in an unbalanced design. For an unbalanced design we have the following identities:

$$\begin{aligned} \text{TSS} &= \text{SST}_{\text{adj}} + \text{SSB} + \text{SSE} \\ &= \text{SST} + \text{SSB}_{\text{adj}} + \text{SSE} \end{aligned}$$

but

$$\text{TSS} \neq \text{SST}_{\text{adj}} + \text{SSB}_{\text{adj}} + \text{SSE}$$

# Exercises

*19.1. Refer to the data of example 19.1 and the SAS computer output shown here. We have notated some items to help you identify quantities in the output.

a. Indicate the complete and reduced models for testing treatments.
b. Construct an analysis of variance table for testing treatments. Give the level of significance for your test and draw conclusions.
c. Indicate the complete and reduced models for testing blocks.
d. Construct an analysis of variance table for testing blocks. Give the levels of significance for your test.

```
STATISTICAL ANALYSIS SYSTEM

DATA COWS;
INPUT Y 1-4 BLOCKS 6 TREATS 8;
CARDS;


   19 OBSERVATIONS IN DATA SET COWS     3 VARIABLES

PROC PRINT;
```

| OBS | Y | BLOCKS | TREATS |
|---|---|---|---|
| 1 | 15.4 | 1 | 1 |
| 2 | 9.6 | 1 | 2 |
| 3 | 9.5 | 1 | 3 |
| 4 | 8.4 | 1 | 4 |
| 5 | 14.8 | 2 | 1 |
| 6 | 9.3 | 2 | 2 |
| 7 | 9.4 | 2 | 3 |
| 8 | 15.9 | 3 | 1 |
| 9 | 9.8 | 3 | 2 |
| 10 | 9.7 | 3 | 3 |
| 11 | 9.3 | 3 | 4 |
| 12 | 15.5 | 4 | 1 |
| 13 | 9.4 | 4 | 2 |
| 14 | 9.2 | 4 | 3 |
| 15 | 8.1 | 4 | 4 |
| 16 | 14.7 | 5 | 1 |
| 17 | 9.2 | 5 | 2 |
| 18 | 9.0 | 5 | 3 |
| 19 | 7.9 | 5 | 4 |

```
PROC REGR;
CLASSES BLOCKS TREATS;
MODEL Y = BLOCKS TREATS;
TITLE 'EXERCISE 19.1';
```

```
*****************************************************************************************************************

PROC REGR : EXERCISE 19.1

DATA SET   : COWS           NUMBER OF VARIABLES = 1  NUMBER OF CLASSES = 2

VARIABLES  : Y

*****************************************************************************************************************

CLASSES      LEVELS              VALUES

BLOCKS         5                 1 2 3 4 5

TREATS         4                 1 2 3 4

*****************************************************************************************************************

EFFECTS                     ASSIGNED

BLOCKS                      DUMMY001 – DUMMY004

TREATS                      DUMMY005 – DUMMY007

*****************************************************************************************************************
```

EXERCISE 19.1

ANALYSIS OF VARIANCE TABLE, REGRESSION COEFFICIENTS, AND STATISTICS OF FIT FOR DEPENDENT VARIABLE Y

| SOURCE | DF | SUM OF SQUARES | MEAN SQUARE | F VALUE | PROB > F | R-SQUARE | C.V. |
|---|---|---|---|---|---|---|---|
| REGRESSION | 7 | 143.41548246 | 20.48792607 | 394.80383 | 0.0001 | 0.99603550 | 2.12065% |
| ERROR | 11 | 0.57083333 | 0.05189394 | | | STD DEV | Y MEAN |
| CORRECTED TOTAL | 18 | 143.98631579 | | | | 0.22780241 | 10.74211 |

| SOURCE | DF | SEQUENTIAL SS | F VALUE | PROB > F | PARTIAL SS | F VALUE | PROB > F |
|---|---|---|---|---|---|---|---|
| BLOCKS | 4 | 2.61464912 | 12.59612 | 0.0007 | 2.11266667 | 10.17781 | 0.0014 |
| TREATS | 3 | 140.80083333 | 904.41411 | 0.0001 | 140.80083333 | 904.41411 | 0.0001 |
| | | SSB ; $SST_{adj}$ | $F$ test for treatments (adj) | | $SSB_{adj}$ | $F$ test for blocks (adj) | |

19.2. Refer to example 19.1. Use the least significant difference criterion for identifying which treatments differ from the others. Use $\alpha = .05$.

19.3. Refer to example 15.1 (p. 415). Suppose that the first observation in block 1 (Plot 1) is missing. Analyze the data by estimating the missing value and then performing an analysis of variance. Use $\alpha = .05$.

19.4. Refer to exercise 19.3. Perform the corresponding analysis by fitting complete and reduced models. Compare your conclusions to those in exercise 19.3.

19.5. Refer to exercise 19.1. Fit the reduced model $y_{ij} = \mu + \alpha_i + \epsilon_{ij}$ to obtain $SSE_2$. The sum of squares drop will be the sum of squares due to blocks, adjusted for treatments. Verify that this computer value for $SSB_{adj}$ is the same as that shown in the partial sum of squares column of the computer output in exercise 19.1.

19.6. Refer to the data of exercise 15.1 (p. 418). Suppose that in the rocket propellant test for the second mixture to be analyzed by Investigator 3, a piece of equipment malfunctioned. Instead of going back to the laboratories to prepare a duplicate mixture, the investigators proceeded to obtain the remaining propellant thrust data.

a. Estimate the missing value.

b. Perform an analysis of variance. Use $\alpha = .05$.

19.7. Refer to exercise 19.6.
   a. Use complete and reduced models to obtain an analysis of variance. Compare your results to those in exercise 19.6.
   b. Indicate how you would analyze the data if the response for Mixture 4 and Investigator 1 was also missing.

## 19.3 A Latin Square Design with Missing Data

Recall that a $t \times t$ Latin square design can be used to compare $t$ treatment means while filtering out two additional sources of variability (rows and columns). The treatments are randomly assigned in such a way that each treatment appears in every row and in every column. In this section we will illustrate the method for performing an analysis of variance in a Latin square design when one observation is missing. Then we will use the general method of fitting complete and reduced models with missing observations, described for the randomized block design in section 19.2, for more complicated designs.

estimating missing value

The formula for estimating a single missing value in a Latin square design is

$$M = \frac{t(T + R + C) - 2G}{(t - 1)(t - 2)}$$

where $T$, $R$, and $C$ represent the treatment, row, and column totals, respectively, corresponding to the missing observation, and $t$ is the number of treatments in the Latin square design.

**EXAMPLE 19.2**

A company has considered the properties (such as strength, elongation, etc.) of many different variations of nylon hose (stockings) in trying to select the experimental hose to be placed in extensive consumer acceptance surveys.

Five versions (A, B, C, D, and E) of the hose have passed the preliminary screening and are scheduled for more extensive testing. As part of the testing, 5 samples of each experimental hose are to be examined for elongation under constant stress by each of 5 investigators on 5 separate days. The analyses are to be performed following the random assignment of a Latin square. The elongation data (in centimeters) are displayed in table 19.5.

**Table 19.5** Elongation data for example 19.2

| *Investigator* | *Day* 1 | | 2 | | 3 | | 4 | | 5 | |
|---|---|---|---|---|---|---|---|---|---|---|
| 1 | B | 22.1 | A | 18.6 | C | 23.0 | E | 24.3 | D | 17.1 |
| 2 | C | 23.5 | D | 16.5 | A | 18.7 | B | 22.0 | E | — |
| 3 | D | 17.4 | E | 23.8 | B | 22.8 | C | 23.9 | A | 20.0 |
| 4 | A | 20.3 | B | 23.4 | E | 25.9 | D | 18.7 | C | 24.2 |
| 5 | E | 25.7 | C | 24.8 | D | 18.9 | A | 20.6 | B | 24.6 |

Note that the measurement on Variety E hose for Investigator 2 is missing and that the experiment was not rerun to obtain an observation. Use the methods of this section to estimate the missing value.

**SOLUTION**

For our data the treatment, row, and column totals corresponding to the missing observations are

$T_E = 99.70 \qquad R_2 = 80.70 \qquad C_5 = 85.90$

Then with $t = r = c = 5$ and $G = 520.80$, we find

$$M = \frac{5(99.7 + 80.7 + 85.9) - 2(520.8)}{4(3)} = 24.2$$

The analysis could now proceed as for a balanced Latin square design, using the shortcut formulas.

Having located a significant effect due to treatments, we can make pairwise treatment comparisons using the formulas below. The least significant difference between the treatment with the missing value and any other treatment is

$$\text{LSD} = t_{\alpha/2}\sqrt{\text{MSE}\left(\frac{2}{t} + \frac{1}{(t-1)(t-2)}\right)}$$

LSD

For any other pair of treatments, the LSD is as before:

$$\text{LSD} = t_{\alpha/2}\sqrt{\frac{2\text{MSE}}{t}}$$

For Latin squares with more than one missing observation, it is perhaps easier to use the method of fitting complete and reduced models to adjust for imbalances caused by the missing values. In general, using the complete model

fitting complete and reduced models

$$y_{ijk} = \mu + \alpha_i + \beta_j + \gamma_k + \epsilon_{ijk}$$

and a computer solution, we would get the analysis of variance table shown in table 19.6. Note that the sum of squares due to rows is unadjusted, the sum of squares for columns is adjusted for rows, and the sum of squares for treatments is adjusted for both rows and columns.

**Table 19.6** AOV table for an unbalanced Latin square design

| *Source* | *SS* | *df* | *MS* | *F* |
|---|---|---|---|---|
| rows | SSR | $t-1$ | — | — |
| columns (adjusted for rows) | $\text{SSC}_{\text{adj}}\text{R}$ | $t-1$ | — | — |
| treatments (adjusted for rows, columns) | $\text{SST}_{\text{adj}}\text{R,C}$ | $t-1$ | $\text{MST}_{\text{adj}}\text{R,C}$ | $\text{MST}_{\text{adj}}\text{R,C/MSE}$ |
| error | SSE | by subtraction | MSE | |
| *Totals* | TSS | $n-1$ | | |

Even though we do not have all the information for the analysis of variance tables, the corresponding tests for either rows or columns can be obtained from the computer output for this same model by using the partial sums of squares column. Here we obtain $\text{SSR}_{\text{adj}}\text{C,T}$ and $\text{SSC}_{\text{adj}}\text{R,T}$. In the computer output the $F$ test and level of significance for these tests are given in the adjacent columns to the right of the partial sums of squares.

# Exercises

19.8. Refer to example 19.2. Perform an analysis of variance, using the estimated value 24.2. Use $\alpha = .05$ to draw your conclusions.

*19.9. Use the SAS computer output shown here to give an analysis of variance table for testing the effect of treatments adjusted for rows (investigators) and columns (days). Indicate the results of testing separately for effects of rows and columns. Use $\alpha = .05$. Compare your results to those of exercise 19.8.

```
STATISTICAL ANALYSIS SYSTEM

DATA NYLON;
INPUT Y 1-4 INV 6 DAY 8 TRT 10;
CARDS;

  24 OBSERVATIONS IN DATA SET NYLON    4 VARIABLES

PROC PRINT;
```

| OBS | Y | INV | DAY | TRT |
|---|---|---|---|---|
| 1 | 22.1 | 1 | 1 | 2 |
| 2 | 23.5 | 2 | 1 | 3 |
| 3 | 17.4 | 3 | 1 | 4 |
| 4 | 20.3 | 4 | 1 | 1 |
| 5 | 25.7 | 5 | 1 | 5 |
| 6 | 18.6 | 1 | 2 | 1 |
| 7 | 16.5 | 2 | 2 | 4 |
| 8 | 23.8 | 3 | 2 | 5 |
| 9 | 23.4 | 4 | 2 | 2 |
| 10 | 24.8 | 5 | 2 | 3 |
| 11 | 23.0 | 1 | 3 | 3 |
| 12 | 18.7 | 2 | 3 | 1 |
| 13 | 22.8 | 3 | 3 | 2 |
| 14 | 25.9 | 4 | 3 | 5 |
| 15 | 18.9 | 5 | 3 | 4 |
| 16 | 24.3 | 1 | 4 | 5 |
| 17 | 22.0 | 2 | 4 | 2 |
| 18 | 23.9 | 3 | 4 | 3 |
| 19 | 18.7 | 4 | 4 | 4 |
| 20 | 20.6 | 5 | 4 | 1 |
| 21 | 17.1 | 1 | 5 | 4 |
| 22 | 20.0 | 3 | 5 | 1 |
| 23 | 24.2 | 4 | 5 | 3 |
| 24 | 24.6 | 5 | 5 | 2 |

```
PROC REGR;
CLASSES INV DAY TRT;
MODEL Y = INV DAT TRT;
TITLE 'EXERCISE 19.9';
```

```
****************************************************************************************************

PROC REGR : EXERCISE 19.9

DATA SET   : NYLON          NUMBER OF VARIABLES = 1  NUMBER OF CLASSES = 3

VARIABLES  : Y

****************************************************************************************************

CLASSES      LEVELS     VALUES

INV             5       1 2 3 4 5

DAY             5       1 2 3 4 5

TRT             5       1 2 3 4 5

****************************************************************************************************
```

| EFFECTS | ASSIGNED |
|---|---|
| INV | DUMMY001 – DUMMY004 |
| DAY | DUMMY005 – DUMMY008 |
| TRT | DUMMY009 – DUMMY012 |

EXERCISE 19.9

ANALYSIS OF VARIANCE TABLE, REGRESSION COEFFICIENTS, AND STATISTICS OF FIT FOR DEPENDENT VARIABLE Y

| SOURCE | DF | SUM OF SQUARES | MEAN SQUARE | F VALUE | PROB > F | R-SQUARE | C.V. |
|---|---|---|---|---|---|---|---|
| REGRESSION | 12 | 189.95683333 | 15.82973611 | 120.65626 | 0.0001 | 0.99245994 | 1.66918 % |
| ERROR | 11 | 1.44316667 | 0.13119697 | | | STD DEV | Y MEAN |
| CORRECTED TOTAL | 23 | 191.40000000 | | | | 0.36221122 | 21.70000 |

| SOURCE | DF | SEQUENTIAL SS | F VALUE | PROB > F | PARTIAL SS | F VALUE | PROB > F |
|---|---|---|---|---|---|---|---|
| INV | 4 | 22.32850000 | 42.54767 | 0.0001 | 14.36883333 | 27.38027 | 0.0001 |
| DAY | 4 | 2.13400000 | 4.06640 | 0.0289 | 0.94283333 | 1.79660 | 0.1994 |
| TRT | 4 | 165.49433333 | 315.35472 | 0.0001 | 165.49433333 | 315.35472 | 0.0001 |

Annotations: SSR (sequential SS for INV); $SSC_{adj}R$ (DAY sequential SS); $SST_{adj}R,C$ (TRT sequential SS); *F* test for treatments (adj); $SSR_{adj}C,T$ (INV partial SS); $SSC_{adj}R,T$ (DAY partial SS); *F* tests for rows and columns

19.10. The data of example 15.3 (p. 426) have been reproduced below. Suppose that a severe accident occurred, knocking out the traffic light at Intersection 3 during the P.M. peak traffic period, thus eliminating any possibility of obtaining a measurement during that period. The investigators decide not to set up the equipment on another day to obtain the measurement.

| *Intersection* | *Time Period* | | | |
|---|---|---|---|---|
| 1 | IV 15.5 | II 33.9 | III 13.2 | I 29.1 |
| 2 | II 16.3 | III 26.6 | I 19.4 | IV 22.8 |
| 3 | III 10.8 | I 31.1 | IV 17.1 | II — |
| 4 | I 14.7 | IV 34.0 | II 19.7 | III 21.6 |

a. Give the investigators an analysis of variance by estimating the missing value. Use $\alpha = .05$.
b. Make treatment comparisons by using Fisher's least significant difference, with $\alpha = .05$.

19.11. Use the method of fitting complete and reduced models to obtain an analysis of variance for the data in exercise 19.10.

# Incomplete Block Designs 19.4

So far we have discussed the analysis for unbalanced designs where the imbalance was not planned but caused by some accident during the collection of the sample data. Sometimes, however, we may be forced to design an experiment in which we must sacrifice some balance to perform the experiment. To illustrate, suppose that a regulatory agency would like to compare the mean potencies for 3 different batches (A, B, C) of the same drug product. Assume for the sake of simplicity that each analyst can do just 2 analyses per day and there are 3 analysts available on a given day. Thus it would be possible to complete a comparison of the 3 batches on a single day if each analyst examines just 2 of the 3 possible batches. One possible design would be the arrangement listed below.

| | *Analyst* | |
|---|---|---|
| *1* | *2* | *3* |
| A | C | B |
| B | A | C |

We can think of this design as a partial (incomplete) randomized block design, where the number of treatments (batches) per block (analyst) is less than $t$, the number of treatments. In fact, a design such as this has become known as an *incomplete block design*.

incomplete block design

There are many different types of incomplete block designs. The one we have constructed belongs to a class of designs that statisticians have called *balanced incomplete block designs*. Although these designs are not balanced as we have defined the term in definition 16.1 (p. 486), the designs do retain some balance. For example, even though all treatments do not appear in the same block, each pair of treatments appears together in a block the same number of times. The pairs of treatments AB, AC, and BC in our design appear together once in a block. We achieved this "balance" by taking all possible combinations of 2 of the 3 treatments for blocks.

balanced incomplete block design

Consider now an extension to the balanced incomplete block design, with 2 treatments per block. Suppose we wish to compare 9 different batches in sets of 3 each. If each analyst can run 3 analyses, how many analysts would be required to maintain a balance similar to that obtained with our previous design?

Following similar logic, we could consider all possible combinations of 3 treatments with one analyst assigned to each different combination. Without too much trouble, we can show that this would require 84 analysts. Obviously, it would be prohibitive for most companies to employ 84 different analysts to compare the 9 treatments in sets of 3. Fortunately, there are other balanced incomplete designs that accomplish the experimental objective. One such design is shown in table 19.7. By employing 12 analysts (blocks), we can compare the 9 treatments.

**Table 19.7** A balanced incomplete block design for comparing 9 treatments with 3 treatments per block

| *Block* | *Treatments* | | | *Block* | *Treatments* | | |
|---|---|---|---|---|---|---|---|
| 1 | A | B | C | 7 | A | E | I |
| 2 | D | E | F | 8 | B | F | G |
| 3 | G | H | I | 9 | C | D | H |
| 4 | A | D | G | 10 | A | F | H |
| 5 | B | E | H | 11 | B | D | I |
| 6 | C | F | I | 12 | C | E | G |

### EXAMPLE 19.3

Identify (by quantity) the parameters listed below for the balanced incomplete block design described in the previous discussion.

$t$: number of treatments

$k$: number of treatments per block

$b$: number of blocks

$r$: number of repetitions of each treatment

$\lambda$: number of times each pair of treatments appears together in a block

### SOLUTION

From table 19.7 we see that

$t = 9 \qquad k = 3 \qquad b = 12 \qquad r = 4$

In addition, after a cursory check we see that $\lambda = 1$ (i.e., each pair of treatments appears once in a block).

Before considering the analysis of variance for balanced incomplete blocks, we should note that a balanced incomplete block design does not exist for all possible values of $t$, $k$, $b$, and $r$. While some researchers in statistics have been concerned with methods for constructing balanced incomplete block designs and other more complicated incomplete block designs, we encourage you to refer to tables of incomplete block designs [see Cochran and Cox (1957)] when searching for a design to satisfy certain experimental objectives.

The analysis of variance for a balanced incomplete block design can be performed either by using specifically developed shortcut formulas or by using the method of fitting complete and reduced models as discussed for unbalanced designs. We will present the shortcut formulas for the analysis of variance table shown in table 19.8.

AOV table

**Table 19.8** Analysis of variance table for a balanced incomplete block design

| *Source* | *SS* | *df* | *MS* | *F* |
|---|---|---|---|---|
| blocks | SSB | $b-1$ | — | — |
| $\text{treatments}_{\text{adj}}$ | $\text{SST}_{\text{adj}}$ | $t-1$ | $\text{MST}_{\text{adj}}$ | $\text{MST}_{\text{adj}}/\text{MSE}$ |
| error | SSE | by subtraction | MSE | |
| *Totals* | TSS | $n-1$ | | |

shortcut formulas

The quantities SSB (the sum of squares unadjusted for treatments) and the total sum of squares are computed as we have done previously.

$$\text{TSS} = \sum y^2 - \frac{G^2}{n}$$

$$\text{SSB} = \sum_{j=1}^{b} \frac{B_j^2}{k} - \frac{G^2}{n}$$

where $B_j$ is the sum of all observations in block $j$. Then if we define

$T_i$ = sum of all observations on treatment $i$

$B_{(i)}$ = sum of all measurements for blocks that contain treatment $i$

the sum of squares for treatments adjusted for blocks is

$$\text{SST}_{\text{adj}} = \frac{t-1}{nk(k-1)} \sum_{i=1}^{t} (kT_i - B_{(i)})^2$$

The sum of squares for error is found by subtraction.

$$\text{SSE} = \text{TSS} - \text{SSB} - \text{SST}_{\text{adj}}$$

As indicated in the analysis of variance table, the test statistic for testing the hypothesis of no difference among the treatment means is $\text{MST}_{\text{adj}}/\text{MSE}$.

**EXAMPLE 19.4**

A large company enlisted the help of a random sample of 20 potential consumers in a given geographical location to compare the physical characteristics (such as firmness, rebound, etc.) of 8 experimental pillows and one presently marketed pillow. Since it was assumed that people would have a difficult time in distinguishing differences among the pillows when presented with all 9 at once, it was decided to employ the balanced incomplete block design shown in table 19.7.

After the pillow types were randomly assigned the letters from A to I, tables were prepared with the appropriate pillow types assigned to each table. For example, pillows G, H, and I were placed on Table 3. Each pillow was sealed in an identical white pillowcase and hence could not be distinguished from the others by color. The only marking on the pillowcase was a 4-digit number, which provided the investigators with an identification code. With all tables in place, the 20 potential consumers were asked to proceed one at a time through the design from Table 1 to Table 12, stopping at each table to compare the 3 pillows. All persons were to record a firmness score for each pillow at each table, based on a 1-to-5-point scale (higher score indicates more firmness). The sums of scores for each pillow at the 12 tables are recorded in table 19.9 (the letters identify the pillow type).

**Table 19.9** Sums of firmness scores, example 19.4

| *Block (table)* | *Treatment (pillow)* | | | *Block Totals* |
|---|---|---|---|---|
| 1 | A 59 | B 26 | C 38 | 123 |
| 2 | D 85 | E 92 | F 69 | 246 |
| 3 | G 74 | H 52 | I 27 | 153 |
| 4 | A 62 | D 70 | G 68 | 200 |
| 5 | B 27 | E 98 | H 59 | 184 |
| 6 | C 31 | F 60 | I 35 | 126 |
| 7 | A 63 | E 85 | I 30 | 178 |
| 8 | B 22 | F 73 | G 75 | 170 |
| 9 | C 45 | D 74 | H 51 | 170 |
| 10 | A 52 | F 76 | H 43 | 171 |
| 11 | B 18 | D 79 | I 41 | 138 |
| 12 | C 41 | E 84 | G 81 | 206 |
| | | | | 2065 |

Use the shortcut formulas of this section to perform an analysis of variance. Use $\alpha = .05$ to test the null hypothesis of no differences among treatment (pillow) means.

## SOLUTION

For an analysis using the shortcut formulas, it is convenient to construct a table of totals, as shown in table 19.10.

From the table we find that

$$\text{SST}_{\text{adj}} = \frac{(t-1)\sum_{i=1}^{t}(kT_i - B_{(i)})^2}{nk(k-1)} = \frac{8(322{,}112.00)}{36(3)(2)} = 11{,}930.07$$

**Table 19.10** Totals for the data of table 19.9

| *Treatment* | $T_i$ | $B_{(i)}$ | $kT_i - B_{(i)}$ |
|---|---|---|---|
| A | 236 | 672 | 36 |
| B | 93 | 615 | −336 |
| C | 155 | 625 | −160 |
| D | 308 | 754 | 170 |
| E | 359 | 814 | 263 |
| F | 278 | 713 | 121 |
| G | 298 | 729 | 165 |
| H | 205 | 678 | −63 |
| I | 133 | 595 | −196 |
| *Total* | 2065 | | |

Similarly, using the block totals from the raw data table, we obtain

$$\text{SSB} = \frac{\sum_{j=1}^{b} B_j^2}{k} - \frac{G^2}{n} = \frac{368{,}991.00}{3} - \frac{(2065)^2}{36}$$

$$= 122{,}997.00 - 118{,}450.69 = 4546.31$$

The total sum of squares is

$$\text{TSS} = \sum y^2 - \frac{G^2}{n} = 135{,}435.00 - 118{,}450.69 = 16{,}984.31$$

The analysis of variance table for testing the hypothesis of no differences among the treatment means is shown in table 19.11.

**Table 19.11** AOV table for the data of example 19.4

| *Source* | *SS* | *df* | *MS* | *F* |
|---|---|---|---|---|
| blocks | 4,546.31 | 11 | — | — |
| $\text{treatments}_{\text{adj}}$ | 11,930.07 | 8 | 1491.26 | 46.97 |
| error | 507.93 | 16 | 31.75 | |
| *Totals* | 16,984.31 | 35 | | |

Since the computed value of $F$, 46.97, exceeds the table value, 2.59, for $df_1 = 8$, $df_2 = 16$, and $a = .05$, we conclude that there are differences among the 9 treatment means.

Following the observation of a significant $F$ test concerning differences among treatment means, we naturally might like to determine which treatment means are significantly different from the others. To do this, we make use of the following notation: $\widehat{\mu}_i$, an estimate of the mean for treatment $i$, given by

comparison among treatment means

$$\widehat{\mu}_i = \bar{y} + \frac{kT_i - B_{(i)}}{t\lambda}$$

where $\bar{y}$ is the overall sample mean. An estimate of the difference between two treatment means $i$ and $j$ is then

$$\hat{\mu}_i - \hat{\mu}_j = \frac{[kT_i - B_{(i)}] - [kT_j - B_{(j)}]}{t\lambda}$$

The least significant difference between any pair of treatment means is

LSD

$$\text{LSD} = t_{\alpha/2}\sqrt{\frac{2k\text{MSE}}{t\lambda}}$$

**EXAMPLE 19.5**

Compute the least significant difference for the 9 treatment means of example 19.4. Determine all pairwise differences, using $\alpha = .05$.

**SOLUTION**

For these data the overall sample mean is $\bar{y} = 2065/36 = 57.36$. Then using the column $kT_i - B_{(i)}$, we have the following estimated treatment means:

| *Treatment* | $\bar{y} + \dfrac{kT_i - B_{(i)}}{t\lambda}$ |
|---|---|
| A | 61.36 |
| B | 20.03 |
| C | 39.58 |
| D | 76.25 |
| E | 86.58 |
| F | 70.80 |
| G | 75.69 |
| H | 50.36 |
| I | 35.58 |

Using MSE = 31.75, based on df = 16, we obtain

$$\text{LSD} = t_{\alpha/2}\sqrt{\frac{2k\text{MSE}}{t\lambda}} = 2.12\sqrt{\frac{2(3)31.75}{9(1)}} = 9.75$$

The 9 treatment means are arranged in ascending order below, with a summary of the significant results. Those treatments underlined by a common line are not significantly different from each other, using the least significant difference criterion (see chapter 14).

| B | I | C | H | A | F | G | D | E |
|---|---|---|---|---|---|---|---|---|
| 20.03 | 35.58 | 39.58 | 50.36 | 61.36 | 70.80 | 75.69 | 76.25 | 86.58 |

## Summary 19.5

In this chapter we have discussed the analysis of variance for some unbalanced designs. We began with a discussion of the analysis for a randomized block design with one missing observation. Two possible analyses were proposed. The first required that we estimate the missing value and then proceed with the usual shortcut formulas developed in chapter 15. While estimating a single missing value is quite easy to do, the procedure becomes more difficult when there is more than one missing value. The second procedure, that of fitting complete and reduced models to obtain adjusted sums of squares, can be used for one or more missing observations.

With the Latin square design, we again showed how to estimate a single missing observation and proceed with the usual analysis. But, as with the randomized block design, the method of analysis by fitting complete and reduced models is more appropriate for more than one missing value.

Finally, we considered another class of unbalanced designs, incomplete block designs. The particular designs we discussed were incomplete randomized block designs in which not all treatments appear in each block. These incomplete block designs retain a certain amount of balance, since all pairs of treatments appear together in a block the same number of times. The analysis for balanced incomplete block designs was illustrated using appropriate shortcut formulas. While no example was given in the chapter to show a computer solution for a balanced incomplete block design, we can obtain the analysis of variance for testing treatment differences following the procedure outlined for a randomized block design with missing observations.

## Exercises

19.12. A physician was interested in comparing the effects of 6 different antihistamines for persons extremely sensitive to a ragweed skin allergy test. To do this, a random sample of 10 allergy patients was selected from the physician's private practice, with treatments (antihistamines) assigned to each patient according to the experimental design shown on the next page. Each person then received injections of the assigned antihistamines in different sections of the right arm. The area of redness surrounding the point of injection was measured after a fixed period of time. The data appear on the next page.

| Person | Treatments | | | | | |
|---|---|---|---|---|---|---|
| 1 | B | 25 | A | 41 | F | 40 |
| 2 | E | 37 | B | 46 | A | 42 |
| 3 | C | 45 | D | 33 | B | 37 |
| 4 | E | 34 | D | 35 | A | 46 |
| 5 | B | 31 | F | 42 | D | 34 |
| 6 | C | 56 | E | 36 | F | 65 |
| 7 | D | 33 | A | 42 | C | 67 |
| 8 | F | 49 | D | 37 | E | 30 |
| 9 | C | 59 | A | 40 | F | 55 |
| 10 | B | 36 | C | 57 | E | 34 |

a. Identify the design.
b. Identify the parameters of the design.

19.13. Refer to the data of exercise 19.12. Do the data indicate differences among the treatment means? Use $\alpha = .05$.

19.14. Refer to exercise 19.13. Use the least significant difference criterion for determining treatment differences. Use $\alpha = .05$.

*19.15. Use a computer program to perform the same analysis as in exercise 19.13. Compare your results.

*19.16. Refer to example 19.4 (p. 603). Use a computer program to perform an analysis of variance. Are your results the same as those found in the example?

19.17. Indicate how you would test for a significant effect due to blocks in a balanced incomplete block design.

# References

Barr, A.J.; Goodnight, J.H.; Sall, J.P.; and Helwig, J.T. 1976. *A user's guide to SAS 76*. Raleigh, N.C.: SAS Institute, Inc.

Cochran, W.G., and Cox, G.M. 1957. *Experimental design*. 2d ed. New York: Wiley.

Davies, O.L. 1956. *The design and analysis of experiments*. New York: Hafner.

Dixon, W.J. 1975. *BMDP, biomedical computer programs*. Berkeley: University of California Press.

Dixon, W.J., and Massey, F.J., Jr. 1969. *Introduction to statistical analysis.* 3d ed. New York: McGraw-Hill.

Draper, N.R., and Smith, H. 1966. *Applied regression analysis.* New York: Wiley.

Fisher, R.A. 1949. *The design of experiments.* Edinburgh: Oliver and Boyd.

Graybill, F.A. 1976. *Theory and application of the linear model.* N. Scituate, Mass.: Duxbury Press.

Harnett, D.L. 1970. *Introduction to statistical methods.* Reading, Mass.: Addison-Wesley.

Kirk, R.E. 1968. *Experimental design: Procedures for the behavioral sciences.* Belmont, Calif.: Brooks/Cole.

Mendenhall, W. 1968. *Introduction to linear models and the design and analysis of experiments.* Belmont, Calif.: Wadsworth.

Nie, N.H.; Hull, C.H.; Jenkins, J.G.; Steinbrenner, K.; and Bent, D.H. 1975. *Statistical package for the social sciences.* 2d ed. New York: McGraw-Hill.

Ostle, B. 1963. *Statistics in research.* 2d ed. Ames, Iowa: Iowa State University Press.

Pearson, E.S., and Hartley, H.O. 1966. *Biometrika tables for statisticians.* 3d ed. Vol. I. London: Cambridge University Press.

Service, J. 1972. *A user's guide to the statistical analysis system.* Raleigh, N.C.: Student Supply Stores, North Carolina State University.

Snedecor, G.W., and Cochran, W.G. 1967. *Statistical methods.* 6th ed. Ames, Iowa: Iowa State University Press.

Sokal, R.R., and Rohlf, F.J. 1969. *Biometry.* San Francisco: W.H. Freeman.

Steel, R.G.D., and Torrie, J.H. 1960. *Principles and procedures of statistics.* New York: McGraw-Hill.

Winer, B.J. 1962. *Statistical principles in experimental design.* New York: McGraw-Hill.

# 20 The Analysis of Covariance

## 20.1 Introduction

The analysis of covariance is a procedure for comparing treatment means that incorporates information on a quantitative variable $x$. This topic appears in most statistical methods textbooks, but it has been our experience that many students become so engrossed in the computational formulas that they rarely understand the problems involved. We will approach the subject in a different way and try to avoid becoming too involved with formulas. Since the analysis of covariance combines features of the analysis of variance and regression, we will make use of a general linear model. We will strive for understanding by making reference to our work with general linear models in preceding chapters. Actually, the topic of covariance analysis can be easily understood if we build on our previous work in this way. We begin our presentation with a completely randomized design.

## 20.2 A Completely Randomized Design with One Covariate

A completely randomized design is used to compare $t$ population means. To do this, we obtain a random sample of $n_i$ observations on the variable $y$ in the

$i$th population ($i = 1, 2, \ldots, t$). Now in addition to measuring the response variable $y$ on each experimental unit, we measure a second variable $x$, often called a *covariable,* or a *covariate*. For example, in studying the effects of different methods of reinforcement on the reading achievement level of eight-year-old children, we could measure not only the final achievement level $y$ for each child but also the prestudy reading performance level $x$. Ultimately we would want to make comparisons among the different methods while taking into account information on both $y$ and $x$.

covariate

It should be noted that while $x$ can be thought of as an independent variable, unlike most situations discussed in previous chapters, we cannot control the value of $x$ (as we controlled settings of temperature or pressure) prior to observing the variable. In spite of this, we may still write a model for the completely randomized design treating the covariate as an independent variable.

For the purposes of illustration we will assume we are interested in comparing $t = 3$ treatments with only one covariate; however, as you will see, the procedure can be applied when more than one covariate is present.

Using the notation of the general linear model, we let $x_1$ denote the covariate. Then we can write the model

model

$$y = \beta_0 + \beta_1 x_1 + \beta_2 x_2 + \beta_3 x_3 + \beta_4 x_1 x_2 + \beta_5 x_1 x_3 + \epsilon$$

where

$$x_2 = 1 \text{ if treatment 2} \qquad x_2 = 0 \text{ otherwise}$$

$$x_3 = 1 \text{ if treatment 3} \qquad x_3 = 0 \text{ otherwise}$$

From our discussions in chapter 16, we recognize this model as a general linear model relating a response $y$ to a quantitative variable (the covariate $x_1$) and a qualitative variable (treatments). The $\beta$'s of the model can be interpreted using table 20.1. As can be seen from the table of expected values, the model defines a straight line for each of the 3 treatments. The intercepts and slopes for the 3 lines are as indicated in table 20.1.

Typically, for a completely randomized design with a single covariate, the

expected values

**Table 20.1** Expected values for the model $y = \beta_0 + \beta_1 x_1 + \beta_2 x_2 + \beta_3 x_3 + \beta_4 x_1 x_2 + \beta_5 x_1 x_3 + \epsilon$

| *Treatment* | *Expected Values of y* |
|---|---|
| 1 | $\beta_0 + \beta_1 x_1$ |
| 2 | $(\beta_0 + \beta_2) + (\beta_1 + \beta_4)x_1$ |
| 3 | $(\beta_0 + \beta_3) + (\beta_1 + \beta_5)x_1$ |

assumptions

response $y$ is assumed to be linearly related to the covariate, and the slope of the straight-line relationship is assumed to be the same for all treatments. While we have already written the model to relate $y$ linearly to $x_1$, we need not be bound by these assumptions, for it is possible to write the model with higher-order terms in $x_1$ (provided, of course, there are enough different values of $x_1$ recorded in the sample data). At present, however, we will illustrate a covariance analysis with a model that is linear in the covariate $x_1$.

tests

The assumption of constant slope for the 3 treatment groups can be tested using the null hypothesis

$$H_0\colon \beta_4 = \beta_5 = 0 \text{ (the slopes are identical)}$$

against the alternative hypothesis that at least one of the slopes differs from the rest. With our original model as the complete model and the model

reduced model

$$y = \beta_0 + \beta_1 x_1 + \beta_2 x_2 + \beta_3 x_3 + \epsilon$$

as the reduced model, the mean square drop based on 2 degrees of freedom would be the numerator in an $F$ test of the null hypothesis

$$H_0\colon \beta_4 = \beta_5 = 0$$

If there is insufficient evidence to reject the hypothesis of equality of the slopes ($\beta_4 = \beta_5 = 0$), we use the reduced model to describe the experimental situation. Under this model, the straight-line relationship between $y$ and the covariate $x_1$ would have expectations

Treatment 1: $\beta_0 + \beta_1 x_1$

Treatment 2: $(\beta_0 + \beta_2) + \beta_1 x_1$

Treatment 3: $(\beta_0 + \beta_3) + \beta_1 x_1$

The model for a covariance analysis in a completely randomized design is not usually written in terms of the parameters in a general linear model. In other sources you might see the model written as

$$y_{ij} = \alpha_i + \beta x + \epsilon_{ij} \qquad (i = 1, 2, \ldots, t;\ j = 1, 2, \ldots, n_i)$$

Note that we assume that the relationship between the response $y$ and the covariate $x$ is linear with the same slope but different intercepts across treatment groups. The reduced model illustrated for our $t = 3$ example,

$$y = \beta_0 + \beta_1 x_1 + \beta_2 x_2 + \beta_3 x_3 + \epsilon$$

is the general linear model analogy to this same situation.

test of means

Referring to our example, when all 3 treatments have the same slope (i.e., $\beta_4 = \beta_5 = 0$ in the complete model), a test of the equality of treatment means can be made using the sum of squares due to treatments adjusted for the covariate. We do this by fitting a complete and a reduced model, using the model

$$y = \beta_0 + \beta_1 x_1 + \beta_2 x_2 + \beta_3 x_3 + \epsilon$$

for the null hypothesis $H_0$: $\beta_2 = \beta_3 = 0$. The sum of squares drop ($SSE_2 - SSE_1$) will be the sum of squares due to treatments adjusted for the covariate. The test statistic $F = MS_{drop}/MSE$ is based on 2 degrees of freedom (the number of $\beta$'s set equal to zero under $H_0$).

**EXAMPLE 20.1**

An investigator is interested in comparing two drug products (A and B) in overweight female volunteers. The experiment calls for 20 randomly selected subjects who are at least 25% overweight. Ten of these women are to be randomly assigned to Product A and the remaining 10 to Product B. The response of interest is a score on a rating scale used to measure the mood of a subject. To obtain a score, a subject must complete a checklist indicating how each of 50 adjectives describes her mood at that time. From this checklist we can obtain an anxiety-tension score, a danger-hostility score, an active score (measuring alertness, energy, etc.), and many others.

On the study day, all 20 volunteers are required to complete the checklist at 8 A.M. Then each subject is given the prescribed medication (A or B). Each subject is required to complete the checklist again at 10 A.M.

Write a model relating a subject's 10-A.M. checklist score $y$ to the two independent variables "drug product" and "8-A.M. (predrug) checklist score." Interpret the $\beta$'s.

**SOLUTION**

For this experimental situation we have one qualitative independent variable (drug) and one quantitative independent variable. Letting $x_1$ denote the checklist score at 8 A.M., the model is

$$y = \beta_0 + \beta_1 x_1 + \beta_2 x_2 + \beta_3 x_1 x_2 + \epsilon$$

where

$x_1$ = checklist score at 8 A.M.

$x_2 = 1$ if Product B $\qquad x_2 = 0$ otherwise

The expected value of $y$ for our model is

$$E(y) = \beta_0 + \beta_1 x_1 + \beta_2 x_2 + \beta_3 x_1 x_2$$

Substituting $x_2 = 0$ and $x_2 = 1$, respectively, for drug products A and B, we have the expected value of $y$, the adjective checklist score at 10 A.M.

Product A: $E(y) = \beta_0 + \beta_1 x_1$

Product B: $E(y) = (\beta_0 + \beta_2) + (\beta_1 + \beta_3)x_1$

Thus $\beta_0$ and $\beta_1$ are the intercept and slope, respectively, defining the linear relationship between $y$ and $x_1$ for Product A. Since $\beta_0 + \beta_2$ and $\beta_1 + \beta_3$ represent the corresponding intercept and slope, respectively, for Product B, $\beta_2$ is the difference between the intercepts for lines A and B, while $\beta_3$ is the difference between the slopes for the two lines.

**EXAMPLE 20.2**

Suppose the 10 anxiety scores for the study of example 20.1 are as shown in table 20.2.

**Table 20.2** Anxiety score data for example 20.2

| *Drug A* | | *Drug B* | |
|---|---|---|---|
| *8 A.M.* $x_1$ | *10 A.M.* $y$ | *8 A.M.* $x_1$ | *10 A.M.* $y$ |
| 5 | 20 | 7 | 19 |
| 10 | 23 | 12 | 26 |
| 12 | 30 | 27 | 33 |
| 9 | 25 | 24 | 35 |
| 23 | 34 | 18 | 30 |
| 21 | 40 | 22 | 31 |
| 14 | 27 | 26 | 34 |
| 18 | 38 | 21 | 28 |
| 6 | 24 | 14 | 23 |
| 13 | 31 | 9 | 22 |

a. Fit the model presented in example 20.1.
b. Test for lack of fit of the linear model. Use $\alpha = .05$.
c. Are the lines for drug products A and B parallel?
d. If so, perform a test to compare the drug product means, adjusting for any differences due to different baseline (8 A.M.) readings. Use $\alpha = .05$.

**SOLUTION**

a. We will run an analysis of covariance by using the two models

$$y = \beta_0 + \beta_1 x_1 + \beta_2 x_2 + \beta_3 x_1 x_2 + \epsilon$$

$$y = \beta_0 + \beta_1 x_1 + \beta_2 x_2 + \epsilon$$

The computer output follows.

```
STATISTICAL ANALYSIS SYSTEM

DATA COVARIATE;
INPUT Y  1-2 X1 4-5 X2 7;
X1X2=X1*X2;
CARDS;


  20 OBSERVATIONS IN DATA SET COVARIAT    4 VARIABLES

PROC PRINT;
```

| OBS | Y | X1 | X2 | X1X2 |
|---|---|---|---|---|
| 1 | 20 | 5 | 0 | 0 |
| 2 | 23 | 10 | 0 | 0 |
| 3 | 30 | 12 | 0 | 0 |
| 4 | 25 | 9 | 0 | 0 |
| 5 | 34 | 23 | 0 | 0 |
| 6 | 40 | 21 | 0 | 0 |
| 7 | 27 | 14 | 0 | 0 |
| 8 | 38 | 18 | 0 | 0 |
| 9 | 24 | 6 | 0 | 0 |
| 10 | 31 | 13 | 0 | 0 |
| 11 | 19 | 7 | 1 | 7 |
| 12 | 26 | 12 | 1 | 12 |
| 13 | 33 | 27 | 1 | 27 |
| 14 | 35 | 24 | 1 | 24 |
| 15 | 30 | 18 | 1 | 18 |
| 16 | 31 | 22 | 1 | 22 |
| 17 | 34 | 26 | 1 | 26 |
| 18 | 28 | 21 | 1 | 21 |
| 19 | 23 | 14 | 1 | 14 |
| 20 | 22 | 9 | 1 | 9 |

```
PROC REGR;
MODEL Y = X1 X2 X1X2;
MODEL Y = X1 X2;
TITLE 'EXAMPLE 20.2';

**************************************************************************************************************

PROC REGR : EXAMPLE 20.2

DATA SET   : COVARIAT   NUMBER OF VARIABLES = 4  NUMBER OF CLASSES = 0

VARIABLES  : Y X1 X2 X1X2

**************************************************************************************************************
```

EXAMPLE 20.2

ANALYSIS OF VARIANCE TABLE, REGRESSION COEFFICIENTS, AND STATISTICS OF FIT FOR DEPENDENT VARIABLE Y

| SOURCE | DF | SUM OF SQUARES | MEAN SQUARE | F VALUE | PROB > F | R-SQUARE | C.V. |
|---|---|---|---|---|---|---|---|
| REGRESSION | 3 | 558.56687443 | 186.18895814 | 27.08619 | 0.0001 | 0.83549005 | 9.15121% |
| ERROR | 16 | 109.98312557 | 6.87394535 | | | STD DEV | Y MEAN |
| CORRECTED TOTAL | 19 | 668.55000000 | | | | 2.62182100 | 28.65000 |

| SOURCE | DF | SEQUENTIAL SS | F VALUE | PROB > F | PARTIAL SS | F VALUE | PROB >F |
|---|---|---|---|---|---|---|---|
| X1 | 1 | 430.92383794 | 62.68945 | 0.0001 | 312.89948313 | 45.51963 | 0.0001 |
| X2 | 1 | 115.30595671 | 16.77435 | 0.0008 | 1.21055917 | 0.17611 | 0.6803 |
| X1X2 | 1 | 12.33707978 | 1.79476 | 0.1991 | 12.33707978 | 1.79476 | 0.1991 |

| SOURCE | B VALUES | T FOR HO:B=0 | PROB > \|T\| | STD ERR B | STD B VALUES |
|---|---|---|---|---|---|
| INTERCEPT | 16.42262086 | 7.94373 | 0.0001 | 2.06736784 | 0.0 |
| X1 | 0.97537245 | 6.74682 | 0.0001 | 0.14456764 | 1.13729607 |
| X2 | −1.31392521 | −0.41965 | 0.6803 | 3.13098272 | −0.11362887 |
| X1X2 | −0.25363332 | −1.33969 | 0.1991 | 0.18932291 | −0.44737430 |

EXAMPLE 20.2

ANALYSIS OF VARIANCE TABLE, REGRESSION COEFFICIENTS, AND STATISTICS OF FIT FOR DEPENDENT VARIABLE Y

| SOURCE | DF | SUM OF SQUARES | MEAN SQUARE | F VALUE | PROB > F | R-SQUARE | C.V. |
|---|---|---|---|---|---|---|---|
| REGRESSION | 2 | 546.22979465 | 273.11489733 | 37.95737 | 0.0001 | 0.81703656 | 9.36268% |
| ERROR | 17 | 122.32020535 | 7.19530620 | | | STD DEV | Y MEAN |
| CORRECTED TOTAL | 19 | 668.55000000 | | | | 2.68240679 | 28.65000 |

| SOURCE | DF | SEQUENTIAL SS | F VALUE | PROB > F | PARTIAL SS | F VALUE | PROB > F |
|---|---|---|---|---|---|---|---|
| X1 | 1 | 430.92383794 | 59.88958 | 0.0001 | 540.17979465 | 75.07391 | 0.0001 |
| X2 | 1 | 115.30595671 | 16.02516 | 0.0009 | 115.30595671 | 16.02516 | 0.0009 |

| SOURCE | B VALUES | T FOR HO: B = 0 | PROB > \|T\| | STD ERR B | STD B VALUES |
|---|---|---|---|---|---|
| INTERCEPT | 18.35999493 | 12.14661 | 0.0001 | 1.51153263 | 0.0 |
| X1 | 0.82748130 | 8.66452 | 0.0001 | 0.09550226 | 0.96485320 |
| X2 | −5.15465839 | −4.00314 | 0.0009 | 1.28765245 | −0.44577729 |

b. Recall that in order to test for lack of fit, we must be able to calculate a sum of squares due to pure experimental error. Since there are no replications of $y$-values at any of the settings of the independent variables $x_1$ and $x_2$, we cannot test for lack of fit of the linear model.

c. Under the assumption that a linear model adequately fits the data for the two drug products, we can test for parallelism of the two lines using the null hypothesis $H_0$: $\beta_3 = 0$. The $t$-value for this test is $-1.33969$, with probability greater than .1991. (These values are read from the computer output.) Since this probability is greater than .05, we have insufficient evidence to reject the null hypothesis of parallelism for the two lines.

d. Having found insufficient evidence to reject the hypothesis of parallelism, we fit a reduced model with $\beta_3 = 0$. In the reduced model ($\beta_3 = 0$), the $t$ test for comparing the mean responses of drugs A and B, adjusted for differences in the covariate $x_1$, is equivalent to testing the null hypothesis $H_0$: $\beta_2 = 0$ (i.e., the intercepts for the two lines are the same). From the computer output, the $t$-value $-4.00314$ has a probability of about .0009. Hence we conclude that there is evidence to indicate a difference in the intercepts for the two curves.

Since the $t$ test on $\beta_1 = 0$ is significant (see the computer output), we conclude that there is a linear relationship between the response $y$ and the covariate $x_1$.

Based on the results of parts c and d, we know that the slope of the line is the same for both drug products but the intercepts are different. Using the computer output, the prediction equations (found by substituting $x_2 = 0$ and 1) are

Drug A: $\hat{y} = 18.36 + .83x_1$

Drug B: $\hat{y} = 18.36 + .83x_1 - 5.15 = 13.21 + .83x_1$

# Exercises

20.1. Consider a completely randomized design for $t = 5$ treatments, with a single covariate $x_1$ and 6 observations per treatment. Write the complete general linear model under the assumption that the response $y$ is linearly related to the covariate $x_1$ for each treatment. Identify the parameters in your model.

20.2. Refer to exercise 20.1. How would you test for parallelism among the straight lines for the 5 treatment groups? Identify how you would obtain the test statistic. What are the degrees of freedom associated with the test statistic?

20.3. Refer to exercise 20.1. Assuming the lines are parallel, give the test for adjusted treatment means. How would you estimate the mean response for treatment 1 with $x_1 = 5$?

20.4. Perform an analysis of covariance. (Hint: When people refer to performing an analysis of covariance, they usually mean to assume a linear relationship, then test for parallelism, and then test for adjusted treatment means assuming parallelism.) The data are given below. Use $\alpha = .05$.

| Treatment | | | | | |
|---|---|---|---|---|---|
| 1 | | 2 | | 3 | |
| $x$ | $y$ | $x$ | $y$ | $x$ | $y$ |
| 26 | 100 | 24 | 118 | 37 | 124 |
| 35 | 150 | 28 | 134 | 31 | 95 |
| 28 | 106 | 29 | 138 | 14 | 60 |
| 21 | 95 | 32 | 147 | 27 | 86 |
| 29 | 113 | 36 | 165 | 18 | 68 |
| 34 | 144 | 35 | 159 | 25 | 81 |

# 20.3 Multiple Covariates and More Complicated Designs

The same procedures discussed in section 20.2 can also be applied to completely randomized designs with one or more covariates. Including more than one covariate in the model merely means that we have more than one quantitative independent variable in our model. For example, we might wish to compare the social status $y$ of several different occupational groups while incorporating information on the number of years $x_1$ of formal education beyond high school and the income level $x_2$ of each individual in a group. As mentioned previously, we need not restrict ourselves to linear terms in the covariate(s). Thus we might have a response related to two covariates ($x_1$ and

$x_2$) and $t = 3$ treatments using the model

$$y = \beta_0 + \beta_1 x_1 + \beta_2 x_1^2 + \beta_3 x_2 + \beta_4 x_3 + \beta_5 x_4 + \beta_6 x_1 x_3 + \beta_7 x_1 x_4 + \beta_8 x_1^2 x_3 + \beta_9 x_1^2 x_4 + \beta_{10} x_2 x_3 + \beta_{11} x_2 x_4 + \epsilon$$

where

$x_3 = 1$ if treatment 2 $\quad x_3 = 0$ otherwise

$x_4 = 1$ if treatment 3 $\quad x_4 = 0$ otherwise

We can readily obtain an interpretation of the $\beta$'s by using a table of expected values similar to table 20.1.

An analysis of covariance for more complicated designs can also be obtained using general linear model methodology. Consider an analysis of covariance for a randomized block design. For simplicity, we will assume that there are 2 blocks, 3 treatments, and one covariate $x_1$.

**EXAMPLE 20.3**

Write the model for the experimental situation described in the previous paragraph, assuming the response is linearly related to $x_1$ for each treatment. Identify the parameters in your model.

**SOLUTION**

The model is

$$y = \beta_0 + \beta_1 x_1 + \beta_2 x_2 + \beta_3 x_3 + \beta_4 x_4 + \beta_5 x_1 x_2 + \beta_6 x_1 x_3 + \beta_7 x_1 x_4 + \epsilon$$

where

$x_2 = 1$ if block 2 $\quad x_2 = 0$ otherwise

$x_3 = 1$ if treatment 2 $\quad x_3 = 0$ otherwise

$x_4 = 1$ if treatment 3 $\quad x_4 = 0$ otherwise

We immediately recognize this as a model relating a response $y$ to a quantitative variable $x_1$ and two qualitative variables, blocks and treatments. An interpretation of the $\beta$'s in the model is readily obtained from the table of expected values shown in table 20.3.

The model we formulated in example 20.3 provides for a linear relationship between $y$ and $x_1$ for each of the treatments in each block, but it allows for differences among intercepts and slopes. If we wanted to test for the equality of the slopes across treatments and blocks, we would use the null hypothesis

$$H_0: \beta_5 = \beta_6 = \beta_7 = 0$$

**Table 20.3** Expected values for the randomized block design with one covariate, example 20.3

| *Block* | *Treatment* | *Expected Values* |
|---|---|---|
| 1 | 1 | $\beta_0 + \beta_1 x_1$ |
| | 2 | $(\beta_0 + \beta_3) + (\beta_1 + \beta_6)x_1$ |
| | 3 | $(\beta_0 + \beta_4) + (\beta_1 + \beta_7)x_1$ |
| 2 | 1 | $(\beta_0 + \beta_2) + (\beta_1 + \beta_5)x_1$ |
| | 2 | $(\beta_0 + \beta_2 + \beta_3) + (\beta_1 + \beta_5 + \beta_6)x_1$ |
| | 3 | $(\beta_0 + \beta_2 + \beta_4) + (\beta_1 + \beta_5 + \beta_7)x_1$ |

If there is insufficient evidence to reject $H_0$, we would proceed with the reduced model (obtained by setting $\beta_5 = \beta_6 = \beta_7 = 0$ in our model).

$$y = \beta_0 + \beta_1 x_1 + \beta_2 x_2 + \beta_3 x_3 + \beta_4 x_4 + \epsilon$$

A test for differences among treatments adjusted for the covariate could be obtained by fitting a complete and a reduced model for the null hypothesis

$$H_0\colon \beta_3 = \beta_4 = 0$$

# Exercises

20.5. Write a model for a $4 \times 4$ Latin square design with one covariate $x_1$. Assume the response is linearly related to the covariate. Identify the parameters in the model.

20.6. Refer to exercise 20.5.

a. Indicate how you would test for parallelism among the different straight lines. How many degrees of freedom would the $F$ test have?
b. Indicate how you would perform a test for the effects of treatments adjusted for the covariate.

20.7. Refer to exercise 20.5. Write a complete model assuming that the response is a second-order function of the covariate $x_1$. Can you identify parameters in the model? How would you test for parallelism of the second-order model?

# Summary | 20.4

In this chapter we have presented a procedure called the analysis of covariance. Here, for each value of $y$, we also observe a value of a concomitant variable $x$. This second variable, called a covariate, is recognized as an uncontrolled quantitative independent variable. Because of this fact we can formulate models using general linear model methodology of previous chapters.

In most situations when reference is made to an analysis of covariance, it is assumed that the response is linearly related to the covariate $x$, with the slope of the line the same for all treatment groups. Then a test for treatments adjusted for the covariate is performed. Actually, many people run analyses of covariance without checking the assumptions of parallelism. Rather than trying to force a particular model onto an experimental situation, it would be much better to postulate a reasonable (not necessarily linear) model relating the response $y$ to the covariate $x$ through the design used. Then by knowing the meanings of the parameters in the model, we can postulate hypotheses concerning the parameters and test these hypotheses by fitting complete and reduced models.

# Exercises

20.8. An investigator studied the effects of 3 different antidepressants (A, B, and C) on patient ratings of depression. To do this, patients were stratified into 6 age-sex combinations. From a random sample of 3 patients from each stratum, the experimenter randomly allocated the 3 antidepressants. On the day the study was to be initiated, a baseline (pretreatment) depression scale rating was obtained from each patient. The assigned therapy was then administered and maintained for one week. At this time, a second rating (posttreatment) was obtained from each patient. The pre- and posttreatment ratings appear below (higher score indicates more depression). Ψ

| | | | *Pretreatment* | | | *Posttreatment* | | |
|---|---|---|---|---|---|---|---|---|
| *Block* | *Sex* | *Age (years)* | *A* | *B* | *C* | *A* | *B* | *C* |
| 1 | F | $<20$ | 48 | 36 | 31 | 21 | 25 | 17 |
| 2 | F | 20–40 | 43 | 31 | 28 | 22 | 21 | 19 |
| 3 | F | $>40$ | 44 | 35 | 29 | 18 | 24 | 18 |
| 4 | M | $<20$ | 42 | 38 | 29 | 26 | 20 | 17 |
| 5 | M | 20–40 | 37 | 34 | 28 | 21 | 24 | 15 |
| 6 | M | $>40$ | 41 | 36 | 26 | 18 | 24 | 19 |

a. Identify the experimental design.
b. Write a first-order model relating the posttreatment response $y$ to the pretreatment rating $x_1$ for each treatment.

*20.9. Refer to exercise 20.8.
a. Use a computer program to fit the model of part b of exercise 20.8. Use $\alpha = .05$.
b. Test for parallelism of the lines.
c. Assuming the lines are parallel, test for differences in treatment means adjusted for the covariate. Use $\alpha = .05$.

*20.10. Refer to exercise 20.8 and 20.9.
a. Assuming parallelism of the response lines, perform a test for block differences adjusted for the covariate. Use $\alpha = .05$.
b. How might you partition the block sum of squares into 5 meaningful single-degree-of-freedom sums of squares?
c. Write a model and perform the tests suggested in part b. Use $\alpha = .05$.

20.11. A random sample of 10 measurements was selected from each of two populations. In addition to measuring a response variable $y$ on each experimental unit, a second variable $x$ was also measured. The sample data appear below.

| *Population 1* | | *Population 2* | |
|---|---|---|---|
| $x$ | $y$ | $x$ | $y$ |
| 30 | 165 | 24 | 180 |
| 27 | 170 | 31 | 169 |
| 20 | 130 | 20 | 171 |
| 21 | 156 | 26 | 161 |
| 33 | 167 | 20 | 180 |
| 29 | 151 | 25 | 170 |
| 27 | 165 | 22 | 169 |
| 25 | 162 | 30 | 160 |
| 28 | 169 | 24 | 178 |
| 32 | 173 | 21 | 182 |

a. Plot the sample data.
b. Write a first-order model relating the response $y$ to the covariate $x$.

*20.12. a. Fit the model of exercise 20.11b.
b. Test for parallelism. Use $\alpha = .05$.
c. What other tests are appropriate?

# References

Barr, A.J.; Goodnight, J.H.; Sall, J.P.; and Helwig, J.T. 1976. *A user's guide to SAS 76.* Raleigh, N.C.: SAS Institute, Inc.

Cochran, W.G., and Cox, G.M. 1957. *Experimental design.* 2d ed. New York: Wiley.

Davies, O.L. 1956. *The design and analysis of experiments.* New York: Hafner.

Dixon, W.J. 1975. *BMDP, biomedical computer programs.* Berkeley: University of California Press.

Dixon, W.J., and Massey, F.J., Jr. 1969. *Introduction to statistical analysis.* 3d ed. New York: McGraw-Hill.

Draper, N.R., and Smith, H. 1966. *Applied regression analysis.* New York: Wiley.

Duncan, A.J. 1959. *Quality control and industrial statistics.* Homewood, Ill.: Richard Irwin, Inc.

Fisher, R.A. 1949. *The design of experiments.* Edinburgh: Oliver and Boyd.

Graybill, F.A. 1976. *Theory and application of the linear model.* N. Scituate, Mass.: Duxbury Press.

Harnett, D.L. 1970. *Introduction to statistical methods.* Reading, Mass.: Addison-Wesley.

Kirk, R.E. 1968. *Experimental design: Procedures for the behavioral sciences.* Belmont, Calif.: Brooks/Cole.

Mendenhall, W. 1968. *Introduction to linear models and the design and analysis of experiments.* Belmont, Calif.: Wadsworth.

Nie, N.H.; Hull, C.H.; Jenkins, J.G.; Steinbrenner, K.; and Bent, D.H. 1975. *Statistical package for the social sciences.* 2d ed. New York: McGraw-Hill.

Ostle, B. 1963. *Statistics in research.* 2d ed. Ames, Iowa: Iowa State University Press.

Pearson, E.S., and Hartley, H.O. 1966. *Biometrika tables for statisticians.* 3d ed. Vol. I. London: Cambridge University Press.

Service, J. 1972. *A user's guide to the statistical analysis system.* Raleigh, N.C.: Student Supply Stores, North Carolina State University.

Siegel, S. 1956. *Nonparametric statistics for the behavioral sciences.* New York: McGraw-Hill.

Snedecor, G.W., and Cochran, W.G. 1967. *Statistical methods.* 6th ed. Ames, Iowa: Iowa State University Press.

Sokal, R.R., and Rohlf, F.J. 1969. *Biometry.* San Francisco: W.H. Freeman.

Steel, R.G.D., and Torrie, J.H. 1960. *Principles and procedures of statistics.* New York: McGraw-Hill.

# 21 Assumptions

## 21.1 Introduction

In previous chapters when statistical tests were presented, we tried to indicate the assumptions under which these tests were based. We did not try to verify these assumptions at that time. Now, however, we must address the subject in more detail to learn how we can perform some simple checks on the underlying assumptions by using the sample data. Since so many of the procedures we use are based on the assumption of normality (the $t$ test, chi-square test, $F$ test, etc.), we begin our discussion with the question of normality.

## 21.2 Normality: Skewness and Kurtosis

Many of the statistical testing and estimation procedures developed so far are based on the assumption that the sample data have been selected from a normal population. Chapter 11 dealt with nonparametric procedures that are appropriate when one or more of the underlying assumptions are violated. At that time we gave no indication about how to check specific assumptions, and as we will find out, there are no clear-cut (no fault) guidelines available. All we can hope to do is to isolate a few tools that will help us in making a rational decision when faced with the possibility that one or more of the assumptions of a parametric test are violated.

Certainly one indication of nonnormality occurs when the relative frequency histogram for the sample data is highly skewed to either the left or the right. A measure of the amount of skewness is given by the quantity $E(y - \mu)^3$, commonly called the third moment about the population mean $\mu$. The meaning of positive and negative values of $E(y - \mu)^3$ is illustrated in figure 21.1.

skewness

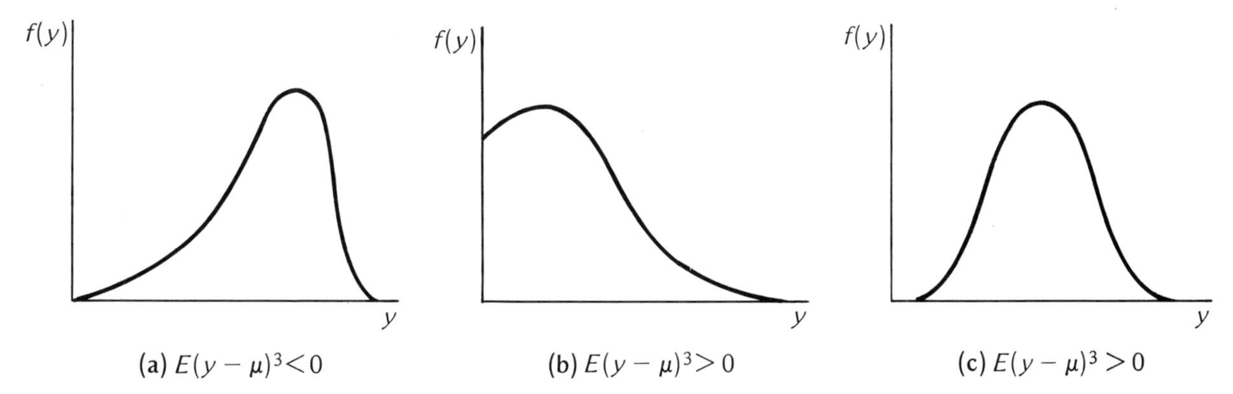


**Figure 21.1** Measuring skewness

Figures 21.1(a) and (b) indicate distributions with a negative and positive skew, respectively. The normal distribution illustrated in figure 21.1(c) has no skew. The actual measure of skewness used by most statisticians is

$$\gamma_1 = \frac{E(y - \mu)^3}{\sigma^3}$$

the *coefficient of skewness*. It is independent of the scale of the measurements and can be estimated from the sample data using the quantity

$\gamma_1$, coefficient of skewness

$$\hat{\gamma}_1 = \left[\frac{m_3^2}{m_2^3}\right]^{1/2} = \frac{m_3}{m_2^{3/2}}$$

where

$$m_3 = \frac{\sum (y - \bar{y})^3}{n} \qquad m_2 = \frac{\sum (y - \bar{y})^2}{n}$$

tests for skewness

A test of the null hypothesis that the sample data are selected from a normal population makes use of the test statistic

$$z = \frac{\hat{\gamma}_1}{\sqrt{6/n}}$$

This test procedure is summarized in the box.

---

**A Test for Skewness**

$H_0$: $\gamma_1 = 0$ (i.e., the data were selected from a symmetrical distribution).

$H_a$: 1. $\gamma_1 > 0$.
2. $\gamma_1 < 0$.
3. $\gamma_1 \neq 0$.

T.S.: $z = \dfrac{\hat{\gamma}_1}{\sqrt{6/n}}$.

R.R.: For a specified value of $\alpha$,
1. reject $H_0$ if $z > z_\alpha$;
2. reject $H_0$ if $z < -z_\alpha$;
3. reject $H_0$ if $|z| > z_{\alpha/2}$.

Note: $n$ must be 150 or more. If it is not, refer to table 16 of the appendix for upper 5% and 1% points for $\hat{\gamma}_1$.

---

A second kind of departure from normality can be detected by examining the kurtosis of a distribution. For a population of measurements, the quantity

$$\gamma_2 = \frac{E(y - \mu)^4}{\sigma^4}$$

$\gamma_2$, coefficient of kurtosis

is called the *coefficient of kurtosis*. Unlike the coefficient of skewness $\gamma_1$, $\gamma_2$ measures the heaviness of the tails of a distribution.

A normal population will have $\gamma_2 = 3$. Those distributions which are lighter-tailed will have more of a pileup near $\mu$, and $\gamma_2 < 3$. The heavier-tailed distributions, such as a $t$ distribution, will have less of a pileup about $\mu$, and $\gamma_2 > 3$. (See figure 21.2.)

The sample data can be used to calculate an estimate of $\gamma_2$:

$$\hat{\gamma}_2 = \frac{m_4}{m_2^2}$$

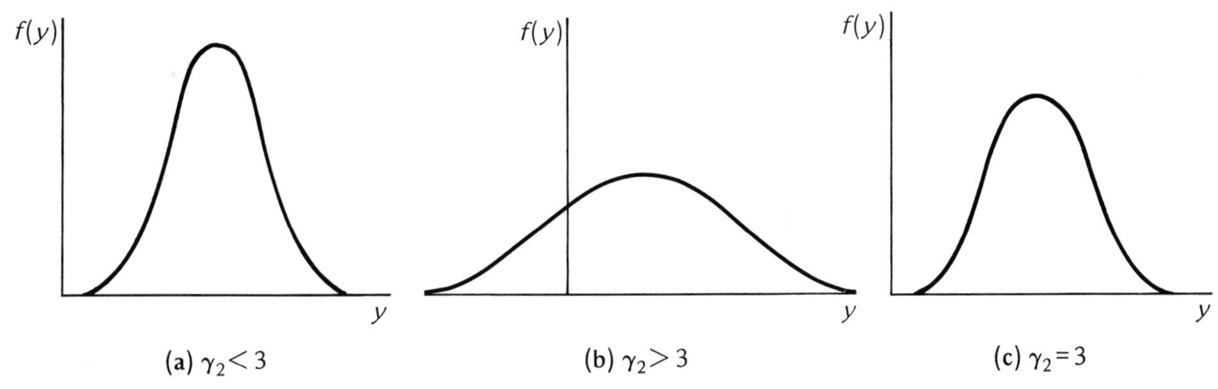


**Figure 21.2** Measuring kurtosis

where

$$m_4 = \frac{\sum (y - \bar{y})^4}{n} \qquad m_2 = \frac{\sum (y - \bar{y})^2}{n}$$

test for kurtosis

For very large sample sizes ($n \geq 1000$), the null hypothesis $\gamma_2 = 3$ can be tested using the test statistic

$$z = \frac{(\hat{\gamma}_2 - 3)}{\sqrt{24/n}}$$

The rejection region for various alternative hypotheses would be found from table 1 in the appendix.

Unfortunately, we rarely have the luxury of sample sizes with 1000 or more measurements, and the test just described has very little utility. For smaller sample sizes, upper and lower percentage points of the distribution of $\hat{\gamma}_2$ have been tabulated and appear in table 17 of the appendix. Thus for $n = 125$, the lower and upper 5% points for $\hat{\gamma}_2$ are 2.40 and 3.71, respectively.

# Exercises

21.1. In chapter 2, exercise 2.7, we discussed an experiment for examining the nitrogen content of apple leaves. To do this, growing tips from 150 leaves were clipped from trees throughout an orchard. The leaves were then ground to form one composite sample. Composite samples were obtained in the same manner from 35 other orchards selected at random from orchards throughout the state. The nitrogen contents (after rounding to the nearest hundredth) for the samples are given below.

| | | | | | |
|---|---|---|---|---|---|
| 2.10 | 2.82 | 2.17 | 1.99 | 2.22 | 3.09 |
| 2.47 | 2.52 | 2.80 | 2.10 | 2.92 | 2.20 |
| 1.75 | 2.77 | 2.82 | 2.67 | 3.05 | 2.93 |
| 2.94 | 1.98 | 2.38 | 2.65 | 2.77 | 1.85 |
| 1.69 | 2.70 | 2.68 | 2.06 | 2.36 | 2.28 |
| 2.75 | 2.43 | 2.39 | 2.55 | 1.80 | 1.96 |

a. Compute the coefficient of skewness.
b. Test the hypothesis $H_0$: $\gamma_1 = 0$ against $H_a$: $\gamma_1 \neq 0$. Use $\alpha = .02$.

21.2. a. Compute the coefficient of kurtosis for the data in exercise 21.1.
b. Can we test the null hypothesis $H_0$: $\gamma_2 = 3$?

21.3. Murder rates in 1971 (per 100,000 inhabitants) for a random sample of 90 cities from the North, South, and West are given in the table on the next page. We used these data in chapter 3.
a. Compute the coefficient of skewness.
b. Conduct a statistical test of $H_0$: $\gamma_1 = 0$ against the alternative $H_a$: $\gamma_1 > 0$. Use $\alpha = .05$.

21.4. Refer to exercise 21.3.
a. Calculate the coefficient of kurtosis and use this value to run a two-sided test of $H_0$: $\gamma_2 = 3$. Choose your own value of $\alpha$.
b. Do you think the sample of 90 murder rates was selected from a normal population?

21.5. Suppose we consider the data of exercise 21.3 as a population of measurements. The data below (adapted from chapter 3, p. 55) represent the means of 50 different random samples of $n = 5$ measurements selected from the population of 90 murder rates. Can we conclude that the 50 sample means were selected from a normal population (the sampling distribution of $\bar{y}$)? Substantiate your conclusion. Note that here we are really trying to verify the Central Limit Theorem for this one set of data.

| | | | | |
|---|---|---|---|---|
| 12.0 | 8.8 | 11.4 | 5.0 | 8.8 |
| 7.0 | 12.4 | 15.0 | 8.6 | 8.6 |
| 14.2 | 7.4 | 10.2 | 8.4 | 7.2 |
| 11.2 | 5.4 | 7.8 | 9.6 | 10.2 |
| 9.2 | 8.8 | 6.2 | 4.4 | 9.0 |
| 11.0 | 6.4 | 6.0 | 10.4 | 8.4 |
| 9.0 | 7.0 | 11.0 | 11.4 | 8.6 |
| 9.8 | 8.2 | 8.2 | 7.4 | 5.4 |
| 7.4 | 10.0 | 13.2 | 7.4 | 8.4 |
| 7.0 | 7.8 | 7.6 | 4.2 | 9.6 |

| *South* | *Rate* | *North* | *Rate* | *West* | *Rate* |
|---|---|---|---|---|---|
| Atlanta | 20 | Albany, NY | 3 | Bakersfield | 8 |
| Augusta, GA | 22 | Allentown | 2 | Boise | 5 |
| Baton Rouge | 10 | Atlantic City | 5 | Colorado Springs | 5 |
| Beaumont, TX | 10 | Canton, OH | 3 | Denver | 8 |
| Birmingham | 14 | Chicago | 13 | Eugene, OR | 4 |
| Charlotte, NC | 25 | Cincinnati | 6 | Fresno | 8 |
| Chattanooga | 15 | Cleveland | 14 | Honolulu | 4 |
| Columbia, SC | 13 | Detroit | 15 | Kansas City, MO | 13 |
| Corpus Christi | 13 | Evansville | 7 | Lawton, OK | 6 |
| Dallas | 18 | Grand Rapids | 3 | Los Angeles | 9 |
| El Paso | 4 | Johnstown, PA | 2 | Modesto, CA | 2 |
| Fort Lauderdale | 14 | Kalamazoo | 4 | Oklahoma City | 6 |
| Greensboro, NC | 11 | Kenosha, WI | 2 | Oxnard, CA | 2 |
| Houston | 17 | Lancaster, PA | 2 | Pueblo, CO | 3 |
| Jackson, MS | 16 | Lansing | 3 | Sacramento | 6 |
| Knoxville | 8 | Lima, OH | 3 | St. Louis | 15 |
| Lexington, KY | 13 | Madison, WI | 2 | Salinas, KS | 6 |
| Lynchburg, VA | 18 | Mansfield, OH | 7 | Salt Lake City | 4 |
| Macon, GA | 13 | Milwaukee | 4 | San Bernardino | 6 |
| Miami | 16 | Newark, NJ | 10 | San Francisco | 8 |
| Monroe, LA | 15 | Paterson, NJ | 3 | San Jose | 2 |
| Nashville | 13 | Philadelphia | 9 | Seattle | 4 |
| Newport News | 8 | Pittsfield, MA | 1 | Sioux City | 3 |
| Orlando | 11 | Racine, WI | 5 | Spokane | 1 |
| Richmond, VA | 15 | Rockford, IL | 4 | Stockton, CA | 9 |
| Roanoke | 10 | South Bend | 6 | Tacoma | 6 |
| Shreveport, LA | 15 | Springfield, IL | 2 | Topeka | 2 |
| Washington, D.C. | 11 | Syracuse | 4 | Tucson | 17 |
| Wichita Falls, TX | 6 | Vineland, NJ | 10 | Vallejo, CA | 4 |
| Wilmington | 7 | Youngstown | 7 | Waterloo, IA | 4 |

*Source: Uniform Crime Reports for the United States: 1970* (Washington, D.C.: Department of Justice), pp. 78–94.

# Equality of Variances 21.3

The assumption of equality of population variances, like the assumption of normality of the populations, has been made throughout the text, such as for the $t$ test when comparing two population means, the analysis of variance $F$ tests in a fixed effects model, the $t$ tests and $F$ tests obtained from fitting a general linear model, to mention just a few.

completely randomized design

Let us consider first a completely randomized design where we wish to compare $t$ population means based on independent random samples from each of the populations. Recall that we assume we are dealing with normal populations with a common variance $\sigma^2$ and possibly different means. If there were just two populations of interest, we could verify the assumption of equality of the two population variances using the $F$ test of chapter 12. However, with $t > 2$, rather than making all pairwise $F$ tests, we seek a single test that can be used to verify the assumption of equality of the population variances.

The one test we will use in this text for the null hypothesis

Hartley's test

$$H_0: \sigma_1^2 = \sigma_2^2 = \cdots = \sigma_t^2$$

was proposed by H. O. Hartley (1940 and 1950) and represents a logical extension to the $F$ test for $t = 2$. If $s_i^2$ denotes the sample variance computed from the $i$th sample, the test statistic is

$F_{max}$

$$F_{max} = \frac{s_{max}^2}{s_{min}^2}$$

where $s_{max}^2$ and $s_{min}^2$ are the largest and smallest of the $s_i^2$'s, respectively. The test procedure is summarized in the box.

**Hartley's Test for Homogeneity of Population Variances**

$H_0$: $\sigma_1^2 = \sigma_2^2 = \cdots = \sigma_t^2$.

$H_a$: Not all population variances are the same.

T.S.: $F_{max} = \dfrac{s_{max}^2}{s_{min}^2}$.

R.R.: For a specified value of $\alpha$, reject $H_0$ if $F_{max}$ exceeds the tabulated $F$-value (table 18) for a $= \alpha$, $df_1 = t$, and $df_2 = n - 1$, where $n$ is the number of observations in each sample.

It should be noted that, theoretically, we require the sample sizes to all be the same. In practice, if the sample sizes are nearly equal, the largest $n$ can be used for running the test of homogeneity. This procedure will result in the probability of a type I error being slightly more than the nominal value $\alpha$.

One other comment should be made. Most practitioners do not routinely run Hartley's test (or any of the other tests mentioned in the references at the end of this chapter). Rather, since the $F$ test in an analysis of variance is valid even with mild departures from the assumption of equality of variances, Hartley's test is used only for the more extreme cases. In these extreme situations, a transformation of the data may be required to stabilize the variances.

A *transformation of the sample data* is defined to be a process whereby the measurements on the original scale are systematically converted to a new scale of measurement. For example, if the original variable is $y$ and the variances associated with the variable across the treatments are heterogeneous, it may be necessary to work with a new variable such as $\sqrt{y}$, $\log y$, or some other transformed variable.

transformation of data

How can we select the appropriate transformation? This is no easy task and often takes a great deal of experience in the experimenter's area of application. In spite of these difficulties, we can consider several guidelines for choosing an appropriate transformation.

Many times the variances across the populations of interest are heterogeneous and seem to vary with the magnitude of the population mean. For example, it may be that the larger the population mean, the larger the population variance. When we are able to identify how the variance varies with the population mean, we can define a suitable transformation from the variable $y$ to a new variable $y_T$. Three specific situations are presented in table 21.1.

guidelines for selecting $y_T$

**Table 21.1** Transformation to achieve uniform variance

| *Relationship Between* $\mu$ and $\sigma^2$ | $y_T$ | *Variance of* $y_T$ *(for a given* $k$*)* |
|---|---|---|
| $\sigma^2 = k\mu$ (when $k = 1$, $y$ is a Poisson variable) | $y_T = \sqrt{y}$ or $\sqrt{y + .375}$ | $\frac{1}{4}$; $(k = 1)$ |
| $\sigma^2 = k\mu^2$ | $y_T = \log y$ or $\log (y + 1)$ | 1; $(k = 1)$ |
| $\sigma^2 = kp(1 - p)$ (when $k = 1/n$, $y$ is a binomial variable) | $y_T = \arcsin \sqrt{y}$ | $1/4n$; $(k = 1/n)$ |

The first row of table 21.1 suggests that if $y$ is a Poisson random variable (refer to chapter 10), the variance of $y$ is equal to the mean of $y$. Thus if the different populations correspond to different Poisson populations, the variances will be heterogeneous provided the means are different. The transformation that will stabilize the variances is $y_T = \sqrt{y}$; or, if the Poisson means are small (under 5), the transformation $y_T = \sqrt{y + .375}$ is better.

$y_T = \sqrt{y}$

**EXAMPLE 21.1**

The mean dissolved oxygen contents (in ppm) of 3 different lakes were to be compared based on independent random samples of 10 observations taken from the center of each lake at a depth of one foot. The sample data are given in table 21.2.

a. Run a test of the equality of the population variances. Use Hartley's test, with $\alpha = .05$.
b. Transform the data using $y_T = \sqrt{y + .375}$.

**Table 21.2** Mean dissolved oxygen contents (in ppm) of 3 lakes, example 21.1

| Lake | | |
|---|---|---|
| *1* | *2* | *3* |
| 0 | 1 | 14 |
| 2 | 3 | 26 |
| 1 | 4 | 25 |
| 3 | 6 | 18 |
| 1 | 8 | 19 |
| 2 | 7 | 22 |
| 3 | 5 | 21 |
| 4 | 3 | 16 |
| 1 | 4 | 20 |
| 5 | 5 | 30 |
| $\bar{y} = 2.2$ | $\bar{y} = 4.6$ | $\bar{y} = 21.1$ |
| $s = 1.55$ | $s = 2.07$ | $s = 4.84$ |

c. Compute the sample means and sample standard deviations for the transformed data.

## SOLUTION

a. The $F$ test for the equality of population variances has

$$F_{max} = \frac{(4.84)^2}{(1.55)^2} = 9.75$$

The critical value of $F_{max}$ for $a = .05$, $df_1 = 3$, and $df_2 = 9$ is 5.34. Since $F_{max}$ is greater than 5.34, we reject the hypothesis of homogeneity of the population variances.

b. The most convenient way to transform the data is to make use of a desk calculator with a square root key or a table of squares and square roots. The square root data appear in table 21.3.

**Table 21.3** Square root transformations ($\sqrt{y + .375}$) of the data of table 21.2

| Lake | | |
|---|---|---|
| *1* | *2* | *3* |
| 0.612 | 1.173 | 3.791 |
| 1.541 | 1.837 | 5.136 |
| 1.173 | 2.092 | 5.037 |
| 1.837 | 2.525 | 4.287 |
| 1.173 | 2.894 | 4.402 |
| 1.541 | 2.716 | 4.730 |
| 1.837 | 2.318 | 4.623 |
| 2.092 | 1.837 | 4.047 |
| 1.173 | 2.092 | 4.514 |
| 2.318 | 2.318 | 5.511 |

c. The sample means and standard deviations for the transformed data are shown in table 21.4. Thus although the original data had heteroge-

**Table 21.4** Sample means and standard deviations for the data in table 21.3

| | Lake | | |
|---|---|---|---|
| | *1* | *2* | *3* |
| *Sample Mean* | 1.53 | 2.18 | 4.61 |
| *Sample Standard Deviation* | .51 | .50 | .52 |

neous variances, the sample variances are all approximately .25, as indicated in table 21.1.

$y_T = \log y$

The second transformation indicated in table 21.1 is for an experimental situation where the population variance is approximately equal to the square of the population mean, or, equivalently, where $\sigma = \mu$. Actually, the logarithmic transformation is appropriate any time the *coefficient of variation* $\sigma_i/\mu_i$ is constant across the populations of interest.

**EXAMPLE 21.2**

Irritable bowel syndrome (IBS) is a nonspecific intestinal disorder characterized by abdominal pain and irregular bowel habits. Each person in a random sample of 24 patients having periodic attacks of IBS was randomly assigned to one of 3 treatment groups, A, B, and C. The number of hours of relief while on therapy is recorded for each patient in table 21.5.

**Table 21.5** Data for hours of relief while on therapy, example 21.2

| | *Treatment* | |
|---|---|---|
| *A* | *B* | C |
| 4.2 | 4.1 | 38.7 |
| 2.3 | 10.7 | 26.3 |
| 6.6 | 14.3 | 5.4 |
| 6.1 | 10.4 | 10.3 |
| 10.2 | 15.3 | 16.9 |
| 11.7 | 11.5 | 43.1 |
| 7.0 | 19.8 | 48.6 |
| 3.6 | 12.6 | 29.5 |
| $\bar{y} = 6.46$ | $\bar{y} = 12.34$ | $\bar{y} = 27.35$ |
| $s = 3.22$ | $s = 4.53$ | $s = 15.66$ |

a. Test for differences among the population variances. Use $\alpha = .05$.
b. Since there are no zero y-values, use the transformation $y_T = \ln y$ ("ln" denotes logarithms to the base e) to try to stabilize the variances.
c. Compute the sample means and the sample standard deviations for the transformed data.

## SOLUTION

a. The $F$ test for a test of the null hypothesis $H_0$: $\sigma_1^2 = \sigma_2^2 = \sigma_3^2$ is

$$F_{max} = \frac{(15.66)^2}{(3.22)^2} = \frac{245.24}{10.37} = 23.65$$

Since the computed value of $F_{max}$ exceeds 8.38, the tabulated value (table 18) for a = .05, $df_1 = 3$, and $df_2 = 7$, we reject $H_0$ and conclude that the population variances are different.

b. We can obtain the natural logarithms (log to the base e) for the sample data using either a desk calculator with an "ln" key or by referring to a table of natural logs (see, for example, the Chemical Rubber Company's *Mathematical Tables,* 1961). The transformed data are shown in table 21.6.

**Table 21.6** Natural logarithms of the data in table 21.5

| *Treatment* | | |
|---|---|---|
| *A* | *B* | *C* |
| 1.435 | 1.411 | 3.656 |
| 0.833 | 2.370 | 3.270 |
| 1.887 | 2.660 | 1.686 |
| 1.808 | 2.342 | 2.332 |
| 2.322 | 2.728 | 2.827 |
| 2.460 | 2.442 | 3.764 |
| 1.946 | 2.986 | 3.884 |
| 1.281 | 2.534 | 3.384 |

c. The sample means and standard deviations for the transformed data are given in table 21.7. Now although the sample variances are not exactly the same, they certainly do not indicate that the corresponding population variances are different.

**Table 21.7** Sample means and standard deviations for the data of table 21.6

| | *Treatment* | | |
|---|---|---|---|
| | *A* | *B* | *C* |
| *Sample Mean* | 1.75 | 2.43 | 3.10 |
| *Sample Standard Deviation* | .54 | .46 | .77 |

$y_T = \arcsin \sqrt{y}$

The third transformation listed in table 21.1 is particularly appropriate for data recorded as percentages or proportions. You will recall that in chapter 10 we introduced the binomial distribution, where $y$ designates the number of successes in $n$ identical trials and $\hat{p} = y/n$ provides an estimate of $p$, the proportion of experimental units in the population possessing the characteristic. Although we may not have mentioned this while studying the binomial, the variance of $\hat{p}$ is given by $pq/n$ (where $q = 1 - p$). Thus if the response variable is the proportion of successes in a random sample of $n$ observations, then the variances for the response variable (the $\hat{p}$'s) will vary, depending on the values of $p$ for the populations from which the samples were drawn. See table 21.8.

**Table 21.8** Variance of $p$, the sample proportion, for several values of $p$ and $n = 20$

| *Values of p* | $pq/n$ |
|---|---|
| .01 | .0005 |
| .05 | .0024 |
| .1 | .0045 |
| .2 | .0080 |
| .3 | .0105 |
| .4 | .0120 |
| .5 | .0125 |

Since the variance of $\hat{p}$ is symmetrical about $p = .5$, the variance of $\hat{p}$ for $p = .7$ and $n = 20$ would be .0105. Similarly, we can determine $pq/n$ for other values of $p > .5$. The important thing to note is that if the populations have values of $p$ in the vicinity of approximately .3 to .5, there is very little difference in the variances for $\hat{p}$. However, the variance of $\hat{p}$ is quite variable for either large or small values of $p$, and for these situations we should consider the possibility of transforming the sample proportions to stabilize the variances.

The transformation we recommend is arcsin $\sqrt{\hat{p}}$ (sometimes written as $\sin^{-1} \sqrt{\hat{p}}$). In words, we are transforming the sample proportion into the angle whose sine is $\sqrt{\hat{p}}$. Some experimenters express these angles in degrees, others in radians. For consistency we will always express our angles in radians. Many hand and desk calculators include a key for the arcsin transformation; table 19* of the appendix provides arcsin computations for various values of $\hat{p}$.

**EXAMPLE 21.3**

In a national public opinion poll, a random sample of 30 registered voters was obtained from each of 24 different standard metropolitan statistical areas (SMSA). Each of the 30 voters in a sample was asked whether he or she favored limiting the FBI director to a fixed term in office (such as 10 years). The data below are the sample proportions for the 24 SMSAs. Transform the data by using $y_T = \text{arcsin} \sqrt{\hat{p}}$. Calculate the sample mean and standard deviation for the transformed data.

| | | | | | |
|---|---|---|---|---|---|
| .13 | .60 | .33 | .03 | .43 | .43 |
| .17 | .70 | .47 | .10 | .60 | .60 |
| .30 | .10 | .57 | .20 | .20 | .67 |
| .53 | .20 | .70 | .33 | .30 | .77 |

*Table 19 of the appendix gives 2 arcsin $\sqrt{\hat{p}}$ values.

**SOLUTION**

Using a calculator or table 19 in the appendix, the transformed data are

| | | | | | |
|---|---|---|---|---|---|
| .37 | .89 | .61 | .17 | .72 | .72 |
| .42 | .99 | .76 | .32 | .89 | .89 |
| .58 | .32 | .86 | .46 | .46 | .96 |
| .82 | .46 | .99 | .61 | .58 | 1.07 |

The sample mean and standard deviation are, respectively, .66 and .25.

when $\hat{p} = 0, 1$

One comment should be made concerning the situation where a sample proportion of 0 or 1 is observed. For these cases we recommend substituting $1/4n$ and $1 - (1/4n)$, respectively, as the corresponding sample proportions to be used in the calculations.

# Exercises

21.6. The data of example 21.3 are shown below. Suppose that the 4 columns represent 4 geographic locations of the country (NE, SE, NW, SW) and that a random sample of 100 voters was obtained from 6 selected SMSAs within each geographic location. Analyze the sample data by using the arcsin transformation to determine if there are differences among the 4 geographic locations. Use $\alpha = .05$.

| *NE* | *SE* | *NW* | *SW* |
|---|---|---|---|
| .13 | .10 | .03 | .20 |
| .17 | .20 | .10 | .30 |
| .30 | .33 | .20 | .43 |
| .53 | .47 | .33 | .60 |
| .60 | .57 | .43 | .67 |
| .70 | .70 | .60 | .77 |

21.7. Refer to exercise 21.6. Suppose that the rows correspond to different socioeconomic levels, so that one SMSA was selected from each socioeconomic level in each of the 4 geographic locations. The sample data then represent the proportion of favorable responses based on independent samples of 100 people for each socioeconomic-geographic location combination.

a. Analyze the transformed data and draw conclusions.

b. Comment on the proposal to take two random samples of size 50 for each socioeconomic-geographic location combination rather than one sample of 100 voters.

21.8. An experiment was conducted to compare the number of major defectives observed along each of 5 production lines during a specific week in which major changes were being instituted. The production was monitored continuously for the week of changes, and the number of major defectives was recorded per day for each line. These data are shown here.

| | *Production Line* | | | | |
|---|---|---|---|---|---|
| *Day* | *1* | *2* | *3* | *4* | *5* |
| Monday | 34 | 54 | 75 | 44 | 80 |
| Tuesday | 44 | 41 | 62 | 43 | 52 |
| Wednesday | 32 | 38 | 45 | 30 | 41 |
| Thursday | 36 | 32 | 10 | 32 | 35 |
| Friday | 51 | 56 | 68 | 55 | 58 |

a. Use the square root transformation to perform an analysis on the sample data.

b. Draw your conclusions concerning differences among the days and the production lines.

# Summary 21.4

In this chapter we have attempted to provide some techniques for checking the assumptions of normality and homogeneous variances. Information obtained from tests for skewness and kurtosis is presented to answer questions of normality. The *F* test for $s^2_{\text{max}}/s^2_{\text{min}}$ is used for comparing the variances of *t* populations.

Sometimes the sample data indicate that the population variances are different. Then when the relationship between the population mean and the population standard deviation is either known or suspected, it is convenient to transform the sample measurement $y$ to new values $y_T$ in order to stabilize the population variances, using the transformations suggested in table 21.1. These transformations include the square root, the logarithmic, and the arcsin transformations, and many others.

The topics in this chapter are certainly not covered in exhaustive detail. However, the material is sufficient for training the beginning researcher to be aware of the assumptions underlying his or her project and to consider either running an alternative analysis (such as using a nonparametric statistical method) when appropriate or applying a transformation to the sample data.

# References

*CRC standard mathematical tables.* 1961. 12th ed. Cleveland, Ohio: The Chemical Rubber Co.

Hartley, H.O. 1940. Testing the homogeneity of a set of variances. *Biometrika* 31: 249–255.

———. 1950. The maximum *F*-ratio as a shortcut for heterogeneity of variance. *Biometrika* 37: 308–312.

Kirk, R.E. 1968. *Experimental design: Procedures for the behavioral sciences.* Belmont, Calif.: Brooks/Cole.

LaValle, I.H. 1970. *An introduction to probability, decision, and inference.* New York: Holt, Rinehart and Winston.

Pearson, E.S., and Hartley, H.O. 1966. *Biometrika tables for statisticians.* 3d ed. Vol. I. London: Cambridge University Press.

Roussas, G. 1973. *A first course in mathematical statistics.* Reading, Mass.: Addison-Wesley.

Snedecor, G.W., and Cochran, W.G. 1967. *Statistical methods.* 6th ed. Ames, Iowa: Iowa State University Press.

Sokal, R.R., and Rohlf, F.J. 1969. *Biometry.* San Francisco: W.H. Freeman.

Yule, G. and Kendall, M.G. 1957. *An introduction to the theory of statistics.* London: Charles Griffin and Company Ltd.

# 22 Sample Survey Design

## 22.1 Introduction

The topic of random sampling was discussed briefly in chapter 3. A sample of $n$ measurements selected from a population of $N$ measurements was said to be random if each possible sample of size $n$ had an equal chance of being selected. In this chapter we will consider an alternative to random sampling, as we have defined it, so that we can estimate a population mean, total, or proportion. The techniques presented here will concentrate on how the sample measurements are selected and will be particularly appropriate for surveys (experiments) conducted in business and the social sciences. Although some of the estimation problems will be similar to those discussed previously, formulas for the bounds on the errors of estimation will be different, due to the fact that we are sampling from a finite (rather than infinite) population of size $N$.

## 22.2 Simple Random Sampling and Stratified Random Sampling

Consider the estimation of a population mean or total using a random sample—which we will now call a *simple random sample*—of units selected from

simple random sample

$\tau$

a population of $N$ experimental units. For purposes of illustration, we will suppose that an auditor is interested in estimating the mean value per account ($\mu$) or the total value ($\tau$) for a population of hospital records. Intuitively we would employ the sample mean $\bar{y}$ as a point estimate of $\mu$. Similarly, if $\tau$ represents the total (sum) of all measurements in the population, then $\tau = N\mu$, and a point estimate of $\tau$ is given by $N\bar{y}$. A summary of the point estimates for $\mu$ and $\tau$ is presented in the box. A formula for a corresponding bound on the error of estimation (BOE) for each is also given.

**Point Estimation of $\mu$ and $\tau$, Simple Random Sampling**

| Parameter: | $\mu$ | $\tau$ |
|---|---|---|
| Point estimate: | $\bar{y}$ | $N\bar{y}$ |
| BOE: | $2\sqrt{\left(\frac{s^2}{n}\right)\left(\frac{N-n}{N}\right)}$ | $2\sqrt{\left(\frac{N^2s^2}{n}\right)\left(\frac{N-n}{N}\right)}$ |

The quantity $(N - n)/N$ is called a finite population correction (fpc) and it accounts for the fact that a random sample of 20 from a population of 40 provides more information than a random sample of 20 from 40,000. For all practical purposes, when $N$ is large relative to $n$ and $(N - n)/N \geq .95$, we can ignore the fpc.

**EXAMPLE 22.1**

An industrial firm is concerned about the time per week spent by researchers on certain trivial tasks. The time log sheets of a simple random sample of 50 employees in one week show that the average amount of time spent on these tasks is 10.31 hours, with a sample variance of 2.25. If the corporation employs 750 researchers, estimate the total number of hours lost per week on these trivial tasks.

**SOLUTION**

We are told that the population consists of $N = 750$ employees, from which a random sample of $n = 50$ time log sheets was obtained. The mean amount of work lost for the sampled employees is 10.31. An estimate of the total amount of time lost per week is

$$N\bar{y} = 750(10.31) = 7732.5 \text{ hours}$$

A bound on the error of estimation is

$$2\sqrt{\left(\frac{N^2s^2}{n}\right)\left(\frac{N-n}{N}\right)} = 2\sqrt{(750)^2\left(\frac{2.25}{50}\right)\left(\frac{750-50}{750}\right)} = 307.4 \text{ hours}$$

Thus we estimate that the total amount of time lost is 7732.5 hours. We are quite confident that the error of estimation is less than 307.4 hours.

Rather than estimating a population mean or total, sample surveys are frequently conducted to estimate the proportion of experimental units in a population that possesses a specified characteristic. For example, suppose we are interested in determining a television program's rating by estimating the proportion of families in a given district who watched the program during a given week. Let $y_i = 0$ if the $i$th family in a simple random sample of $n$ does not possess the characteristic of interest (did not watch the TV show) and $y_i = 1$ if the $i$th family does possess the characteristic of interest. Then the proportion of elements in the sample possessing the specified characteristic also represents the sample mean of the $y_i$'s. With this convention, $\hat{p}$ is actually $\bar{y}$ and $p$ can be thought of as the mean for the entire population of 0s and 1s. Thus formulas developed for estimating $\mu$ can also be used for estimating $p$.

**EXAMPLE 22.2**

A simple random sample of 150 high school seniors were interviewed to determine those students who had held a part-time job over the past year. There are 1000 seniors in the district sampled and 60 of the 150 indicated they had held some part-time job over the past year. Use this information to estimate the proportion of all the high school students who worked on a part-time job during the last year. Place a bound on the error of estimation.

**SOLUTION**

The point estimate of the population proportion is $\hat{p} = 60/150 = .40$. Since the sample proportion is also the sample mean of the 0s and 1s, the bound on the error of estimation is

$$2\sqrt{\left(\frac{s^2}{n}\right)\left(\frac{N-n}{N}\right)}$$

Now, for the sample data of 1s and 0s corresponding to those who did hold and those who did not hold a part-time job in the last year,

$$\sum_{i=1}^{150} y_i = \sum_{i=1}^{150} y_i^2 = 60$$

Hence

$$s^2 = \frac{1}{n-1}\left[\sum y^2 - \frac{\left(\sum y\right)^2}{n}\right] = \frac{1}{149}\left[60 - \frac{(60)^2}{150}\right] = .2416$$

The BOE is then

$$2\sqrt{\left(\frac{.2416}{150}\right)\left(\frac{1000-150}{1000}\right)} = .074$$

Thus we estimate that .40 of all the high school seniors have held a part-time job at some time over the past year. We are quite confident that the point estimate (.40) is within .074 of the actual proportion of seniors who had held a part-time job during the last year.

stratification

Stratification offers an alternative to simple random sampling, and, in many instances, stratification increases the quantity of information available for estimating $\mu$ or $\tau$.

---

**Definition 22.1**

A *stratified random sample* is a sample obtained by dividing the population of experimental units into nonoverlapping groups (called strata) and then selecting a simple random sample from each stratum.

---

The first step in selecting a stratified random sample is to clearly specify the strata, making certain that each experimental unit can be classified into only one stratum. For example, in a local election survey we might wish to stratify registered voters according to 1 of 5 voting precincts. If precinct boundary lines are clearly defined and the voter registration lists are up to date, there should be no problem in placing registered voters into the appropriate precincts (strata).

After the experimental units are divided into strata, we select a simple random sample of units from each stratum, using the techniques discussed earlier in this section.

notation

Before presenting the formulas for the estimates of $\mu$ (or $p$) and $\tau$, we need the following notation.

$\ell$: the number of strata.

$N_i$: the number of elements in stratum $i$ ($i = 1, 2, \ldots, \ell$).

$n_i$: the sample size in stratum $i$.

$N$: the total population size; $N = \Sigma_{i=1}^{\ell} N_i$.

$n$: the total sample size; $n = \Sigma_{i=1}^{\ell} n_i$.

$\bar{y}_i$: the sample mean in stratum $i$.

$s_i^2$: the sample variance in stratum $i$.

The point estimation procedures for $\mu$ and $\tau$ are given in the box.

---

**Point Estimation of $\mu$ and $\tau$, Stratified Random Sampling**

| | | |
|---|---|---|
| Parameter: | $\mu$ | $\tau$ |
| Point estimate: | $\bar{y}_{\mathrm{ST}} = \dfrac{\sum_{i=1}^{\ell} N_i \bar{y}_i}{N}$ | $N\bar{y}_{\mathrm{ST}}$ |

| | | |
|---|---|---|
| BOE: | $2\sqrt{\dfrac{1}{N^2}\sum_{i=1}^{\ell} N_i^2\left(\dfrac{N_i - n_i}{N_i}\right)\left(\dfrac{s_i^2}{n_i}\right)}$ | $2\sqrt{\sum_{i=1}^{\ell} N_i^2\left(\dfrac{N_i - n_i}{N_i}\right)\left(\dfrac{s_i^2}{n_i}\right)}$ |

### EXAMPLE 22.3

A wholesale food distributor in a large metropolitan area would like to know if demand is great enough to justify adding a new product to his stock. To aid in making his decision, he plans to add this product to a sample of the stores he services to estimate average monthly sales. Since he only services 4 large chains in the metropolitan area, it is administratively convenient to stratify the stores, with each chain serving as a stratum. There are 24 stores in stratum 1, 36 in stratum 2, 30 in stratum 3, and 30 in stratum 4. The wholesaler decides that he has enough money to obtain data in a total of 20 retail stores. If we allocate the total sample size among the strata with the stratum sample sizes proportional to the stratum sizes, we obtain sample sizes of 4, 6, 5, and 5 for the 4 strata, respectively. Thus the product is introduced into 4 stores chosen at random from Chain 1, 6 stores from Chain 2, and 5 stores each from chains 3 and 4. The sales (in hundreds of dollars, after a one-month trial period) appear in table 22.1.

**Table 22.1** Sales data for example 22.3

| | *Stratum (Chain)* | | | |
|---|---|---|---|---|
| | *1* | *2* | *3* | *4* |
| | 84 | 91 | 108 | 102 |
| | 80 | 99 | 96 | 120 |
| | 92 | 93 | 100 | 104 |
| | 100 | 105 | 93 | 101 |
| | | 111 | 93 | 123 |
| | | 101 | | |
| *Sample Mean* | 89 | 100 | 98 | 110 |
| *Sample Variance* | 78.67 | 55.60 | 39.50 | 112.50 |

Estimate the average sales for the month and place a bound on the error of estimation.

### SOLUTION

The point estimate of $\mu$, the average montly sales for all stores across the 4 chains, is

$$\bar{y}_{ST} = \frac{\sum_i N_i\bar{y}_i}{N} = \frac{24(89) + 36(100) + 30(98) + 30(110)}{120} = 99.8$$

The corresponding bound on the error of estimation can be found by substituting into the formula given in the box.

$$\text{BOE} = 2\left\{\frac{1}{(120)^2}\left[(24)^2\left(\frac{24-4}{24}\right)\left(\frac{78.67}{4}\right) + (36)^2\left(\frac{36-6}{36}\right)\left(\frac{55.60}{6}\right) + (30)^2\left(\frac{30-5}{30}\right)\left(\frac{39.50}{5}\right) + (30)^2\left(\frac{30-5}{30}\right)\left(\frac{112.50}{5}\right)\right]\right\}^{1/2}$$

$$= 2\sqrt{2.9339} = 3.4257$$

To summarize, we estimate the average monthly sales to be \$9,980, with a bound on the error of estimation of \$343.

advantages

There are several reasons why stratified random sampling often results in an increase in information for a given cost. First, the data should be more homogeneous within each stratum than in the population as a whole. Taking advantage of the reduced variability within each stratum, we obtain estimates that have smaller BOEs than comparable estimates from a simple random sample of the same size. Second, the cost of conducting a stratified random sample tends to be less than that for a simple random sample. The elements in each stratum are usually located within a smaller geographic area, and separate teams of interviewers can be sent to the strata for collection of the sample data. Third, separate estimates of population parameters for each stratum can be obtained without additional sampling.

# Exercises

22.1. A group of college students conducted a survey to determine the average number of college hours (credits) an undergraduate student must earn for various majors to obtain a bachelor's degree from a large university. To do this, the departments of the university were stratified by colleges. Use the sample data shown below to estimate $\mu$, the average number of credit hours required for a bachelor's degree. Place a bound on the error of estimation.

| *College* | *Number of Departments* | *Sample Data* |
|---|---|---|
| architecture and fine arts | 7 | 192, 199, 188, 191 |
| arts and sciences | 40 | 186, 195, 186, 189, 186, 192, 193, 195, 200, 183, 187, 192 |
| business administration | 6 | 193, 186, 180, 182 |
| education | 8 | 197, 198, 188, 196 |
| engineering | 20 | 202, 203, 213, 202, 204, 206, 210, 206 |

22.2. Use the data of exercise 22.1 to estimate the average number of hours for all departments within the college of arts and sciences. Place a bound on the error of estimation.

22.3. Refer to example 22.3. Use the sales data from all 20 stores as if it were obtained from a single simple random sample. Estimate the mean sales and place a bound on the error of estimation. Compare your results to those of example 22.3. Does it appear that stratification has helped?

22.4. A student newspaper recently conducted a survey to estimate the proportion of faculty members in the state university system who favor collective bargaining for faculty members on matters of salary and fringe benefits. Of the 157 professors interviewed, 86 stated they favored collective bargaining. Use these data to estimate the proportion of all faculty members who favor collective bargaining, and place a bound on the error of estimation. (Hint: Ignore the fpc.)

22.5. The alumni association of a university recently sent a survey to a sample of former students. The set of all alumni was stratified according to class into the following strata:

| *Stratum* | *Class* | *Size* |
|---|---|---|
| 1 | 1900–1930 | 100 |
| 2 | 1931–1950 | 500 |
| 3 | 1951–1960 | 1200 |
| 4 | 1961–1970 | 1800 |
| 5 | 1971–present | 900 |

It was decided that the total of 1000 interviews should be distributed among the strata as follows: 20 from stratum 1, 110 from stratum 2, 270 from stratum 3, 400 from stratum 4, and 200 from stratum 5. Based on the records from the alumni association of the university, simple random samples of these sizes were selected from the strata above. Each person to be included in the survey was contacted and asked to respond to a number of questions, one of which related to their attitudes towards annual support drives. For the question, "The annual fund should provide money to help the Athletic Association," the following proportions of persons responded positively:

stratum 1: .40

stratum 2: .45

stratum 3: .62

stratum 4: .48

stratum 5: .37

Use these data to estimate the proportion of all alumni favoring annual fund support for the Athletic Association. Place a bound on the error of estimation.

22.6. In the same survey referred to in exercise 22.5, each person was asked to state the number of years since his or her last visit to the campus. These data are summarized below.

| *Stratum* | *Sample Mean* | *Sample Standard Deviation* |
|---|---|---|
| 1 | 35.6 | 15.2 |
| 2 | 22.1 | 12.8 |
| 3 | 10.3 | 7.5 |
| 4 | 6.3 | 4.6 |
| 5 | 1.9 | 2.1 |

Estimate the mean number of years since the last visit for all alumni. Place a bound on the error of estimation.

22.7. Rather than obtaining an overall estimate of $\mu$ in exercise 22.6, estimate the mean length of time since the last visit for each stratum. Place bounds on the errors of estimation.

22.8. A campus survey was conducted to determine the average occupancy per car for staff and faculty members entering the university during the 7:30–9:00 A.M. rush hour period for a representative midweek day. There are 4 legal entrances to the university and it was assumed that data would be collected from these checkpoints. The day chosen for study was a Wednesday in the middle of the semester. One person was stationed at each checkpoint and instructed to sample every second car. For each car included in the sample, the person first checked the sticker on the front of the car to verify that it was a faculty or staff member's car. Then the number of occupants (including the driver) was recorded. Assuming the 4 checkpoints represent 4 strata and the method of sampling simulates simple random sampling of each stratum, use the data on the next page to estimate the average occupancy per car. Place a bound on the error of estimation.

| Stratum (entrance) | $N_i$ | $n_i$ | $\sum_i y_i$ | $s_i^2$ |
|---|---|---|---|---|
| north | 294 | 147 | 185 | .24 |
| south | 150 | 75 | 89 | .26 |
| east | 230 | 115 | 144 | .26 |
| west | 322 | 161 | 203 | .18 |

22.9. Would simple random sampling have been an alternative to stratified random sampling for estimating $\mu$ in exercise 22.8? Explain.

22.10. Refer to exercise 22.8. Estimate $\tau$, the total number of occupants entering the campus during the 7:30–9:00 A.M. period. Place a bound on the error of estimation.

22.11. Each meter in a simple random sample of 100 water meters within a community was monitored to estimate the average daily water consumption per day over a specified dry spell. The sample mean and sample standard deviation were found to be 40.2 gallons and 36.8 gallons, respectively. If there are 5500 homes in the community, estimate $\mu$, the average consumption for all homes in the community. Place a bound on the error of estimation.

e

22.12. Refer to exercise 22.11. Estimate the total amount of water used by the 5500 homes in the community and place a bound on the error of estimation.

## Cluster Sampling 22.3

cluster sample

The third type of sampling procedure, called *cluster sampling*, will sometimes give more information per unit cost than either simple or stratified random sampling. The population of experimental units are divided into groups, called clusters.

**Definition 22.2**

A *cluster sample* is a sample obtained by taking a simple random sample of clusters, with an observation obtained from each unit in the sampled clusters.

Note that cluster sampling is similar to stratified random sampling in that we first divide the population of experimental units into groups. However, rather than obtaining a simple random sample of units from each group, we take a simple random sample of groups and sample all units in the chosen groups.

advantages

Cluster sampling can be less costly than either simple random sampling or stratified random sampling if the cost of obtaining observations increases as the distance separating experimental units increases or if the cost of obtaining a list of experimental units is high. For example, suppose we wish to estimate the average income per household in a large city. How should we choose the sample? If we use simple random sampling, we will need a list of all households in the city, and this might be costly or even impossible to obtain. Even with stratified random sampling, we would need a list of all households within each stratum. However, by dividing the city into regions (perhaps blocks), we could obtain a simple random sample of blocks and then we could interview each household in the sampled blocks. Even if a list of households did exist, it still might be better to use cluster sampling, especially if sampling scattered households increases the travel costs and interviewing time.

Before presenting methods for estimating a population mean or total, we need the following notation:

notation

$N$: the number of clusters.

$n$: the number of clusters selected in a simple random sample.

$m_i$: the number of elements in cluster $i$ ($i = 1, 2, \ldots, N$).

$\bar{m}$: the average cluster size for the sampled clusters; $\bar{m} = \Sigma_{i=1}^{n} m_i/n$.

$M$: the number of elements in the population; $\bar{M} = \Sigma_{i=1}^{N} m_i$.

$\bar{M}$: the average cluster size for the population; $\bar{M} = M/N$.

$y_i$: the total for all observations in the $i$th cluster.

The estimation procedures for $\mu$ and $\tau$ are presented in the box.

**Point Estimation of $\mu$ or $\tau$, Cluster Sampling**

| Parameter: | $\mu$ | $\tau$ |
|---|---|---|
| Point estimate: | $\bar{y}_c = \dfrac{\sum_{i=1}^{n} y_i}{\sum_{i=1}^{n} m_i}$ | $M\bar{y}_c$ |
| BOE: | $2\sqrt{\left(\dfrac{N-n}{Nn\bar{M}^2}\right)\left(\dfrac{\sum_{i=1}^{n}(y_i - \bar{y}m_i)^2}{n-1}\right)}$ | $2\sqrt{N^2\left(\dfrac{N-n}{Nn}\right)\left(\dfrac{\sum_{i=1}^{n}(y_i - \bar{y}m_i)^2}{n-1}\right)}$ |

**EXAMPLE 22.4**

Interviews are conducted in each of 25 blocks sampled in a city. The data on income for adult males are presented in table 22.2. Use the data to estimate the average income per adult male in the city and place a bound on the error of estimation. (Note: $N = 415$.)

**Table 22.2** Incomes for adult males, example 22.4

| Cluster $i$ | Number of Adult Males $m_i$ | Total Income Per Cluster $y_i$ | Cluster $i$ | Number of Adult Males $m_i$ | Total Income Per Cluster $y_i$ |
|---|---|---|---|---|---|
| 1 | 8 | \$ 96,000 | 14 | 10 | \$49,000 |
| 2 | 12 | 121,000 | 15 | 9 | 53,000 |
| 3 | 4 | 42,000 | 16 | 3 | 50,000 |
| 4 | 5 | 65,000 | 17 | 6 | 32,000 |
| 5 | 6 | 52,000 | 18 | 5 | 22,000 |
| 6 | 6 | 40,000 | 19 | 5 | 45,000 |
| 7 | 7 | 75,000 | 20 | 4 | 37,000 |
| 8 | 5 | 65,000 | 21 | 6 | 51,000 |
| 9 | 8 | 45,000 | 22 | 8 | 30,000 |
| 10 | 3 | 50,000 | 23 | 7 | 39,000 |
| 11 | 2 | 85,000 | 24 | 3 | 47,000 |
| 12 | 6 | 43,000 | 25 | 8 | 41,000 |
| 13 | 5 | 54,000 | | | |

$$\sum_{i=1}^{25} m_i = 151 \qquad \sum_{i=1}^{25} y_i = \$1{,}329{,}000$$

**SOLUTION**

The best estimate of the population mean $\mu$ is

$$\bar{y} = \frac{\sum_{i=1}^{n} y_i}{\sum_{i=1}^{n} m_i} = \frac{\$1{,}329{,}000}{151} = \$8801$$

To calculate the BOE, we must compute

$$\sum_{i=1}^{n} (y_i - \bar{y}m_i)^2 = \sum_{i=1}^{n} y_i^2 - 2\bar{y}\sum_{i=1}^{n} y_i m_i + \bar{y}^2 \sum_{i=1}^{n} m_i^2$$

Thus we have

$$\sum_{i=1}^{25} y_i^2 = y_1^2 + y_2^2 + \cdots + y_{25}^2$$

$$= (96{,}000)^2 + (121{,}000)^2 + \cdots + (41{,}000)^2 = 82{,}039{,}000{,}000$$

$$\sum_{i=1}^{25} m_i^2 = m_1^2 + m_2^2 + \cdots + m_{25}^2$$

$$= (8)^2 + (12)^2 + \cdots + (8)^2 = 1047$$

$$\sum_{i=1}^{25} y_i m_i = y_1 m_1 + y_2 m_2 + \cdots + y_{25} m_{25}$$

$$= (96{,}000)(8) + (121{,}000)(12) + \cdots + (41{,}000)(8) = 8{,}403{,}000$$

and hence

$$\sum_{i=1}^{25} (y_i - \bar{y}m_i)^2 = 82{,}039{,}000{,}000 - 2(8801)(8{,}403{,}000) + (8801)^2(1047)$$

$$= 15{,}227{,}502{,}247$$

Since $M$ is not known, the $\bar{M}$ appearing in the BOE must be estimated by $\bar{m}$, where

$$\bar{m} = \frac{\sum_{i=1}^{n} m_i}{n} = \frac{151}{25} = 6.04$$

Then

$$\left(\frac{N-n}{Nn\bar{M}^2}\right)\left(\frac{\sum_{i=1}^{n} (y_i - \bar{y}m_i)^2}{n-1}\right) = \left(\frac{415-25}{(415)(25)(6.04)^2}\right)\left(\frac{15{,}227{,}502{,}247}{24}\right)$$

$$= 653{,}785$$

and the bound on the error of estimation is

$$\text{BOE} = 2\sqrt{653{,}785} = 1617$$

The best estimate of the average income for adult males in the city is \$8801, with a bound on the error of estimation of \$1617. This is a rather large BOE and it could be reduced by sampling more clusters and thereby increasing the sample size.

# 22.4 Summary

The first and simplest type of sampling procedure discussed was simple random sampling. We obtain a simple random sample of $n$ elements if each sample of size $n$ from the $N$ elements in the population has the same probability of being selected. For estimating a population mean $\mu$ or total $\tau$, we use the sample mean $\bar{y}$ or sample total $N\bar{y}$, respectively. The procedure for estimating a population proportion can be thought of as a variation on estimating a population mean. By assigning a 1 to each element possessing the characteristic of interest and a 0 to all others, the population and sample proportions $p$ and $\hat{p}$ are also the population and sample mean $\mu$ and $\bar{y}$, respectively.

The second sampling procedure we presented was stratified random sampling. We obtain a stratified random sample by separating the population of elements into groups (strata) such that each element belongs to one and only one stratum. A simple random sample is then selected from each of the strata. Stratified random sampling has three major advantages over simple random sampling. First, the variance of estimation procedures for $\mu$ and $\tau$ are usually reduced because the variance of observations within each stratum is usually smaller than the overall population variance. Second, the cost of collecting observations is often reduced due to separating the population of elements into smaller groups. And finally, separate estimates of parameters in each stratum can also be computed from the same sample data.

The last sampling procedure we presented was cluster sampling. In this procedure the elements of the population are separated into groups called clusters, and the experimenter selects a simple random sample of clusters. An observation is obtained from each element in the sampled clusters. Cluster sampling may provide more information per unit cost than simple random sampling or stratified random sampling either when a list of the elements in the population is not available or when the cost of obtaining observations increases as the distance between elements increases.

The three procedures presented in this chapter are certainly not exhaustive, but they do give you an initial exposure to the concepts and methods of survey sampling. You can find more in-depth coverage in the references at the end of this chapter.

# Exercises

22.13. A study was conducted to determine the mobility of residents in a small community. The community was divided into 20 geographic areas (clusters) and a simple random sample of 4 clusters was selected. Each home in the selected samples was observed to determine the number of years each family lived in that particular home. The results of the survey are listed on the next page.

| Cluster | Cluster Size | Sample Data |
|---|---|---|
| 1 | 10 | 2, 1, 3, 4, 5, 3, 4, 2, 1, 3 |
| 2 | 15 | 1, 3, 5, 7, 6, 8, 9, 10, 12, 10, 11, 14, 2, 3, 15 |
| 3 | 12 | 7, 5, 6, 8, 12, 15, 18, 30, 22, 11, 13, 6 |
| 4 | 11 | 1, 2, 4, 3, 4, 1, 1, 2, 3, 2, 3 |

If there are 200 families in the community, estimate $\mu$, the average number of years families have spent in their present home, and place a bound on the error of estimation.

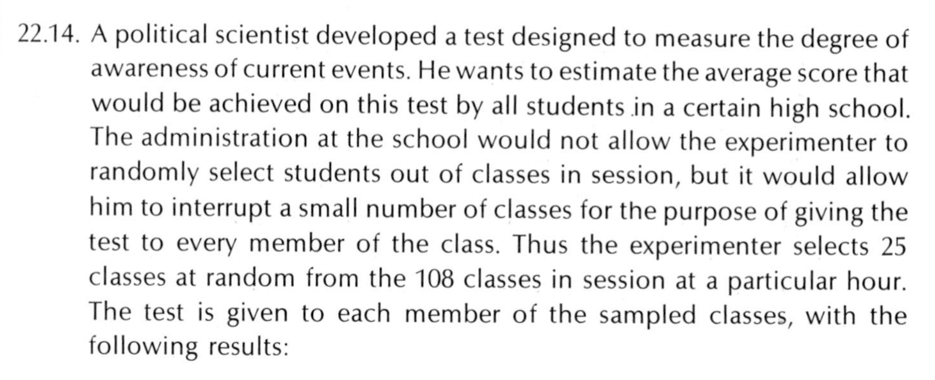

22.14. A political scientist developed a test designed to measure the degree of awareness of current events. He wants to estimate the average score that would be achieved on this test by all students in a certain high school. The administration at the school would not allow the experimenter to randomly select students out of classes in session, but it would allow him to interrupt a small number of classes for the purpose of giving the test to every member of the class. Thus the experimenter selects 25 classes at random from the 108 classes in session at a particular hour. The test is given to each member of the sampled classes, with the following results:

| Class | Number of Students | Total Score | Class | Number of Students | Total Score |
|---|---|---|---|---|---|
| 1 | 31 | 1590 | 14 | 40 | 1980 |
| 2 | 29 | 1510 | 15 | 38 | 1990 |
| 3 | 25 | 1490 | 16 | 28 | 1420 |
| 4 | 35 | 1610 | 17 | 17 | 900 |
| 5 | 15 | 800 | 18 | 22 | 1080 |
| 6 | 31 | 1720 | 19 | 41 | 2010 |
| 7 | 22 | 1310 | 20 | 32 | 1740 |
| 8 | 27 | 1427 | 21 | 35 | 1750 |
| 9 | 25 | 1290 | 22 | 19 | 890 |
| 10 | 19 | 860 | 23 | 29 | 1470 |
| 11 | 30 | 1620 | 24 | 18 | 910 |
| 12 | 18 | 710 | 25 | 31 | 1740 |
| 13 | 21 | 1140 | | | |

Estimate the average score that would be achieved on this test by all students in the school. Place a bound on the error of estimation.

$ 22.15. An economic survey is designed to estimate the average amount spent on utilities for households in a city. Since no list of households is available, cluster sampling is used with divisions (wards) forming the clusters. A simple random sample of 20 wards is selected from the 60 wards of the city. Interviewers then obtain the cost of utilities from each household within the sampled wards; the total costs are tabulated below.

| *Sampled Ward* | *Number of Households* | *Total Amount Spent on Utilities* | *Sampled Ward* | *Number of Households* | *Total Amount Spent on Utilities* |
|---|---|---|---|---|---|
| 1 | 55 | $2210 | 11 | 73 | $2930 |
| 2 | 60 | 2390 | 12 | 64 | 2470 |
| 3 | 63 | 2430 | 13 | 69 | 2830 |
| 4 | 58 | 2380 | 14 | 58 | 2370 |
| 5 | 71 | 2760 | 15 | 63 | 2390 |
| 6 | 78 | 3110 | 16 | 75 | 2870 |
| 7 | 69 | 2780 | 17 | 78 | 3210 |
| 8 | 58 | 2370 | 18 | 51 | 2430 |
| 9 | 52 | 1990 | 19 | 67 | 2730 |
| 10 | 71 | 2810 | 20 | 70 | 2880 |

Estimate the average amount a household in the city spends on utilities, and place a bound on the error of estimation.

22.16. An inspector wants to estimate the average weight of fill for cereal boxes packaged in a certain factory. The cereal is available to him in cartons containing 12 boxes each. The inspector randomly selects 5 cartons and measures the weight of fill for every box in the sampled cartons, with the following results (in ounces): $

| *Carton* | *Ounces to Fill* |
|---|---|
| 1 | 16.1, 15.9, 16.1, 16.2, 15.9, 15.8, 16.1, 16.2, 16.0, 15.9, 15.8, 16.0 |
| 2 | 15.9, 16.2, 15.8, 16.0, 16.3, 16.1, 15.8, 15.9, 16.0, 16.1, 16.1, 15.9 |
| 3 | 16.2, 16.0, 15.7, 16.3, 15.8, 16.0, 15.9, 16.0, 16.1, 16.0, 15.9, 16.1 |
| 4 | 15.9, 16.1, 16.2, 16.1, 16.1, 16.3, 15.9, 16.1, 15.9, 15.9, 16.0, 16.0 |
| 5 | 16.0, 15.8, 16.3, 15.7, 16.1, 15.9, 16.0, 16.1, 15.8, 16.0, 16.1, 15.9 |

Estimate the average weight of fill for boxes packaged by this factory, and place a bound on the error of estimation. Assume that the total number of cartons packaged by the factory is large enough for the finite population correction to be ignored.

22.17. Refer to exercise 22.15. Use the sample data to estimate $\tau$, the total amount of money spent on utilities. Place a bound on the error of estimation. (Hint: The total number of households in the city is 3950).

# References

Cochran, W.G. 1963. *Sampling techniques*. 2d ed. New York: Wiley.
Kish, L. 1965. *Survey sampling*. New York: Wiley.
Mendenhall, W.; Ott, L.; and Scheaffer, R.L. 1971. *Elementary survey sampling*. Belmont, Calif.: Wadsworth.

# Glossary of Statistical Estimation and Test Procedures

| | |
|---|---|
| Chapter 9, p. 243 | Sample sizes required for two-sided tests for $\mu$ or $\mu_1 - \mu_2$ |
| Chapter 9, pp. 249-252 | Test for $\mu_d = \mu_1 - \mu_2$ using paired data |
| Chapter 9, p. 252 | Confidence interval for $\mu_d = \mu_1 - \mu_2$ using paired data |
| Chapter 10, p. 270 | Chi-square goodness-of-fit test |
| Chapter 10, pp. 278-279 | Point estimation of the binomial parameter $p$ |
| Chapter 10, p. 280 | Confidence interval for the binomial parameter $p$ |
| Chapter 10, p. 281 | Test for the binomial parameter $p$ |
| Chapter 10, pp. 290-291 | Test for the Poisson parameter $\mu$ |
| Chapter 10, pp. 294-297 | Chi-square test of independence for an $r \times c$ contingency table |
| Chapter 11, p. 308 | Sign test for comparing two populations using paired data |
| Chapter 11, p. 310 | Sign test for $H_0$: $\mu = \mu_0$ |
| Chapter 11, pp. 313-314 | Wilcoxon's signed-rank test for comparing two populations using paired data: large-sample results |
| Chapter 11, p. 315 | Wilcoxon's signed-rank test for comparing two populations using paired data: small-sample results |
| Chapter 11, pp. 317-318 | Wilcoxon's rank sum test for comparing two populations |
| Chapter 11, pp. 320-321 | Kruskal-Wallis test for comparing more than two populations |
| Chapter 11, pp. 325-327 | Small-sample runs test |
| Chapter 11, p. 329 | Tukey-Duckworth two-sample test |
| Chapter 11, pp. 330-332 | Quadrant sum test for association |
| Chapter 11, p. 334 | An approxiation for the least squares equation $\hat{y} = \hat{\beta}_0 + \hat{\beta}_1 x$ |
| Chapter 12, p. 341 | Confidence interval for $\sigma^2$ |
| Chapter 12, p. 342 | Test for $\sigma^2$ |
| Chapter 12, p. 346 | Confidence interval for $\sigma_1^2/\sigma_2^2$ |
| Chapter 12, p. 348 | Test for comparing two variances |
| Chapter 13, pp. 356-363 | One-way analysis of variance |
| Chapter 14, p. 384 | Fisher's least significant difference for comparing $t$ treatment means |
| Chapter 14, pp. 389-390 | Tukey's $W$ procedure for comparing $t$ treatment means |
| Chapter 14, pp. 393-394 | Duncan's new multiple range test for comparing $t$ treatment means |
| Chapter 14, p. 396 | Scheffé's $S$ method for comparing $t$ treatment means |
| Chapter 14, p. 400 | Waller-Duncan $k$ ratio procedure for comparing t treatment means |
| Chapter 15, pp. 408-415 | Analysis of variance for a randomized block design |
| Chapter 15, pp. 421-426 | Analysis of variance for a Latin square design |
| Chapter 15, pp. 430-442 | Analysis of a $k$-way classification (factorial experiment) |
| Chapter 16, pp. 469-472 | Testing a set of $\beta$'s in the general linear model |
| Chapter 16, pp. 487-489 | Comparing the slopes of two regression lines |
| Chapter 16, pp. 499-513 | Analyzing multidimensional contingency tables |
| Chapter 18, p. 561 | Analysis of variance for a one-way classification: random effects model |
| Chapter 18, p. 566 | Analysis of variance for a two-way classification: random effects model |

# Appendix: Statistical Tables

Table 1 Normal curve areas

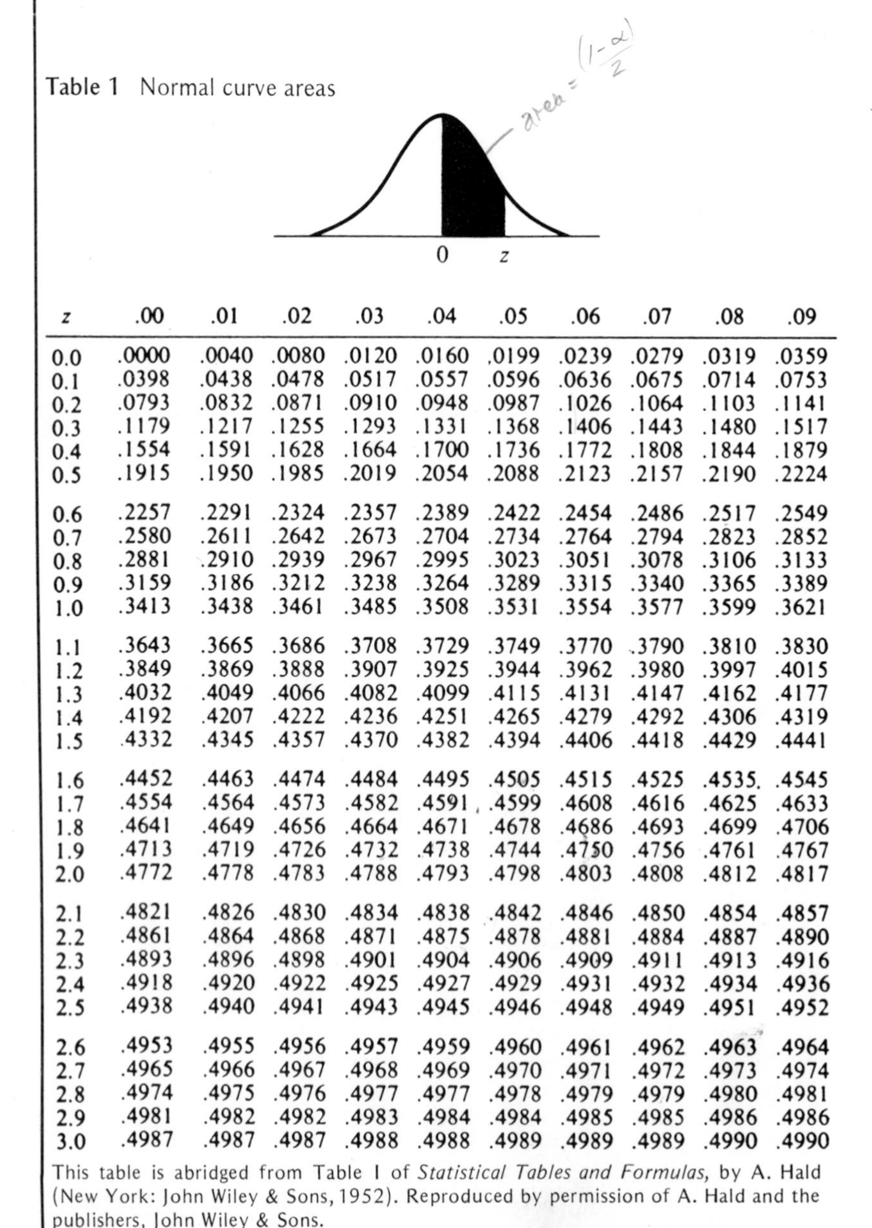


| z | .00 | .01 | .02 | .03 | .04 | .05 | .06 | .07 | .08 | .09 |
|---|---|---|---|---|---|---|---|---|---|---|
| 0.0 | .0000 | .0040 | .0080 | .0120 | .0160 | .0199 | .0239 | .0279 | .0319 | .0359 |
| 0.1 | .0398 | .0438 | .0478 | .0517 | .0557 | .0596 | .0636 | .0675 | .0714 | .0753 |
| 0.2 | .0793 | .0832 | .0871 | .0910 | .0948 | .0987 | .1026 | .1064 | .1103 | .1141 |
| 0.3 | .1179 | .1217 | .1255 | .1293 | .1331 | .1368 | .1406 | .1443 | .1480 | .1517 |
| 0.4 | .1554 | .1591 | .1628 | .1664 | .1700 | .1736 | .1772 | .1808 | .1844 | .1879 |
| 0.5 | .1915 | .1950 | .1985 | .2019 | .2054 | .2088 | .2123 | .2157 | .2190 | .2224 |
| 0.6 | .2257 | .2291 | .2324 | .2357 | .2389 | .2422 | .2454 | .2486 | .2517 | .2549 |
| 0.7 | .2580 | .2611 | .2642 | .2673 | .2704 | .2734 | .2764 | .2794 | .2823 | .2852 |
| 0.8 | .2881 | .2910 | .2939 | .2967 | .2995 | .3023 | .3051 | .3078 | .3106 | .3133 |
| 0.9 | .3159 | .3186 | .3212 | .3238 | .3264 | .3289 | .3315 | .3340 | .3365 | .3389 |
| 1.0 | .3413 | .3438 | .3461 | .3485 | .3508 | .3531 | .3554 | .3577 | .3599 | .3621 |
| 1.1 | .3643 | .3665 | .3686 | .3708 | .3729 | .3749 | .3770 | .3790 | .3810 | .3830 |
| 1.2 | .3849 | .3869 | .3888 | .3907 | .3925 | .3944 | .3962 | .3980 | .3997 | .4015 |
| 1.3 | .4032 | .4049 | .4066 | .4082 | .4099 | .4115 | .4131 | .4147 | .4162 | .4177 |
| 1.4 | .4192 | .4207 | .4222 | .4236 | .4251 | .4265 | .4279 | .4292 | .4306 | .4319 |
| 1.5 | .4332 | .4345 | .4357 | .4370 | .4382 | .4394 | .4406 | .4418 | .4429 | .4441 |
| 1.6 | .4452 | .4463 | .4474 | .4484 | .4495 | .4505 | .4515 | .4525 | .4535 | .4545 |
| 1.7 | .4554 | .4564 | .4573 | .4582 | .4591 | .4599 | .4608 | .4616 | .4625 | .4633 |
| 1.8 | .4641 | .4649 | .4656 | .4664 | .4671 | .4678 | .4686 | .4693 | .4699 | .4706 |
| 1.9 | .4713 | .4719 | .4726 | .4732 | .4738 | .4744 | .4750 | .4756 | .4761 | .4767 |
| 2.0 | .4772 | .4778 | .4783 | .4788 | .4793 | .4798 | .4803 | .4808 | .4812 | .4817 |
| 2.1 | .4821 | .4826 | .4830 | .4834 | .4838 | .4842 | .4846 | .4850 | .4854 | .4857 |
| 2.2 | .4861 | .4864 | .4868 | .4871 | .4875 | .4878 | .4881 | .4884 | .4887 | .4890 |
| 2.3 | .4893 | .4896 | .4898 | .4901 | .4904 | .4906 | .4909 | .4911 | .4913 | .4916 |
| 2.4 | .4918 | .4920 | .4922 | .4925 | .4927 | .4929 | .4931 | .4932 | .4934 | .4936 |
| 2.5 | .4938 | .4940 | .4941 | .4943 | .4945 | .4946 | .4948 | .4949 | .4951 | .4952 |
| 2.6 | .4953 | .4955 | .4956 | .4957 | .4959 | .4960 | .4961 | .4962 | .4963 | .4964 |
| 2.7 | .4965 | .4966 | .4967 | .4968 | .4969 | .4970 | .4971 | .4972 | .4973 | .4974 |
| 2.8 | .4974 | .4975 | .4976 | .4977 | .4977 | .4978 | .4979 | .4979 | .4980 | .4981 |
| 2.9 | .4981 | .4982 | .4982 | .4983 | .4984 | .4984 | .4985 | .4985 | .4986 | .4986 |
| 3.0 | .4987 | .4987 | .4987 | .4988 | .4988 | .4989 | .4989 | .4989 | .4990 | .4990 |

This table is abridged from Table I of *Statistical Tables and Formulas*, by A. Hald (New York: John Wiley & Sons, 1952). Reproduced by permission of A. Hald and the publishers, John Wiley & Sons.

**Table 2** Percentage points of the $t$ distribution

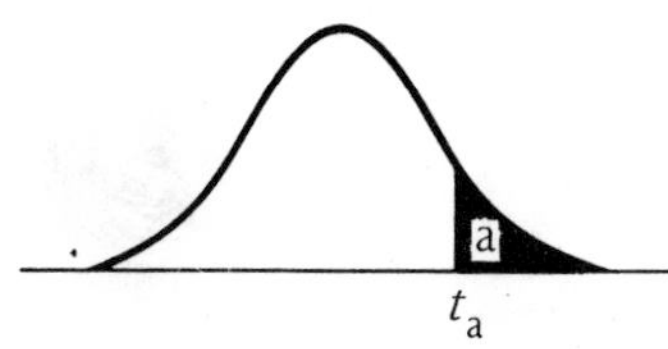


| df | a = .10 | a = .05 | a = .025 | a = .010 | a = .005 |
|---|---|---|---|---|---|
| 1 | 3.078 | 6.314 | 12.706 | 31.821 | 63.657 |
| 2 | 1.886 | 2.920 | 4.303 | 6.965 | 9.925 |
| 3 | 1.638 | 2.353 | 3.182 | 4.541 | 5.841 |
| 4 | 1.533 | 2.132 | 2.776 | 3.747 | 4.604 |
| 5 | 1.476 | 2.015 | 2.571 | 3.365 | 4.032 |
| 6 | 1.440 | 1.943 | 2.447 | 3.143 | 3.707 |
| 7 | 1.415 | 1.895 | 2.365 | 2.998 | 3.499 |
| 8 | 1.397 | 1.860 | 2.306 | 2.896 | 3.355 |
| 9 | 1.383 | 1.833 | 2.262 | 2.821 | 3.250 |
| 10 | 1.372 | 1.812 | 2.228 | 2.764 | 3.169 |
| 11 | 1.363 | 1.796 | 2.201 | 2.718 | 3.106 |
| 12 | 1.356 | 1.782 | 2.179 | 2.681 | 3.055 |
| 13 | 1.350 | 1.771 | 2.160 | 2.650 | 3.012 |
| 14 | 1.345 | 1.761 | 2.145 | 2.624 | 2.977 |
| 15 | 1.341 | 1.753 | 2.131 | 2.602 | 2.947 |
| 16 | 1.337 | 1.746 | 2.120 | 2.583 | 2.921 |
| 17 | 1.333 | 1.740 | 2.110 | 2.567 | 2.898 |
| 18 | 1.330 | 1.734 | 2.101 | 2.552 | 2.878 |
| 19 | 1.328 | 1.729 | 2.093 | 2.539 | 2.861 |
| 20 | 1.325 | 1.725 | 2.086 | 2.528 | 2.845 |
| 21 | 1.323 | 1.721 | 2.080 | 2.518 | 2.831 |
| 22 | 1.321 | 1.717 | 2.074 | 2.508 | 2.819 |
| 23 | 1.319 | 1.714 | 2.069 | 2.500 | 2.807 |
| 24 | 1.318 | 1.711 | 2.064 | 2.492 | 2.797 |
| 25 | 1.316 | 1.708 | 2.060 | 2.485 | 2.787 |
| 26 | 1.315 | 1.706 | 2.056 | 2.479 | 2.779 |
| 27 | 1.314 | 1.703 | 2.052 | 2.473 | 2.771 |
| 28 | 1.313 | 1.701 | 2.048 | 2.467 | 2.763 |
| 29 | 1.311 | 1.699 | 2.045 | 2.462 | 2.756 |
| inf. | 1.282 | 1.645 | 1.960 | 2.326 | 2.576 |

Table 3 Percentage points of the chi-square distribution

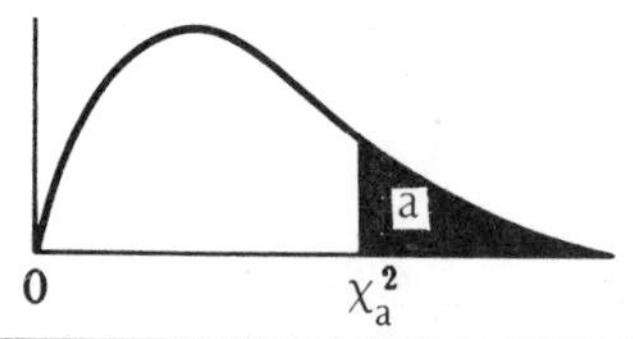


| df | a = .995 | a = .990 | a = .975 | a = .950 | a = .900 |
|---|---|---|---|---|---|
| 1 | 0.0000393 | 0.0001571 | 0.0009821 | 0.0039321 | 0.0157908 |
| 2 | 0.0100251 | 0.0201007 | 0.0506356 | 0.102587 | 0.210720 |
| 3 | 0.0717212 | 0.114832 | 0.215795 | 0.351846 | 0.584375 |
| 4 | 0.206990 | 0.297110 | 0.484419 | 0.710721 | 1.063623 |
| 5 | 0.411740 | 0.554300 | 0.831211 | 1.145476 | 1.61031 |
| 6 | 0.675727 | 0.872085 | 1.237347 | 1.63539 | 2.20413 |
| 7 | 0.989265 | 1.239043 | 1.68987 | 2.16735 | 2.83311 |
| 8 | 1.344419 | 1.646482 | 2.17973 | 2.73264 | 3.48954 |
| 9 | 1.734926 | 2.087912 | 2.70039 | 3.32511 | 4.16816 |
| 10 | 2.15585 | 2.55821 | 3.24697 | 3.94030 | 4.86518 |
| 11 | 2.60321 | 3.05347 | 3.81575 | 4.57481 | 5.57779 |
| 12 | 3.07382 | 3.57056 | 4.40379 | 5.22603 | 6.30380 |
| 13 | 3.56503 | 4.10691 | 5.00874 | 5.89186 | 7.04150 |
| 14 | 4.07468 | 4.66043 | 5.62872 | 6.57063 | 7.78953 |
| 15 | 4.60094 | 5.22935 | 6.26214 | 7.26094 | 8.54675 |
| 16 | 5.14224 | 5.81221 | 6.90766 | 7.96164 | 9.31223 |
| 17 | 5.69724 | 6.40776 | 7.56418 | 8.67176 | 10.0852 |
| 18 | 6.26481 | 7.01491 | 8.23075 | 9.39046 | 10.8649 |
| 19 | 6.84398 | 7.63273 | 8.90655 | 10.1170 | 11.6509 |
| 20 | 7.43386 | 8.26040 | 9.59083 | 10.8508 | 12.4426 |
| 21 | 8.03366 | 8.89720 | 10.28293 | 11.5913 | 13.2396 |
| 22 | 8.64272 | 9.54249 | 10.9823 | 12.3380 | 14.0415 |
| 23 | 9.26042 | 10.19567 | 11.6885 | 13.0905 | 14.8479 |
| 24 | 9.88623 | 10.8564 | 12.4011 | 13.8484 | 15.6587 |
| 25 | 10.5197 | 11.5240 | 13.1197 | 14.6114 | 16.4734 |
| 26 | 11.1603 | 12.1981 | 13.8439 | 15.3791 | 17.2919 |
| 27 | 11.8076 | 12.8786 | 14.5733 | 16.1513 | 18.1138 |
| 28 | 12.4613 | 13.5648 | 15.3079 | 16.9279 | 18.9392 |
| 29 | 13.1211 | 14.2565 | 16.0471 | 17.7083 | 19.7677 |
| 30 | 13.7867 | 14.9535 | 16.7908 | 18.4926 | 20.5992 |
| 40 | 20.7065 | 22.1643 | 24.4331 | 26.5093 | 29.0505 |
| 50 | 27.9907 | 29.7067 | 32.3574 | 34.7642 | 37.6886 |
| 60 | 35.5346 | 37.4848 | 40.4817 | 43.1879 | 46.4589 |
| 70 | 43.2752 | 45.4418 | 48.7576 | 51.7393 | 55.3290 |
| 80 | 51.1720 | 53.5400 | 57.1532 | 60.3915 | 64.2778 |
| 90 | 59.1963 | 61.7541 | 65.6466 | 69.1260 | 73.2912 |
| 100 | 67.3276 | 70.0648 | 74.2219 | 77.9295 | 82.3581 |

From "Tables of the Percentage Points of the $\chi^2$-Distribution." *Biometrika,* Vol. 32 (1941), pp. 188–189, by Catherine M. Thompson. Reproduced by permission of the *Biometrika* Trustees.

Table 3 (*continued*)

| a = .10 | a = .05 | a = .025 | a = .010 | a = .005 | df |
|---|---|---|---|---|---|
| 2.70554 | 3.84146 | 5.02389 | 6.63490 | 7.87944 | 1 |
| 4.60517 | 5.99147 | 7.37776 | 9.21034 | 10.5966 | 2 |
| 6.25139 | 7.81473 | 9.34840 | 11.3449 | 12.8381 | 3 |
| 7.77944 | 9.48773 | 11.1433 | 13.2767 | 14.8602 | 4 |
| 9.23635 | 11.0705 | 12.8325 | 15.0863 | 16.7496 | 5 |
| 10.6446 | 12.5916 | 14.4494 | 16.8119 | 18.5476 | 6 |
| 12.0170 | 14.0671 | 16.0128 | 18.4753 | 20.2777 | 7 |
| 13.3616 | 15.5073 | 17.5346 | 20.0902 | 21.9550 | 8 |
| 14.6837 | 16.9190 | 19.0228 | 21.6660 | 23.5893 | 9 |
| 15.9871 | 18.3070 | 20.4831 | 23.2093 | 25.1882 | 10 |
| 17.2750 | 19.6751 | 21.9200 | 24.7250 | 26.7569 | 11 |
| 18.5494 | 21.0261 | 23.3367 | 26.2170 | 28.2995 | 12 |
| 19.8119 | 22.3621 | 24.7356 | 27.6883 | 29.8194 | 13 |
| 21.0642 | 23.6848 | 26.1190 | 29.1413 | 31.3193 | 14 |
| 22.3072 | 24.9958 | 27.4884 | 30.5779 | 32.8013 | 15 |
| 23.5418 | 26.2962 | 28.8454 | 31.9999 | 34.2672 | 16 |
| 24.7690 | 27.5871 | 30.1910 | 33.4087 | 35.7185 | 17 |
| 25.9894 | 28.8693 | 31.5264 | 34.8053 | 37.1564 | 18 |
| 27.2036 | 30.1435 | 32.8523 | 36.1908 | 38.5822 | 19 |
| 28.4120 | 31.4104 | 34.1696 | 37.5662 | 39.9968 | 20 |
| 29.6151 | 32.6705 | 35.4789 | 38.9321 | 41.4010 | 21 |
| 30.8133 | 33.9244 | 36.7807 | 40.2894 | 42.7956 | 22 |
| 32.0069 | 35.1725 | 38.0757 | 41.6384 | 44.1813 | 23 |
| 33.1963 | 36.4151 | 39.3641 | 42.9798 | 45.5585 | 24 |
| 34.3816 | 37.6525 | 40.6465 | 44.3141 | 46.9278 | 25 |
| 35.5631 | 38.8852 | 41.9232 | 45.6417 | 48.2899 | 26 |
| 36.7412 | 40.1133 | 43.1944 | 46.9630 | 49.6449 | 27 |
| 37.9159 | 41.3372 | 44.4607 | 48.2782 | 50.9933 | 28 |
| 39.0875 | 42.5569 | 45.7222 | 49.5879 | 52.3356 | 29 |
| 40.2560 | 43.7729 | 46.9792 | 50.8922 | 53.6720 | 30 |
| 51.8050 | 55.7585 | 59.3417 | 63.6907 | 66.7659 | 40 |
| 63.1671 | 67.5048 | 71.4202 | 76.1539 | 79.4900 | 50 |
| 74.3970 | 79.0819 | 83.2976 | 88.3794 | 91.9517 | 60 |
| 85.5271 | 90.5312 | 95.0231 | 100.425 | 104.215 | 70 |
| 96.5782 | 101.879 | 106.629 | 112.329 | 116.321 | 80 |
| 107.565 | 113.145 | 118.136 | 124.116 | 128.299 | 90 |
| 118.498 | 124.342 | 129.561 | 135.807 | 140.169 | 100 |

Table 4 Percentage points of the $F$ distribution

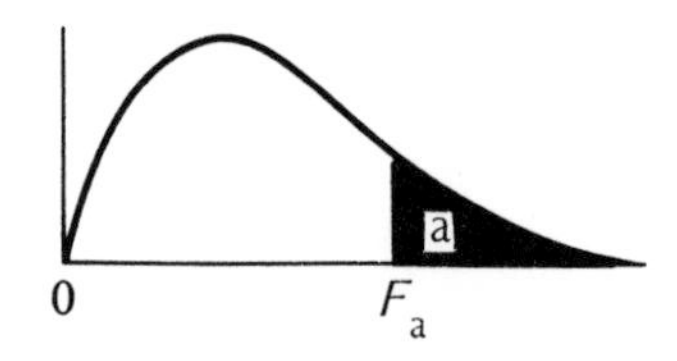


Degrees of freedom (a = .05)

| $df_1$ / $df_2$ | 1 | 2 | 3 | 4 | 5 | 6 | 7 | 8 | 9 |
|---|---|---|---|---|---|---|---|---|---|
| 1 | 161.4 | 199.5 | 215.7 | 224.6 | 230.2 | 234.0 | 236.8 | 238.9 | 240.5 |
| 2 | 18.51 | 19.00 | 19.16 | 19.25 | 19.30 | 19.33 | 19.35 | 19.37 | 19.38 |
| 3 | 10.13 | 9.55 | 9.28 | 9.12 | 9.01 | 8.94 | 8.89 | 8.85 | 8.81 |
| 4 | 7.71 | 6.94 | 6.59 | 6.39 | 6.26 | 6.16 | 6.09 | 6.04 | 6.00 |
| 5 | 6.61 | 5.79 | 5.41 | 5.19 | 5.05 | 4.95 | 4.88 | 4.82 | 4.77 |
| 6 | 5.99 | 5.14 | 4.76 | 4.53 | 4.39 | 4.28 | 4.21 | 4.15 | 4.10 |
| 7 | 5.59 | 4.74 | 4.35 | 4.12 | 3.97 | 3.87 | 3.79 | 3.73 | 3.68 |
| 8 | 5.32 | 4.46 | 4.07 | 3.84 | 3.69 | 3.58 | 3.50 | 3.44 | 3.39 |
| 9 | 5.12 | 4.26 | 3.86 | 3.63 | 3.48 | 3.37 | 3.29 | 3.23 | 3.18 |
| 10 | 4.96 | 4.10 | 3.71 | 3.48 | 3.33 | 3.22 | 3.14 | 3.07 | 3.02 |
| 11 | 4.84 | 3.98 | 3.59 | 3.36 | 3.20 | 3.09 | 3.01 | 2.95 | 2.90 |
| 12 | 4.75 | 3.89 | 3.49 | 3.26 | 3.11 | 3.00 | 2.91 | 2.85 | 2.80 |
| 13 | 4.67 | 3.81 | 3.41 | 3.18 | 3.03 | 2.92 | 2.83 | 2.77 | 2.71 |
| 14 | 4.60 | 3.74 | 3.34 | 3.11 | 2.96 | 2.85 | 2.76 | 2.70 | 2.65 |
| 15 | 4.54 | 3.68 | 3.29 | 3.06 | 2.90 | 2.79 | 2.71 | 2.64 | 2.59 |
| 16 | 4.49 | 3.63 | 3.24 | 3.01 | 2.85 | 2.74 | 2.66 | 2.59 | 2.54 |
| 17 | 4.45 | 3.59 | 3.20 | 2.96 | 2.81 | 2.70 | 2.61 | 2.55 | 2.49 |
| 18 | 4.41 | 3.55 | 3.16 | 2.93 | 2.77 | 2.66 | 2.58 | 2.51 | 2.46 |
| 19 | 4.38 | 3.52 | 3.13 | 2.90 | 2.74 | 2.63 | 2.54 | 2.48 | 2.42 |
| 20 | 4.35 | 3.49 | 3.10 | 2.87 | 2.71 | 2.60 | 2.51 | 2.45 | 2.39 |
| 21 | 4.32 | 3.47 | 3.07 | 2.84 | 2.68 | 2.57 | 2.49 | 2.42 | 2.37 |
| 22 | 4.30 | 3.44 | 3.05 | 2.82 | 2.66 | 2.55 | 2.46 | 2.40 | 2.34 |
| 23 | 4.28 | 3.42 | 3.03 | 2.80 | 2.64 | 2.53 | 2.44 | 2.37 | 2.32 |
| 24 | 4.26 | 3.40 | 3.01 | 2.78 | 2.62 | 2.51 | 2.42 | 2.36 | 2.30 |
| 25 | 4.24 | 3.39 | 2.99 | 2.76 | 2.60 | 2.49 | 2.40 | 2.34 | 2.28 |
| 26 | 4.23 | 3.37 | 2.98 | 2.74 | 2.59 | 2.47 | 2.39 | 2.32 | 2.27 |
| 27 | 4.21 | 3.35 | 2.96 | 2.73 | 2.57 | 2.46 | 2.37 | 2.31 | 2.25 |
| 28 | 4.20 | 3.34 | 2.95 | 2.71 | 2.56 | 2.45 | 2.36 | 2.29 | 2.24 |
| 29 | 4.18 | 3.33 | 2.93 | 2.70 | 2.55 | 2.43 | 2.35 | 2.28 | 2.22 |
| 30 | 4.17 | 3.32 | 2.92 | 2.69 | 2.53 | 2.42 | 2.33 | 2.27 | 2.21 |
| 40 | 4.08 | 3.23 | 2.84 | 2.61 | 2.45 | 2.34 | 2.25 | 2.18 | 2.12 |
| 60 | 4.00 | 3.15 | 2.76 | 2.53 | 2.37 | 2.25 | 2.17 | 2.10 | 2.04 |
| 120 | 3.92 | 3.07 | 2.68 | 2.45 | 2.29 | 2.17 | 2.09 | 2.02 | 1.96 |
| ∞ | 3.84 | 3.00 | 2.60 | 2.37 | 2.21 | 2.10 | 2.01 | 1.94 | 1.88 |

From "Tables of Percentage Points of the Inverted Beta ($F$)-Distribution," *Biometrika*, Vol. 33 (1943), pp. 73–88, by Maxine Merrington and Catherine M. Thompson. Reproduced by permission of the *Biometrika* Trustees.

Table 4 (*continued*)

| 10 | 12 | 15 | 20 | 24 | 30 | 40 | 60 | 120 | ∞ | $df_1$ / $df_2$ |
|---|---|---|---|---|---|---|---|---|---|---|
| 241.9 | 243.9 | 245.9 | 248.0 | 249.1 | 250.1 | 251.1 | 252.2 | 253.3 | 254.3 | 1 |
| 19.40 | 19.41 | 19.43 | 19.45 | 19.45 | 19.46 | 19.47 | 19.48 | 19.49 | 19.50 | 2 |
| 8.79 | 8.74 | 8.70 | 8.66 | 8.64 | 8.62 | 8.59 | 8.57 | 8.55 | 8.53 | 3 |
| 5.96 | 5.91 | 5.86 | 5.80 | 5.77 | 5.75 | 5.72 | 5.69 | 5.66 | 5.63 | 4 |
| 4.74 | 4.68 | 4.62 | 4.56 | 4.53 | 4.50 | 4.46 | 4.43 | 4.40 | 4.36 | 5 |
| 4.06 | 4.00 | 3.94 | 3.87 | 3.84 | 3.81 | 3.77 | 3.74 | 3.70 | 3.67 | 6 |
| 3.64 | 3.57 | 3.51 | 3.44 | 3.41 | 3.38 | 3.34 | 3.30 | 3.27 | 3.23 | 7 |
| 3.35 | 3.28 | 3.22 | 3.15 | 3.12 | 3.08 | 3.04 | 3.01 | 2.97 | 2.93 | 8 |
| 3.14 | 3.07 | 3.01 | 2.94 | 2.90 | 2.86 | 2.83 | 2.79 | 2.75 | 2.71 | 9 |
| 2.98 | 2.91 | 2.85 | 2.77 | 2.74 | 2.70 | 2.66 | 2.62 | 2.58 | 2.54 | 10 |
| 2.85 | 2.79 | 2.72 | 2.65 | 2.61 | 2.57 | 2.53 | 2.49 | 2.45 | 2.40 | 11 |
| 2.75 | 2.69 | 2.62 | 2.54 | 2.51 | 2.47 | 2.43 | 2.38 | 2.34 | 2.30 | 12 |
| 2.67 | 2.60 | 2.53 | 2.46 | 2.42 | 2.38 | 2.34 | 2.30 | 2.25 | 2.21 | 13 |
| 2.60 | 2.53 | 2.46 | 2.39 | 2.35 | 2.31 | 2.27 | 2.22 | 2.18 | 2.13 | 14 |
| 2.54 | 2.48 | 2.40 | 2.33 | 2.29 | 2.25 | 2.20 | 2.16 | 2.11 | 2.07 | 15 |
| 2.49 | 2.42 | 2.35 | 2.28 | 2.24 | 2.19 | 2.15 | 2.11 | 2.06 | 2.01 | 16 |
| 2.45 | 2.38 | 2.31 | 2.23 | 2.19 | 2.15 | 2.10 | 2.06 | 2.01 | 1.96 | 17 |
| 2.41 | 2.34 | 2.27 | 2.19 | 2.15 | 2.11 | 2.06 | 2.02 | 1.97 | 1.92 | 18 |
| 2.38 | 2.31 | 2.23 | 2.16 | 2.11 | 2.07 | 2.03 | 1.98 | 1.93 | 1.88 | 19 |
| 2.35 | 2.28 | 2.20 | 2.12 | 2.08 | 2.04 | 1.99 | 1.95 | 1.90 | 1.84 | 20 |
| 2.32 | 2.25 | 2.18 | 2.10 | 2.05 | 2.01 | 1.96 | 1.92 | 1.87 | 1.81 | 21 |
| 2.30 | 2.23 | 2.15 | 2.07 | 2.03 | 1.98 | 1.94 | 1.89 | 1.84 | 1.78 | 22 |
| 2.27 | 2.20 | 2.13 | 2.05 | 2.01 | 1.96 | 1.91 | 1.86 | 1.81 | 1.76 | 23 |
| 2.25 | 2.18 | 2.11 | 2.03 | 1.98 | 1.94 | 1.89 | 1.84 | 1.79 | 1.73 | 24 |
| 2.24 | 2.16 | 2.09 | 2.01 | 1.96 | 1.92 | 1.87 | 1.82 | 1.77 | 1.71 | 25 |
| 2.22 | 2.15 | 2.07 | 1.99 | 1.95 | 1.90 | 1.85 | 1.80 | 1.75 | 1.69 | 26 |
| 2.20 | 2.13 | 2.06 | 1.97 | 1.93 | 1.88 | 1.84 | 1.79 | 1.73 | 1.67 | 27 |
| 2.19 | 2.12 | 2.04 | 1.96 | 1.91 | 1.87 | 1.82 | 1.77 | 1.71 | 1.65 | 28 |
| 2.18 | 2.10 | 2.03 | 1.94 | 1.90 | 1.85 | 1.81 | 1.75 | 1.70 | 1.64 | 29 |
| 2.16 | 2.09 | 2.01 | 1.93 | 1.89 | 1.84 | 1.79 | 1.74 | 1.68 | 1.62 | 30 |
| 2.08 | 2.00 | 1.92 | 1.84 | 1.79 | 1.74 | 1.69 | 1.64 | 1.58 | 1.51 | 40 |
| 1.99 | 1.92 | 1.84 | 1.75 | 1.70 | 1.65 | 1.59 | 1.53 | 1.47 | 1.39 | 60 |
| 1.91 | 1.83 | 1.75 | 1.66 | 1.61 | 1.55 | 1.50 | 1.43 | 1.35 | 1.25 | 120 |
| 1.83 | 1.75 | 1.67 | 1.57 | 1.52 | 1.46 | 1.39 | 1.32 | 1.22 | 1.00 | ∞ |

**Table 5** Percentage points of the $F$ distribution

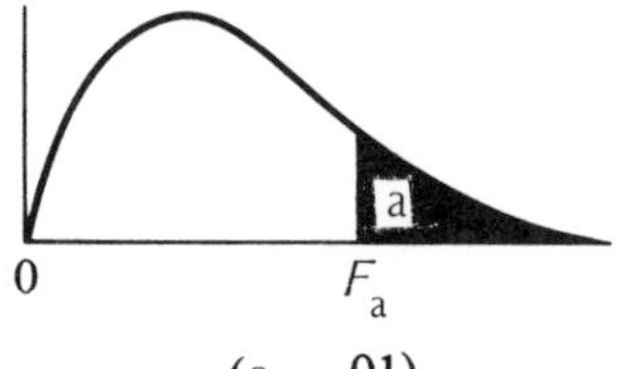


Degrees of freedom (a = .01)

| $df_2$ \ $df_1$ | 1 | 2 | 3 | 4 | 5 | 6 | 7 | 8 | 9 |
|---|---|---|---|---|---|---|---|---|---|
| 1 | 4052 | 4999.5 | 5403 | 5625 | 5764 | 5859 | 5928 | 5982 | 6022 |
| 2 | 98.50 | 99.00 | 99.17 | 99.25 | 99.30 | 99.33 | 99.36 | 99.37 | 99.39 |
| 3 | 34.12 | 30.82 | 29.46 | 28.71 | 28.24 | 27.91 | 27.67 | 27.49 | 27.35 |
| 4 | 21.20 | 18.00 | 16.69 | 15.98 | 15.52 | 15.21 | 14.98 | 14.80 | 14.66 |
| 5 | 16.26 | 13.27 | 12.06 | 11.39 | 10.97 | 10.67 | 10.46 | 10.29 | 10.16 |
| 6 | 13.75 | 10.92 | 9.78 | 9.15 | 8.75 | 8.47 | 8.26 | 8.10 | 7.98 |
| 7 | 12.25 | 9.55 | 8.45 | 7.85 | 7.46 | 7.19 | 6.99 | 6.84 | 6.72 |
| 8 | 11.26 | 8.65 | 7.59 | 7.01 | 6.63 | 6.37 | 6.18 | 6.03 | 5.91 |
| 9 | 10.56 | 8.02 | 6.99 | 6.42 | 6.06 | 5.80 | 5.61 | 5.47 | 5.35 |
| 10 | 10.04 | 7.56 | 6.55 | 5.99 | 5.64 | 5.39 | 5.20 | 5.06 | 4.94 |
| 11 | 9.65 | 7.21 | 6.22 | 5.67 | 5.32 | 5.07 | 4.89 | 4.74 | 4.63 |
| 12 | 9.33 | 6.93 | 5.95 | 5.41 | 5.06 | 4.82 | 4.64 | 4.50 | 4.39 |
| 13 | 9.07 | 6.70 | 5.74 | 5.21 | 4.86 | 4.62 | 4.44 | 4.30 | 4.19 |
| 14 | 8.86 | 6.51 | 5.56 | 5.04 | 4.69 | 4.46 | 4.28 | 4.14 | 4.03 |
| 15 | 8.68 | 6.36 | 5.42 | 4.89 | 4.56 | 4.32 | 4.14 | 4.00 | 3.89 |
| 16 | 8.53 | 6.23 | 5.29 | 4.77 | 4.44 | 4.20 | 4.03 | 3.89 | 3.78 |
| 17 | 8.40 | 6.11 | 5.18 | 4.67 | 4.34 | 4.10 | 3.93 | 3.79 | 3.68 |
| 18 | 8.29 | 6.01 | 5.09 | 4.58 | 4.25 | 4.01 | 3.84 | 3.71 | 3.60 |
| 19 | 8.18 | 5.93 | 5.01 | 4.50 | 4.17 | 3.94 | 3.77 | 3.63 | 3.52 |
| 20 | 8.10 | 5.85 | 4.94 | 4.43 | 4.10 | 3.87 | 3.70 | 3.56 | 3.46 |
| 21 | 8.02 | 5.78 | 4.87 | 4.37 | 4.04 | 3.81 | 3.64 | 3.51 | 3.40 |
| 22 | 7.95 | 5.72 | 4.82 | 4.31 | 3.99 | 3.76 | 3.59 | 3.45 | 3.35 |
| 23 | 7.88 | 5.66 | 4.76 | 4.26 | 3.94 | 3.71 | 3.54 | 3.41 | 3.30 |
| 24 | 7.82 | 5.61 | 4.72 | 4.22 | 3.90 | 3.67 | 3.50 | 3.36 | 3.26 |
| 25 | 7.77 | 5.57 | 4.68 | 4.18 | 3.85 | 3.63 | 3.46 | 3.32 | 3.22 |
| 26 | 7.72 | 5.53 | 4.64 | 4.14 | 3.82 | 3.59 | 3.42 | 3.29 | 3.18 |
| 27 | 7.68 | 5.49 | 4.60 | 4.11 | 3.78 | 3.56 | 3.39 | 3.26 | 3.15 |
| 28 | 7.64 | 5.45 | 4.57 | 4.07 | 3.75 | 3.53 | 3.36 | 3.23 | 3.12 |
| 29 | 7.60 | 5.42 | 4.54 | 4.04 | 3.73 | 3.50 | 3.33 | 3.20 | 3.09 |
| 30 | 7.56 | 5.39 | 4.51 | 4.02 | 3.70 | 3.47 | 3.30 | 3.17 | 3.07 |
| 40 | 7.31 | 5.18 | 4.31 | 3.83 | 3.51 | 3.29 | 3.12 | 2.99 | 2.89 |
| 60 | 7.08 | 4.98 | 4.13 | 3.65 | 3.34 | 3.12 | 2.95 | 2.82 | 2.72 |
| 120 | 6.85 | 4.79 | 3.95 | 3.48 | 3.17 | 2.96 | 2.79 | 2.66 | 2.56 |
| ∞ | 6.63 | 4.61 | 3.78 | 3.32 | 3.02 | 2.80 | 2.64 | 2.51 | 2.41 |

From "Tables of Percentage Points of the Inverted Beta ($F$)-Distribution," *Biometrika,* Vol. 33 (1943), pp. 73–88, by Maxine Merrington and Catherine M. Thompson. Reproduced by permission of the *Biometrika* Trustees.

Table 5 (*continued*)

| 10 | 12 | 15 | 20 | 24 | 30 | 40 | 60 | 120 | ∞ | $df_1$ / $df_2$ |
|---|---|---|---|---|---|---|---|---|---|---|
| 6056 | 6106 | 6157 | 6209 | 6235 | 6261 | 6287 | 6313 | 6339 | 6366 | 1 |
| 99.40 | 99.42 | 99.43 | 99.45 | 99.46 | 99.47 | 99.47 | 99.48 | 99.49 | 99.50 | 2 |
| 27.23 | 27.05 | 26.87 | 26.69 | 26.60 | 26.50 | 26.41 | 26.32 | 26.22 | 26.13 | 3 |
| 14.55 | 14.37 | 14.20 | 14.02 | 13.93 | 13.84 | 13.75 | 13.65 | 13.56 | 13.46 | 4 |
| 10.05 | 9.89 | 9.72 | 9.55 | 9.47 | 9.38 | 9.29 | 9.20 | 9.11 | 9.02 | 5 |
| 7.87 | 7.72 | 7.56 | 7.40 | 7.31 | 7.23 | 7.14 | 7.06 | 6.97 | 6.88 | 6 |
| 6.62 | 6.47 | 6.31 | 6.16 | 6.07 | 5.99 | 5.91 | 5.82 | 5.74 | 5.65 | 7 |
| 5.81 | 5.67 | 5.52 | 5.36 | 5.28 | 5.20 | 5.12 | 5.03 | 4.95 | 4.86 | 8 |
| 5.26 | 5.11 | 4.96 | 4.81 | 4.73 | 4.65 | 4.57 | 4.48 | 4.40 | 4.31 | 9 |
| 4.85 | 4.71 | 4.56 | 4.41 | 4.33 | 4.25 | 4.17 | 4.08 | 4.00 | 3.91 | 10 |
| 4.54 | 4.40 | 4.25 | 4.10 | 4.02 | 3.94 | 3.86 | 3.78 | 3.69 | 3.60 | 11 |
| 4.30 | 4.16 | 4.01 | 3.86 | 3.78 | 3.70 | 3.62 | 3.54 | 3.45 | 3.36 | 12 |
| 4.10 | 3.96 | 3.82 | 3.66 | 3.59 | 3.51 | 3.43 | 3.34 | 3.25 | 3.17 | 13 |
| 3.94 | 3.80 | 3.66 | 3.51 | 3.43 | 3.35 | 3.27 | 3.18 | 3.09 | 3.00 | 14 |
| 3.80 | 3.67 | 3.52 | 3.37 | 3.29 | 3.21 | 3.13 | 3.05 | 2.96 | 2.87 | 15 |
| 3.69 | 3.55 | 3.41 | 3.26 | 3.18 | 3.10 | 3.02 | 2.93 | 2.84 | 2.75 | 16 |
| 3.59 | 3.46 | 3.31 | 3.16 | 3.08 | 3.00 | 2.92 | 2.83 | 2.75 | 2.65 | 17 |
| 3.51 | 3.37 | 3.23 | 3.08 | 3.00 | 2.92 | 2.84 | 2.75 | 2.66 | 2.57 | 18 |
| 3.43 | 3.30 | 3.15 | 3.00 | 2.92 | 2.84 | 2.76 | 2.67 | 2.58 | 2.49 | 19 |
| 3.37 | 3.23 | 3.09 | 2.94 | 2.86 | 2.78 | 2.69 | 2.61 | 2.52 | 2.42 | 20 |
| 3.31 | 3.17 | 3.03 | 2.88 | 2.80 | 2.72 | 2.64 | 2.55 | 2.46 | 2.36 | 21 |
| 3.26 | 3.12 | 2.98 | 2.83 | 2.75 | 2.67 | 2.58 | 2.50 | 2.40 | 2.31 | 22 |
| 3.21 | 3.07 | 2.93 | 2.78 | 2.70 | 2.62 | 2.54 | 2.45 | 2.35 | 2.26 | 23 |
| 3.17 | 3.03 | 2.89 | 2.74 | 2.66 | 2.58 | 2.49 | 2.40 | 2.31 | 2.21 | 24 |
| 3.13 | 2.99 | 2.85 | 2.70 | 2.62 | 2.54 | 2.45 | 2.36 | 2.27 | 2.17 | 25 |
| 3.09 | 2.96 | 2.81 | 2.66 | 2.58 | 2.50 | 2.42 | 2.33 | 2.23 | 2.13 | 26 |
| 3.06 | 2.93 | 2.78 | 2.63 | 2.55 | 2.47 | 2.38 | 2.29 | 2.20 | 2.10 | 27 |
| 3.03 | 2.90 | 2.75 | 2.60 | 2.52 | 2.44 | 2.35 | 2.26 | 2.17 | 2.06 | 28 |
| 3.00 | 2.87 | 2.73 | 2.57 | 2.49 | 2.41 | 2.33 | 2.23 | 2.14 | 2.03 | 29 |
| 2.98 | 2.84 | 2.70 | 2.55 | 2.47 | 2.39 | 2.30 | 2.21 | 2.11 | 2.01 | 30 |
| 2.80 | 2.66 | 2.52 | 2.37 | 2.29 | 2.20 | 2.11 | 2.02 | 1.92 | 1.80 | 40 |
| 2.63 | 2.50 | 2.35 | 2.20 | 2.12 | 2.03 | 1.94 | 1.84 | 1.73 | 1.60 | 60 |
| 2.47 | 2.34 | 2.19 | 2.03 | 1.95 | 1.86 | 1.76 | 1.66 | 1.53 | 1.38 | 120 |
| 2.32 | 2.18 | 2.04 | 1.88 | 1.79 | 1.70 | 1.59 | 1.47 | 1.32 | 1.00 | ∞ |

**Table 6** Squares and square roots

| $n$ | $n^2$ | $\sqrt{n}$ | $\sqrt{10n}$ | $n$ | $n^2$ | $\sqrt{n}$ | $\sqrt{10n}$ |
|---|---|---|---|---|---|---|---|
| | | | | 35 | 1 225 | 5.916 08 | 18.70829 |
| 1 | 1 | 1.000 00 | 3.162 28 | 36 | 1 296 | 6.000 00 | 18.97367 |
| 2 | 4 | 1.414 21 | 4.472 14 | 37 | 1 369 | 6.082 76 | 19.23538 |
| 3 | 9 | 1.732 05 | 5.477 23 | 38 | 1 444 | 6.164 41 | 19.49359 |
| 4 | 16 | 2.000 00 | 6.324 56 | 39 | 1 521 | 6.245 00 | 19.74842 |
| 5 | 25 | 2.236 07 | 7.071 07 | 40 | 1 600 | 6.324 56 | 20.00000 |
| 6 | 36 | 2.449 49 | 7.745 97 | 41 | 1 681 | 6.403 12 | 20.24846 |
| 7 | 49 | 2.645 75 | 8.366 60 | 42 | 1 764 | 6.480 74 | 20.49390 |
| 8 | 64 | 2.828 43 | 8.944 27 | 43 | 1 849 | 6.557 44 | 20.73644 |
| 9 | 81 | 3.000 00 | 9.486 83 | 44 | 1 936 | 6.633 25 | 20.97618 |
| 10 | 100 | 3.162 28 | 10.00000 | 45 | 2 025 | 6.708 20 | 21.21320 |
| 11 | 121 | 3.316 63 | 10.48809 | 46 | 2 116 | 6.782 33 | 21.44761 |
| 12 | 144 | 3.464 10 | 10.95445 | 47 | 2 209 | 6.855 66 | 21.67948 |
| 13 | 169 | 3.605 55 | 11.40175 | 48 | 2 304 | 6.928 20 | 21.90890 |
| 14 | 196 | 3.741 66 | 11.83216 | 49 | 2 401 | 7.000 00 | 22.13594 |
| 15 | 225 | 3.872 98 | 12.24745 | 50 | 2 500 | 7.071 07 | 22.36068 |
| 16 | 256 | 4.000 00 | 12.64911 | 51 | 2 601 | 7.141 43 | 22.58318 |
| 17 | 289 | 4.123 11 | 13.03840 | 52 | 2 704 | 7.211 10 | 22.80351 |
| 18 | 324 | 4.242 64 | 13.41641 | 53 | 2 809 | 7.280 11 | 23.02173 |
| 19 | 361 | 4.358 90 | 13.78405 | 54 | 2 916 | 7.348 47 | 23.23790 |
| 20 | 400 | 4.472 14 | 14.14214 | 55 | 3 025 | 7.416 20 | 23.45208 |
| 21 | 441 | 4.582 58 | 14.49138 | 56 | 3 136 | 7.483 32 | 23.66432 |
| 22 | 484 | 4.690 42 | 14.83240 | 57 | 3 249 | 7.549 83 | 23.87467 |
| 23 | 529 | 4.795 83 | 15.16575 | 58 | 3 364 | 7.615 77 | 24.08319 |
| 24 | 576 | 4.898 98 | 15.49193 | 59 | 3 481 | 7.618 15 | 24.28992 |
| 25 | 625 | 5.000 00 | 15.81139 | 60 | 3 600 | 7.745 97 | 24.49490 |
| 26 | 676 | 5.099 02 | 16.12452 | 61 | 3 721 | 7.810 25 | 24.69818 |
| 27 | 729 | 5.196 15 | 16.43168 | 62 | 3 844 | 7.874 01 | 24.89980 |
| 28 | 784 | 5.291 50 | 16.73320 | 63 | 3 969 | 7.937 25 | 25.09980 |
| 29 | 841 | 5.385 17 | 17.02939 | 64 | 4 096 | 8.000 00 | 25.29822 |
| 30 | 900 | 5.477 23 | 17.32051 | 65 | 4 225 | 8.062 26 | 25.49510 |
| 31 | 961 | 5.567 76 | 17.60682 | 66 | 4 356 | 8.124 04 | 25.69047 |
| 32 | 1 024 | 5.656 85 | 17.88854 | 67 | 4 489 | 8.185 35 | 25.88436 |
| 33 | 1 089 | 5.744 56 | 18.16590 | 68 | 4 624 | 8.246 21 | 26.07681 |
| 34 | 1 156 | 5.830 95 | 18.43909 | 69 | 4 761 | 8.306 62 | 26.26785 |

Computed by J. Huang, Department of Statistics, University of Florida.

Table 6 (*continued*)

| $n$ | $n^2$ | $\sqrt{n}$ | $\sqrt{10n}$ | $n$ | $n^2$ | $\sqrt{n}$ | $\sqrt{10n}$ |
|---|---|---|---|---|---|---|---|
| 70 | 4 900 | 8.366 60 | 26.45751 | 105 | 11 025 | 10.24695 | 32.40370 |
| 71 | 5 041 | 8.426 15 | 26.64583 | 106 | 11 236 | 10.29563 | 32.55764 |
| 72 | 5 184 | 8.485 28 | 26.83282 | 107 | 11 449 | 10.34408 | 32.71085 |
| 73 | 5 329 | 8.544 00 | 27.01851 | 108 | 11 664 | 10.39230 | 32.86335 |
| 74 | 5 476 | 8.602 33 | 27.20294 | 109 | 11 881 | 10.44031 | 33.01515 |
| 75 | 5 625 | 8.660 25 | 27.38613 | 110 | 12 100 | 10.48809 | 33.16625 |
| 76 | 5 776 | 8.717 80 | 27.56810 | 111 | 12 321 | 10.53565 | 33.31666 |
| 77 | 5 929 | 8.774 96 | 27.74887 | 112 | 12 544 | 10.58301 | 33.46640 |
| 78 | 6 084 | 8.831 76 | 27.92848 | 113 | 12 769 | 10.63015 | 33.61547 |
| 79 | 6 241 | 8.888 19 | 28.10694 | 114 | 12 996 | 10.67708 | 33.76389 |
| 80 | 6 400 | 8.944 27 | 28.28427 | 115 | 13 225 | 10.72381 | 33.91165 |
| 81 | 6 561 | 9.000 00 | 28.46050 | 116 | 13 456 | 10.77033 | 34.05877 |
| 82 | 6 724 | 9.055 39 | 28.63564 | 117 | 13 689 | 10.81665 | 34.20526 |
| 83 | 6 889 | 9.110 43 | 28.80972 | 118 | 13 924 | 10.86278 | 34.35113 |
| 84 | 7 056 | 9.165 15 | 28.98275 | 119 | 14 161 | 10.90871 | 34.49638 |
| 85 | 7 225 | 9.219 54 | 29.15476 | 120 | 14 400 | 10.95445 | 34.64102 |
| 86 | 7 396 | 9.273 62 | 29.32576 | 121 | 14 641 | 11.00000 | 34.78505 |
| 87 | 7 569 | 9.327 38 | 29.49576 | 122 | 14 884 | 11.04536 | 34.92850 |
| 88 | 7 744 | 9.380 83 | 29.66479 | 123 | 15 129 | 11.09054 | 35.07136 |
| 89 | 7 921 | 9.433 98 | 29.83287 | 124 | 15 376 | 11.13553 | 35.21363 |
| 90 | 8 100 | 9.486 83 | 30.00000 | 125 | 15 625 | 11.18034 | 35.35534 |
| 91 | 8 281 | 9.539 39 | 30.16621 | 126 | 15 876 | 11.22497 | 35.49648 |
| 92 | 8 464 | 9.591 66 | 30.33150 | 127 | 16 129 | 11.26943 | 35.63706 |
| 93 | 8 649 | 9.643 65 | 30.49590 | 128 | 16 384 | 11.31371 | 35.77709 |
| 94 | 8 836 | 9.695 36 | 30.65942 | 129 | 16 641 | 11.35782 | 35.91657 |
| 95 | 9 025 | 9.746 79 | 30.82207 | 130 | 16 900 | 11.40175 | 36.05551 |
| 96 | 9 216 | 9.797 96 | 30.98387 | 131 | 17 161 | 11.44552 | 36.19392 |
| 97 | 9 409 | 9.848 86 | 31.14482 | 132 | 17 424 | 11.48913 | 36.33180 |
| 98 | 9 604 | 9.899 50 | 31.30495 | 133 | 17 689 | 11.53256 | 36.46917 |
| 99 | 9 801 | 9.949 87 | 31.46427 | 134 | 17 956 | 11.57584 | 36.60601 |
| 100 | 10 000 | 10.00000 | 31.62278 | 135 | 18 225 | 11.61895 | 36.74235 |
| 101 | 10 201 | 10.04998 | 31.78050 | 136 | 18 496 | 11.66190 | 36.87818 |
| 102 | 10 404 | 10.09950 | 31.93744 | 137 | 18 769 | 11.70470 | 37.01351 |
| 103 | 10 609 | 10.14889 | 32.09361 | 138 | 19 044 | 11.74734 | 37.14835 |
| 104 | 10 816 | 10.19804 | 32.24903 | 139 | 19 321 | 11.78983 | 37.28270 |

Table 6 (*continued*)

| $n$ | $n^2$ | $\sqrt{n}$ | $\sqrt{10n}$ | $n$ | $n^2$ | $\sqrt{n}$ | $\sqrt{10n}$ |
|---|---|---|---|---|---|---|---|
| 140 | 19 600 | 11.83216 | 37.41657 | 175 | 30 625 | 13.22876 | 41.83300 |
| 141 | 19 881 | 11.87434 | 37.54997 | 176 | 30 976 | 13.26650 | 41.95235 |
| 142 | 20 164 | 11.91638 | 37.68289 | 177 | 31 329 | 13.30413 | 42.07137 |
| 143 | 20 449 | 11.95826 | 37.81534 | 178 | 31 684 | 13.34166 | 42.19005 |
| 144 | 20 736 | 12.00000 | 37.94733 | 179 | 32 041 | 13.37909 | 42.30829 |
| 145 | 21 025 | 12.04159 | 38.07887 | 180 | 32 400 | 13.41641 | 42.42641 |
| 146 | 21 316 | 12.08305 | 38.20995 | 181 | 32 761 | 13.45362 | 42.54409 |
| 147 | 21 609 | 12.12436 | 38.34058 | 182 | 33 124 | 13.49074 | 42.66146 |
| 148 | 21 904 | 12.16553 | 38.47077 | 183 | 33 489 | 13.52775 | 42.77850 |
| 149 | 22 201 | 12.20656 | 38.60052 | 184 | 33 856 | 13.56466 | 42.89522 |
| 150 | 22 500 | 12.24745 | 38.72983 | 185 | 34 225 | 13.60147 | 43.01163 |
| 151 | 22 801 | 12.28821 | 38.85872 | 186 | 34 596 | 13.63818 | 43.12772 |
| 152 | 23 104 | 12.32883 | 38.98718 | 187 | 34 969 | 13.67479 | 43.24350 |
| 153 | 23 409 | 12.36932 | 39.11521 | 188 | 35 344 | 13.71131 | 43.35897 |
| 154 | 23 716 | 12.40967 | 39.24283 | 189 | 35 721 | 13.74773 | 43.47413 |
| 155 | 24 025 | 12.44990 | 39.37004 | 190 | 36 100 | 13.78405 | 43.58899 |
| 156 | 24 336 | 12.49000 | 39.49684 | 191 | 36 481 | 13.82027 | 43.70355 |
| 157 | 24 649 | 12.52996 | 39.62323 | 192 | 36 864 | 13.85641 | 43.81780 |
| 158 | 24 964 | 12.56981 | 39.74921 | 193 | 37 249 | 13.89244 | 43.93177 |
| 159 | 25 281 | 12.60952 | 39.87480 | 194 | 37 636 | 13.92839 | 44.04543 |
| 160 | 25 600 | 12.64911 | 40.00000 | 195 | 38 025 | 13.96424 | 44.15880 |
| 161 | 25 921 | 12.68858 | 40.12481 | 196 | 38 416 | 14.00000 | 44.27189 |
| 162 | 26 244 | 12.72792 | 40.24922 | 197 | 38 809 | 14.03567 | 44.38468 |
| 163 | 26 569 | 12.76715 | 40.37326 | 198 | 39 204 | 14.07125 | 44.49719 |
| 164 | 26 806 | 12.80625 | 40.49691 | 199 | 39 601 | 14.10674 | 44.60942 |
| 165 | 27 225 | 12.84523 | 40.62019 | 200 | 40 000 | 14.14214 | 44.72136 |
| 166 | 27 556 | 12.88410 | 40.74310 | 201 | 40 401 | 14.17745 | 44.83302 |
| 167 | 27 889 | 12.92285 | 40.86563 | 202 | 40 804 | 14.21267 | 44.94441 |
| 168 | 28 224 | 12.96148 | 40.98780 | 203 | 41 209 | 14.24781 | 45.05552 |
| 169 | 28 561 | 13.00000 | 41.10961 | 204 | 41 616 | 14.28286 | 45.16636 |
| 170 | 28 900 | 13.03840 | 41.23106 | 205 | 42 025 | 14.31782 | 45.27693 |
| 171 | 29 241 | 13.07670 | 41.35215 | 206 | 42 436 | 14.35270 | 45.38722 |
| 172 | 29 584 | 13.11488 | 41.47288 | 207 | 42 849 | 14.38749 | 45.49725 |
| 173 | 29 929 | 13.15295 | 41.59327 | 208 | 43 264 | 14.42221 | 45.60702 |
| 174 | 30 276 | 13.19091 | 41.71331 | 209 | 43 681 | 14.45683 | 45.71652 |

Table 6 (*continued*)

| $n$ | $n^2$ | $\sqrt{n}$ | $\sqrt{10n}$ | $n$ | $n^2$ | $\sqrt{n}$ | $\sqrt{10n}$ |
|---|---|---|---|---|---|---|---|
| 210 | 44 100 | 14.49138 | 45.82576 | 245 | 60 025 | 15.65248 | 49.49747 |
| 211 | 44 521 | 14.52584 | 45.93474 | 246 | 60 516 | 15.68439 | 49.59839 |
| 212 | 44 944 | 14.56022 | 46.04346 | 247 | 61 009 | 15.71623 | 49.69909 |
| 213 | 45 369 | 14.59452 | 46.15192 | 248 | 61 504 | 15.74902 | 49.79960 |
| 214 | 45 796 | 14.62874 | 46.26013 | 249 | 62 001 | 15.77973 | 49.89990 |
| 215 | 46 225 | 14.66288 | 46.36809 | 250 | 62 500 | 15.81139 | 50.00000 |
| 216 | 46 656 | 14.69694 | 46.47580 | 251 | 63 001 | 15.84298 | 50.09990 |
| 217 | 47 089 | 14.73092 | 46.58326 | 252 | 63 504 | 15.87451 | 50.19960 |
| 218 | 47 524 | 14.76482 | 46.69047 | 253 | 64 009 | 15.90597 | 50.29911 |
| 219 | 47 961 | 14.79865 | 46.79744 | 254 | 64 516 | 15.93738 | 50.39841 |
| 220 | 48 400 | 14.83240 | 46.90416 | 255 | 65 025 | 15.96872 | 50.49752 |
| 221 | 48 841 | 14.86607 | 47.01064 | 256 | 65 536 | 16.00000 | 50.59644 |
| 222 | 49 284 | 14.89966 | 47.11688 | 257 | 66 049 | 16.03122 | 50.69517 |
| 223 | 49 729 | 14.93318 | 47.22288 | 258 | 66 564 | 16.06238 | 50.79370 |
| 224 | 50 176 | 14.96663 | 47.32864 | 259 | 67 081 | 16.09348 | 50.89204 |
| 225 | 50 625 | 15.00000 | 47.43416 | 260 | 67 600 | 16.12452 | 50.99020 |
| 226 | 51 076 | 15.03330 | 47.53946 | 261 | 68 121 | 16.15549 | 51.08816 |
| 227 | 51 529 | 15.06652 | 47.64452 | 262 | 68 644 | 16.18641 | 51.18594 |
| 228 | 51 984 | 15.09967 | 47.74935 | 263 | 69 169 | 16.21727 | 51.28353 |
| 229 | 52 441 | 15.13275 | 47.85394 | 264 | 69 696 | 16.24808 | 51.38093 |
| 230 | 52 900 | 15.16575 | 47.95832 | 265 | 70 225 | 16.27882 | 51.47815 |
| 231 | 53 361 | 15.19868 | 48.06246 | 266 | 70 756 | 16.30951 | 51.57519 |
| 232 | 53 824 | 15.23155 | 48.16638 | 267 | 71 289 | 16.34013 | 51.67204 |
| 233 | 54 289 | 15.26434 | 48.27007 | 268 | 71 824 | 16.37071 | 51.76872 |
| 234 | 54 756 | 15.29706 | 48.37355 | 269 | 72 361 | 16.40122 | 51.86521 |
| 235 | 55 225 | 15.32971 | 48.47680 | 270 | 72 900 | 16.43168 | 51.96152 |
| 236 | 55 696 | 15.36229 | 48.57983 | 271 | 73 441 | 16.46208 | 52.05766 |
| 237 | 56 169 | 15.39480 | 48.68265 | 272 | 73 984 | 16.49242 | 52.15362 |
| 238 | 56 644 | 15.42725 | 48.78524 | 273 | 74 529 | 16.52271 | 52.24940 |
| 239 | 57 121 | 15.45962 | 48.88763 | 274 | 75 076 | 16.55295 | 52.34501 |
| 240 | 57 600 | 15.49193 | 48.98979 | 275 | 75 625 | 16.58312 | 52.44044 |
| 241 | 58 081 | 15.52417 | 49.09175 | 276 | 76 176 | 16.61235 | 52.53570 |
| 242 | 58 564 | 15.55635 | 49.19350 | 277 | 76 729 | 16.64332 | 52.63079 |
| 243 | 59 049 | 15.58846 | 49.29503 | 278 | 77 284 | 16.67333 | 52.72571 |
| 244 | 59 536 | 15.62050 | 49.39636 | 279 | 77 841 | 16.70329 | 52.82045 |

Table 6 (*continued*)

| $n$ | $n^2$ | $\sqrt{n}$ | $\sqrt{10n}$ | $n$ | $n^2$ | $\sqrt{n}$ | $\sqrt{10n}$ |
|---|---|---|---|---|---|---|---|
| 280 | 78 400 | 16.73320 | 52.91503 | 315 | 99 225 | 17.74824 | 56.12486 |
| 281 | 78 961 | 16.76305 | 53.00943 | 316 | 99 856 | 17.77639 | 56.21388 |
| 282 | 79 524 | 16.79286 | 53.10367 | 317 | 100 489 | 17.80449 | 56.30275 |
| 283 | 80 089 | 16.82260 | 53.19774 | 318 | 101 124 | 17.83255 | 56.39149 |
| 284 | 80 656 | 16.85230 | 53.29165 | 319 | 101 761 | 17.86057 | 56.48008 |
| 285 | 81 225 | 16.88194 | 53.38539 | 320 | 102 400 | 17.88854 | 56.56854 |
| 286 | 81 796 | 16.91153 | 53.47897 | 321 | 103 041 | 17.91647 | 56.65686 |
| 287 | 82 369 | 16.94107 | 53.57238 | 322 | 103 684 | 17.94436 | 56.74504 |
| 288 | 82 944 | 16.97056 | 53.66563 | 323 | 104 329 | 17.97220 | 56.83309 |
| 289 | 83 521 | 17.00000 | 53.75872 | 324 | 104 976 | 18.00000 | 56.92100 |
| 290 | 84 100 | 17.02939 | 53.85165 | 325 | 105 625 | 18.02776 | 57.00877 |
| 291 | 84 681 | 17.05872 | 53.94442 | 326 | 106 276 | 18.05547 | 57.09641 |
| 292 | 85 264 | 17.08801 | 54.03702 | 327 | 106 929 | 18.08314 | 57.18391 |
| 293 | 85 849 | 17.11724 | 54.12947 | 328 | 107 584 | 18.11077 | 57.27128 |
| 294 | 86 436 | 17.14643 | 54.22177 | 329 | 108 241 | 18.13836 | 57.35852 |
| 295 | 87 025 | 17.17556 | 54.31390 | 330 | 108 900 | 18.16590 | 57.44563 |
| 296 | 87 616 | 17.20465 | 54.40588 | 331 | 109 561 | 18.19341 | 57.53260 |
| 297 | 88 209 | 17.23369 | 54.49771 | 332 | 110 224 | 18.22087 | 57.61944 |
| 298 | 88 804 | 17.26268 | 54.58938 | 333 | 110 889 | 18.24829 | 57.70615 |
| 299 | 89 401 | 17.29162 | 54.68089 | 334 | 111 556 | 18.27567 | 57.79273 |
| 300 | 90 000 | 17.32051 | 54.77226 | 335 | 112 225 | 18.30301 | 57.87918 |
| 301 | 90 601 | 17.34935 | 54.86347 | 336 | 112 896 | 18.33030 | 57.96551 |
| 302 | 91 204 | 17.37815 | 54.95453 | 337 | 113 569 | 18.35756 | 58.05170 |
| 303 | 91 809 | 17.40690 | 55.04544 | 338 | 114 244 | 18.38478 | 58.13777 |
| 304 | 92 416 | 17.43560 | 55.13620 | 339 | 114 921 | 18.41195 | 58.22371 |
| 305 | 93 025 | 17.46425 | 55.22681 | 340 | 115 600 | 18.43909 | 58.30952 |
| 306 | 93 636 | 17.49286 | 55.31727 | 341 | 116 281 | 18.46619 | 58.39521 |
| 307 | 94 249 | 17.52142 | 55.40758 | 342 | 116 964 | 18.49324 | 58.48077 |
| 308 | 94 864 | 17.54993 | 55.49775 | 343 | 117 649 | 18.52026 | 58.56620 |
| 309 | 95 481 | 17.57840 | 55.58777 | 344 | 118 336 | 18.54724 | 58.65151 |
| 310 | 96 100 | 17.60682 | 55.67764 | 345 | 119 025 | 18.57418 | 58.73670 |
| 311 | 96 721 | 17.63519 | 55.76737 | 346 | 119 716 | 18.60108 | 58.82176 |
| 312 | 97 344 | 17.66352 | 55.85696 | 347 | 120 409 | 18.62794 | 58.90671 |
| 313 | 97 969 | 17.69181 | 55.94640 | 348 | 121 104 | 18.65476 | 58.99152 |
| 314 | 98 596 | 17.72005 | 56.03570 | 349 | 121 801 | 18.68154 | 59.07622 |

Table 6 (*continued*)

| $n$ | $n^2$ | $\sqrt{n}$ | $\sqrt{10n}$ | $n$ | $n^2$ | $\sqrt{n}$ | $\sqrt{10n}$ |
|---|---|---|---|---|---|---|---|
| 350 | 122 500 | 18.70829 | 59.16080 | 385 | 148 225 | 19.62142 | 62.04837 |
| 351 | 123 201 | 18.73499 | 59.24525 | 386 | 148 996 | 19.64688 | 62.12890 |
| 352 | 123 904 | 18.76166 | 59.32959 | 387 | 149 769 | 19.67232 | 62.20932 |
| 353 | 124 609 | 18.78829 | 59.41380 | 388 | 150 544 | 19.69772 | 62.28965 |
| 354 | 125 316 | 18.81489 | 59.49790 | 389 | 151 321 | 19.72308 | 62.36986 |
| 355 | 126 025 | 18.84144 | 59.58188 | 390 | 152 100 | 19.74842 | 62.44998 |
| 356 | 126 736 | 18.86796 | 59.66574 | 391 | 152 881 | 19.77372 | 62.52999 |
| 357 | 127 449 | 18.89444 | 59.74948 | 392 | 153 664 | 19.79899 | 62.60990 |
| 358 | 128 164 | 18.92089 | 59.83310 | 393 | 154 449 | 19.82423 | 62.68971 |
| 359 | 128 881 | 18.94730 | 59.91661 | 394 | 155 236 | 19.84943 | 62.76942 |
| 360 | 129 600 | 18.97367 | 60.00000 | 395 | 156 025 | 19.87461 | 62.84903 |
| 361 | 130 321 | 19.00000 | 60.08328 | 396 | 156 816 | 19.89975 | 62.92853 |
| 362 | 131 044 | 19.02630 | 60.16644 | 397 | 157 609 | 19.92486 | 63.00794 |
| 363 | 131 769 | 19.05256 | 60.24948 | 398 | 158 404 | 19.94994 | 63.08724 |
| 364 | 132 496 | 19.07878 | 60.33241 | 399 | 159 201 | 19.97498 | 63.16645 |
| 365 | 133 225 | 19.10497 | 60.41523 | 400 | 160 000 | 20.00000 | 63.24555 |
| 366 | 133 956 | 19.13113 | 60.49793 | 401 | 160 801 | 20.02498 | 63.32456 |
| 367 | 134 689 | 19.15724 | 60.58052 | 402 | 161 604 | 20.04994 | 63.40347 |
| 368 | 135 424 | 19.18333 | 60.66300 | 403 | 162 409 | 20.07486 | 63.48228 |
| 369 | 136 161 | 19.20937 | 60.74537 | 404 | 163 216 | 20.09975 | 63.56099 |
| 370 | 136 900 | 19.23538 | 60.82763 | 405 | 164 025 | 20.12461 | 63.63961 |
| 371 | 137 641 | 19.26136 | 60.90977 | 406 | 164 836 | 20.14944 | 63.71813 |
| 372 | 138 384 | 19.28730 | 60.99180 | 407 | 165 649 | 20.17424 | 63.79655 |
| 373 | 139 129 | 19.31321 | 61.07373 | 408 | 166 464 | 20.19901 | 63.87488 |
| 374 | 139 876 | 19.33908 | 61.15554 | 409 | 167 281 | 20.22375 | 63.95311 |
| 375 | 140 625 | 19.36492 | 61.23724 | 410 | 168 100 | 20.24864 | 64.03124 |
| 376 | 141 376 | 19.39072 | 61.31884 | 411 | 168 921 | 20.27313 | 64.10928 |
| 377 | 142 129 | 19.41649 | 61.40033 | 412 | 169 744 | 20.29778 | 64.18723 |
| 378 | 142 884 | 19.44222 | 61.48170 | 413 | 170 569 | 20.32240 | 64.26508 |
| 379 | 143 641 | 19.46792 | 61.56298 | 414 | 171 396 | 20.34699 | 64.34283 |
| 380 | 144 400 | 19.49359 | 61.64414 | 415 | 172 225 | 20.37155 | 64.42049 |
| 381 | 145 161 | 19.51922 | 61.72520 | 416 | 173 056 | 20.39608 | 64.49806 |
| 382 | 145 924 | 19.54482 | 61.80615 | 417 | 173 889 | 20.42058 | 64.57554 |
| 383 | 146 689 | 19.57039 | 61.88699 | 418 | 174 724 | 20.44505 | 64.65292 |
| 384 | 147 456 | 19.59592 | 61.96773 | 419 | 175 561 | 20.46949 | 64.73021 |

Table 6 (*continued*)

| $n$ | $n^2$ | $\sqrt{n}$ | $\sqrt{10n}$ | $n$ | $n^2$ | $\sqrt{n}$ | $\sqrt{10n}$ |
|---|---|---|---|---|---|---|---|
| 420 | 176 400 | 20.49390 | 64.80741 | 455 | 207 025 | 21.33073 | 67.45369 |
| 421 | 177 241 | 20.51828 | 64.88451 | 456 | 207 936 | 21.35416 | 67.52777 |
| 422 | 178 084 | 20.54264 | 64.96153 | 457 | 208 849 | 21.37756 | 67.60178 |
| 423 | 178 929 | 20.56696 | 65.03845 | 458 | 209 764 | 21.40093 | 67.67570 |
| 424 | 179 776 | 20.59126 | 65.11528 | 459 | 210 681 | 21.42429 | 67.74954 |
| 425 | 180 625 | 20.61553 | 65.19202 | 460 | 211 600 | 21.44761 | 67.82330 |
| 426 | 181 476 | 20.63977 | 65.26868 | 461 | 212 521 | 21.47091 | 67.89698 |
| 427 | 182 329 | 20.66398 | 65.34524 | 462 | 213 444 | 21.49419 | 67.97058 |
| 428 | 183 184 | 20.68816 | 65.42171 | 463 | 214 369 | 21,51743 | 68.04410 |
| 429 | 184 041 | 20.71232 | 65.49809 | 464 | 215 296 | 21.54066 | 68.11755 |
| 430 | 184 900 | 20.73644 | 65.57439 | 465 | 216 225 | 21.56386 | 68.19091 |
| 431 | 185 761 | 20.76054 | 65.65059 | 466 | 217 156 | 21.58703 | 68.26419 |
| 432 | 186 624 | 20.78461 | 65.72671 | 467 | 218 089 | 21.61018 | 68.33740 |
| 433 | 187 489 | 20.80865 | 65.80274 | 468 | 219 024 | 21.63331 | 68.41053 |
| 434 | 188 356 | 20.83267 | 65.87868 | 469 | 219 961 | 21.65641 | 68.48957 |
| 435 | 189 225 | 20.85665 | 65.95453 | 470 | 220 900 | 21.67948 | 68.55655 |
| 436 | 190 096 | 20.88061 | 66.03030 | 471 | 221 841 | 21.70253 | 68.62944 |
| 437 | 190 969 | 20.90454 | 66.10598 | 472 | 222 784 | 21.72556 | 68.70226 |
| 438 | 191 844 | 20.92845 | 66.18157 | 473 | 223 729 | 21.74856 | 68.77500 |
| 439 | 192 721 | 20.95233 | 66.25708 | 474 | 224 676 | 21.77154 | 68.84766 |
| 440 | 193 600 | 20.97618 | 66.33250 | 475 | 225 625 | 21.79449 | 68.92024 |
| 441 | 194 481 | 21.00000 | 66.40783 | 476 | 226 576 | 21.81742 | 68.99275 |
| 442 | 195 364 | 21.02380 | 66.48308 | 477 | 227 529 | 21.84033 | 69.06519 |
| 443 | 196 249 | 21.04757 | 66.55825 | 478 | 228 484 | 21.86321 | 69.13754 |
| 444 | 197 136 | 21.07131 | 66.63332 | 479 | 229 441 | 21.88607 | 69.20983 |
| 445 | 198 025 | 21.09502 | 66.70832 | 480 | 230 400 | 21.90890 | 69.28203 |
| 446 | 198 916 | 21.11871 | 66.78323 | 481 | 231 361 | 21.93171 | 69.35416 |
| 447 | 199 809 | 21.14237 | 66.85806 | 482 | 232 324 | 21.95450 | 69.42622 |
| 448 | 200 704 | 21.16601 | 66.93280 | 483 | 233 289 | 21.97726 | 69.49820 |
| 449 | 201 601 | 21.18962 | 67.00746 | 484 | 234 256 | 22.00000 | 69.57011 |
| 450 | 202 500 | 21.21320 | 67.08204 | 485 | 235 225 | 22.02272 | 69.64194 |
| 451 | 203 401 | 21.23676 | 67.15653 | 486 | 236 196 | 22.04541 | 69.71370 |
| 452 | 204 304 | 21.26029 | 67.23095 | 487 | 237 169 | 22.06808 | 69.78539 |
| 453 | 205 209 | 21.28380 | 67.30527 | 488 | 238 144 | 22.09072 | 69.85700 |
| 454 | 206 116 | 21.30728 | 67.37952 | 489 | 239 121 | 22.11334 | 69.92853 |

Table 6 *(continued)*

| $n$ | $n^2$ | $\sqrt{n}$ | $\sqrt{10n}$ | $n$ | $n^2$ | $\sqrt{n}$ | $\sqrt{10n}$ |
|---|---|---|---|---|---|---|---|
| 490 | 240 100 | 22.13594 | 70.00000 | 525 | 275 625 | 22.91288 | 72.45688 |
| 491 | 241 081 | 22.15852 | 70.07139 | 526 | 276 676 | 22.93469 | 72.52586 |
| 492 | 242 064 | 22.18107 | 70.14271 | 527 | 277 729 | 22.95648 | 72.59477 |
| 493 | 243 049 | 22.20360 | 70.21396 | 528 | 278 784 | 22.97825 | 72.66361 |
| 494 | 244 036 | 22.22611 | 70.28513 | 529 | 279 841 | 23.00000 | 72.73239 |
| 495 | 245 025 | 22.24860 | 70.35624 | 530 | 280 900 | 23.02173 | 72.80110 |
| 496 | 246 016 | 22.27106 | 70.42727 | 531 | 281 961 | 23.04344 | 72.86975 |
| 497 | 247 009 | 22.29350 | 70.49823 | 532 | 283 024 | 23.06513 | 72.93833 |
| 498 | 248 004 | 22.31591 | 70.56912 | 533 | 284 089 | 23.08679 | 73.00685 |
| 499 | 249 001 | 22.33831 | 70.63993 | 534 | 285 156 | 23.10844 | 73.07530 |
| 500 | 250 000 | 22.36068 | 70.71068 | 535 | 286 225 | 23.13007 | 73.14369 |
| 501 | 251 001 | 22.38303 | 70.78135 | 536 | 287 296 | 23.15167 | 73.21202 |
| 502 | 252 004 | 22.40536 | 70.85196 | 537 | 288 369 | 23.17326 | 73.28028 |
| 503 | 253 009 | 22.42766 | 70.92249 | 538 | 289 444 | 23.19483 | 73.34848 |
| 504 | 254 016 | 22.44994 | 70.99296 | 539 | 290 521 | 23.21637 | 73.41662 |
| 505 | 255 025 | 22.47221 | 71.06335 | 540 | 291 600 | 23.23790 | 73.48469 |
| 506 | 256 036 | 22.49444 | 71.13368 | 541 | 292 681 | 23.25941 | 73.55270 |
| 507 | 257 049 | 22.51666 | 71.20393 | 542 | 293 764 | 23.28089 | 73.62065 |
| 508 | 258 064 | 22.53886 | 71.27412 | 543 | 294 849 | 23.30236 | 73.68853 |
| 509 | 259 081 | 22.56103 | 71.34424 | 544 | 295 936 | 23.32381 | 73.75636 |
| 510 | 260 100 | 22.58318 | 71.41428 | 545 | 297 025 | 23.34524 | 73.82412 |
| 511 | 261 121 | 22.60531 | 71.48426 | 546 | 298 116 | 23.36664 | 73.89181 |
| 512 | 262 144 | 22.62742 | 71.55418 | 547 | 299 209 | 23.38803 | 73.95945 |
| 513 | 263 169 | 22.64950 | 71.62402 | 548 | 300 304 | 23.40940 | 74.02702 |
| 514 | 264 196 | 22.67157 | 71.69379 | 549 | 301 401 | 23.43075 | 74.09453 |
| 515 | 265 225 | 22.69361 | 71.76350 | 550 | 302 500 | 23.45208 | 74.16198 |
| 516 | 266 256 | 22.71563 | 71.83314 | 551 | 303 601 | 23.47339 | 74.22937 |
| 517 | 267 289 | 22.73763 | 71.90271 | 552 | 304 704 | 23.49468 | 74.29670 |
| 518 | 268 324 | 22.75961 | 71.97222 | 553 | 305 809 | 23.51595 | 74.36397 |
| 519 | 269 361 | 22.78157 | 72.04165 | 554 | 306 916 | 23.53720 | 74.43118 |
| 520 | 270 400 | 22.80351 | 72.11103 | 555 | 308 025 | 23.55844 | 74.49832 |
| 521 | 271 441 | 22.82542 | 72.18033 | 556 | 309 136 | 23.57965 | 74.56541 |
| 522 | 272 484 | 22.84732 | 72.24957 | 557 | 310 249 | 23.60085 | 74.63243 |
| 523 | 273 529 | 22.86919 | 72.31874 | 558 | 311 364 | 23.62202 | 74.69940 |
| 524 | 274 576 | 22.89105 | 72.38784 | 559 | 312 481 | 23.64318 | 74.76630 |

Table 6 (*continued*)

| $n$ | $n^2$ | $\sqrt{n}$ | $\sqrt{10n}$ | $n$ | $n^2$ | $\sqrt{n}$ | $\sqrt{10n}$ |
|---|---|---|---|---|---|---|---|
| 560 | 313 600 | 23.66432 | 74.83315 | 595 | 354 025 | 24.39262 | 77.13624 |
| 561 | 314 721 | 23.68544 | 74.89993 | 596 | 355 216 | 24.41311 | 77.20104 |
| 562 | 315 844 | 23.70654 | 74.96666 | 597 | 356 409 | 24.43358 | 77.26578 |
| 563 | 316 969 | 23.72762 | 75.03333 | 598 | 357 604 | 24.45404 | 77.33046 |
| 564 | 318 096 | 23.74868 | 75.09993 | 599 | 358 801 | 24.47448 | 77.39509 |
| 565 | 319 225 | 23.76973 | 75.16648 | 600 | 360 000 | 24.49490 | 77.45967 |
| 566 | 320 356 | 23.79075 | 75.23297 | 601 | 361 201 | 24.51530 | 77.52419 |
| 567 | 321 489 | 23.81176 | 75.29940 | 602 | 362 404 | 24.53569 | 77.58866 |
| 568 | 322 624 | 23.83275 | 75.36577 | 603 | 363 609 | 24.55606 | 77.65307 |
| 569 | 323 761 | 23.85372 | 75.43209 | 604 | 364 816 | 24.57641 | 77.71744 |
| 570 | 324 900 | 23.87467 | 75.49834 | 605 | 366 025 | 24.59675 | 77.78175 |
| 571 | 326 041 | 23.89561 | 75.56454 | 606 | 367 236 | 24.61707 | 77.84600 |
| 572 | 327 184 | 23.91652 | 75.63068 | 607 | 368 449 | 24.63737 | 77.91020 |
| 573 | 328 329 | 23.93742 | 75.69676 | 608 | 369 664 | 24.65766 | 77.97435 |
| 574 | 329 476 | 23.95830 | 75.76279 | 609 | 370 881 | 24.67793 | 78.03845 |
| 575 | 330 625 | 23.97916 | 75.82875 | 610 | 372 100 | 24.69818 | 78.10250 |
| 576 | 331 776 | 24.00000 | 75.89466 | 611 | 373 321 | 24.71841 | 78.16649 |
| 577 | 332 929 | 24.02082 | 75.96052 | 612 | 374 544 | 24.73863 | 78.23043 |
| 578 | 334 084 | 24.04163 | 76.02631 | 613 | 375 769 | 24.75884 | 78.29432 |
| 579 | 335 241 | 24.06242 | 76.09205 | 614 | 376 996 | 24.77902 | 78.35815 |
| 580 | 336 400 | 24.08319 | 76.15773 | 615 | 378 225 | 24.79919 | 78.42194 |
| 581 | 337 561 | 24.10394 | 76.22336 | 616 | 379 456 | 24.81935 | 78.48567 |
| 582 | 338 724 | 24.12468 | 76.28892 | 617 | 380 689 | 24.83948 | 78.54935 |
| 583 | 339 889 | 24.14539 | 76.35444 | 618 | 381 924 | 24.85961 | 78.61298 |
| 584 | 341 056 | 24.16609 | 76.41989 | 619 | 383 161 | 24.87971 | 78.67655 |
| 585 | 342 225 | 24.18677 | 76.48529 | 620 | 384 400 | 24.89980 | 78.74008 |
| 586 | 343 396 | 24.20744 | 76.55064 | 621 | 385 641 | 24.91987 | 78.80355 |
| 587 | 344 569 | 24.22808 | 76.61593 | 622 | 386 884 | 24.93993 | 78.86698 |
| 588 | 345 744 | 24.24871 | 76.68116 | 623 | 388 129 | 24.95997 | 78.93035 |
| 589 | 346 921 | 24.26932 | 76.74634 | 624 | 389 376 | 24.97999 | 78.99367 |
| 590 | 348 100 | 24.28992 | 76.81146 | 625 | 390 625 | 25.00000 | 79.05694 |
| 591 | 349 281 | 24.31049 | 76.87652 | 626 | 391 876 | 25.01999 | 79.12016 |
| 592 | 350 464 | 24.33105 | 76.94154 | 627 | 393 129 | 25.03997 | 79.18333 |
| 593 | 351 649 | 24.35159 | 77.00649 | 628 | 394 384 | 25.05993 | 79.24645 |
| 594 | 352 836 | 24.37212 | 77.07140 | 629 | 395 641 | 25.07987 | 79.30952 |

Table 6 (*continued*)

| $n$ | $n^2$ | $\sqrt{n}$ | $\sqrt{10n}$ | $n$ | $n^2$ | $\sqrt{n}$ | $\sqrt{10n}$ |
|---|---|---|---|---|---|---|---|
| 630 | 396 900 | 25.09980 | 79.37254 | 665 | 442 225 | 25.78759 | 81.54753 |
| 631 | 398 161 | 25.11971 | 79.43551 | 666 | 443 556 | 25.80698 | 81.60882 |
| 632 | 399 424 | 25.13961 | 79.49843 | 667 | 444 889 | 25.82634 | 81.67007 |
| 633 | 400 689 | 25.15949 | 79.56130 | 668 | 446 224 | 25.84570 | 81.73127 |
| 634 | 401 956 | 25.17936 | 79.62412 | 669 | 447 561 | 25.86503 | 81.79242 |
| 635 | 403 225 | 25.19921 | 79.68689 | 670 | 448 900 | 25.88436 | 81.85353 |
| 636 | 404 496 | 25.21904 | 79.74961 | 671 | 450 241 | 25.90367 | 81.91459 |
| 637 | 405 769 | 25.23886 | 79.81228 | 672 | 451 584 | 25.92296 | 81.97561 |
| 638 | 407 044 | 25.25866 | 79.87490 | 673 | 452 929 | 25.94224 | 82.03658 |
| 639 | 408 321 | 25.27845 | 79.93748 | 674 | 454 276 | 25.96151 | 82.09750 |
| 640 | 409 600 | 25.29822 | 80.00000 | 675 | 455 625 | 25.98076 | 82.15838 |
| 641 | 410 881 | 25.31798 | 80.06248 | 676 | 456 976 | 26.00000 | 82.21922 |
| 642 | 412 164 | 25.33772 | 80.12490 | 677 | 458 329 | 26.01922 | 82.28001 |
| 643 | 413 449 | 25.35744 | 80.18728 | 678 | 459 684 | 26.03843 | 82.34076 |
| 644 | 414 736 | 25.37716 | 80.24961 | 679 | 461 041 | 26.05763 | 82.40146 |
| 645 | 416 025 | 25.39685 | 80.31189 | 680 | 462 400 | 26.07681 | 82.46211 |
| 646 | 417 316 | 25.41653 | 80.37413 | 681 | 463 761 | 26.09598 | 82.52272 |
| 647 | 418 609 | 25.43619 | 80.43631 | 682 | 465 124 | 26.11513 | 82.58329 |
| 648 | 419 904 | 25.45584 | 80.49845 | 683 | 466 489 | 26.13427 | 82.64381 |
| 649 | 421 201 | 25.47548 | 80.56054 | 684 | 467 856 | 26.15339 | 82.70429 |
| 650 | 422 500 | 25.49510 | 80.62258 | 685 | 469 225 | 26.17250 | 82.76473 |
| 651 | 423 801 | 25.51470 | 80.68457 | 686 | 470 596 | 26.19160 | 82.82512 |
| 652 | 425 104 | 25.53429 | 80.74652 | 687 | 471 969 | 26.21068 | 82.88546 |
| 653 | 426 409 | 25.55386 | 80.80842 | 688 | 473 344 | 26.22975 | 82.94577 |
| 654 | 427 716 | 25.57342 | 80.87027 | 689 | 474 721 | 26.24881 | 83.00602 |
| 655 | 429 025 | 25.59297 | 80.93207 | 690 | 476 100 | 26.26785 | 83.06624 |
| 656 | 430 336 | 25.61250 | 80.99383 | 691 | 477 481 | 26.28688 | 83.12641 |
| 657 | 431 649 | 25.63201 | 81.05554 | 692 | 478 864 | 26.30589 | 83.18654 |
| 658 | 432 964 | 25.65151 | 81.11720 | 693 | 480 249 | 26.32489 | 83.24662 |
| 659 | 434 281 | 25.67100 | 81.17881 | 694 | 481 636 | 26.34388 | 83.30666 |
| 660 | 435 600 | 25.69047 | 81.24038 | 695 | 483 025 | 26.36285 | 83.36666 |
| 661 | 436 921 | 25.70992 | 81.30191 | 696 | 484 416 | 26.38181 | 83.42661 |
| 662 | 438 244 | 25.72936 | 81.36338 | 697 | 485 809 | 26.40076 | 83.48653 |
| 663 | 439 569 | 25.74879 | 81.42481 | 698 | 487 204 | 26.41969 | 83.54639 |
| 664 | 440 896 | 25.76820 | 81.48620 | 699 | 488 601 | 26.43861 | 83.60622 |

**Table 6** (*continued*)

| $n$ | $n^2$ | $\sqrt{n}$ | $\sqrt{10n}$ | $n$ | $n^2$ | $\sqrt{n}$ | $\sqrt{10n}$ |
|---|---|---|---|---|---|---|---|
| 700 | 490 000 | 26.45751 | 83.66600 | 735 | 540 225 | 27.11088 | 85.73214 |
| 701 | 491 401 | 26.47640 | 83.72574 | 736 | 541 696 | 27.12932 | 85.79044 |
| 702 | 492 804 | 26.49528 | 83.78544 | 737 | 543 169 | 27.14774 | 85.84870 |
| 703 | 494 209 | 26.51415 | 83.84510 | 738 | 544 644 | 27.16616 | 85.90693 |
| 704 | 495 616 | 26.53300 | 83.90471 | 739 | 546 121 | 27.18455 | 85.96511 |
| 705 | 497 025 | 26.55184 | 83.96428 | 740 | 547 600 | 27.20294 | 86.02325 |
| 706 | 498 436 | 26.57066 | 84.02381 | 741 | 549 081 | 27.22132 | 86.08136 |
| 707 | 499 849 | 26.58947 | 84.08329 | 742 | 550 564 | 27.23968 | 86.13942 |
| 708 | 501 264 | 26.60827 | 84.14274 | 743 | 552 049 | 27.25803 | 86.19745 |
| 709 | 502 681 | 26.62705 | 84.20214 | 744 | 553 536 | 27.27636 | 86.25543 |
| 710 | 504 100 | 26.64583 | 84.26150 | 745 | 555 025 | 27.29469 | 86.31338 |
| 711 | 505 521 | 26.66458 | 84.32082 | 746 | 556 516 | 27.31300 | 86.37129 |
| 712 | 506 944 | 26.68333 | 84.38009 | 747 | 558 009 | 27.33130 | 86.42916 |
| 713 | 508 369 | 26.70206 | 84.43933 | 748 | 559 504 | 27.34959 | 86.48699 |
| 714 | 509 796 | 26.72078 | 84.49852 | 749 | 561 001 | 27.36786 | 86.54479 |
| 715 | 511 225 | 26.73948 | 84.55767 | 750 | 562 500 | 27.38613 | 86.60254 |
| 716 | 512 656 | 26.75818 | 84.61678 | 751 | 564 001 | 27.40438 | 86.66026 |
| 717 | 514 089 | 26.77686 | 84.67585 | 752 | 565 504 | 27.42262 | 86.71793 |
| 718 | 515 524 | 26.79552 | 84.73488 | 753 | 567 009 | 27.44085 | 86.77557 |
| 719 | 516 961 | 26.81418 | 84.79387 | 754 | 568 516 | 27.45906 | 86.83317 |
| 720 | 518 400 | 26.83282 | 84.85281 | 755 | 570 025 | 27.47726 | 86.89074 |
| 721 | 519 841 | 26.85144 | 84.91172 | 756 | 571 536 | 27.49545 | 86.94826 |
| 722 | 521 284 | 26.87006 | 84.97058 | 757 | 573 049 | 27.51363 | 87.00575 |
| 723 | 522 729 | 26.88866 | 85.02941 | 758 | 574 564 | 27.53180 | 87.06320 |
| 724 | 524 176 | 26.90725 | 85.08819 | 759 | 576 081 | 27.54995 | 87.12061 |
| 725 | 525 625 | 26.92582 | 85.14693 | 760 | 577 600 | 27.56810 | 87.17798 |
| 726 | 527 076 | 26.94439 | 85.20563 | 761 | 579 121 | 27.58623 | 87.23531 |
| 727 | 528 529 | 26.96294 | 85.26429 | 762 | 580 644 | 27.60435 | 87.29261 |
| 728 | 529 984 | 26.98148 | 85.32292 | 763 | 582 169 | 27.62245 | 87.34987 |
| 729 | 531 441 | 27.00000 | 85.38150 | 764 | 583 696 | 27.64055 | 87.40709 |
| 730 | 532 900 | 27.01851 | 85.44004 | 765 | 585 225 | 27.65863 | 87.46428 |
| 731 | 534 361 | 27.03701 | 85.49854 | 766 | 586 756 | 27.67671 | 87.52143 |
| 732 | 535 824 | 27.05550 | 85.55700 | 767 | 588 289 | 27.69476 | 87.57854 |
| 733 | 537 289 | 27.07397 | 85.61542 | 768 | 589 824 | 27.71281 | 87.63561 |
| 734 | 538 756 | 27.09243 | 85.67380 | 769 | 591 361 | 27.73085 | 87.69265 |

Table 6 (*continued*)

| $n$ | $n^2$ | $\sqrt{n}$ | $\sqrt{10n}$ | $n$ | $n^2$ | $\sqrt{n}$ | $\sqrt{10n}$ |
|---|---|---|---|---|---|---|---|
| 770 | 592 900 | 27.74887 | 87.74964 | 805 | 648 025 | 28.37252 | 89.72179 |
| 771 | 594 441 | 27.76689 | 87.80661 | 806 | 649 636 | 28.39014 | 89.77750 |
| 772 | 595 984 | 27.78489 | 87.86353 | 807 | 651 249 | 28.40775 | 89.83318 |
| 773 | 597 529 | 27.80288 | 87.92042 | 808 | 652 864 | 28.42534 | 89.88882 |
| 774 | 599 076 | 27.82086 | 87.97727 | 809 | 654 481 | 28.44293 | 89.94443 |
| 775 | 600 625 | 27.83882 | 88.03408 | 810 | 656 100 | 28.46050 | 90.00000 |
| 776 | 602 176 | 27.85678 | 88.09086 | 811 | 657 721 | 28.47806 | 90.05554 |
| 777 | 603 729 | 27.87472 | 88.14760 | 812 | 659 344 | 28.49561 | 90.11104 |
| 778 | 605 284 | 27.89265 | 88.20431 | 813 | 660 969 | 28.51315 | 90.16651 |
| 779 | 606 841 | 27.91057 | 88.26098 | 814 | 662 596 | 28.53069 | 90.22195 |
| 780 | 608 400 | 27.92848 | 88.31761 | 815 | 664 225 | 28.54820 | 90.27735 |
| 781 | 609 961 | 27.94638 | 88.37420 | 816 | 665 856 | 28.56571 | 90.33272 |
| 782 | 611 524 | 27.96426 | 88.43076 | 817 | 667 489 | 28.58321 | 90.38805 |
| 783 | 613 089 | 27.98214 | 88.48729 | 818 | 669 124 | 28.60070 | 90.44335 |
| 784 | 614 656 | 28.00000 | 88.54377 | 819 | 670 761 | 28.61818 | 90.49862 |
| 785 | 616 225 | 28.01785 | 88.60023 | 820 | 672 400 | 28.63564 | 90.55385 |
| 786 | 617 796 | 28.03569 | 88.65664 | 821 | 674 041 | 28.65310 | 90.60905 |
| 787 | 619 369 | 28.05352 | 88.71302 | 822 | 675 684 | 28.67054 | 90.66422 |
| 788 | 620 944 | 28.07134 | 88.76936 | 823 | 677 329 | 28.68798 | 90.71935 |
| 789 | 622 521 | 28.08914 | 88.82567 | 824 | 678 976 | 28.70540 | 90.77445 |
| 790 | 624 100 | 28.10694 | 88.88194 | 825 | 680 625 | 28.72281 | 90.82951 |
| 791 | 625 681 | 28.12472 | 88.93818 | 826 | 682 726 | 28.74022 | 90.88454 |
| 792 | 627 264 | 28.14249 | 88.99428 | 827 | 683 929 | 28.75761 | 90.93954 |
| 793 | 628 849 | 28.16026 | 89.05055 | 828 | 685 584 | 28.77499 | 90.99451 |
| 794 | 630 436 | 28.17801 | 89.10668 | 829 | 687 241 | 28.79236 | 91.04944 |
| 795 | 632 025 | 28.19574 | 89.16277 | 830 | 688 900 | 28.80972 | 91.10434 |
| 796 | 633 616 | 28.21347 | 89.21883 | 831 | 690 561 | 28.82707 | 91.15920 |
| 797 | 635 209 | 28.23119 | 89.27486 | 832 | 692 224 | 28.84441 | 91.21403 |
| 798 | 636 804 | 28.24889 | 89.33085 | 833 | 693 889 | 28.86174 | 91.26883 |
| 799 | 638 401 | 28.26659 | 89.38680 | 834 | 695 556 | 28.87906 | 91.32360 |
| 800 | 640 000 | 28.28472 | 89.44272 | 835 | 697 225 | 28.89637 | 91.37833 |
| 801 | 641 601 | 28.30194 | 89.49860 | 836 | 698 896 | 28.91366 | 91.43304 |
| 802 | 643 204 | 28.31960 | 89.55445 | 837 | 700 569 | 28.93095 | 91.48770 |
| 803 | 644 809 | 28.33725 | 89.61027 | 838 | 702 244 | 28.94823 | 91.54234 |
| 804 | 646 416 | 28.35489 | 89.66605 | 839 | 703 921 | 28.96550 | 91.59694 |

Table 6 (*continued*)

| $n$ | $n^2$ | $\sqrt{n}$ | $\sqrt{10n}$ | $n$ | $n^2$ | $\sqrt{n}$ | $\sqrt{10n}$ |
|---|---|---|---|---|---|---|---|
| 840 | 705 600 | 28.98275 | 91.65151 | 875 | 765 625 | 29.58040 | 93.54143 |
| 841 | 707 281 | 29.00000 | 91.70605 | 876 | 767 376 | 29.59730 | 93.59487 |
| 842 | 708 964 | 29.01724 | 91.76056 | 877 | 769 129 | 29.61419 | 93.64828 |
| 843 | 710 649 | 29.03446 | 91.81503 | 878 | 770 884 | 29.63106 | 93.70165 |
| 844 | 712 336 | 29.05168 | 91.86947 | 879 | 772 641 | 29.64793 | 93.75500 |
| 845 | 714 025 | 29.06888 | 91.92388 | 880 | 774 400 | 29.66479 | 93.80832 |
| 846 | 715 716 | 29.08608 | 91.97826 | 881 | 776 161 | 29.68164 | 93.86160 |
| 847 | 717 409 | 29.10326 | 92.03260 | 882 | 777 924 | 29.69848 | 93.91486 |
| 848 | 719 104 | 29.12044 | 92.08692 | 883 | 779 689 | 29.71532 | 93.96808 |
| 849 | 720 801 | 29.13760 | 92.14120 | 884 | 781 456 | 29.73214 | 94.02127 |
| 850 | 722 500 | 29.15476 | 92.19544 | 885 | 783 225 | 29.74895 | 94.07444 |
| 851 | 724 201 | 29.17190 | 92.24966 | 886 | 784 996 | 29.76575 | 94.12757 |
| 852 | 725 904 | 29.18904 | 92.30385 | 887 | 786 769 | 29.78255 | 94.18068 |
| 853 | 727 609 | 29.20616 | 92.35800 | 888 | 788 544 | 29.79933 | 94.23375 |
| 854 | 729 316 | 29.22328 | 92.41212 | 889 | 790 321 | 29.81610 | 94.28680 |
| 855 | 731 025 | 29.24038 | 92.46621 | 890 | 792 100 | 29.83287 | 94.33981 |
| 856 | 732 736 | 29.25748 | 92.52027 | 891 | 793 881 | 29.84962 | 94.39280 |
| 857 | 734 449 | 29.27456 | 92.57429 | 892 | 795 664 | 29.86637 | 94.44575 |
| 858 | 736 164 | 29.29164 | 92.62829 | 893 | 797 449 | 29.88311 | 94.49868 |
| 859 | 737 881 | 29.30870 | 92.68225 | 894 | 799 236 | 29.89983 | 94.55157 |
| 860 | 739 600 | 29.32576 | 92.73618 | 895 | 801 025 | 29.91655 | 94.60444 |
| 861 | 741 321 | 29.34280 | 92.79009 | 896 | 802 816 | 29.93326 | 94.65728 |
| 862 | 743 044 | 29.35984 | 92.84396 | 897 | 804 609 | 29.94996 | 94.71008 |
| 863 | 744 769 | 29.37686 | 92.89779 | 898 | 806 404 | 29.96665 | 94.76286 |
| 864 | 746 496 | 29.39388 | 92.95160 | 899 | 808 201 | 29.98333 | 94.81561 |
| 865 | 748 225 | 29.41088 | 93.00538 | 900 | 810 000 | 30.00000 | 94.86833 |
| 866 | 749 956 | 29.42788 | 93.05912 | 901 | 811 801 | 30.01666 | 94.92102 |
| 867 | 751 689 | 29.44486 | 93.11283 | 902 | 813 604 | 30.03331 | 94.97368 |
| 868 | 753 424 | 29.46184 | 93.16652 | 903 | 815 409 | 30.04996 | 95.02631 |
| 869 | 755 161 | 29.47881 | 93.22017 | 904 | 817 216 | 30.06659 | 95.07891 |
| 870 | 756 900 | 29.49576 | 93.27379 | 905 | 819 025 | 30.08322 | 95.13149 |
| 871 | 758 641 | 29.51271 | 93.32738 | 906 | 820 836 | 30.09983 | 95.18403 |
| 872 | 760 384 | 29.52965 | 93.38094 | 907 | 822 649 | 30.11644 | 95.23655 |
| 873 | 762 129 | 29.54657 | 93.43447 | 908 | 824 464 | 30.13304 | 95.28903 |
| 874 | 763 876 | 29.56349 | 93.48797 | 909 | 826 281 | 30.14963 | 95.34149 |

Table 6 (*continued*)

| $n$ | $n^2$ | $\sqrt{n}$ | $\sqrt{10n}$ | $n$ | $n^2$ | $\sqrt{n}$ | $\sqrt{10n}$ |
|---|---|---|---|---|---|---|---|
| 910 | 828 100 | 30.16621 | 95.39392 | 945 | 893 025 | 30.74085 | 97.21111 |
| 911 | 829 921 | 30.18278 | 95.44632 | 946 | 894 916 | 30.75711 | 97.26253 |
| 912 | 831 744 | 30.19934 | 95.49869 | 947 | 896 809 | 30.77337 | 97.31393 |
| 913 | 833 569 | 30.21589 | 95.55103 | 948 | 898 704 | 30.78961 | 97.36529 |
| 914 | 835 396 | 30.23243 | 95.60335 | 949 | 900 601 | 30.80584 | 97.41663 |
| 915 | 837 225 | 30.24897 | 95.65563 | 950 | 902 500 | 30.82207 | 97.46794 |
| 916 | 839 056 | 30.26549 | 95.70789 | 951 | 904 401 | 30.83829 | 97.51923 |
| 917 | 840 889 | 30.28201 | 95.76012 | 952 | 906 304 | 30.85450 | 97.57049 |
| 918 | 842 724 | 30.29851 | 95.81232 | 953 | 908 209 | 30.87070 | 97.62172 |
| 919 | 844 561 | 30.31501 | 95.86449 | 954 | 910 116 | 30.88689 | 97.67292 |
| 920 | 846 400 | 30.33150 | 95.91663 | 955 | 912 025 | 30.90307 | 97.72410 |
| 921 | 848 241 | 30.34798 | 95.96874 | 956 | 913 936 | 30.91925 | 97.77525 |
| 922 | 850 084 | 30.36445 | 96.02083 | 957 | 915 849 | 30.93542 | 97.82638 |
| 923 | 851 929 | 30.38092 | 96.07289 | 958 | 917 764 | 30.95158 | 97.87747 |
| 924 | 853 776 | 30.39737 | 96.12492 | 959 | 919 681 | 30.96773 | 97.92855 |
| 925 | 855 625 | 30.41381 | 96.17692 | 960 | 921 600 | 30.98387 | 97.97959 |
| 926 | 857 476 | 30.43025 | 96.22889 | 961 | 923 521 | 31.00000 | 98.03061 |
| 927 | 859 329 | 30.44667 | 96.28084 | 962 | 925 444 | 31.01612 | 98.08160 |
| 928 | 861 184 | 30.46309 | 96.33276 | 963 | 927 369 | 31.03224 | 98.13256 |
| 929 | 863 041 | 30.47950 | 96.38465 | 964 | 929 296 | 31.04835 | 98.18350 |
| 930 | 864 900 | 30.49590 | 96.43651 | 965 | 931 225 | 31.06445 | 98.23441 |
| 931 | 866 761 | 30.51229 | 96.48834 | 966 | 933 156 | 31.08054 | 98.28530 |
| 932 | 868 624 | 30.52868 | 96.54015 | 967 | 935 089 | 31.09662 | 98.33616 |
| 933 | 870 489 | 30.54505 | 96.59193 | 968 | 937 024 | 31.11270 | 98.38699 |
| 934 | 872 356 | 30.56141 | 96.64368 | 969 | 938 961 | 31.12876 | 98.43780 |
| 935 | 874 225 | 30.57777 | 96.69540 | 970 | 940 900 | 31.14482 | 98.48858 |
| 936 | 876 096 | 30.59412 | 96.74709 | 971 | 942 841 | 31.16087 | 98.53933 |
| 937 | 877 969 | 30.61046 | 96.79876 | 972 | 944 784 | 31.17691 | 98.59006 |
| 938 | 879 844 | 30.62679 | 96.85040 | 973 | 946 729 | 31.19295 | 98.64076 |
| 939 | 881 721 | 30.64311 | 96.90201 | 974 | 948 676 | 31.20897 | 98.69144 |
| 940 | 883 600 | 30.65942 | 96.95360 | 975 | 950 625 | 31.22499 | 98.74209 |
| 941 | 885 481 | 30.67572 | 97.00515 | 976 | 952 576 | 31.24100 | 98.79271 |
| 942 | 887 364 | 30.69202 | 97.05668 | 977 | 954 529 | 31.25700 | 98.84331 |
| 943 | 889 249 | 30.70831 | 97.10819 | 978 | 956 484 | 31.27299 | 98.89388 |
| 944 | 891 136 | 30.72458 | 97.15966 | 979 | 958 441 | 31.28898 | 98.94443 |

Table 6 (*continued*)

| $n$ | $n^2$ | $\sqrt{n}$ | $\sqrt{10n}$ | $n$ | $n^2$ | $\sqrt{n}$ | $\sqrt{10n}$ |
|---|---|---|---|---|---|---|---|
| 980 | 960 400 | 31.30495 | 98.99495 | 990 | 980 100 | 31.46427 | 99.49874 |
| 981 | 962 361 | 31.32092 | 99.04544 | 991 | 982 081 | 31.48015 | 99.54898 |
| 982 | 964 324 | 31.33688 | 99.09591 | 992 | 984 064 | 31.49603 | 99.59920 |
| 983 | 966 289 | 31.35283 | 99.14636 | 993 | 986 049 | 31.51190 | 99.64939 |
| 984 | 968 256 | 31.36877 | 99.19677 | 994 | 988 036 | 31.52777 | 99.69955 |
| 985 | 970 225 | 31.38471 | 99.24717 | 995 | 990 025 | 31.54362 | 99.74969 |
| 986 | 972 196 | 31.40064 | 99.29753 | 996 | 992 016 | 31.55947 | 99.79980 |
| 987 | 974 169 | 31.41656 | 99.34787 | 997 | 994 009 | 31.57531 | 99.84989 |
| 988 | 976 144 | 31.43247 | 99.39819 | 998 | 996 004 | 31.59114 | 99.89995 |
| 989 | 978 121 | 31.44837 | 99.44848 | 999 | 998 001 | 31.60696 | 99.94999 |
| | | | | 1000 | 1000 000 | 31.62278 | 100.00000 |

**Table 7** Individual terms of the Poisson distribution

| y | μ 0·1 | 0·2 | 0·3 | 0·4 | 0·5 | 0·6 | 0·7 | 0·8 | 0·9 | 1·0 | y |
|---|---|---|---|---|---|---|---|---|---|---|---|
| 0 | ·904837 | ·818731 | ·740818 | ·670320 | ·606531 | ·548812 | ·496585 | ·449329 | ·406570 | ·367879 | 0 |
| 1 | ·090484 | ·163746 | ·222245 | ·268128 | ·303265 | ·329287 | ·347610 | ·359463 | ·365913 | ·367879 | 1 |
| 2 | ·004524 | ·016375 | ·033337 | ·053626 | ·075816 | ·098786 | ·121663 | ·143785 | ·164661 | ·183940 | 2 |
| 3 | ·000151 | ·001092 | ·003334 | ·007150 | ·012636 | ·019757 | ·028388 | ·038343 | ·049398 | ·061313 | 3 |
| 4 | ·000004 | ·000055 | ·000250 | ·000715 | ·001580 | ·002964 | ·004968 | ·007669 | ·011115 | ·015328 | 4 |
| 5 | — | ·000002 | ·000015 | ·000057 | ·000158 | ·000356 | ·000696 | ·001227 | ·002001 | ·003066 | 5 |
| 6 | — | — | ·000001 | ·000004 | ·000013 | ·000036 | ·000081 | ·000164 | ·000300 | ·000511 | 6 |
| 7 | — | — | — | — | ·000001 | ·000003 | ·000008 | ·000019 | ·000039 | ·000073 | 7 |
| 8 | — | — | — | — | — | — | ·000001 | ·000002 | ·000004 | ·000009 | 8 |
| 9 | — | — | — | — | — | — | — | — | — | ·000001 | 9 |

| y | 1·1 | 1·2 | 1·3 | 1·4 | 1·5 | 1·6 | 1·7 | 1·8 | 1·9 | 2·0 | y |
|---|---|---|---|---|---|---|---|---|---|---|---|
| 0 | ·332871 | ·301194 | ·272532 | ·246597 | ·223130 | .201897 | ·182684 | ·165299 | ·149569 | ·135335 | 0 |
| 1 | ·366158 | ·361433 | ·354291 | ·345236 | ·334695 | .323034 | ·310562 | ·297538 | ·284180 | ·270671 | 1 |
| 2 | ·201387 | ·216860 | ·230289 | ·241665 | ·251021 | .258428 | ·263978 | ·267784 | ·269971 | ·270671 | 2 |
| 3 | ·073842 | ·086744 | ·099792 | ·112777 | ·125510 | ·137828 | ·149587 | ·160671 | ·170982 | ·180447 | 3 |
| 4 | ·020307 | ·026023 | ·032432 | ·039472 | ·047067 | ·055131 | ·063575 | ·072302 | ·081216 | ·090224 | 4 |
| 5 | ·004467 | ·006246 | ·008432 | ·011052 | ·014120 | ·017642 | ·021615 | ·026029 | ·030862 | ·036089 | 5 |
| 6 | ·000819 | ·001249 | ·001827 | ·002579 | ·003530 | ·004705 | ·006124 | ·007809 | ·009773 | ·012030 | 6 |
| 7 | ·000129 | ·000214 | ·000339 | ·000516 | ·000756 | ·001075 | ·001487 | ·002008 | ·002653 | ·003437 | 7 |
| 8 | ·000018 | ·000032 | ·000055 | ·000090 | ·000142 | ·000215 | ·000316 | ·000452 | ·000630 | ·000859 | 8 |
| 9 | ·000002 | ·000004 | ·000008 | ·000014 | ·000024 | ·000038 | ·000060 | ·000090 | ·000133 | ·000191 | 9 |
| 10 | — | ·000001 | ·000001 | ·000002 | ·000004 | ·000006 | ·000010 | ·000016 | ·000025 | ·000038 | 10 |
| 11 | — | — | — | — | — | ·000001 | ·000002 | ·000003 | ·000004 | ·000007 | 11 |
| 12 | — | — | — | — | — | — | — | — | ·000001 | ·000001 | 12 |

| y | 2·1 | 2·2 | 2·3 | 2·4 | 2·5 | 2·6 | 2·7 | 2·8 | 2·9 | 3·0 | y |
|---|---|---|---|---|---|---|---|---|---|---|---|
| 0 | ·122456 | ·110803 | ·100259 | ·090718 | ·082085 | ·074274 | ·067206 | ·060810 | ·055023 | ·049787 | 0 |
| 1 | ·257159 | ·243767 | ·230595 | ·217723 | ·205212 | ·193111 | ·181455 | ·170268 | ·159567 | ·149361 | 1 |
| 2 | ·270016 | ·268144 | ·265185 | ·261268 | ·256516 | ·251045 | ·244964 | ·238375 | ·231373 | ·224042 | 2 |
| 3 | ·189012 | ·196639 | ·203308 | ·209014 | ·213763 | ·217572 | ·220468 | ·222484 | ·223660 | ·224042 | 3 |
| 4 | ·099231 | ·108151 | ·116902 | ·125409 | ·133602 | ·141422 | ·148816 | ·155739 | ·162154 | ·168031 | 4 |
| 5 | ·041677 | ·047587 | ·053775 | ·060196 | ·066801 | ·073539 | ·080360 | ·087214 | ·094049 | ·100819 | 5 |
| 6 | ·014587 | ·017448 | ·020614 | ·024078 | ·027834 | ·031867 | ·036162 | ·040700 | ·045457 | ·050409 | 6 |
| 7 | ·004376 | ·005484 | ·006773 | ·008255 | ·009941 | ·011836 | ·013948 | ·016280 | ·018832 | ·021604 | 7 |
| 8 | ·001149 | ·001508 | ·001947 | ·002477 | ·003106 | ·003847 | ·004708 | ·005698 | ·006827 | ·008102 | 8 |
| 9 | ·000268 | ·000369 | ·000498 | ·000660 | ·000863 | ·001111 | ·001412 | ·001773 | ·002200 | ·002701 | 9 |
| 10 | ·000056 | ·000081 | ·000114 | ·000158 | ·000216 | ·000289 | ·000381 | ·000496 | ·000638 | ·000810 | 10 |
| 11 | ·000011 | ·000016 | ·000024 | ·000035 | ·000049 | ·000068 | ·000094 | ·000126 | ·000168 | ·000221 | 11 |
| 12 | ·000002 | ·000003 | ·000005 | ·000007 | ·000010 | ·000015 | ·000021 | ·000029 | ·000041 | ·000055 | 12 |
| 13 | — | ·000001 | ·000001 | ·000001 | ·000002 | ·000003 | ·000004 | ·000006 | ·000009 | ·000013 | 13 |
| 14 | — | — | — | — | — | ·000001 | ·000001 | ·000001 | ·000002 | ·000003 | 14 |
| 15 | — | — | — | — | — | — | — | — | — | ·000001 | 15 |

Table 7 *(continued)*

| $y$ | $\mu$ | | | | | | | | | | $y$ |
|---|---|---|---|---|---|---|---|---|---|---|---|
| | 3·1 | 3·2 | 3·3 | 3·4 | 3·5 | 3·6 | 3·7 | 3·8 | 3·9 | 4·0 | |
| 0 | ·045049 | ·040762 | ·036883 | ·033373 | ·030197 | ·027324 | ·024724 | ·022371 | ·020242 | ·018316 | 0 |
| 1 | ·139653 | ·130439 | ·121714 | ·113469 | ·105691 | ·098365 | ·091477 | ·085009 | ·078943 | ·073263 | 1 |
| 2 | ·216461 | ·208702 | ·200829 | ·192898 | ·184959 | ·177058 | ·169233 | ·161517 | ·153940 | ·146525 | 2 |
| 3 | ·223677 | ·222616 | ·220912 | ·218617 | ·215785 | ·212469 | ·208720 | ·204588 | ·200122 | ·195367 | 3 |
| 4 | ·173350 | ·178093 | ·182252 | ·185825 | ·188812 | ·191222 | ·193066 | ·194359 | ·195119 | ·195367 | 4 |
| 5 | ·107477 | ·113979 | ·120286 | ·126361 | ·132169 | ·137680 | ·142869 | ·147713 | ·152193 | ·156293 | 5 |
| 6 | ·055530 | ·060789 | ·066158 | ·071604 | ·077098 | ·082608 | ·088102 | ·093551 | ·098925 | ·104196 | 6 |
| 7 | ·024592 | ·027789 | ·031189 | ·034779 | ·038549 | ·042484 | ·046568 | ·050785 | ·055115 | ·059540 | 7 |
| 8 | ·009529 | ·011116 | ·012865 | ·014781 | ·016865 | ·019118 | ·021538 | ·024123 | ·026869 | ·029770 | 8 |
| 9 | ·003282 | ·003952 | ·004717 | ·005584 | ·006559 | ·007647 | ·008854 | ·010185 | ·011643 | ·013231 | 9 |
| 10 | ·001018 | ·001265 | ·001557 | ·001899 | ·002296 | ·002753 | ·003276 | ·003870 | ·004541 | ·005292 | 10 |
| 11 | ·000287 | ·000368 | ·000467 | ·000587 | ·000730 | ·000901 | ·001102 | ·001337 | ·001610 | ·001925 | 11 |
| 12 | ·000074 | ·000098 | ·000128 | ·000166 | ·000213 | ·000270 | ·000340 | ·000423 | ·000523 | ·000642 | 12 |
| 13 | ·000018 | ·000024 | ·000033 | ·000043 | ·000057 | ·000075 | ·000097 | ·000124 | ·000157 | ·000197 | 13 |
| 14 | ·000004 | ·000006 | ·000008 | ·000011 | ·000014 | ·000019 | ·000026 | ·000034 | ·000044 | ·000056 | 14 |
| 15 | ·000001 | ·000001 | ·000002 | ·000002 | ·000003 | ·000005 | ·000006 | ·000009 | ·000011 | ·000015 | 15 |
| 16 | — | — | — | ·000001 | ·000001 | ·000001 | ·000001 | ·000002 | ·000003 | ·000004 | 16 |
| 17 | — | — | — | — | — | — | — | — | ·000001 | ·000001 | 17 |
| | 4·1 | 4·2 | 4·3 | 4·4 | 4·5 | 4·6 | 4·7 | 4·8 | 4·9 | 5·0 | |
| 0 | ·016573 | ·014996 | ·013569 | ·012277 | ·011109 | ·010052 | ·009095 | ·008230 | ·007447 | ·006738 | 0 |
| 1 | ·067948 | ·062981 | ·058345 | ·054020 | ·049990 | ·046238 | ·042748 | ·039503 | ·036488 | ·033690 | 1 |
| 2 | ·139293 | ·132261 | ·125441 | ·118845 | ·112479 | ·106348 | ·100457 | ·094807 | ·089396 | ·084224 | 2 |
| 3 | ·190368 | ·185165 | ·179799 | ·174305 | ·168718 | ·163068 | ·157383 | ·151691 | ·146014 | ·140374 | 3 |
| 4 | ·195127 | ·194424 | ·193284 | ·191736 | ·189808 | ·187528 | ·184925 | ·182029 | ·178867 | ·175467 | 4 |
| 5 | ·160004 | ·163316 | ·166224 | ·168728 | ·170827 | ·172525 | ·173830 | ·174748 | ·175290 | ·175467 | 5 |
| 6 | ·109336 | ·114321 | ·119127 | ·123734 | ·128120 | ·132270 | ·136167 | ·139798 | ·143153 | ·146223 | 6 |
| 7 | ·064040 | ·068593 | ·073178 | ·077775 | ·082363 | ·086920 | ·091426 | ·095862 | ·100207 | ·104445 | 7 |
| 8 | ·032820 | ·036011 | ·039333 | ·042776 | ·046329 | ·049979 | ·053713 | ·057517 | ·061377 | ·065278 | 8 |
| 9 | ·014951 | ·016805 | ·018793 | ·020913 | ·023165 | ·025545 | ·028050 | ·030676 | ·033416 | ·036266 | 9 |
| 10 | ·006130 | ·007058 | ·008081 | ·009202 | ·010424 | ·011751 | ·013184 | ·014724 | ·016374 | ·018133 | 10 |
| 11 | ·002285 | ·002695 | ·003159 | ·003681 | ·004264 | ·004914 | ·005633 | ·006425 | ·007294 | ·008242 | 11 |
| 12 | ·000781 | ·000943 | ·001132 | ·001350 | ·001599 | ·001884 | ·002206 | ·002570 | ·002978 | ·003434 | 12 |
| 13 | ·000246 | ·000305 | ·000374 | ·000457 | ·000554 | ·000667 | ·000798 | ·000949 | ·001123 | ·001321 | 13 |
| 14 | ·000072 | ·000091 | ·000115 | ·000144 | ·000178 | ·000219 | ·000268 | ·000325 | ·000393 | ·000472 | 14 |
| 15 | ·000020 | ·000026 | ·000033 | ·000042 | ·000053 | ·000067 | ·000084 | ·000104 | ·000128 | ·000157 | 15 |
| 16 | ·000005 | ·000007 | ·000009 | ·000012 | ·000015 | ·000019 | ·000025 | ·000031 | ·000039 | ·000049 | 16 |
| 17 | ·000001 | ·000902 | ·000002 | ·000003 | ·000004 | ·000005 | ·000007 | ·000009 | ·000011 | ·000014 | 17 |
| 18 | — | — | ·000001 | ·000001 | ·000001 | ·000001 | ·000002 | ·000002 | ·000003 | ·000004 | 18 |
| 19 | — | — | — | — | — | — | — | ·000001 | ·000001 | ·000001 | 19 |
| | 5·1 | 5·2 | 5·3 | 5·4 | 5·5 | 5·6 | 5·7 | 5·8 | 5·9 | 6·0 | |
| 0 | ·006097 | ·005517 | ·004992 | ·004517 | ·004087 | ·003698 | ·003346 | ·003028 | ·002739 | ·002479 | 0 |
| 1 | ·031093 | ·028686 | ·026455 | ·024390 | ·022477 | ·020708 | ·019072 | ·017560 | ·016163 | ·014873 | 1 |
| 2 | ·079288 | ·074584 | ·070107 | ·065852 | ·061812 | ·057982 | ·054355 | ·050923 | ·047680 | ·044618 | 2 |
| 3 | ·134790 | ·129279 | ·123856 | ·118533 | ·113323 | ·108234 | ·103275 | ·098452 | ·093771 | ·089235 | 3 |

Table 7 *(continued)*

| $y$ | $\mu$ | | | | | | | | | | $y$ |
|---|---|---|---|---|---|---|---|---|---|---|---|
| | 5·1 | 5·2 | 5·3 | 5·4 | 5·5 | 5·6 | 5·7 | 5·8 | 5·9 | 6·0 | |
| 4 | ·171857 | ·168063 | ·164109 | ·160020 | ·155819 | ·151528 | ·147167 | ·142755 | ·138312 | ·133853 | 4 |
| 5 | ·175294 | ·174785 | ·173955 | ·172821 | ·171401 | ·169711 | ·167770 | ·165596 | ·163208 | ·160623 | 5 |
| 6 | ·149000 | ·151480 | ·153660 | ·155539 | ·157117 | ·158397 | ·159382 | ·160076 | ·160488 | ·160623 | 6 |
| 7 | ·108557 | ·112528 | ·116343 | ·119987 | ·123449 | ·126717 | ·129782 | ·132635 | ·135268 | ·137677 | 7 |
| 8 | ·069205 | ·073143 | ·077077 | ·080991 | ·084871 | ·088702 | ·092470 | ·096160 | ·099760 | ·103258 | 8 |
| 9 | ·039216 | ·042261 | ·045390 | ·048595 | ·051866 | ·055192 | ·058564 | ·061970 | ·065398 | ·068838 | 9 |
| 10 | ·020000 | ·021976 | ·024057 | ·026241 | ·028526 | ·030908 | ·033382 | ·035943 | ·038585 | ·041303 | 10 |
| 11 | ·009273 | ·010388 | ·011591 | ·012882 | ·014263 | ·015735 | ·017298 | ·018952 | ·020696 | ·022529 | 11 |
| 12 | ·003941 | ·004502 | ·005119 | ·005797 | ·006537 | ·007343 | ·008216 | ·009160 | ·010175 | ·011264 | 12 |
| 13 | ·001546 | ·001801 | ·002087 | ·002408 | ·002766 | ·003163 | ·003603 | ·004087 | ·004618 | ·005199 | 13 |
| 14 | ·000563 | ·000669 | ·000790 | ·000929 | ·001087 | ·001265 | ·001467 | ·001693 | ·001946 | ·002228 | 14 |
| 15 | ·000191 | ·000232 | ·000279 | ·000334 | ·000398 | ·000472 | ·000557 | ·000655 | ·000766 | ·000891 | 15 |
| 16 | ·000061 | ·000075 | ·000092 | ·000113 | ·000137 | ·000165 | ·000199 | ·000237 | ·000282 | ·000334 | 16 |
| 17 | ·000018 | ·000023 | ·000029 | ·000036 | ·000044 | ·000054 | ·000067 | ·000081 | ·000098 | ·000118 | 17 |
| 18 | ·000005 | ·000007 | ·000008 | ·000011 | ·000014 | ·000017 | ·000021 | ·000026 | ·000032 | ·000039 | 18 |
| 19 | ·000001 | ·000002 | ·000002 | ·000003 | ·000004 | ·000005 | ·000006 | ·000008 | ·000010 | ·000012 | 19 |
| 20 | — | — | ·000001 | ·000001 | ·000001 | ·000001 | ·000002 | ·000002 | ·000003 | ·000004 | 20 |
| 21 | — | — | — | — | — | — | — | ·000001 | ·000001 | ·000001 | 21 |

| $y$ | 6·1 | 6·2 | 6·3 | 6·4 | 6·5 | 6·6 | 6·7 | 6·8 | 6·9 | 7·0 | $y$ |
|---|---|---|---|---|---|---|---|---|---|---|---|
| 0 | ·002243 | ·002029 | ·001836 | ·001662 | ·001503 | ·001360 | ·001231 | ·001114 | ·001008 | ·000912 | 0 |
| 1 | ·013682 | ·012582 | ·011569 | ·010634 | ·009772 | ·008978 | ·008247 | ·007574 | ·006954 | ·006383 | 1 |
| 2 | ·041729 | ·039006 | ·036441 | ·034029 | ·031760 | ·029629 | ·027628 | ·025751 | ·023990 | ·022341 | 2 |
| 3 | ·084848 | ·080612 | ·076527 | ·072595 | ·068814 | ·065183 | ·061702 | ·058368 | ·055178 | ·052129 | 3 |
| 4 | ·129393 | ·124948 | ·120530 | ·116151 | ·111822 | ·107553 | ·103351 | ·099225 | ·095182 | ·091226 | 4 |
| 5 | ·157860 | ·154936 | ·151868 | ·148674 | ·145369 | ·141969 | ·138490 | ·134946 | ·131351 | ·127717 | 5 |
| 6 | ·160491 | ·160100 | ·159461 | ·158585 | ·157483 | ·156166 | ·154648 | ·152939 | ·151053 | ·149003 | 6 |
| 7 | ·139856 | ·141803 | ·143515 | ·144992 | ·146234 | ·147243 | ·148020 | ·148569 | ·148895 | ·149003 | 7 |
| 8 | ·106640 | ·109897 | ·113018 | ·115994 | ·118815 | ·121475 | ·123967 | ·126284 | ·128422 | ·130377 | 8 |
| 9 | ·072278 | ·075707 | ·079113 | ·082484 | ·085811 | ·089082 | ·092286 | ·095415 | ·098457 | ·101405 | 9 |
| 10 | ·044090 | ·046938 | ·049841 | ·052790 | ·055777 | ·058794 | ·061832 | ·064882 | ·067935 | ·070983 | 10 |
| 11 | ·024450 | ·026456 | ·028545 | ·030714 | ·032959 | ·035276 | ·037661 | ·040109 | ·042614 | ·045171 | 11 |
| 12 | ·012429 | ·013669 | ·014986 | ·016381 | ·017853 | ·019402 | ·021028 | ·022728 | ·024503 | ·026350 | 12 |
| 13 | ·005832 | ·006519 | ·007263 | ·008064 | ·008926 | ·009850 | ·010837 | ·011889 | ·013005 | ·014188 | 13 |
| 14 | ·002541 | ·002887 | ·003268 | ·003687 | ·004144 | ·004644 | ·005186 | ·005774 | ·006410 | ·007094 | 14 |
| 15 | ·001033 | ·001193 | ·001373 | ·001573 | ·001796 | ·002043 | ·002317 | ·002618 | ·002949 | ·003311 | 15 |
| 16 | ·000394 | ·000462 | ·000540 | ·000629 | ·000730 | ·000843 | ·000970 | ·001113 | ·001272 | ·001448 | 16 |
| 17 | ·000141 | ·000169 | ·000200 | ·000237 | ·000279 | ·000327 | ·000382 | ·000445 | ·000516 | ·000596 | 17 |
| 18 | ·000048 | ·000058 | ·000070 | ·000084 | ·000101 | ·000120 | ·000142 | ·000168 | ·000198 | ·000232 | 18 |
| 19 | ·000015 | ·000019 | ·000023 | ·000028 | ·000034 | ·000042 | ·000050 | ·000060 | ·000072 | ·000085 | 19 |
| 20 | ·000005 | ·000006 | ·000007 | ·000009 | ·000011 | ·000014 | ·000017 | ·000020 | ·000025 | ·000030 | 20 |
| 21 | ·000001 | ·000002 | ·000002 | ·000003 | ·000003 | ·000004 | ·000005 | ·000007 | ·000008 | ·000010 | 21 |
| 22 | — | — | ·000001 | ·000001 | ·000001 | ·000001 | ·000002 | ·000002 | ·000003 | ·000003 | 22 |
| 23 | — | — | — | — | — | — | — | ·000001 | ·000001 | ·000001 | 23 |

Table 7 *(continued)*

| y | μ 7·1 | 7·2 | 7·3 | 7·4 | 7·5 | 7·6 | 7·7 | 7·8 | 7·9 | 8·0 | y |
|---|---|---|---|---|---|---|---|---|---|---|---|
| 0 | ·000825 | ·000747 | ·000676 | ·000611 | ·000553 | ·000500 | ·000453 | ·000410 | ·000371 | ·000335 | 0 |
| 1 | ·005858 | ·005375 | ·004931 | ·004523 | ·004148 | ·003803 | ·003487 | ·003196 | ·002929 | ·002684 | 1 |
| 2 | ·020797 | ·019352 | ·018000 | ·016736 | ·015555 | ·014453 | ·013424 | ·012464 | ·011569 | ·010735 | 2 |
| 3 | ·049219 | ·046444 | ·043799 | ·041282 | ·038889 | ·036614 | ·034455 | ·032407 | ·030465 | ·028626 | 3 |
| 4 | ·087364 | ·083598 | ·079934 | ·076372 | ·072916 | ·069567 | ·066326 | ·063193 | ·060169 | ·057252 | 4 |
| 5 | ·124057 | ·120382 | ·116703 | ·113031 | ·109375 | ·105742 | ·102142 | ·098581 | ·095067 | ·091604 | 5 |
| 6 | ·146800 | ·144458 | ·141989 | ·139405 | ·136718 | ·133940 | ·131082 | ·128156 | ·125171 | ·122138 | 6 |
| 7 | ·148897 | ·148586 | ·148074 | ·147371 | ·146484 | ·145421 | ·144191 | ·142802 | ·141264 | ·139587 | 7 |
| 8 | ·132146 | ·133727 | ·135118 | ·136318 | ·137329 | ·138150 | ·138783 | ·139232 | ·139499 | ·139587 | 8 |
| 9 | ·104249 | ·106982 | ·109596 | ·112084 | ·114440 | ·116660 | ·118737 | ·120668 | ·122449 | ·124077 | 9 |
| 10 | ·074017 | ·077027 | ·080005 | ·082942 | ·085830 | ·088661 | ·091427 | ·094121 | ·096735 | ·099262 | 10 |
| 11 | ·047774 | ·050418 | ·053094 | ·055797 | ·058521 | ·061257 | ·063999 | ·066740 | ·069473 | ·072190 | 11 |
| 12 | ·028267 | ·030251 | ·032299 | ·034408 | ·036575 | ·038796 | ·041066 | ·043381 | ·045736 | ·048127 | 12 |
| 13 | ·015438 | ·016754 | ·018137 | ·019586 | ·021101 | ·022681 | ·024324 | ·026029 | ·027794 | ·029616 | 13 |
| 14 | ·007829 | ·008616 | ·009457 | ·010353 | ·011304 | ·012312 | ·013378 | ·014502 | ·015684 | ·016924 | 14 |
| 15 | ·003706 | ·004136 | ·004603 | ·005107 | ·005652 | ·006238 | ·006867 | ·007541 | ·008260 | ·009026 | 15 |
| 16 | ·001644 | ·001861 | ·002100 | ·002362 | ·002649 | ·002963 | ·003305 | ·003676 | ·004078 | ·004513 | 16 |
| 17 | ·000687 | ·000788 | ·000902 | ·001028 | ·001169 | ·001325 | ·001497 | ·001687 | ·001895 | ·002124 | 17 |
| 18 | ·000271 | ·000315 | ·000366 | ·000423 | ·000487 | ·000559 | ·000640 | ·000731 | ·000832 | ·000944 | 18 |
| 19 | ·000101 | ·000119 | ·000141 | ·000165 | ·000192 | ·000224 | ·000259 | ·000300 | ·000346 | ·000397 | 19 |
| 20 | ·000036 | ·000043 | ·000051 | ·000061 | ·000072 | ·000085 | ·000100 | ·000117 | ·000137 | ·000159 | 20 |
| 21 | ·000012 | ·000015 | ·000018 | ·000021 | ·000026 | ·000031 | ·000037 | ·000043 | ·000051 | ·000061 | 21 |
| 22 | ·000004 | ·000005 | ·000006 | ·000007 | ·000009 | ·000011 | ·000013 | ·000015 | ·000018 | ·000022 | 22 |
| 23 | ·000001 | ·000002 | ·000002 | ·000002 | ·000003 | ·000004 | ·000004 | ·000005 | ·000006 | ·000008 | 23 |
| 24 | — | — | ·000001 | ·000001 | ·000001 | ·000001 | ·000001 | ·000002 | ·000002 | ·000003 | 24 |
| 25 | — | — | — | — | — | — | — | ·000001 | ·000001 | ·000001 | 25 |

| y | 8·1 | 8·2 | 8·3 | 8·4 | 8·5 | 8·6 | 8·7 | 8·8 | 8·9 | 9·0 | y |
|---|---|---|---|---|---|---|---|---|---|---|---|
| 0 | ·000304 | ·000275 | ·000249 | ·000225 | ·000203 | ·000184 | ·000167 | ·000151 | ·000136 | ·000123 | 0 |
| 1 | ·002459 | ·002252 | ·002063 | ·001889 | ·001729 | ·001583 | ·001449 | ·001326 | ·001214 | ·001111 | 1 |
| 2 | ·009958 | ·009234 | ·008560 | ·007933 | ·007350 | ·006808 | ·006304 | ·005836 | ·005402 | ·004998 | 2 |
| 3 | ·026885 | ·025239 | ·023683 | ·022213 | ·020826 | ·019517 | ·018283 | ·017120 | ·016025 | ·014994 | 3 |
| 4 | ·054443 | ·051740 | ·049142 | ·046648 | ·044255 | ·041961 | ·039765 | ·037664 | ·035656 | ·033737 | 4 |
| 5 | ·088198 | ·084854 | ·081576 | ·078368 | ·075233 | ·072174 | ·069192 | ·066289 | ·063467 | ·060727 | 5 |
| 6 | ·119067 | ·115967 | ·112847 | ·109716 | ·106581 | ·103449 | ·100328 | ·097224 | ·094143 | ·091090 | 6 |
| 7 | ·137778 | ·135848 | ·133805 | ·131659 | ·129419 | ·127094 | ·124693 | ·122224 | ·119696 | ·117116 | 7 |
| 8 | ·139500 | ·139244 | ·138823 | ·138242 | ·137508 | ·136626 | ·135604 | ·134446 | ·133161 | ·131756 | 8 |
| 9 | ·125550 | ·126866 | ·128025 | ·129026 | ·129869 | ·130554 | ·131084 | ·131459 | ·131682 | ·131756 | 9 |
| 10 | ·101696 | ·104031 | ·106261 | ·108382 | ·110388 | ·112277 | ·114043 | ·115684 | ·117197 | ·118580 | 10 |
| 11 | ·074885 | ·077550 | ·080179 | ·082764 | ·085300 | ·087780 | ·090197 | ·092547 | ·094823 | ·097020 | 11 |
| 12 | ·050547 | ·052993 | ·055457 | ·057935 | ·060421 | ·062909 | ·065393 | ·067868 | ·070327 | ·072765 | 12 |
| 13 | ·031495 | ·033426 | ·035407 | ·037435 | ·039506 | ·041617 | ·043763 | ·045941 | ·048147 | ·050376 | 13 |
| 14 | ·018222 | ·019578 | ·020991 | ·022461 | ·023986 | ·025565 | ·027196 | ·028877 | ·030608 | ·032384 | 14 |
| 15 | ·009840 | ·010703 | ·011615 | ·012578 | ·013592 | ·014657 | ·015773 | ·016941 | ·018161 | ·019431 | 15 |
| 16 | ·004981 | ·005485 | ·006025 | ·006604 | ·007221 | ·007878 | ·008577 | ·009318 | ·010102 | ·010930 | 16 |
| 17 | ·002373 | ·002646 | ·002942 | ·003263 | ·003610 | ·003985 | ·004389 | ·004823 | ·005289 | ·005786 | 17 |
| 18 | ·001068 | ·001205 | ·001356 | ·001523 | ·001705 | ·001904 | ·002121 | ·002358 | ·002615 | ·002893 | 18 |
| 19 | ·000455 | ·000520 | ·000593 | ·000673 | ·000763 | ·000862 | ·000971 | ·001092 | ·001225 | ·001370 | 19 |
| 20 | ·000184 | ·000213 | ·000246 | ·000283 | ·000324 | ·000371 | ·000423 | ·000481 | ·000545 | ·000617 | 20 |

**Table 7** *(continued)*

| y | μ | | | | | | | | | | y |
|---|---|---|---|---|---|---|---|---|---|---|---|
| | 8·1 | 8·2 | 8·3 | 8·4 | 8·5 | 8·6 | 8·7 | 8·8 | 8·9 | 9·0 | |
| 21 | ·000071 | ·000083 | ·000097 | ·000113 | ·000131 | ·000152 | ·000175 | ·000201 | ·000231 | ·000264 | 21 |
| 22 | ·000026 | ·000031 | ·000037 | ·000043 | ·000051 | ·000059 | ·000069 | ·000081 | ·000093 | ·000108 | 22 |
| 23 | ·000009 | ·000011 | ·000013 | ·000016 | ·000019 | ·000022 | ·000026 | ·000031 | ·000036 | ·000042 | 23 |
| 24 | ·000003 | ·000004 | ·000005 | ·000006 | ·000007 | ·000008 | ·000009 | ·000011 | ·000013 | ·000016 | 24 |
| 25 | ·000001 | ·000001 | ·000002 | ·000002 | ·000002 | ·000003 | ·000003 | ·000004 | ·000005 | ·000006 | 25 |
| 26 | — | — | — | ·000001 | ·000001 | ·000001 | ·000001 | ·000001 | ·000002 | ·000002 | 26 |
| 27 | — | — | — | — | — | — | — | — | ·000001 | ·000001 | 27 |
| | 9·1 | 9·2 | 9·3 | 9·4 | 9·5 | 9·6 | 9·7 | 9·8 | 9·9 | 10·0 | |
| 0 | ·000112 | ·000101 | ·000091 | ·000083 | ·000075 | ·000068 | ·000061 | ·000055 | ·000050 | ·000045 | 0 |
| 1 | ·001016 | ·000930 | ·000850 | ·000778 | ·000711 | ·000650 | ·000594 | ·000543 | ·000497 | ·000454 | 1 |
| 2 | ·004624 | ·004276 | ·003954 | ·003655 | ·003378 | ·003121 | ·002883 | ·002663 | ·002459 | ·002270 | 2 |
| 3 | ·014025 | ·013113 | ·012256 | ·011452 | ·010696 | ·009987 | ·009322 | ·008698 | ·008114 | ·007567 | 3 |
| 4 | ·031906 | ·030160 | ·028496 | ·026911 | ·025403 | ·023969 | ·022606 | ·021311 | ·020082 | ·018917 | 4 |
| 5 | ·058069 | ·055494 | ·053002 | ·050593 | ·048266 | ·046020 | ·043855 | ·041770 | ·039763 | ·037833 | 5 |
| 6 | ·088072 | ·085091 | ·082154 | ·079262 | ·076421 | ·073632 | ·070899 | ·068224 | ·065609 | ·063055 | 6 |
| 7 | ·114493 | ·111834 | ·109147 | ·106438 | ·103714 | ·100981 | ·098246 | ·095514 | ·092790 | ·090079 | 7 |
| 8 | ·130236 | ·128609 | ·126883 | ·125065 | ·123160 | ·121178 | ·119123 | ·117004 | ·114827 | ·112599 | 8 |
| 9 | ·131683 | ·131467 | ·131113 | ·130623 | ·130003 | ·129256 | ·128388 | ·127405 | ·126310 | ·125110 | 9 |
| 10 | ·119832 | ·120950 | ·121935 | ·122786 | ·123502 | ·124086 | ·124537 | ·124857 | ·125047 | ·125110 | 10 |
| 11 | ·099133 | ·101158 | ·103090 | ·104926 | ·106661 | ·108293 | ·109819 | ·111236 | ·112542 | ·113736 | 11 |
| 12 | ·075176 | ·077555 | ·079895 | ·082192 | ·084440 | ·086634 | ·088770 | ·090843 | ·092847 | ·094780 | 12 |
| 13 | ·052623 | ·054885 | ·057156 | ·059431 | ·061706 | ·063976 | ·066236 | ·068481 | ·070707 | ·072908 | 13 |
| 14 | ·034205 | ·036067 | ·037968 | ·039904 | ·041872 | ·043869 | ·045892 | ·047937 | ·050000 | ·052077 | 14 |
| 15 | ·020751 | ·022121 | ·023540 | ·025006 | ·026519 | ·028076 | ·029677 | ·031319 | ·033000 | ·034718 | 15 |
| 16 | ·011802 | ·012720 | ·013683 | ·014691 | ·015746 | ·016846 | ·017992 | ·019183 | ·020419 | ·021699 | 16 |
| 17 | ·006318 | ·006884 | ·007485 | ·008123 | ·008799 | ·009513 | ·010266 | ·011058 | ·011891 | ·012764 | 17 |
| 18 | ·003194 | ·003518 | ·003867 | ·004242 | ·004644 | ·005074 | ·005532 | ·006021 | ·006540 | ·007091 | 18 |
| 19 | ·001530 | ·001704 | ·001893 | ·002099 | ·002322 | ·002563 | ·002824 | ·003105 | ·003408 | ·003732 | 19 |
| 20 | ·000696 | ·000784 | ·000880 | ·000986 | ·001103 | ·001230 | ·001370 | ·001522 | ·001687 | ·001866 | 20 |
| 21 | ·000302 | ·000343 | ·000390 | ·000442 | ·000499 | ·000563 | ·000633 | ·000710 | ·000795 | ·000889 | 21 |
| 22 | ·000125 | ·000144 | ·000165 | ·000189 | ·000215 | ·000245 | ·000279 | ·000316 | ·000358 | ·000404 | 22 |
| 23 | ·000049 | ·000057 | ·000067 | ·000077 | ·000089 | ·000102 | ·000118 | ·000135 | ·000154 | ·000176 | 23 |
| 24 | ·000019 | ·000022 | ·000026 | ·000030 | ·000035 | ·000041 | ·000048 | ·000055 | ·000064 | ·000073 | 24 |
| 25 | ·000007 | ·000008 | ·000010 | ·000011 | ·000013 | ·000016 | ·000018 | ·000022 | ·000025 | ·000029 | 25 |
| 26 | ·000002 | ·000003 | ·000003 | ·000004 | ·000005 | ·000006 | ·000007 | ·000008 | ·000010 | ·000011 | 26 |
| 27 | ·000001 | ·000001 | ·000001 | ·000001 | ·000002 | ·000002 | ·000002 | ·000003 | ·000004 | ·000004 | 27 |
| 28 | — | — | — | — | ·000001 | ·000001 | ·000001 | ·000001 | ·000001 | ·000001 | 28 |
| 29 | — | — | — | — | — | — | — | — | — | ·000001 | 29 |
| | 10·1 | 10·2 | 10·3 | 10·4 | 10·5 | 10·6 | 10·7 | 10·8 | 10·9 | 11·0 | |
| 0 | ·000041 | ·000037 | ·000034 | ·000030 | ·000028 | ·000025 | ·000023 | ·000020 | ·000018 | ·000017 | 0 |
| 1 | ·000415 | ·000379 | ·000346 | ·000317 | ·000289 | ·000264 | ·000241 | ·000220 | ·000201 | ·000184 | 1 |
| 2 | ·002095 | ·001934 | ·001784 | ·001646 | ·001518 | ·001400 | ·001291 | ·001190 | ·001097 | ·001010 | 2 |
| 3 | ·007054 | ·006574 | ·006125 | ·005705 | ·005313 | ·004946 | ·004603 | ·004283 | ·003984 | ·003705 | 3 |

Table 7 *(continued)*

| $y$ | $\mu$ | | | | | | | | | | $y$ |
|---|---|---|---|---|---|---|---|---|---|---|---|
| | 10·1 | 10·2 | 10·3 | 10·4 | 10·5 | 10·6 | 10·7 | 10·8 | 10·9 | 11·0 | |
| 4 | ·017811 | ·016764 | ·015773 | ·014834 | ·013946 | ·013107 | ·012313 | ·011564 | ·010856 | ·010189 | 4 |
| 5 | ·035979 | ·034199 | ·032492 | ·030855 | ·029287 | ·027786 | ·026350 | ·024978 | ·023667 | ·022415 | 5 |
| 6 | ·060565 | ·058139 | ·055777 | ·053482 | ·051252 | ·049089 | ·046991 | ·044960 | ·042995 | ·041095 | 6 |
| 7 | ·087387 | ·084716 | ·082072 | ·079458 | ·076878 | ·074334 | ·071830 | ·069367 | ·066949 | ·064577 | 7 |
| 8 | ·110326 | ·108013 | ·105668 | ·103296 | ·100902 | ·098493 | ·096072 | ·093646 | ·091218 | ·088794 | 8 |
| 9 | ·123810 | ·122415 | ·120931 | ·119364 | ·117720 | ·116003 | ·114219 | ·112375 | ·110475 | ·108526 | 9 |
| 10 | ·125048 | ·124863 | ·124559 | ·124139 | ·123606 | ·122963 | ·122215 | ·121365 | ·120418 | ·119378 | 10 |
| 11 | ·114817 | ·115782 | ·116633 | ·117368 | ·117987 | ·118492 | ·118882 | ·119159 | ·119323 | ·119378 | 11 |
| 12 | ·096637 | ·098415 | ·100110 | ·101719 | ·103239 | ·104667 | ·106003 | ·107243 | ·108386 | ·109430 | 12 |
| 13 | ·075080 | ·077218 | ·079318 | ·081375 | ·083385 | ·085344 | ·087248 | ·089094 | ·090877 | ·092595 | 13 |
| 14 | ·054165 | ·056259 | ·058355 | ·060450 | ·062539 | ·064618 | ·066683 | ·068730 | ·070754 | ·072753 | 14 |
| 15 | ·036471 | ·038256 | ·040071 | ·041912 | ·043777 | ·045663 | ·047567 | ·049485 | ·051415 | ·053352 | 15 |
| 16 | ·023022 | ·024388 | ·025795 | ·027243 | ·028729 | ·030252 | ·031810 | ·033403 | ·035026 | ·036680 | 16 |
| 17 | ·013678 | ·014633 | ·015629 | ·016666 | ·017744 | ·018863 | ·020022 | ·021220 | ·022458 | ·023734 | 17 |
| 18 | ·007675 | ·008292 | ·008943 | ·009629 | ·010351 | ·011108 | ·011902 | ·012732 | ·013600 | ·014504 | 18 |
| 19 | ·004080 | ·004451 | ·004848 | ·005271 | ·005720 | ·006197 | ·006703 | ·007237 | ·007802 | ·008397 | 19 |
| 20 | ·002060 | ·002270 | ·002497 | ·002741 | ·003003 | ·003285 | ·003586 | ·003908 | ·004252 | ·004618 | 20 |
| 21 | ·000991 | ·001103 | ·001225 | ·001357 | ·001502 | ·001658 | ·001827 | ·002010 | ·002207 | ·002419 | 21 |
| 22 | ·000455 | ·000511 | ·000573 | ·000642 | ·000717 | ·000799 | ·000889 | ·000987 | ·001093 | ·001210 | 22 |
| 23 | ·000200 | ·000227 | ·000257 | ·000290 | ·000327 | ·000368 | ·000413 | ·000463 | ·000518 | ·000578 | 23 |
| 24 | ·000084 | ·000096 | ·000110 | ·000126 | ·000143 | ·000163 | ·000184 | ·000208 | ·000235 | ·000265 | 24 |
| 25 | ·000034 | ·000039 | ·000045 | ·000052 | ·000060 | ·000069 | ·000079 | ·000090 | ·000103 | ·000117 | 25 |
| 26 | ·000013 | ·000015 | ·000018 | ·000021 | ·000024 | ·000028 | ·000032 | ·000037 | ·000043 | ·000049 | 26 |
| 27 | ·000005 | ·000006 | ·000007 | ·000008 | ·000009 | ·000011 | ·000013 | ·000015 | ·000017 | ·000020 | 27 |
| 28 | ·000002 | ·000002 | ·000003 | ·000003 | ·000004 | ·000004 | ·000005 | ·000006 | ·000007 | ·000008 | 28 |
| 29 | ·000001 | ·000001 | ·000001 | ·000001 | ·000001 | ·000002 | ·000002 | ·000002 | ·000003 | ·000003 | 29 |
| 30 | — | — | — | — | — | ·000001 | ·000001 | ·000001 | ·000001 | ·000001 | 30 |
| | 11·1 | 11·2 | 11·3 | 11·4 | 11·5 | 11·6 | 11·7 | 11·8 | 11·9 | 12·0 | |
| 0 | ·000015 | ·000014 | ·000012 | ·000011 | ·000010 | ·000009 | ·000008 | ·000008 | ·000007 | ·000006 | 0 |
| 1 | ·000168 | ·000153 | ·000140 | ·000128 | ·000116 | ·000106 | ·000097 | ·000089 | ·000081 | ·000074 | 1 |
| 2 | ·000931 | ·000858 | ·000790 | ·000727 | ·000670 | ·000617 | ·000568 | ·000522 | ·000481 | ·000442 | 2 |
| 3 | ·003445 | ·003202 | ·002976 | ·002764 | ·002568 | ·002385 | ·002214 | ·002055 | ·001907 | ·001770 | 3 |
| 4 | ·009559 | ·008965 | ·008406 | ·007879 | ·007382 | ·006915 | ·006476 | ·006062 | ·005674 | ·005309 | 4 |
| 5 | ·021221 | ·020082 | ·018997 | ·017963 | ·016979 | ·016043 | ·015153 | ·014307 | ·013504 | ·012741 | 5 |
| 6 | ·039259 | ·037487 | ·035778 | ·034130 | ·032544 | ·031017 | ·029549 | ·028137 | ·026782 | ·025481 | 6 |
| 7 | ·062253 | ·059979 | ·057755 | ·055584 | ·053465 | ·051400 | ·049388 | ·047432 | ·045530 | ·043682 | 7 |
| 8 | ·086376 | ·083970 | ·081579 | ·079206 | ·076856 | ·074529 | ·072231 | ·069962 | ·067725 | ·065523 | 8 |
| 9 | ·106531 | ·104496 | ·102427 | ·100328 | ·098204 | ·096060 | ·093900 | ·091728 | ·089548 | ·087364 | 9 |
| 10 | ·118249 | ·117036 | ·115743 | ·114374 | ·112935 | ·111430 | ·109863 | ·108239 | ·106562 | ·104837 | 10 |
| 11 | ·119324 | ·119164 | ·118899 | ·118533 | ·118068 | ·117508 | ·116854 | ·116110 | ·115281 | ·114368 | 11 |
| 12 | ·110375 | ·111220 | ·111964 | ·112607 | ·113149 | ·113591 | ·113933 | ·114175 | ·114320 | ·114363 | 12 |
| 13 | ·094243 | ·095820 | ·097322 | ·098747 | ·100093 | ·101358 | ·102539 | ·103636 | ·104647 | ·105570 | 13 |
| 14 | ·074721 | ·076656 | ·078553 | ·080409 | ·082219 | ·083982 | ·085694 | ·087350 | ·088950 | ·090489 | 14 |
| 15 | ·055294 | ·057236 | ·059177 | ·061110 | ·063035 | ·064946 | ·066841 | ·068716 | ·070567 | ·072391 | 15 |
| 16 | ·038360 | ·040065 | ·041793 | ·043541 | ·045306 | ·047086 | ·048877 | ·050678 | ·052484 | ·054293 | 16 |
| 17 | ·025047 | ·026396 | ·027780 | ·029198 | ·030648 | ·032129 | ·033639 | ·035176 | ·036739 | ·038325 | 17 |
| 18 | ·015446 | ·016424 | ·017440 | ·018492 | ·019581 | ·020706 | ·021865 | ·023060 | ·024288 | ·025550 | 18 |
| 19 | ·009023 | ·009682 | ·010372 | ·011095 | ·011852 | ·012641 | ·013465 | ·014322 | ·015212 | ·016137 | 19 |

Table 7 *(continued)*

| $y$ | $\mu$ | | | | | | | | | | $y$ |
|---|---|---|---|---|---|---|---|---|---|---|---|
| | *11·1* | *11·2* | *11·3* | *11·4* | *11·5* | *11·6* | *11·7* | *11·8* | *11·9* | *12·0* | |
| *20* | ·005008 | ·005422 | ·005860 | ·006324 | ·006815 | ·007332 | ·007877 | ·008450 | ·009051 | ·009682 | *20* |
| *21* | ·002647 | ·002892 | ·003153 | ·003433 | ·003732 | ·004050 | ·004388 | ·004748 | ·005129 | ·005533 | *21* |
| *22* | ·001336 | ·001472 | ·001620 | ·001779 | ·001951 | ·002136 | ·002334 | ·002547 | ·002774 | ·003018 | *22* |
| *23* | ·000645 | ·000717 | ·000796 | ·000882 | ·000975 | ·001077 | ·001187 | ·001307 | ·001435 | ·001575 | *23* |
| *24* | ·000298 | ·000335 | ·000375 | ·000419 | ·000467 | ·000521 | ·000579 | ·000642 | ·000712 | ·000787 | *24* |
| *25* | ·000132 | ·000150 | ·000169 | ·000191 | ·000215 | ·000242 | ·000271 | ·000303 | ·000339 | ·000378 | *25* |
| *26* | ·000057 | ·000065 | ·000074 | ·000084 | ·000095 | ·000108 | ·000122 | ·000138 | ·000155 | ·000174 | *26* |
| *27* | ·000023 | ·000027 | ·000031 | ·000035 | ·000041 | ·000046 | ·000053 | ·000060 | ·000068 | ·000078 | *27* |
| *28* | ·000009 | ·000011 | ·000012 | ·000014 | ·000017 | ·000019 | ·000022 | ·000025 | ·000029 | ·000033 | *28* |
| *29* | ·000004 | ·000004 | ·000005 | ·000006 | ·000007 | ·000008 | ·000009 | ·000010 | ·000012 | ·000014 | *29* |
| *30* | ·000001 | ·000002 | ·000002 | ·000002 | ·000003 | ·000003 | ·000003 | ·000004 | ·000005 | ·000005 | *30* |
| *31* | — | ·000001 | ·000001 | ·000001 | ·000001 | ·000001 | ·000001 | ·000002 | ·000002 | ·000002 | *31* |
| *32* | — | — | — | — | — | — | — | ·000001 | ·000001 | ·000001 | *32* |
| | *12·1* | *12·2* | *12·3* | *12·4* | *12·5* | *12·6* | *12·7* | *12·8* | *12·9* | *13·0* | |
| *0* | ·000006 | ·000005 | ·000005 | ·000004 | ·000004 | ·000003 | ·000003 | ·000003 | ·000002 | ·000002 | *0* |
| *1* | ·000067 | ·000061 | ·000056 | ·000051 | ·000047 | ·000042 | ·000039 | ·000035 | ·000032 | ·000029 | *1* |
| *2* | ·000407 | ·000374 | ·000344 | ·000317 | ·000291 | ·000268 | ·000246 | ·000226 | ·000208 | ·000191 | *2* |
| *3* | ·001641 | ·001522 | ·001412 | ·001309 | ·001213 | ·001124 | ·001042 | ·000965 | ·000894 | ·000828 | *3* |
| *4* | ·004966 | ·004643 | ·004341 | ·004057 | ·003791 | ·003541 | ·003307 | ·003088 | ·002882 | ·002690 | *4* |
| *5* | ·012017 | ·011330 | ·010679 | ·010062 | ·009477 | ·008924 | ·008400 | ·007905 | ·007436 | ·006994 | *5* |
| *6* | ·024233 | ·023037 | ·021892 | ·020794 | ·019744 | ·018740 | ·017781 | ·016864 | ·015988 | ·015153 | *6* |
| *7* | ·041889 | ·040151 | ·038467 | ·036836 | ·035258 | ·033733 | ·032259 | ·030837 | ·029464 | ·028141 | *7* |
| *8* | ·063358 | ·061230 | ·059142 | ·057095 | ·055091 | ·053129 | ·051212 | ·049339 | ·047511 | ·045730 | *8* |
| *9* | ·085181 | ·083000 | ·080828 | ·078665 | ·076515 | ·074381 | ·072266 | ·070171 | ·068100 | ·066054 | *9* |
| *10* | ·103069 | ·101261 | ·099418 | ·097544 | ·095644 | ·093720 | ·091777 | ·089819 | ·087849 | ·085870 | *10* |
| *11* | ·113376 | ·112308 | ·111168 | ·109959 | ·108686 | ·107352 | ·105961 | ·104516 | ·103023 | ·101483 | *11* |
| *12* | ·114321 | ·114180 | ·113947 | ·113624 | ·113215 | ·112720 | ·112142 | ·111484 | ·110749 | ·109940 | *12* |
| *13* | ·106406 | ·107153 | ·107811 | ·108380 | ·108860 | ·109251 | ·109554 | ·109769 | ·109897 | ·109940 | *13* |
| *14* | ·091965 | ·093376 | ·094720 | ·095994 | ·097197 | ·098326 | ·099381 | ·100360 | ·101263 | ·102087 | *14* |
| *15* | ·074185 | ·075946 | ·077670 | ·079355 | ·080997 | ·082594 | ·084143 | ·085641 | ·087086 | ·088475 | *15* |
| *16* | ·056103 | ·057909 | ·059709 | ·061500 | ·063279 | ·065043 | ·066788 | ·068513 | ·070213 | ·071886 | *16* |
| *17* | ·039932 | ·041558 | ·043201 | ·044859 | ·046529 | ·048208 | ·049895 | ·051586 | ·053279 | ·054972 | *17* |
| *18* | ·026843 | ·028167 | ·029521 | ·030903 | ·032312 | ·033746 | ·035204 | ·036683 | ·038183 | ·039702 | *18* |
| *19* | ·017095 | ·018086 | ·019111 | ·020168 | ·021258 | ·022379 | ·023531 | ·024713 | ·025925 | ·027164 | *19* |
| *20* | ·010342 | ·011033 | ·011753 | ·012504 | ·013286 | ·014099 | ·014942 | ·015816 | ·016721 | ·017657 | *20* |
| *21* | ·005959 | ·006409 | ·006884 | ·007383 | ·007908 | ·008459 | ·009036 | ·009640 | ·010272 | ·010930 | *21* |
| *22* | ·003278 | ·003554 | ·003849 | ·004162 | ·004493 | ·004845 | ·005216 | ·005609 | ·006023 | ·006459 | *22* |
| *23* | ·001724 | ·001885 | ·002058 | ·002244 | ·002442 | ·002654 | ·002880 | ·003122 | ·003378 | ·003651 | *23* |
| *24* | ·000869 | ·000958 | ·001055 | ·001159 | ·001272 | ·001393 | ·001524 | ·001665 | ·001816 | ·001977 | *24* |
| *25* | ·000421 | ·000468 | ·000519 | ·000575 | ·000636 | ·000702 | ·000774 | ·000852 | ·000937 | ·001028 | *25* |
| *26* | ·000196 | ·000219 | ·000246 | ·000274 | ·000306 | ·000340 | ·000378 | ·000420 | ·000465 | ·000514 | *26* |
| *27* | ·000088 | ·000099 | ·000112 | ·000126 | ·000142 | ·000159 | ·000178 | ·000199 | ·000222 | ·000248 | *27* |
| *28* | ·000038 | ·000043 | ·000049 | ·000056 | ·000063 | ·000071 | ·000081 | ·000091 | ·000102 | ·000115 | *28* |
| *29* | ·000016 | ·000018 | ·000021 | ·000024 | ·000027 | ·000031 | ·000035 | ·000040 | ·000046 | ·000052 | *29* |
| *30* | ·000006 | ·000007 | ·000009 | ·000010 | ·000011 | ·000013 | ·000015 | ·000017 | ·000020 | ·000022 | *30* |
| *31* | ·000002 | ·000003 | ·000003 | ·000004 | ·000005 | ·000005 | ·000006 | ·000007 | ·000008 | ·000009 | *31* |
| *32* | ·000001 | ·000001 | ·000001 | ·000002 | ·000002 | ·000002 | ·000002 | ·000003 | ·000003 | ·000004 | *32* |
| *33* | — | — | — | ·000001 | ·000001 | ·000001 | ·000001 | ·000001 | ·000001 | ·000002 | *33* |
| *34* | — | — | — | — | — | — | — | — | — | ·000001 | *34* |

Table 7 *(continued)*

| y | μ | | | | | | | | | | y |
|---|---|---|---|---|---|---|---|---|---|---|---|
| | 13·1 | 13·2 | 13·3 | 13·4 | 13·5 | 13·6 | 13·7 | 13·8 | 13·9 | 14·0 | |
| 0 | ·000002 | ·000002 | ·000002 | ·000002 | ·000001 | ·000001 | ·000001 | ·000001 | ·000001 | ·000001 | 0 |
| 1 | ·000027 | ·000024 | ·000022 | ·000020 | ·000019 | ·000017 | ·000015 | ·000014 | ·000013 | ·000012 | 1 |
| 2 | ·000175 | ·000161 | ·000148 | ·000136 | ·000125 | ·000115 | ·000105 | ·000097 | ·000089 | ·000081 | 2 |
| 3 | ·000766 | ·000709 | ·000657 | ·000608 | ·000562 | ·000520 | ·000481 | ·000445 | ·000411 | ·000380 | 3 |
| 4 | ·002510 | ·002341 | ·002183 | ·002035 | ·001897 | ·001768 | ·001648 | ·001535 | ·001429 | ·001331 | 4 |
| 5 | ·006575 | ·006180 | ·005807 | ·005455 | ·005123 | ·004810 | ·004514 | ·004236 | ·003974 | ·003727 | 5 |
| 6 | ·014356 | ·013596 | ·012872 | ·012183 | ·011526 | ·010902 | ·010308 | ·009743 | ·009206 | ·008696 | 6 |
| 7 | ·026867 | ·025639 | ·024458 | ·023322 | ·022230 | ·021181 | ·020173 | ·019207 | ·018280 | ·017392 | 7 |
| 8 | ·043994 | ·042304 | ·040661 | ·039064 | ·037512 | ·036007 | ·034547 | ·033132 | ·031762 | ·030435 | 8 |
| 9 | ·064036 | ·062046 | ·060088 | ·058161 | ·056269 | ·054410 | ·052588 | ·050802 | ·049054 | ·047344 | 9 |
| 10 | ·083887 | ·081901 | ·079916 | ·077936 | ·075963 | ·073998 | ·072046 | ·070107 | ·068185 | ·066282 | 10 |
| 11 | ·099901 | ·098281 | ·096626 | ·094940 | ·093227 | ·091489 | ·089730 | ·087953 | ·086162 | ·084359 | 11 |
| 12 | ·109059 | ·108109 | ·107094 | ·106017 | ·104880 | ·103687 | ·102441 | ·101146 | ·099804 | ·098418 | 12 |
| 13 | ·109898 | ·109773 | ·109566 | ·109279 | ·108914 | ·108473 | ·107957 | ·107370 | ·106713 | ·105989 | 13 |
| 14 | ·102833 | ·103500 | ·104087 | ·104595 | ·105024 | ·105373 | ·105644 | ·105836 | ·105951 | ·105989 | 14 |
| 15 | ·089807 | ·091080 | ·092291 | ·093439 | ·094522 | ·095539 | ·096488 | ·097369 | ·098181 | ·098923 | 15 |
| 16 | ·073530 | ·075141 | ·076717 | ·078255 | ·079753 | ·081208 | ·082618 | ·083981 | ·085295 | ·086558 | 16 |
| 17 | ·056661 | ·058345 | ·060019 | ·061683 | ·063333 | ·064966 | ·066580 | ·068173 | ·069741 | ·071283 | 17 |
| 18 | ·041237 | ·042786 | ·044348 | ·045920 | ·047500 | ·049086 | ·050675 | ·052266 | ·053856 | ·055442 | 18 |
| 19 | ·028432 | ·029725 | ·031043 | ·032385 | ·033750 | ·035135 | ·036539 | ·037962 | ·039400 | ·040852 | 19 |
| 20 | ·018623 | ·019619 | ·020644 | ·021698 | ·022781 | ·023892 | ·025030 | ·026193 | ·027383 | ·028597 | 20 |
| 21 | ·011617 | ·012332 | ·013074 | ·013846 | ·014645 | ·015473 | ·016329 | ·017213 | ·018125 | ·019064 | 21 |
| 22 | ·006917 | ·007399 | ·007904 | ·008433 | ·008987 | ·009565 | ·010168 | ·010797 | ·011452 | ·012132 | 22 |
| 23 | ·003940 | ·004246 | ·004571 | ·004913 | ·005275 | ·005656 | ·006057 | ·006478 | ·006921 | ·007385 | 23 |
| 24 | ·002151 | ·002336 | ·002533 | ·002743 | ·002967 | ·003205 | ·003457 | ·003725 | ·004008 | ·004308 | 24 |
| 25 | ·001127 | ·001233 | ·001348 | ·001470 | ·001602 | ·001744 | ·001895 | ·002056 | ·002229 | ·002412 | 25 |
| 26 | ·000568 | ·000626 | ·000689 | ·000758 | ·000832 | ·000912 | ·000998 | ·001091 | ·001191 | ·001299 | 26 |
| 27 | ·000275 | ·000306 | ·000340 | ·000376 | ·000416 | ·000459 | ·000507 | ·000558 | ·000613 | ·000674 | 27 |
| 28 | ·000129 | ·000144 | ·000161 | ·000180 | ·000201 | ·000223 | ·000248 | ·000275 | ·000305 | ·000337 | 28 |
| 29 | ·000058 | ·000066 | ·000074 | ·000083 | ·000093 | ·000105 | ·000117 | ·000131 | ·000146 | ·000163 | 29 |
| 30 | ·000025 | ·000029 | ·000033 | ·000037 | ·000042 | ·000047 | ·000053 | ·000060 | ·000068 | ·000076 | 30 |
| 31 | ·000011 | ·000012 | ·000014 | ·000016 | ·000018 | ·000021 | ·000024 | ·000027 | ·000030 | ·000034 | 31 |
| 32 | ·000004 | ·000005 | ·000006 | ·000007 | ·000008 | ·000009 | ·000010 | ·000012 | ·000013 | ·000015 | 32 |
| 33 | ·000002 | ·000002 | ·000002 | ·000003 | ·000003 | ·000004 | ·000004 | ·000005 | ·000006 | ·000006 | 33 |
| 34 | ·000001 | ·000001 | ·000001 | ·000001 | ·000001 | ·000001 | ·000002 | ·000002 | ·000002 | ·000003 | 34 |
| 35 | — | — | — | — | — | ·000001 | ·000001 | ·000001 | ·000001 | ·000001 | 35 |
| | 14·1 | 14·2 | 14·3 | 14·4 | 14·5 | 14·6 | 14·7 | 14·8 | 14·9 | 15·0 | |
| 0 | ·000001 | ·000001 | ·000001 | ·000001 | ·000001 | — | — | — | — | — | 0 |
| 1 | ·000011 | ·000010 | ·000009 | ·000008 | ·000007 | ·000007 | ·000006 | ·000006 | ·000005 | ·000005 | 1 |
| 2 | ·000075 | ·000069 | ·000063 | ·000058 | ·000053 | ·000049 | ·000045 | ·000041 | ·000038 | ·000034 | 2 |
| 3 | ·000352 | ·000325 | ·000300 | ·000277 | ·000256 | ·000237 | ·000219 | ·000202 | ·000186 | ·000172 | 3 |
| 4 | ·001239 | ·001153 | ·001073 | ·000999 | ·000929 | ·000864 | ·000803 | ·000747 | ·000694 | ·000645 | 4 |
| 5 | ·003494 | ·003275 | ·003070 | ·002876 | ·002694 | ·002523 | ·002362 | ·002211 | ·002069 | ·001936 | 5 |
| 6 | ·008212 | ·007752 | ·007316 | ·006902 | ·006510 | ·006139 | ·005787 | ·005454 | ·005138 | ·004839 | 6 |
| 7 | ·016541 | ·015726 | ·014946 | ·014199 | ·013486 | ·012804 | ·012152 | ·011530 | ·010937 | ·010370 | 7 |
| 8 | ·029153 | ·027913 | ·026715 | ·025559 | ·024443 | ·023367 | ·022330 | ·021331 | ·020370 | ·019444 | 8 |
| 9 | ·045673 | ·044040 | ·042447 | ·040894 | ·039380 | ·037907 | ·036472 | ·035078 | ·033723 | ·032407 | 9 |
| 10 | ·064399 | ·062537 | ·060700 | ·058887 | ·057101 | ·055343 | ·053614 | ·051915 | ·050247 | ·048611 | 10 |
| 11 | ·082547 | ·080730 | ·078910 | ·077089 | ·075270 | ·073456 | ·071648 | ·069850 | ·068062 | ·066287 | 11 |

Table 7 *(continued)*

| $y$ | $\mu$ | | | | | | | | | | $y$ |
|---|---|---|---|---|---|---|---|---|---|---|---|
| | 14·1 | 14·2 | 14·3 | 14·4 | 14·5 | 14·6 | 14·7 | 14·8 | 14·9 | 15·0 | |
| 12 | ·096993 | ·095530 | ·094034 | ·092507 | ·090951 | ·089371 | ·087769 | ·086148 | ·084510 | ·082859 | 12 |
| 13 | ·105200 | ·104349 | ·103437 | ·102469 | ·101446 | ·100371 | ·099247 | ·098076 | ·096862 | ·095607 | 13 |
| 14 | ·105951 | ·105839 | ·105654 | ·105396 | ·105069 | ·104672 | ·104209 | ·103681 | ·103089 | ·102436 | 14 |
| 15 | ·099594 | ·100195 | ·100723 | ·101181 | ·101567 | ·101881 | ·102125 | ·102298 | ·102402 | ·102436 | 15 |
| 16 | ·087768 | ·088923 | ·090021 | ·091063 | ·092045 | ·092967 | ·093827 | ·094626 | ·095361 | ·096034 | 16 |
| 17 | ·072795 | ·074277 | ·075724 | ·077135 | ·078509 | ·079842 | ·081133 | ·082380 | ·083581 | ·084736 | 17 |
| 18 | ·057023 | ·058596 | ·060158 | ·061708 | ·063243 | ·064761 | ·066259 | ·067735 | ·069187 | ·070613 | 18 |
| 19 | ·042317 | ·043793 | ·045277 | ·046768 | ·048264 | ·049763 | ·051263 | ·052762 | ·054257 | ·055747 | 19 |
| 20 | ·029834 | ·031093 | ·032373 | ·033673 | ·034992 | ·036327 | ·037678 | ·039044 | ·040422 | ·041810 | 20 |
| 21 | ·020031 | ·021025 | ·022045 | ·023090 | ·024161 | ·025256 | ·026375 | ·027517 | ·028680 | ·029865 | 21 |
| 22 | ·012838 | ·013570 | ·014329 | ·015114 | ·015924 | ·016761 | ·017623 | ·018511 | ·019424 | ·020362 | 22 |
| 23 | ·007870 | ·008378 | ·008909 | ·009462 | ·010039 | ·010640 | ·011264 | ·011911 | ·012584 | ·013280 | 23 |
| 24 | ·004624 | ·004957 | ·005308 | ·005677 | ·006065 | ·006472 | ·006899 | ·007345 | ·007812 | ·008300 | 24 |
| 25 | ·002608 | ·002816 | ·003036 | ·003270 | ·003518 | ·003780 | ·004057 | ·004348 | ·004656 | ·004980 | 25 |
| 26 | ·001414 | ·001538 | ·001670 | ·001811 | ·001962 | ·002123 | ·002294 | ·002475 | ·002668 | ·002873 | 26 |
| 27 | ·000739 | ·000809 | ·000884 | ·000966 | ·001054 | ·001148 | ·001249 | ·001357 | ·001473 | ·001596 | 27 |
| 28 | ·000372 | ·000410 | ·000452 | ·000497 | ·000546 | ·000598 | ·000656 | ·000717 | ·000784 | ·000855 | 28 |
| 29 | ·000181 | ·000201 | ·000223 | ·000247 | ·000273 | ·000301 | ·000332 | ·000366 | ·000403 | ·000442 | 29 |
| 30 | ·000085 | ·000095 | ·000106 | ·000118 | ·000132 | ·000147 | ·000163 | ·000181 | ·000200 | ·000221 | 30 |
| 31 | ·000039 | ·000044 | ·000049 | ·000055 | ·000062 | ·000069 | ·000077 | ·000086 | ·000096 | ·000107 | 31 |
| 32 | ·000017 | ·000019 | ·000022 | ·000025 | ·000028 | ·000032 | ·000035 | ·000040 | ·000045 | ·000050 | 32 |
| 33 | ·000007 | ·000008 | ·000009 | ·000011 | ·000012 | ·000014 | ·000016 | ·000018 | ·000020 | ·000023 | 33 |
| 34 | ·000003 | ·000003 | ·000004 | ·000005 | ·000005 | ·000006 | ·000007 | ·000008 | ·000009 | ·000010 | 34 |
| 35 | ·000001 | ·000001 | ·000002 | ·000002 | ·000002 | ·000002 | ·000003 | ·000003 | ·000004 | ·000004 | 35 |
| 36 | — | ·000001 | ·000001 | ·000001 | ·000001 | ·000001 | ·000001 | ·000001 | ·000002 | ·000002 | 36 |
| 37 | — | — | — | — | — | — | — | ·000001 | ·000001 | ·000001 | 37 |

**Table 8** Random numbers

| *Line/Col.* | *(1)* | *(2)* | *(3)* | *(4)* | *(5)* | *(6)* | *(7)* | *(8)* | *(9)* | *(10)* | *(11)* | *(12)* | *(13)* | *(14)* |
|---|---|---|---|---|---|---|---|---|---|---|---|---|---|---|
| 1 | 10480 | 15011 | 01536 | 02011 | 81647 | 91646 | 69179 | 14194 | 62590 | 36207 | 20969 | 99570 | 91291 | 90700 |
| 2 | 22368 | 46573 | 25595 | 85393 | 30995 | 89198 | 27982 | 53402 | 93965 | 34095 | 52666 | 19174 | 39615 | 99505 |
| 3 | 24130 | 48360 | 22527 | 97265 | 76393 | 64809 | 15179 | 24830 | 49340 | 32081 | 30680 | 19655 | 63348 | 58629 |
| 4 | 42167 | 93093 | 06243 | 61680 | 07856 | 16376 | 39440 | 53537 | 71341 | 57004 | 00849 | 74917 | 97758 | 16379 |
| 5 | 37570 | 39975 | 81837 | 16656 | 06121 | 91782 | 60468 | 81305 | 49684 | 60672 | 14110 | 06927 | 01263 | 54613 |
| 6 | 77921 | 06907 | 11008 | 42751 | 27756 | 53498 | 18602 | 70659 | 90655 | 15053 | 21916 | 81825 | 44394 | 42880 |
| 7 | 99562 | 72905 | 56420 | 69994 | 98872 | 31016 | 71194 | 18738 | 44013 | 48840 | 63213 | 21069 | 10634 | 12952 |
| 8 | 96301 | 91977 | 05463 | 07972 | 18876 | 20922 | 94595 | 56869 | 69014 | 60045 | 18425 | 84903 | 42508 | 32307 |
| 9 | 89579 | 14342 | 63661 | 10281 | 17453 | 18103 | 57740 | 84378 | 25331 | 12566 | 58678 | 44947 | 05585 | 56941 |
| 10 | 85475 | 36857 | 53342 | 53988 | 53060 | 59533 | 38867 | 62300 | 08158 | 17983 | 16439 | 11458 | 18593 | 64952 |
| 11 | 28918 | 69578 | 88231 | 33276 | 70997 | 79936 | 56865 | 05859 | 90106 | 31595 | 01547 | 85590 | 91610 | 78188 |
| 12 | 63553 | 40961 | 48235 | 03427 | 49626 | 69445 | 18663 | 72695 | 52180 | 20847 | 12234 | 90511 | 33703 | 90322 |
| 13 | 09429 | 93969 | 52636 | 92737 | 88974 | 33488 | 36320 | 17617 | 30015 | 08272 | 84115 | 27156 | 30613 | 74952 |
| 14 | 10365 | 61129 | 87529 | 85689 | 48237 | 52267 | 67689 | 93394 | 01511 | 26358 | 85104 | 20285 | 29975 | 89868 |
| 15 | 07119 | 97336 | 71048 | 08178 | 77233 | 13916 | 47564 | 81056 | 97735 | 85977 | 29372 | 74461 | 28551 | 90707 |
| 16 | 51085 | 12765 | 51821 | 51259 | 77452 | 16308 | 60756 | 92144 | 49442 | 53900 | 70960 | 63990 | 75601 | 40719 |
| 17 | 02368 | 21382 | 52404 | 60268 | 89368 | 19885 | 55322 | 44819 | 01188 | 65255 | 64835 | 44919 | 05944 | 55157 |
| 18 | 01011 | 54092 | 33362 | 94904 | 31273 | 04146 | 18594 | 29852 | 71585 | 85030 | 51132 | 01915 | 92747 | 64951 |
| 19 | 52162 | 53916 | 46369 | 58586 | 23216 | 14513 | 83149 | 98736 | 23495 | 64350 | 94738 | 17752 | 35156 | 35749 |
| 20 | 07056 | 97628 | 33787 | 09998 | 42698 | 06691 | 76988 | 13602 | 51851 | 46104 | 88916 | 19509 | 25625 | 58104 |
| 21 | 48663 | 91245 | 85828 | 14346 | 09172 | 30168 | 90229 | 04734 | 59193 | 22178 | 30421 | 61666 | 99904 | 32812 |
| 22 | 54164 | 58492 | 22421 | 74103 | 47070 | 25306 | 76468 | 26384 | 58151 | 06646 | 21524 | 15227 | 96909 | 44592 |
| 23 | 32639 | 32363 | 05597 | 24200 | 13363 | 38005 | 94342 | 28728 | 35806 | 06912 | 17012 | 64161 | 18296 | 22851 |
| 24 | 29334 | 27001 | 87637 | 87308 | 58731 | 00256 | 45834 | 15398 | 46557 | 41135 | 10367 | 07684 | 36188 | 18510 |
| 25 | 02488 | 33062 | 28834 | 07351 | 19731 | 92420 | 60952 | 61280 | 50001 | 67658 | 32586 | 86679 | 50720 | 94953 |

**Table 9** Critical values for the Wilcoxon signed-rank test

$n = 5(1)50$

| *One-sided* | *Two-sided* | $n = 5$ | $n = 6$ | $n = 7$ | $n = 8$ | $n = 9$ | $n = 10$ | $n = 11$ | $n = 12$ | $n = 13$ | $n = 14$ | $n = 15$ | $n = 16$ |
|---|---|---|---|---|---|---|---|---|---|---|---|---|---|
| .05 | .10 | 1 | 2 | 4 | 6 | 8 | 11 | 14 | 17 | 21 | 26 | 30 | 36 |
| .025 | .05 | | 1 | 2 | 4 | 6 | 8 | 11 | 14 | 17 | 21 | 25 | 30 |
| .01 | .02 | | | 0 | 2 | 3 | 5 | 7 | 10 | 13 | 16 | 20 | 24 |
| .005 | .01 | | | | 0 | 2 | 3 | 5 | 7 | 10 | 13 | 16 | 19 |

| *One-sided* | *Two-sided* | $n = 17$ | $n = 18$ | $n = 19$ | $n = 20$ | $n = 21$ | $n = 22$ | $n = 23$ | $n = 24$ | $n = 25$ | $n = 26$ | $n = 27$ | $n = 28$ |
|---|---|---|---|---|---|---|---|---|---|---|---|---|---|
| .05 | .10 | 41 | 47 | 54 | 60 | 68 | 75 | 83 | 92 | 101 | 110 | 120 | 130 |
| .025 | .05 | 35 | 40 | 46 | 52 | 59 | 66 | 73 | 81 | 90 | 98 | 107 | 117 |
| .01 | .02 | 28 | 33 | 38 | 43 | 49 | 56 | 62 | 69 | 77 | 85 | 93 | 102 |
| .005 | .01 | 23 | 28 | 32 | 37 | 43 | 49 | 55 | 61 | 68 | 76 | 84 | 92 |

| *One-sided* | *Two-sided* | $n = 29$ | $n = 30$ | $n = 31$ | $n = 32$ | $n = 33$ | $n = 34$ | $n = 35$ | $n = 36$ | $n = 37$ | $n = 38$ | $n = 39$ |
|---|---|---|---|---|---|---|---|---|---|---|---|---|
| .05 | .10 | 141 | 152 | 163 | 175 | 188 | 201 | 214 | 228 | 242 | 256 | 271 |
| .025 | .05 | 127 | 137 | 148 | 159 | 171 | 183 | 195 | 208 | 222 | 235 | 250 |
| .01 | .02 | 111 | 120 | 130 | 141 | 151 | 162 | 174 | 186 | 198 | 211 | 224 |
| .005 | .01 | 100 | 109 | 118 | 128 | 138 | 149 | 160 | 171 | 183 | 195 | 208 |

| *One-sided* | *Two-sided* | $n = 40$ | $n = 41$ | $n = 42$ | $n = 43$ | $n = 44$ | $n = 45$ | $n = 46$ | $n = 47$ | $n = 48$ | $n = 49$ | $n = 50$ |
|---|---|---|---|---|---|---|---|---|---|---|---|---|
| .05 | .10 | 287 | 303 | 319 | 336 | 353 | 371 | 389 | 408 | 427 | 446 | 466 |
| .025 | .05 | 264 | 279 | 295 | 311 | 327 | 344 | 361 | 379 | 397 | 415 | 434 |
| .01 | .02 | 238 | 252 | 267 | 281 | 297 | 313 | 329 | 345 | 362 | 380 | 398 |
| .005 | .01 | 221 | 234 | 248 | 262 | 277 | 292 | 307 | 323 | 339 | 356 | 373 |

**Table 10** Distribution of the total number of runs $r$ in samples of size $(n_1, n_2)$; $P[r \leqslant \ell]$

| | $\ell$ | | | | | | | | |
|---|---|---|---|---|---|---|---|---|---|
| $(n_1, n_2)$ | 2 | 3 | 4 | 5 | 6 | 7 | 8 | 9 | 10 |
| (2,3) | .200 | .500 | .900 | 1.000 | | | | | |
| (2,4) | .133 | .400 | .800 | 1.000 | | | | | |
| (2,5) | .095 | .333 | .714 | 1.000 | | | | | |
| (2,6) | .071 | .286 | .643 | 1.000 | | | | | |
| (2,7) | .056 | .250 | .583 | 1.000 | | | | | |
| (2,8) | .044 | .222 | .533 | 1.000 | | | | | |
| (2,9) | .036 | .200 | .491 | 1.000 | | | | | |
| (2,10) | .030 | .182 | .455 | 1.000 | | | | | |
| (3,3) | .100 | .300 | .700 | .900 | 1.000 | | | | |
| (3,4) | .057 | .200 | .543 | .800 | .971 | 1.000 | | | |
| (3,5) | .036 | .143 | .429 | .714 | .929 | 1.000 | | | |
| (3,6) | .024 | .107 | .345 | .643 | .881 | 1.000 | | | |
| (3,7) | .017 | .083 | .283 | .583 | .833 | 1.000 | | | |
| (3,8) | .012 | .067 | .236 | .533 | .788 | 1.000 | | | |
| (3,9) | .009 | .055 | .200 | .491 | .745 | 1.000 | | | |
| (3,10) | .007 | .045 | .171 | .455 | .706 | 1.000 | | | |
| (4,4) | .029 | .114 | .371 | .629 | .886 | .971 | 1.000 | | |
| (4,5) | .016 | .071 | .262 | .500 | .786 | .929 | .992 | 1.000 | |
| (4,6) | .010 | .048 | .190 | .405 | .690 | .881 | .976 | 1.000 | |
| (4,7) | .006 | .033 | .142 | .333 | .606 | .833 | .954 | 1.000 | |
| (4,8) | .004 | .024 | .109 | .279 | .533 | .788 | .929 | 1.000 | |
| (4,9) | .003 | .018 | .085 | .236 | .471 | .745 | .902 | 1.000 | |
| (4,10) | .002 | .014 | .068 | .203 | .419 | .706 | .874 | 1.000 | |
| (5,5) | .008 | .040 | .167 | .357 | .643 | .833 | .960 | .992 | 1.000 |
| (5,6) | .004 | .024 | .110 | .262 | .522 | .738 | .911 | .976 | .998 |
| (5,7) | .003 | .015 | .076 | .197 | .424 | .652 | .854 | .955 | .992 |
| (5,8) | .002 | .010 | .054 | .152 | .347 | .576 | .793 | .929 | .984 |
| (5,9) | .001 | .007 | .039 | .119 | .287 | .510 | .734 | .902 | .972 |
| (5,10) | .001 | .005 | .029 | .095 | .239 | .455 | .678 | .874 | .958 |
| (6,6) | .002 | .013 | .067 | .175 | .392 | .608 | .825 | .933 | .987 |
| (6,7) | .001 | .008 | .043 | .121 | .296 | .500 | .733 | .879 | .966 |
| (6,8) | .001 | .005 | .028 | .086 | .226 | .413 | .646 | .821 | .937 |
| (6,9) | .000 | .003 | .019 | .063 | .175 | .343 | .566 | .762 | .902 |
| (6,10) | .000 | .002 | .013 | .047 | .137 | .288 | .497 | .706 | .864 |
| (7,7) | .001 | .004 | .025 | .078 | .209 | .383 | .617 | .791 | .922 |
| (7,8) | .000 | .002 | .015 | .051 | .149 | .296 | .514 | .704 | .867 |
| (7,9) | .000 | .001 | .010 | .035 | .108 | .231 | .427 | .622 | .806 |
| (7,10) | .000 | .001 | .006 | .024 | .080 | .182 | .355 | .549 | .743 |
| (8,8) | .000 | .001 | .009 | .032 | .100 | .214 | .405 | .595 | .786 |
| (8,9) | 000 | .001 | .005 | .020 | .069 | .157 | .319 | .500 | .702 |
| (8,10) | .000 | .000 | .003 | .013 | .048 | .117 | .251 | .419 | .621 |
| (9,9) | .000 | .000 | .003 | .012 | .044 | .109 | .238 | .399 | .601 |
| (9,10) | .000 | .000 | .002 | .008 | .029 | .077 | .179 | .319 | .510 |
| (10,10) | .000 | .000 | .001 | .004 | .019 | .051 | .128 | .242 | .414 |

From "Tables for Testing Randomness of Grouping in a Sequence of Alternatives," C. Eisenhart and F. Swed, *Annals of Mathematical Statistics*, Volume 14 (1943). Reproduced with the kind permission of the Editor, *Annals of Mathematical Statistics*.

Table 10 *(continued)*

| | $\ell$ | | | | | | | | | |
|---|---|---|---|---|---|---|---|---|---|---|
| $(n_1, n_2)$ | 11 | 12 | 13 | 14 | 15 | 16 | 17 | 18 | 19 | 20 |
| (2,3) | | | | | | | | | | |
| (2,4) | | | | | | | | | | |
| (2,5) | | | | | | | | | | |
| (2,6) | | | | | | | | | | |
| (2,7) | | | | | | | | | | |
| (2,8) | | | | | | | | | | |
| (2,9) | | | | | | | | | | |
| (2,10) | | | | | | | | | | |
| (3,3) | | | | | | | | | | |
| (3,4) | | | | | | | | | | |
| (3,5) | | | | | | | | | | |
| (3,6) | | | | | | | | | | |
| (3,7) | | | | | | | | | | |
| (3,8) | | | | | | | | | | |
| (3,9) | | | | | | | | | | |
| (3,10) | | | | | | | | | | |
| (4,4) | | | | | | | | | | |
| (4,5) | | | | | | | | | | |
| (4,6) | | | | | | | | | | |
| (4,7) | | | | | | | | | | |
| (4,8) | | | | | | | | | | |
| (4,9) | | | | | | | | | | |
| (4,10) | | | | | | | | | | |
| (5,5) | | | | | | | | | | |
| (5,6) | 1.000 | | | | | | | | | |
| (5,7) | 1.000 | | | | | | | | | |
| (5,8) | 1.000 | | | | | | | | | |
| (5,9) | 1.000 | | | | | | | | | |
| (5,10) | 1.000 | | | | | | | | | |
| (6,6) | .998 | 1.000 | | | | | | | | |
| (6,7) | .992 | .999 | 1.000 | | | | | | | |
| (6,8) | .984 | .998 | 1.000 | | | | | | | |
| (6,9) | .972 | .994 | 1.000 | | | | | | | |
| (6,10) | .958 | .990 | 1.000 | | | | | | | |
| (7,7) | .975 | .996 | .999 | 1.000 | | | | | | |
| (7,8) | .949 | .988 | .998 | 1.000 | 1.000 | | | | | |
| (7,9) | .916 | .975 | .994 | .999 | 1.000 | | | | | |
| (7,10) | .879 | .957 | .990 | .998 | 1.000 | | | | | |
| (8,8) | .900 | .968 | .991 | .999 | 1.000 | 1.000 | | | | |
| (8,9) | .843 | .939 | .980 | .996 | .999 | 1.000 | 1.000 | | | |
| (8,10) | .782 | .903 | .964 | .990 | .998 | 1.000 | 1.000 | | | |
| (9,9) | .762 | .891 | .956 | .988 | .997 | 1.000 | 1.000 | 1.000 | | |
| (9,10) | .681 | .834 | .923 | .974 | .992 | .999 | 1.000 | 1.000 | 1.000 | |
| (10,10) | .586 | .758 | .872 | .949 | .981 | .996 | .999 | 1.000 | 1.000 | 1.000 |

**Table 11** Percentage points of the studentized range

| Error df | $\alpha$ | *t* = *number of treatment means* 2 | 3 | 4 | 5 | 6 | 7 | 8 | 9 | 10 | 11 |
|---|---|---|---|---|---|---|---|---|---|---|---|
| 5 | .05 | 3.64 | 4.60 | 5.22 | 5.67 | 6.03 | 6.33 | 6.58 | 6.80 | 6.99 | 7.17 |
| | .01 | 5.70 | 6.98 | 7.80 | 8.42 | 8.91 | 9.32 | 9.67 | 9.97 | 10.24 | 10.48 |
| 6 | .05 | 3.46 | 4.34 | 4.90 | 5.30 | 5.63 | 5.90 | 6.12 | 6.32 | 6.49 | 6.65 |
| | .01 | 5.24 | 6.33 | 7.03 | 7.56 | 7.97 | 8.32 | 8.61 | 8.87 | 9.10 | 9.30 |
| 7 | .05 | 3.34 | 4.16 | 4.68 | 5.06 | 5.36 | 5.61 | 5.82 | 6.00 | 6.16 | 6.30 |
| | .01 | 4.95 | 5.92 | 6.54 | 7.01 | 7.37 | 7.68 | 7.94 | 8.17 | 8.37 | 8.55 |
| 8 | .05 | 3.26 | 4.04 | 4.53 | 4.89 | 5.17 | 5.40 | 5.60 | 5.77 | 5.92 | 6.05 |
| | .01 | 4.75 | 5.64 | 6.20 | 6.62 | 6.96 | 7.24 | 7.47 | 7.68 | 7.86 | 8.03 |
| 9 | .05 | 3.20 | 3.95 | 4.41 | 4.76 | 5.02 | 5.24 | 5.43 | 5.59 | 5.74 | 5.87 |
| | .01 | 4.60 | 5.43 | 5.96 | 6.35 | 6.66 | 6.91 | 7.13 | 7.33 | 7.49 | 7.65 |
| 10 | .05 | 3.15 | 3.88 | 4.33 | 4.65 | 4.91 | 5.12 | 5.30 | 5.46 | 5.60 | 5.72 |
| | .01 | 4.48 | 5.27 | 5.77 | 6.14 | 6.43 | 6.67 | 6.87 | 7.05 | 7.21 | 7.36 |
| 11 | .05 | 3.11 | 3.82 | 4.26 | 4.57 | 4.82 | 5.03 | 5.30 | 5.35 | 5.49 | 5.61 |
| | .01 | 4.39 | 5.15 | 5.62 | 5.97 | 6.25 | 6.48 | 6.67 | 6.84 | 6.99 | 7.13 |
| 12 | .05 | 3.08 | 3.77 | 4.20 | 4.52 | 4.75 | 4.95 | 5.12 | 5.27 | 5.39 | 5.51 |
| | .01 | 4.32 | 5.05 | 5.50 | 5.84 | 6.10 | 6.32 | 6.51 | 6.67 | 6.81 | 6.94 |
| 13 | .05 | 3.06 | 3.73 | 4.15 | 4.45 | 4.69 | 4.88 | 5.05 | 5.19 | 5.32 | 5.43 |
| | .01 | 4.26 | 4.96 | 5.40 | 5.73 | 5.98 | 6.19 | 6.37 | 6.53 | 6.67 | 6.79 |
| 14 | .05 | 3.03 | 3.70 | 4.11 | 4.41 | 4.64 | 4.83 | 4.99 | 5.13 | 5.25 | 5.36 |
| | .01 | 4.21 | 4.89 | 5.32 | 5.63 | 5.88 | 6.08 | 6.26 | 6.41 | 6.54 | 6.66 |
| 15 | .05 | 3.01 | 3.67 | 4.08 | 4.37 | 4.59 | 4.78 | 4.94 | 5.08 | 5.20 | 5.31 |
| | .01 | 4.17 | 4.84 | 5.25 | 5.56 | 5.80 | 5.99 | 6.16 | 6.31 | 6.44 | 6.55 |
| 16 | .05 | 3.00 | 3.65 | 4.05 | 4.33 | 4.56 | 4.74 | 4.90 | 5.03 | 5.15 | 5.26 |
| | .01 | 4.13 | 4.79 | 5.19 | 5.49 | 5.72 | 5.92 | 6.08 | 6.22 | 6.35 | 6.46 |
| 17 | .05 | 2.98 | 3.63 | 4.02 | 4.30 | 4.52 | 4.70 | 4.86 | 4.99 | 5.11 | 5.21 |
| | .01 | 4.10 | 4.74 | 5.14 | 5.43 | 5.66 | 5.85 | 6.01 | 6.15 | 6.27 | 6.38 |
| 18 | .05 | 2.97 | 3.61 | 4.00 | 4.28 | 4.49 | 4.67 | 4.82 | 4.96 | 5.07 | 5.17 |
| | .01 | 4.07 | 4.70 | 5.09 | 5.38 | 5.60 | 5.79 | 5.94 | 6.08 | 6.20 | 6.31 |
| 19 | .05 | 2.96 | 3.59 | 3.98 | 4.25 | 4.47 | 4.65 | 4.79 | 4.92 | 5.04 | 5.14 |
| | .01 | 4.05 | 4.67 | 5.05 | 5.33 | 5.55 | 5.73 | 5.89 | 6.02 | 6.14 | 6.25 |
| 20 | .05 | 2.95 | 3.58 | 3.96 | 4.23 | 4.45 | 4.62 | 4.77 | 4.90 | 5.01 | 5.11 |
| | .01 | 4.02 | 4.64 | 5.02 | 5.29 | 5.51 | 5.69 | 5.84 | 5.97 | 6.09 | 6.19 |
| 24 | .05 | 2.92 | 3.53 | 3.90 | 4.17 | 4.37 | 4.54 | 4.68 | 4.81 | 3.92 | 5.01 |
| | .01 | 3.96 | 4.55 | 4.91 | 5.17 | 5.37 | 5.54 | 5.69 | 5.81 | 5.92 | 6.02 |
| 30 | .05 | 2.89 | 3.49 | 3.85 | 4.10 | 4.30 | 4.46 | 4.60 | 4.72 | 4.82 | 4.92 |
| | .01 | 3.89 | 4.45 | 4.80 | 5.05 | 5.24 | 5.40 | 5.54 | 5.65 | 5.76 | 5.85 |
| 40 | .05 | 2.86 | 3.44 | 3.79 | 4.04 | 4.23 | 4.39 | 4.52 | 4.63 | 4.73 | 4.82 |
| | .01 | 3.82 | 4.37 | 4.70 | 4.93 | 5.11 | 5.26 | 5.39 | 5.50 | 5.60 | 5.69 |
| 60 | .05 | 2.83 | 3.40 | 3.74 | 3.98 | 4.16 | 4.31 | 4.44 | 4.55 | 4.65 | 4.73 |
| | .01 | 3.76 | 4.28 | 4.59 | 4.82 | 4.99 | 5.13 | 5.25 | 5.36 | 5.45 | 5.53 |
| 120 | .05 | 2.80 | 3.36 | 3.68 | 3.92 | 4.10 | 4.24 | 4.36 | 4.47 | 4.56 | 4.64 |
| | .01 | 3.70 | 4.20 | 4.50 | 4.71 | 4.87 | 5.01 | 5.12 | 5.21 | 5.30 | 5.37 |
| $\infty$ | .05 | 2.77 | 3.31 | 3.63 | 3.86 | 4.03 | 4.17 | 4.29 | 4.39 | 4.47 | 4.55 |
| | .01 | 3.64 | 4.12 | 4.40 | 4.60 | 4.76 | 4.88 | 4.99 | 5.08 | 5.16 | 5.23 |

This table is abridged from Table 29 in *Biometrika Tables for Statisticians*, Vol. 1, 2d ed. New York: Cambridge, 1958. Edited by E.S. Pearson and H.O. Hartley. Reproduced with the kind permission of the editors and the trustees of *Biometrika*.

Table 11 *(continued)*

| 12 | 13 | 14 | 15 | 16 | 17 | 18 | 19 | 20 | α | Error df |
|---|---|---|---|---|---|---|---|---|---|---|
| | | | | *t* = number of treatment means | | | | | | |
| 7.32 | 7.47 | 7.60 | 7.72 | 7.83 | 7.93 | 8.03 | 8.12 | 8.21 | .05 | 5 |
| 10.70 | 10.89 | 11.08 | 11.24 | 11.40 | 11.55 | 11.68 | 11.81 | 11.93 | .01 | |
| 6.79 | 6.92 | 7.03 | 7.14 | 7.24 | 7.34 | 7.43 | 7.51 | 7.59 | .05 | 6 |
| 9.48 | 9.65 | 9.81 | 9.95 | 10.08 | 10.21 | 10.32 | 10.43 | 10.54 | .01 | |
| 6.43 | 6.55 | 6.66 | 6.76 | 6.85 | 6.94 | 7.02 | 7.10 | 7.17 | .05 | 7 |
| 8.71 | 8.86 | 9.00 | 9.12 | 9.24 | 9.35 | 9.46 | 9.55 | 9.65 | .01 | |
| 6.18 | 6.29 | 6.39 | 6.48 | 6.57 | 6.65 | 6.73 | 6.80 | 6.87 | .05 | 8 |
| 8.18 | 8.31 | 8.44 | 8.55 | 8.66 | 8.76 | 8.85 | 8.94 | 9.03 | .01 | |
| 5.98 | 6.09 | 6.19 | 6.28 | 6.36 | 6.44 | 6.51 | 6.58 | 6.64 | .05 | 9 |
| 7.78 | 7.91 | 8.03 | 8.13 | 8.23 | 8.33 | 8.41 | 8.49 | 8.57 | .01 | |
| 5.83 | 5.93 | 6.03 | 6.11 | 6.19 | 6.27 | 6.34 | 6.40 | 6.47 | .05 | 10 |
| 7.49 | 7.60 | 7.71 | 7.81 | 7.91 | 7.99 | 8.08 | 8.15 | 8.23 | .01 | |
| 5.71 | 5.81 | 5.90 | 5.98 | 6.06 | 6.13 | 6.20 | 6.27 | 6.33 | .05 | 11 |
| 7.25 | 7.36 | 7.46 | 7.56 | 7.65 | 7.73 | 7.81 | 7.88 | 7.95 | .01 | |
| 5.61 | 5.71 | 5.80 | 5.88 | 5.95 | 6.02 | 6.09 | 6.15 | 6.21 | .05 | 12 |
| 7.06 | 7.17 | 7.26 | 7.36 | 7.44 | 7.52 | 7.59 | 7.66 | 7.73 | .01 | |
| 5.53 | 5.63 | 5.71 | 5.79 | 5.86 | 5.93 | 5.99 | 6.05 | 6.11 | .05 | 13 |
| 6.90 | 7.01 | 7.10 | 7.19 | 7.27 | 7.35 | 7.42 | 7.48 | 7.55 | .01 | |
| 5.46 | 5.55 | 5.64 | 5.71 | 5.79 | 5.85 | 5.91 | 5.97 | 6.03 | .05 | 14 |
| 6.77 | 6.87 | 6.96 | 7.05 | 7.13 | 7.20 | 7.27 | 7.33 | 7.39 | .01 | |
| 5.40 | 5.49 | 5.57 | 5.65 | 5.72 | 5.78 | 5.85 | 5.90 | 5.96 | .05 | 15 |
| 6.66 | 6.76 | 6.84 | 6.93 | 7.00 | 7.07 | 7.14 | 7.20 | 7.26 | .01 | |
| 5.35 | 5.44 | 5.52 | 5.59 | 5.66 | 5.73 | 5.79 | 5.84 | 5.90 | .05 | 16 |
| 6.56 | 6.66 | 6.74 | 6.82 | 6.90 | 6.97 | 7.03 | 7.09 | 7.15 | .01 | |
| 5.31 | 5.39 | 5.47 | 5.54 | 5.61 | 5.67 | 5.73 | 5.79 | 5.84 | .05 | 17 |
| 6.48 | 6.57 | 6.66 | 6.73 | 6.81 | 6.87 | 6.94 | 7.00 | 7.05 | .01 | |
| 5.27 | 5.35 | 5.43 | 5.50 | 5.57 | 5.63 | 5.69 | 5.74 | 5.79 | .05 | 18 |
| 6.41 | 6.50 | 6.58 | 6.65 | 6.73 | 6.79 | 6.85 | 6.91 | 6.97 | .01 | |
| 5.23 | 5.31 | 5.39 | 5.46 | 5.53 | 5.59 | 5.65 | 5.70 | 5.75 | .05 | 19 |
| 6.34 | 6.43 | 6.51 | 6.58 | 6.65 | 6.72 | 6.78 | 6.84 | 6.89 | .01 | |
| 5.20 | 5.28 | 5.36 | 5.43 | 5.49 | 5.55 | 5.61 | 5.66 | 5.71 | .05 | 20 |
| 6.28 | 6.37 | 6.45 | 6.52 | 6.59 | 6.65 | 6.71 | 6.77 | 6.82 | .01 | |
| 5.10 | 5.18 | 5.25 | 5.32 | 5.38 | 5.44 | 5.49 | 5.55 | 5.59 | .05 | 24 |
| 6.11 | 6.19 | 6.26 | 6.33 | 6.39 | 6.45 | 6.51 | 6.56 | 6.61 | .01 | |
| 5.00 | 5.08 | 5.15 | 5.21 | 5.27 | 5.33 | 5.38 | 5.43 | 5.47 | .05 | 30 |
| 5.93 | 6.01 | 6.08 | 6.14 | 6.20 | 6.26 | 6.31 | 6.36 | 6.41 | .01 | |
| 4.90 | 4.98 | 5.04 | 5.11 | 5.16 | 5.22 | 5.27 | 5.31 | 5.36 | .05 | 40 |
| 5.76 | 5.83 | 5.90 | 5.96 | 6.02 | 6.07 | 6.12 | 6.16 | 6.21 | .01 | |
| 4.81 | 4.88 | 4.94 | 5.00 | 5.06 | 5.11 | 5.15 | 5.20 | 5.24 | .05 | 60 |
| 5.60 | 5.67 | 5.73 | 5.78 | 5.84 | 5.89 | 5.93 | 5.97 | 6.01 | .01 | |
| 4.71 | 4.78 | 4.84 | 4.90 | 4.95 | 5.00 | 5.04 | 5.09 | 5.13 | .05 | 120 |
| 5.44 | 5.50 | 5.56 | 5.61 | 5.66 | 5.71 | 5.75 | 5.79 | 5.83 | .01 | |
| 4.62 | 4.68 | 4.74 | 4.80 | 4.85 | 4.89 | 4.93 | 4.97 | 5.01 | .05 | ∞ |
| 5.29 | 5.35 | 5.40 | 5.45 | 5.49 | 5.54 | 5.57 | 5.61 | 5.65 | .01 | |

## Table 12 Percentage points of the Duncan new multiple range test

| Error df | α | r = number of ordered steps between means 2 | 3 | 4 | 5 | 6 | 7 | 8 | 9 | 10 | 12 | 14 | 16 | 18 | 20 |
|---|---|---|---|---|---|---|---|---|---|---|---|---|---|---|---|
| 1 | .05 | 18.0 | 18.0 | 18.0 | 18.0 | 18.0 | 18.0 | 18.0 | 18.0 | 18.0 | 18.0 | 18.0 | 18.0 | 18.0 | 18.0 |
| | .01 | 90.0 | 90.0 | 90.0 | 90.0 | 90.0 | 90.0 | 90.0 | 90.0 | 90.0 | 90.0 | 90.0 | 90.0 | 90.0 | 90.0 |
| 2 | .05 | 6.09 | 6.09 | 6.09 | 6.09 | 6.09 | 6.09 | 6.09 | 6.09 | 6.09 | 6.09 | 6.09 | 6.09 | 6.09 | 6.09 |
| | .01 | 14.0 | 14.0 | 14.0 | 14.0 | 14.0 | 14.0 | 14.0 | 14.0 | 14.0 | 14.0 | 14.0 | 14.0 | 14.0 | 14.0 |
| 3 | .05 | 4.50 | 4.50 | 4.50 | 4.50 | 4.50 | 4.50 | 4.50 | 4.50 | 4.50 | 4.50 | 4.50 | 4.50 | 4.50 | 4.50 |
| | .01 | 8.26 | 8.5 | 8.6 | 8.7 | 8.8 | 8.9 | 8.9 | 9.0 | 9.0 | 9.0 | 9.1 | 9.2 | 9.3 | 9.3 |
| 4 | .05 | 3.93 | 4.01 | 4.02 | 4.02 | 4.02 | 4.02 | 4.02 | 4.02 | 4.02 | 4.02 | 4.02 | 4.02 | 4.02 | 4.02 |
| | .01 | 6.51 | 6.8 | 6.9 | 7.0 | 7.1 | 7.1 | 7.2 | 7.2 | 7.3 | 7.3 | 7.4 | 7.4 | 7.5 | 7.5 |
| 5 | .05 | 3.64 | 3.74 | 3.79 | 3.83 | 3.83 | 3.83 | 3.83 | 3.83 | 3.83 | 3.83 | 3.83 | 3.83 | 3.83 | 3.83 |
| | .01 | 5.70 | 5.96 | 6.11 | 6.18 | 6.26 | 6.33 | 6.40 | 6.44 | 6.5 | 6.6 | 6.6 | 6.7 | 6.7 | 6.8 |
| 6 | .05 | 3.46 | 3.58 | 3.64 | 3.68 | 3.68 | 3.68 | 3.68 | 3.68 | 3.68 | 3.68 | 3.68 | 3.68 | 3.68 | 3.68 |
| | .01 | 5.24 | 5.51 | 5.65 | 5.73 | 5.81 | 5.88 | 5.95 | 6.00 | 6.0 | 6.1 | 6.2 | 6.2 | 6.3 | 6.3 |
| 7 | .05 | 3.35 | 3.47 | 3.54 | 3.58 | 3.60 | 3.61 | 3.61 | 3.61 | 3.61 | 3.61 | 3.61 | 3.61 | 3.61 | 3.61 |
| | .01 | 4.95 | 5.22 | 5.37 | 5.45 | 5.53 | 5.61 | 5.69 | 5.73 | 5.8 | 5.8 | 5.9 | 5.9 | 6.0 | 6.0 |
| 8 | .05 | 3.26 | 3.39 | 3.47 | 3.52 | 3.55 | 3.56 | 3.56 | 3.56 | 3.56 | 3.56 | 3.56 | 3.56 | 3.56 | 3.56 |
| | .01 | 4.74 | 5.00 | 5.14 | 5.23 | 5.32 | 5.40 | 5.47 | 5.51 | 5.5 | 5.6 | 5.7 | 5.7 | 5.8 | 5.8 |
| 9 | .05 | 3.20 | 3.34 | 3.41 | 3.47 | 3.50 | 3.52 | 3.52 | 3.52 | 3.52 | 3.52 | 3.52 | 3.52 | 3.52 | 3.52 |
| | .01 | 4.60 | 4.86 | 4.99 | 5.08 | 5.17 | 5.25 | 5.32 | 5.36 | 5.4 | 5.5 | 5.5 | 5.6 | 5.7 | 5.7 |
| 10 | .05 | 3.15 | 3.30 | 3.37 | 3.43 | 3.46 | 3.47 | 3.47 | 3.47 | 3.47 | 3.47 | 3.47 | 3.47 | 3.47 | 3.48 |
| | .01 | 4.48 | 4.73 | 4.88 | 4.96 | 5.06 | 5.13 | 5.20 | 5.24 | 5.28 | 5.36 | 5.42 | 5.48 | 5.54 | 5.55 |
| 11 | .05 | 3.11 | 3.27 | 3.35 | 3.39 | 3.43 | 3.44 | 3.45 | 3.46 | 3.46 | 3.46 | 3.46 | 3.46 | 3.47 | 3.48 |
| | .01 | 4.39 | 4.63 | 4.77 | 4.86 | 4.94 | 5.01 | 5.06 | 5.12 | 5.15 | 5.24 | 5.28 | 5.34 | 5.38 | 5.39 |
| 12 | .05 | 3.08 | 3.23 | 3.33 | 3.36 | 3.40 | 3.42 | 3.44 | 3.44 | 3.46 | 3.46 | 3.46 | 3.46 | 3.47 | 3.48 |
| | .01 | 4.32 | 4.55 | 4.68 | 4.76 | 4.84 | 4.92 | 4.96 | 5.02 | 5.07 | 5.13 | 5.17 | 5.22 | 5.23 | 5.26 |
| 13 | .05 | 3.06 | 3.21 | 3.30 | 3.35 | 3.38 | 3.41 | 3.42 | 3.44 | 3.45 | 3.45 | 3.46 | 3.46 | 3.47 | 3.47 |
| | .01 | 4.26 | 4.48 | 4.62 | 4.69 | 4.74 | 4.84 | 4.88 | 4.94 | 4.98 | 5.04 | 5.08 | 5.13 | 5.14 | 5.15 |
| 14 | .05 | 3.03 | 3.18 | 3.27 | 3.33 | 3.37 | 3.39 | 3.41 | 3.42 | 3.44 | 3.45 | 3.46 | 3.46 | 3.47 | 3.47 |
| | .01 | 4.21 | 4.42 | 4.55 | 4.63 | 4.70 | 4.78 | 4.83 | 4.87 | 4.91 | 4.96 | 5.00 | 5.04 | 5.06 | 5.07 |
| 15 | .05 | 3.01 | 3.16 | 3.25 | 3.31 | 3.36 | 3.38 | 3.40 | 3.42 | 3.43 | 3.44 | 3.45 | 3.46 | 3.47 | 3.47 |
| | .01 | 4.17 | 4.37 | 4.50 | 4.58 | 4.64 | 4.72 | 4.77 | 4.81 | 4.84 | 4.90 | 4.94 | 4.97 | 4.99 | 5.00 |
| 16 | .05 | 3.00 | 3.15 | 3.23 | 3.30 | 3.34 | 3.37 | 3.39 | 3.41 | 3.43 | 3.44 | 3.45 | 3.46 | 3.47 | 3.47 |
| | .01 | 4.13 | 4.34 | 4.45 | 4.54 | 4.60 | 4.67 | 4.72 | 4.76 | 4.79 | 4.84 | 4.88 | 4.91 | 4.93 | 4.94 |
| 17 | .05 | 2.98 | 3.13 | 3.22 | 3.28 | 3.33 | 3.36 | 3.38 | 3.40 | 3.42 | 3.44 | 3.45 | 3.46 | 3.47 | 3.47 |
| | .01 | 4.10 | 4.30 | 4.41 | 4.50 | 4.56 | 4.63 | 4.68 | 4.72 | 4.75 | 4.80 | 4.83 | 4.86 | 4.88 | 4.89 |
| 18 | .05 | 2.97 | 3.12 | 3.21 | 3.27 | 3.32 | 3.35 | 3.37 | 3.39 | 3.41 | 3.43 | 3.45 | 3.46 | 3.47 | 3.47 |
| | .01 | 4.07 | 4.27 | 4.38 | 4.46 | 4.53 | 4.59 | 4.64 | 4.68 | 4.71 | 4.76 | 4.79 | 4.82 | 4.84 | 4.85 |
| 19 | .05 | 2.96 | 3.11 | 3.19 | 3.26 | 3.31 | 3.35 | 3.37 | 3.39 | 3.41 | 3.43 | 3.44 | 3.46 | 3.47 | 3.47 |
| | .01 | 4.05 | 4.24 | 4.35 | 4.43 | 4.50 | 4.56 | 4.61 | 4.64 | 4.67 | 4.72 | 4.76 | 4.79 | 4.81 | 4.82 |
| 20 | .05 | 2.95 | 3.10 | 3.18 | 3.25 | 3.30 | 3.34 | 3.36 | 3.38 | 3.40 | 3.43 | 3.44 | 3.46 | 3.46 | 3.47 |
| | .01 | 4.02 | 4.22 | 4.33 | 4.40 | 4.47 | 4.53 | 4.58 | 4.61 | 4.65 | 4.69 | 4.73 | 4.76 | 4.78 | 4.79 |
| 22 | .05 | 2.93 | 3.08 | 3.17 | 3.24 | 3.29 | 3.32 | 3.35 | 3.37 | 3.39 | 3.42 | 3.44 | 3.45 | 3.46 | 3.47 |
| | .01 | 3.99 | 4.17 | 4.28 | 4.36 | 4.42 | 4.48 | 4.53 | 4.57 | 4.60 | 4.65 | 4.68 | 4.71 | 4.74 | 4.75 |
| 24 | .05 | 2.92 | 3.07 | 3.15 | 3.22 | 3.28 | 3.31 | 3.34 | 3.37 | 3.38 | 3.41 | 3.44 | 3.45 | 3.46 | 3.47 |
| | .01 | 3.96 | 4.14 | 4.24 | 4.33 | 4.39 | 4.44 | 4.49 | 4.53 | 4.57 | 4.62 | 4.64 | 4.67 | 4.70 | 4.72 |
| 26 | .05 | 2.91 | 3.06 | 3.14 | 3.21 | 3.27 | 3.30 | 3.34 | 3.36 | 3.38 | 3.41 | 3.43 | 3.45 | 3.46 | 3.47 |
| | .01 | 3.93 | 4.11 | 4.21 | 4.30 | 4.36 | 4.41 | 4.46 | 4.50 | 4.53 | 4.58 | 4.62 | 4.65 | 4.67 | 4.69 |
| 28 | .05 | 2.90 | 3.04 | 3.13 | 3.20 | 3.26 | 3.30 | 3.33 | 3.35 | 3.37 | 3.40 | 3.43 | 3.45 | 3.46 | 3.47 |
| | .01 | 3.91 | 3.08 | 4.18 | 4.28 | 4.34 | 4.39 | 4.43 | 4.47 | 4.51 | 4.56 | 4.60 | 4.62 | 4.65 | 4.67 |
| 30 | .05 | 2.89 | 3.04 | 3.12 | 3.20 | 3.25 | 3.29 | 3.32 | 3.35 | 3.37 | 3.40 | 3.43 | 3.44 | 3.46 | 3.47 |
| | .01 | 3.89 | 4.06 | 4.16 | 4.22 | 4.32 | 4.36 | 4.41 | 4.45 | 4.48 | 4.54 | 4.58 | 4.61 | 4.63 | 4.65 |
| 40 | .05 | 2.86 | 3.01 | 3.10 | 3.17 | 3.22 | 3.27 | 3.30 | 3.33 | 3.35 | 3.39 | 3.42 | 3.44 | 3.46 | 3.47 |
| | .01 | 3.82 | 3.99 | 4.10 | 4.17 | 4.24 | 4.30 | 4.34 | 4.37 | 4.41 | 4.46 | 4.51 | 4.54 | 4.57 | 4.59 |
| 60 | .05 | 2.83 | 2.98 | 3.08 | 3.14 | 3.20 | 3.24 | 3.28 | 3.31 | 3.33 | 3.37 | 3.40 | 3.43 | 3.45 | 3.47 |
| | .01 | 3.76 | 3.92 | 4.03 | 4.12 | 4.17 | 4.23 | 4.27 | 4.31 | 4.34 | 4.39 | 4.44 | 4.47 | 4.50 | 4.53 |
| 100 | .05 | 2.80 | 2.95 | 3.05 | 3.12 | 3.18 | 3.22 | 3.26 | 3.29 | 3.32 | 3.36 | 3.40 | 3.42 | 3.45 | 3.47 |
| | .01 | 3.71 | 3.86 | 3.93 | 4.06 | 4.11 | 4.17 | 4.21 | 4.25 | 4.29 | 4.35 | 4.38 | 4.42 | 4.45 | 4.48 |
| ∞ | .05 | 2.77 | 2.92 | 3.02 | 3.09 | 3.15 | 3.19 | 3.23 | 3.26 | 3.29 | 3.34 | 3.38 | 3.41 | 3.44 | 3.47 |
| | .01 | 3.64 | 3.80 | 3.90 | 3.98 | 4.04 | 4.09 | 4.14 | 4.17 | 4.20 | 4.26 | 4.31 | 4.34 | 4.38 | 4.41 |

Reproduced from: D.B. Duncan, Multiple Range and Multiple *F* Tests. *Biometrics*, 11: 1–42, 1955. With permission from the Biometric Society and the author.

**Table 13** Waller-Duncan $k$ ratio test
Values of $t_c$ for $k = 100$

$F = 1.2$ (a = .913, $b$ = 2.449)

| $df_1$ \ $df_2$ | 6 | 8 | 10 | 12 | 14 | 16 | 18 | 20 | 24 | 30 | 40 | 60 | 120 |
|---|---|---|---|---|---|---|---|---|---|---|---|---|---|
| 2-6 | * | * | * | * | * | * | * | * | * | * | * | * | * |
| 8 | 2.91 | 2.94 | 2.96 | 2.97 | 2.98 | 2.99 | 2.99 | 2.99 | 3.00 | 3.00 | 3.00 | 3.00 | 3.00 |
| 10 | 2.93 | 2.98 | 3.01 | 3.04 | 3.05 | 3.06 | 3.07 | 3.08 | 3.09 | 3.10 | 3.10 | 3.11 | 3.12 |
| 12 | 2.95 | 3.01 | 3.05 | 3.08 | 3.10 | 3.12 | 3.13 | 3.14 | 3.16 | 3.17 | 3.19 | 3.20 | 3.21 |
| 14 | 2.96 | 3.03 | 3.08 | 3.12 | 3.14 | 3.16 | 3.18 | 3.19 | 3.21 | 3.23 | 3.25 | 3.27 | 3.29 |
| 16 | 2.97 | 3.05 | 3.11 | 3.15 | 3.18 | 3.20 | 3.22 | 3.24 | 3.26 | 3.28 | 3.31 | 3.33 | 3.36 |
| 20 | 2.99 | 3.08 | 3.14 | 3.19 | 3.23 | 3.26 | 3.28 | 3.30 | 3.33 | 3.37 | 3.40 | 3.44 | 3.47 |
| 40 | 3.02 | 3.13 | 3.22 | 3.29 | 3.35 | 3.39 | 3.43 | 3.47 | 3.52 | 3.58 | 3.64 | 3.72 | 3.79 |
| 100 | 3.04 | 3.17 | 3.28 | 3.36 | 3.44 | 3.50 | 3.55 | 3.59 | 3.67 | 3.76 | 3.86 | 3.98 | 4.11 |
| ∞ | 3.05 | 3.20 | 3.32 | 3.42 | 3.50 | 3.58 | 3.64 | 3.70 | 3.80 | 3.91 | 4.06 | 4.24 | 4.45 |

$F = 1.4$ (a = .845, $b$ = 1.871)

| $df_1$ \ $df_2$ | 6 | 8 | 10 | 12 | 14 | 16 | 18 | 20 | 24 | 30 | 40 | 60 | 120 |
|---|---|---|---|---|---|---|---|---|---|---|---|---|---|
| 2-4 | * | * | * | * | * | * | * | * | * | * | * | * | * |
| 6 | 2.85 | 2.84 | 2.83 | 2.82 | 2.82 | 2.81 | 2.80 | 2.80 | 2.79 | 2.78 | 2.77 | 2.75 | 2.74 |
| 8 | 2.88 | 2.89 | 2.90 | 2.90 | 2.90 | 2.89 | 2.89 | 2.89 | 2.88 | 2.88 | 2.87 | 2.86 | 2.85 |
| 10 | 2.90 | 2.93 | 2.94 | 2.95 | 2.95 | 2.96 | 2.96 | 2.96 | 2.95 | 2.95 | 2.95 | 2.94 | 2.93 |
| 12 | 2.92 | 2.95 | 2.98 | 2.99 | 3.00 | 3.00 | 3.01 | 3.01 | 3.01 | 3.01 | 3.01 | 3.00 | 2.99 |
| 14 | 2.93 | 2.97 | 3.00 | 3.02 | 3.03 | 3.04 | 3.04 | 3.05 | 3.05 | 3.06 | 3.06 | 3.05 | 3.05 |
| 16 | 2.94 | 2.99 | 3.02 | 3.04 | 3.06 | 3.07 | 3.08 | 3.08 | 3.09 | 3.09 | 3.10 | 3.10 | 3.09 |
| 20 | 2.95 | 3.01 | 3.05 | 3.08 | 3.10 | 3.11 | 3.12 | 3.13 | 3.14 | 3.15 | 3.16 | 3.16 | 3.16 |
| 40 | 2.98 | 3.06 | 3.12 | 3.16 | 3.19 | 3.22 | 3.24 | 3.25 | 3.28 | 3.30 | 3.31 | 3.32 | 3.32 |
| 100 | 2.99 | 3.09 | 3.16 | 3.22 | 3.26 | 3.29 | 3.32 | 3.34 | 3.38 | 3.41 | 3.43 | 3.45 | 3.42 |
| ∞ | 3.01 | 3.12 | 3.20 | 3.26 | 3.31 | 3.35 | 3.39 | 3.42 | 3.46 | 3.50 | 3.53 | 3.54 | 3.46 |

$F = 1.7$ (a = .767, $b$ = 1.558)

| $df_1$ \ $df_2$ | 6 | 8 | 10 | 12 | 14 | 16 | 18 | 20 | 24 | 30 | 40 | 60 | 120 |
|---|---|---|---|---|---|---|---|---|---|---|---|---|---|
| 2 | * | * | * | * | * | * | * | * | * | * | * | * | * |
| 4 | * | * | * | * | * | 2.61 | 2.59 | 2.58 | 2.56 | 2.54 | 2.52 | 2.50 | 2.48 |
| 6 | 2.82 | 2.79 | 2.76 | 2.74 | 2.72 | 2.71 | 2.70 | 2.69 | 2.67 | 2.65 | 2.63 | 2.61 | 2.58 |
| 8 | 2.84 | 2.83 | 2.81 | 2.80 | 2.78 | 2.77 | 2.76 | 2.75 | 2.74 | 2.72 | 2.70 | 2.68 | 2.65 |
| 10 | 2.86 | 2.86 | 2.85 | 2.84 | 2.83 | 2.82 | 2.81 | 2.80 | 2.79 | 2.77 | 2.75 | 2.73 | 2.70 |
| 12 | 2.87 | 2.88 | 2.88 | 2.87 | 2.86 | 2.85 | 2.84 | 2.84 | 2.82 | 2.81 | 2.79 | 2.76 | 2.73 |
| 14 | 2.88 | 2.90 | 2.90 | 2.89 | 2.89 | 2.88 | 2.87 | 2.86 | 2.85 | 2.83 | 2.81 | 2.79 | 2.75 |
| 16 | 2.89 | 2.91 | 2.91 | 2.91 | 2.90 | 2.90 | 2.89 | 2.89 | 2.87 | 2.86 | 2.84 | 2.81 | 2.77 |
| 20 | 2.90 | 2.93 | 2.93 | 2.94 | 2.93 | 2.93 | 2.92 | 2.92 | 2.91 | 2.89 | 2.87 | 2.84 | 2.80 |
| 40 | 2.93 | 2.97 | 2.99 | 3.00 | 3.00 | 3.00 | 3.00 | 2.99 | 2.98 | 2.97 | 2.94 | 2.89 | 2.83 |
| 100 | 2.94 | 2.99 | 3.02 | 3.04 | 3.05 | 3.05 | 3.05 | 3.05 | 3.04 | 3.02 | 2.98 | 2.92 | 2.83 |
| ∞ | 2.95 | 3.01 | 3.05 | 3.07 | 3.08 | 3.09 | 3.09 | 3.08 | 3.07 | 3.05 | 3.01 | 2.93 | 2.81 |

$F = 2.0$ (a = .707, $b$ = 1.414)

| $df_1$ \ $df_2$ | 6 | 8 | 10 | 12 | 14 | 16 | 18 | 20 | 24 | 30 | 40 | 60 | 120 |
|---|---|---|---|---|---|---|---|---|---|---|---|---|---|
| 2 | * | * | * | * | * | * | * | * | * | * | * | * | * |
| 4 | 2.74 | 2.67 | 2.63 | 2.59 | 2.56 | 2.54 | 2.52 | 2.51 | 2.49 | 2.46 | 2.44 | 2.41 | 2.39 |
| 6 | 2.79 | 2.74 | 2.70 | 2.67 | 2.64 | 2.62 | 2.60 | 2.59 | 2.57 | 2.54 | 2.52 | 2.49 | 2.46 |
| 8 | 2.81 | 2.77 | 2.74 | 2.71 | 2.69 | 2.67 | 2.65 | 2.64 | 2.62 | 2.59 | 2.56 | 2.53 | 2.49 |
| 10 | 2.83 | 2.80 | 2.77 | 2.74 | 2.72 | 2.70 | 2.69 | 2.67 | 2.65 | 2.62 | 2.59 | 2.56 | 2.52 |
| 12 | 2.84 | 2.82 | 2.79 | 2.77 | 2.75 | 2.73 | 2.71 | 2.70 | 2.67 | 2.64 | 2.61 | 2.57 | 2.53 |
| 14 | 2.85 | 2.83 | 2.81 | 2.79 | 2.77 | 2.75 | 2.73 | 2.72 | 2.69 | 2.66 | 2.63 | 2.59 | 2.54 |
| 16 | 2.85 | 2.84 | 2.82 | 2.80 | 2.78 | 2.76 | 2.74 | 2.73 | 2.70 | 2.67 | 2.64 | 2.59 | 2.54 |
| 20 | 2.86 | 2.85 | 2.84 | 2.82 | 2.80 | 2.78 | 2.77 | 2.75 | 2.72 | 2.69 | 2.65 | 2.61 | 2.55 |
| 40 | 2.88 | 2.89 | 2.88 | 2.86 | 2.85 | 2.83 | 2.81 | 2.80 | 2.77 | 2.73 | 2.68 | 2.62 | 2.55 |
| 100 | 2.89 | 2.91 | 2.90 | 2.89 | 2.88 | 2.86 | 2.84 | 2.82 | 2.79 | 2.75 | 2.69 | 2.62 | 2.53 |
| ∞ | 2.90 | 2.92 | 2.92 | 2.91 | 2.90 | 2.88 | 2.86 | 2.85 | 2.81 | 2.76 | 2.69 | 2.61 | 2.52 |

$F = 2.4$ (a = .645, $b$ = 1.309)

| $df_1$ \ $df_2$ | 6 | 8 | 10 | 12 | 14 | 16 | 18 | 20 | 24 | 30 | 40 | 60 | 120 |
|---|---|---|---|---|---|---|---|---|---|---|---|---|---|
| 2 | * | * | * | * | * | * | * | * | * | * | * | * | 2.18 |
| 4 | 2.71 | 2.63 | 2.57 | 2.53 | 2.49 | 2.47 | 2.44 | 2.43 | 2.40 | 2.37 | 2.34 | 2.31 | 2.28 |
| 6 | 2.75 | 2.68 | 2.63 | 2.58 | 2.55 | 2.52 | 2.50 | 2.48 | 2.46 | 2.42 | 2.39 | 2.36 | 2.32 |
| 8 | 2.77 | 2.71 | 2.66 | 2.62 | 2.59 | 2.56 | 2.54 | 2.52 | 2.49 | 2.45 | 2.42 | 2.38 | 2.34 |
| 10 | 2.79 | 2.73 | 2.68 | 2.64 | 2.61 | 2.58 | 2.56 | 2.54 | 2.50 | 2.47 | 2.43 | 2.39 | 2.34 |
| 12 | 2.79 | 2.74 | 2.70 | 2.66 | 2.62 | 2.60 | 2.57 | 2.55 | 2.52 | 2.48 | 2.44 | 2.39 | 2.35 |
| 14 | 2.80 | 2.75 | 2.71 | 2.67 | 2.64 | 2.61 | 2.58 | 2.56 | 2.53 | 2.49 | 2.44 | 2.40 | 2.35 |
| 16 | 2.81 | 2.76 | 2.72 | 2.68 | 2.65 | 2.62 | 2.59 | 2.57 | 2.53 | 2.49 | 2.45 | 2.40 | 2.34 |
| 20 | 2.82 | 2.77 | 2.73 | 2.69 | 2.66 | 2.63 | 2.60 | 2.58 | 2.54 | 2.50 | 2.45 | 2.40 | 2.34 |
| 40 | 2.83 | 2.80 | 2.76 | 2.72 | 2.69 | 2.66 | 2.63 | 2.60 | 2.56 | 2.51 | 2.46 | 2.39 | 2.33 |
| 100 | 2.84 | 2.81 | 2.78 | 2.74 | 2.71 | 2.67 | 2.64 | 2.62 | 2.57 | 2.51 | 2.45 | 2.39 | 2.32 |
| ∞ | 2.85 | 2.83 | 2.79 | 2.76 | 2.72 | 2.68 | 2.65 | 2.62 | 2.57 | 2.51 | 2.45 | 2.38 | 2.31 |

$F = 3.0$ (a = .577, $b$ = 1.225)

| $df_1$ \ $df_2$ | 6 | 8 | 10 | 12 | 14 | 16 | 18 | 20 | 24 | 30 | 40 | 60 | 120 |
|---|---|---|---|---|---|---|---|---|---|---|---|---|---|
| 2 | * | * | 2.41 | 2.36 | 2.32 | 2.29 | 2.27 | 2.25 | 2.22 | 2.20 | 2.17 | 2.14 | 2.11 |
| 4 | 2.68 | 2.57 | 2.50 | 2.45 | 2.41 | 2.38 | 2.35 | 2.33 | 2.30 | 2.27 | 2.24 | 2.20 | 2.17 |
| 6 | 2.71 | 2.61 | 2.54 | 2.49 | 2.44 | 2.41 | 2.39 | 2.36 | 2.33 | 2.29 | 2.26 | 2.22 | 2.18 |
| 8 | 2.72 | 2.63 | 2.56 | 2.51 | 2.47 | 2.43 | 2.40 | 2.38 | 2.34 | 2.31 | 2.27 | 2.22 | 2.18 |
| 10 | 2.74 | 2.65 | 2.58 | 2.52 | 2.48 | 2.44 | 2.41 | 2.39 | 2.35 | 2.31 | 2.27 | 2.22 | 2.18 |
| 12 | 2.74 | 2.66 | 2.59 | 2.53 | 2.49 | 2.45 | 2.42 | 2.40 | 2.36 | 2.31 | 2.27 | 2.22 | 2.18 |
| 14 | 2.75 | 2.66 | 2.60 | 2.54 | 2.49 | 2.46 | 2.43 | 2.40 | 2.36 | 2.32 | 2.27 | 2.22 | 2.17 |
| 16 | 2.75 | 2.67 | 2.60 | 2.55 | 2.50 | 2.46 | 2.43 | 2.40 | 2.36 | 2.32 | 2.27 | 2.22 | 2.17 |
| 20 | 2.76 | 2.68 | 2.61 | 2.55 | 2.51 | 2.47 | 2.43 | 2.41 | 2.36 | 2.32 | 2.27 | 2.22 | 2.17 |
| 40 | 2.77 | 2.70 | 2.63 | 2.57 | 2.52 | 2.48 | 2.44 | 2.41 | 2.37 | 2.32 | 2.26 | 2.21 | 2.16 |
| 100 | 2.78 | 2.71 | 2.64 | 2.58 | 2.53 | 2.49 | 2.45 | 2.42 | 2.37 | 2.31 | 2.26 | 2.21 | 2.16 |
| ∞ | 2.79 | 2.71 | 2.65 | 2.59 | 2.53 | 2.49 | 2.45 | 2.42 | 2.37 | 2.31 | 2.26 | 2.20 | 2.15 |

$F = 4.0$ (a = .500, $b$ = 1.155)

| $df_1$ \ $df_2$ | 6 | 8 | 10 | 12 | 14 | 16 | 18 | 20 | 24 | 30 | 40 | 60 | 120 |
|---|---|---|---|---|---|---|---|---|---|---|---|---|---|
| 2 | 2.58 | 2.44 | 2.35 | 2.29 | 2.25 | 2.22 | 2.20 | 2.18 | 2.15 | 2.12 | 2.09 | 2.06 | 2.03 |
| 4 | 2.63 | 2.50 | 2.41 | 2.35 | 2.30 | 2.27 | 2.24 | 2.22 | 2.18 | 2.15 | 2.12 | 2.08 | 2.05 |
| 6 | 2.65 | 2.52 | 2.43 | 2.37 | 2.32 | 2.28 | 2.25 | 2.23 | 2.19 | 2.16 | 2.12 | 2.08 | 2.04 |
| 10 | 2.67 | 2.55 | 2.46 | 2.39 | 2.34 | 2.30 | 2.26 | 2.24 | 2.20 | 2.16 | 2.12 | 2.08 | 2.04 |
| 20 | 2.69 | 2.57 | 2.47 | 2.40 | 2.35 | 2.30 | 2.27 | 2.24 | 2.20 | 2.15 | 2.11 | 2.07 | 2.03 |
| ∞ | 2.71 | 2.59 | 2.49 | 2.42 | 2.36 | 2.31 | 2.27 | 2.24 | 2.19 | 2.15 | 2.11 | 2.06 | 2.02 |

$F = 6.0$ (a = .408, $b$ = 1.095)

| $df_1$ \ $df_2$ | 6 | 8 | 10 | 12 | 14 | 16 | 18 | 20 | 24 | 30 | 40 | 60 | 120 |
|---|---|---|---|---|---|---|---|---|---|---|---|---|---|
| 2 | 2.53 | 2.37 | 2.27 | 2.21 | 2.16 | 2.13 | 2.10 | 2.08 | 2.05 | 2.02 | 1.99 | 1.96 | 1.93 |
| 4 | 2.56 | 2.40 | 2.30 | 2.23 | 2.18 | 2.14 | 2.12 | 2.09 | 2.06 | 2.02 | 1.99 | 1.96 | 1.93 |
| 6 | 2.58 | 2.42 | 2.31 | 2.24 | 2.19 | 2.15 | 2.12 | 2.09 | 2.06 | 2.02 | 1.99 | 1.95 | 1.92 |
| 10 | 2.59 | 2.43 | 2.32 | 2.24 | 2.19 | 2.15 | 2.12 | 2.09 | 2.06 | 2.02 | 1.99 | 1.95 | 1.92 |
| 20 | 2.60 | 2.44 | 2.32 | 2.25 | 2.19 | 2.15 | 2.12 | 2.09 | 2.05 | 2.02 | 1.98 | 1.95 | 1.92 |
| ∞ | 2.61 | 2.44 | 2.33 | 2.25 | 2.19 | 2.15 | 2.12 | 2.09 | 2.05 | 2.02 | 1.98 | 1.95 | 1.92 |

$F = 10.0$ (a = .316, $b$ = 1.054)

| $df_1$ \ $df_2$ | 6 | 8 | 10 | 12 | 14 | 16 | 18 | 20 | 24 | 30 | 40 | 60 | 120 |
|---|---|---|---|---|---|---|---|---|---|---|---|---|---|
| 2 | 2.48 | 2.30 | 2.19 | 2.12 | 2.07 | 2.04 | 2.01 | 1.99 | 1.96 | 1.93 | 1.90 | 1.87 | 1.85 |
| 4 | 2.49 | 2.31 | 2.20 | 2.13 | 2.08 | 2.04 | 2.01 | 1.99 | 1.96 | 1.93 | 1.90 | 1.87 | 1.84 |
| 6 | 2.50 | 2.31 | 2.20 | 2.13 | 2.08 | 2.04 | 2.01 | 1.99 | 1.96 | 1.93 | 1.90 | 1.87 | 1.84 |
| ⩾ 10 | 2.51 | 2.32 | 2.20 | 2.13 | 2.08 | 2.04 | 2.01 | 1.99 | 1.96 | 1.93 | 1.90 | 1.87 | 1.84 |

$F = 25.0$ (a = .200, $b$ = 1.021)

| $df_1$ \ $df_2$ | 6 | 8 | 10 | 12 | 14 | 16 | 18 | 20 | 24 | 30 | 40 | 60 | 120 |
|---|---|---|---|---|---|---|---|---|---|---|---|---|---|
| 2-4 | 2.40 | 2.20 | 2.10 | 2.03 | 1.99 | 1.95 | 1.93 | 1.91 | 1.88 | 1.86 | 1.83 | 1.80 | 1.78 |
| ⩾ 6 | 2.41 | 2.21 | 2.10 | 2.03 | 1.99 | 1.95 | 1.93 | 1.91 | 1.88 | 1.86 | 1.83 | 1.80 | 1.78 |

$F = \infty$ (a = 0, $b$ = 1)

| $df_1$ \ $df_2$ | 6 | 8 | 10 | 12 | 14 | 16 | 18 | 20 | 24 | 30 | 40 | 60 | 120 |
|---|---|---|---|---|---|---|---|---|---|---|---|---|---|
| ⩾ 2 | 2.33 | 2.13 | 2.03 | 1.97 | 1.93 | 1.90 | 1.88 | 1.86 | 1.84 | 1.81 | 1.79 | 1.76 | 1.74 |

*All differences not significant. $a = 1/F^{1/2}$; $b = [F/(F-1)]^{1/2}$.
From "Corrigenda" by R. Waller, and D. Duncan. *Journal of the American Statistical Association*, 67, 1972, pp. 254-55, Table A2. Reproduced by permission of the American Statistical Association.

**Table 14** Waller-Duncan $k$ ratio test
Values of $t_c$ for $k = 500$

| | df$_2$ | | | | | | | | | | | | |
|---|---|---|---|---|---|---|---|---|---|---|---|---|---|
| df$_1$ | 6 | 8 | 10 | 12 | 14 | 16 | 18 | 20 | 24 | 30 | 40 | 60 | 120 |
| $F = 1.2$ (a = .913, $b$ = 2.449) | | | | | | | | | | | | | |
| 2-16 | * | * | * | * | * | * | * | * | * | * | * | * | * |
| 20 | 4.70 | 4.82 | 4.89 | * | * | * | * | * | * | * | * | * | * |
| 40 | 4.75 | 4.91 | 5.03 | 5.12 | 5.20 | 5.25 | 5.30 | 5.34 | 5.41 | 5.48 | 5.55 | 5.61 | 5.67 |
| 100 | 4.79 | 4.98 | 5.13 | 5.25 | 5.34 | 5.43 | 5.50 | 5.56 | 5.65 | 5.76 | 5.89 | 6.02 | 6.13 |
| ∞ | 4.81 | 5.03 | 5.20 | 5.34 | 5.46 | 5.56 | 5.65 | 5.73 | 5.86 | 6.02 | 6.20 | 6.41 | 6.56 |
| $F = 1.4$ (a = .845, $b$ = 1.871) | | | | | | | | | | | | | |
| 2-14 | * | * | * | * | * | * | * | * | * | * | * | * | * |
| 16 | 4.61 | 4.66 | 4.68 | 4.69 | 4.69 | 4.69 | 4.69 | 4.68 | 4.67 | 4.65 | 4.62 | 4.58 | 4.53 |
| 20 | 4.64 | 4.70 | 4.73 | 4.75 | 4.76 | 4.77 | 4.77 | 4.76 | 4.76 | 4.74 | 4.72 | 4.68 | 4.62 |
| 40 | 4.68 | 4.78 | 4.85 | 4.89 | 4.92 | 4.94 | 4.96 | 4.96 | 4.97 | 4.97 | 4.95 | 4.90 | 4.81 |
| ∞ | 4.74 | 4.88 | 4.99 | 5.06 | 5.12 | 5.17 | 5.20 | 5.23 | 5.26 | 5.28 | 5.26 | 5.16 | 4.82 |
| $F = 1.7$ (a = .767, $b$ = 1.558) | | | | | | | | | | | | | |
| 2-8 | * | * | * | * | * | * | * | * | * | * | * | * | * |
| 10 | * | * | * | * | * | * | * | * | * | 4.08 | 4.02 | 3.95 | 3.87 |
| 12 | 4.50 | 4.46 | 4.42 | 4.38 | 4.34 | 4.30 | 4.27 | 4.24 | 4.19 | 4.14 | 4.07 | 3.99 | 3.90 |
| 20 | 4.55 | 4.54 | 4.52 | 4.49 | 4.46 | 4.43 | 4.40 | 4.37 | 4.32 | 4.26 | 4.18 | 4.08 | 3.95 |
| 40 | 4.59 | 4.61 | 4.61 | 4.60 | 4.57 | 4.55 | 4.52 | 4.49 | 4.44 | 4.36 | 4.26 | 4.12 | 3.93 |
| ∞ | 4.64 | 4.69 | 4.71 | 4.72 | 4.71 | 4.69 | 4.66 | 4.63 | 4.57 | 4.46 | 4.31 | 4.07 | 3.76 |
| $F = 2.0$ (a = .707, $b$ = 1.414) | | | | | | | | | | | | | |
| 2-6 | * | * | * | * | * | * | * | * | * | * | * | * | * |
| 8 | * | * | * | * | * | 3.98 | 3.93 | 3.89 | 3.83 | 3.76 | 3.69 | 3.60 | 3.51 |
| 10 | 4.41 | 4.31 | 4.22 | 4.15 | 4.08 | 4.03 | 3.98 | 3.94 | 3.88 | 3.80 | 3.72 | 3.63 | 3.53 |
| 20 | 4.48 | 4.41 | 4.34 | 4.27 | 4.21 | 4.16 | 4.10 | 4.06 | 3.98 | 3.89 | 3.78 | 3.65 | 3.51 |
| 40 | 4.51 | 4.47 | 4.41 | 4.35 | 4.29 | 4.23 | 4.17 | 4.12 | 4.03 | 3.92 | 3.78 | 3.62 | 3.44 |
| ∞ | 4.55 | 4.53 | 4.49 | 4.43 | 4.37 | 4.31 | 4.25 | 4.19 | 4.07 | 3.93 | 3.75 | 3.54 | 3.33 |
| $F = 2.4$ (a = .645, $b$ = 1.309) | | | | | | | | | | | | | |
| 2-4 | * | * | * | * | * | * | * | * | * | * | * | * | * |
| 6 | * | * | * | * | 3.77 | 3.71 | 3.65 | 3.61 | 3.54 | 3.47 | 3.39 | 3.30 | 3.22 |
| 8 | 4.31 | 4.14 | 4.01 | 3.91 | 3.83 | 3.76 | 3.70 | 3.66 | 3.58 | 3.50 | 3.41 | 3.32 | 3.22 |
| 10 | 4.33 | 4.18 | 4.05 | 3.95 | 3.87 | 3.79 | 3.73 | 3.68 | 3.60 | 3.51 | 3.42 | 3.31 | 3.21 |
| 20 | 4.39 | 4.26 | 4.14 | 4.04 | 3.95 | 3.87 | 3.80 | 3.74 | 3.64 | 3.53 | 3.41 | 3.28 | 3.15 |
| ∞ | 4.45 | 4.35 | 4.25 | 4.14 | 4.03 | 3.94 | 3.85 | 3.78 | 3.64 | 3.50 | 3.34 | 3.18 | 3.04 |
| $F = 3.0$ (a = .577, $b$ = 1.225) | | | | | | | | | | | | | |
| 2 | * | * | * | * | * | * | * | * | * | * | * | * | * |
| 4 | * | * | * | * | * | 3.43 | 3.38 | 3.33 | 3.26 | 3.19 | 3.12 | 3.04 | 2.97 |
| 6 | 4.19 | 3.95 | 3.79 | 3.66 | 3.56 | 3.49 | 3.43 | 3.37 | 3.30 | 3.21 | 3.13 | 3.04 | 2.95 |
| 10 | 4.24 | 4.02 | 3.85 | 3.72 | 3.62 | 3.53 | 3.46 | 3.40 | 3.31 | 3.21 | 3.12 | 3.02 | 2.92 |
| 20 | 4.28 | 4.08 | 3.91 | 3.77 | 3.65 | 3.56 | 3.48 | 3.41 | 3.31 | 3.20 | 3.09 | 2.98 | 2.87 |
| ∞ | 4.33 | 4.15 | 3.97 | 3.82 | 3.69 | 3.57 | 3.48 | 3.40 | 3.28 | 3.15 | 3.03 | 2.92 | 2.82 |

*All differences not significant. $a = 1/F^{1/2}$; $b = [F/(F-1)]^{1/2}$.

From "Corrigenda" by R. Waller and D. Duncan. *Journal of the American Statistical Association,* 67, 1972, p. 255, Table A3. Reproduced by permission of the American Statistical Association.

Table 14 *(continued)*

| $df_1$ | 6 | 8 | 10 | 12 | 14 | 16 | 18 | 20 | 24 | 30 | 40 | 60 | 120 |
|---|---|---|---|---|---|---|---|---|---|---|---|---|---|
| | | | | | | | $df_2$ | | | | | | |
| $F = 4.0$ (a = .500, $b$ = 1.155) | | | | | | | | | | | | | |
| 2 | * | * | * | * | * | * | * | * | * | * | * | 2.81 | 2.75 |
| 4 | * | 3.74 | 3.54 | 3.40 | 3.30 | 3.22 | 3.16 | 3.11 | 3.04 | 2.96 | 2.89 | 2.81 | 2.74 |
| 6 | 4.08 | 3.78 | 3.58 | 3.43 | 3.32 | 3.24 | 3.17 | 3.12 | 3.04 | 2.95 | 2.87 | 2.79 | 2.71 |
| 10 | 4.12 | 3.83 | 3.62 | 3.46 | 3.34 | 3.25 | 3.17 | 3.11 | 3.03 | 2.94 | 2.85 | 2.77 | 2.69 |
| 20 | 4.15 | 3.86 | 3.64 | 3.48 | 3.35 | 3.25 | 3.17 | 3.10 | 3.01 | 2.92 | 2.83 | 2.74 | 2.66 |
| ∞ | 4.19 | 3.90 | 3.67 | 3.49 | 3.35 | 3.24 | 3.15 | 3.09 | 2.99 | 2.89 | 2.80 | 2.72 | 2.65 |
| $F = 6.0$ (a = .408, $b$ = 1.095) | | | | | | | | | | | | | |
| 2 | * | * | 3.28 | 3.14 | 3.04 | 2.97 | 2.91 | 2.87 | 2.81 | 2.74 | 2.68 | 2.62 | 2.56 |
| 4 | 3.90 | 3.54 | 3.32 | 3.17 | 3.06 | 2.98 | 2.92 | 2.87 | 2.80 | 2.73 | 2.66 | 2.60 | 2.53 |
| 6 | 3.93 | 3.57 | 3.33 | 3.18 | 3.06 | 2.98 | 2.91 | 2.86 | 2.79 | 2.72 | 2.65 | 2.58 | 2.52 |
| 10 | 3.95 | 3.59 | 3.34 | 3.18 | 3.06 | 2.97 | 2.91 | 2.85 | 2.78 | 2.71 | 2.64 | 2.57 | 2.51 |
| 20 | 3.97 | 3.60 | 3.35 | 3.18 | 3.06 | 2.97 | 2.90 | 2.84 | 2.77 | 2.70 | 2.63 | 2.56 | 2.51 |
| ∞ | 3.99 | 3.62 | 3.36 | 3.18 | 3.05 | 2.96 | 2.89 | 2.83 | 2.76 | 2.69 | 2.62 | 2.56 | 2.50 |
| $F = 10.0$ (a = .316, $b$ = 1.054) | | | | | | | | | | | | | |
| 2 | 3.72 | 3.33 | 3.10 | 2.96 | 2.86 | 2.79 | 2.74 | 2.70 | 2.64 | 2.58 | 2.52 | 2.47 | 2.42 |
| 4 | 3.75 | 3.35 | 3.11 | 2.96 | 2.86 | 2.79 | 2.73 | 2.69 | 2.63 | 2.57 | 2.51 | 2.46 | 2.41 |
| 10 | 3.78 | 3.36 | 3.11 | 2.96 | 2.85 | 2.78 | 2.72 | 2.68 | 2.62 | 2.56 | 2.50 | 2.45 | 2.40 |
| 20 | 3.79 | 3.36 | 3.11 | 2.96 | 2.85 | 2.78 | 2.72 | 2.68 | 2.62 | 2.56 | 2.50 | 2.45 | 2.40 |
| ∞ | 3.80 | 3.37 | 3.11 | 2.95 | 2.85 | 2.77 | 2.72 | 2.67 | 2.61 | 2.56 | 2.50 | 2.45 | 2.40 |
| $F = 25.0$ (a = .200, $b$ = 1.021) | | | | | | | | | | | | | |
| 2 | 3.55 | 3.14 | 2.92 | 2.79 | 2.70 | 2.64 | 2.59 | 2.56 | 2.51 | 2.46 | 2.41 | 2.36 | 2.32 |
| 10 | 3.57 | 3.14 | 2.92 | 2.79 | 2.70 | 2.64 | 2.59 | 2.55 | 2.50 | 2.45 | 2.41 | 2.36 | 2.32 |
| ∞ | 3.57 | 3.14 | 2.92 | 2.78 | 2.70 | 2.63 | 2.59 | 2.55 | 2.50 | 2.45 | 2.41 | 2.36 | 2.32 |
| $F = \infty$ (a = 0, $b$ = 1) | | | | | | | | | | | | | |
| ⩾ 2 | 3.39 | 3.00 | 2.80 | 2.69 | 2.61 | 2.55 | 2.51 | 2.48 | 2.44 | 2.39 | 2.35 | 2.31 | 2.27 |

## Table 15 Orthogonal polynomials

For $x$ assuming the values 1, 2, 3, . . . , $p$

$$x_1^* = \lambda_1 (x - \bar{x})$$

$$x_2^* = \lambda_2 \left\{ (x - \bar{x})^2 - \frac{p^2 - 1}{12} \right\}$$

$$x_3^* = \lambda_3 \left\{ (x - \bar{x})^3 - (x - \bar{x}) \left( \frac{3p^2 - 7}{20} \right) \right\}$$

$$x_4^* = \lambda_4 \left\{ (x - \bar{x})^4 - (x - \bar{x})^2 \left( \frac{3p^2 - 13}{14} \right) + \frac{3(p^2 - 1)(p^2 - 9)}{560} \right\}$$

| | $p = 3$ | | $p = 4$ | | | $p = 5$ | | | | $p = 6$ | | | | $p = 7$ | | | |
|---|---|---|---|---|---|---|---|---|---|---|---|---|---|---|---|---|---|
| | $x_1^*$ | $x_2^*$ | $x_1^*$ | $x_2^*$ | $x_3^*$ | $x_1^*$ | $x_2^*$ | $x_3^*$ | $x_4^*$ | $x_1^*$ | $x_2^*$ | $x_3^*$ | $x_4^*$ | $x_1^*$ | $x_2^*$ | $x_3^*$ | $x_4^*$ |
| | -1 | 1 | -3 | 1 | -1 | -2 | 2 | -1 | 1 | -5 | 5 | -5 | 1 | -3 | 5 | -1 | 3 |
| | 0 | -2 | -1 | -1 | 3 | -1 | -1 | 2 | -4 | -3 | -1 | 7 | -3 | -2 | 0 | 1 | -7 |
| | 1 | 1 | 1 | -1 | -3 | 0 | -2 | 0 | 6 | -1 | -4 | 4 | 2 | -1 | -3 | 1 | 1 |
| | | | 3 | 1 | 1 | 1 | -1 | -2 | -4 | 1 | -4 | -4 | 2 | 0 | -4 | 0 | 6 |
| | | | | | | 2 | 2 | 1 | 1 | 3 | -1 | -7 | -3 | 1 | -3 | -1 | 1 |
| | | | | | | | | | | 5 | 5 | 5 | 1 | 2 | 0 | -1 | -7 |
| | | | | | | | | | | | | | | 3 | 5 | 1 | 3 |
| $\Sigma x_i^{*2}$ | 2 | 6 | 20 | 4 | 20 | 10 | 14 | 10 | 70 | 70 | 84 | 180 | 28 | 28 | 84 | 6 | 154 |
| $\lambda_i$ | 1 | 3 | 2 | 1 | $\frac{10}{3}$ | 1 | 1 | $\frac{5}{6}$ | $\frac{35}{12}$ | 2 | $\frac{3}{2}$ | $\frac{5}{3}$ | $\frac{7}{12}$ | 1 | 1 | $\frac{1}{6}$ | $\frac{7}{12}$ |

| | $p = 8$ | | | | $p = 9$ | | | | $p = 10$ | | | |
|---|---|---|---|---|---|---|---|---|---|---|---|---|
| | $x_1^*$ | $x_2^*$ | $x_3^*$ | $x_4^*$ | $x_1^*$ | $x_2^*$ | $x_3^*$ | $x_4^*$ | $x_1^*$ | $x_2^*$ | $x_3^*$ | $x_4^*$ |
| | -7 | 7 | -7 | 7 | -4 | 28 | -14 | 14 | -9 | 6 | -42 | 18 |
| | -5 | 1 | 5 | -13 | -3 | 7 | 7 | -21 | -7 | 2 | 14 | -22 |
| | -3 | -3 | 7 | -3 | -2 | -8 | 13 | -11 | -5 | -1 | 35 | -17 |
| | -1 | -5 | 3 | 9 | -1 | -17 | 9 | 9 | -3 | -3 | 31 | 3 |
| | 1 | -5 | -3 | 9 | 0 | -20 | 0 | 18 | -1 | -4 | 12 | 18 |
| | 3 | -3 | -7 | -3 | | | | | 1 | -4 | -12 | 18 |
| | 5 | 1 | -5 | -13 | 1 | -17 | -9 | 9 | 3 | -3 | -31 | 3 |
| | 7 | 7 | 7 | 7 | 2 | -8 | -13 | -11 | 5 | -1 | -35 | -17 |
| | | | | | 3 | 7 | -7 | -21 | 7 | 2 | -14 | -22 |
| | | | | | 4 | 28 | 14 | 14 | 9 | 6 | 42 | 18 |
| $\Sigma x_i^{*2}$ | 168 | 168 | 264 | 616 | 60 | 2,772 | 990 | 2,002 | 330 | 132 | 8,580 | 2,860 |
| $\lambda_i$ | 2 | 1 | $\frac{2}{3}$ | $\frac{7}{12}$ | 1 | 3 | $\frac{5}{6}$ | $\frac{7}{12}$ | 2 | $\frac{1}{2}$ | $\frac{5}{3}$ | $\frac{5}{12}$ |

Abridged from *Biometrika Tables for Statisticians*, Vol. I, edited by E.S. Pearson and H.O. Hartley (New York: Cambridge University Press, 1966) Table 47, p. 236. Reproduced by permission of the *Biometrika* Trustees.

**Table 16** Testing for skewness
Upper-tail values* for $\hat{\gamma}_1 = m_3/m_2^{3/2}$

| Sample Size $n$ | a = .05 | a = .01 |
|---|---|---|
| 25 | 0.711 | 1.061 |
| 30 | 0.662 | 0.986 |
| 35 | 0.621 | 0.923 |
| 40 | 0.587 | 0.870 |
| 45 | 0.558 | 0.825 |
| 50 | 0.534 | 0.787 |
| 60 | 0.492 | 0.723 |
| 70 | 0.459 | 0.673 |
| 80 | 0.432 | 0.631 |
| 90 | 0.409 | 0.596 |
| 100 | 0.389 | 0.567 |
| 100 | 0.389 | 0.567 |
| 125 | 0.350 | 0.508 |
| 150 | 0.321 | 0.464 |

*Upper-tail values suffice since the distribution of $\hat{\gamma}_1$ is symmetrical about zero.
Abridged from *Biometrika Tables for Statisticians,* Vol. I, edited by E.S. Pearson and H.O. Hartley (New York: Cambridge University Press, 1966), Table 34B, p. 207. Reproduced by permission of the *Biometrika* Trustees.

**Table 17** Testing for kurtosis
Upper- and lower-tail values for $\hat{\gamma}_2 = m_4/m_2^2$

| Size of Sample $n$ | Upper 1% | Upper 5% | Lower 5% | Lower 1% | Size of Sample $n$ | Upper 1% | Upper 5% | Lower 5% | Lower 1% |
|---|---|---|---|---|---|---|---|---|---|
| 50 | 4.88 | 3.99 | 2.15 | 1.95 | 600 | 3.54 | 3.34 | 2.70 | 2.60 |
| 75 | 4.59 | 3.87 | 2.27 | 2.08 | 650 | 3.52 | 3.33 | 2.71 | 2.61 |
| 100 | 4.39 | 3.77 | 2.35 | 2.18 | 700 | 3.50 | 3.31 | 2.72 | 2.62 |
| 125 | 4.24 | 3.71 | 2.40 | 2.24 | 750 | 3.48 | 3.30 | 2.73 | 2.64 |
| 150 | 4.13 | 3.65 | 2.45 | 2.29 | 800 | 3.46 | 3.29 | 2.74 | 2.65 |
| | | | | | 850 | 3.45 | 3.28 | 2.74 | 2.66 |
| 200 | 3.98 | 3.57 | 2.51 | 2.37 | 900 | 3.43 | 3.28 | 2.75 | 2.66 |
| 250 | 3.87 | 3.52 | 2.55 | 2.42 | 950 | 3.42 | 3.27 | 2.76 | 2.67 |
| 300 | 3.79 | 3.47 | 2.59 | 2.46 | 1000 | 3.41 | 3.26 | 2.76 | 2.68 |
| 350 | 3.72 | 3.44 | 2.62 | 2.50 | | | | | |
| 400 | 3.67 | 3.41 | 2.64 | 2.52 | | | | | |
| 450 | 3.63 | 3.39 | 2.66 | 2.55 | | | | | |
| 500 | 3.60 | 3.37 | 2.67 | 2.57 | | | | | |
| 550 | 3.57 | 3.35 | 2.69 | 2.58 | | | | | |
| 600 | 3.54 | 3.34 | 2.70 | 2.60 | | | | | |

Abridged from *Biometrika Tables for Statisticians,* Vol I, edited by E.S. Pearson and H.O. Hartley (New York: Cambridge University Press, 1966) Table 34C, p. 208. Reproduced by permission of the *Biometrika* Trustees.

**Table 18** Percentage points of $F_{max} = s^2_{max}/s^2_{min}$

Upper 5% points

| $df_2$ \ $t$ | 2 | 3 | 4 | 5 | 6 | 7 | 8 | 9 | 10 | 11 | 12 |
|---|---|---|---|---|---|---|---|---|---|---|---|
| 2 | 39.0 | 87.5 | 142 | 202 | 266 | 333 | 403 | 475 | 550 | 626 | 704 |
| 3 | 15.4 | 27.8 | 39.2 | 50.7 | 62.0 | 72.9 | 83.5 | 93.9 | 104 | 114 | 124 |
| 4 | 9.60 | 15.5 | 20.6 | 25.2 | 29.5 | 33.6 | 37.5 | 41.1 | 44.6 | 48.0 | 51.4 |
| 5 | 7.15 | 10.8 | 13.7 | 16.3 | 18.7 | 20.8 | 22.9 | 24.7 | 26.5 | 28.2 | 29.9 |
| 6 | 5.82 | 8.38 | 10.4 | 12.1 | 13.7 | 15.0 | 16.3 | 17.5 | 18.6 | 19.7 | 20.7 |
| 7 | 4.99 | 6.94 | 8.44 | 9.70 | 10.8 | 11.8 | 12.7 | 13.5 | 14.3 | 15.1 | 15.8 |
| 8 | 4.43 | 6.00 | 7.18 | 8.12 | 9.03 | 9.78 | 10.5 | 11.1 | 11.7 | 12.2 | 12.7 |
| 9 | 4.03 | 5.34 | 6.31 | 7.11 | 7.80 | 8.41 | 8.95 | 9.45 | 9.91 | 10.3 | 10.7 |
| 10 | 3.72 | 4.85 | 5.67 | 6.34 | 6.92 | 7.42 | 7.87 | 8.28 | 8.66 | 9.01 | 9.34 |
| 12 | 3.28 | 4.16 | 4.79 | 5.30 | 5.72 | 6.09 | 6.42 | 6.72 | 7.00 | 7.25 | 7.48 |
| 15 | 2.86 | 3.54 | 4.01 | 4.37 | 4.68 | 4.95 | 5.19 | 5.40 | 5.59 | 5.77 | 5.93 |
| 20 | 2.46 | 2.95 | 3.29 | 3.54 | 3.76 | 3.94 | 4.10 | 4.24 | 4.37 | 4.49 | 4.59 |
| 30 | 2.07 | 2.40 | 2.61 | 2.78 | 2.91 | 3.02 | 3.12 | 3.21 | 3.29 | 3.36 | 3.39 |
| 60 | 1.67 | 1.85 | 1.96 | 2.04 | 2.11 | 2.17 | 2.22 | 2.26 | 2.30 | 2.33 | 2.36 |
| ∞ | 1.00 | 1.00 | 1.00 | 1.00 | 1.00 | 1.00 | 1.00 | 1.00 | 1.00 | 1.00 | 1.00 |

Upper 1% points

| $df_2$ \ $t$ | 2 | 3 | 4 | 5 | 6 | 7 | 8 | 9 | 10 | 11 | 12 |
|---|---|---|---|---|---|---|---|---|---|---|---|
| 2 | 199 | 448 | 729 | 1036 | 1362 | 1705 | 2063 | 2432 | 2813 | 3204 | 3605 |
| 3 | 47.5 | 85 | 120 | 151 | 184 | 21(6) | 24(9) | 28(1) | 31(0) | 33(7) | 36(1) |
| 4 | 23.2 | 37 | 49 | 59 | 69 | 79 | 89 | 97 | 106 | 113 | 120 |
| 5 | 14.9 | 22 | 28 | 33 | 38 | 42 | 46 | 50 | 54 | 57 | 60 |
| 6 | 11.1 | 15.5 | 19.1 | 22 | 25 | 27 | 30 | 32 | 34 | 36 | 37 |
| 7 | 8.89 | 12.1 | 14.5 | 16.5 | 18.4 | 20 | 22 | 23 | 24 | 26 | 27 |
| 8 | 7.50 | 9.9 | 11.7 | 13.2 | 14.5 | 15.8 | 16.9 | 17.9 | 18.9 | 19.8 | 21 |
| 9 | 6.54 | 8.5 | 9.9 | 11.1 | 12.1 | 13.1 | 13.9 | 14.7 | 15.3 | 16.0 | 16.6 |
| 10 | 5.85 | 7.4 | 8.6 | 9.6 | 10.4 | 11.1 | 11.8 | 12.4 | 12.9 | 13.4 | 13.9 |
| 12 | 4.91 | 6.1 | 6.9 | 7.6 | 8.2 | 8.7 | 9.1 | 9.5 | 9.9 | 10.2 | 10.6 |
| 15 | 4.07 | 4.9 | 5.5 | 6.0 | 6.4 | 6.7 | 7.1 | 7.3 | 7.5 | 7.8 | 8.0 |
| 20 | 3.32 | 3.8 | 4.3 | 4.6 | 4.9 | 5.1 | 5.3 | 5.5 | 5.6 | 5.8 | 5.9 |
| 30 | 2.63 | 3.0 | 3.3 | 3.4 | 3.6 | 3.7 | 3.8 | 3.9 | 4.0 | 4.1 | 4.2 |
| 60 | 1.96 | 2.2 | 2.3 | 2.4 | 2.4 | 2.5 | 2.5 | 2.6 | 2.6 | 2.7 | 2.7 |
| ∞ | 1.00 | 1.0 | 1.0 | 1.0 | 1.0 | 1.0 | 1.0 | 1.0 | 1.0 | 1.0 | 1.0 |

$s^2_{max}$ is the largest and $s^2_{min}$ the smallest in a set of $t$ independent mean squares, each based on $df_2 = n - 1$ degrees of freedom.

Values in the column $t = 2$ and in the rows $df_2 = 2$ and ∞ are exact. Elsewhere the third digit may be in error by a few units for the 5% points and several units for the 1% points. The third digit figures in brackets for $df_2 = 3$ are the most uncertain.

**Table 19** Values of 2 arcsin $\sqrt{\hat{p}}$

| $\hat{p}$ | | $\hat{p}$ | | $\hat{p}$ | | $\hat{p}$ | | $\hat{p}$ | |
|---|---|---|---|---|---|---|---|---|---|
| .001 | .0633 | .041 | .4078 | .36 | 1.2870 | .76 | 2.1177 | .971 | 2.7993 |
| .002 | .0895 | .042 | .4128 | .37 | 1.3078 | .77 | 2.1412 | .972 | 2.8053 |
| .003 | .1096 | .043 | .4178 | .38 | 1.3284 | .78 | 2.1652 | .973 | 2.8115 |
| .004 | .1266 | .044 | .4227 | .39 | 1.3490 | .79 | 2.1895 | .974 | 2.8177 |
| .005 | .1415 | .045 | .4275 | .40 | 1.3694 | .80 | 2.2143 | .975 | 2.8240 |
| .006 | .1551 | .046 | .4323 | .41 | 1.3898 | .81 | 2.2395 | .976 | 2.8305 |
| .007 | .1675 | .047 | .4371 | .42 | 1.4101 | .82 | 2.2653 | .977 | 2.8371 |
| .008 | .1791 | .048 | .4418 | .43 | 1.4303 | .83 | 2.2916 | .978 | 2.8438 |
| .009 | .1900 | .049 | .4464 | .44 | 1.4505 | .84 | 2.3186 | .979 | 2.8507 |
| .010 | .2003 | .050 | .4510 | .45 | 1.4706 | .85 | 2.3462 | .980 | 2.8578 |
| .011 | .2101 | .06 | .4949 | .46 | 1.4907 | .86 | 2.3746 | .981 | 2.8650 |
| .012 | .2195 | .07 | .5355 | .47 | 1.5108 | .87 | 2.4039 | .982 | 2.8725 |
| .013 | .2285 | .08 | .5735 | .48 | 1.5308 | .88 | 2.4341 | .983 | 2.8801 |
| .014 | .2372 | .09 | .6094 | .49 | 1.5508 | .89 | 2.4655 | .984 | 2.8879 |
| .015 | .2456 | .10 | .6435 | .50 | 1.5708 | .90 | 2.4981 | .985 | 2.8960 |
| .016 | .2537 | .11 | .6761 | .51 | 1.5908 | .91 | 2.5322 | .986 | 2.9044 |
| .017 | .2615 | .12 | .7075 | .52 | 1.6108 | .92 | 2.5681 | .987 | 2.9131 |
| .018 | .2691 | .13 | .7377 | .53 | 1.6308 | .93 | 2.6062 | .988 | 2.9221 |
| .019 | .2766 | .14 | .7670 | .54 | 1.6509 | .94 | 2.6467 | .989 | 2.9315 |
| .020 | .2838 | .15 | .7954 | .55 | 1.6710 | .95 | 2.6906 | .990 | 2.9413 |
| .021 | .2909 | .16 | .8230 | .56 | 1.6911 | .951 | 2.6952 | .991 | 2.9516 |
| .022 | .2978 | .17 | .8500 | .57 | 1.7113 | .952 | 2.6998 | .992 | 2.9625 |
| .023 | .3045 | .18 | .8763 | .58 | 1.7315 | .953 | 2.7045 | .993 | 2.9741 |
| .024 | .3111 | .19 | .9021 | .59 | 1.7518 | .954 | 2.7093 | .994 | 2.9865 |
| .025 | .3176 | .20 | .9273 | .60 | 1.7722 | .955 | 2.7141 | .995 | 3.0001 |
| .026 | .3239 | .21 | .9521 | .61 | 1.7926 | .956 | 2.7189 | .996 | 3.0150 |
| .027 | .3301 | .22 | .9764 | .62 | 1.8132 | .957 | 2.7238 | .997 | 3.0320 |
| .028 | .3363 | .23 | 1.0004 | .63 | 1.8338 | .958 | 2.7288 | .998 | 3.0521 |
| .029 | .3423 | .24 | 1.0239 | .64 | 1.8338 | .959 | 2.7338 | .999 | 3.0783 |
| .030 | .3482 | .25 | 1.0472 | .65 | 1.8546 | .960 | 2.7389 | | |
| .031 | .3540 | .26 | 1.0701 | .66 | 1.8965 | .961 | 2.7440 | | |
| .032 | .3597 | .27 | 1.0928 | .67 | 1.9177 | .962 | 2.7492 | | |
| .033 | .3654 | .28 | 1.1152 | .68 | 1.9391 | .963 | 2.7545 | | |
| .034 | .3709 | .29 | 1.1374 | .69 | 1.9606 | .964 | 2.7598 | | |
| .035 | .3764 | .30 | 1.1593 | .70 | 1.9823 | .965 | 2.7652 | | |
| .036 | .3818 | .31 | 1.1810 | .71 | 2.0042 | .966 | 2.7707 | | |
| .037 | .3871 | .32 | 1.2025 | .72 | 2.0264 | .967 | 2.7762 | | |
| .038 | .3924 | .33 | 1.2239 | .73 | 2.0488 | .968 | 2.7819 | | |
| .039 | .3976 | .34 | 1.2451 | .74 | 2.0715 | .969 | 2.7876 | | |
| .040 | .4027 | .35 | 1.2661 | .75 | 2.0944 | .970 | 2.7934 | | |

**Table 20** Values of $d_n$ used in control limits for $\mu$

| $n$ | $d_n$ |
|---|---|
| 2 | 1.128 |
| 3 | 1.693 |
| 4 | 2.059 |
| 5 | 2.326 |
| 6 | 2.534 |
| 7 | 2.704 |
| 8 | 2.847 |
| 9 | 2.970 |
| 10 | 3.078 |
| 11 | 3.173 |
| 12 | 3.258 |

**Table 21** Values of $D_n$ and $D'_n$ used in control limits for product variability

| *Number of Observations in Sample, n* | $D_n$ | $D'_n$ |
|---|---|---|
| 2 | 0 | 3.267 |
| 3 | 0 | 2.575 |
| 4 | 0 | 2.282 |
| 5 | 0 | 2.115 |
| 6 | 0 | 2.004 |
| 7 | 0.076 | 1.924 |
| 8 | 0.136 | 1.864 |
| 9 | 0.184 | 1.816 |
| 10 | 0.223 | 1.777 |
| 11 | 0.256 | 1.744 |
| 12 | 0.284 | 1.716 |
| 13 | 0.308 | 1.692 |
| 14 | 0.329 | 1.671 |
| 15 | 0.348 | 1.652 |
| 16 | 0.364 | 1.636 |
| 17 | 0.379 | 1.621 |
| 18 | 0.392 | 1.608 |
| 19 | 0.404 | 1.596 |
| 20 | 0.414 | 1.586 |
| 21 | 0.425 | 1.575 |
| 22 | 0.434 | 1.566 |
| 23 | 0.443 | 1.557 |
| 24 | 0.452 | 1.548 |
| 25 | 0.459 | 1.541 |
| Over 25 | $3/\sqrt{n}$ | $3/\sqrt{n}$ |

# Selected Answers

## CHAPTER 2

2.2. a. Not particularly, since a pie chart is used to display the percentage of a whole assigned to each of several categories.

2.3. a. range = 1.05 - .72 = .33

b.

| Class | Class Intervals | Class Frequency | Relative Frequency |
|---|---|---|---|
| 1 | .705- .755 | 2 | 2/25 |
| 2 | .755- .805 | 4 | 4/25 |
| 3 | .805- .855 | 8 | 8/25 |
| 4 | .855- .905 | 4 | 4/25 |
| 5 | .905- .955 | 4 | 4/25 |
| 6 | .955-1.005 | 2 | 2/25 |
| 7 | 1.005-1.055 | 1 | 1/25 |

d. $\frac{7}{25}$

2.6. a. Median = (7.5 + 9.5)/2 = 8.5 b. mean = 8.77. No, only for symmetrical distributions.

2.7. a. The rounded measurements are

| | | | |
|---|---|---|---|
| 2.10 | 1.98 | 1.99 | 2.77 |
| 2.47 | 2.70 | 2.10 | 2.36 |
| 1.75 | 2.43 | 2.67 | 1.80 |
| 2.94 | 2.17 | 2.65 | 3.09 |
| 1.69 | 2.80 | 2.06 | 2.20 |
| 2.75 | 2.82 | 2.55 | 2.93 |
| 2.82 | 2.38 | 2.22 | 1.85 |
| 2.52 | 2.68 | 2.92 | 2.28 |
| 2.77 | 2.39 | 3.05 | 1.96 |

b. mode = 2.10, 2.77, 2.82 c. median = (2.43 + 2.47)/2 = 2.45 d. mean = 2.43

2.8. Since 10 of the 15 measurements are less than the sample mean, the distribution appears to be skewed to the right. The sample median is a more appropriate measure of central tendency for these data; median = 61.61.

2.10. a. $n$ = 90, mean = 425.395
b. smallest = 91, largest = 949

2.11. b. $s^2 = \frac{1}{4}[37 - (11)^2/5] = 3.20; s = \sqrt{3.20} = 1.79$

2.12. a. $s^2 = .158; s = \sqrt{.158} = .397$
b. $s \approx$ range/4 = (3.09 - 1.69)/4 = .35
c. $\bar{y} \pm s$, 2.03 to 2.83, 67% (24/36) of measurements in interval
$\bar{y} \pm 2s$, 1.64 to 3.22, 100% of measurements in interval
$\bar{y} \pm 3s$, 1.24 to 3.62, 100% of measurements in interval

2.13. The 50 times are arranged in order of magnitude below:
4,4,5,5,5,5,5,6,6,7,8,9,9,9,10,11,11,12,12,12,13,14,14, 15,15,15,16,16,16,17,17,17,18,19,21,21,21,22,23,25,25, 25,26,27,27,29,29,33,33,34
a. mode = 5; median = 15; mean = 15.96
b. $s \approx$ range/4 = (34 - 4)/4 = 7.5
c. $s$ = 8.39 d. No, they are not mound-shaped.

2.14.

| Class | Class Frequency | Fre-quency $f$ | Mid-point, $y$ | $y_c$ | $fy_c$ | $fy_c^2$ |
|---|---|---|---|---|---|---|
| 1 | 3.55-3.65 | 1 | 3.6 | -7 | -7 | 49 |
| 2 | 3.65-3.75 | 1 | 3.7 | -6 | -6 | 36 |
| 3 | 3.75-3.85 | 6 | 3.8 | -5 | -30 | 150 |
| 4 | 3.85-3.95 | 6 | 3.9 | -4 | -24 | 96 |
| 5 | 3.95-4.05 | 10 | 4.0 | -3 | -30 | 90 |
| 6 | 4.05-4.15 | 10 | 4.1 | -2 | -20 | 40 |
| 7 | 4.15-4.25 | 13 | 4.2 | -1 | -13 | 13 |
| 8 | 4.25-4.35 | 11 | 4.3 | 0 | 0 | 0 |
| 9 | 4.35-4.45 | 13 | 4.4 | 1 | 13 | 13 |
| 10 | 4.45-4.55 | 7 | 4.5 | 2 | 14 | 28 |
| 11 | 4.55-4.65 | 6 | 4.6 | 3 | 18 | 54 |
| 12 | 4.65-4.75 | 7 | 4.7 | 4 | 28 | 112 |
| 13 | 4.75-4.85 | 5 | 4.8 | 5 | 25 | 125 |
| 14 | 4.85-4.95 | 4 | 4.9 | 6 | 24 | 144 |
| *Totals* | | | | | -8 | 950 |

$\bar{y}_c = -8/100 = -.08$; substituting, $\bar{y} = w\bar{y}_c + m = .1(-.08) + 4.3 = 4.29$

$s_c^2 = 1/99\,[950 - (-.08)^2/100] = 9.60$ and $s_c = \sqrt{9.60} = 3.10$

Then $s = ws_c = .1(3.10) = .31$.

2.15, 2.16. Refer to the solution to exercise 2.14, using $m = 4.4$. $\bar{y}$ and $s$ should be identical to those values in exercise 2.14.

2.17. a. mean = .65; median = .55; mode = .5

b. mean = 1.21; median = .55; mode = .5; median and mode are same as in part a.

2.18. a. range = 1.2 - .3 = .9; $s$ = .28  b. $s \approx$ range/4 = .9/4 = .23

c. range = 6.5; $s$ = 1.97

2.19. The coded measurements are 7,5,5,6,5,4,3,9,12,9

The mean and standard deviation for the coded measurements are, respectively, 6.5 and 2.76. Then $\bar{y} = \bar{y}_c/10 = .65$ and $s = s_c/10 = .276$ or .28.

2.21. a. mode = 2.5; median = $L + (w/fn)(.5n - cf_b) = 5.5 + (2/13)(45 - 35) = 7.04$

b. mean = $\Sigma fy/n = 747/90 = 8.3$

c. Since the distribution is skewed to the right, the median provides a better measure of the center of the distribution.

2.22. a. $P_{75} = 11.5 + (2/7)(67.50 - 65) = 12.21$; $P_{25} = 3.5 + (2/15)(22.50 - 20) = 3.83$; interquartile range = 8.38

2.24. $P_{25} = -.5 + (1/170)(125 - 0) = .24$; $P_{75} = 1.5 + (1/75)(375 - 355) = 1.77$

2.25. a. median = .5 + (1/185)(250 - 170) = .93; mode = 1.0

b. mean = $\Sigma fy/500 = 663/500 = 1.33$  c. The distribution is skewed to the right.

2.26. a.

| Class | $f$ | $y$ | $y_c$ | $fy_c$ | $fy_c^2$ |
|---|---|---|---|---|---|
| 1 | 170 | 0 | -2 | -340 | 680 |
| 2 | 185 | 1 | -1 | -185 | 185 |
| 3 | 75 | 2 | 0 | 0 | 0 |
| 4 | 25 | 3 | 1 | 25 | 25 |
| 5 | 15 | 4 | 2 | 30 | 60 |
| 6 | 10 | 5 | 3 | 30 | 90 |
| 7 | 8 | 6 | 4 | 32 | 128 |
| 8 | 5 | 7 | 5 | 25 | 125 |
| 9 | 4 | 8 | 6 | 24 | 144 |
| 10 | 2 | 9 | 7 | 14 | 98 |
| 11 | 1 | 10 | 8 | 8 | 64 |
| *Totals* | | | | -337 | 1599 |

$s_c^2 = 1/499\,[1599 - (-337)^2/500] = 2.75$ and $s_c = 1.66$. Then $s = ws_c = 1(1.66) = 1.66$.

b. Since the distribution is skewed, the Empirical Rule would not give a good description of the set of measurements.

2.27. mean = 19.12; standard deviation = 4.25

2.31. $s = 183.536$; $s/\sqrt{n} = 19.3464$

2.32. a. $\bar{y} = 16{,}354.938$, $n = 60$  c. $s = 4436.070$

d. 15,350

## CHAPTER 3

3.1.

| | | | | |
|---|---|---|---|---|
| 150 | 729 | 611 | 584 | 255 |
| 465 | 143 | 127 | 323 | 225 |
| 483 | 368 | 213 | 270 | 062 |
| 399 | 695 | 540 | 330 | 110 |
| 069 | 409 | 539 | 015 | 564 |

3.2. 054, 636, 533, 482, 526

3.4. The sampling distribution of sample means will be approximately normal, with mean = 60 and standard deviation $5/\sqrt{16} = 1.25$. We would expect that $\bar{y}$ would lie in the interval $60 \pm 2(1.25)$, or 57.50 to 62.50, approximately 95% of the time in repeated sampling.

3.5. The sampling distribution of the sample sum will be approximately normal, with mean = 16(60) = 960 and standard deviation = $5\sqrt{16} = 20$. Observing a measurement more than 70 units (3.5 standard deviations) away from the mean (960) of a normal distribution is very improbable.

3.7. a. .4332  b. .4641

3.8. a. .95  b. .9802

3.9. 1.645, -1.645

3.10. a. 2.33  b. -1.28

3.11. a. .025  b. .0136  c. .0021  d. .2327

3.12. a. .1151  b. .0044  c. .9671  d. .9850

3.13. a. .0336

b. Since 55 is 2.67 standard deviations above $\mu = 39$ and the probability of observing a value of 55 or greater is .0038, we would probably conclude that the voucher had been lost or misplaced.

3.14. a. .7462 b. .1587 c. .0188
3.15. a. 1096.4 b. 1017.75 - 842.25 = 175.5
3.16. a. 125 ± 32 should contain approximately 68% of the weeks; 125 ± 64 should contain approximately 95% of the weeks; 125 ± 96 should contain approximately all of the weeks.
b. .1379
3.17. 178,200 3.18. a. ,01 b. .0038
3.19. a. 3.21 - 4.19 b. 4.19
3.20. a. approximately normal with mean = 3.7 and standard deviation = .06
b. approximately normal with mean = 3.7 and standard deviation = .03
c. essentially 0
3.21. a. approximately normal with mean = 65 and standard deviation = 1.79 b. .9974
3.22. a. approximately normal with mean = 1300 and standard deviation = 35.78
b. 25th percentile = 1275.85; 75th percentile = 1324.15
3.23. a. .0708 b. approximately normal with mean = 2250 and standard deviation = 2.63
3.24. 2.63 3.25. essentially 0
3.26. No, it has a mean breaking strength higher than the old fabric.

## CHAPTER 4

4.1. point estimate = 850; BOE = 25.82
4.2. point estimate = 22; BOE = 1.06
4.3. point estimate = 98.4; BOE = .05 4.4. a. 98.4 ± .05
b. 98.4 ± .06
4.5. 9.02 ± .29 4.6. 430 ± 17.12 4.7. 22 ± 1.37
4.8. This is a one-tailed test. Reject $H_0: \mu = 760; z = 4.21$.
4.9. a. $H_0: \mu = 5.2$. $H_a: \mu < 5.2$. T.S.: $z = (5.0 - 5.2)/(.70/\sqrt{50}) = -2.02$. R.R.: For $\alpha = .05$, reject $H_0$ if $z < -1.645$.
b. Since $z < -1.645$, we reject $H_0$ and conclude that the mean dissolved oxygen count is less than 5.2 ppm.
4.10. .8212, .0455 4.11. $z = -2.36, p = .0091$
4.12. $H_0: \mu = .145$. $H_a: \mu < .145$. $z = -2.12$. Reject $H_0$.
4.13. 430 ± 22.53 4.14. .13 ± .018
4.15. a. 30.51 ± 4.09 b. 30.51 ± 5.39
4.16. $n = 587$ 4.17. $z = .72, p \approx .24$ 4.18. $H_a: \mu \neq 29$. $p \approx .48$
4.19. .76 ± .03 4.20. .76 ± .04 4.21. 75 ± 4.16
4.22. 58 ± 4.08
4.23. $\beta < .001$. If it is important to detect $\mu = 40$, then we would accept $H_0$ because $\beta$ is so small.
4.24. $\beta = 0$ for $\mu = 40$. $\beta = .0301$ for $\mu = 38$. $\beta = .3192$ for $\mu = 36$. $\beta = .8264$ for $\mu = .34$.
4.26. a. 0.76200 b. .98
c. 0.72044 to 0.80356. We are 98% confident that the population mean proportion of patients per hospital with group medical insurance lies in the interval .72044 to .80356.
4.27. 35 ± 1.95 4.28. $z = 2.39; p = .0168$

## CHAPTER 5

5.1. a. 9.7 ± 2.21
b. We are 95% confident that this interval captures the population mean speed for all fourth-grade students.
c. 9.7 ± 2.76
5.2. $t = 1.09, p > .10$. There is very little evidence to indicate the mean comprehension score for all fourth graders is greater than 80.
5.3. $H_0: \mu = 80$. $H_a: \mu \neq 80$. T.S.: $t = (84.2 - .80)/(12.22/\sqrt{10}) = 1.09$. $p > .20$.
5.5. $t = -.31$. There is insufficient evidence to indicate $\mu < 5.0$.
5.6. The process is still in control; however the process must be monitored closely to make certain the upward drift does not continue beyond the UCL.
5.7. $t = 10.71$; reject $H_0: \mu_1 - \mu_2 = 0$. 5.8. $t = -5.85$; reject $H_0: \mu_1 - \mu_2 = 0$.
5.9. $s = 4.87, t = 1.91, .025 < p < .05$
5.11. It would be difficult to justify using a $t$ interval since the sample mean is less than .5 standard deviations away from zero.
5.12. Since the sample mean is further away from zero than for exercise 5.11, we will use the confidence interval based on $t$; 17 ± 7.33.
5.13. $t = .83, p > .20$ 5.14. 1.3 ± 3.35 5.15. $t = 12.28$; reject $H_0: \mu_1 - \mu_2 = 0$.
5.16. 6.16 ± .88 5.17. a. $t = 3.11$; reject $H_0: \mu_1 - \mu_2 = 0$.
b. Same conclusion as in a.
5.18. 4.9 ± 2.59 5.19. 3.2 ± 3.88 5.20. $t = 2.12$; reject $H_0: \mu_1 - \mu_2 = 0$.
5.21. $t = 3.47$; reject $H_0: \mu_1 - \mu_2 = 0$. 5.22. 8.1 ± 5.43
5.23. $-t = 4.09, p < .01$
5.24. A 95% C.I. interval on $\mu_1 - \mu_2$ is -23.40 ± 12.02.
5.25. $t = -4.47, p < .005$ 5.27. $t = 1.75, p < .10$

## CHAPTER 6

6.1. a. $y = \beta_0 + \beta_1 x_1 + \beta_2 x_2 + \epsilon$
b. $y = \beta_0 + \beta_1 x_1 + \beta_2 x_1^2 + \beta_3 x_2 + \beta_4 x_2^2 + \epsilon$
$y = \beta_0 + \beta_1 x_1 + \beta_2 x_2 + \beta_3 x_1 x_2 + \epsilon$
$y = \beta_0 + \beta_1 x_1 + \beta_2 x_1^2 + \beta_3 x_2 + \beta_4 x_2^2 + \beta_5 x_1 x_2 + \epsilon$

6.2.

| | |
|---|---|
| $\beta_1 x_1$ | 1st degree |
| $\beta_2 x_1^2$ | 2d degree |
| $\beta_3 x_2$ | 1st degree |
| $\beta_4 x_1 x_2$ | 2d degree |
| $\beta_5 x_1^2 x_2$ | 3d degree |

6.7. For a deterministic model a setting of the independent variables determines the exact value of $y$. For a probabilistic model a setting of the independent variables determines a distribution of possible $y$-values.
6.8. a. $\beta_1 x_1$ 1st degree

$\beta_2 x_1^2$ 2d degree

$\beta_3 x_1^3$ 3d degree

$\beta_4 x_1^4$ 4th degree

b. 4th order

6.9. $y = \beta_0 + \beta_1 x_1 + \beta_2 x_2 + \beta_3 x_3 + \beta_4 x_1^2 + \beta_5 x_2^2 + \beta_6 x_3^2 + \beta_7 x_1 x_2 + \beta_8 x_1 x_3 + \beta_9 x_2 x_3 + \epsilon$

6.11. a. $\beta_1 x_1$ 1st degree

$\beta_2 x_1^2$ 2d degree

$\beta_3 x_2$ 1st degree

$\beta_4 x_1 x_2$ 2d degree

b. Neither; the first-order model for $x_1$ and $x_2$ is $y = \beta_0 + \beta_1 x_1 + \beta_2 x_2 + \epsilon$ and the second-order model is

$$y = \beta_0 + \beta_1 x_1 + \beta_2 x_1^2 + \beta_3 x_2 + \beta_4 x_2^2 + \beta_5 x_1 x_2 + \epsilon$$

6.12. $y = \beta_0 + \beta_1 x_1 + \beta_2 x_1^2 + \beta_3 x_2 + \beta_4 x_2^2 + \beta_5 x_3 + \beta_6 x_3^2 + \beta_7 x_4 + \beta_8 x_4^2 + \beta_9 x_1 x_2 + \beta_{10} x_1 x_3 + \beta_{11} x_1 x_4 + \beta_{12} x_2 x_3 + \beta_{13} x_2 x_4 + \beta_{14} x_3 x_4 + \epsilon$

6.15. $y = \beta_0 + \beta_1 x + \epsilon$ 6.17. $y = \beta_0 + \beta_1 x + \beta_2 x^2 + \epsilon$

6.18. either 2 peaks and 1 valley or 1 peak and 2 valleys

6.20. $y = \beta_0 + \beta_1 x + \beta_2 x^2 + \epsilon$

## CHAPTER 7

7.1. b. $\hat{y} = 48.93 + 10.33x$ 7.2. 59.26

7.3. b. $\hat{y} = 8.46 - .14x$ 7.4. $\hat{y} = 2.54 + 1.58x$ 7.5. 7.06

7.6. a. $\begin{bmatrix} 4 & 2 \\ 4 & 3 \end{bmatrix}$ b. by inspection c. $\begin{bmatrix} 5 & 3 \\ 8 & 5 \end{bmatrix}$

7.7. a. $\mathbf{A}' = \begin{bmatrix} 2 & 3 \\ 1 & 2 \end{bmatrix}$ $\mathbf{B}' = \begin{bmatrix} 2 & 1 \\ 1 & 1 \end{bmatrix}$ b. $\mathbf{A}^{-1} = \begin{bmatrix} 2 & -1 \\ -3 & 2 \end{bmatrix}$

7.8. $\begin{bmatrix} 5 & -3 \\ -8 & 5 \end{bmatrix}$ 7.9. a. $\begin{bmatrix} 3 & 0 & 2 \\ 0 & 2 & 0 \\ 2 & 0 & 2 \end{bmatrix}$ b. 4

7.10. $\mathbf{A}^{-1} = \begin{bmatrix} 1 & 0 & -1 \\ 0 & .5 & 0 \\ -1 & 0 & 1.5 \end{bmatrix}$ 7.11. a. $\begin{bmatrix} 1 & 1 & 1 \\ 1 & 2 & 3 \end{bmatrix}$

b. $\begin{bmatrix} 3 & 6 \\ 6 & 14 \end{bmatrix}$

7.12. $\begin{bmatrix} 7/3 & -1 \\ -1 & 1/2 \end{bmatrix}$ 7.13. $\begin{bmatrix} 4/3 \\ 1 \end{bmatrix}$

7.14. a. $\begin{bmatrix} 2 & -1 & 1 \\ 2 & 1 & 1 \\ 3 & 1 & 4 \end{bmatrix}$ b. $\begin{bmatrix} 1 & -1 & 1 \\ 2 & 0 & 1 \\ 3 & 1 & 3 \end{bmatrix}$

7.15. a. 4 b. $\begin{bmatrix} -\frac{1}{4} & 1 & -\frac{1}{4} \\ -\frac{3}{4} & 0 & \frac{1}{4} \\ \frac{2}{4} & -1 & \frac{2}{4} \end{bmatrix}$

7.16. a. $\mathbf{X} = \begin{bmatrix} 1 & .0 \\ 1 & .0 \\ 1 & 1.5 \\ 1 & 1.5 \\ 1 & 3.0 \\ 1 & 3.0 \end{bmatrix}$ $\mathbf{Y} = \begin{bmatrix} 50.5 \\ 46.8 \\ 62.3 \\ 67.7 \\ 80.1 \\ 79.2 \end{bmatrix}$

b. $\mathbf{X'X} = \begin{bmatrix} 6 & 9 \\ 9 & 22.5 \end{bmatrix}$ $\mathbf{X'X}^{-1} = \begin{bmatrix} .42 & -.17 \\ -.17 & .11 \end{bmatrix}$

$\mathbf{X'Y} = \begin{bmatrix} 386.60 \\ 672.90 \end{bmatrix}$

7.17. $\hat{y} = 48.93 + 10.33x$

7.18. b. $\mathbf{X} = \begin{bmatrix} 1 & 1.8 \\ 1 & 2.6 \\ 1 & 4.2 \\ 1 & 5.0 \\ 1 & 4.8 \\ 1 & 3.4 \end{bmatrix}$ $\mathbf{Y} = \begin{bmatrix} 4.0 \\ 5.2 \\ 8.5 \\ 11.6 \\ 10.1 \\ 6.3 \end{bmatrix}$ $\mathbf{X'X} = \begin{bmatrix} 6 & 21.8 \\ 21.8 & 87.24 \end{bmatrix}$

$\mathbf{X'Y} = \begin{bmatrix} 45.70 \\ 184.32 \end{bmatrix}$

7.19. a. $(\mathbf{X'X})^{-1} = \begin{bmatrix} 1.80996 & -.45228 \\ -.45228 & .12448 \end{bmatrix}$, $\hat{y} = -.65 + 2.28x$

b. 6.19

7.22. c. 4.94

7.23. b. $y = 8.67 + .58x$; $y = 4.48 + 1.51x - .027x^2$

7.24. a.

| *Dose Level* | *Log Dose* |
|---|---|
| 2 | |
| 4 | |
| 8 | |
| 16 | |
| 32 | |

b. $\hat{y} = 1.20 + 16.17x_1$

7.25. b. $\hat{y} = 63.14 - .71x + .003x^2$

7.26. $\hat{y} = 60.48 - .71x_1 + .003x_1^2 + 8.88x_2$

$\hat{y} = 42.28 - .42x_1 + .002x_1^2 + 69.54x_2 - .95x_1x_2 + .003x_1^2x_2$

7.27. $\hat{y} = -19.41 + 29.10x_1 + 3.29x_2$

$\hat{y} = -23.81 + 31.30x_1 + 3.84x_2 - .28x_1x_2$

7.28. $\hat{y} = -13.74 + 33.26x_1 + .23x_1^2 - 1.48x_2 + .40x_2^2 + 2.48x_1x_2 - .254x_1x_2^2 - .875x_1^2x_2 + .075x_1^2x_2^2$

7.29. b. $\hat{y} = 38.957 - 6.28x + .107x^2$

7.30. a. $\hat{y} = 3.51 + 1.45x$ b. 45.56

## CHAPTER 8

8.1. a. $\hat{y} = 48.93 + 10.33x$ b. $0.1111\sigma^2$ c. 5.6983

8.2. a. .6331 b. $t = 12.9864$; reject $H_0$

8.3. a.

| Biological Recovery (%) | log (%) | Biological Recovery (%) | log (%) |
|---|---|---|---|
| 70.6 | 1.85 | 10.0 | 1.00 |
| 52.0 | 1.72 | 9.1 | .96 |
| 33.4 | 1.52 | 8.3 | .92 |
| 22.0 | 1.34 | 7.9 | .90 |
| 18.3 | 1.26 | 7.7 | .89 |
| 15.1 | 1.18 | 7.7 | .89 |
| 13.0 | 1.11 | | |

8.4. a. $\hat{y} = 1.854 - .036x + .0003x^2$ b. .0006
c. .0170, .0013, .00002
8.5. $t = 15.60$; reject $H_0$: $\beta_2 = 0$ 8.6. $1.854 \pm .038$
8.7. b. $\hat{y} = 12.509 + 35.828x$
8.8. a. $45.266\sigma^2$ b. 1.069 c. $t = 5.15$; reject $H_0$: $\beta_1 = 0$ d. 6.957
8.9. a. Since there is a nonzero fixed percentage of lysine mixed into all feed, the mean response for no lysine eaten, $\beta_0$, should be zero.
b. $\hat{y} = 106.524x$, $\hat{\sigma}^2 = 11.681$ with 11 degrees of freedom
8.10. 1.059 to 1.103 8.11. 18.553 to 20.079
8.12. 16.889 to 21.743
8.13. 1.024 to 1.138 8.15. $\hat{y} = 1.474 + .797x$; $t = 7.53$; reject $H_0$: $\beta_1 = 0$
8.16. a. $\hat{x} = 14.47$ b. 11.08 to 17.63
8.17. $\hat{y} = -4.505 + 3.075x$; $t = 15.59$; reject $H_0$:$\beta_1 = 0$
8.20. a. $r = .948$ b. $t = 8.43, p < .005$
8.21. a. $\hat{y} = 11.700 - .793x + .023x^2$ b. $R^2 = .98$
8.22. $t = 3.242$; $p = .0176$
8.23. a. $\hat{y} = -1.234 + 23.404x$ b. $r = .971$
8.24. a. $t = 12.898$; reject $H_0$: $\beta_1 = 0$ b. $\hat{y} = 16.319$
8.25. c. $t = 4.789$; reject $H_0$
8.27. a. $r_{y1} = .000$; $r_{y2} = .571$; $r_{12} = .000$
b. $r_{y1.2} = r_{y1} = .000$; $r_{y2.1} = r_{y2} = .571$

## CHAPTER 9

9.1. 400 9.2. 100 9.3. a. 54 b. 44
9.4. 100 9.5. 492 9.6. $t = 4.91$ 9.7. $t = 4.95$
9.8. 865 9.9. 3 X 4 factorial experiment
9.10. 4 X 5 factorial experiment
9.11. a. yes b. $t = 5.31, p < .001$

## CHAPTER 10

10.1. $\chi^2 = 6.95$; insufficient evidence to reject $H_0$
10.2. $\chi^2 = 181.65$; reject $H_0$
10.3. $\chi^2 = 6.00$; reject $H_0$
10.4. $\chi^2 = 120.71$; reject $H_0$ 10.5. $\chi^2 = 123.60$; $p < .005$
10.6. $\chi^2 = 27.75$; $p < .005$ 10.7. $\chi^2 = 2.40$; insufficient to reject $H_0$
10.8. a. yes; $n$ is greater than 5/.096 = 52.08 b. $z = -5.02$; reject $H_0$
10.9. a. $\hat{p} = .37$; half-width of a 95% confidence interval is .024
b. 8955
10.10. $\hat{p} = .4$ with a BOE = .219
10.14. Yes, none of the percentages were outside the range 0 to 30.
10.15. $\chi^2 = 8.11$; insufficient evidence to reject $H_0$
10.16. $\chi^2 = 26.80$; reject $H_0$
10.17. $\chi^2 = 8.48$; insufficient evidence to reject $H_0$
10.18. $\chi^2 = 13.57$; reject $H_0$
10.19. $\chi^2 = 6.43$; insufficient evidence to reject $H_0$
10.22. $z = -1.98$; probability of $y \leq 3$ for $n = 40$ and $p = .2$ is approximately .0239.
10.23. a. $z = 1.78$; probability of $y \leq 3.5$ is approximately .0375.
b. Part a is more precise. Since the probability associated with $y = 3$ is represented by the area under a rectangle from $y = 2.5$ to $y = 3.5$, part a is more precise. Even so, the method of exercise 10.22 is still a good approximation that can be used under most circumstances.
10.24. $\chi^2 = 17.07$
10.26. $z = .85$; insufficient evidence to reject $H_0$: $p = .3$

## CHAPTER 11

11.1. $z = 4.02$; reject $H_0$ 11.2. $z = 2.24$; reject $H_0$
11.3. $z = 2.14$; reject $H_0$
11.4. a. $t = -2.43$; reject $H_0$ b. $z = -2$; reject $H_0$
c. Yes; they may be different when there are approximately the same number of plus and minus signs but the magnitude of differences associated with the plus (minus) signs is larger than the magnitude associated with the minus (plus) signs.
11.5. $z = 2.11$; reject $H_0$ 11.6. $p < .02$
11.7. A $t$ test of $H_0$:$\mu = 82$ would appear to be appropriate for these data since the variability of the sample observations is small. Under these circumstances the assumption of normality is not violated.
11.8. $T = 51$; reject $H_0$; $p < .01$. Wilcoxon signed-rank test is more powerful.
11.9. $T = 30$; reject $H_0$ 11.10. $T = 4$; reject $H_0$
11.11. $z = 2.71$; reject $H_0$. Same conclusion as in exercise 11.10.
11.12. a. $z = 2.92$; reject $H_0$
b. two-sample $t$ test if 2 populations have a common variance
11.13. $z = 2.92$; reject $H_0$
11.14. a. $T = 29$; reject $H_0$.
b. The conclusions from all three tests is to reject $H_0$.
11.15. $T = 7.5$; reject $H_0$ 11.16. $H^* = 10.01$; $.02 < p < .05$
11.17. a. $z = -1.55$; cannot reject $H_0$
b. Nonrandom sequence:
SSSSSSSSSSSSSSSSSSSSSSSSSSSFFF
This may indicate a malfunctioning machine and would be valuable to the manufacturer.
11.18. $z = -1.54$; cannot reject $H_0$ 11.19. $C = 14$; reject $H_0$

11.21. $C = 10$; reject $H_0$ at the .01 level. Conclusion agrees with exercise 11.12.

## CHAPTER 12

12.1. $.0017 < \sigma^2 < .0038$

12.2. $\chi^2_{49} = .245$; reject $H_0$

12.3. $2.0434 < \sigma^2 < 15.7067$

12.4. $F_{9,9} = 1.81$; cannot reject $H_0$

12.5. $F_{13,13} = 7.83$; reject $H_0$. A $t$ test would have been inappropriate because the two samples did not have a common variance.

12.6. a. $.7276 < \sigma_1^2/\sigma_2^2 < 185.8072$ b. $F_{4,4} = 11.63$; reject $H_0$

12.7. $10.4203 < \sigma^2 < 58.7436$

12.8. $\chi^2_{19} = 11.55$; cannot reject $H_0$

12.9. a. 22.5 mg to 27.5 mg
b. For testing $H_0: \sigma^2 = (27.5 - 22.5)^2/4 = 1.5625$, $\chi^2_{29} = 40.058$; cannot reject $H_0$

12.10. $F_{8,8} = 3.95$; reject $H_0$

## CHAPTER 13

13.1. $F_{2,15} = 39.14; p < .05$. Reject $H_0$ that the means are equal.

13.2. a. $F_{2,15} = 27.96; p < .05$ b. reject $H_0$ that the means are equal

13.3. a. $\chi^2_3 = 26.62; p < .05$
b. $F_{3,28} = 55.67; p < .05$. Both tests indicate rejection of the null hypothesis that all means are equal.

13.4. $\chi^2_3 = 21.15; p < .05$

13.5 a. $y_{ij} = \mu + \alpha_i + \epsilon_{ij}; i = 1, 2, 3; j = 1, 2, \ldots, n_i; n_1 = 6, n_2 = 5, n_3 = 4$
$y_{ij}$ = $j$th sample measurement from $i$th nutrient
$\mu$ = overall mean
$\alpha_i$ = effect of $i$th nutrient
$\epsilon_{ij}$ = random error associated with $j$th observation from $i$th nutrient
b. $F_{2,12} = 24.31; p < .05$

13.6. a. $y_{ij} = \mu + \alpha_i + \epsilon_{ij}; i = 1, 2, 3; j = 1, 2, \ldots, n_i; n_1 = 12, n_2 = 14, n_3 = 11$
b. $F_{2,34} = 39.805; p < .01$

13.7. a. $y_{ij} = \mu + \alpha_i + \epsilon_{ij}; i = 1, 2, 3, 4; j = 1, 2, \ldots, 8$
b. $F_{3,28} = 11.05; p < .05$; reject $H_0$

13.8. $\chi^2_3 = 16.56; p < .05$. Both tests rejected the null hypothesis that all means are equal.

13.9. $F_{2,33} = 6.65; p < .05$

13.10. $F_{2,42} = 5.038; p < .05$

13.16. $\chi^2_2 = 9.996; p < .05$. Both tests rejected the null hypothesis that all means are equal at $\alpha = .05$.

## CHAPTER 14

14.1. $\hat{\ell}_1$ and $\hat{\ell}_2$ are linear contrasts but they are not orthogonal.

14.2. $\hat{\ell}_1 = -3\bar{y}_1 - \bar{y}_2 + \bar{y}_3 + 3\bar{y}_4$
$\hat{\ell}_2 = \bar{y}_1 - \bar{y}_2 - \bar{y}_3 + \bar{y}_4$
$\hat{\ell}_3 = -\bar{y}_1 + 3\bar{y}_2 - 3\bar{y}_3 + \bar{y}_4$

14.3. $\hat{\ell}_1 = -2\bar{y}_1 - \bar{y}_2 + \bar{y}_4 + 2\bar{y}_5$
$\hat{\ell}_2 = 2\bar{y}_1 - \bar{y}_2 - 2\bar{y}_3 - \bar{y}_4 + 2\bar{y}_5$
$\hat{\ell}_3 = -\bar{y}_1 + 2\bar{y}_2 - 2\bar{y}_4 + \bar{y}_5$
$\hat{\ell}_4 = \bar{y}_1 - 4\bar{y}_2 + 6\bar{y}_3 - 4\bar{y}_4 + \bar{y}_5$

14.5. $F_{4,45} = 15.069; p < .05$; reject $H_0$ that all means are equal

14.9. a. $\hat{\ell} = -2.11$; reject $H_0$ that $\ell = 0$
b. $\hat{\ell} = -.12$; cannot reject $H_0: \ell = 0$ at $\alpha = .05$
c. $\hat{\ell} = .46$; cannot reject $H_0: \ell = 0$ at $\alpha = .05$
d. $\hat{\ell} = .90$; cannot reject $H_0: \ell = 0$ at $\alpha = .05$

14.10. a. $-3.24 < -2.11 < -.98$ b. $-1.55 < -.12 < 1.31$
c. $-.78 < .46 < 1.69$ d. $-.27 < .90 < 2.06$

14.11. a. LSD = 4.708: A B C
b. $W = 5.734$: A B C. Conclusions are the same.

14.12. c. LSD = 4.33: A B C
d. All three tests gave the same results: A B C.

14.13. $W_2 = 3.538$; $W_3 = 3.714$: 3 2 1

14.14. when $k = 100$, $t_c = 1.86$; when $k = 500$, $t_c = 2.45$

14.15. LSD = .83
4 3 2 1 5

14.16. $W_2 = .46$, $W_3 = .48$, $W_4 = .49$
4 1 2 3

14.17. Fisher's LSD procedure. Summary: 4 3 2 1. Per-comparison error rate is controlled.

14.18. 95% confidence limits

| *Population Difference,* $(\mu_i - \mu_j)$ | *Mean Difference,* $\bar{y}_i - \bar{y}_j$ | *95% Confidence Limits* *Upper* | *Lower* |
|---|---|---|---|
| $\mu_1 - \mu_4$ | 5.74 | 7.32 | 4.16 |
| $\mu_2 - \mu_4$ | 4.26 | 5.84 | 2.68 |
| $\mu_3 - \mu_4$ | 1.43 | 3.01 | -.15 |
| $\mu_1 - \mu_3$ | 4.31 | 5.89 | 2.73 |
| $\mu_2 - \mu_3$ | 2.83 | 4.41 | 1.26 |
| $\mu_1 - \mu_2$ | 1.48 | 3.06 | -.10 |

4 3 2 1

14.19. LSD = 12.72. Summary: 1 2 3.

## CHAPTER 15

15.1. a. Blocks are investigators and treatments are mixtures.
b. Mixtures are applied at random to experimental units within an investigator, with the stipulation that each mixture appears once in an investigator.

15.2. a.

| *Source* | *SS* | *df* | *MS* | *F* | *p* |
|---|---|---|---|---|---|
| mixtures | 261260.95 | 3 | 87086.983 | 1264.73 | <.01 |
| investigators | 452.50 | 4 | 113.125 | 1.64 | .23 |
| error | 826.30 | 12 | 68.858 | | |
| *Totals* | 262539.75 | 19 | | | |

b. The "best" mixture appears to be mixture 2. A multiple comparison procedure (Section 15.5) can be used to see whether mixture 2 is significantly better than the other mixtures.

c. 1.135 = relative efficiency

15.3.

| *Source* | *SS* | *df* | *MS* | *F* | *P* |
|---|---|---|---|---|---|
| year | .220 | 1 | .220 | 6.47 | .03 |
| curricula | 21.260 | 10 | 2.126 | 62.53 | <.01 |
| error | .340 | 10 | .034 | | |
| *Totals* | 21.820 | 21 | | | |

15.5. a. $y_{ijk} = \mu + \alpha_i + \beta_j + \gamma_k + \epsilon_{ijk}$; $i = 1, 2, 3, 4$; $j = 1, 2, 3, 4$; $k = 1, 2, 3, 4$

b.

| *Source* | *SS* | *df* | *MS* | *F* | *p* |
|---|---|---|---|---|---|
| rows | .00085 | 3 | .000283 | 2.26 | .18 |
| columns | .01235 | 3 | .004117 | 32.94 | <.01 |
| treatments | .48015 | 3 | .16005 | 1280.4 | <.01 |
| error | .00075 | 6 | .000125 | | |
| *Totals* | .49410 | 15 | | | |

We cannot reject the null hypothesis that rows are equal but we can reject the hypotheses that columns and treatments are equal.

15.6.

| *Source* | *SS* | *df* | *MS* | *F* | *p* |
|---|---|---|---|---|---|
| rows | .075 | 3 | .025 | 1.88 | .23 |
| columns | 1.260 | 3 | .420 | 31.58 | <.01 |
| treatments | 28.435 | 3 | 9.4783 | 712.65 | <.01 |
| error | .080 | 6 | .0133 | | |
| *Totals* | 29.850 | 15 | | | |

We cannot reject the null hypothesis that rows are equal but we reject the null hypotheses that columns and treatments are equal.

15.8. $\bar{\epsilon}_4 - \bar{\epsilon}_2$ 15.9. a. 0 b. 8

15.11. a. $y_{ij} = \mu + \alpha_i + \beta_j + \gamma_{ij} + \epsilon_{ij}$; $i = 1, 2, 3, 4$; $j = 1, 2, 3, 4$

15.12. a.

| *Source* | *SS* | *df* | *MS* |
|---|---|---|---|
| copper | 171856.19 | 3 | 57285.396 |
| manganese | 3745496.19 | 3 | 1248498.7 |
| interaction | 9994.56 | 9 | 1110.5069 |
| *Totals* | 3927346.94 | 15 | |

We cannot test for interaction.

b. $y = \beta_0 + \beta_1 x_1 + \beta_2 x_1^2 + \beta_3 x_1^3 + \beta_4 x_2 + \beta_5 x_2^2 + \beta_6 x_2^3 + \epsilon$

15.13. $y_{ijk} = \mu + \alpha_i + \beta_j + \gamma_k + \alpha\beta_{ij} + \alpha\gamma_{ik} + \beta\gamma_{jk} + \epsilon_{ijk}$; $i = 1, 2, 3, 4$; $j = 1, 2, 3$; $k = 1, 2, 3, 4, 5, 6$

| *Source* | *SS* | *df* | *MS* | *F* |
|---|---|---|---|---|
| Main Effects | | | | |
| A | SSA | 3 | SSA/3 | MSA/MSE |
| B | SSB | 2 | SSB/2 | MSB/MSE |
| C | SSC | 5 | SSC/5 | MSC/MSE |
| Interaction | | | | |
| AB | SSAB | 6 | SSAB/6 | MSAB/MSE |
| AC | SSAC | 15 | SSAC/15 | MSAC/MSE |
| BC | SSBC | 10 | SSBC/10 | MSBC/MSE |
| Error | SSE | 30 | SSE/30 | |
| *Totals* | TSS | 71 | | |

15.14. a. $y_{ijk} = \mu + \alpha_i + \beta_j + \gamma_{ij} + \epsilon_{ijk}$; $i = 1, 2, 3, 4$; $j = 1, 2, 3$; $k = 1, 2, 3$

b.

| *Source* | *SS* | *df* | *MS* | *F* | *p* |
|---|---|---|---|---|---|
| pH | 4.4608 | 3 | 1.4869 | 21.93 | <.01 |
| calcium | 1.4672 | 2 | .7336 | 10.82 | <.01 |
| interaction | 3.2550 | 6 | .5425 | 8.00 | <.01 |
| error | 1.6267 | 24 | .0678 | | |
| *Totals* | 10.8097 | 35 | | | |

15.15. treatment: 4 3 2 1 LSD = .02
1.32 1.42 1.68 1.74

15.16. Summary: Mn 20 50 80 110
1506.5 1948.25 2470 2767.5

15.17. Summary: pH 4.0 7.0 5.0 6.0
6.5 7.0 7.3 7.4
Ca 300 100 200
6.87 6.96 7.33

15.18. a. $y_{ijk} = \mu + \alpha_i + \beta_j + \gamma_{ij} + \epsilon_{ijk}$; $\mu$ = overall mean; $\alpha_i$ = effect of investigator $i$, $i = 1, 2, 3$; $\beta_j$ = effect of treatment $j$, $j = 1, 2, \ldots, 9$; $\gamma_{ij}$ = interaction between investigator $i$ and treatment $j$; $\epsilon_{ijk}$ = error, $k = 1, 2, \ldots, 5$

b.

| *Source* | *SS* | *df* | *MS* | *F* | *P* |
|---|---|---|---|---|---|
| investigators | .3628 | 2 | .1814 | 7.37 | <.01 |
| treatments | 3.5117 | 8 | .4390 | 17.85 | <.01 |
| interaction | .5012 | 16 | .0313 | 1.27 | .23 |
| error | 2.6520 | 108 | .0246 | | |
| *Totals* | 7.0277 | 134 | | | |

15.19. See AOV table in exercise 15.18.

15.20. a. Completely randomized design; $y_{ij} = \mu + \alpha_i + \epsilon_{ij}$; $i = 1, 2, 3, 4$; $j = 1, 2, ..., 7$

b.

| Source | SS | df | MS | F | p |
|---|---|---|---|---|---|
| class | 892.7143 | 3 | 297.5714 | 11.055 | <.01 |
| error | 646.0000 | 24 | 26.9167 | | |
| Totals | 1538.7143 | 27 | | | |

15.21. a. We could use the 4 excess weight classifications of exercise 15.20 as well as 4 age classifications and 2 sex classifications. This would be a 4 X 4 X 2 factorial experiment. The number of observations per factor-level combination might depend on the amount of money available for this study.

b. $y$ = fatigue time, $x_1$ = percentage overweight, $x_2$ = age; $y = \beta_0 + \beta_1 x_1 + \beta_2 x_2 + \beta_3 x_1 x_2 + \epsilon$

15.22. a. Randomized block design; $y_{ij} = \mu + \alpha_i + \beta_j + \epsilon_{ij}$; $i = 1, 2, ..., 5$; $j = 1, 2, ..., 5$

b.

| Source | SS | df | MS | F | p |
|---|---|---|---|---|---|
| temperature | 16.4216 | 4 | 4.1054 | 42.92 | <.01 |
| pane | 7.1176 | 4 | 1.7794 | 18.60 | <.01 |
| error | 1.5304 | 16 | .09565 | | |
| Totals | 25.0696 | 24 | | | |

15.23. Summary:

| A | B | C | D | E |
|---|---|---|---|---|
| 9.38 | 9.66 | 9.98 | 10.04 | 10.96 |

15.24. a. 4 X 3 X 3 factorial

b. $y_{ijk} = \mu + \alpha_i + \beta_j + \gamma_k + \alpha\beta_{ij} + \alpha\gamma_{ik} + \beta\gamma_{jk} + \epsilon_{ijk}$; $i = 1, 2, 3, 4$; $j = 1, 2, 3$; $k = 1, 2, 3$

c.

| Source | SS | df | MS | F | p |
|---|---|---|---|---|---|
| Main Effects | | | | | |
| pH | 4.4608 | 3 | 1.4869 | 28.48 | <.01 |
| calcium | 1.4672 | 2 | .7336 | 14.05 | <.01 |
| grove | .0089 | 2 | .0044 | .08 | .92 |
| Interaction | | | | | |
| pH-calcium | 3.2550 | 6 | .5425 | 10.39 | <.01 |
| pH-grove | .4000 | 6 | .0667 | 1.28 | .34 |
| calcium-grove | .5911 | 4 | .1478 | 2.83 | .07 |
| Error | .6267 | 12 | .0522 | | |
| Totals | 10.8097 | 35 | | | |

## CHAPTER 16

16.1. $y = \beta_0 + \beta_1 x_1 + \beta_2 x_2 + \beta_3 x_3 + \beta_4 x_4 + \beta_5 x_5 + \beta_6 x_6 + \epsilon$

$x_1$ = 1 if block 2    $x_4$ = 1 if treatment 3
$x_1$ = 0 otherwise    $x_4$ = 0 otherwise
$x_2$ = 1 if block 3    $x_5$ = 1 if treatment 4
$x_2$ = 0 otherwise    $x_5$ = 0 otherwise
$x_3$ = 1 if treatment 2    $x_6$ = 1 if treatment 5
$x_3$ = 0 otherwise    $x_6$ = 0 otherwise

16.2. $y = \beta_0 + \beta_1 x_1 + \beta_2 x_2 + \beta_3 x_3 + \beta_4 x_4 + \beta_5 x_5 + \beta_6 x_6 + \epsilon$

$x_1$ = 1 if row 2    $x_4$ = 1 if column 3
$x_1$ = 0 otherwise    $x_4$ = 0 otherwise
$x_2$ = 1 if row 3    $x_5$ = 1 if treatment 2
$x_2$ = 0 otherwise    $x_5$ = 0 otherwise
$x_3$ = 1 if column 2    $x_6$ = 1 if treatment 3
$x_3$ = 0 otherwise    $x_6$ = 0 otherwise

16.3. $y = \beta_0 + \beta_1 x_1 + \beta_2 x_1^2 + \beta_3 x_2 + \beta_4 x_3 + \beta_5 x_1 x_2 + \beta_6 x_1 x_3 + \beta_7 x_1^2 x_2 + \beta_8 x_1^2 x_3 + \epsilon$

$x_1$ = rate    $x_3$ = 1 if Fertilizer C
$x_2$ = 1 if Fertilizer B    $x_3$ = 0 otherwise
$x_2$ = 0 otherwise

A: $\beta_0 + \beta_1 x_1 + \beta_2 x_1^2$

B: $\beta_0 + \beta_3 + (\beta_1 + \beta_5)x_1 + (\beta_2 + \beta_7)x_1^2$

C: $\beta_0 + \beta_4 + (\beta_1 + \beta_6)x_1 + (\beta_2 + \beta_8)x_1^2$

16.4. $y = \beta_0 + \beta_1 x_1 + \beta_2 x_2 + \beta_3 x_3 + \beta_4 x_4 + \beta_5 x_5 + \beta_6 x_6 + \beta_7 x_7 + \epsilon$

$x_1$ = 1 if Mixture 2    $x_5$ = 1 if Investigator 3
$x_1$ = 0 otherwise    $x_5$ = 0 otherwise
$x_2$ = 1 if Mixture 3    $x_6$ = 1 if Investigator 4
$x_2$ = 0 otherwise    $x_6$ = 0 otherwise
$x_3$ = 1 if Mixture 4    $x_7$ = 1 if Investigator 5
$x_3$ = 0 otherwise    $x_7$ = 0 otherwise
$x_4$ = 1 if Investigator 2
$x_4$ = 0 otherwise

$\beta_0$ = mean response for Mixture 1, Investigator I
$\beta_1 = \mu_2 - \mu_1$    $\beta_5 = \mu_{III} - \mu_I$
$\beta_2 = \mu_3 - \mu_1$    $\beta_6 = \mu_{IV} - \mu_I$
$\beta_3 = \mu_4 - \mu_1$    $\beta_7 = \mu_V - \mu_I$
$\beta_4 = \mu_{II} - \mu_I$

16.5. $y = \beta_0 + \beta_1 x_1 + \beta_2 x_2 + \beta_3 x_3 + \beta_4 x_4 + \beta_5 x_5 + \beta_6 x_6 + \beta_7 x_7 + \beta_8 x_8 + \beta_9 x_9 + \epsilon$

$\beta_0$ = mean response for row 1, column A, treatment I
$\beta_1 = \mu_2 - \mu_1$    $\beta_5 = \mu_C - \mu_A$
$\beta_2 = \mu_3 - \mu_1$    $\beta_6 = \mu_D - \mu_A$
$\beta_3 = \mu_4 - \mu_1$    $\beta_7 = \mu_{II} - \mu_I$
$\beta_4 = \mu_B - \mu_A$    $\beta_8 = \mu_{III} - \mu_I$
$\beta_9 = \mu_{IV} - \mu_I$

16.6. $F_{2,8} = 30.795$; reject $H_0$ that $\beta_2 = \beta_3 = 0$ at $\alpha = .05$

16.7. a. $F_{1,8} = 279.916$; reject $H_0$ that $\beta_1 = 0$ at $\alpha = .05$

b. The $t$ test for $H_0: \beta_1 = 0$ in the complete model of part a would give you the same result.

16.8. $F_{2,6} = 44.063$; reject $H_0$ that $\beta_1 = \beta_2 = 0$ at $\alpha = .05$

16.9.

| Source of Variation | SS |
|---|---|
| rows (intersections) | 5.897 |

columns (times) 736.912
treatments (devices) 108.982

16.11. a. $y = \beta_0 + \beta_1 x_1 + \beta_2 x_2 + \beta_3 x_3 + \beta_4 x_1 x_2 + \beta_5 x_1 x_3 + \epsilon$

$x_1$ = log of drug dose; $\beta_0$ = $y$-intercept for Product A regression line

$x_2$ = 1 if Product B; $\beta_1$ = slope of Product A regression line

$x_2$ = 0 otherwise; $\beta_2$ = difference in $y$-intercepts for products B and A

$x_3$ = 1 if Product C; $\beta_3$ = difference in $y$-intercepts for products C and A

$x_3$ = 0 otherwise; $\beta_4$ = difference in slopes for products B and A

$\beta_5$ = difference in slopes for products C and A

b. $y = \beta_0 + \beta_1 x_1 + \beta_4 x_1 x_2 + \beta_5 x_1 x_3 + \epsilon$

16.12. a. $F_{2,6} = 28.09$; reject $H_0$ that $\beta_4 = \beta_5 = 0$ at $\alpha = .05$
b. no c. Fit a reduced model with $\beta_2 = \beta_3 = \beta_0$.

16.13. $y = \beta_0 + \beta_1 x + \beta_2 x^2 + \beta_3 x^3 + \epsilon$

16.14. a. $\hat{\beta} = \begin{bmatrix} -119.333 \\ 24.344 \\ -1.447 \\ .028 \end{bmatrix}$

b. We cannot test for lack of fit. At most, we can fit a cubic, as there are four levels of $x$.

16.15. a. $\hat{\beta} = \begin{bmatrix} -91.000 \\ 19.217 \\ -1.163 \\ .023 \end{bmatrix}$

b. In neither model could the lack of fit be tested because the highest-order polynomial that could be fitted was a cubic. There is some variation in the coefficients between the two exercises.

16.16. a. 2 X 2 factorial experiment with augmented center points

b. $\hat{\beta} = \begin{bmatrix} 100.167 \\ -2.825 \\ 4.750 \\ 1.425 \end{bmatrix}$

16.17. a. no
b. $F_{1,1} = 1.127$; cannot reject $H_0$ that the assumed model is correct at $\alpha = .05$.

16.18. a. [0 0 0 0 0 0 1 0 -1] b. [0 0 0 0 0 0 2 1 1]
c. [0 0 0 0 0 0 1 2 1] d. [0 0 0 0 0 0 1 1 2]

16.19.

| Comparison | $\chi_1^2$ | $p$ |
|---|---|---|
| A vs. B | 1.7884 | .1811 |
| A vs. C | .1118 | .7381 |
| A vs. D | 1.0444 | .3068 |
| B vs. D | 5.5914 | .0180 |
| C vs. D | 1.8671 | .1718 |

16.20. a. no

b. $y = \beta_0 + \beta_1 x_1 + \beta_2 x_1^2 + \beta_3 x_2 + \beta_4 x_2^2 + \beta_5 x_2^3 + \beta_6 x_2^4 + \beta_7 x_1 x_2 + \beta_8 x_1 x_2^2 + \beta_9 x_1 x_2^3 + \beta_{10} x_1 x_2^4 + \beta_{11} x_1^2 x_2 + \beta_{12} x_1^2 x_2^2 + \beta_{13} x_1^2 x_2^3 + \beta_{14} x_1^2 x_2^4 + \epsilon$

16.21. a. $y = \beta_0 + \beta_1 x + \beta_2 x^2 + \epsilon$; $\hat{\beta} = \begin{bmatrix} 44.1786 \\ -.4701 \\ .0012 \end{bmatrix}$

b. $F_{3,12} = 2.043$; cannot reject $H_0$ that the assumed model is correct.

16.22. a. $\hat{\beta} = \begin{bmatrix} 44.1824 \\ -.4940 \\ .0014 \end{bmatrix}$

b. $F_{3,9} = 1.20$; cannot reject $H_0$ that the assumed model is correct.

16.23. a. $F_{6,36} = 2.35$; reject $H_0$ that the assumed model is correct.

b. $y = \beta_0 + \beta_1 x_1 + \beta_2 x_1^2 + \beta_3 x_1^3 + \beta_4 x_1^4 + \beta_5 x_1^5 + \beta_6 x_2 + \beta_7 x_1 x_2 + \beta_8 x_1^2 x_2 + \beta_9 x_1^3 x_2 + \beta_{10} x_1^4 x_2 + \beta_{11} x_1^5 x_2 + \epsilon$

16.24. $F_{2,10} = 5.452$; reject $H_0$ that the assumed model is correct.

16.25. A model was fitted relating the probability of a yes response to a linear function of a destination area and number of passengers effect. The interaction and main effects were tested and the results were as follows:

1. Interaction: $\chi_9^2 = 18.15$; $p = .0335$. There was a significant difference in the yes responders in some combinations of destination area and number of passengers.
2. Number of passengers: $\chi_3^2 = 31.64$; $p = 0$. There was a significant difference in yes responders among the four groups of number of passengers.
3. Destination area: $\chi_3^2 = 73.88$; $p = 0$. There was a significant difference in yes responders among the four destination areas.

16.26. test for $\beta_3 = 0$: $t = 4.69$; $p = .0001$

## CHAPTER 17

17.1. a. $\beta_0 = \mu + \alpha_1$; $\beta_1 = \alpha_2 - \alpha_1$; $\beta_2 = \alpha_3 - \alpha_1$

b. $\begin{bmatrix} \beta_0 \\ \beta_1 \\ \beta_2 \end{bmatrix} = \begin{bmatrix} \mu + \alpha_1 \\ \alpha_2 - \alpha_1 \\ \alpha_3 - \alpha_1 \end{bmatrix}$

17.2. a. $\beta_0 = \mu + \alpha_1 + \gamma_1$; $\beta_1 = \alpha_2 - \alpha_1$; $\beta_2 = \alpha_3 - \alpha_1$; $\beta_3 = \alpha_4 - \alpha_1$; $\beta_4 = \gamma_2 - \gamma_1$; $\beta_5 = \gamma_3 - \gamma_1$

b. $\begin{bmatrix} \beta_0 \\ \beta_1 \\ \beta_2 \\ \beta_3 \\ \beta_4 \\ \beta_5 \end{bmatrix} = \begin{bmatrix} \mu + \alpha_1 + \gamma_1 \\ \alpha_2 - \alpha_1 \\ \alpha_3 - \alpha_1 \\ \alpha_4 - \alpha_1 \\ \gamma_2 - \gamma_1 \\ \gamma_3 - \gamma_1 \end{bmatrix}$

17.3. $\beta_0 = \mu + \alpha_1 + \gamma_1 + \delta_1$ $\quad \beta_4 = \gamma_3 - \gamma_1$
$\beta_1 = \alpha_2 - \alpha_1$ $\quad \beta_5 = \delta_2 - \delta_1$
$\beta_2 = \alpha_3 - \alpha_1$ $\quad \beta_6 = \delta_3 - \delta_1$
$\beta_3 = \gamma_2 - \gamma_1$

17.4. a.

$$\mathbf{X}^* = \begin{bmatrix} 1 & 1 & 1 & 1 & 1 & 1 \\ 1 & 1 & 1 & -1 & 1 & 1 \\ 1 & 1 & 1 & 0 & -2 & 1 \\ 1 & 1 & 1 & 0 & 0 & -3 \\ 1 & -1 & 1 & 1 & 1 & 1 \\ 1 & -1 & 1 & -1 & 1 & 1 \\ 1 & -1 & 1 & 0 & -2 & 1 \\ 1 & -1 & 1 & 0 & 0 & -3 \\ 1 & 0 & -2 & 1 & 1 & 1 \\ 1 & 0 & -2 & -1 & 1 & 1 \\ 1 & 0 & -2 & 0 & -2 & 1 \\ 1 & 0 & -2 & 0 & 0 & -3 \end{bmatrix}$$

b.

| *Source* | *SS* | *df* | *MS* | *F* | *p* |
|---|---|---|---|---|---|
| Insecticides | 1925.17 | 2 | 962.58 | | |
| $\ell_1$ | 1770.13 | 1 | 1770.13 | 451.56 | $<.01$ |
| $\ell_2$ | 155.04 | 1 | 155.04 | 39.55 | $<.01$ |
| Plots | 386.25 | 3 | 128.75 | | |
| $\ell_3$ | 73.5 | 1 | 73.5 | 18.75 | $<.01$ |
| $\ell_4$ | 234.72 | 1 | 234.72 | 59.88 | $<.01$ |
| $\ell_5$ | 78.03 | 1 | 78.03 | 19.91 | $<.01$ |
| Error | 23.50 | 6 | 3.92 | | |
| *Totals* | 2334.92 | 11 | | | |

17.5. (1) $SS_{\text{insecticides}} = (\Sigma T_i^2/n_i - G^2/n)$; where $T_1$ = total for Insecticide 1, $n_1$ = sample size; $I_2$= total for Insecticide 2, $n_2$ = sample size; $T_3$ = total for Insecticide 3, $n_3$ = sample size; $G = T_1 + T_2 + T_3$; $n = n_1 + n_2 + n_3$

(2) Complete model:

$$y = \beta_0 + \beta_1 x_1 + \beta_2 x_2 + \beta_3 x_3 + \beta_4 x_4 + \beta_5 x_5$$

Reduced model:

$$y = \beta_0 + \beta_3 x_3 + \beta_4 x_4 + \beta_5 x_5$$

$$SS_{\text{drop}} = SSE_{\text{reduced model}} - SSE_{\text{complete model}}$$

17.6.

$$\mathbf{X}^* = \begin{bmatrix} 1 & 1 & 1 & 1 & 1 & 1 & 1 & 0 & 0 & -3 \\ 1 & 1 & 1 & 1 & -1 & 1 & 1 & -1 & 1 & 1 \\ 1 & 1 & 1 & 1 & 0 & -2 & 1 & 0 & -2 & 1 \\ 1 & 1 & 1 & 1 & 0 & 0 & -3 & 1 & 1 & 1 \\ 1 & -1 & 1 & 1 & 1 & 1 & 1 & -1 & 1 & 1 \\ 1 & -1 & 1 & 1 & -1 & 1 & 1 & 0 & -2 & 1 \\ 1 & -1 & 1 & 1 & 0 & -2 & 1 & 1 & 1 & 1 \\ 1 & -1 & 1 & 1 & 0 & 0 & -3 & 0 & 0 & -3 \\ 1 & 0 & -2 & 1 & 1 & 1 & 1 & 0 & -2 & 1 \\ 1 & 0 & -2 & 1 & -1 & 1 & 1 & 1 & 1 & 1 \\ 1 & 0 & -2 & 1 & 0 & -2 & 1 & 0 & 0 & -3 \\ 1 & 0 & -2 & 1 & 0 & 0 & -3 & -1 & 1 & 1 \\ 1 & 0 & 0 & -3 & 1 & 1 & 1 & 1 & 1 & 1 \\ 1 & 0 & 0 & -3 & -1 & 1 & 1 & 0 & 0 & -3 \\ 1 & 0 & 0 & -3 & 0 & -2 & 1 & -1 & 1 & 1 \\ 1 & 0 & 0 & -3 & 0 & 0 & -3 & 0 & -2 & 1 \end{bmatrix}$$

17.7.

| *Source* | *SS* | *df* | *MS* | *F* | *p* |
|---|---|---|---|---|---|
| Time | 736.91 | 3 | | | |
| $\ell_1$ | 583.11 | 1 | 583.11 | 146.88 | $<.01$ |
| $\ell_2$ | 81.03 | 1 | 81.03 | 20.41 | $<.01$ |
| $\ell_3$ | 72.77 | 1 | 72.77 | 18.33 | $<.01$ |
| Intersections | 5.90 | 3 | | | |
| $\ell_4$ | 5.44 | 1 | 5.44 | 1.37 | $>.05$ |
| $\ell_5$ | .14 | 1 | .14 | .04 | $>.05$ |
| $\ell_6$ | .32 | 1 | .32 | .08 | $>.05$ |
| Devices | 108.98 | 3 | | | |
| $\ell_7$ | 4.35 | 1 | 4.35 | 1.10 | $>.05$ |
| $\ell_8$ | 104.58 | 1 | 104.58 | 26.34 | $<.01$ |
| $\ell_9$ | .05 | 1 | .05 | .01 | $>.05$ |
| Error | 23.81 | 6 | 3.97 | | |
| *Totals* | 875.6 | 15 | | | |

17.9.

$$\mathbf{X}^* = \begin{bmatrix} 1 & 1 & 1 & 1 & 1 & 1 & 1 & 1 & 1 & 1 & 1 & 1 & 1 & 1 & 1 & 1 \\ 1 & 1 & 1 & 1 & 1 & 1 & 1 & 1 & 1 & 1 & 1 & 1 & 1 & 1 & 1 & 1 \\ 1 & -1 & 1 & 1 & 1 & 1 & 1 & -1 & -1 & -1 & 1 & 1 & 1 & 1 & 1 & 1 \\ 1 & -1 & 1 & 1 & 1 & 1 & 1 & -1 & -1 & -1 & 1 & 1 & 1 & 1 & 1 & 1 \\ 1 & 0 & -2 & 1 & 1 & 1 & 1 & 0 & 0 & 0 & -2 & -2 & -2 & 1 & 1 & 1 \\ 1 & 0 & -2 & 1 & 1 & 1 & 1 & 0 & 0 & 0 & -2 & -2 & -2 & 1 & 1 & 1 \\ 1 & 0 & 0 & -3 & 1 & 1 & 1 & 0 & 0 & 0 & 0 & 0 & 0 & -3 & -3 & -3 \\ 1 & 0 & 0 & -3 & 1 & 1 & 1 & 0 & 0 & 0 & 0 & 0 & 0 & -3 & -3 & -3 \\ 1 & 1 & 1 & 1 & -1 & 1 & 1 & -1 & 1 & 1 & -1 & 1 & 1 & -1 & 1 & 1 \\ 1 & 1 & 1 & 1 & -1 & 1 & 1 & -1 & 1 & 1 & -1 & 1 & 1 & -1 & 1 & 1 \\ 1 & -1 & 1 & 1 & -1 & 1 & 1 & 1 & -1 & -1 & -1 & 1 & 1 & -1 & 1 & 1 \\ 1 & -1 & 1 & 1 & -1 & 1 & 1 & 1 & -1 & -1 & -1 & 1 & 1 & -1 & 1 & 1 \\ 1 & 0 & -2 & 1 & -1 & 1 & 1 & 0 & 0 & 0 & 2 & -2 & -2 & -1 & 1 & 1 \\ 1 & 0 & -2 & 1 & -1 & 1 & 1 & 0 & 0 & 0 & 2 & -2 & -2 & -1 & 1 & 1 \\ 1 & 0 & 0 & -3 & -1 & 1 & 1 & 0 & 0 & 0 & 0 & 0 & 0 & 3 & -3 & -3 \\ 1 & 0 & 0 & -3 & -1 & 1 & 1 & 0 & 0 & 0 & 0 & 0 & 0 & 3 & -3 & -3 \\ 1 & 1 & 1 & 1 & 0 & -2 & 1 & 0 & -2 & 1 & 0 & -2 & 1 & 0 & -2 & 1 \\ 1 & 1 & 1 & 1 & 0 & -2 & 1 & 0 & -2 & 1 & 0 & -2 & 1 & 0 & -2 & 1 \\ 1 & -1 & 1 & 1 & 0 & -2 & 1 & 0 & 2 & -1 & 0 & -2 & 1 & 0 & -2 & 1 \\ 1 & -1 & 1 & 1 & 0 & -2 & 1 & 0 & 2 & -1 & 0 & -2 & 1 & 0 & -2 & 1 \\ 1 & 0 & -2 & 1 & 0 & -2 & 1 & 0 & 0 & 0 & 0 & 4 & -2 & 0 & -2 & 1 \\ 1 & 0 & -2 & 1 & 0 & -2 & 1 & 0 & 0 & 0 & 0 & 4 & -2 & 0 & -2 & 1 \\ 1 & 0 & 0 & -3 & 0 & -2 & 1 & 0 & 0 & 0 & 0 & 0 & 0 & 0 & 6 & -3 \\ 1 & 0 & 0 & -3 & 0 & -2 & 1 & 0 & 0 & 0 & 0 & 0 & 0 & 0 & 6 & -3 \\ 1 & 1 & 1 & 1 & 0 & 0 & -3 & 0 & 0 & -3 & 0 & 0 & -3 & 0 & 0 & -3 \\ 1 & 1 & 1 & 1 & 0 & 0 & -3 & 0 & 0 & -3 & 0 & 0 & -3 & 0 & 0 & -3 \\ 1 & -1 & 1 & 1 & 0 & 0 & -3 & 0 & 0 & 3 & 0 & 0 & -3 & 0 & 0 & -3 \\ 1 & -1 & 1 & 1 & 0 & 0 & -3 & 0 & 0 & 3 & 0 & 0 & -3 & 0 & 0 & -3 \\ 1 & 0 & -2 & 1 & 0 & 0 & -3 & 0 & 0 & 0 & 0 & 0 & 6 & 0 & 0 & -3 \\ 1 & 0 & -2 & 1 & 0 & 0 & -3 & 0 & 0 & 0 & 0 & 0 & 6 & 0 & 0 & -3 \\ 1 & 0 & 0 & -3 & 0 & 0 & -3 & 0 & 0 & 0 & 0 & 0 & 0 & 0 & 0 & 9 \\ 1 & 0 & 0 & -3 & 0 & 0 & -3 & 0 & 0 & 0 & 0 & 0 & 0 & 0 & 0 & 9 \end{bmatrix}$$

17.8.

$$
\mathbf{X} = \begin{bmatrix}
1 & 0 & 0 & 0 & 0 & 0 & 0 & 0 & 0 & 1 \\
1 & 0 & 0 & 0 & 1 & 0 & 0 & 1 & 0 & 0 \\
1 & 0 & 0 & 0 & 0 & 1 & 0 & 0 & 1 & 0 \\
1 & 0 & 0 & 0 & 0 & 0 & 1 & 0 & 0 & 0 \\
1 & 1 & 0 & 0 & 0 & 0 & 0 & 1 & 0 & 0 \\
1 & 1 & 0 & 0 & 1 & 0 & 0 & 0 & 1 & 0 \\
1 & 1 & 0 & 0 & 0 & 1 & 0 & 0 & 0 & 0 \\
1 & 1 & 0 & 0 & 0 & 0 & 1 & 0 & 0 & 1 \\
1 & 0 & 1 & 0 & 0 & 0 & 0 & 0 & 1 & 0 \\
1 & 0 & 1 & 0 & 1 & 0 & 0 & 0 & 0 & 0 \\
1 & 0 & 1 & 0 & 0 & 1 & 0 & 0 & 0 & 1 \\
1 & 0 & 1 & 0 & 0 & 0 & 1 & 1 & 0 & 0 \\
1 & 0 & 0 & 1 & 0 & 0 & 0 & 0 & 0 & 0 \\
1 & 0 & 0 & 1 & 1 & 0 & 0 & 0 & 0 & 1 \\
1 & 0 & 0 & 1 & 0 & 1 & 0 & 1 & 0 & 0 \\
1 & 0 & 0 & 1 & 0 & 0 & 1 & 0 & 1 & 0
\end{bmatrix}
$$

$$\beta = (\mathbf{X}'\mathbf{X})^{-1}\mathbf{X}'\mathbf{X}^*\beta^*$$

*Ed. Note:* Positioning of 17.9 before 17.8 was necessary because of space limitations.

17.10.

| *Source* | *SS* | *df* | *MS* | *F* | *p* |
|---|---|---|---|---|---|
| Machines | 2797.094 | 3 | 932.365 | 75.92 | $<.01$ |
| $\ell_1$ | 6.25 | 1 | 6.25 | .51 | $>.05$ |
| $\ell_2$ | 2790.75 | 1 | 2790.75 | 227.24 | $<.01$ |
| $\ell_3$ | .094 | 1 | .094 | .01 | $>.05$ |
| Mode | 4989.094 | 3 | 1663.031 | 135.41 | $<.01$ |
| $\ell_4$ | 2401.00 | 1 | 2401.00 | 195.50 | $<.01$ |
| $\ell_5$ | 1900.083 | 1 | 1900.083 | 154.72 | $<.01$ |
| $\ell_6$ | 688.010 | 1 | 688.010 | 56.02 | $<.01$ |
| Interaction | 1684.281 | 9 | 187.142 | 15.24 | $<.01$ |
| $\ell_7$ | 420.5 | 1 | 420.5 | 34.24 | $<.01$ |
| $\ell_8$ | 88.167 | 1 | 88.167 | 7.18 | $<.025$ |
| $\ell_9$ | 140.083 | 1 | 140.083 | 11.41 | $<.01$ |
| $\ell_{10}$ | 240.667 | 1 | 240.667 | 19.60 | $<.01$ |
| $\ell_{11}$ | 304.222 | 1 | 304.222 | 24.77 | $<.01$ |
| $\ell_{12}$ | 250.694 | 1 | 250.694 | 20.41 | $<.01$ |
| $\ell_{13}$ | 120.333 | 1 | 120.333 | 9.80 | $<.01$ |
| $\ell_{14}$ | 34.028 | 1 | 34.028 | 2.77 | $>.05$ |
| $\ell_{15}$ | 85.587 | 1 | 85.587 | 6.97 | $<.025$ |
| Error | 196.500 | 16 | 12.281 | | |
| *Totals* | 9666.969 | 31 | | | |

17.11.

| *Source* | *SS* | *df* | *MS* | *F* | *p* |
|---|---|---|---|---|---|
| machines | 2797.094 | 3 | 932.365 | 75.92 | $<.01$ |
| modes | 4989.094 | 3 | 1663.031 | 135.41 | $<.01$ |
| interaction | 1684.281 | 9 | 187.142 | 15.24 | $<.01$ |
| error | 196.500 | 16 | 12.281 | | |
| *Totals* | 9666.969 | 31 | | | |

17.12. $\mathbf{X} = \begin{bmatrix}
1 & 0 & 0 & 0 & 0 & 0 & 0 & 0 & 0 & 0 & 0 & 0 & 0 & 0 & 0 & 0 \\
1 & 0 & 0 & 0 & 0 & 0 & 0 & 0 & 0 & 0 & 0 & 0 & 0 & 0 & 0 & 0 \\
1 & 1 & 0 & 0 & 0 & 0 & 0 & 0 & 0 & 0 & 0 & 0 & 0 & 0 & 0 & 0 \\
1 & 1 & 0 & 0 & 0 & 0 & 0 & 0 & 0 & 0 & 0 & 0 & 0 & 0 & 0 & 0 \\
1 & 0 & 1 & 0 & 0 & 0 & 0 & 0 & 0 & 0 & 0 & 0 & 0 & 0 & 0 & 0 \\
1 & 0 & 1 & 0 & 0 & 0 & 0 & 0 & 0 & 0 & 0 & 0 & 0 & 0 & 0 & 0 \\
1 & 0 & 0 & 1 & 0 & 0 & 0 & 0 & 0 & 0 & 0 & 0 & 0 & 0 & 0 & 0 \\
1 & 0 & 0 & 1 & 0 & 0 & 0 & 0 & 0 & 0 & 0 & 0 & 0 & 0 & 0 & 0 \\
1 & 0 & 0 & 0 & 1 & 0 & 0 & 0 & 0 & 0 & 0 & 0 & 0 & 0 & 0 & 0 \\
1 & 0 & 0 & 0 & 1 & 0 & 0 & 0 & 0 & 0 & 0 & 0 & 0 & 0 & 0 & 0 \\
1 & 1 & 0 & 0 & 1 & 0 & 0 & 1 & 0 & 0 & 0 & 0 & 0 & 0 & 0 & 0 \\
1 & 1 & 0 & 0 & 1 & 0 & 0 & 1 & 0 & 0 & 0 & 0 & 0 & 0 & 0 & 0 \\
1 & 0 & 1 & 0 & 1 & 0 & 0 & 0 & 0 & 0 & 1 & 0 & 0 & 0 & 0 & 0 \\
1 & 0 & 1 & 0 & 1 & 0 & 0 & 0 & 0 & 0 & 1 & 0 & 0 & 0 & 0 & 0 \\
1 & 0 & 0 & 1 & 1 & 0 & 0 & 0 & 0 & 0 & 0 & 0 & 0 & 1 & 0 & 0 \\
1 & 0 & 0 & 1 & 1 & 0 & 0 & 0 & 0 & 0 & 0 & 0 & 0 & 1 & 0 & 0 \\
1 & 0 & 0 & 0 & 0 & 1 & 0 & 0 & 0 & 0 & 0 & 0 & 0 & 0 & 0 & 0 \\
1 & 0 & 0 & 0 & 0 & 1 & 0 & 0 & 0 & 0 & 0 & 0 & 0 & 0 & 0 & 0 \\
1 & 1 & 0 & 0 & 0 & 1 & 0 & 0 & 1 & 0 & 0 & 0 & 0 & 0 & 0 & 0 \\
1 & 1 & 0 & 0 & 0 & 1 & 0 & 0 & 1 & 0 & 0 & 0 & 0 & 0 & 0 & 0 \\
1 & 0 & 1 & 0 & 0 & 1 & 0 & 0 & 0 & 0 & 0 & 1 & 0 & 0 & 0 & 0 \\
1 & 0 & 1 & 0 & 0 & 1 & 0 & 0 & 0 & 0 & 0 & 1 & 0 & 0 & 0 & 0 \\
1 & 0 & 0 & 1 & 0 & 1 & 0 & 0 & 0 & 0 & 0 & 0 & 0 & 0 & 1 & 0 \\
1 & 0 & 0 & 1 & 0 & 1 & 0 & 0 & 0 & 0 & 0 & 0 & 0 & 0 & 1 & 0 \\
1 & 0 & 0 & 0 & 0 & 0 & 1 & 0 & 0 & 0 & 0 & 0 & 0 & 0 & 0 & 0 \\
1 & 0 & 0 & 0 & 0 & 0 & 1 & 0 & 0 & 0 & 0 & 0 & 0 & 0 & 0 & 0 \\
1 & 1 & 0 & 0 & 0 & 0 & 1 & 0 & 0 & 1 & 0 & 0 & 0 & 0 & 0 & 0 \\
1 & 1 & 0 & 0 & 0 & 0 & 1 & 0 & 0 & 1 & 0 & 0 & 0 & 0 & 0 & 0 \\
1 & 0 & 1 & 0 & 0 & 0 & 1 & 0 & 0 & 0 & 0 & 0 & 1 & 0 & 0 & 0 \\
1 & 0 & 1 & 0 & 0 & 0 & 1 & 0 & 0 & 0 & 0 & 0 & 1 & 0 & 0 & 0 \\
1 & 0 & 0 & 1 & 0 & 0 & 1 & 0 & 0 & 0 & 0 & 0 & 0 & 0 & 0 & 1 \\
1 & 0 & 0 & 1 & 0 & 0 & 1 & 0 & 0 & 0 & 0 & 0 & 0 & 0 & 0 & 1
\end{bmatrix}$

$\beta = (\mathbf{X}'\mathbf{X})^{-1}\mathbf{X}'\mathbf{X}^*\beta^*$

17.13. b. $y = \beta_0 + \beta_1 x_1 + \beta_2 x_1^2 + \epsilon$ c. $\mathbf{X}^* = \begin{bmatrix} 1 & -7 & 7 \\ 1 & -5 & 1 \\ 1 & -3 & -3 \\ 1 & -1 & -5 \\ 1 & 1 & -5 \\ 1 & 3 & -3 \\ 1 & 5 & 1 \\ 1 & 7 & 7 \end{bmatrix}$

17.14. a. $\beta_0 = \beta_0^* - 2\beta_1^*\bar{x} + \beta_2^*[\bar{x}^2 - (63/12)]$

$\beta_1 = 2\beta_1^* - 2\beta_2^*\bar{x}$

$\beta_2 = \beta_2^*$

b. $\hat{\beta}^* = \begin{bmatrix} 136.250 \\ 9.583 \\ -9.940 \end{bmatrix}$ c. $\hat{\beta} = \begin{bmatrix} 99.098 \\ 108.626 \\ -9.940 \end{bmatrix}$

17.15. $\mathbf{X}^* = \begin{bmatrix}
1 & 1 & 1 & 1 & -3 & 1 & -1 & -3 & 1 & -1 & -3 & 1 & -1 & -3 & 1 & -1 \\
1 & 1 & 1 & 1 & -3 & 1 & -1 & -3 & 1 & -1 & -3 & 1 & -1 & -3 & 1 & -1 \\
1 & -1 & 1 & 1 & -3 & 1 & -1 & 3 & -1 & 1 & -3 & 1 & -1 & -3 & 1 & -1 \\
1 & -1 & 1 & 1 & -3 & 1 & -1 & 3 & -1 & 1 & -3 & 1 & -1 & -3 & 1 & -1 \\
1 & 0 & -2 & 1 & -3 & 1 & -1 & 0 & 0 & 0 & 6 & -2 & 2 & -3 & 1 & -1 \\
1 & 0 & -2 & 1 & -3 & 1 & -1 & 0 & 0 & 0 & 6 & -2 & 2 & -3 & 1 & -1 \\
1 & 0 & 0 & -3 & -3 & 1 & -1 & 0 & 0 & 0 & 0 & 0 & 0 & 9 & -3 & 3 \\
1 & 0 & 0 & -3 & -3 & 1 & -1 & 0 & 0 & 0 & 0 & 0 & 0 & 9 & -3 & 3 \\
1 & 1 & 1 & 1 & -1 & -1 & 3 & -1 & -1 & 3 & -1 & -1 & 3 & -1 & -1 & 3 \\
1 & 1 & 1 & 1 & -1 & -1 & 3 & -1 & -1 & 3 & -1 & -1 & 3 & -1 & -1 & 3 \\
1 & -1 & 1 & 1 & -1 & -1 & 3 & 1 & 1 & -3 & -1 & -1 & 3 & -1 & -1 & 3 \\
1 & -1 & 1 & 1 & -1 & -1 & 3 & 1 & 1 & -3 & -1 & -1 & 3 & -1 & -1 & 3 \\
1 & 0 & -2 & 1 & -1 & -1 & 3 & 0 & 0 & 0 & 2 & 2 & -6 & -1 & -1 & 3 \\
1 & 0 & -2 & 1 & -1 & -1 & 3 & 0 & 0 & 0 & 2 & 2 & -6 & -1 & -1 & 3 \\
1 & 0 & 0 & -3 & -1 & -1 & 3 & 0 & 0 & 0 & 0 & 0 & 0 & 3 & 3 & -9 \\
1 & 0 & 0 & -3 & -1 & -1 & 3 & 0 & 0 & 0 & 0 & 0 & 0 & 3 & 3 & -9 \\
1 & 1 & 1 & 1 & 1 & -1 & -3 & 1 & -1 & -3 & 1 & -1 & -3 & 1 & -1 & -3 \\
1 & 1 & 1 & 1 & 1 & -1 & -3 & 1 & -1 & -3 & 1 & -1 & -3 & 1 & -1 & -3 \\
1 & -1 & 1 & 1 & 1 & -1 & -3 & -1 & 1 & 3 & 1 & -1 & -3 & 1 & -1 & -3 \\
1 & -1 & 1 & 1 & 1 & -1 & -3 & -1 & 1 & 3 & 1 & -1 & -3 & 1 & -1 & -3 \\
1 & 0 & -2 & 1 & 1 & -1 & -3 & 0 & 0 & 0 & -2 & 2 & 6 & 1 & -1 & -3 \\
1 & 0 & -2 & 1 & 1 & -1 & -3 & 0 & 0 & 0 & -2 & 2 & 6 & 1 & -1 & -3 \\
1 & 0 & 0 & -3 & 1 & -1 & -3 & 0 & 0 & 0 & 0 & 0 & 0 & -3 & 3 & 9 \\
1 & 0 & 0 & -3 & 1 & -1 & -3 & 0 & 0 & 0 & 0 & 0 & 0 & -3 & 3 & 9 \\
1 & 1 & 1 & 1 & 3 & 1 & 1 & 3 & 1 & 1 & 3 & 1 & 1 & 3 & 1 & 1 \\
1 & 1 & 1 & 1 & 3 & 1 & 1 & 3 & 1 & 1 & 3 & 1 & 1 & 3 & 1 & 1 \\
1 & -1 & 1 & 1 & 3 & 1 & 1 & -3 & -1 & -1 & 3 & 1 & 1 & 3 & 1 & 1 \\
1 & -1 & 1 & 1 & 3 & 1 & 1 & -3 & -1 & -1 & 3 & 1 & 1 & 3 & 1 & 1 \\
1 & 0 & -2 & 1 & 3 & 1 & 1 & 0 & 0 & 0 & -6 & -2 & -2 & 3 & 1 & 1 \\
1 & 0 & -2 & 1 & 3 & 1 & 1 & 0 & 0 & 0 & -6 & -2 & -2 & 3 & 1 & 1 \\
1 & 0 & 0 & -3 & 3 & 1 & 1 & 0 & 0 & 0 & 0 & 0 & 0 & -9 & -3 & -3 \\
1 & 0 & 0 & -3 & 3 & 1 & 1 & 0 & 0 & 0 & 0 & 0 & 0 & -9 & -3 & -3
\end{bmatrix}$

17.16. a.

| *Source* | *SS* | *df* | *MS* | *F* | *p* |
|---|---|---|---|---|---|
| Machines | 2797.094 | 3 | 932.365 | 75.92 | $<.05$ |
| $\ell_1$ | 6.25 | 1 | 6.25 | .51 | $>.05$ |
| $\ell_2$ | 2790.75 | 1 | 2790.75 | 227.24 | $<.05$ |
| $\ell_3$ | .094 | 1 | .094 | .01 | $>.05$ |
| Mode | 4989.093 | 3 | 1663.031 | 135.41 | $<.05$ |
| $\ell_4$ | 3562.656 | 1 | 3562.656 | 290.09 | $<.05$ |
| $\ell_5$ | 1391.281 | 1 | 1391.281 | 113.29 | $<.05$ |
| $\ell_6$ | 35.156 | 1 | 35.156 | 2.86 | $>.05$ |
| Interaction | 1684.281 | 9 | 187.142 | 15.24 | $<.05$ |
| $\ell_7$ | 432.450 | 1 | 432.450 | 35.21 | $<.05$ |
| $\ell_8$ | 132.250 | 1 | 132.250 | 10.77 | $<.05$ |
| $\ell_9$ | 84.050 | 1 | 84.050 | 6.84 | $<.05$ |
| $\ell_{10}$ | 714.150 | 1 | 714.150 | 58.15 | $<.05$ |
| $\ell_{11}$ | 80.083 | 1 | 80.083 | 6.52 | $<.05$ |
| $\ell_{12}$ | 1.350 | 1 | 1.350 | .11 | $>.05$ |
| $\ell_{13}$ | 191.269 | 1 | 191.269 | 15.57 | $<.05$ |
| $\ell_{14}$ | 23.010 | 1 | 23.010 | 1.87 | $>.05$ |
| $\ell_{15}$ | 25.669 | 1 | 25.669 | 2.09 | $>.05$ |
| Error | 196.500 | 16 | 12.281 | | |
| *Totals* | 9666.968 | 31 | | | |

b. $\boldsymbol{\beta} = (\mathbf{X}'\mathbf{X})^{-1}\mathbf{X}'\mathbf{X}^*\boldsymbol{\beta}^*$; $\mathbf{X}$ = matrix derived in exercise 17.9; $\mathbf{X}^*$ = matrix derived in exercise 17.15

$$\hat{\boldsymbol{\beta}}^* = \begin{bmatrix} 42.969 \\ .625 \\ 7.625 \\ .031 \\ 4.719 \\ -6.594 \\ .469 \\ 2.325 \\ -2.875 \\ 1.025 \\ 1.725 \\ -1.292 \\ .075 \\ .631 \\ -.490 \\ .231 \end{bmatrix} \qquad \hat{\boldsymbol{\beta}} = \begin{bmatrix} 42.969 \\ .625 \\ 7.625 \\ .031 \\ -12.250 \\ -6.292 \\ -2.677 \\ -7.250 \\ -1.917 \\ -1.708 \\ -3.167 \\ -2.056 \\ -1.319 \\ -1.583 \\ -.486 \\ -.545 \end{bmatrix}$$

17.17. $\boldsymbol{\beta} = (\mathbf{X}'\mathbf{X})^{-1}\mathbf{X}'\mathbf{X}^*\boldsymbol{\beta}^*$

$$\mathbf{X} = \begin{bmatrix} 1 & 1 & 0 & 1 & 0 & 0 \\ 1 & 1 & 0 & 0 & 1 & 0 \\ 1 & 1 & 0 & 0 & 0 & 1 \\ 1 & 1 & 0 & 0 & 0 & 0 \\ 1 & 0 & 1 & 1 & 0 & 0 \\ 1 & 0 & 1 & 0 & 1 & 0 \\ 1 & 0 & 1 & 0 & 0 & 1 \\ 1 & 0 & 1 & 0 & 0 & 0 \\ 1 & 0 & 0 & 1 & 0 & 0 \\ 1 & 0 & 0 & 0 & 1 & 0 \\ 1 & 0 & 0 & 0 & 0 & 1 \\ 1 & 0 & 0 & 0 & 0 & 0 \end{bmatrix} \qquad \mathbf{X}^* = \begin{bmatrix} 1 & 1 & 1 & 1 & 1 & 1 \\ 1 & 1 & 1 & -1 & 1 & 1 \\ 1 & 1 & 1 & 0 & -2 & 1 \\ 1 & 1 & 1 & 0 & 0 & -3 \\ 1 & -1 & 1 & 1 & 1 & 1 \\ 1 & -1 & 1 & -1 & 1 & 1 \\ 1 & -1 & 1 & 0 & -2 & 1 \\ 1 & -1 & 1 & 0 & 0 & -3 \\ 1 & 0 & -2 & 1 & 1 & 1 \\ 1 & 0 & -2 & -1 & 1 & 1 \\ 1 & 0 & -2 & 0 & -2 & 1 \\ 1 & 0 & -2 & 0 & 0 & -3 \end{bmatrix}$$

$$\hat{\boldsymbol{\beta}}^* = \begin{bmatrix} 74.917 \\ -14.875 \\ -2.542 \\ 3.500 \\ -3.611 \\ -1.472 \end{bmatrix} \quad \hat{\boldsymbol{\beta}} = \begin{bmatrix} 74.917 \\ -17.417 \\ 12.333 \\ -1.583 \\ -8.583 \\ 5.750 \end{bmatrix} = \begin{bmatrix} \hat{\beta}_0 \\ \hat{\beta}_1 \\ \hat{\beta}_2 \\ \hat{\beta}_3 \\ \hat{\beta}_4 \\ \hat{\beta}_5 \end{bmatrix}$$

$$\hat{\boldsymbol{\beta}}^* = \begin{bmatrix} \hat{\beta}_0 \\ \frac{1}{2}(\hat{\beta}_1 - \hat{\beta}_2) \\ \frac{1}{2}(\hat{\beta}_1 + \hat{\beta}_2) \\ \frac{1}{2}(\hat{\beta}_3 - \hat{\beta}_4) \\ \frac{1}{6}(\hat{\beta}_3 + \hat{\beta}_4 - 2\hat{\beta}_5) \\ \frac{1}{3}(\hat{\beta}_1 + \hat{\beta}_2 + \hat{\beta}_3) \end{bmatrix} = \begin{bmatrix} 74.917 \\ -14.875 \\ -2.542 \\ 3.500 \\ -3.611 \\ -1.472 \end{bmatrix}$$

17.18. a. $\hat{\beta} = \begin{bmatrix} 171.111 \\ 20.833 \\ -3.611 \end{bmatrix}$

b.

| *Source* | *SS* | *df* | *MS* | *F* | *p* |
|---|---|---|---|---|---|
| $x_1^*$ | 2604.1667 | 1 | 2604.1667 | 8.93 | $<.05$ |
| $x_2^*$ | 234.7222 | 1 | 234.7222 | .80 | $>.05$ |
| error | 1750 | 6 | 291.6667 | | |
| *Totals* | 4588.8889 | 8 | | | |

17.19. a. $\hat{\beta} = \begin{bmatrix} 93.3333 \\ .6417 \\ -.0011 \end{bmatrix}$

b.

| *Source* | *SS* | *df* | *MS* | *F* | *p* |
|---|---|---|---|---|---|
| linear | 2604.1667 | 1 | 2604.1667 | 8.93 | $<.05$ |
| quadratic | 234.7222 | 1 | 234.7222 | .80 | $>.05$ |
| error | 1750 | 6 | 291.6667 | | |
| *Totals* | 4588.8889 | 8 | | | |

c. From both exercises, the conclusions are identical:
(1) Reject $H_0$ that $\beta_1 = 0$; there is a significant linear effect.
(2) Cannot reject $H_0$ that $\beta_2 = 0$; there is no quadratic effect.

17.20.

| *Source* | *SS* | *df* | *MS* | *F* | *p* |
|---|---|---|---|---|---|
| Calcium | 1.467 | 2 | | | |
| $\ell_1$ | .844 | 1 | .844 | 12.45 | $<.05$ |
| $\ell_2$ | .623 | 1 | .623 | 9.20 | $<.05$ |
| pH | 4.461 | 3 | | | |
| $\ell_3$ | 2.961 | 1 | 2.961 | 43.68 | $<.05$ |
| $\ell_4$ | 1.467 | 1 | 1.467 | 21.64 | $<.05$ |
| $\ell_5$ | .033 | 1 | .033 | .49 | $>.05$ |
| Interaction | 3.254 | 6 | | | |
| $\ell_6$ | 1.920 | 1 | 1.920 | 28.33 | $<.05$ |
| $\ell_7$ | .250 | 1 | .250 | 3.69 | $>.05$ |
| $\ell_8$ | .661 | 1 | .661 | 9.76 | $<.05$ |
| $\ell_9$ | .054 | 1 | .054 | .80 | $>.05$ |
| $\ell_{10}$ | .095 | 1 | .095 | 1.40 | $>.05$ |
| $\ell_{11}$ | .274 | 1 | .274 | 4.05 | $>.05$ |
| Error | 1.627 | 24 | .0678 | | |
| *Totals* | 10.809 | 35 | | | |

17.21. a. $\hat{\beta} = \begin{bmatrix} 7.0528 \\ .0458 \\ -.1403 \\ .0794 \\ .3028 \\ -.1008 \end{bmatrix}$ b. $\hat{y} = 7.22286$

17.22. $F_{6,24} = 5.062; p < .01$; reject $H_0$ that the assumed model is correct.

17.23. a. $\begin{bmatrix} 11.3333 \\ 1.1000 \\ 0 \\ 1.0333 \end{bmatrix}$

b. $y = 11.82$; confidence limits are $\hat{y} \pm t_{\alpha/2,8}\, s\sqrt{1 + \mathbf{a}'(\mathbf{X}'\mathbf{X})^{-1}\mathbf{a}}$, where $\hat{y} = 11.82$, $s = \sqrt{\text{MSE}} = 1.5$[illegible]

$$\mathbf{a} = \begin{bmatrix} 1 \\ -.2 \\ -1.24 \\ .68 \end{bmatrix} \quad (\mathbf{X}'\mathbf{X})^{-1} = \begin{bmatrix} .0833 & 0 & 0 & 0 \\ 0 & .0167 & 0 & 0 \\ 0 & 0 & .0833 & 0 \\ 0 & 0 & 0 & .0167 \end{bmatrix}$$

c. $\hat{\beta}_0 = \hat{\beta}_0^* - \hat{\beta}_1^*\bar{x}(\frac{2}{5}) + \hat{\beta}_2^*(\frac{\bar{x}^2}{25} - \frac{15}{12}) + \hat{\beta}_3^*\bar{x}(\frac{-2\bar{x}^2}{75} + \frac{41}{30}) = -119.333$

$\hat{\beta}_1 = \frac{2}{5}\hat{\beta}_1^* - \hat{\beta}_2^*\bar{x}(\frac{2}{25}) + \hat{\beta}_3^*(\frac{2\bar{x}^2}{25} - \frac{41}{30}) = 24.344$

$\hat{\beta}_2 = \hat{\beta}_2^*(\frac{1}{25}) - \hat{\beta}_3^*(\frac{2\bar{x}}{25}) = -1.447$

$\hat{\beta}_3 = \frac{2}{75}\hat{\beta}_3^* = .028$

## CHAPTER 18

18.1. a. $y_{ij} = \mu + \alpha_i + \epsilon_{ij}; i = 1, 2, ..., 5; j = 1, 2, 3, 4$
$y_{ij}$ = $j$th observation from Vat $i$
$\mu$ = overall unknown mean
$\alpha_i$ = random effect due to the $i$th vat; $\alpha_i$ is normally distributed with mean 0 and variance $\sigma_\alpha^2$
$\epsilon_{ij}$ = random error associated with response $j$ on $i$th vat

b.

| *Source* | *SS* | *df* | *MS* | *EMS* | *F* |
|---|---|---|---|---|---|
| vats | 11.948 | 4 | 2.987 | $\sigma^2 + 4\sigma_\alpha^2$ | 32.53 |
| error | 1.3775 | 15 | .09183 | $\sigma^2$ | |
| *Totals* | 13.3255 | 19 | | | |

18.2. a. 3.135 b. .773

18.3. a. $y_{ij} = \mu + \alpha_i + \beta_j + \epsilon_{ij}; i = 1, 2, 3; j = 1, 2, 3$
$y_{ij}$ = observation from $j$th subject in Analyst $i$
$\mu$ = overall unknown concentration mean
$\alpha_i$ = random effect due to the $i$th analyst
$\beta_j$ = random effect due to the $j$th subject
$\epsilon_{ij}$ = random error associated with response in Analyst $i$, Subject $j$

b.

| *Source* | *EMS* |
|---|---|
| analysts | $\sigma^2 + 3\sigma_\alpha^2$ |
| subjects | $\sigma^2 + 3\sigma_\beta^2$ |
| error | $\sigma^2$ |

18.4.

| Source | SS | df | MS | EMS | F | p |
|---|---|---|---|---|---|---|
| analysts | .8822 | 2 | .4411 | $\sigma^2 + 3\sigma_\alpha^2$ | 19.36 | $<.05$ |
| subjects | 33.2356 | 2 | 16.6178 | $\sigma^2 + 3\sigma_\beta^2$ | 729.5 | $<.05$ |
| error | .0911 | 4 | .02278 | $\sigma^2$ | | |
| *Totals* | 34.2089 | 8 | | | | |

18.5. a. $y_{ijk} = \mu + \alpha_i + \beta_j + \gamma_{ij} + \epsilon_{ijk}$; $i = 1, 2, 3, 4$; $j = 1, 2, 3$

$y_{ijk}$ = response (sales volume) for the $k$th observation at the $i$th level of SMSA and $j$th level of chain store

$\mu$ = overall unknown sales volume mean

$\alpha_i$ = random effect due to the $i$th SMSA

$\beta_j$ = random effect due to the $j$th store

$\gamma_{ij}$ = random effect due to the $i$th level of SMSA and the $j$th level of store

$\epsilon_{ijk}$ = random error associated with $k$th response at SMSA $i$, Store $j$

b.

| Source | SS | df | MS | EMS | F | p |
|---|---|---|---|---|---|---|
| SMSA | 136644.125 | 3 | 45548.042 | $\sigma^2 + 2\sigma_\gamma^2 + 6\sigma_\alpha^2$ | 5.049 | $<.05$ |
| store | 62973.000 | 2 | 31486.5 | $\sigma^2 + 2\sigma_\gamma^2 + 8\sigma_\beta^2$ | 3.490 | $>.05$ |
| interaction | 54127.000 | 6 | 9021.1667 | $\sigma^2 + 2\sigma_\gamma^2$ | 22.03 | $<.05$ |
| error | 4913.500 | 12 | 409.4583 | $\sigma^2$ | | |
| *Totals* | 258657.625 | 23 | | | | |

18.6. a. $y_{ijk} = \mu + \alpha_i + \beta_j + \gamma_{ij} + \epsilon_{ijk}$; $i = 1, 2, \ldots, 5$; $j = 1, 2, 3, 4$

$y_{ijk}$ = response (degree of cure) for the $k$th observation at the $i$th level of physician and the $j$th level of patient

$\mu$ = overall unknown degree of cure mean

$\alpha_i$ = fixed effect due to the $i$th physician

$\beta_j$ = random effect due to the $j$th patient

$\gamma_{ij}$ = random effect due to the $i$th level of physician and the $j$th level of patient

$\epsilon_{ijk}$ = random error associated with the $k$th response at Physician $i$, Patient $j$

b.

| Source | EMS |
|---|---|
| physician | $\sigma^2 + 2\sigma_\gamma^2 + \theta_A$ |
| patients | $\sigma^2 + 10\sigma_B^2$ |
| interaction | $\sigma^2 + 2\sigma_\gamma^2$ |
| error | $\sigma^2$ |

18.7.

| Source | SS | df | MS | EMS | F | p |
|---|---|---|---|---|---|---|
| physicians | 3.8115 | 4 | .9529 | $\sigma^2 + 2\sigma_\gamma^2 + \theta_A$ | .71 | $>.05$ |
| patients | 180.13275 | 3 | 60.0442 | $\sigma^2 + 10\sigma_B^2$ | 173.41 | $<.05$ |
| interaction | 16.1585 | 12 | 1.3465 | $\sigma^2 + 2\sigma_\gamma^2$ | 3.89 | $<.05$ |
| error | 6.925 | 20 | .34625 | $\sigma^2$ | | |
| *Totals* | 207.02775 | 39 | | | | |

18.8. a.

| Source | EMS | Source | EMS |
|---|---|---|---|
| A | $\sigma^2 + \theta_A$ | AC | $\sigma^2 + \theta_{AC}$ |
| B | $\sigma^2 + \theta_B$ | BC | $\sigma^2 + \theta_{BC}$ |
| C | $\sigma^2 + \theta_C$ | ABC | $\sigma^2 + \theta_{ABC}$ |
| AB | $\sigma^2 + \theta_{AB}$ | error | $\sigma^2$ |

b. The test for all sources—main effects (3), two-way interactions (3), and one three-way interaction—consists of the $MS_{source}$ divided by $MS_{error}$ with $df_1$ = degrees of freedom in source and $df_2$ = degrees of freedom in error.

18.9. a.

| Source | EMS |
|---|---|
| A | $\sigma^2 + r\sigma^2_{ABC} + cr\sigma^2_{AB} + br\sigma^2_{AC} + bcr\sigma^2_A$ |
| B | $\sigma^2 + r\sigma^2_{ABC} + cr\sigma^2_{AB} + ar\sigma^2_{BC} + acr\sigma^2_B$ |
| C | $\sigma^2 + r\sigma^2_{ABC} + br\sigma^2_{AC} + ar\sigma^2_{BC} + abr\sigma^2_C$ |
| AB | $\sigma^2 + r\sigma^2_{ABC} + cr\sigma^2_{AB}$ |
| AC | $\sigma^2 + r\sigma^2_{ABC} + br\sigma^2_{AC}$ |
| BC | $\sigma^2 + r\sigma^2_{ABC} + ar\sigma^2_{BC}$ |
| ABC | $\sigma^2 + r\sigma^2_{ABC}$ |
| error | $\sigma^2$ |

b.

| Source | F Test | $df_1$ | $df_2$ |
|---|---|---|---|
| ABC | MSABC/MSE | $(a-1)(b-1)(c-1)$ | $abc(r-1)$ |
| BC | MSBC/MSABC | $(b-1)(c-1)$ | $(a-1)(b-1)(c-1)$ |
| AC | MSAC/MSABC | $(a-1)(c-1)$ | $(a-1)(b-1)(c-1)$ |
| AB | MSAB/MSABC | $(a-1)(b-1)$ | $(a-1)(b-1)(c-1)$ |

There are no exact $F$ tests for $\sigma^2_A$, $\sigma^2_B$, and $\sigma^2_C$.

18.10. a.

| Source | EMS |
|---|---|
| A | $\sigma^2 + cr\sigma^2_{AB} + br\sigma^2_{AC} + r\sigma^2_{ABC} + \theta_A$ |
| B | $\sigma^2 + r\sigma^2_{ABC} + ar\sigma^2_{BC} + acr\sigma^2_B$ |
| C | $\sigma^2 + r\sigma^2_{ABC} + ar\sigma^2_{BC} + abr\sigma^2_C$ |
| AB | $\sigma^2 + r\sigma^2_{ABC} + cr\sigma^2_{AB}$ |
| AC | $\sigma^2 + r\sigma^2_{ABC} + br\sigma^2_{AC}$ |
| BC | $\sigma^2 + ar\sigma^2_{BC}$ |
| ABC | $\sigma^2 + r\sigma^2_{ABC}$ |
| error | $\sigma^2$ |

b.

| Source | F Test | $df_1$ | $df_2$ |
|---|---|---|---|
| ABC | MSABC/MSE | $(a-1)(b-1)(c-1)$ | $abc(r-1)$ |
| BC | MSBC/MSE | $(b-1)(c-1)$ | $abc(r-1)$ |
| AC | MSAC/MSABC | $(a-1)(c-1)$ | $(a-1)(b-1)(c-1)$ |
| AB | MSAB/MSABC | $(a-1)(b-1)$ | $(a-1)(b-1)(c-1)$ |

18.11. a. $y_{ij} = \mu + \alpha_i + \epsilon_{ij}$; $i = 1, 2, ..., 6$; $j = 1, 2$
$y_{ij}$ = $j$th observation from week $i$
$\mu$ = overall unknown mean
$\alpha_i$ = random effect due to the $i$th week
$\epsilon_{ij}$ = random error associated with response $j$ at week $i$

b.

| Source | SS | df | MS | EMS | F | p |
|---|---|---|---|---|---|---|
| weeks | 255089.42 | 5 | 51017.88 | $\sigma^2 + 2\sigma^2_\alpha$ | 26.76 | $<.05$ |
| error | 11439.5 | 6 | 1906.58 | $\sigma^2$ | | |
| Totals | 266528.92 | 11 | | | | |

18.12. $\bar{y} = 525.083$; BOE = 130.41

18.13. a. $y_{ijk} = \mu + \alpha_i + \beta_j + \gamma_{ij} + \epsilon_{ijk}$; $i = 1, 2, ..., 5$; $j = 1, 2, 3, 4$; $k = 1, 2$; in this model $\alpha_i$ is a random effect due to the $i$th physician, while in exercise 18.6 it was a fixed effect. In this model $\alpha_i$, $\beta_j$, and $\gamma_{ij}$ are all independent.

b.

| Source | SS | df | MS | EMS | F | p |
|---|---|---|---|---|---|---|
| physicians | 3.8115 | 4 | .9529 | $\sigma^2 + 2\sigma^2_\gamma + 8\sigma^2_\alpha$ | .71 | $>.05$ |
| patients | 180.1328 | 3 | 60.0442 | $\sigma^2 + 2\sigma^2_\gamma + 10\sigma^2_\beta$ | 44.59 | $<.05$ |
| interaction | 16.1585 | 12 | 1.3465 | $\sigma^2 + 2\sigma^2_\gamma$ | 3.89 | |
| error | 6.9250 | 20 | .34625 | $\sigma^2$ | | |
| Totals | 207.0278 | 39 | | | | |

c. In the practical situation, the model with physicians as fixed effects and patients as random effects is more appropriate because the physicians who are tops in their field have been asked to do the study and they have consented to do the experiment.

d. Yes, if you consider your population to consist of all and the only ones available for treatment.

18.14. a. $y_{ij} = \mu + \alpha_i + \beta_j + \epsilon_{ij}$; $i = 1, 2, 3, 4$; $j = 1, 2, ..., 5$
$y_{ij}$ = observation from $j$th investigator in Mixture $i$
$\mu$ = overall unknown mean
$\alpha_i$ = fixed effect due to $i$th mixture
$\beta_j$ = random effect due to $j$th investigator; $\beta_j$ is normally distributed with mean 0 and variance $\sigma^2_\beta$; the $\beta$'s are independent
$\epsilon_{ij}$ = random error associated with response in Mixture $i$, Investigator $j$; the $\epsilon_{ij}$ are independent

b.

| Source | SS | df | MS | EMS | F | p |
|---|---|---|---|---|---|---|
| mixtures | 261260.95 | 3 | 87086.983 | $\sigma^2 + \theta_A$ | 1264.73 | $<.05$ |
| investigators | 452.50 | 4 | 113.125 | $\sigma^2 + 4\sigma^2_\beta$ | 1.64 | $>.05$ |
| error | 826.30 | 12 | 68.858 | $\sigma^2$ | | |
| Totals | 262539.75 | | | | | |

c. In exercise 15.2 $H_0$ was $\beta_1 = \beta_2 = \beta_3 = \beta_4 = \beta_5 = 0$; in exercise 18.14 $H_0$ is $\sigma^2_\beta = 0$. In exercise 15.2 we could not reject $H_0$ and concluded that the sampled $\beta$'s (which were the only $\beta$'s) were identically 0. In exercise 18.14 we cannot reject $H_0$ and we conclude that the sampled $\beta$'s as well as all other $\beta$'s in the population are 0.

18.15.

| *Source* | *EMS* |
|---|---|
| rows (intersections) | $\sigma^2 + 4\sigma_\alpha^2$ |
| columns (time) | $\sigma^2 + \theta_B$ |
| treatments (devices) | $\sigma^2 + \theta_C$ |
| error | $\sigma^2$ |

18.16.

| *Source* | *EMS* |
|---|---|
| rows | $\sigma^2 + 4\sigma_\alpha^2$ |
| columns | $\sigma^2 + 4\sigma_\beta^2$ |
| treatments | $\sigma^2 + \theta_C$ |
| error | $\sigma^2$ |

18.17. a.

| *Source* | *EMS* |
|---|---|
| investigators | $\sigma^2 + 45\sigma^2$ |
| treatments | $\sigma^2 + 5\sigma_\gamma^2 + \theta_B$ |
| interaction | $\sigma^2 + 5\sigma_\gamma^2$ |
| error | $\sigma^2$ |

b. In the $F$ test for differences among treatments in this exercise, the T.S. is $MST/MS_{interaction}$ with $df_1 = 8$ and $df_2 = 16$. In exercise 15.18 the T.S. was MST/MSE with $df_1 = 8$ and $df_2 = 108$. If the test for investigators was significant in this exercise, we would conclude that $\sigma_\alpha^2$ is greater than 0, whereas in exercise 15.18 the conclusion would be that the means are unequal.

18.18. a. $\hat{\sigma}_\gamma^2 = (MSAB - MSE)/2$; $\hat{\sigma}_\alpha^2 = (MSA - MSAB)/6$; $\hat{\sigma}_\beta^2 = (MSB - MSAB)/6$; $\hat{\sigma}^2 = MSE$

b. Since MSAB is so small relative to MSE, we can assume $\sigma_\gamma^2 = 0$ and we will pool SSAB and SSE to form a combined estimate of $\sigma^2$ as we did in section 18.3. The pooled mean square is .16. Since $\sigma_\gamma^2 = 0$, the EMS for factors $A$ and $B$ are $\sigma^2 + 6\sigma_\alpha^2$ and $\sigma^2 + 6\sigma_\beta^2$, respectively. Solving for $\sigma_\alpha^2$ we get $(MSA - MSE)/6 = 4.62$ and similarly $\hat{\sigma}_\beta^2 = (MSB - MSE)/6 = 0$.

18.19. $\hat{\sigma}^2 = 409.46$; $\hat{\sigma}_\beta^2 = 2808.17$; $\hat{\sigma}_\gamma^2 = 4305.85$; $\hat{\sigma}_\alpha^2 = 6087.81$

## CHAPTER 19

19.2.

| $\bar{T}_4$ | $\bar{T}_3$ | $\bar{T}_2$ | $\bar{T}_1$ |
|---|---|---|---|
| 8.38 | 9.36 | 9.46 | 15.26 |

Note: The mean of $\bar{T}_4$ includes the estimate of the missing value.

19.3.

| *Source* | *SS* | *df* | *MS* | *F* | *p* |
|---|---|---|---|---|---|
| plots | 386.89 | 3 | 128.96 | 27.44 | <.01 |
| insecticides | 1932.14 | 2 | 966.07 | 205.55 | <.01 |
| error | 23.49 | 5 | 4.70 | | |
| *Totals* | 2342.52 | 10 | | | |

19.4. AOV for testing insecticides:

| *Source* | *SS* | *df* | *MS* | *F* | *p* |
|---|---|---|---|---|---|
| plots | 446.545 | 3 | | | |
| insecticides (adj.) | 1474.514 | 2 | 737.257 | 156.96 | <.01 |
| error | 23.486 | 5 | 4.697 | | |
| *Totals* | 1944.545 | 10 | | | |

AOV for testing plots:

| *Source* | *SS* | *df* | *MS* | *F* | *p* |
|---|---|---|---|---|---|
| plots (adj.) | 383.264 | 3 | 127.755 | 27.20 | <.01 |
| insecticides | 1537.795 | 2 | | | |
| error | 23.486 | 5 | 4.697 | | |
| *Totals* | 1944.545 | 10 | | | |

19.6. a. 2654.5

b.

| *Source* | *SS* | *df* | *MS* | *F* | *p* |
|---|---|---|---|---|---|
| investigators | 359.05 | 4 | 89.76 | 1.30 | >.05 |
| mixtures | 257,299.04 | 3 | 85,766.35 | 1241.19 | <.01 |
| error | 760.15 | 11 | 69.10 | | |
| *Totals* | 258,418.24 | 18 | | | |

19.7. a. AOV for testing mixtures:

| *Source* | *SS* | *df* | *MS* | *F* | *p* |
|---|---|---|---|---|---|
| investigators | 8910.03 | 4 | | | |
| mixtures (adj.) | 210,236.35 | 3 | 70,078.78 | 1014.16 | <.01 |
| error | 760.15 | 11 | 69.10 | | |
| *Totals* | 219,906.53 | 18 | | | |

AOV for testing investigators:

| Source | SS | df | MS | F | p |
|---|---|---|---|---|---|
| investigators (adj.) | 344.60 | 4 | 86.15 | 1.25 | > .05 |
| mixtures | 218,801.78 | 3 | | | |
| error | 760.15 | 11 | 69.10 | | |
| *Totals* | 219,906.53 | 18 | | | |

b. Use complete and reduced models excluding this observation also.

19.8.

| Source | SS | df | MS | F | p |
|---|---|---|---|---|---|
| investigator | 15.368 | 4 | 3.842 | 29.33 | < .01 |
| day | 1.032 | 4 | .258 | 1.97 | > .05 |
| treatment | 179.556 | 4 | 44.889 | 342.66 | < .01 |
| error | 1.444 | 11 | .131 | | |
| *Totals* | 197.400 | 23 | | | |

19.10. a. $26.3 = M$

| Source | SS | df | MS | F | p |
|---|---|---|---|---|---|
| intersection | 8.3469 | 3 | 2.7823 | 0.78 | > .05 |
| time | 710.3619 | 3 | 236.7873 | 66.11 | < .01 |
| device | 89.6319 | 3 | 29.8773 | 8.34 | < .05 |
| error | 17.9087 | 5 | 3.5817 | | |
| *Totals* | 826.2494 | 14 | | | |

b. III IV I II
18.05 22.35 23.58 24.05

19.11.

| Source | SS | df | MS | F | p |
|---|---|---|---|---|---|
| intersection (adj. for time, device) | 7.492 | 3 | 2.497 | 0.70 | > .05 |
| time (adj. for intersection, device) | 695.087 | 3 | 231.696 | 64.68 | < .01 |
| device (adj. for intersection, time) | 82.297 | 3 | 27.432 | 7.66 | > .05 |
| error | 17.908 | 5 | 3.582 | | |

19.12. a. balanced incomplete block design
b. $t = 6$; $b = 10$; $\lambda = 2$; $k = 3$; $r = 5$

19.13.

| Source | SS | df | MS | F | p |
|---|---|---|---|---|---|
| persons | 1034.8 | 9 | | | |
| treatments (adj.) | 1747.056 | 5 | 349.411 | 11.55 | < .01 |
| error | 453.611 | 15 | 30.241 | | |
| *Totals* | 3235.467 | 29 | | | |

19.14. E D B A F C
31.55 35.55 37.55 41.383 51.55 55.216

19.15. same answer as exercise 19.13

19.16.

| Source | SS | df | MS | F | p |
|---|---|---|---|---|---|
| blocks | 4546.31 | 11 | | | |
| treatments (adj.) | 11930.07 | 8 | 1491.26 | 46.98 | |
| error | 507.93 | 16 | 31.75 | | |
| *Totals* | 16984.31 | 35 | | | |

9.17. Use the partial sum of squares for blocks.

## :HAPTER 20

'0.1. $y = \beta_0 + \beta_1 x_1 + \beta_2 x_2 + \beta_3 x_3 + \beta_4 x_4 + \beta_5 x_5 + \beta_6 x_1 x_2 + \beta_7 x_1 x_3 + \beta_8 x_1 x_4 + \beta_9 x_1 x_5 + \epsilon$

$x_1$ = covariate
$x_2 = 1$ if treatment 2 is applied
$x_2 = 0$ otherwise
$x_3 = 1$ if treatment 3 is applied
$x_3 = 0$ otherwise
$x_4 = 1$ if treatment 4 is applied
$x_4 = 0$ otherwise
$x_5 = 1$ if treatment 5 is applied
$x_5 = 0$ otherwise

20.2. The assumption of constant slope for the 5 treatment groups can be tested using the following null hypothesis:
$$H_0: \beta_6 = \beta_7 = \beta_8 = \beta_9 = 0$$
The complete model is given in exercise 20.1 and the reduced model is $y = \beta_0 + \beta_1 x_1 + \beta_2 x_2 + \beta_3 x_3 + \beta_4 x_4 + \beta_5 x_5 + \epsilon$. The test statistic is $F = \text{MS}_{\text{drop}}/\text{MSE}_1$, with $df_1 = 4$ and $df_2 = 20$.

20.3. Complete model: $y = \beta_0 + \beta_1 x_1 + \beta_2 x_2 + \beta_3 x_3 + \beta_4 x_4 + \beta_5 x_5 + \epsilon$

Reduced model: $y = \beta_0 + \beta_1 x_1 + \epsilon$

$F_{4,24} = \text{MS}_{\text{drop}}/\text{MSE}_1$ and $\hat{y} = \hat{\beta}_0 + 5\hat{\beta}_1$

20.4. Test for parallelism: $F_{2,12} = 4.133$. The test for adjusted treatment means is not valid since the slopes are not constant for the three groups.

20.5.

$$y = \beta_0 + \beta_1x_1 + \beta_2x_2 + \beta_3x_3 + \beta_4x_4 + \beta_5x_5 + \beta_6x_6 + \beta_7x_7 + \beta_8x_8 + \beta_9x_9 + \beta_{10}x_{10} + \beta_{11}x_1x_2 + \beta_{12}x_1x_3 + \beta_{13}x_1x_4 + \beta_{14}x_1x_5 + \beta_{15}x_1x_6 + \beta_{16}x_1x_7 + \beta_{17}x_1x_8 + \beta_{18}x_1x_9 + \beta_{19}x_1x_{10} + \epsilon$$

| | |
|---|---|
| $x_1$ = covariate | $x_6$ = 1 if row 3 |
| $x_2$ = 1 if column 2 | $x_6$ = 0 otherwise |
| $x_2$ = 0 otherwise | $x_7$ = 1 if row 4 |
| $x_3$ = 1 if column 3 | $x_7$ = 0 otherwise |
| $x_3$ = 0 otherwise | $x_8$ = 1 if treatment 2 |
| $x_4$ = 1 if column 4 | $x_8$ = 0 otherwise |
| $x_4$ = 0 otherwise | $x_9$ = 1 if treatment 3 |
| $x_5$ = 1 if row 2 | $x_9$ = 0 otherwise |
| $x_5$ = 0 otherwise | $x_{10}$ = 1 if treatment 4 |
| | $x_{10}$ = 0 otherwise |

20.6. a. Obtain $SSE_1$ from the complete model described in exercise 20.5. Obtain $SSE_2$ from the reduced model:

$$y = \beta_0 + \beta_1x_1 + \beta_2x_2 + \beta_3x_3 + \beta_4x_4 + \beta_5x_5 + \beta_6x_6 + \beta_7x_7 + \beta_8x_8 + \beta_9x_9 + \beta_{10}x10 + \epsilon$$

$SSE_{drop} = SSE_2 - SSE_1$; $MS_{drop} = SS_{drop}/9$; $F = MS_{drop}/MSE_1$, with $df_1 = 9$ and $df_2$ = degrees of freedom for error in the complete model.

b. Use the reduced model described in exercise 20.6a as the complete model and set $\beta_8 = \beta_9 = \beta_{10}$ equal to 0 for the reduced model.

20.7.

$$y = \beta_0 + \beta_1x_1 + \beta_2x_1^2 + \beta_3x_2 + \beta_4x_3 + \beta_5x_4 + \beta_6x_5 + \beta_7x_6 + \beta_8x_7 + \beta_9x_8 + \beta_{10}x_9 + \beta_{11}x_{10} + \beta_{12}x_1x_2 + \beta_{13}x_1x_3 + \beta_{14}x_1x_4 + \beta_{15}x_1x_5 + \beta_{16}x_1x_6 + \beta_{17}x_1x_7 + \beta_{18}x_1x_8 + \beta_{19}x_1x_9 + \beta_{20}x_1x_{10} + \beta_{21}x_1^2x_2 + \beta_{22}x_1^2x_3 + \beta_{23}x_1^2x_4 + \beta_{24}x_1^2x_5 + \beta_{25}x_1^2x_6 + \beta_{26}x_1^2x_7 + \beta_{27}x_1^2x_8 + \beta_{28}x_1^2x_9 + \beta_{29}x_1^2x_{10} + \epsilon$$

| | |
|---|---|
| $x_1$ = covariate | $x_6$ = 1 if column 3 |
| $x_2$ = 1 if row 2 | $x_6$ = 0 otherwise |
| $x_2$ = 0 otherwise | $x_7$ = 1 if column 4 |
| $x_3$ = 1 if row 3 | $x_7$ = 0 otherwise |
| $x_3$ = 0 otherwise | $x_8$ = 1 if treatment 2 |
| $x_4$ = 1 if row 4 | $x_8$ = 0 otherwise |
| $x_4$ = 0 otherwise | $x_9$ = 1 if treatment 3 |
| $x_5$ = 1 if column 2 | $x_9$ = 0 otherwise |
| $x_5$ = 0 otherwise | $x_{10}$ = 1 if treatment 4 |
| | $x_{10}$ = 0 otherwise |

Run a complete model as described above and a reduced model in which you set $\beta_{12} = \beta_{13} = \cdots = \beta_{29} = 0$. $F = MS_{drop}/MSE_1$, with $df_1 = 18$ and $df_2$ = degrees of freedom in error term in the complete model.

20.8. a. Randomized block design with antidepressants as treatments and age-sex combinations as blocks.

b. $y = \beta_0 + \beta_1x_1 + \beta_2x_2 + \beta_3x_3 + \beta_4x_4 + \beta_5x_5 + \beta_6x_6 + \beta_7x_7 + \beta_8x_8 + \beta_9x_1x_2 + \beta_{10}x_1x_3 + \beta_{11}x_1x_4 + \beta_{12}x_1x_5 + \beta_{13}x_1x_6 + \beta_{14}x_1x_7 + \beta_{15}x_1x_8 + \epsilon$

20.9. a.

| *Source* | *SS* | *df* | *MS* |
|---|---|---|---|
| covariate | 25.799 | 1 | 25.799 |
| blocks | 2.043 | 5 | .409 |
| treatments | 72.077 | 2 | 36.039 |
| interaction (with covariate) | 28.633 | 7 | 4.090 |
| error | 39.948 | 2 | 19.974 |
| *Totals* | 168.500 | 17 | |

b. $F_{7,2} = .205$; cannot reject $H_0$ that the lines are parallel.

c. $F_{2,9} = 4.729$; reject $H_0$ that the three treatments are equal.

20.10. a. $F_{5,9} = .174$; cannot reject $H_0$ that the blocks are equal.

b. Use 5 orthogonal contrasts:

(1) males vs. females
(2) $< 20$ vs. 20–40
(3) $< 20$ and 20–40 vs. 2($> 40$)
(4) interaction between sex and first 2 age groups
(5) interaction between sex and all age groups

c. $y = \beta_0 + \beta_1x_1 + \beta_2x_2 + \beta_3x_3 + \beta_4x_4 + \beta_5x_5 + \beta_6x_6 + \beta_7x_4x_5 + \beta_8x_4x_6 + \epsilon$

| | |
|---|---|
| $x_1$ = covariate | $x_5$ = 1 if less than 20 |
| $x_2$ = 1 if treatment B | $x_5$ = −1 if 20–40 |
| $x_2$ = 0 otherwise | $x_5$ = 0 if greater than 40 |
| $x_3$ = 1 if treatment C | $x_6$ = 1 if less than 20 or 20–40 |
| $x_3$ = 0 otherwise | $x_6$ = −2 if greater than 40 |
| $x_4$ = 1 if female | |
| $x_4$ = −1 if male | |

| *Source* | *SS* | *df* |
|---|---|---|
| $x_1$ (covariate) | 25.799 | 1 |
| $x_2$ (treatments) | 59.260 | 1 |
| $x_3$ (treatments) | 8.224 | 1 |
| $x_4$ (sex) | .183 | 1 |
| $x_5$ (age) | 4.66 | 1 |
| $x_6$ (age) | 1.292 | 1 |
| $x_7$ (age × sex) | .114 | 1 |
| $x_8$ (age × sex) | .386 | 1 |
| error | 68.581 | 9 |
| *Totals* | 168.500 | 17 |

Although the single-degree-of-freedom sums of squares are given in the AOV table, no additional tests need be performed since the test for blocks (sex, age, age × sex) of part a failed to show any significant effects.

20.11. b. $y = \beta_0 + \beta_1 x_1 + \beta_2 x_2 + \beta_3 x_1 x_2 + \epsilon$

20.12. a.

| Source | SS | df | MS | F | p |
|---|---|---|---|---|---|
| $x_1$ | .294 | 1 | .294 | .004 | $>.05$ |
| $x_2$ | 726.744 | 1 | 726.744 | 10.865 | $<.01$ |
| $x_1 x_2$ | 825.577 | 1 | 825.577 | 12.343 | $<.01$ |
| error | 1070.185 | 16 | 66.887 | | |
| *Totals* | 2622.8 | | | | |

b. $F_{1,16} = 12.343$; reject $H_0$ that the lines are parallel.
c. The test for a difference between the 2 populations is not valid because the lines are not parallel. The assumptions for a covariance analysis are not met.

## CHAPTER 21

21.1. a. $\hat{\gamma}_1 = -.197$ b. cannot reject $H_0$: $\gamma_1 = 0$
21.2. a. $\hat{\gamma}_2 = 1.904$ b. no, $n < 50$
21.3. a. $\hat{\gamma}_1 = .757$ b. reject $H_0$ that $\gamma_1 = 0$
21.4. a. $\hat{\gamma}_2 = 2.837$ b. cannot reject $H_0$ that $\gamma_2 = 3$
c. No, the distribution is skewed to the right.
21.5. $\hat{\gamma}_1 = .435$; cannot reject $H_0$ that $\gamma_1 = 0$. $\hat{\gamma}_2 = 3.166$, cannot reject $H_0$ that $\gamma_2 = 3$. Therefore, we can conclude that the 50 sample means were selected from a normal population.

21.6.

| Source | SS | df | MS | F | p |
|---|---|---|---|---|---|
| location | .1924 | 3 | .0641 | 1.005 | $>.05$ |
| error | 1.2755 | 20 | .0638 | | |
| *Totals* | 1.4679 | 23 | | | |

21.7. a.

| Source | SS | df | MS | F | p |
|---|---|---|---|---|---|
| location | .1924 | 3 | .0641 | 91.57 | $<.01$ |
| level | 1.2653 | 5 | .2531 | 361.57 | $<.01$ |
| error | .0102 | 15 | .0007 | | |
| *Totals* | 1.4679 | 23 | | | |

21.8. a.

| Source | SS | df | MS | F | p |
|---|---|---|---|---|---|
| line | 3.2469 | 4 | .8117 | 1.028 | $>.05$ |
| day | 19.2188 | 4 | 4.8047 | 6.086 | $<.01$ |
| error | 12.6322 | 16 | .7895 | | |
| *Totals* | 35.0979 | 24 | | | |

b. We would conclude that the production lines are equal but that at least one of the days differs from the others.

## CHAPTER 22

22.1. $\bar{y}_{st} = 194.4$ credits, $B = 1.3886$
22.2. $\bar{y} = 190.3$ credits, $B = 2.4051$
22.3. $\bar{y} = \$9980$, $B = \$432$
22.4. $\hat{p} = .55$, $B = .0797$
22.5. $\hat{p}_{st} = .49$, $B = .0275$
22.6. $\bar{y}_{st} = 8.89$ years, $B = .3881$

22.7.

| Stratum | $\bar{y}$ | B |
|---|---|---|
| 1 | 35.6 | 6.08 |
| 2 | 22.1 | 2.1557 |
| 3 | 10.3 | .8036 |
| 4 | 6.3 | .4057 |
| 5 | 1.9 | .2619 |

22.8. $\bar{y}_{st} = 1.25$, $B = .0303$
22.10. $\hat{\tau} = 1242$, $B = 30.1543$
22.11. $\bar{y} = 40.2$, $B = 7.36$
22.12. $\hat{\tau} = 221{,}100$ gallons, $B = 40{,}480$ gallons
22.13. $\bar{y}_c = 6.73$ years, $B = 4.98$
22.14. $\bar{y}_c = 51.56$, $B = 1.344$
22.15. $\bar{y}_c = 40.1688$, $B = .6406$
22.16. $\bar{y}_c = 16.005$, $B = .0215$
22.17. $\hat{\tau} = \$158{,}667$, $B = \$2504$

# Index